Measurement Error Models

Measurement Error Models

WAYNE A. FULLER
Iowa State University
Ames, Iowa

A JOHN WILEY & SONS, INC., PUBLICATION

Published by John Wiley & Sons, Inc., Hoboken, New Jersey.
Published simultaneously in Canada.

For general information on our other products and services or for technical support, please contact our Customer Care Department within the United States at (800) 762-2974, outside the United States at (317) 572-3993 or fax (317) 572-4002.

Wiley also publishes its books in a variety of electronic formats. Some content that appears in print may not be available in electronic format. For information about Wiley products, visit our web site at www.wiley.com.

Library of Congress Cataloging-in-Publication Data is available.

ISBN-13 978-0-470-09571-3
ISBN-10 0-470-09571-7

10 9 8 7 6 5 4 3 2 1

To Doug and Bret

Preface

The study of regression models wherein the independent variables are measured with error predates the twentieth century. There has been a continuing interest in the problem among statisticians and there is considerable literature on the subject. Also, for over 80 years, studies have documented the presence of sizable measurement error in data collected from human respondents. Despite these two lines of research, only a fraction of the statistical studies appearing in the literature use procedures designed for explanatory variables measured with error.

This book is an outgrowth of research on the measurement error, also called response error, in data collected from human respondents. The book was written with the objective of increasing the use of statistical techniques explicitly recognizing the presence of measurement error. To this end, a number of real examples have been included in the text. An attempt has been made to choose examples from a variety of areas of application, but the reader will understand if many of the examples have an agricultural aspect.

The book may be used as a text for a graduate course concentrating on statistical analyses in the presence of measurement error. It is hoped that it will also find use as an auxiliary text in courses on statistical methodology that heretofore have ignored, or given cursory treatment to, the problems associated with measurement error. Chapter 1 was developed to provide an introduction to techniques for a range of simple models. While the models of Chapter 1 are special cases of models discussed in later chapters, it is felt that the concepts are better communicated with the small models. There is some flexibility in the order in which the material can be covered. One can move from a section in Chapter 1 to the corresponding section in Chapter 2 or Chapter 4. To facilitate flexible use, Sections 1.2, 1.3, and 1.4 are largely self-supporting. As a result, there is some duplication in the treatment of topics such as prediction. Some repetition seems advantageous because the models of this book differ from those typically encountered by students in courses on regression estimation.

The proofs of most of the theorems require a background of statistical theory. One will be comfortable with the proofs only if one has an understanding of large sample theory. Also, the treatment assumes a background in ordinary linear regression methods. In attempting to make the book useful to those interested in the methods, as well as to those interested in an introduction to the theory, derivations are concentrated in the proofs of theorems. It is hoped that the text material, the statements of the theorems, and the examples will serve the person interested in applications.

Computer programs are required for any extensive application of the methods of this book. Perhaps the most general program for normal distribution linear models is LISREL® VI by Jöreskog and Sörbom. LISREL VI is available in SPSS^{X}™ and can be used for a wide range of models of the factor type. A program with similar capabilities, which can also perform some least squares fitting of the type discussed in Section 4.2, is EQS developed by Bentler. EQS is available from BMDP® Statistical Software, Inc. Dan Schnell has placed the procedures of Chapter 2 and Section 3.1 in a program for the IBM® Personal Computer AT. This program, called EV CARP, is available from the Statistical Laboratory, Iowa State University. The packages SAS® and BMDP contain algorithms for simple factor analysis. A program, ISU Factor, written with Proc MATRIX of SAS by Sastry Pantula, Department of Statistics, North Carolina State University, can be used to estimate the factor model, to estimate multivariate models with known error variances, and to estimate the covariance matrix of the factor estimates. A program for nonlinear models, written with Proc MATRIX of SAS by Dan Schnell, is available from Iowa State University.

I have been fortunate to work with a number of graduate students on topics related to those of this text. Each has contributed to my understanding of the field, but none is to be held responsible for remaining shortcomings. I express my sincere thanks to each of them. In chronological order they are James S. DeGracie, Angel Martinez-Garza, George E. Battese, A. Ronald Gallant, Gordon D. Booth, Kirk M. Wolter, Michael A. Hidiroglou, Randy Lee Carter, P. Fred Dahm, Fu-hua Yu, Ronald Mowers, Yasuo Amemiya, Sastry Pantula, Tin-Chiu Chua, Hsien-Ming Hung, Daniel Schnell, Stephen Miller, Nancy Hasabelnaby, Edina Miazaki, Neerchal Nagaraj, and John Eltinge. I owe a particular debt to Yasuo Amemiya for proofs of many theorems and for reading and repair of much of the manuscript. I thank Sharon Loubert, Clifford Spiegelman, and Leonard Stefanski for useful comments. I also express my appreciation to the United Kingdom Science and Engineering Research Council and the U.S. Army European Research Office for supporting the "Workshop on Functional and Structural Relationships and Factor Analysis" held at Dundee, Scotland, August 24 through September 9, 1983. Material presented at that stimulating conference had an influence on several sections of this book. I am grateful to Jane Stowe, Jo Ann Hershey, and Christine Olson for repeated typings of the manuscript. A part of the research for this book was supported by joint statistical agreements with the United States Bureau of the Census and by cooperative research agreements with the Statistical Reporting Service of the United States Department of Agriculture.

WAYNE A. FULLER

Ames, Iowa
February 1987

Contents

List of Examples

Number	*Topic*

List of Principal Results

List of Figures

Figure	*Title*

CHAPTER 1

A Single Explanatory Variable

In Section 1 of this chapter we introduce the linear model containing measurement error and investigate the effects of measurement error on the ordinary least squares estimators. Estimation for a simple model is considered.

In Sections 2 and 3 we study the two-variable measurement error model for two types of information about the measurement error variances. Because the objective is to introduce the reader to models and estimation procedures, we concentrate on the normal bivariate model. Sections 4 and 5 treat situations in which three variables are observed. The estimation methods employed in Sections 2–5 are closely related to maximum likelihood estimation. Because of the simple nature of the models, most estimators can be obtained by the method of moments. Some specialized models and methods are considered in Section 6.

Later chapters will be devoted to models of higher dimension and to an expanded treatment of models with fixed independent variables measured with error. In particular, Sections 2.1–2.4 of Chapter 2 extend the results of Sections 1.1–1.4 of this chapter to vector explanatory variables.

1.1. INTRODUCTION

1.1.1. Ordinary Least Squares and Measurement Error

The classical linear regression model with one independent variable is defined by

$$Y_t = \beta_0 + \beta_1 x_t + e_t, \qquad t = 1, 2, \ldots, n, \tag{1.1.1}$$

where $(x_1, x_2, \ldots, x_n)$ is fixed in repeated sampling and the e_t are independent $N(0, \sigma_{ee})$ random variables. We shall use the convention of identifying the mean and variance of a random variable within parentheses and shall use N to identify the normal distribution. It is well known that the least squares estimator

$$\hat{\beta}_1 = \left[\sum_{t=1}^{n} (x_t - \bar{x})^2\right]^{-1} \sum_{t=1}^{n} (x_t - \bar{x})(Y_t - \bar{Y})$$

is unbiased for β_1 and has the smallest variance of unbiased linear estimators.

In a second form of the regression model, the x_t are assumed to be independent drawings from a $N(\mu_x, \sigma_{xx})$ distribution. In the second model it is also assumed that the vector $(e_1, e_2, \ldots, e_n)$ is independent of the vector $(x_1, x_2, \ldots, x_n)$. The estimator $\hat{\beta}_1$ is the maximum likelihood estimator for β_1 and is unbiased for β_1 under both models.

We shall study models of the regression type where one is unable to observe x_t directly. Instead of observing x_t, one observes the sum

$$X_t = x_t + u_t, \tag{1.1.2}$$

where u_t is a $(0, \sigma_{uu})$ random variable. The observed variable X_t is sometimes called the *manifest* variable or the *indicator* variable. The unobserved variable x_t is called a *latent* variable in certain areas of application. Models with fixed x_t are called *functional* models, while models with random x_t are called *structural* models. To aid in remembering the meaning of functional and structural, note that "F" is the first letter of the words fixed and functional, while "S" is the first letter of the words stochastic and structural.

As an example of a situation where x_t cannot be observed, consider the relationship between the yield of corn and available nitrogen in the soil. Assume that (1.1.1) is an adequate approximation to the relationship between yield and nitrogen. The coefficient β_1 is the amount that yield is increased when soil nitrogen increases one unit. To estimate the available soil nitrogen, it is necessary to sample the soil of the experimental plot and to perform a laboratory analysis on the selected sample. As a result of the sampling and of the laboratory analysis, we do not observe x_t but observe an estimate of x_t. Therefore, we represent the observed nitrogen by X_t, where X_t satisfies (1.1.2) and u_t is the measurement error introduced by sampling and laboratory analysis.

The description of the collection of the soil nitrogen data permits two interpretations of the true values x_t. First, assume that the fields are a set of experimental fields managed by the experiment station in ways that produce different levels of soil nitrogen in the different fields. For example, the application of varying rates of fertilizer and the growing of different fractions of legumes in the rotations would produce different levels of soil nitrogen. In such a situation, one would treat the true, but unknown, nitrogen levels in the different fields as fixed. On the other hand, if the fields were a random sample of farmers' fields in the state of Iowa, the true values of soil nitrogen could be treated as random variables.

Let us investigate the effect of measurement error on the least squares coefficient in the simple model (1.1.1) and (1.1.2), under the assumption that the x_t are random variables with $\sigma_{xx} > 0$. We assume

$$(x_t, e_t, u_t)' \sim \mathrm{NI}[(\mu_x, 0, 0)', \operatorname{diag}(\sigma_{xx}, \sigma_{ee}, \sigma_{uu})], \tag{1.1.3}$$

where $\sim$NI is an abbreviation for "distributed normally and independently," and $\operatorname{diag}(\sigma_{xx}, \sigma_{ee}, \sigma_{uu})$ is a diagonal matrix with the given elements on the diagonal.

It follows from the structural model (1.1.3) that the vector $(Y_t, X_t)'$, where Y_t is defined by (1.1.1) and X_t is defined in (1.1.2), is distributed as a bivariate normal vector with mean vector

$$E\{(Y, X)\} = (\mu_Y, \mu_X) = (\beta_0 + \beta_1 \mu_x, \mu_x)$$

and covariance matrix

$$\begin{bmatrix} \sigma_{YY} & \sigma_{XY} \\ \sigma_{XY} & \sigma_{XX} \end{bmatrix} = \begin{bmatrix} \beta_1^2 \sigma_{xx} + \sigma_{ee} & \beta_1 \sigma_{xx} \\ \beta_1 \sigma_{xx} & \sigma_{xx} + \sigma_{uu} \end{bmatrix}. \tag{1.1.4}$$

Let

$$\hat{\gamma}_{1\ell} = \left[\sum_{t=1}^{n} (X_t - \bar{X})^2 \right]^{-1} \sum_{t=1}^{n} (X_t - \bar{X})(Y_t - \bar{Y}) \tag{1.1.5}$$

be the regression coefficient computed from the observed variables. By the properties of the bivariate normal distribution,

$$E\{\hat{\gamma}_{1\ell}\} = \sigma_{XX}^{-1} \sigma_{XY} = \beta_1 (\sigma_{xx} + \sigma_{uu})^{-1} \sigma_{xx} = \gamma_1 \tag{1.1.6}$$

We conclude that, for the bivariate model with independent measurement error in X, the least squares regression coefficient is biased toward zero. It is important to remember that Equation (1.1.6) was derived under the assumption that the measurement error in X_t is independent of the true values, x_t, and of the errors, e_t.

One way to describe the effect of measurement error displayed in (1.1.6) is to say that the regression coefficient has been *attenuated* by the measurement error. There are a number of names for the ratio $\kappa_{xx} = \sigma_{XX}^{-1} \sigma_{xx}$ that defines the degree of attenuation. The ratio is called the reliability of X_t in the social science literature, but because reliability is a heavily used word in statistics, we call κ_{xx} the *reliability ratio*.

The reliability ratio is called *heritability* in genetics. An observed characteristic of a plant or animal, the X value, is called the phenotype and the unobserved true genetic makeup of the individual, the x value, is called the genotype. The phenotype is the sum of the genotype and the environment effect, where the environment effect is the measurement error.

Because the bias in $\hat{\gamma}_{1\ell}$ as an estimator of β_1 is multiplicative, the test of the hypothesis that $\beta_1 = 0$ remains valid in the presence of independent measurement error. That is, if model (1.1.1)–(1.1.5) holds with $\beta_1 = 0$ and $\sigma_{eu} = 0$, the population correlation between X_t and Y_t is zero for all values of σ_{uu}. It follows that the usual regression "t test" of the hypothesis that $\beta_1 = 0$ has Student's t distribution when $\beta_1 = 0$, whether or not the explanatory variable is measured with error. The use of the t distribution for hypotheses other than H_0: $\beta_1 = 0$ leads to biased tests in the presence of measurement error. Also, the presence of measurement error will reduce the power of the test of $\beta_1 = 0$. See Exercise 1.1.

For the classical regression model with fixed x_t, the estimator

$$\left[\sum_{t=1}^{n} (x_t - \bar{x})^2\right]^{-1} \sum_{t=1}^{n} (x_t - \bar{x})(Y_t - \bar{Y})$$

is called a linear estimator. The adjective "linear" is a modifier for the random variables Y_t and means that the estimator is linear in Y_t. The term linear can also be interpreted to mean that the error in the estimator is a linear function of the random variables e_t.

If there are errors of measurement in the explanatory variable, the error in the least squares estimator is not linear in the full set of random errors. When $X_t = x_t + u_t$, where u_t is measurement error, the error in the least squares estimator is

$$\hat{\gamma}_{1\ell} - \beta_1 = \left[\sum_{t=1}^{n} \{(x_t - \bar{x})^2 + 2(x_t - \bar{x})(u_t - \bar{u}) + (u_t - \bar{u})^2\}\right]^{-1}$$
$$\times \left[\sum_{t=1}^{n} \{(x_t - \bar{x})(v_t - \bar{v}) + (u_t - \bar{u})(v_t - \bar{v})\}\right],$$

where $v_t = e_t - u_t\beta_1$. The measurement errors u_t enter as squares and as products with x_t in the denominator, and as squares and as products with e_t in the numerator of $\hat{\gamma}_{1\ell} - \beta_1$. The covariance between u_t and v_t produces the bias characterized in (1.1.6). It is possible for the measurement errors u_t to be correlated with the true values x_t and with e_t. In such cases the terms $\sum_{t=1}^{n} (x_t - \bar{x})(u_t - \bar{u})$ and $\sum_{t=1}^{n} (u_t - \bar{u})(e_t - \bar{e})$ will also make a contribution to the bias.

The population squared correlation between x_t and Y_t is

$$R_{xY}^2 = (\sigma_{xx}\sigma_{YY})^{-1}\sigma_{xY}^2 = \sigma_{YY}^{-1}\beta_1^2\sigma_{xx},$$

while the population squared correlation between X_t and Y_t is

$$R_{XY}^2 = (\sigma_{XX}\sigma_{YY})^{-1}\sigma_{XY}^2 = \kappa_{xx}R_{xY}^2.$$

Thus, the introduction of independent measurement error leads to a reduction in the squared correlation, where the factor by which the correlation is reduced is the factor by which the regression coefficient is biased toward zero. As with the regression coefficient, it is said that the correlation has been attenuated by the presence of measurement error.

1.1.2. Estimation with Known Reliability Ratio

We have shown that the expected value of the least squares estimator of β_1 is the true β_1 multiplied by κ_{xx}. Therefore, if we know the ratio $\sigma_{XX}^{-1}\sigma_{xx}$, it is possible to construct an unbiased estimator of β_1. There are a number of situations, particularly in psychology, sociology, and survey sampling, where information about κ_{xx} is available.

Assume we are interested in a population of individuals possessing a trait x. Examples of such traits are intelligence, community loyalty, social consciousness, willingness to adopt new practices, managerial ability, and the probability that the individual will vote in a particular election. It is impossible to observe directly the value of the trait for a particular individual. One can obtain only an estimate for the individual. This estimate, or response, for an individual is obtained frequently as a score constructed from the answers to a number of questions. The battery of questions is sometimes called an *instrument*. The score of an IQ test is, perhaps, the best known example. The difference between the individual's value for the trait and the response is the measurement error. Often the instrument has been studied and the ratio κ_{xx} is so well estimated that it may be treated as known. The reliability ratio κ_{xx} is published as a part of the supporting material for a number of standard instruments, such as IQ tests.

Suppose that we draw a simple random sample of individuals from a population for which the ratio κ_{xx} is known. An unbiased estimator of the structural regression coefficient β_1 relating a characteristic Y to the true value of a trait x of model (1.1.1) is given by

$$\hat{\beta}_1 = \kappa_{xx}^{-1}\hat{\gamma}_{1\ell}, \tag{1.1.7}$$

where $\hat{\gamma}_{1\ell}$ is the least squares coefficient defined in (1.1.5). The coefficient (1.1.7) is sometimes called the regression coefficient corrected for attenuation.

Because (Y_t, X_t) is distributed as a bivariate normal, the conditional distribution of $\hat{\gamma}_{1\ell}$ is normal and the conditional mean and variance of $\hat{\gamma}_{1\ell}$, given $\mathbf{X} = (X_1, X_2, \ldots, X_n)'$, are

$$E\{\hat{\gamma}_{1\ell} \mid \mathbf{X}\} = \gamma_1, \tag{1.1.8}$$

$$V\{\hat{\gamma}_{1\ell} \mid \mathbf{X}\} = \left[\sum_{t=1}^{n} (X_t - \bar{X})^2\right]^{-1} (\sigma_{YY} - \gamma_1\sigma_{XY}). \tag{1.1.9}$$

An unbiased estimator of (1.1.9) (for $n > 2$) is

$$\hat{V}\{\hat{\gamma}_{1\ell} \mid \mathbf{X}\} = \left[\sum_{t=1}^{n} (X_t - \bar{X})^2\right]^{-1} s_\ell^2, \tag{1.1.10}$$

where

$$s_\ell^2 = (n-2)^{-1} \sum_{t=1}^{n} [Y_t - \bar{Y} - \hat{\gamma}_{1\ell}(X_t - \bar{X})]^2.$$

The unconditional variance of $\hat{\gamma}_{1\ell}$ is the variance of the conditional expectation plus the expected value of the conditional variance. Because the estimator $\hat{\gamma}_{1\ell}$ is conditionally unbiased for γ_1, the unconditional variance is the expected value of the conditional variance. The quantity $\sigma_{XX}^{-1} \sum_{t=1}^{n} (X_t - \bar{X})^2$ is distributed as a chi-square random variable with $n-1$ degrees of freedom, and the unconditional variance of $\hat{\gamma}_{1\ell}$ is obtained by evaluating the expectation of (1.1.9). For $n > 3$,

$$V\{\hat{\gamma}_{1\ell}\} = [(n-3)\sigma_{XX}]^{-1}(\sigma_{YY} - \gamma_1 \sigma_{XY}). \tag{1.1.11}$$

The estimated conditional variance (1.1.10) is an unbiased estimator of the unconditional variance because $\hat{\gamma}_{1\ell}$ is conditionally unbiased. Therefore, an unbiased estimator of the conditional variance of $\hat{\beta}_1$ of (1.1.7) is

$$\hat{V}\{\hat{\beta}_1 \mid \mathbf{X}\} = \kappa_{xx}^{-2}\left[\sum_{t=1}^{n} (X_t - \bar{X})^2\right]^{-1} s_\ell^2. \tag{1.1.12}$$

The conditional and unconditional distributions of $\hat{\beta}_1$ are closely related and the estimated variance of the conditional distribution is also an estimator of the variance of the unconditional distribution. This is not true for

$$\hat{\beta}_0 = \bar{Y} - \hat{\beta}_1 \bar{X} \tag{1.1.13}$$

because the conditional expected value of $\hat{\beta}_0$ is a function of $\mathbf{X}$. The limiting distribution of the estimator of β_0 is established by noting that

$$\hat{\beta}_0 = \beta_0 + \bar{v} - (\hat{\beta}_1 - \beta_1)\bar{X}, \tag{1.1.14}$$

where $\bar{v} = n^{-1} \sum_{t=1}^{n} v_t$ and $v_t = e_t - u_t \beta_1$. It follows that $n^{1/2}(\hat{\beta}_0 - \beta_0, \hat{\beta}_1 - \beta_1)$ is normally distributed in the limit and that a consistent estimator of the covariance matrix of $(\hat{\beta}_0, \hat{\beta}_1)'$ is

$$\hat{\mathbf{V}}\{(\hat{\beta}_0, \hat{\beta}_1)'\} = \begin{bmatrix} n^{-1}s_{vv} + \bar{X}^2 \hat{V}\{\hat{\beta}_1 \mid \mathbf{X}\} & -\bar{X}\hat{V}\{\hat{\beta}_1 \mid \mathbf{X}\} \\ -\bar{X}\hat{V}\{\hat{\beta}_1 \mid \mathbf{X}\} & \hat{V}\{\hat{\beta}_1 \mid \mathbf{X}\} \end{bmatrix}, \tag{1.1.15}$$

where

$$s_{vv} = (n-2)^{-1} \sum_{t=1}^{n} (Y_t - \hat{\beta}_0 - \hat{\beta}_1 X_t)^2$$

and $\hat{V}\{\hat{\beta}_1 \mid \mathbf{X}\}$ is defined in (1.1.12).

Also, by standard regression results,

$$t = \left[\kappa_{xx}^{2} s_{\ell}^{-2} \sum_{t=1}^{n} (X_t - \bar{X})^2\right]^{1/2} (\hat{\beta}_1 - \beta_1)$$

$$= \left[s_{\ell}^{-2} \sum_{t=1}^{n} (\bar{X}_t - \bar{X})^2\right]^{1/2} (\hat{\gamma}_{1\ell} - \gamma_1) \qquad (1.1.16)$$

is distributed as Student's t with $n - 2$ degrees of freedom. Any linear hypothesis about β_1 can be transformed into a hypothesis about γ_1 by using the reliability ratio. Therefore, in the bivariate situation, knowledge of the ratio $\sigma_{XX}^{-1}\sigma_{xx}$ permits one to construct an unbiased estimator of the parameter β_1 and to apply the usual normal theory for hypothesis testing and confidence interval construction. Unfortunately, these simple results do not extend to the vector-**x** case. See Section 3.1.3.

To construct the estimator (1.1.7), we must assume that the X values are from a distribution where the reliability ratio is known. Generally, this means that the X values of the current study must be a random sample from the population that generated the random sample used to estimate the ratio. For example, assume that the population of graduate students scores higher on IQ tests than does the general population. Then the ratio $\sigma_{XX}^{-1}\sigma_{xx}$ of an IQ test computed for the general population is not applicable for a sample of graduate students.

Recalling that

$$R_{XY}^2 = (\sigma_{XX}\sigma_{YY})^{-1}\sigma_{XY}^2 = \kappa_{xx}R_{xY}^2,$$

an estimator of the squared correlation between x and Y is

$$\hat{R}_{xY}^2 = \kappa_{xx}^{-1}\hat{R}_{XY}^2, \qquad (1.1.17)$$

where $\hat{R}_{XY}^2 = (m_{XX}m_{YY})^{-1}m_{XY}^2$ and (m_{YY}, m_{XY}, m_{XX}) is the sample estimator of $(\sigma_{YY}, \sigma_{XY}, \sigma_{XX})$. The estimator (1.1.17) is said to be the squared correlation corrected for attenuation.

It is possible for $\hat{R}_{xY}^2$ defined by (1.1.17) to exceed one. In such cases, the maximum likelihood estimator of R_{xY}^2 is one and the maximum likelihood estimator of β_1 *is* $\hat{\beta}_1 = (\operatorname{sgn} m_{XY})[m_{XX}^{-1}\kappa_{xx}^{-1}m_{YY}]^{1/2}$.

We have studied the effect of measurement error on ordinary regression statistics and have demonstrated one method of correcting for that effect. We now give an indication of the order of magnitude of such effects. Table 1.1.1 contains estimates of the effect of measurement error on the regression coefficient for a number of socioeconomic variables. The attenuation coefficients, denoted by κ_{xx}, of the table are the correlations between two determinations on the same characteristic. For continuous variables, such as income, the attenuation coefficient is $\sigma_{XX}^{-1}\sigma_{xx}$. For zero–one variables,

TABLE 1.1.1. Attenuation coefficients for selected variables

Variable	κ_{xx}	Sample Size, $R^2_{XY} = 0.5$	
		50/50	MEM Superior
Sex[a]	0.98	2157	97
Age[a]	0.99	6976	171
Age[a] (45–49)(0–1)	0.92	132	27
Education[b]	0.88	58	19
Income[b]	0.85	35	15
Unemployed[c] (0–1)	0.77	14	11
Poverty status[d] (0–1)	0.58	5	7

[a] All persons.
[b] Persons 14 and over.
[c] Persons 16 and over.
[d] All families. Unemployed attenuation coefficient is from Fuller and Chua (1984). All other coefficients calculated from data in U.S. Department of Commerce (1975).

such as sex, the correlation between two independent, identically distributed, determinations defines the multiplicative effect of measurement error on the regression coefficient of the latent class model. See Section 3.4.

The estimates of the table were constructed from repeated interview studies conducted by the United States Bureau of the Census. In such studies, the same data are collected in two different interviews. Most of the estimates of Table 1.1.1 come from a comparison of responses in the 1970 Census with the same data collected in the Current Population Survey. In survey sampling the measurement error in data collected from human respondents is usually called *response error*.

The increase in response error as one moves down the table is associated with a corresponding increase in the complexity of the concept being measured. Sex and age are relatively well-defined characteristics, while income, poverty, and unemployment are all relatively complex. In fact, whether or not a family is above the poverty level, or an individual is unemployed, is determined by responses to a number of questions.

Age illustrates how subdividing the population can increase the effect of measurement error. For the population as a whole about 1% of the observed variation in age is due to measurement error. However, few studies investigate the effect of age for the entire population. If one is interested in a single 5-year category, such as ages 45–49, the effect of measurement error on the estimated regression coefficient increases to about 8%.

Measurement error variance is about 15% of total variation for income. This level of variation is typical of many related variables, such as occupational status and socioeconomic status. As with age, the percent of variation due to measurement error increases if we restrict the population. For example, $\kappa_{xx} = 0.82$ for persons with some income.

The 50/50 sample size given in the table is the sample size for which 50% of the mean square error of an ordinary least squares regression coefficient is squared measurement error bias, given that R_{XY}^2 for the observed variables is 0.5. If the sample size is larger than that of the table, more than half of the mean square error is due to squared bias.

The last column of Table 1.1.1 contains the sample size for which the correction-for-attenuation estimator (1.1.7) has the same mean square error as the ordinary least squares estimator in a population with $R_{XY}^2 = 0.5$. For any larger sample size, the corrected coefficient of (1.1.7) has smaller mean square error. The sample size is determined by solving the equation

$$(n-3)^{-1}(1-R_{XY}^2) + (1-\kappa_{xx})^2\kappa_{xx}^{-2}R_{XY}^2 = (n-3)^{-1}\kappa_{xx}^{-2}(1-R_{XY}^2) \quad (1.1.18)$$

for n, where the left side of the equality is the mean square error of the ordinary least squares coefficient and the right side of the equality is the variance of the estimator (1.1.7), both in standardized units. This equation was derived under the assumption of bivariate normality, but it is a useful approximation for a wide range of distributions.

References containing discussions of measurement error and its effects include Morgenstern (1963), Cochran (1968), Hunter (1980), and Pierce (1981). Dalenius (1977a–c) has given a bibliography for response errors in surveys.

1.1.3. Identification

The model (1.1.1)–(1.1.3) can be used to illustrate the idea of *identification*. Identification is a concept closely related to the ability to estimate the parameters of a model from a sample generated by the model. For the purposes of our discussion we consider a *model* to be a specification of:

(a) Variables and parameters of interest.
(b) Relationships among the variables.
(c) Assumptions about the stochastic properties of the random variables.

It is understood that the parameters of interest are specified by a vector $\boldsymbol{\theta} \in \Theta$, where Θ is the space of possible parameter values and the dimension of $\boldsymbol{\theta}$ is the minimum required to fully define the model. The observable random vectors have a sampling distribution defined on a probability space.

Definition 1.1.1. Let $\mathbf{Z}$ be the vector of observable random variables and let $F_{\mathbf{Z}}(\mathbf{a}: \boldsymbol{\theta})$ be the distribution function of $\mathbf{Z}$ for parameter $\boldsymbol{\theta}$ evaluated at $\mathbf{Z} = \mathbf{a}$. The parameter $\boldsymbol{\theta}$ is identified if, for any $\boldsymbol{\theta}_1 \in \Theta$ and $\boldsymbol{\theta}_2 \in \Theta$, $\boldsymbol{\theta}_1 \neq \boldsymbol{\theta}_2$

implies that

$$F_{\mathbf{Z}}(\mathbf{a}: \boldsymbol{\theta}_1) \neq F_{\mathbf{Z}}(\mathbf{a}: \boldsymbol{\theta}_2)$$

for some $\mathbf{a}$. If the vector $\boldsymbol{\theta}$ is identified, we also say that the model is identified.

Definition 1.1.2. The parameter θ_i, where θ_i is the ith element of $\boldsymbol{\theta}$, is identified if no two values of $\boldsymbol{\theta} \in \Theta$, for which θ_i differ, lead to the same sampling distribution of the observable random variables. The model is identified if and only if every element of $\boldsymbol{\theta}$ is identified.

For the model of Section 1.1.1, Equations (1.1.1) and (1.1.2) give the algebraic relationship among the variables x_t, e_t, u_t, X_t, and Y_t, where only Y_t and X_t can be observed. For the normal structural model, the random vector (x_t, u_t, e_t) is distributed as a multivariate normal with $\sigma_{eu} = \sigma_{xu} = \sigma_{ex} = 0$ and $\sigma_{xx} > 0$. This exhausts our prior information about the distribution of the variables. The vector of unknown parameters for the model is $\boldsymbol{\theta} = (\mu_x, \sigma_{xx}, \sigma_{ee}, \sigma_{uu}, \beta_0, \beta_1)$. Under the assumptions, the distribution of the observations (Y_t, X_t) is bivariate normal. Hence, the distribution of (Y_t, X_t) is characterized completely by the elements of its mean vector and its covariance matrix, a total of five parameters. Because the model contains six parameters, there are many different parametric configurations (different $\boldsymbol{\theta}$) that lead to the same distribution of the observations. Therefore, the model is not identified. For example, the parameter sets

$$\boldsymbol{\theta}_1 = (\mu_x, \sigma_{xx}, \sigma_{ee}, \sigma_{uu}, \beta_0, \beta_1) = (1, 1, 1, 1, 1, 1)$$

and

$$\boldsymbol{\theta}_2 = (\mu_x, \sigma_{xx}, \sigma_{ee}, \sigma_{uu}, \beta_0, \beta_1) = (1, 2, 1.5, 0, 1.5, 0.5)$$

are both such that

$$\begin{bmatrix} Y_t \\ X_t \end{bmatrix} \sim \text{NI}\left(\begin{bmatrix} 2 \\ 1 \end{bmatrix}, \begin{bmatrix} 2 & 1 \\ 1 & 2 \end{bmatrix}\right).$$

While the normal structural model is not identified, one of the parameters is identified. The mean of x is equal to the mean of X. Thus, the parameter μ_x is identified because, given the sample distribution, the parameter vector $\boldsymbol{\theta}$ is restricted to a subspace where μ_x is equal to the mean of X.

For a model that is not identified, it is possible for the sample distribution to contain some information about the parameters. We have seen that one of the parameters, the mean of x, is identified even though the other parameters are not identified. In fact, the sample distribution associated with model (1.1.1)–(1.1.3) contains additional information about the parameters. We note that we are able to detect the situation $\beta_1 = 0$ from the distribution of

(X_t, Y_t). That is, given $\sigma_{eu} = 0$, we have $\beta_1 = 0$ if and only if $\sigma_{XY} = 0$. However, we are still unable to determine σ_{uu} and σ_{xx} when $\beta_1 = 0$.

In general, given $\sigma_{eu} = 0$, the distribution of (X_t, Y_t) permits one to establish bounds for σ_{uu}. Because $\sigma_{uu} \geqslant 0$, we need only establish an upper bound for σ_{uu}. From the equations following (1.1.4), we can obtain an equation for σ_{ee} in terms of σ_{uu} and an equation for σ_{uu} in terms of σ_{ee}. These equations are

$$\sigma_{ee} = \sigma_{YY} - \sigma_{XY}^2(\sigma_{XX} - \sigma_{uu})^{-1},$$
$$\sigma_{uu} = \sigma_{XX} - \sigma_{XY}^2(\sigma_{YY} - \sigma_{ee})^{-1}.$$

A maximum value for σ_{uu} is obtained by setting $\sigma_{ee} = 0$ and a maximum value for σ_{ee} is obtained by setting $\sigma_{uu} = 0$. These bounds lead to the bounds for the parameters displayed in Table 1.1.2. When $\beta_1 \neq 0$, the bounds for β_1 are the population regression of Y on X and the inverse of the population regression of X on Y.

The reader should remember that the assumptions $\sigma_{xx} > 0$ and $\sigma_{eu} = 0$ were used in this development. It is not posssible to set bounds for the parameters if σ_{eu} is also unknown.

To summarize, the sampling distribution of (X_t, Y_t) tells us something about the measurement error model (1.1.1)–(1.1.3). However, the model is not identified because it is not possible to find a unique relationship between the parameter vector of the distribution of (X_t, Y_t) and the parameter vector $\boldsymbol{\theta}$. If we are to construct a consistent estimator for the vector $\boldsymbol{\theta}$, our model must contain a specification of additional information. In Section 1.1.2, knowledge of $\sigma_{XX}^{-1}\sigma_{xx}$ enabled us to estimate the remaining parameters.

In the following sections we use the method of moments and the method of maximum likelihood, methods that are often equivalent under normality, to construct estimators of the unknown parameter vector $\boldsymbol{\theta}$. The properties of the estimators depend on the type of information that is used to identify the model.

TABLE 1.1.2. Bounds for the parameters of model (1.1.1)–(1.1.3) obtained from the distribution[a] of (X_t, Y_t)

$\sigma_{XY} = 0$	$\sigma_{XY} > 0$	$\sigma_{XY} < 0$
$\beta_1 = 0$	$\sigma_{XX}^{-1}\sigma_{XY} \leqslant \beta_1 \leqslant \sigma_{XY}^{-1}\sigma_{YY}$	$\sigma_{XY}^{-1}\sigma_{YY} \leqslant \beta_1 \leqslant \sigma_{XX}^{-1}\sigma_{XY}$
$\sigma_{ee} = \sigma_{YY}$	$0 \leqslant \sigma_{ee} \leqslant \sigma_{YY} - \sigma_{XX}^{-1}\sigma_{XY}^2$	$0 \leqslant \sigma_{ee} \leqslant \sigma_{YY} - \sigma_{XX}^{-1}\sigma_{XY}^2$
$0 \leqslant \sigma_{uu} \leqslant \sigma_{XX}$	$0 \leqslant \sigma_{uu} \leqslant \sigma_{XX} - \sigma_{YY}^{-1}\sigma_{XY}^2$	$0 \leqslant \sigma_{uu} \leqslant \sigma_{XX} - \sigma_{YY}^{-1}\sigma_{XY}^2$
$\mu_x = \mu_X$	$\mu_x = \mu_X$	$\mu_x = \mu_X$
$\beta_0 = \mu_Y$	$\mu_Y - \sigma_{XY}^{-1}\sigma_{YY}\mu_X \leqslant \beta_0$	$\mu_Y - \sigma_{XX}^{-1}\sigma_{XY}\mu_X \leqslant \beta_0$
	$\leqslant \mu_Y - \sigma_{XX}^{-1}\sigma_{XY}\mu_X$	$\leqslant \mu_Y - \sigma_{XY}^{-1}\sigma_{YY}\mu_X$

[a] The covariance σ_{eu} is zero and $\sigma_{xx} > 0$.

REFERENCES

Allen (1939), Cochran (1968), Fisher (1966), Moran (1971), Spiegelman (1982), U.S. Department of Commerce (1975).

EXERCISES

1. (Section 1.1.1) Assume that model (1.1.1)–(1.1.3) holds with $\sigma_{xx} = 1$, $\sigma_{ee} = 0.1$, and $n = 100$. Compute the power of the regression t test (0.05 level) of the hypothesis $\beta_1 = 0$ against H_A: $\beta_1 \neq 0$ at $\beta_1 = 0.063$ for $\sigma_{uu} = 0$ and for $\sigma_{uu} = 0.5$. You may approximate the noncentral t distribution by the normal distribution with mean $E\{\hat{\gamma}_{1\ell}\}[n^{-1}E\{s_\ell^2\}]^{-1/2}$ and variance 1.

2. (Section 1.1.1) Let

$$Y_t = \beta_0 + \beta_1 x_t + e_t, \qquad X_t = x_t + u_t,$$

$$(x_t, e_t, u_t)' \sim \mathrm{NI}[(\mu_x, 0, 0)', \mathrm{diag}(\sigma_{xx}, \sigma_{ee}, \sigma_{uu})],$$

$$a_t = Y_t - E\{Y_t\} - \gamma_1(X_t - E\{X_t\}), \quad \text{and} \quad \gamma_1 = \sigma_{XX}^{-1}\sigma_{XY}.$$

(a) Show that $V\{a_t\} = \sigma_{ee} + \gamma_1^2\sigma_{uu} + (\beta_1 - \gamma_1)^2\sigma_{xx}$.

(b) Let $\hat{\gamma}_{1\ell} = m_{XX}^{-1}m_{XY}$. Give the unconditional variance of $\hat{\gamma}_{1\ell}$. Is it possible for the unconditional variance of $\hat{\gamma}_{1\ell}$ to be smaller than that of $m_{xx}^{-1}m_{xY}$?

3. (Section 1.1.3) In discussing the identification of models we often compared the number of unknown parameters of the model with the dimension of the minimal sufficient statistic. If a model is to be identified, it is necessary that the number of parameters be no greater than the dimension of the minimal sufficient statistic. Construct an example to demonstrate that not all parameters need be identified if the number of unknown parameters is less than or equal to the dimension of the minimal sufficient statistic.

4. (Section 1.1.3) Assume the model $y_t = \beta_0 + x_t\beta_1$, with

$$(x_t, e_t, u_t)' \sim \mathrm{NI}[(\mu_x, 0, 0)', \text{block diag}(\sigma_{xx}, \boldsymbol{\Sigma}_{\epsilon\epsilon})]$$

and $(Y_t, X_t) = (y_t, x_t) + (e_t, u_t)$. Discuss carefully the identification status for cases (a), (b), and (c), giving the parameter space in each case. In all cases all parameters except those specified are unknown.

(a) The parameters β_0, σ_{eu}, and σ_{uu} are known.
(b) The parameters σ_{ee} and σ_{uu} are known.
(c) The parameters β_0, σ_{ee}, and σ_{uu} are known.

5. (Section 1.1.2) Assume that the twelve (Y_t, X_t) pairs (3.6, 3.9), (2.5, 2.9), (3.9, 4.4), (5.0, 5.9), (4.9, 5.4), (4.5, 4.2), (2.9, 2.3), (5.2, 4.5), (2.7, 3.5), (5.8, 6.0), (4.1, 3.3), and (5.1, 4.1) satisfy model (1.1.1)–(1.1.3) with $\kappa_{xx} = 0.85$. Estimate (β_0, β_1). Estimate the covariance matrix of your estimators.

6. (Section 1.1.3) Let the following model hold:

$$Y_t = \beta_0 + \sum_{j=1}^{p} x_{tj}\beta_j + e_t,$$

$$(e_t, u_t, \mathbf{x}_t)' \sim \mathrm{NI}[(0, 0, \boldsymbol{\mu}_x)', \text{block diag}(\sigma_{ee}, \sigma_{uu}, \boldsymbol{\Sigma}_{xx})],$$

where $X_{t1} = x_{t1} + u_t$. Obtain an expression for the expected value of the coefficient for X_{t1} in the ordinary least squares regression of Y_t on $X_{t1}, x_{t2}, x_{t3}, \ldots, x_{tp}$. Establish bounds for β_1 in terms of the population covariance matrix of $(Y_t, \mathbf{X}_t)$.

7. (Section 1.1.2, Appendix 1.B) Let

$$[(y_t, x_t), (e_t, u_t)]' \sim \text{NI}[(1, 0, 0, 0)', \text{block diag}(\boldsymbol{\Sigma}_{zz}, 0.2\mathbf{I})],$$

where $\sigma_{xx} = \sigma_{yy} = 1$, $\sigma_{xy} = \rho$, and $(Y_t, X_t) = (y_t, x_t) + (e_t, u_t)$.

(a) What is the reliability ratio of X_t? Of $X_t + Y_t$?

(b) What is the reliability ratio of Y_t^2? Of X_t^2? Of $X_t Y_t$?

(c) What is the reliability ratio of Y_t^3? Of X_t^3? Of $X_t^2 Y_t$?

8. (Section 1.1.2) Verify the equation used to construct the last column of Table 1.1.1 by showing that the left side of Equation (1.1.18) is the standardized mean squared error of $\hat{\gamma}_{1\ell}$ and that the right side of (1.1.18) is the standardized mean squared error of $\hat{\beta}_1$, both as estimators of β_1.

9. (Section 1.1.3) Assume the model

$$Y_t = \beta_0 + \beta_1 x_t + e_t, \qquad X_t = x_t + u_t,$$
$$(x_t, e_t, u_t)' \sim \text{NI}[(\mu_x, 0, 0)', \text{block diag}(\sigma_{xx}, \boldsymbol{\Sigma}_{\epsilon\epsilon})],$$

with σ_{xx}, σ_{ee}, σ_{eu}, and σ_{uu} unknown and $\sigma_{xx} > 0$. Show that, given any positive definite covariance matrix for (Y_t, X_t), the interval $(-\infty, \infty)$ is the set of possible values for β_1.

10. (Section 1.1.2, Appendix 1.A) Show that $\hat{\beta}_1$ of (1.1.7) satisfies

$$\hat{\beta}_1 - \beta_1 = (n-1)^{-1}\sigma_{xx}^{-1} \sum_{t=1}^{n} (X_t - \bar{X})(r_t - \bar{r}) + O_p(n^{-1}),$$

where $r_t = Y_t - \gamma_0 - \gamma_1 X_t$ and $\gamma_0 = E\{Y_t - \gamma_1 X_t\}$. Hence, $\bar{v}$ and $\hat{\beta}_1 - \beta_1$, where $\bar{v}$ is defined in (1.1.14), are independent in the limit.

11. (Section 1.1.2, Appendix 1.A) In Section 1.1.2 it was assumed that $\kappa_{xx} = \sigma_{XX}^{-1}\sigma_{xx}$ was known. Prove the following theorem.

Theorem. Let model (1.1.1)–(1.1.3) hold and let $\hat{\kappa}_{xx}$ be an estimator of κ_{xx}, where $\hat{\kappa}_{xx}$ is independent of (Y_t, X_t), $t = 1, 2, \ldots$, and

$$n^{1/2}(\hat{\kappa}_{xx} - \kappa_{xx}) \xrightarrow{L} N(0, \sigma_{\kappa\kappa}).$$

Let $\hat{\beta}_1 = \hat{\kappa}_{xx}^{-1}\hat{\gamma}_{1\ell}$. Then

$$n^{1/2}(\hat{\beta}_1 - \beta_1) \xrightarrow{L} N[0, \kappa_{xx}^{-2}\{\sigma_{XX}^{-1}(\sigma_{YY} - \gamma_1\sigma_{XY}) + \kappa_{xx}^{-2}\gamma_1^2\sigma_{\kappa\kappa}\}].$$

1.2. MEASUREMENT VARIANCE KNOWN

1.2.1. Introduction and Estimators

In this section, we retain the normal distribution model introduced in Section 1.1.1. That is, we assume

$$Y_t = \beta_0 + \beta_1 x_t + e_t, \qquad X_t = x_t + u_t, \tag{1.2.1}$$
$$(x_t, e_t, u_t)' \sim \text{NI}[(\mu_x, 0, 0)', \text{diag}(\sigma_{xx}, \sigma_{ee}, \sigma_{uu})].$$

The first equation of (1.2.1) is a classical regression specification, but the true explanatory variable x_t is not observed directly. The observed measure of x_t, denoted by X_t, may be obtained by asking people questions, by reading an imperfect instrument, or by performing a laboratory analysis. It is assumed that the variance of the measurement error, σ_{uu}, has been determined, perhaps by making a large number of independent repeated measurements. Estimators of the remaining parameters will be derived under the assumption that σ_{uu} is known.

Because $\mathbf{Z}_t = (Y_t, X_t)$ is bivariate normal, the sample mean $\bar{\mathbf{Z}} = (\bar{Y}, \bar{X})$ and sample covariances (m_{YY}, m_{XY}, m_{XX}), where, for example,

$$m_{XY} = (n-1)^{-1} \sum_{t=1}^{n} (X_t - \bar{X})(Y_t - \bar{Y}),$$

form a set of sufficient statistics for estimation of the parameters. If the parameter vector is identified, the maximum likelihood estimator will be a function of these statistics. See Kendall and Stuart (1979 Vol. 2, Chaps. 17 and 23). If there are no parametric restrictions on the covariance matrix of $\mathbf{Z}_t$, then $n^{-1}(n-1)\mathbf{m}_{ZZ}$ is the maximum likelihood estimator of the covariance matrix of $\mathbf{Z}_t$, where

$$\mathbf{m}_{ZZ} = (n-1)^{-1} \sum_{t=1}^{n} (\mathbf{Z}_t - \bar{\mathbf{Z}})'(\mathbf{Z}_t - \bar{\mathbf{Z}}).$$

We shall use the unbiased estimator $\mathbf{m}_{ZZ}$ in our discussion. We call $\mathbf{m}_{ZZ}$ the maximum likelihood estimator adjusted for degrees of freedom.

Recall that, under model (1.2.1), the population moments of (Y_t, X_t) satisfy

$$(\sigma_{YY}, \sigma_{XY}, \sigma_{XX}) = (\beta_1^2\sigma_{xx} + \sigma_{ee}, \beta_1\sigma_{xx}, \sigma_{xx} + \sigma_{uu})$$

and

$$(\mu_Y, \mu_X) = (\beta_0 + \beta_1\mu_x, \mu_x). \tag{1.2.2}$$

See (1.1.4). We create estimators of the unknown parameters by replacing the unknown population moments on the left side of (1.2.2) with their sample estimators to obtain a system of equations in the unknown parameters. Solving, we have

$$\begin{aligned} \hat{\beta}_1 &= (m_{XX} - \sigma_{uu})^{-1} m_{XY}, \qquad (1.2.3) \\ (\hat{\sigma}_{xx}, \hat{\sigma}_{ee}) &= (m_{XX} - \sigma_{uu}, m_{YY} - \hat{\beta}_1 m_{XY}), \\ (\hat{\mu}_x, \hat{\beta}_0) &= (\bar{X}, \bar{Y} - \hat{\beta}_1\bar{X}). \end{aligned}$$

The knowledge of σ_{uu} has enabled us to construct a one-to-one mapping from the minimal sufficient statistic to the vector

$$(\hat{\mu}_x, \hat{\sigma}_{xx}, \hat{\beta}_0, \hat{\beta}_1, \hat{\sigma}_{ee}).$$

For the quantities defined in (1.2.3) to be proper estimators (i.e., to be in the parameter space), $\hat{\sigma}_{xx}$ and $\hat{\sigma}_{ee}$ must be nonnegative. The estimators of σ_{xx} and σ_{ee} in (1.2.3) will be positive if and only if

$$m_{YY}(m_{XX} - \sigma_{uu}) - m_{YX}^2 > 0. \tag{1.2.4}$$

Estimators for samples in which (1.2.4) is violated are $\hat{\sigma}_{ee} = 0$,

$$\hat{\beta}_1 = m_{XY}^{-1} m_{YY}, \quad \text{and} \quad \hat{\sigma}_{xx} = m_{YY}^{-1} m_{XY}^2,$$

with $(\hat{\mu}_x, \hat{\beta}_0)$ defined by the last equation of (1.2.3).

1.2.2. Sampling Properties of the Estimators

The sampling behavior of the estimator $\hat{\beta}_1$ defined in (1.2.3) is not obtained easily. While the properties of the normal distribution can be used to derive the distribution of the estimator conditional on the observed X values, the expressions are not particularly useful because the conditional mean is a function of the X_t and the X_t are functions of the measurement errors. (See Exercise 1.13.) Therefore, it seems necessary to use large sample theory to develop an approximation to the distribution of $\hat{\beta}_1$.

The sample moments (m_{YY}, m_{XY}, m_{XX}) are unbiased for the population moments and all have variances that are decreasing at the rate n^{-1}. These properties enable us to obtain the limiting distribution of the estimators.

Theorem 1.2.1. Let model (1.2.1) hold with σ_{uu} known, $\sigma_{ee} > 0$, and $\sigma_{xx} > 0$. Then the vector $n^{1/2}[(\hat{\beta}_0 - \beta_0), (\hat{\beta}_1 - \beta_1)]$, where $(\hat{\beta}_0, \hat{\beta}_1)$ is defined in (1.2.3), converges in distribution to a normal vector random variable with zero mean and covariance matrix

$$\mathbf{\Gamma} = \begin{bmatrix} \mu_x^2 \sigma_{xx}^{-2}(\sigma_{XX}\sigma_{vv} + \sigma_{Xv}^2) + \sigma_{vv} & -\mu_x \sigma_{xx}^{-2}(\sigma_{XX}\sigma_{vv} + \sigma_{Xv}^2) \\ -\mu_x \sigma_{xx}^{-2}(\sigma_{XX}\sigma_{vv} + \sigma_{Xv}^2) & \sigma_{xx}^{-2}(\sigma_{XX}\sigma_{vv} + \sigma_{Xv}^2) \end{bmatrix}, \tag{1.2.5}$$

where $v_t = Y_t - \beta_0 - X_t\beta_1 = e_t - u_t\beta_1$ and $\sigma_{Xv} = \sigma_{uv} = -\beta_1\sigma_{uu}$. Furthermore, $n\hat{\mathbf{V}}\{(\hat{\beta}_0, \hat{\beta}_1)'\}$ converges in probability to $\mathbf{\Gamma}$, where

$$\hat{\mathbf{V}}\{(\hat{\beta}_0, \hat{\beta}_1)'\} = \begin{bmatrix} \bar{X}^2\hat{V}\{\hat{\beta}_1\} + n^{-1}s_{vv} & -\bar{X}\hat{V}\{\hat{\beta}_1\} \\ -\bar{X}\hat{V}\{\hat{\beta}_1\} & \hat{V}\{\hat{\beta}_1\} \end{bmatrix}, \tag{1.2.6}$$

$$\hat{V}\{\hat{\beta}_1\} = (n-1)^{-1}\hat{\sigma}_{xx}^{-2}(m_{XX}s_{vv} + \hat{\beta}_1^2\sigma_{uu}^2), \tag{1.2.7}$$

$$s_{vv} = (n-2)^{-1}\sum_{t=1}^{n}[Y_t - \bar{Y} - (X_t - \bar{X})\hat{\beta}_1]^2,$$

$\hat{\sigma}_{xx} = m_{XX} - \sigma_{uu}$, and $\hat{\sigma}_{uv} = -\hat{\beta}_1\sigma_{uu}$.

Proof. Under the normality assumption, m_{XY} is unbiased for σ_{XY} and

$$V\{m_{XY}\} = (n-1)^{-1}(\sigma_{XX}\sigma_{YY} + \sigma_{XY}^2).$$

(See Appendix 1.B for the moments of the normal distribution.) Because the sample moments are converging in probability to the population moments, we can expand $\hat{\beta}_1$ in a Taylor series about the population values to obtain

$$\begin{aligned}\hat{\beta}_1 &= (\sigma_{XX} - \sigma_{uu})^{-1}\sigma_{XY} + (\sigma_{XX} - \sigma_{uu})^{-1}(m_{XY} - \sigma_{XY}) \\ &\quad -(\sigma_{XX} - \sigma_{uu})^{-2}\sigma_{XY}(m_{XX} - \sigma_{XX}) + O_p(n^{-1}).\end{aligned}$$

(See Appendix 1.A.) After algebraic simplification, this expression can be written

$$n^{1/2}(\hat{\beta}_1 - \beta_1) = n^{1/2}\sigma_{xx}^{-1}(m_{Xv} - \sigma_{Xv}) + O_p(n^{-1/2}),$$

where $m_{Xv} = (n-1)^{-1}\sum_{t=1}^{n}(X_t - \bar{X})v_t$. It follows that the limiting distribution of $n^{1/2}(\hat{\beta}_1 - \beta_1)$ is the same as the limiting distribution of

$$n^{1/2}\sigma_{xx}^{-1}(m_{Xv} - \sigma_{uv}) = n^{1/2}\sigma_{xx}^{-1}(m_{xv} + m_{uv} - \sigma_{uv}).$$

In a similar manner

$$\begin{aligned}\hat{\beta}_0 &= \bar{Y} - \hat{\beta}_1\bar{X} = \beta_0 + \beta_1\bar{x} + \bar{e} - \hat{\beta}_1(\bar{x} + \bar{u}) \\ &= \beta_0 - (\hat{\beta}_1 - \beta_1)\mu_x + \bar{v} + O_p(n^{-1}),\end{aligned}$$

where $\bar{v} = \bar{e} - \beta_1\bar{u}$, and the limiting distribution of $n^{1/2}(\hat{\beta}_0 - \beta_0)$ is the same as that of

$$n^{1/2}[\bar{v} - (\hat{\beta}_1 - \beta_1)\mu_x].$$

By the properties of the normal distribution, the covariance between m_{Xv} and $\bar{v}$ is zero and, by Corollary 1.C.1 of Appendix 1.C, the limiting distribution of $n^{1/2}(m_{Xv} - \sigma_{uv}, \bar{v})$ is bivariate normal. The limiting distribution of $n^{1/2}[(\hat{\beta}_0, \hat{\beta}_1)' - (\beta_0, \beta_1)']$ follows because, to the order required, the error in the estimators is a linear function of $(m_{Xv} - \sigma_{uv}, \bar{v})$.

By the normal moment properties, $\hat{\sigma}_{xx} = \sigma_{xx} + O_p(n^{-1/2})$, and we have shown that $\hat{\beta}_1 - \beta_1 = O_p(n^{-1/2})$. Therefore,

$$\begin{aligned}s_{vv} &= (n-2)^{-1}\sum_{t=1}^{n}(v_t - \bar{v})^2 + O_p(n^{-1/2}) \\ &= \sigma_{vv} + O_p(n^{-1/2}).\end{aligned}$$

It follows that

$$n\hat{\mathbf{V}}\{\hat{\beta}_1\} = \sigma_{xx}^{-1}\sigma_{vv} + \sigma_{xx}^{-2}(\sigma_{uu}\sigma_{vv} + \beta_1^2\sigma_{uu}^2) + O_p(n^{-1/2}).$$

Similar results hold for the remaining two entries of (1.2.6). □

The random variable v_t, introduced in Theorem 1.2.1, will be central in our study of measurement error models. Its role is analogous to that of the deviation from the population regression line in ordinary fixed-x regression models. This analogy can be made more apparent if we substitute $x_t = X_t - u_t$ into the first equation of (1.2.1) to obtain

$$Y_t = \beta_0 + \beta_1 X_t + v_t.$$

The v_t differs from the error in the ordinary fixed-x regression model because X_t and v_t are correlated.

In Theorem 1.2.1 it is assumed that the true values x_t are normally distributed. The theorem holds if (e_t, u_t) is independent of x_t and the x_t satisfy mild regularity conditions. Normality of the error vector (e_t, u_t) permits us to give an explicit expression for the covariance matrix of the estimators in terms of second moments of the distribution of the original variables. The approximations based on normality remain useful for error distributions displaying modest departures from normality, provided the error variances are small relative to the variance of x. Also see Theorem 2.2.1 and Section 3.1.

The variance of the limiting distribution of $n^{1/2}(\hat{\beta}_1 - \beta_1)$ is considerably larger than the corresponding variance for the ordinary least squares regression coefficient constructed with X_t. The difference is due to several sources. The divisor, $(m_{XX} - \sigma_{uu})$, has a smaller expectation than the divisor of the ordinary least squares estimator and the variance of v_t is larger than the variance about the ordinary least squares line.

The quantity $n^{1/2}$ was used to standardize the estimators in obtaining the limiting distributions. We choose to use the divisor $(n - 1)$ in the estimated variance (1.2.7) because the variance estimator so defined reduces to the ordinary least squares variance estimator when $\sigma_{uu} = 0$. For the same reason we used the divisor $(n - 2)$ in the definition of s_{vv}. The use of the divisor $(n - 2)$ in the estimator of σ_{vv} leads to an internal inconsistency with the estimator of σ_{ee} defined in (1.2.3) because $\sigma_{vv} = \sigma_{ee} + \beta_1^2\sigma_{uu}$.

The estimator $n\hat{\mathbf{V}}\{(\hat{\beta}_0, \hat{\beta}_1)'\}$ is an estimator of the covariance matrix of the limiting distribution of $n^{1/2}(\hat{\beta}_0 - \beta_0, \hat{\beta}_1 - \beta_1)$. Because the expectation of $\hat{\beta}_1$ is not defined, it is not technically correct to speak of (1.2.7) as an estimator of the variance of $\hat{\beta}_1$. It is an estimator of the variance of the approximating distribution.

Because $n\hat{\mathbf{V}}\{(\hat{\beta}_0, \hat{\beta}_1)'\}$ is a consistent estimator of (1.2.5), it follows that

$$t = [\hat{V}\{\hat{\beta}_1\}]^{-1/2}(\hat{\beta}_1 - \beta_1) \tag{1.2.8}$$

is approximately distributed as a $N(0, 1)$ random variable. In practice it seems reasonable to approximate the distribution of (1.2.8) with the distribution of Student's t with $n - 2$ degrees of freedom. Care is required when using this approximation. Throughout, we have assumed $\sigma_{xx} > 0$. If σ_{xx} is small relative to the variance of $m_{XX} - \sigma_{uu}$, the approximations of Theorem 1.2.1 will not perform well. A test of the hypothesis that $\sigma_{xx} = 0$ is given by the statistic

$$\chi^2 = \sigma_{uu}^{-1} \sum_{t=1}^{n} (X_t - \bar{X})^2. \tag{1.2.9}$$

If $\sigma_{xx} = 0$, the distribution of χ^2 is that of a chi-square random variable with $n - 1$ degrees of freedom. For the approximations of Theorem 1.2.1 and Equation (1.2.8) to perform well, the population analogue of (1.2.9) should be large. That is, $(n-1)\sigma_{uu}^{-1}\sigma_{XX}$ should be large. As a rule of thumb, one might judge Student's t distribution with $n-2$ degrees of freedom an adequate approximation to the distribution of (1.2.8) if

$$(n-1)^{-1}(m_{XX}-\sigma_{uu})^{-2}\sigma_{uu}^2 < 0.001. \tag{1.2.10}$$

This means that the approximation will perform well if n is large and (or) σ_{uu} is small. A partial theoretical justification for the rule of thumb is developed in Exercise 1.59. If the χ^2 value of (1.2.9) is small compared to the tabular value of the chi-square distribution with $n-1$ degrees of freedom, other testing and confidence interval methods for β_1 should be considered. See Section 2.5.1.

Example 1.2.1. To illustrate the computation of estimates when σ_{uu} is known, we use a small data set adapted from Voss (1969) and presented by DeGracie and Fuller (1972). The data given in Table 1.2.1 are yields of corn and determinations of available soil nitrogen collected at 11 sites on Marshall soil in Iowa. The estimates of soil nitrogen contain measurement error arising from two sources. First, only a small sample of soil is selected from each plot and, as a result, there is the sampling error associated with the use of the sample to represent the whole. Second, there is a measurement error associated with the chemical analysis used to determine the level of nitrogen in the soil sample. The variance arising from these two sources has been estimated to be $\sigma_{uu}=57$. For the purposes of our example, we assume σ_{uu} is known and that model (1.2.1) holds.

The statistics associated with the data of Table 1.2.1 are

$$(\bar{Y}, \bar{X}) = (97.4545, 70.6364),$$

$$\begin{bmatrix} m_{YY} & m_{YX} \\ m_{XY} & m_{XX} \end{bmatrix} = \begin{bmatrix} 87.6727 & 104.8818 \\ 104.8818 & 304.8545 \end{bmatrix}. \tag{1.2.11}$$

TABLE 1.2.1. Yields of corn on Marshall soil in Iowa

Site	Yield (Y)	Soil Nitrogen (X)	Site	Yield (Y)	Soil Nitrogen (X)
1	86	70	7	99	50
2	115	97	8	96	70
3	90	53	9	99	94
4	86	64	10	104	69
5	110	95	11	96	51
6	91	64			

To investigate the magnitude of the variance of the unobserved x_t, we compute

$$\sigma_{uu}^{-1} m_{XX} = (57)^{-1} 304.8545 = 5.35. \tag{1.2.12}$$

If $\sigma_{xx} = 0$, this ratio is distributed as Snedecor's F with 10 and infinity degrees of freedom. Because of the large value of the ratio relative to the tabular value, we retain our assumption that $\sigma_{xx} > 0$. The determinant

$$\begin{vmatrix} m_{YY} & m_{YX} \\ m_{XY} & m_{XX} - \sigma_{uu} \end{vmatrix} = \begin{vmatrix} 87.6727 & 104.8818 \\ 104.8818 & 247.8545 \end{vmatrix} = 10729.88$$

and the estimators (1.2.3) are in the parameter space. We have

$$\begin{aligned}
\hat{\sigma}_{xx} &= m_{XX} - \sigma_{uu} = 304.8545 - 57 = 247.8545, \\
\hat{\beta}_1 &= [m_{XX} - \sigma_{uu}]^{-1} m_{XY} = (247.8545)^{-1} 104.8818 = 0.4232, \\
\hat{\sigma}_{ee} &= m_{YY} - \hat{\beta}_1^2 \hat{\sigma}_{xx} = 87.6727 - (0.4232)^2 (247.8545) = 43.2910, \\
\hat{\beta}_0 &= \bar{Y} - \hat{\beta}_1 \bar{X} = 97.4545 - (0.4232)(70.6364) = 67.5613.
\end{aligned}$$

The slope of the measurement error line is $\hat{\beta}_1$ and the line passes through the sample mean $(\bar{Y}, \bar{X})$.

The estimated covariance matrix (1.2.6) is

$$\hat{\mathbf{V}}\{(\hat{\beta}_0, \hat{\beta}_1)'\} = \begin{bmatrix} 157.3153 & -2.1506 \\ -2.1506 & 0.0304 \end{bmatrix}, \tag{1.2.13}$$

where $s_{vv} = 59.4440$. Because the sample is small and $m_{XX}^{-1}\sigma_{uu}$ is of moderate size, the distribution of the statistic in (1.2.8) may deviate considerably from that of Student's t. Nevertheless, we use Theorem 1.2.1 and (1.2.8) to establish (0.0288, 0.8176) as an approximate 95% confidence interval for β_1. By analogy to ordinary least squares theory, we used $t = 2.262$ from the table of Student's t with nine degrees of freedom to construct the interval.

Because our distribution theory is approximate, no simple relationship exists between the test of the hypothesis H_0: $\beta_1 = 0$ and the approximate standard error of the estimate of β_1. Recall that the ordinary least squares coefficient can be used to test the hypothesis that $\beta_1 = 0$. For these data the least squares test statistic for the hypothesis $\beta_1 = 0$ is

$$t_\ell = (0.0188)^{-1/2} 0.3440 = 2.51, \tag{1.2.14}$$

where 0.3440 is the ordinary least squares regression coefficient. If we use $\hat{\beta}_1$ and the standard error computed from $\hat{\mathbf{V}}\{(\hat{\beta}_0, \hat{\beta}_1)'\}$, we have

$$t_a = (0.0304)^{-1/2} 0.4232 = 2.43. \tag{1.2.15}$$

The two "t statistics" are similar but not identical. The least squares statistic is the preferred method for testing the hypothesis that $\beta_1 = 0$ but is appropriate for that hypothesis only. The statistic (1.2.8) or the likelihood ratio statistic discussed in Section 2.5 should be used to set approximate confidence intervals for β_1 or to test H_0: $\beta_1 = \beta_1^0$, when $\beta_1^0 \neq 0$.

The computed statistics enable one to estimate the correlation between yield Y_t and true soil nitrogen x_t by

$$\hat{R}_{xY}^2 = (\hat{\sigma}_{xx} m_{YY})^{-1} m_{XY}^2 = 0.5062. \qquad \square\,\square$$

1.2.3. Estimation of True x Values

In this section we construct an estimator of the x value that generated the vector (Y_t, X_t). The prediction of Y_t, given an observation on X_t, is discussed in Section 1.6. With knowledge of the parameters of the structural relationship and an observation (Y_t, X_t), it is possible to construct an estimator of x_t superior to X_t. In constructing an estimator of x_t, the unknown x_t can be treated as fixed or random, and the two assumptions lead to different procedures.

We begin by treating the x_t as fixed unknown constants. Let the error assumptions of model (1.2.1) hold and let β_0, β_1, σ_{ee}, and σ_{uu} be known. Then we can write

$$\begin{bmatrix} Y_t - \beta_0 \\ X_t \end{bmatrix} = \begin{bmatrix} \beta_1 \\ 1 \end{bmatrix} x_t + \begin{bmatrix} e_t \\ u_t \end{bmatrix}. \tag{1.2.16}$$

Equation (1.2.16) is given in the form of the classical regression model, where x_t is the unknown parameter to be estimated. Therefore, if we treat x_t as a fixed unknown constant, the best linear unbiased estimator of x_t is given by the generalized least squares estimator

$$\ddot{x}_t = [(\beta_1, 1)\,\Sigma_{\varepsilon\varepsilon}^{-1}(\beta_1, 1)']^{-1}(\beta_1, 1)\Sigma_{\varepsilon\varepsilon}^{-1}(Y_t - \beta_0, X_t)', \tag{1.2.17}$$

where $\Sigma_{\varepsilon\varepsilon}$ is the covariance matrix of $\varepsilon_t = (e_t, u_t)$. If $\Sigma_{\varepsilon\varepsilon} = \text{diag}(\sigma_{ee}, \sigma_{uu})$, the expression for $\ddot{x}_t$ reduces to

$$\ddot{x}_t = [\beta_1^2\sigma_{uu} + \sigma_{ee}]^{-1}[\sigma_{uu}(Y_t - \beta_0)\beta_1 + \sigma_{ee}X_t].$$

The variance of estimator (1.2.17) is

$$V\{\ddot{x}_t - x_t\} = [(\beta_1, 1)\Sigma_{\varepsilon\varepsilon}^{-1}(\beta_1, 1)']^{-1}, \tag{1.2.18}$$

which, for diagonal $\Sigma_{\varepsilon\varepsilon}$, reduces to

$$V\{\ddot{x}_t - x_t\} = [\beta_1^2\sigma_{ee}^{-1} + \sigma_{uu}^{-1}]^{-1}.$$

A representation for $\ddot{x}_t$ in terms of the variable v_t will be very useful. The quantity $\ddot{x}_t$ is a linear combination of Y_t and X_t, say

$$\ddot{x}_t = a_1(Y_t - \beta_0) + a_2 X_t, \tag{1.2.19}$$

with the properties that

$$E\{\ddot{x}_t | x_t\} = x_t, \quad \text{for all } x_t, \tag{1.2.20}$$

and $V\{\ddot{x}_t | x_t\}$ is a minimum for estimators of the form (1.2.19). Now v_t is also a linear combination of Y_t and X_t, so that it must be possible to write

$$\ddot{x}_t = c_1 X_t + c_2 v_t + c_3. \tag{1.2.21}$$

To satisfy (1.2.20) for all x_t, we must have $c_1 = 1$ and $c_3 = 0$ because $E\{v_t | x_t\} = 0$. The variance of (1.2.21) is

$$V\{X_t + c_2 v_t | x_t\} = V\{u_t + c_2 v_t\}, \tag{1.2.22}$$

which is minimized for $c_2 = -\sigma_{vv}^{-1}\sigma_{uv}$, where $-c_2$ is the regression coefficient in the population regression of u_t on v_t. Therefore, we can write

$$\ddot{x}_t = X_t - \sigma_{vv}^{-1}\sigma_{uv} v_t. \tag{1.2.23}$$

Using this regression form of the estimator, we have

$$V\{\ddot{x}_t - x_t\} = \sigma_{uu} - \sigma_{vv}^{-1}\sigma_{uv}^2. \tag{1.2.24}$$

In most situations, β_0, β_1, and σ_{ee} will be estimated. Let $\hat{x}_t$ denote the estimator

$$\hat{x}_t = [(\hat{\beta}_1, 1)\hat{\Sigma}_{\varepsilon\varepsilon}^{-1}(\hat{\beta}_1, 1)']^{-1}(\hat{\beta}_1, 1)\hat{\Sigma}_{\varepsilon\varepsilon}^{-1}(Y_t - \hat{\beta}_0, X_t)', \tag{1.2.25}$$

where $\hat{\Sigma}_{\varepsilon\varepsilon} = \text{diag}(\hat{\sigma}_{ee}, \sigma_{uu})$ and $(\hat{\beta}_0, \hat{\beta}_1, \hat{\sigma}_{ee})$ is defined in (1.2.3). The error in the estimator $(\hat{\beta}_0, \hat{\beta}_1, \hat{\sigma}_{ee})$ is $O_p(n^{-1/2})$, and

$$\hat{x}_t - x_t = [(\beta_1, 1)\Sigma_{\varepsilon\varepsilon}^{-1}(\beta_1, 1)']^{-1}(\beta_1, 1)\Sigma_{\varepsilon\varepsilon}^{-1}(e_t, u_t)' + O_p(n^{-1/2}).$$

It follows that (1.2.18) can be used to approximate the variance of the estimation error in $\hat{x}_t$, and the estimator

$$\hat{V}\{\hat{x}_t - x_t\} = [(\hat{\beta}_1, 1)\hat{\Sigma}_{\varepsilon\varepsilon}^{-1}(\hat{\beta}_1, 1)']^{-1} \tag{1.2.26}$$

is a consistent estimator of the variance of $\hat{x}_t$. Using (1.2.23), the estimator $\hat{x}_t$ can be written

$$\hat{x}_t = X_t - \hat{\sigma}_{vv}^{-1}\hat{\sigma}_{uv}\hat{v}_t, \tag{1.2.27}$$

where $\hat{v}_t = Y_t - \hat{\beta}_0 - X_t\hat{\beta}_1$, $\hat{\sigma}_{uv} = \sigma_{eu} - \hat{\beta}_1\sigma_{uu}$, and

$$\hat{\sigma}_{vv} = m_{YY} - 2\hat{\beta}_1 m_{XY} + \hat{\beta}_1^2 m_{XX} = (n-1)^{-1}\sum_{t=1}^{n} \hat{v}_t^2.$$

In (1.2.27) we defined $\hat{x}_t$ using $\hat{\sigma}_{vv}$, the estimator of σ_{vv} constructed with a divisor of $(n-1)$, instead of s_{vv}, which has a divisor of $(n-2)$. The forms (1.2.25) and (1.2.27) are identical with $\hat{\sigma}_{vv}$ as the divisor. An alternative form for the estimator of variance that is equivalent to the one in (1.2.26) is

$$\hat{V}\{\hat{x}_t - x_t\} = \sigma_{uu} - \hat{\sigma}_{vv}^{-1}\hat{\sigma}_{uv}^2. \tag{1.2.28}$$

It follows from (1.2.23) that $\ddot{x}_t$ and v_t are uncorrelated. This property is mirrored in the estimators in that

$$\sum_{t=1}^{n} \hat{x}_t \hat{v}_t = 0. \tag{1.2.29}$$

The estimation procedure transforms the original vector (Y_t, X_t) into two uncorrelated parts $\hat{x}_t$ and $\hat{v}_t$. This is analogous to the result that the explanatory variables and the deviations from fit are uncorrelated in ordinary least squares regression. In the presence of measurement error it is necessary to estimate x_t and it is the estimated x_t that is uncorrelated with the residual.

In the preceding discussion of the estimation of x_t, the quantity x_t was treated as fixed. One also can predict x_t under the assumption that the x_t values are a random sample chosen from a population of x values. Under the assumption of random x_t, the population covariance matrix of (Y_t, X_t, x_t) is

$$\begin{bmatrix} \beta_1^2\sigma_{xx} + \sigma_{ee} & \beta_1\sigma_{xx} + \sigma_{eu} & \beta_1\sigma_{xx} \\ \beta_1\sigma_{xx} + \sigma_{eu} & \sigma_{xx} + \sigma_{uu} & \sigma_{xx} \\ \beta_1\sigma_{xx} & \sigma_{xx} & \sigma_{xx} \end{bmatrix}. \tag{1.2.30}$$

Therefore, under the assumption of normality, the expected value of x_t given $\mathbf{Z}_t$ is

$$\begin{aligned} E\{x_t \mid \mathbf{Z}_t\} &= \mu_x + \gamma_1(Y_t - \mu_y) + \gamma_2(X_t - \mu_x) \\ &= \gamma_0 + \gamma_1 Y_t + \gamma_2 X_t, \end{aligned} \tag{1.2.31}$$

where $\mathbf{Z}_t = (Y_t, X_t)$, $\gamma_0 = (1 - \gamma_2)\mu_x - \gamma_1\mu_y$, $\mu_y = \beta_0 + \beta_1\mu_x$, and

$$(\gamma_1, \gamma_2)' = \mathbf{\Sigma}_{ZZ}^{-1}(\beta_1\sigma_{xx}, \sigma_{xx})'.$$

Using the fact that $\mathbf{\Sigma}_{ZZ} = (\beta_1, 1)'\sigma_{xx}(\beta_1, 1) + \mathbf{\Sigma}_{\varepsilon\varepsilon}$, the expected value (1.2.31) can be written

$$E\{x_t \mid \mathbf{Z}_t\} = X_t - (\mathbf{Z}_t - \boldsymbol{\mu}_Z)\mathbf{\Sigma}_{ZZ}^{-1}\mathbf{\Sigma}_{Zu}, \tag{1.2.32}$$

where $\mathbf{\Sigma}_{Zu} = \mathbf{\Sigma}_{\varepsilon u} = (\sigma_{eu}, \sigma_{uu})'$ and $\boldsymbol{\varepsilon}_t = (e_t, u_t)$. The conditional variance of x_t is

$$V\{x_t \mid \mathbf{Z}_t\} = \sigma_{xx} - (\gamma_1, \gamma_2)(\beta_1\sigma_{xx}, \sigma_{xx})'.$$

It is interesting to compare the estimator (1.2.23) that treats x_t as fixed with the predictor (1.2.32) constructed by treating x_t as random. In both situations the estimator (predictor) is obtained by subtracting a predictor of u_t from X_t. If x_t is fixed, the predictor of u_t is constructed from the random variable v_t that contains no fixed part. If x_t is random, the best predictor of u_t is constructed using the random vector (Y_t, X_t).

Treating x_t as random and replacing parameters with estimators, the predictor of x_t, given (Y_t, X_t), is

$$\begin{aligned} \tilde{x}_t &= \tilde{\gamma}_0 + \tilde{\gamma}_1 Y_t + \tilde{\gamma}_2 X_t \\ &= X_t - (\mathbf{Z}_t - \bar{Z})\mathbf{m}_{ZZ}^{-1}(\sigma_{eu}, \sigma_{uu})', \end{aligned} \tag{1.2.33}$$

where $\tilde{\gamma}_0 = (1 - \tilde{\gamma}_2)\bar{X} - \tilde{\gamma}_1 \bar{Y}$,

$$(\tilde{\gamma}_1, \tilde{\gamma}_2)' = \mathbf{m}_{ZZ}^{-1}[\mathbf{m}_{ZX} - (\sigma_{eu}, \sigma_{uu})'],$$

$$(\mathbf{m}_{ZZ}, \mathbf{m}_{ZX}) = (n - 1)^{-1} \sum_{t=1}^{n} (\mathbf{Z}_t - \bar{Z})'(\mathbf{Z}_t - \bar{Z}, X_t - \bar{X}).$$

As before, the use of estimators in constructing the predictor introduces an additional error that is $O_p(n^{-1/2})$. If one ignores this additional component, an estimator of the variance of the prediction error is

$$\hat{V}\{\tilde{x}_t - x_t | \mathbf{Z}_t\} = \sigma_{uu} - (\sigma_{eu}, \sigma_{uu})\mathbf{m}_{ZZ}^{-1}(\sigma_{eu}, \sigma_{uu})'. \tag{1.2.34}$$

Example 1.2.2. We use the estimates from Example 1.2.1 to construct an improved estimate of the soil nitrogen at each site. From Example 1.2.1, $\sigma_{uu} = 57$, $\hat{\sigma}_{ee} = 43.29$, $\hat{\beta}_0 = 67.56$, and $\hat{\beta}_1 = 0.4232$. Treating x_t as a fixed unknown constant, the estimator (1.2.25) is

$$\hat{x}_t = 0.4510(Y_t - 67.5613) + 0.8092X_t$$

and the estimated variance (1.2.26) is

$$\hat{V}\{\hat{x}_t - x_t\} = [\hat{\beta}_1^2 \hat{\sigma}_{ee}^{-1} + \sigma_{uu}^{-1}]^{-1} = 46.12.$$

Using the information available in Y_t reduces the variance of our estimate of soil nitrogen by about 20%, to a variance of 46 from a variance of 57. Table 1.2.2 contains the $\hat{x}_t$ and $\hat{v}_t$. The reader may verify that

$$(n - 1)^{-1} \sum_{t=1}^{n} (Y_t - \bar{Y})\hat{v}_t = 43.29 = \hat{\sigma}_{ee},$$

$$(n - 1)^{-1} \sum_{t=1}^{n} (X_t - \bar{X})\hat{v}_t = -24.13 = -\hat{\beta}_1 \sigma_{uu},$$

$$(n - 1)^{-1} \sum_{t=1}^{n} (\hat{x}_t - \bar{X})\hat{v}_t = 0.$$

We now construct the predictor of x_t under the assumption that the x_t are a random sample from a normal distribution. The estimated mean vector is

$$(\hat{\mu}_y, \hat{\mu}_x) = (\bar{Y}, \bar{X}) = (97.4545, 70.6364)$$

TABLE 1.2.2. Predicted soil nitrogen, estimated soil nitrogen, and deviations from fit

Site	Observed Yield Y_t	Observed Nitrogen X_t	$\hat{v}_t$	Estimated (Fixed x) $\hat{x}_t$	Predicted (Random x) $\tilde{x}_t$
1	86	70	−11.18	64.96	65.85
2	115	97	6.39	99.88	95.29
3	90	53	0.01	53.00	55.77
4	86	64	−8.65	60.10	61.75
5	110	95	2.24	96.01	92.03
6	91	64	−3.65	62.36	63.66
7	99	50	10.28	54.64	57.14
8	96	70	−1.18	69.47	69.65
9	99	94	−8.34	90.24	87.16
10	104	69	7.24	72.26	72.01
11	96	51	6.86	54.09	56.69

and by (1.2.33),

$$\begin{bmatrix} \tilde{\gamma}_1 \\ \tilde{\gamma}_2 \end{bmatrix} = \begin{bmatrix} 87.6727 & 104.8818 \\ 104.8818 & 304.8545 \end{bmatrix}^{-1} \begin{bmatrix} 104.8818 \\ 247.8545 \end{bmatrix} = \begin{bmatrix} 0.3801 \\ 0.6822 \end{bmatrix}.$$

It follows that the estimated prediction equation for x_t, given (Y_t, X_t) and treating x_t as random, is

$$\tilde{x}_t = -14.5942 + 0.3801 Y_t + 0.6822 X_t.$$

By (1.2.34), the estimated variance of the prediction error is

$$\hat{V}\{\tilde{x}_t - x_t \mid (Y_t, X_t)\} = 38.90.$$

If we are willing to treat the x_t as a random sample from a normal distribution and to condition on the observed $\mathbf{Z}_t$, the variance of the prediction error is about 84% of that of the estimate constructed treating x_t as fixed. The values of $\tilde{x}_t$ are compared to the values $\hat{x}_t$, where the $\hat{x}_t$ were computed treating x_t as fixed, in Table 1.2.2. Note that the $\tilde{x}_t$ values are all slightly closer to the mean of X than are the $\hat{x}_t$ values. See Exercise 1.18. □ □

In Example 1.2.2 we constructed estimators of the x_t values under the two assumptions, fixed x_t and random x_t. In fact, when we considered the x_t to be random, we assumed them to be normally distributed. In the fixed-x case, the x values are the same for every possible sample that we are considering. Therefore, if $\mathbf{x} = (x_1, x_2, \ldots, x_n)$ is fixed and one is constructing unbiased estimators of **x**, one is asking that the average of the estimator be **x**, where the average is over all samples in the population of samples of interest. In the soil nitrogen example the population of fixed-x samples are those samples always selected from the same set of 11 fields (sites).

It is possible to think of a sampling scheme wherein the fields of Example 1.2.2 are selected at random and then the soil sampled in each selected field. If we consider the total sampling scheme consisting of random selection of fields combined with random selection of soil within fields, we have a population of samples in which the x values are random. However, it is also legitimate to *consider* the subset of all possible samples that contain the 11 fields actually selected. That is, we may restrict our attention to the conditional distribution of sample outcomes conditional on the particular set of 11 fields that was selected. When we restrict attention to the subpopulation of samples of soil selected from the 11 fields, we are treating the x values for those 11 fields as fixed.

When the data are collected in an operation where fields are randomly selected, it is legitimate, in one context, to treat the x values as random—for example, to estimate the variance of x—and in another context to treat the x values as fixed—for example, to estimate the individual x values for the 11 fields. In the first context the population of interest is the overall population of fields from which the sample was randomly selected. In the second context the population of interest is the set of 11 fields.

1.2.4. Model Checks

In ordinary regression analysis it is good practice to plot the residuals from the fitted regression equation. For examples of such plots, see Draper and Smith (1981, Chap. 3). The construction of plots remains good practice when fitting measurement error models. The form of the plots for ordinary least squares and for measurement error models will differ somewhat because of the different nature of the measurement error problem. For an ordinary least squares problem, a common plot is that of the residuals against the independent variables. This plot often will give an indication of nonlinearity in the regression, of lack of homogeneity of the error variances, of nonnormality of the errors, or of outlier observations.

For a fitted measurement error model, the quantites that correspond most closely to the residual and to the independent variable of ordinary least squares are $\hat{v}_t$ and $\hat{x}_t$, respectively. A measurement error model such as that given by (1.2.1) postulates constant variance for e_t and u_t. Therefore, $v_t = e_t - \beta_1 u_t$ will also have constant variance. Because (e_t, u_t) is independent of x_t, the expected value of v_t given x_t is zero. The best estimator of x_t is $\ddot{x}_t$

defined in (1.2.17) and (1.2.23). Under normality, v_t and $\ddot{x}_t$ are independent. Therefore, the mean of v_t is zero for all $\ddot{x}_t$ and the variance of v_t is σ_{vv} for all $\ddot{x}_t$.

We are able to observe only $\hat{v}_t$ and $\hat{x}_t$, not v_t and $\ddot{x}_t$, but the properties of the estimators should be similar to the properties of the true variables. It is suggested that the plot of $\hat{v}_t$ against $\hat{x}_t$ be used in the same manner as the analogous ordinary least squares plot of deviations from fit against the explanatory variable. That is, the plot can be used to check for outliers, for nonlinearity, and for lack of variance homogeneity. Also, for samples of reasonable size, tests of normality can be applied to the $\hat{v}_t$. Estimators for models with heterogeneous error variances, for models with nonnormal errors, and for nonlinear models are discussed in Sections 3.1 and 3.2.

As in ordinary least squares, the variances of the $\hat{v}_t$ are not all the same. Because β_0 and β_1 are estimated, the variances of the $\hat{v}_t$ will differ by a quantity of order n^{-1}. An estimator of the approximate covariance matrix of $(\hat{v}_1, \hat{v}_2, \ldots, \hat{v}_n)$ is given in Section 2.2.3. Using the estimated variances of that section, one could plot the standardized quantities $[\hat{V}\{\hat{v}_t\}]^{-1/2}\hat{v}_t$ against $\hat{x}_t$. However, for most purposes, the plot of $\hat{v}_t$ against $\hat{x}_t$ is adequate.

Example 1.2.3. Figure 1.2.1 is a plot of the $\hat{v}$ values against the $\hat{x}$ values for the corn yield–soil nitrogen data, where the plotted values are taken from Table 1.2.2. This plot is for a small sample, but the plot contains no obvious anomalies. The mean and the range of the $\hat{v}$ values are similar for large, medium, and small $\hat{x}$ values. □ □

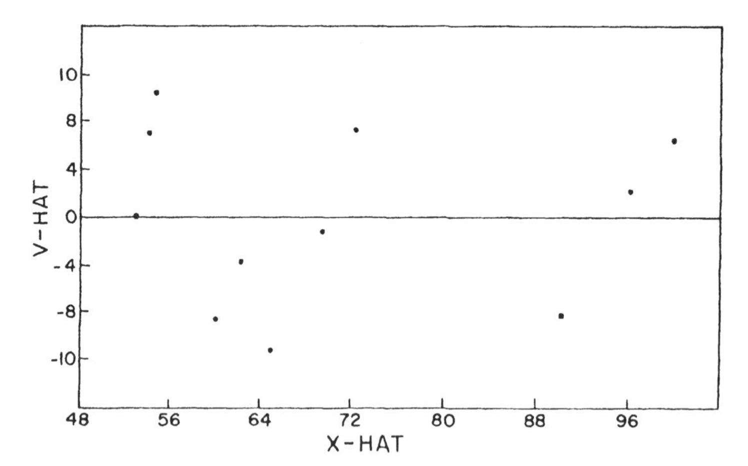

FIGURE 1.2.1. Residual plot for the corn–nitrogen data.

REFERENCES

Birch (1964), DeGracie and Fuller (1972), Fuller (1980), Kendall and Stuart (1979), Lord (1960), Madansky (1959), Neyman (1951), Neyman and Scott (1948).

EXERCISES

12. (Sections 1.1.3, 1.2)
   (a) Assume model (1.2.1) holds with $\beta_1 \neq 0$ and no information available on σ_{uu} and σ_{ee}. Extend the results of Section 1.1.3 by constructing a pair of estimators $\hat{\beta}_{0\ell}$ and $\hat{\beta}_{0L}$ such that the parameter β_0 is contained in the closed interval formed by plim $\hat{\beta}_{0\ell}$ and plim $\hat{\beta}_{0L}$.
   (b) Assume that the data below were generated by model (1.2.1).

t	X_t	Y_t	t	X_t	Y_t
1	3.58	3.94	11	3.88	3.32
2	2.52	2.44	12	5.93	5.70
3	3.85	3.95	13	5.84	5.25
4	5.02	5.93	14	4.50	4.24
5	4.94	5.37	15	5.81	4.61
6	4.49	4.55	16	3.67	3.63
7	2.57	2.60	17	5.94	4.20
8	5.05	3.81	18	3.54	4.03
9	1.84	1.75	19	4.34	4.22
10	5.82	6.04	20	5.25	4.33

   Using these data construct an interval in β_1 space such that the probability of β_1 being contained in a region so constructed is greater than or equal to 0.95.
   (c) Assume that the data listed below were generated by model (1.2.1). Construct a region in β_1 space such that the probability of β_1 being contained in a region so constructed is greater than or equal to 0.95.

t	X_t	Y_t	t	X_t	Y_t
1	3.62	1.67	11	2.46	0.93
2	1.64	0.45	12	0.85	1.05
3	3.72	4.43	13	3.59	3.30
4	3.28	1.39	14	2.76	1.71
5	4.24	1.08	15	4.51	4.74
6	2.67	2.41	16	3.18	2.75
7	4.73	1.79	17	4.17	1.55
8	2.76	2.94	18	4.06	2.61
9	4.16	1.92	19	1.53	2.42
10	2.50	2.12	20	2.38	0.70

13. (Section 1.2.2)
   (a) Let model (1.2.1) hold and let $\hat{\beta}_1$ be defined by (1.2.3). For samples with $\hat{\sigma}_{xx}$ and $\hat{\sigma}_{ee}$ positive, show that the conditional mean and variance of $\hat{\beta}_1$, given the sample values of X_t, are

$$E\{\hat{\beta}_1|\mathbf{X}\} = \sigma_{XX}^{-1}\sigma_{XY}(m_{XX} - \sigma_{uu})^{-1}m_{XX},$$
$$V\{\hat{\beta}_1|\mathbf{X}\} = (m_{XX} - \sigma_{uu})^{-2}m_{XX}^2[(n-1)m_{XX}]^{-1}(\sigma_{YY} - \sigma_{XX}^{-1}\sigma_{XY}^2),$$

   where $\mathbf{X} = (X_1, X_2, \ldots, X_n)$.
   (b) Show that

$$\int_0^\infty (w - c)^{-1} dF(w),$$

   where $F(w)$ is the distribution function of a chi-square random variable and c is a positive number, is not defined. Hence, show that $E\{\hat{\beta}_1\}$ is not defined.
14. (Section 1.2.2, Section 1.2.3) Reilly and Patino-Leal (1981) give the following data:

Condition (i)	**Replicate (j)**	Y_{ij}	X_{ij}
1	1	3.44	16.61
	2	3.78	19.73
	3	3.33	17.23
2	1	1.11	9.19
	2	1.78	11.89
	3	2.19	10.37
	4	1.85	10.43
3	1	6.86	20.55
	2	5.00	17.33

Assume the data satisfy the model

$$Y_{ij} = \beta_0 + \beta_1 x_{ij} + e_{ij}, \qquad X_{ij} = x_{ij} + u_{ij},$$
$$(x_{ij}, e_{ij}, u_{ij})' \sim \text{NI}[(\mu_x, 0, 0)', \text{diag}(\sigma_{xx}, \sigma_{ee}, 1.0)].$$

   (a) Using the nine (Y_{ij}, X_{ij}) observations, estimate $(\beta_0, \beta_1, \sigma_{xx}, \sigma_{ee})$. Using Theorem 1.2.1, estimate the covariance matrix of the approximate distribution of $(\hat{\beta}_0, \hat{\beta}_1)$. Estimate the x values treating them as fixed. Give the estimated variance of the estimated x values. Plot $\hat{v}_{ij}$ against $\hat{x}_{ij}$.
   (b) Assume that $\sigma_{eu} = -0.4$ and $\sigma_{uu} = 0.8$. Transform the data so that the covariance of the measurement errors in the transformed variables is zero. Estimate the transformed parameters using equations (1.2.3). Estimate the covariance matrix of the transformed parameters. Then, using the inverse transformation, estimate the original parameters.
15. (Section 1.2.1) Let $\mathbf{X}_t = (X_{t0}, X_{t1}) = (1, X_{t1})$, and let (β_0, β_1) of model (1.2.1) be estimated by

$$(\hat{\beta}_{0M}, \hat{\beta}_{1M})' = (\mathbf{M}_{XX} - \mathbf{\Sigma}_{uu})^{-1}\mathbf{M}_{XY},$$

where $(\mathbf{M}_{XY}, \mathbf{M}_{XX}) = n^{-1}\sum_{t=1}^n \mathbf{X}_t'(Y_t, \mathbf{X}_t)$ and $\mathbf{\Sigma}_{uu} = \text{diag}(0, \sigma_{uu})$. How do these estimators differ from those defined in Equations (1.2.3)?
16. (Section 1.2.1) Show that $\hat{\sigma}_{ee}$ of (1.2.3) satisfies

$$\hat{\sigma}_{ee} = m_{Yv} - (\hat{\beta}_1 - \beta_1)\sigma_{XY} + O_p(n^{-1}) = m_{vv} - \beta_1^2\sigma_{uu} + O_p(n^{-1}).$$

17. (Section 1.2) Suppose that the relationship between phosphorus content in the leaves of corn (Y) and the concentration of inorganic phosphorus (x) in the soil is defined by the linear model

$$Y_t = \beta_0 + \beta_1 x_t + e_t, \qquad X_t = x_t + u_t,$$
$$(x_t, e_t, u_t)' \sim \mathrm{NI}[(\mu_x, 0, 0)', \mathrm{diag}(\sigma_{xx}, \sigma_{ee}, 0.25)],$$

where X_t is the observed estimate of inorganic phosphorus in the soil. The chemical determinations for 18 soils are as follows:

t	Y_t	X_t	t	Y_t	X_t
1	64	1.18	10	51	3.69
2	60	1.18	11	76	3.45
3	71	2.02	12	96	4.91
4	61	1.26	13	77	4.91
5	54	2.39	14	93	4.75
6	77	1.64	15	95	4.91
7	81	3.22	16	54	1.70
8	93	3.33	17	68	5.27
9	93	3.55	18	99	5.56

(a) Calculate estimates of the parameters of the model.
(b) Estimate the covariance matrix of the approximate distribution of $(\hat{\beta}_0, \hat{\beta}_1)$.
(c) Estimate the true x_t values treating the x_t as fixed. Estimate the variance of these estimates.
(d) Plot $\hat{v}_t$ against $\hat{x}_t$.

18. (Section 1.2.3)
(a) Using the $\hat{v}_t$ values of Table 1.2.2, regress X_t on $\hat{v}_t$ using an ordinary regression program. Do not include an intercept in the regression. Verify that the deviations from fit of this regression are the $\hat{x}_t$ values of Table 1.2.2. How would you interpret the "Y-hat" values obtained from a regression of X_t on v_t?
(b) Verify that the $\tilde{x}_t$ values of Example 1.2.2 satisfy

$$\tilde{x}_t - \bar{X} = [\hat{\sigma}_{vv}\hat{\sigma}_{xx} + \hat{\sigma}_{ee}\sigma_{uu}]^{-1}\hat{\sigma}_{vv}\hat{\sigma}_{xx}(\hat{x}_t - \bar{X}),$$

where $\hat{x}_t$ is defined in (1.2.25) and $\tilde{x}_t$ is defined in (1.2.33). Show that

$$\tilde{x}_t = \bar{X} + \hat{g}(\hat{x}_t - \bar{X}),$$

where $\hat{g} = [\hat{V}\{\eta_t\}]^{-1}\hat{C}\{\hat{x}_t, x_t\}$, $\eta_t = X_t - v_t\sigma_{vv}^{-1}\sigma_{uv}$, and $\hat{V}\{\eta_t\}$ is an estimator of the variance of η_t.
(c) Use $v_t = (e_t, u_t)(1, -\beta_1)'$ and the equality

$$\ddot{x}_t - x_t = [(\beta_1, 1)\Sigma_{\varepsilon\varepsilon}^{-1}(\beta_1, 1)']^{-1}(\beta_1, 1)\Sigma_{\varepsilon\varepsilon}^{-1}(e_t, u_t)'$$

to show that $E\{(\ddot{x}_t - x_t)v_t\} = 0$.
(d) Compute $\sum_{t=1}^{11} \hat{x}_t^2$ for Example 1.2.1. Express this quantity as a function of the estimated parameters $\hat{\beta}_1$, $\hat{\sigma}_{xx}$, $\hat{\sigma}_{ee}$ and $\hat{\mu}_x$.

19. (Section 1.2.3)
(a) Verify Equation (1.2.32).
(b) Show that the vector form of Equation (1.2.34) is

$$\hat{\mathbf{V}}\{\tilde{\mathbf{z}}_t' \mid \mathbf{Z}_t'\} = \mathbf{\Sigma}_{\varepsilon\varepsilon} - \mathbf{\Sigma}_{\varepsilon\varepsilon}\mathbf{m}_{ZZ}^{-1}\mathbf{\Sigma}_{\varepsilon\varepsilon},$$

where $\tilde{\mathbf{z}}_t = (\tilde{y}_t, \tilde{x}_t)$.

1.3. RATIO OF MEASUREMENT VARIANCES KNOWN

1.3.1. Introduction

In this section we shall study parameter estimation for the measurement error model, given information on the relative magnitude of the error variances associated with the two variables in the model. We write the model as

$$\begin{aligned} y_t &= \beta_0 + \beta_1 x_t, \\ (Y_t, X_t) &= (y_t, x_t) + (e_t, u_t), \end{aligned} \tag{1.3.1}$$

where (Y_t, X_t) is observed, y_t is the true value of the dependent variable, x_t is the true value of the independent variable, and (e_t, u_t) is the vector of measurement errors. We have retained the terms "independent variable" and "dependent variable" although the model is symmetric in x and y for $\beta_1 \neq 0$.

The model (1.3.1) with $\sigma_{eu} = 0$ and the ratio

$$\delta = \sigma_{uu}^{-1}\sigma_{ee}$$

known, can be called the classical errors-in-variables model. R. J. Adcock (1877, 1878) considered estimation for the model under the (implicit) assumption that $\delta = 1$ and K. Pearson (1901) suggested the same estimator obtained by Adcock. C. H. Kummell (1879) considered the problem for a general δ, though he did not formulate the model in precisely our manner. The model was treated in considerable detail by Koopmans (1937). Others discussing the model include Tintner (1945), Lindley (1947), Madansky (1959). Anderson (1951b), Barnett (1967), Moran (1971), Kendall and Stuart (1979, Chap. 29), and Fuller (1980).

We devote considerable space to this model for two reasons. First, it is an important model in its own right, furnishing an approximation to real world situations. Second, a careful examination of this simple model helps one to understand the theoretical underpinnings of methods for other models, including those of Sections 1.4 and 1.5 and Chapters 3 and 4.

1.3.2. Method of Moments Estimators

To introduce estimation for the model we assume that (x_t, e_t, u_t) are normally and independently distributed,

$$(x_t, e_t, u_t)' \sim \text{NI}[(\mu_x, 0, 0)', \text{diag}(\sigma_{xx}, \sigma_{ee}, \sigma_{uu})], \tag{1.3.2}$$

where $\sigma_{ee} = \delta\sigma_{uu}$. Then, under the specification (1.3.1) and (1.3.2),

$$\begin{bmatrix} Y_t \\ X_t \end{bmatrix} \sim \text{NI}\left(\begin{bmatrix} \beta_0 + \beta_1\mu_x \\ \mu_x \end{bmatrix}, \begin{bmatrix} \beta_1^2\sigma_{xx} + \delta\sigma_{uu} & \beta_1\sigma_{xx} \\ \beta_1\sigma_{xx} & \sigma_{xx} + \sigma_{uu} \end{bmatrix}\right).$$

We assume that a sample of n observations on (Y_t, X_t) is available. To obtain estimators of β_0 and β_1, we recall that the sample mean and covariance matrix are minimal sufficient statistics for the normal distribution estimation problem. Let the sample mean be $(\bar{Y}, \bar{X})$ and let the sample covariance matrix be

$$\mathbf{m}_{ZZ} = (n-1)^{-1} \sum_{t=1}^{n} (Y_t - \bar{Y}, X_t - \bar{X})'(Y_t - \bar{Y}, X_t - \bar{X}),$$

where $\mathbf{Z}_t = (Y_t, X_t)$ and the elements of $\mathbf{m}_{ZZ}$ are m_{YY}, m_{XY}, and m_{XX}. We define estimators by equating the sample moments to their expectations. Under the normality assumption the estimators so obtained are the maximum likelihood estimators adjusted for degrees of freedom. That is, the estimators of σ_{xx} and σ_{uu} differ from the maximum likelihood estimators by a factor of $n(n-1)^{-1}$. We have

$$\begin{aligned} m_{YY} &= \hat{\beta}_1^2 \hat{\sigma}_{xx} + \delta \hat{\sigma}_{uu}, \\ (m_{XY}, m_{XX}) &= (\hat{\beta}_1 \hat{\sigma}_{xx}, \hat{\sigma}_{xx} + \hat{\sigma}_{uu}), \end{aligned} \tag{1.3.3}$$

$\hat{\mu}_x = \bar{X}$, $\hat{\beta}_0 = \bar{Y} - \hat{\beta}_1 \bar{X}$, and $\hat{\sigma}_{ee} = \delta \hat{\sigma}_{uu}$. For samples with $m_{XY} \neq 0$, it follows from (1.3.3) that

$$m_{YY} - \delta m_{XX} = \hat{\beta}_1^2 \hat{\sigma}_{xx} - \delta \hat{\sigma}_{xx}, \tag{1.3.4}$$

$$\hat{\beta}_1 \hat{\sigma}_{xx} = m_{XY}, \tag{1.3.5}$$

and

$$\hat{\beta}_1^2 m_{XY} - \hat{\beta}_1 (m_{YY} - \delta m_{XX}) - \delta m_{XY} = 0. \tag{1.3.6}$$

There are two solutions to (1.3.6) and the two roots are real, one positive and one negative. Because $m_{XY} = \hat{\beta}_1 \hat{\sigma}_{xx}$, $\hat{\beta}_1$ will have the same sign as m_{XY}. Therefore, we choose the root

$$\hat{\beta}_1 = \frac{m_{YY} - \delta m_{XX} + [(m_{YY} - \delta m_{XX})^2 + 4\delta m_{XY}^2]^{1/2}}{2m_{XY}} \tag{1.3.7}$$

as our estimator of β_1. Given $\hat{\beta}_1$, the estimators of the other parameters are given by (1.3.3) and (1.3.5).

To obtain an explicit expression for $\hat{\sigma}_{xx}$, we can solve (1.3.5) for $\hat{\beta}_1$ and substitute the expression for $\hat{\beta}_1$ into (1.3.4) to obtain

$$\delta \hat{\sigma}_{xx}^2 + (m_{YY} - \delta m_{XX}) \hat{\sigma}_{xx} - m_{XY}^2 = 0.$$

The quadratic has a positive and a negative root. The estimator of σ_{xx} is the positive root

$$\hat{\sigma}_{xx} = (2\delta)^{-1} \{[(m_{YY} - \delta m_{XX})^2 + 4\delta m_{XY}^2]^{1/2} - (m_{YY} - \delta m_{XX})\}. \tag{1.3.8}$$

By (1.3.7) and (1.3.8),

$$\hat{\beta}_1 = \hat{\sigma}_{xx}^{-1} m_{XY}. \tag{1.3.9}$$

We can construct an explicit expression for $\hat{\sigma}_{uu}$ by substituting the expressions for $\hat{\beta}_1\hat{\sigma}_{xx}$ and $\hat{\sigma}_{xx}$ from the m_{XX} and m_{XY} equations of (1.3.3) into the m_{YY} equation to obtain

$$\delta\hat{\sigma}_{uu}^2 - (m_{YY} + \delta m_{XX})\hat{\sigma}_{uu} + m_{YY}m_{XX} - m_{XY}^2 = 0.$$

It follows that

$$\hat{\sigma}_{uu} = (2\delta)^{-1}\{m_{YY} + \delta m_{XX} - [(m_{YY} - \delta m_{XX})^2 + 4\delta m_{XY}^2]^{1/2}\}, \tag{1.3.10}$$

where we choose the smallest root of the quadratic equation so that $\hat{\sigma}_{uu} < m_{XX}$.

If $m_{XY} = 0$ and $m_{YY} < \delta m_{XX}$, it is natural to take $\hat{\beta}_1 = 0$ and $\hat{\sigma}_{xx} = m_{XX} - \delta^{-1}m_{YY}$. If $m_{XY} = 0$ and $m_{YY} > \delta m_{XX}$, the assumption that σ_{xx} is positive is suspect and it is reasonable to consider $\hat{\beta}_1$ to be undefined. For random variables with a density, the probability that $m_{XY} = 0$ is zero.

Because the estimators are continuous differentiable functions of the sample moments, the methods of Appendix 1.A can be used to establish the limiting distribution of the estimators.

Theorem 1.3.1. Let the model defined by (1.3.1) and (1.3.2) hold with $\sigma_{uu} > 0$ and $\sigma_{xx} > 0$, where $\delta > 0$ is known. Let $\boldsymbol{\theta} = (\beta_0, \beta_1, \sigma_{uu})'$ and let $\hat{\boldsymbol{\theta}} = (\hat{\beta}_0, \hat{\beta}_1, \hat{\sigma}_{uu})'$ be defined by (1.3.3), (1.3.7), and (1.3.10). Then, as $n \to \infty$,

$$n^{1/2}(\hat{\boldsymbol{\theta}} - \boldsymbol{\theta}) \xrightarrow{L} N(\mathbf{0}, \boldsymbol{\Gamma}_{\theta\theta}),$$

where

$$\boldsymbol{\Gamma}_{\theta\theta} = \begin{bmatrix} \sigma_{vv} + \mu_x^2\Gamma_{22} & -\mu_x\Gamma_{22} & 0 \\ -\mu_x\Gamma_{22} & \Gamma_{22} & 0 \\ 0 & 0 & 2\sigma_{uu}^2 \end{bmatrix},$$

$$\Gamma_{22} = \sigma_{xx}^{-2}(\sigma_{xx}\sigma_{vv} + \sigma_{uu}\sigma_{vv} - \sigma_{uv}^2),$$

$\sigma_{vv} = \sigma_{ee} + \beta_1^2\sigma_{uu}$, $\sigma_{uv} = -\beta_1\sigma_{uu}$, and the symbol $\xrightarrow{L}$ is used to denote convergence in distribution, also called convergence in law.

Proof. The estimators are functions of the vector $(\bar{Y}, \bar{X}, m_{YY}, m_{XY}, m_{XX}) = \hat{\mathbf{b}}$. This vector differs from the sample mean of the vectors $[Y_t, X_t, (Y_t - \mu_Y)^2, (X_t - \mu_X)(Y_t - \mu_Y), (X_t - \mu_X)^2]$ by a quantity that is $O_p(n^{-1})$. In fact, for normal (Y_t, X_t), it is known that $(\bar{Y}, \bar{X})$ is independent of (m_{YY}, m_{XY}, m_{XX}) and that (m_{YY}, m_{XY}, m_{XX}) can be expressed as the mean of $(n - 1)$ independent random vectors. By the assumptions that $\sigma_{xx} > 0$ and $\sigma_{uu} > 0$, the first and second derivatives of $\hat{\boldsymbol{\theta}}$ are continuous at the point $(\mu_Y, \mu_X, \sigma_{YY}, \sigma_{XY}, \sigma_{XX}) = \mathbf{b} = E\{\hat{\mathbf{b}}\}$. Therefore, Corollary 1.A.1 of Appendix 1.A can be used to obtain the limiting distribution of $n^{1/2}(\hat{\boldsymbol{\theta}} - \boldsymbol{\theta})$. Letting $\hat{\boldsymbol{\theta}} = \mathbf{g}(\hat{\mathbf{b}})$, we have, for example,

$$\frac{\partial \hat{\beta}_1}{\partial m_{YY}} = \frac{\partial g_2(\mathbf{b})}{\partial b_3} = \beta_1[\sigma_{xx}(\beta_1^2 + \delta)]^{-1},$$

$$\frac{\partial \hat{\beta}_1}{\partial m_{XY}} = \frac{\partial g_2(\mathbf{b})}{\partial b_4} = \sigma_{xx}^{-1} - 2\beta_1^2[\sigma_{xx}(\beta_1^2 + \delta)]^{-1},$$

$$\frac{\partial \hat{\beta}_1}{\partial m_{XX}} = \frac{\partial g_2(\mathbf{b})}{\partial b_5} = -\delta\beta_1[\sigma_{xx}(\beta_1^2 + \delta)]^{-1},$$

where $g_2(\mathbf{b}) = \beta_1$ and the partial derivatives are evaluated at the parameter vector $\mathbf{b}$. Using the derivatives, we have

$$\begin{aligned} \hat{\beta}_0 - \beta_0 &= \bar{v} - (\hat{\beta}_1 - \beta_1)\mu_x + O_p(n^{-1}), \\ \hat{\beta}_1 - \beta_1 &= \sigma_{xx}^{-1}[m_{Xv} - \sigma_{vv}^{-1} m_{vv}\sigma_{uv}] + O_p(n^{-1}), \\ \hat{\sigma}_{uu} - \sigma_{uu} &= (\beta_1^2 + \delta)^{-1}(m_{vv} - \sigma_{vv}) + O_p(n^{-1}), \end{aligned} \tag{1.3.11}$$

where $m_{vv} = (n-1)^{-1}\sum_{t=1}^{n}(v_t - \bar{v})^2$, $\sigma_{vv} = (\delta + \beta_1^2)\sigma_{uu}$, and $\sigma_{Xv} = \sigma_{uv} = -\beta_1\sigma_{uu}$. Equations (1.3.11) correspond to Equation (1.A.3) of Appendix 1.A. The result follows by applying Corollary 1.A.1 and evaluating the covariance matrix of the leading terms of (1.3.11). □

The assumption of normality was used in the proof of Theorem 1.3.1 only to obtain expressions for the variances of the sample moments. The estimators $(\hat{\beta}_0, \hat{\beta}_1)$ that define the line are often of the most interest and the covariance matrix for these estimators is that of Theorem 1.3.1 for normal (e_t, u_t) under a wide range of assumptions for x_t.

Comparing the distributions of Theorem 1.2.1 and of Theorem 1.3.1, we see that the variance of the limiting distribution of $\hat{\beta}_1$ computed with $\delta = \sigma_{uu}^{-1}\sigma_{ee}$ known, is smaller than the variance of the limiting distribution of $\hat{\beta}_1$ computed with σ_{uu} known. The difference between the two variances is the quantity

$$2\sigma_{xx}^{-2}\sigma_{uv}^2 = 2\sigma_{xx}^{-2}\beta_1^2\sigma_{uu}^2.$$

An estimator of the covariance matrix of the approximate distribution of $(\hat{\beta}_0, \hat{\beta}_1)$ of Theorem 1.3.1 is

$$\hat{\mathbf{V}}\{(\hat{\beta}_0, \hat{\beta}_1)'\} = \begin{bmatrix} n^{-1}s_{vv} + \bar{X}^2\hat{V}\{\hat{\beta}_1\} & -\bar{X}\hat{V}\{\hat{\beta}_1\} \\ -\bar{X}\hat{V}\{\hat{\beta}_1\} & \hat{V}\{\hat{\beta}_1\} \end{bmatrix}, \tag{1.3.12}$$

where

$$\hat{V}\{\hat{\beta}_1\} = (n-1)^{-1}\hat{\sigma}_{xx}^{-2}[\hat{\sigma}_{xx}s_{vv} + \hat{\sigma}_{uu}s_{vv} - \hat{\sigma}_{uv}^2],$$

$$s_{vv} = (n-2)^{-1}\sum_{t=1}^{n}[Y_t - \bar{Y} - \hat{\beta}_1(X_t - \bar{X})]^2$$

$$= (n-2)^{-1}(n-1)(\delta + \hat{\beta}_1^2)\hat{\sigma}_{uu},$$

$\hat{\sigma}_{uv} = -\hat{\beta}_1\hat{\sigma}_{uu}$, and $\hat{\sigma}_{xx}$ is defined in (1.3.8).

Because $n^{1/2}(\hat{\beta}_1 - \beta_1)$ is approximately normally distributed and $n\hat{V}\{\hat{\beta}_1\}$ is a consistent estimator of the variance of the limiting distribution, the quantity

$$t = [\hat{V}\{\hat{\beta}_1\}]^{-1/2}(\hat{\beta}_1 - \beta_1) \tag{1.3.13}$$

is approximately distributed as a $N(0, 1)$ random variable. It is suggested that, in small samples, the distribution of the statistic defined in (1.3.13) be approximated by Student's t distribution with $n - 2$ degrees of freedom. We also suggest that the distribution of the statistic $\sigma_{uu}^{-1}(n-1)\hat{\sigma}_{uu}$ be approximated by that of a chi-square random variable with $n - 2$ degrees of freedom.

Example 1.3.1. To illustrate the estimation of the parameters of model (1.3.1), we use some data collected by the Iowa Conservation Commission. Table 1.3.1 contains indexes of the number of hen pheasants in Iowa at two times during the year. These indexes are based on the average number of

TABLE 1.3.1. Measures of the number of hen pheasants in Iowa

Year	August Hens Y_t	Spring Hens X_t	Year	August Hens Y_t	Spring Hens X_t
1976	8.0	9.0	1968	8.1	10.9
1975	6.0	6.6	1967	8.7	10.4
1974	9.8	12.3	1966	8.7	10.2
1973	10.8	11.9	1965	7.4	7.4
1972	9.7	11.9	1964	10.1	11.0
1971	9.3	12.0	1963	10.0	11.8
1970	9.2	9.6	1962	7.3	8.2
1969	6.9	7.5			

Source: Data from Iowa Conservation Commission. Data for 1962–1968 have been adjusted.

birds sighted by trained observers traveling a number of specific routes in late April and early May, and again in August. Both measures are subject to error for two reasons. First, the routes are a sample of all possible routes in Iowa. Second, observers cannot be expected to sight all pheasants along the route. In August, it is relatively easy to sight hen pheasants because most hens are accompanied by a brood of chicks at that time. On the basis of other analyses, it has been estimated that the error variance for the spring count is about six times that for August. For the purposes of the present example, we treat this ratio as known. We do not believe that the number of hen pheasants in August is a perfect linear function of the number in the spring, but such a specification is considered a reasonable approximation. Although we proceed with this model, we must remember that our estimated error variances will contain components due to variations in survival rates as well as that due to observational and sampling error. The working model is

$$y_t = \beta_0 + \beta_1 x_t,$$
$$(Y_t, X_t) = (y_t, x_t) + (e_t, u_t),$$
$$(x_t, e_t, u_t)' \sim \text{NI}[(\mu_x, \mathbf{0})', \text{diag}(\sigma_{xx}, 6^{-1}\sigma_{uu}, \sigma_{uu})].$$

The sample means are $(\bar{Y}, \bar{X}) = (8.6667, 10.0467)$ and the sample covariances are

$$(m_{YY}, m_{YX}, m_{XX}) = (1.84952, 2.35167, 3.62124).$$

With $\delta = 6^{-1}$, Equation (1.3.6) becomes

$$2.35167\hat{\beta}_1^2 - 1.24598\hat{\beta}_1 - 0.39195 = 0$$

and the estimated parameters of the line are

$$(\hat{\beta}_0, \hat{\beta}_1) = (1.1158, 0.7516).$$

The data and the estimated structural line are plotted in Figure 1.3.1. If, at first glance, the line does not appear to be drawn correctly, recall that the ratio of σ_{ee} to σ_{uu} is 6^{-1}. This means that horizontal deviations are six times more important in determining the line than are vertical distances.

The estimated covariance matrix of the approximate distribution of $(\hat{\beta}_0, \hat{\beta}_1)$ is

$$\hat{\mathbf{V}}\{(\hat{\beta}_0, \hat{\beta}_1)'\} = \begin{bmatrix} 0.95929 & -0.09291 \\ -0.09291 & 0.00925 \end{bmatrix},$$

where $\hat{\sigma}_{xx} = 3.1290$, $\hat{\sigma}_{uu} = 0.4923$, $\hat{\sigma}_{uv} = 0.3700$, and

$$s_{vv} = (n-2)^{-1}(n-1)(\delta + \hat{\beta}_1^2)\hat{\sigma}_{uu} = 0.38783.$$

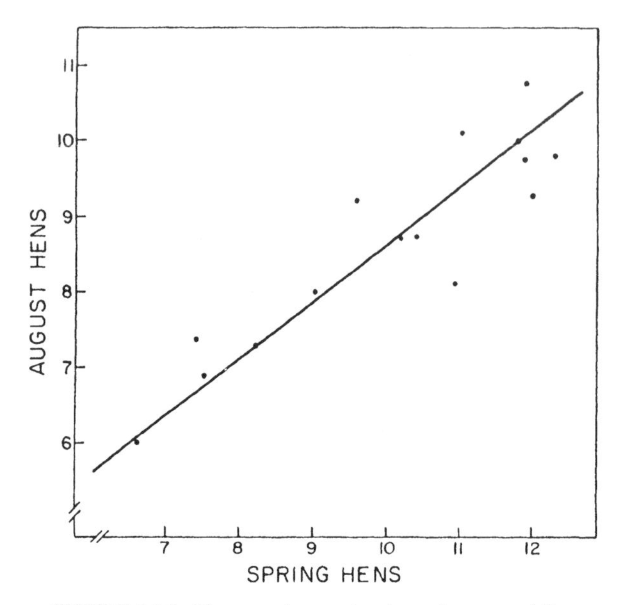

FIGURE 1.3.1. Pheasant data and estimated structural line.

For illustrative purposes we have chosen a rather small data set. However, even in this small example, different conclusions can be drawn if the measurement error is ignored. The ordinary least squares regression of Y_t on X_t is

$$\hat{Y}_t = \underset{(0.845)}{2.142} + \underset{(0.083)}{0.649} X_t,$$

where the numbers in parentheses are the ordinary least squares estimated standard errors. The estimated line constructed under the assumption $6\sigma_{ee} = \sigma_{uu}$ has an intercept of 1.116 with a standard error of 0.979. Therefore, the ordinary least squares line has a significant positive intercept while one can easily accept the hypothesis of a zero intercept for the structural line.

□ □

1.3.3. Least Squares Estimation

We now derive an estimator of β_1 for model (1.3.1) by the method of least squares. In the classical least squares problem the estimator of (β_0, β_1) is the $(\hat{\beta}_0, \hat{\beta}_1)$ for which the sum of squared vertical deviations of the points (Y_t, x_t) from the line $y_t = \beta_0 + \beta_1 x_t$ is minimized. The vertical distance is the proper statistical distance to consider because the x values are fixed in repeated sampling; the observation vector deviates from the true line by the distance e_t in the vertical direction.

In the errors-in-variables situation an observation can deviate from the true line in both the horizontal and vertical direction. That is, the point on the line is $(\beta_0 + \beta_1 x_t, x_t)$ and the observed point is $(Y_t, X_t) = (\beta_0 + \beta_1 x_t + e_t, x_t + u_t)$. The squared Euclidean distance from the observation to the point on the line that generated the point is

$$[Y_t - (\beta_0 + \beta_1 x_t)]^2 + (X_t - x_t)^2 = e_t^2 + u_t^2. \tag{1.3.14}$$

If e_t and u_t are independent random variables with unit variance, the Euclidean distance is the proper statistical distance to consider. If e_t and u_t are independent, but with different variances, the squared statistical distance is

$$[\text{statistical distance}]^2 = \sigma_{ee}^{-1} e_t^2 + \sigma_{uu}^{-1} u_t^2. \tag{1.3.15}$$

If the covariance matrix of (e_t, u_t) is $\Sigma_{\varepsilon\varepsilon}$, where $\Sigma_{\varepsilon\varepsilon}$ is nonsingular, then the squared statistical distance from (Y_t, X_t) to the point on the line associated with x_t is

$$(Y_t - \beta_0 - \beta_1 x_t, X_t - x_t)\Sigma_{\varepsilon\varepsilon}^{-1}(Y_t - \beta_0 - \beta_1 x_t, X_t - x_t)'. \tag{1.3.16}$$

Using the least squares criterion we shall choose as our estimators of β_0 and β_1 those values $(\hat{\beta}_0, \hat{\beta}_1)$ for which the sum of squared statistical distances of the observed points from the estimated line is a minimum. That is, we find $(\hat{\beta}_0, \hat{\beta}_1, \hat{x}_1, \hat{x}_2, \ldots, \hat{x}_n)$ to minimize the sum of (1.3.16) over the n observations.

Given (β_0, β_1), the x value for the point on the line that is the smallest statistical distance from (Y_t, X_t) was obtained in (1.2.17) of Section 1.2.3. That x value is

$$\begin{aligned} \ddot{x}_t &= [(\beta_1, 1)\Sigma_{\varepsilon\varepsilon}^{-1}(\beta_1, 1)']^{-1}(\beta_1, 1)\Sigma_{\varepsilon\varepsilon}^{-1}(Y_t - \beta_0, X_t)' \\ &= X_t - \sigma_{vv}^{-1}\sigma_{uv}v_t, \end{aligned} \tag{1.3.17}$$

where $\sigma_{vv} = \sigma_{ee} - 2\beta_1\sigma_{eu} + \beta_1^2\sigma_{uu}$, $\sigma_{uv} = \sigma_{eu} - \beta_1\sigma_{uu}$, and $v_t = Y_t - \beta_0 - \beta_1 X_t$. The corresponding y value is

$$\ddot{y}_t = \beta_0 + \beta_1 \ddot{x}_t. \tag{1.3.18}$$

If we substitute (1.3.17) and (1.3.18) into (1.3.16), we find that the minimum squared statistical distance from (X_t, Y_t) to the line is

$$(\sigma_{ee} - 2\beta_1\sigma_{eu} + \beta_1^2\sigma_{uu})^{-1}(Y_1 - \beta_0 - \beta_1 X_t)^2. \tag{1.3.19}$$

Thus, the least squares (minimum statistical distance) estimator of (β_0, β_1) for model (1.3.1), with general error covariance matrix, is the $(\hat{\beta}_0, \hat{\beta}_1)$ which minimizes

$$(\sigma_{ee} - 2\beta_1\sigma_{eu} + \beta_1^2\sigma_{uu})^{-1} \sum_{t=1}^{n} (Y_t - \beta_0 - \beta_1 X_t)^2. \tag{1.3.20}$$

The partial derivative of (1.3.20) with respect to β_0 is zero for

$$\hat{\beta}_0 = \bar{Y} - \beta_1 \bar{X}. \tag{1.3.21}$$

Substituting (1.3.21) into (1.3.20) and dividing by $(n - 1)$, we obtain

$$(\sigma_{ee} - 2\beta_1\sigma_{eu} + \beta_1^2\sigma_{uu})^{-1}(m_{YY} - 2\beta_1 m_{XY} + \beta_1^2 m_{XX}). \tag{1.3.22}$$

If we set $\sigma_{eu} = 0$ and $\sigma_{uu}^{-1}\sigma_{ee} = \delta$ and equate the derivative of (1.3.22) with respect to β_1 to zero, we obtain (1.3.6) as the equation defining the estimator of β_1. Therefore, the method of moments estimator and the least squares estimator of β_1 are the same.

We consider (1.3.22) further to obtain an alternative form for the estimator. If we set $\beta_1 = -\alpha_2\alpha_1^{-1}$, we can write (1.3.22) as

$$(\alpha_1^2\sigma_{ee} + 2\alpha_1\alpha_2\sigma_{eu} + \alpha_2^2\sigma_{uu})^{-1}(\alpha_1^2 m_{YY} + 2\alpha_1\alpha_2 m_{XY} + \alpha_2^2 m_{XX}). \tag{1.3.23}$$

Minimizing the ratio (1.3.23) is formally equivalent to minimizing the Lagrangian

$$\alpha_1^2 m_{YY} + 2\alpha_1\alpha_2 m_{XY} + \alpha_2^2 m_{XX} - \lambda(\alpha_1^2\sigma_{ee} + 2\alpha_1\alpha_2\sigma_{eu} + \alpha_2^2\sigma_{uu} - A) \tag{1.3.24}$$

with respect to α_1 and α_2, where A is any positive constant. Setting the partial derivatives with respect to α_1 and α_2 equal to zero, we obtain

$$\begin{aligned} 2\hat{\alpha}_1 m_{YY} + 2\hat{\alpha}_2 m_{XY} - 2\lambda\hat{\alpha}_1\sigma_{ee} - 2\lambda\hat{\alpha}_2\sigma_{eu} &= 0, \\ 2\hat{\alpha}_1 m_{XY} + 2\hat{\alpha}_2 m_{XX} - 2\lambda\hat{\alpha}_1\sigma_{eu} - 2\lambda\hat{\alpha}_2\sigma_{uu} &= 0. \end{aligned} \tag{1.3.25}$$

This system of equations has a nontrivial solution only if

$$\left| \begin{bmatrix} m_{YY} & m_{XY} \\ m_{XY} & m_{XX} \end{bmatrix} - \lambda \begin{bmatrix} \sigma_{ee} & \sigma_{eu} \\ \sigma_{eu} & \sigma_{uu} \end{bmatrix} \right| = 0. \tag{1.3.26}$$

We choose the smallest root $\hat{\lambda}$ of the determinantal equation (1.3.26) to minimize the original ratio (1.3.23). Then the estimator of $\beta_1 = -\alpha_2\alpha_1^{-1}$ is given by the solution to

$$(m_{XX} - \hat{\lambda}\sigma_{uu})\hat{\beta}_1 = m_{XY} - \hat{\lambda}\sigma_{eu}. \tag{1.3.27}$$

While not obvious, this estimator of β_1 is identical to that of (1.3.7) when $\sigma_{eu} = 0$ and $\sigma_{ee} = \delta\sigma_{uu}$.

Given an estimator $(\hat{\beta}_0, \hat{\beta}_1)$, one can replace (β_0, β_1) with $(\hat{\beta}_0, \hat{\beta}_1)$ in (1.3.17) and (1.3.18) to obtain an estimator of (x_t, y_t),

$$\begin{aligned} \hat{x}_t &= [(\hat{\beta}_1, 1)\boldsymbol{\Sigma}_{\varepsilon\varepsilon}^{-1}(\hat{\beta}_1, 1)']^{-1}(\hat{\beta}_1, 1)\boldsymbol{\Sigma}_{\varepsilon\varepsilon}^{-1}(Y_t - \hat{\beta}_0, X_t)' \\ &= X_t - \hat{\sigma}_{vv}^{-1}\hat{\sigma}_{uv}\hat{v}_t, \end{aligned} \tag{1.3.28}$$

$$\hat{y}_t = \hat{\beta}_0 + \hat{\beta}_1\hat{x}_t, \tag{1.3.29}$$

where $\hat{\sigma}_{vv} = \sigma_{ee} - 2\hat{\beta}_1\sigma_{eu} + \hat{\beta}_1^2\sigma_{uu}$, $\hat{\sigma}_{uv} = \sigma_{eu} - \hat{\beta}_1\sigma_{uu}$, and $\hat{v}_t = Y_t - \hat{\beta}_0 - \hat{\beta}_1 X_t$. It is interesting that the estimator of x_t, treating x_t as fixed, can be constructed by subtracting the best predictor of u_t, treating u_t as random, from X_t. That is, expression (1.3.28) is the sample estimator of

$$\ddot{x}_t = X_t - \ddot{u}_t = X_t - v_t \sigma_{vv}^{-1} \sigma_{uv},$$

where the best predictor of u_t, given v_t, is

$$\ddot{u}_t = v_t \sigma_{vv}^{-1} \sigma_{uv},$$

and $\sigma_{vv}^{-1} \sigma_{uv}$ is the population regression coefficient for the regression of u_t on v_t. Also see Equation (1.2.23) of Section 1.2.3.

Because the error made in estimating the parameters is small relative to the errors (e_t, u_t), the variance expression $V\{\ddot{x}_t - x_t\}$ is generally adequate for $\hat{x}_t$. For an illustration of the exception, see Example 2.3.4 of Section 2.3.4. An estimator of the variance of $\hat{x}_t - x_t$ is

$$\begin{aligned} \hat{V}\{\hat{x}_t - x_t\} &= [(\hat{\beta}_1, 1)\boldsymbol{\Sigma}_{\varepsilon\varepsilon}^{-1}(\hat{\beta}_1, 1)']^{-1} \\ &= \sigma_{uu} - \hat{\sigma}_{vv}^{-1} \hat{\sigma}_{uv}^2. \end{aligned} \tag{1.3.30}$$

The second expressions of (1.3.28) and (1.3.30) are valid for singular $\boldsymbol{\Sigma}_{\varepsilon\varepsilon}$.

In the derivation of this section we treated the unknown x_t values as fixed. Just as the estimator of β_1 was the same as that obtained in Section 1.3.2 under the random-x assumption, the covariance matrix of the approximate distribution given in Theorem 1.3.1 is appropriate with the understanding that σ_{xx} stands for the mean squares of the true x values.

If the entire error covariance matrix is known, instead of being known up to a multiple, the estimated variance of $\hat{\beta}_1$ takes a different form. The estimated variance is

$$\hat{V}\{\hat{\beta}_1\} = (n-1)^{-1} \hat{m}_{xx}^{-2} [\hat{m}_{xx} \hat{\sigma}_{vv} + \sigma_{uu} \hat{\sigma}_{vv} - \hat{\sigma}_{uv}^2], \tag{1.3.31}$$

where $\hat{\sigma}_{uv} = \sigma_{ue} - \hat{\beta} \sigma_{uu}$, $\hat{\sigma}_{vv} = (1, -\hat{\beta}_1)\boldsymbol{\Sigma}_{\varepsilon\varepsilon}(1, -\hat{\beta}_1)'$,

$$\begin{aligned} \hat{m}_{xx} &= \hat{\mathbf{H}}_2'(\mathbf{m}_{ZZ} - \boldsymbol{\Sigma}_{\varepsilon\varepsilon})\hat{\mathbf{H}}_2, \\ \hat{\mathbf{H}}_2 &= [(0, 1)' - (1, -\hat{\beta}_1)'\hat{\sigma}_{vv}^{-1} \hat{\sigma}_{vu}]. \end{aligned}$$

The estimator $\hat{\sigma}_{vv}$ is superior to s_{vv} because it contains no variation due to variation in the v_t^2, only variation due to estimating β_1. The estimator $\hat{m}_{xx}$ is the maximum likelihood estimator for σ_{xx} of the structural model with known $\boldsymbol{\Sigma}_{\varepsilon\varepsilon}$, provided the largest root of (1.3.26) is greater than one. It is the mean square of the $\hat{x}_t$ reduced by an estimator of the variance of $\hat{x}_t$ and equals the estimator (1.3.8) for $\hat{\lambda} = 1$. See Exercise 1.24. The estimator $\hat{m}_{xx}$ is derived in Section 4.1.

In Section 1.3.2 the model specified $\sigma_{eu} = 0$ and $\sigma_{ee} = \delta \sigma_{uu}$, where δ is known, but σ_{ee} and σ_{uu} are unknown. This is equivalent to the specification

$$\boldsymbol{\Sigma}_{\varepsilon\varepsilon} = \boldsymbol{\Upsilon}_{\varepsilon\varepsilon}\sigma^2 = \sigma_{uu} \operatorname{diag}(\delta, 1), \tag{1.3.32}$$

where $\sigma^2 = \sigma_{uu}$ is unknown. Then the smallest root of the determinantal equation (1.3.26), with diag(δ, 1) replacing $\mathbf{\Sigma}_{\varepsilon\varepsilon}$, is

$$\hat{\lambda} = (2\delta)^{-1}\{m_{YY} + \delta m_{XX} - [(m_{YY} - \delta m_{XX})^2 + 4\delta m_{XY}^2]^{1/2}\}. \quad (1.3.33)$$

Thus, $\hat{\lambda} = \hat{\sigma}_{uu}$, where $\hat{\sigma}_{uu}$ is defined in (1.3.10) and the smallest root $\hat{\lambda}$ is an estimator of σ^2. If $n = 2$, the estimated line passes through the two observed points and $\hat{\lambda} = 0$. Therefore, it is reasonable to define the estimator of σ^2 for $n > 2$ by

$$\hat{\sigma}^2 = (n-2)^{-1}(n-1)\hat{\lambda}. \quad (1.3.34)$$

This definition of $\hat{\sigma}^2$ is also consistent with the approximate chi-square distribution suggested for $\hat{\sigma}_{uu}$.

For unknown σ^2 there are three sample covariances (m_{YY}, m_{XY}, m_{XX}) available for the estimation of three parameters (σ_{xx}, β_1, σ^2). For known σ^2 (known $\mathbf{\Sigma}_{\varepsilon\varepsilon}$) there are three sample covariances available for the estimation of two parameters (σ_{xx}, β_1). Therefore, knowledge of the full covariance matrix $\mathbf{\Sigma}_{\varepsilon\varepsilon}$ permits us to check for model validity. The smallest root of the determinantal equation (1.3.26) satisfies

$$\hat{\lambda} = (\sigma_{ee} - 2\hat{\beta}_1\sigma_{eu} + \hat{\beta}_1^2\sigma_{uu})^{-1}(m_{YY} - 2\hat{\beta}_1 m_{XY} + \hat{\beta}_1^2 m_{XX}). \quad (1.3.35)$$

The numerator of this statistic is a multiple of the estimator of σ_{vv} constructed with deviations from the fitted line. The denominator is an estimator of σ_{vv} constructed with the known $\mathbf{\Sigma}_{\varepsilon\varepsilon}$. If $\hat{\beta}_1$ were replaced by β_1, the ratio would be distributed as Snedecor's F with $n - 1$ and infinity degrees of freedom. In the next chapter it is demonstrated that $(n-2)^{-1}(n-1)\hat{\lambda}$ is approximately distributed as Snedecor's F with $n - 2$ and infinity degrees of freedom. Therefore, when σ_{ee}, σ_{eu}, and σ_{uu} are known, the statistic $(n-2)^{-1}(n-1)\hat{\lambda}$ can be used as a test of the model. If $(n-2)^{-1}(n-1)\hat{\lambda}$ is large relative to the tabular value of Snedecor's F with $n - 2$ and infinity degrees of freedom, the model is suspect.

Our derivation of Equation (1.3.27) demonstrates that the knowledge of σ^2 does not change the estimator of β_1 for the functional model. But knowledge of the error variance permits us to construct a test of model adequacy. This can be compared to the case of ordinary fixed-x regression. Knowledge of the error variance does not change the form of the fixed-x regression estimator of β_1, but knowledge of the error variance permits construction of a test for lack of fit.

Example 1.3.2. We analyze some data of Cohen, D'Eustachio, and Edelman (1977). See also, Cohen and D'Eustachio (1978) and Fuller (1978). The basic data, displayed in Table 1.3.2, are the numbers of two types of

TABLE 1.3.2. Numbers of two types of cells in a fraction of the spleens of fetal mice

Individual	Number of Cells Forming Rosettes	Number of Nucleated Cells	Y	X
1	52	337	7.211	18.358
2	6	141	2.449	11.874
3	14	177	3.742	13.304
4	5	116	2.236	10.770
5	5	88	2.236	9.381

Source: Cohen and D'Eustachio (1978).

cells in a specified fraction (aliquot) of the spleens of fetal mice. Cohen and D'Eustachio (1978) argued, on the basis of sampling, that it is reasonable to assume the original counts to be Poisson random variables. Therefore, the square roots of the counts, given in the last two columns of the table, will have, approximately, constant variance equal to one-fourth. The postulated model is

$$y_t = \beta_0 + \beta_1 x_t,$$

where $(Y_t, X_t) = (y_t, x_t) + (e_t, u_t)$, Y_t is the square root of the number of cells forming rosettes for the tth individual, and X_t is the square root of the number of nucleated cells for the tth individual. On the basis of the sampling, the (e_t, u_t) have a covariance matrix that is, approximately, $\boldsymbol{\Sigma}_{\varepsilon\varepsilon} = \text{diag}(0.25, 0.25)$. The square roots of the counts cannot be exactly normally distributed, but we assume the distribution is close enough to normal to permit the use of the formulas based on normality. Thus, our operating assumption is

$$(e_t, u_t)' \sim \text{NI}(\mathbf{0}, 0.25\mathbf{I})$$

and we assume (e_t, u_t) is independent of x_j for all t and j.

The statistics associated with the data of Table 1.3.2 are $(\bar{Y}, \bar{X}) = (3.5748, 12.7374)$ and

$$(m_{YY}, m_{XY}, m_{XX}) = (4.5255, 7.1580, 11.9484).$$

The smallest root of

$$|\mathbf{m}_{ZZ} - \lambda(0.25)\mathbf{I}| = 0$$

is $\hat{\lambda} = 0.69605$. Using (1.3.21) and (1.3.27), the estimated parameters of the line are

$$(\hat{\beta}_0, \hat{\beta}_1) = (-4.1686, 0.6079).$$

TABLE 1.3.3. Estimated numbers of two types of cells

t	Observed Y_t	Observed X_t	$\hat{y}_t$	$\hat{x}_t$	$\hat{v}_t$
1	7.211	18.358	7.0508	18.4563	0.2194
2	2.449	11.874	2.8878	11.6078	−0.6009
3	3.742	13.304	3.8714	13.2260	−0.1772
4	2.236	10.770	2.3403	10.7071	−0.1428
5	2.236	9.381	1.7237	9.6929	0.7015

The estimates of the x_t and y_t are, by (1.3.28) and (1.3.29),

$$\hat{x}_t = 1.8503 + 0.4439Y_t + 0.7302X_t,$$
$$\hat{y}_t = -3.0438 + 0.2698Y_t + 0.4439X_t.$$

The estimated true values are given in Table 1.3.3 and are plotted in Figure 1.3.2 as the dots on the estimated line. The original observations are plotted as crosses. In this example the covariance matrix of the measurement errors is proportional to the identity matrix. Therefore, the statistical distance is proportional to the Euclidean distance and the lines formed by joining (Y_t, X_t) and $(\hat{y}_t, \hat{x}_t)$ are perpendicular to the estimated functional line. The statistical distance from (Y_t, X_t) to $(\hat{y}_t, \hat{x}_t)$ is, by (1.3.16) and (1.3.19),

$$\text{statistical distance} = \left[\frac{(Y_t - \hat{\beta}_0 - \hat{\beta}_1 X_t)^2}{(0.25)(1 + \hat{\beta}_1^2)}\right]^{1/2}.$$

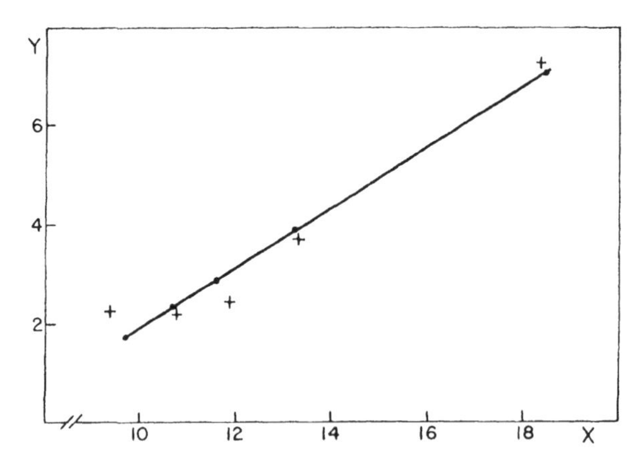

FIGURE 1.3.2. Estimated minimum distance line and estimated true values for two types of cells.

Therefore, the statistical distance is proportional to the absolute value of $\hat{v}_t$. The $\hat{v}_t$ are given in Table 1.3.3. As in Example 1.2.2, the sample correlation between $\hat{x}_t$ and $\hat{v}_t$ is zero. The estimation procedure has transformed the observed vector (Y_t, X_t) into a new vector $(\hat{v}_t, \hat{x}_t)$, where the sample covariance between $\hat{v}_t$ and $\hat{x}_t$ is zero. By (1.3.30), the approximate variance of $\hat{x}_t$ is

$$\hat{V}\{\hat{x}_t - x_t\} = [(0.6079, 1)4\mathbf{I}(0.6079, 1)']^{-1} = 0.1825.$$

If the model is true, $(n-2)^{-1}(n-1)\hat{\lambda}$ is approximately distributed as Snedecor's F with three and infinity degrees of freedom. Because

$$F = (3^{-1})4(0.69605) = 0.9281$$

is approximately equal to one, the data and the model are compatible. Figure 1.3.2 might lead one to conclude that there is a nonlinear relationship between y_t and x_t. This interpretation should be tempered by two facts. First there are only five observations in the plot. Second, as the F statistic indicates, the deviations are not large relative to the independent estimate of the standard deviation of v_t, where $\hat{\sigma}_{vv}^{1/2} = 0.5(1 + \hat{\beta}_1^2)^{1/2} = 0.59$. Also see Exercise 3.10.

We can check the assumption that $\sum(x_t - \bar{x})^2 > 0$ by computing

$$\sigma_{uu}^{-1} m_{XX} = 4(11.9484) = 47.79.$$

If $m_{xx} = (n-1)^{-1}\sum(x_t - \bar{x})^2$ were zero, the ratio would be distributed as Snedecor's F with four and infinity degrees of freedom. It seems clear that m_{xx} is positive and that the model is identified. By (1.3.31),

$$\hat{\mathbf{V}}\{(\hat{\beta}_0, \hat{\beta}_1)'\} = \begin{bmatrix} 1.2720 & -0.0945 \\ -0.0945 & 0.0074 \end{bmatrix},$$

where $\hat{\sigma}_{uv} = -\hat{\beta}_1\sigma_{uu} = -0.1520$, $\hat{\sigma}_{vv} = 0.3424$,

$$\hat{m}_{xx} = \hat{\mathbf{H}}_2'\mathbf{m}_{ZZ}\hat{\mathbf{H}}_2 - \hat{V}\{\hat{x}_t - x_t\} = 11.7192,$$
$$\hat{\mathbf{H}}_2' = (1 + \hat{\beta}_1^2)^{-1}(\hat{\beta}_1, 1) = (0.4439, 0.7302).$$

A hypothesis of interest in the original study is the hypothesis that β_0 is zero. On the basis of the approximate distribution of Theorem 1.3.1,

$$t = (1.2608)^{-1/2}(-4.1686) = -3.71$$

is approximately distributed as a $N(0, 1)$ random variable when β_0 is zero. Thus, the hypothesis that the intercept is zero would be rejected at the 0.001 level.

An alternative method of testing the hypothesis that $\beta_0 = 0$ is to estimate β_1 using the uncorrected sums of squares and products. Under the null hypothesis that $\beta_0 = 0$, an estimator of β_1 is

$$(M_{XX} - \hat{\lambda}\sigma_{uu})^{-1}M_{XY},$$

where $(M_{XY}, M_{XX}) = n^{-1} \sum_{t=1}^{n} X_t(Y_t, X_t)$ and $\hat{\lambda}$ is the smallest root of

$$|\mathbf{M}_{ZZ} - \lambda(0.25)\mathbf{I}| = 0.$$

For these data the smallest root is $\hat{\lambda} = 4.0573$. If the model is true, then $(n-1)^{-1}n\hat{\lambda}$ is approximately distributed as Snedecor's F with four and infinity degrees of freedom. Because $F = 5.07$, the hypothesis that $\beta_0 = 0$ is rejected at the 0.001 level. The t test is a test of the hypothesis that $y_t = \beta_1 x_1$ given that the alternative model for y_t is $y_t = \beta_0 + \beta_1 x_t$. One conjectures that the $\hat{\lambda}$ test will have smaller power than the t test against the alternative of the affine model, because the λ test has power against a wide range of alternatives. □ □

One cannot hope to specify parametric configurations completely such that the approximate distribution of Theorem 1.3.1 will be appropriate in practice. However, some guidance is possible. One rule of thumb is to consider the approximations adequate if

$$(n-1)^{-1}\hat{m}_{xx}^{-2}\hat{\sigma}_{uu}^2 < 0.001. \tag{1.3.36}$$

This is roughly equivalent to requiring the coefficient of variation of $(m_{XX} - \sigma_{uu})^{-1}m_{XX}$ as an estimator of $m_{xx}^{-1}(m_{xx} + \sigma_{uu})$ to be less than 5%. This suggestion is supported by Monte Carlo studies such as that of Martinez-Garza (1970) and Miller (1984), by higher-order expansions such as those described in Cochran (1977, p. 162), DeGracie and Fuller (1972), and Anderson (1974, 1976, 1977), and by the exact distributional theory of Sawa (1969) and Mariano and Sawa (1972). Also see Exercises 1.59 and 1.60.

For the pheasant data of Example 1.3.1,

$$(n-1)^{-1}\hat{m}_{xx}^{-2}\hat{\sigma}_{uu}^2 = (14)^{-1}[(3.12895)^{-1}(0.49229)]^2 = 0.0018.$$

For the cell data of Example 1.3.2,

$$(n-1)^{-1}\hat{m}_{xx}^{-2}\sigma_{uu}^2 = (4)^{-1}[(11.7744)^{-1}0.25]^2 = 0.0001.$$

Thus, although the sample size, n, for the pheasant example is larger, we expect the approximations to perform better for the cell example because $\hat{m}_{xx}^{-1}\sigma_{uu}$ is smaller for the cell example. This is confirmed when we compare exact and approximate confidence intervals for the two examples. See Exercise 1.31 and Example 1.3.3.

1.3.4. Tests of Hypotheses for the Slope

While we are unable to establish the exact distribution of $\hat{\beta}_1$, we can construct some exact tests for the model. Let model (1.3.1)–(1.3.2) hold with $\delta = \sigma_{uu}^{-1}\sigma_{ee}$

known and $\sigma_{eu} = 0$. We consider the null hypothesis that $\beta_1 = \beta_1^0$. If $\beta_1 = \beta_1^0$, then

$$Y_t - \bar{Y} - \beta_1^0(X_t - \bar{X}) = v_t - \bar{v},$$

where $v_t = e_t - \beta_1^0 u_t$, and the variance of v_t is

$$\sigma_{vv} = \sigma_{ee} + (\beta_1^0)^2\sigma_{uu} = [\delta + (\beta_1^0)^2]\sigma_{uu}.$$

Because we know δ and because (e_t, u_t) is normally distributed, we can construct a second random variable, say h_t, that will be independent of v_t if $\beta_1 = \beta_1^0$. We define

$$h_t = a_1 Y_t + a_2 X_t = a_1(\beta_0 + \beta_1 x_t) + a_2 x_t + a_1 e_t + a_2 u_t,$$

where a_1 and a_2 $(a_1^2 + a_2^2 \neq 0)$ are chosen so that, under the null, the covariance between h_t and v_t is zero. Because

$$C\{h_t, v_t\} = (a_1\delta - a_2\beta_1^0)\sigma_{uu},$$

h_t defined with $a_1 = \delta^{-1}\beta_1^0$ and $a_2 = 1$ will be uncorrelated with v_t under the null. If $\beta_1^0 \neq 0$, the definitions of h_t and v_t are symmetric. That is, the hypothesis that $\beta_1 = -\delta(\beta_1^0)^{-1}$ leads to a pair that is a multiple of the pair (v_t, h_t), but with the identification interchanged. This symmetry means that we must consider two types of alternative hypothesis. First, assume that we know the sign of β_1. Then the hypotheses are

$$\begin{aligned} &H_0: \beta_1 = \beta_1^0, \\ &H_A: \beta_1 \geqslant 0, \quad \text{and} \quad \beta_1 \neq \beta_1^0, \end{aligned} \tag{1.3.37}$$

where we have chosen nonnegative values as the parameter space for β_1, with no loss of generality. Under alternative hypothesis (1.3.37), the null hypothesis $\beta_1 = \beta_1^0$ corresponds to the hypothesis that the correlation between v_t and h_t is zero. A t test or an F test can be used to test for zero correlation. The null hypothesis that $\beta_1 = \beta_1^0$ is accepted if the hypothesis of zero correlation is accepted. Furthermore, the set of β_1 with $\beta_1 \in [0, \infty)$ for which the hypothesis of zero correlation is accepted at level α constitutes a $1 - \alpha$ confidence set for β_1.

We now consider the hypotheses,

$$\begin{aligned} &H_0: \beta_1 = \beta_1^0, \qquad \beta_1^0 \neq 0 \\ &H_A: \beta_1 \neq \beta_1^0. \end{aligned}$$

With this alternative hypothesis the symmetry in the definition of v_t and h_t must be recognized. The acceptance of the hypothesis of zero correlation is equivalent to accepting the hypothesis that $\beta_1 = \beta_1^0$ or that $\beta_1 = -\delta(\beta_1^0)^{-1}$.

However, if $\beta_1 = \beta_1^0$ and $\sigma_{xx} > 0$, then

$$V\{h_t\} = [\delta^{-1}(\beta_1^0)^2 + 1]^2\sigma_{xx} + [\delta^{-1}(\beta_1^0)^2 + 1]\sigma_{uu},$$
$$V\{v_t\} = \sigma_{ee} + (\beta_1^0)^2\sigma_{uu} = [1 + (\beta_1^0)^2\delta^{-1}]\delta\sigma_{uu}.$$

Therefore, the hypothesis that $\beta_1 = \beta_1^0$ corresponds to the composite hypothesis

$$H_0\colon C\{h_t, v_t\} = 0 \quad \text{and} \quad V\{h_t\} \geqslant \delta^{-1}V\{v_t\}.$$

Under the null hypothesis the three mean squares

$$\text{ms}_{(1)} = (n-2)^{-1}\left\{\sum_{t=1}^{n}(v_t - \bar{v})^2 - \left[\sum_{t=1}^{n}(h_t - \bar{h})^2\right]^{-1}\left[\sum_{t=1}^{n}(h_t - \bar{h})v_t\right]^2\right\},$$

$$\text{ms}_{(2)} = \left[\sum_{t=1}^{n}(h_t - \bar{h})^2\right]^{-1}\left[\sum_{t=1}^{n}(h_t - \bar{h})v_t\right]^2,$$

$$m_{hh} = (n-1)^{-1}\sum_{t=1}^{n}(h_t - \bar{h})^2$$

are mutually independent. This is because $\text{ms}_{(1)}$ is the residual mean square computed from the regression of v_t on h_t and $\text{ms}_{(2)}$ is the mean square due to regression. Under the null, h_t and v_t are independent and, hence, m_{hh} is independent of Σv_t^2. Because the regression residual mean square $\text{ms}_{(1)}$ is independent of m_{hh}, it follows that $\text{ms}_{(2)}$ is independent of m_{hh}. Furthermore,

$$E\{\text{ms}_{(1)}\} = E\{\text{ms}_{(2)}\} = \sigma_{vv}.$$

Also, the sum

$$(n-1)m_{vv} = (n-2)\text{ms}_{(1)} + \text{ms}_{(2)}$$

is independent of the ratio $[\text{ms}_{(1)}]^{-1}\text{ms}_{(2)}$. Therefore, the F test of the hypothesis that $C\{h_t, v_t\} = 0$, given by

$$F_{n-2}^{1} = [\text{ms}_{(1)}]^{-1}\text{ms}_{(2)}, \tag{1.3.38}$$

is independent of the F test of the hypothesis that $\sigma_{hh} \geqslant \delta^{-1}\sigma_{vv}$, given by

$$F_{n-1}^{n-1} = \delta^{-1}m_{hh}^{-1}m_{vv}. \tag{1.3.39}$$

These two tests may be combined in a number of different ways to obtain a test of the composite hypothesis. One procedure is composed of two steps.

Step 1. Test at level α_1 the hypothesis that the correlation between h_t and v_t is zero using the F test

$$F_{n-2}^{1} = [\text{ms}_{(1)}]^{-1}\text{ms}_{(2)}.$$

If the hypothesis of zero correlation is rejected, reject the hypothesis that $\beta_1 = \beta_1^0$. If the hypothesis of zero correlation is accepted, proceed to Step 2.

Step 2. Test at level α_2 the hypothesis that $\sigma_{hh} \geq \delta^{-1}\sigma_{vv}$ against the alternative that $\delta^{-1}\sigma_{vv} > \sigma_{hh}$ using the F test

$$F_{n-1}^{n-1} = \delta^{-1} m_{hh}^{-1} m_{vv}.$$

If the hypothesis that $\delta^{-1}\sigma_{vv} \leq \sigma_{hh}$ is accepted, accept the null hypothesis that $\beta_1 = \beta_1^0$. Otherwise, reject the hypothesis that $\beta_1 = \beta_1^0$.

The probability of rejecting the hypothesis $\beta_1 = \beta_1^0$ when it is true is

$$P\{\text{reject } \beta_1^0 | \beta_1\} = \alpha_1 + (1 - \alpha_1)P\{\text{reject } \delta^{-1}\sigma_{vv} \leq \sigma_{hh} | \beta_1 = \beta_1^0\}.$$

The probability of rejecting the hypothesis that $\delta^{-1}\sigma_{vv} \leq \sigma_{hh}$ is a function of the true σ_{xx}. This probability achieves its maximum of α_2 for $\sigma_{xx} = 0$. Therefore,

$$\alpha_1 \leq P\{\text{reject } \beta_1^0 | \beta_1 = \beta_1^0\} \leq \alpha_1 + (1 - \alpha_1)\alpha_2.$$

Given a test, it is possible to construct the associated confidence set. Therefore, a confidence set for β_1 with coverage probability greater than or equal to $\alpha_1 + (1 - \alpha_1)\alpha_2$ is the set of β_1 for which

$$[\text{ms}_{(1)}]^{-1}\text{ms}_{(2)} \leq F_{n-2}^{1}(\alpha_1)$$

and

$$\delta^{-1} m_{hh}^{-1} m_{vv} \leq F_{n-1}^{n-1}(\alpha_2).$$

The confidence sets constructed by the methods of this section will not always be a single interval and can be the whole real line. For composite hypotheses any β_1 on the real line is consistent with the data when the quadratic in β_1 defined by (1.3.39) has no real root. This is roughly equivalent to the data failing to reject the hypothesis that σ_{xx} is zero. Also, it is possible that the hypothesis of zero covariance will be accepted for all β_1.

In our treatment we have assumed that $\sigma_{eu} = 0$. This simplifies the discussion and represents no loss of generality. One can always transform the problem to one with equal-variance uncorrelated errors. For example, if $\boldsymbol{\Upsilon}_{\varepsilon\varepsilon}$ is known, we define

$$Y_t^* = (\Upsilon_{ee} - \Upsilon_{eu}\Upsilon_{uu}^{-1}\Upsilon_{ue})^{-1/2}(Y_t - X_t\Upsilon_{uu}^{-1}\Upsilon_{ue}),$$
$$X_t^* = \Upsilon_{uu}^{-1/2}X_t.$$

Then the error covariance matrix for (Y_t^*, X_t^*) is a multiple of the identity matrix. The transformed model is

$$Y_t^* = \beta_0^* + \beta_1^* x_t + e_t, \tag{1.3.40}$$

where the hypothesized value for β_1^* is $(\Upsilon_{ee} - \Upsilon_{eu}\Upsilon_{uu}^{-1}\Upsilon_{ue})^{1/2}\Upsilon_{uu}^{1/2}\beta_1^0 - \Upsilon_{uu}^{-1/2}\Upsilon_{ue}$.

Example 1.3.3. We illustrate the methods of this section using the data of Example 1.3.2. We construct a test of the hypothesis

$$H_0: \beta_1 = 0.791$$

against the alternative, $H_A: \beta_1 \geqslant 0$ and $\beta_1 \neq 0.791$. Under the null hypothesis $Y_t - 0.791X_t$ has variance

$$[1 + (0.791)^2]0.25 = 0.4064$$

and is uncorrelated with $h_t = 0.791Y_t + X_t$. Because we assume we know that $\beta_1 \geqslant 0$, the null hypothesis that $\beta_1 = \beta_1^0$ is equivalent to the hypothesis that $\gamma_1 = 0$ in the regression equation

$$Y_t - 0.791X_t = \gamma_0 + \gamma_1 h_t + v_t,$$

where $v_t \sim \text{NI}(0, 0.4064)$. Computing the regression of $Y_t - 0.791X_t$ on h_t, we obtain

$$t = \left[0.4064\left\{\sum_{t=1}^{5}(h_t - \bar{h})^2\right\}^{-1}\right]^{-1/2}\hat{\gamma}_1 = -1.96.$$

Thus, if $\beta_1 = 0.791$, the probability of getting a $\hat{\gamma}_1$ this small or smaller is 0.025. Because we know σ_{vv} under the null hypothesis, the distribution of the "t statistic" is that of a $N(0, 1)$ random variable. We chose the value 0.791 to yield a value of -1.96 for the test statistic. It can be verified that the hypothesis $H_0: \beta_1 = 0.450$ will yield a test statistic of 1.96. Therefore, an exact 95% confidence interval for β_1 is the interval (0.450, 0.791). That is, all values in the interval are accepted by the test.

If we use the approximate distribution theory and the estimated variance $\hat{V}\{\hat{\beta}_1\} = 0.007418$ calculated in Example 1.3.3, we obtain the approximate 95% confidence interval whose end points are $0.6079 \pm 1.96(0.0861)$. This interval is (0.439, 0.777), which is very close to the exact interval (0.450, 0.791). In this example the approximate theory works well because σ_{uu} is small relative to m_{xx}. □ □

REFERENCES

Sections 1.3.1–1.3.3. Adcock (1877, 1878), Amemiya (1980), Anderson (1951b, 1976, 1977, 1984), Birch (1964), Fuller (1980), Kendall (1951, 1952), Kendall and Stuart (1979), Koopmans (1937), Kummell (1879), Pearson (1901), Sprent (1966), Tintner (1945), Villegas (1961).

Section 1.3.4. Creasy (1956), Fieller (1954), Gleser and Hwang (1985), Williams (1955).

EXERCISES

20. (Sections 1.3.2, 1.3.3) The data below are 10 pairs of observations on hectares of corn for 10 area segments as determined by aerial photography (Y_t) and by personal interview (X_t).

$$\mathbf{Y} = (97.1, 89.8, 84.2, 88.2, 87.0, 93.1, 99.6, 94.7, 83.4, 78.5),$$
$$\mathbf{X} = (96.3, 87.4, 88.6, 88.6, 88.6, 93.5, 92.9, 99.0, 77.7, 76.1).$$

Assume that the data satisfy model (1.3.1), (1.3.2) with $\sigma_{ee} = \sigma_{uu}$.
(a) Estimate $(\beta_0, \beta_1, \sigma_{uu})$ and the covariance matrix of the estimators.
(b) Treating x_t as fixed, compute the estimates of (x_t, y_t).
(c) Plot the data, the line, and $(\hat{x}_t, \hat{y}_t)$. Plot $\hat{v}_t$ against $\hat{x}_t$.
(d) Compute an estimate of the variance of $\hat{x}_t$.

21. (Sections 1.3.2, 1.3.3)
(a) Verify, for $\sigma_{eu} = 0$, that the estimators (1.3.27) and (1.3.7) are algebraically equivalent.
(b) Verify Equation (1.3.19).

22. (Section 1.3.2) Estimate the parameters of the model studied in Example 1.3.1 under the assumption that $\delta = 3$. Estimate the covariance matrix of your estimators.

23. (Section 1.3.2) Let $y_t = \beta_0 + \beta_1 x_t$,

$$(x_t, e_t, u_t)' \sim \mathrm{NI}[(\mu_x, 0, 0)', \mathrm{diag}(\sigma_{xx}, \sigma_{ee}, \sigma_{uu})],$$

and $(Y_t, X_t) = (y_t, x_t) + (e_t, u_t)$. Let $\gamma_{yy} = \sigma_{ee}\sigma_{yy}^{-1}$ and $\gamma_{xx} = \sigma_{uu}\sigma_{xx}^{-1}$, where $\sigma_{yy} = \beta_1^2\sigma_{xx}$. The ratio $\gamma_{xx} = \sigma_{uu}\sigma_{xx}^{-1}$ sometimes is called the *noise-to-signal ratio* of X. Give the variance expression Γ_{22} of Theorem 1.3.1 in terms of γ_{yy} and γ_{xx}.

24. (Sections 1.3.2, 1.3.3) Let $\hat{\beta}_1$, $\hat{v}_t$, and $\hat{x}_t$ be defined by (1.3.27), (1.3.28), and (1.3.29).
(a) Show that $\sum_{t=1}^{n} \hat{v}_t \hat{x}_t = 0$.
(b) Show that

$$\hat{\beta}_1 = \left[\sum_{t=1}^{n} (\hat{x}_t - \bar{X})^2\right]^{-1} \sum_{t=1}^{n} (\hat{x}_t - \bar{X})(Y_t - \bar{Y}).$$

(c) Let $\hat{m}_{xx}$, $\hat{\sigma}_{vv}$, and $\hat{\sigma}_{uv}$ be defined by (1.3.31). Show that

$$\hat{m}_{xx} = m_{XX} - \hat{\lambda}\sigma_{uu} + (\hat{\lambda} - 1)(\sigma_{uu} - \hat{\sigma}_{vv}^{-1}\hat{\sigma}_{vu}^2),$$

where $\hat{\lambda}$ is the smallest root of (1.3.27).

25. (Sections 1.3.2, 1.3.3) Estimate the true x and y values for Example 1.3.1, assuming the x values to be a random sample from a normal population. Give the approximate covariance matrix of your estimators.

26. (Section 1.3.4) Construct a 95% confidence interval for the β_1 of Example 1.3.1 given that the parameter space is the entire real line.

27. (Section 1.3.2) Show the equivalence of the two expressions for s_{vv} defined following (1.3.12).

28. (Sections 1.3.2, 1.3.3) Let

$$y_t = \beta x_t, \qquad (Y_t, X_t) = (y_t, x_t) + (e_t, u_t),$$

for $t = 1, 2, \ldots, n$, where $E\{(e_t, u_t)\} = (0, 0)$. Let $u_t \equiv 0$ so that

$$E\{(e_t, u_t)'(e_t, u_t)\} = \sigma_{ee}\,\mathrm{diag}(1, 0).$$

Use (1.3.27) to obtain an estimator of β. Show that this estimator is the ordinary least squares regression coefficient obtained by regressing Y_t on x_t. Use formula (1.3.12) to construct an estimator of the variance of the estimator.

29. (Section 1.3.3) Treating $\hat{\beta}_0$ and $\hat{\beta}_1$ as known parameters, compute the estimated variance of $\hat{y}_t - y_t$ for Example 1.3.2. Why is the estimated variance of $\hat{y}_t - y_t$ smaller than the estimated variance of $\hat{x}_t - x_t$?

30. (Section 1.3.3) Let $\mathbf{X}_t = (1, X_{t2})$,

$$\mathbf{M} = n^{-1} \sum_{t=1}^{n} (Y_t, \mathbf{X}_t)'(Y_t, \mathbf{X}_t),$$

$$\mathbf{m} = (n-1)^{-1} \sum_{t=1}^{n} (Y_t - \bar{Y}, X_{t2} - \bar{X}_2)'(Y_t - \bar{Y}, X_{t2} - \bar{X}_2),$$

where $\bar{X}_2 = n^{-1} \sum_{t=1}^{n} X_{t2}$. Let $\hat{\lambda}$ and $\hat{\gamma}$ be the smallest roots of

$$|\mathbf{M} - \lambda\mathbf{\Sigma}| = 0 \quad \text{and} \quad |\mathbf{m} - \gamma\mathbf{I}_2| = 0,$$

respectively, where $\mathbf{\Sigma} = \text{diag}(1, 0, 1)$ and $\mathbf{I}_2$ is the identity matrix of dimension two. What is the relationship between $\hat{\lambda}$ and $\hat{\gamma}$?

31. (Section 1.3.4) Using the methods of Section 1.3.4, construct an exact 95% confidence interval for β_1 for the data of Example 1.3.1. You may assume that it is known that $\beta_1 \geqslant 0$. Compare the exact and approximate intervals.

32. (Section 1.3.4) Using the data of Example 1.3.1, construct a test of the hypothesis H_0: $(\beta_0, \beta_1) = (0.10, 0.95)$ against the alternative, H_A: $(\beta_0, \beta_1) \neq (0.10, 0.95)$, $\beta_1 > 0$.

33. (Section 1.3.3) Let $\hat{\lambda}_1 \geqslant \hat{\lambda}_2$ be the two roots of

$$|\mathbf{m}_{ZZ} - \lambda\mathbf{\Upsilon}_{\varepsilon\varepsilon}| = 0,$$

where (Y_t, X_t) satisfies model (1.3.2), with $\mathbf{\Sigma}_{\varepsilon\varepsilon} = \mathbf{\Upsilon}_{\varepsilon\varepsilon}\sigma^2$, and σ^2 is unknown. Show that

$$\hat{\sigma}_{xx} = [(\hat{\beta}_1, 1)\mathbf{\Upsilon}_{\varepsilon\varepsilon}^{-1}(\hat{\beta}_1, 1)']^{-1}(\hat{\lambda}_1 - \hat{\lambda}_2),$$

where $\hat{\sigma}_{xx}$ is defined in (1.3.8).

34. (Sections 1.3.2, 1.3.3) Assume that the data of Exercise 1.5 satisfy model (1.3.1) with $\delta = 0.5$. Estimate $(\beta_0, \beta_1, \sigma_{uu})$. Estimate the covariance matrix of the approximate distribution of $(\hat{\beta}_0, \hat{\beta}_1)$. Plot $\hat{v}_t$ against $\hat{x}_t$.

1.4. INSTRUMENTAL VARIABLE ESTIMATION

In Sections 1.1.2, 1.2, and 1.3, we constructed estimators of the parameters of structural models that were identified by knowledge about the error variances. In this section we consider the use of a different type of auxiliary information.

Assume that the model of interest specifies

$$Y_t = \beta_0 + \beta_1 x_t + e_t, \qquad X_t = x_t + u_t, \tag{1.4.1}$$

for $t = 1, 2, \ldots, n$, where the e_t are independent $(0, \sigma_{ee})$ random variables. The vector (Y_t, X_t) is observed, where u_t is the measurement error in X_t. In addition, we have available a third variable, denoted by W_t, known to be correlated with x_t. For example, we might conduct an agronomic experiment where X_t is the observed nitrogen in the leaves of the plant and Y_t is the dry weight of the plant. We would expect the true nitrogen in the leaves, x_t, to be correlated with nitrogen fertilizer, W_t, applied to the experimental plot. Furthermore, it is reasonable to assume that both u_t and e_t are independent of W_t. We give the definition of a variable such as W_t.

Definition 1.4.1. Let model (1.4.1) hold and let a variable W_t satisfy

$$\text{(i) } E\left\{n^{-1}\sum_{t=1}^{n}(W_t-\bar{W})(e_t, u_t)\right\}=(0, 0), \tag{1.4.2}$$

$$\text{(ii) } E\left\{n^{-1}\sum_{t=1}^{n}(W_t-\bar{W})x_t\right\}\neq 0, \tag{1.4.3}$$

where $\bar{W}=n^{-1}\sum_{t=1}^{n}W_t$. Then W_t is called an *instrumental variable* for x_t of model (1.4.1).

It is convenient to have a parametric expression for the fact that x_t and W_t are related, and we use the parameters of the population regression of x_t on W_t to quantify the relationship. Let

$$\pi_{22}=\left[E\left\{\sum_{t=1}^{n}(W_t-\bar{W})^2\right\}\right]^{-1}E\left\{\sum_{t=1}^{n}(W_t-\bar{W})x_t\right\}, \tag{1.4.4}$$

$$\pi_{12}=E\{\bar{x}-\pi_{22}\bar{W}\}. \tag{1.4.5}$$

Condition (ii) of Definition 1.4.1 is equivalent to specifying $\pi_{22}\neq 0$. The π coefficients are defined with double subscripts so that they will be consistent with the notation of higher-order models. Using the π coefficients, we can write

$$x_t=\pi_{12}+\pi_{22}W_t+r_t, \qquad t=1, 2, \ldots, n, \tag{1.4.6}$$

where r_t is the failure of x_t to be perfectly linearly related to W_t. By the regression method of construction, r_t has zero correlation with W_t. Equation (1.4.6) and the definition of X_t yield

$$X_t=\pi_{12}+\pi_{22}W_t+a_{t2}, \tag{1.4.7}$$

where $a_{t2}=r_t+u_t$ and $E\{\sum_{t=1}^{n}W_t a_{t2}\}=0$.

In model (1.4.1) and Definition 1.4.1, x_t can be fixed, random, or a sum of fixed and random components. Likewise, the variable W_t can be fixed, random, or a sum of random and fixed components. If W_t and x_t contain fixed components, π_{22} and π_{12} could properly be subscripted with n because the expected value will be a function of the fixed components of the n observations. If both x_t and W_t are fixed, condition (ii) of Definition 1.4.1 reduces to

$$n^{-1}\sum_{t=1}^{n}(W_t-\bar{W})x_t\neq 0. \tag{1.4.8}$$

One possible choice for W_t is a measurement of x_t obtained by an independent method. If x_t is fixed and if W_t is a measure of x_t containing error, then W_t is a sum of fixed and random components.

We defined an instrumental variable in terms of model (1.4.1). If it is known that $\beta_0 = 0$, then the conditions (i) and (ii) become

$$\text{(i}'\text{)} \quad E\left\{n^{-1} \sum_{t=1}^{n} W_t(e_t, u_t)\right\} = (0, 0).$$

$$\text{(ii}'\text{)} \quad E\left\{n^{-1} \sum_{t=1}^{n} W_t x_t\right\} \neq 0.$$

Therefore, if the mean of x_t is not zero, the variable that is identically one becomes a possible instrumental variable for the model with structural line passing through the origin.

The idea behind instrumental variable estimation is seen easily when all variables are normally distributed. Therefore, to introduce the estimators, we assume

$$\begin{bmatrix} x_t \\ e_t \\ u_t \\ W_t \end{bmatrix} \sim \mathrm{NI}\left(\begin{bmatrix} \mu_x \\ 0 \\ 0 \\ \mu_W \end{bmatrix}, \begin{bmatrix} \sigma_{xx} & \sigma_{xe} & \sigma_{xu} & \sigma_{xW} \\ \sigma_{xe} & \sigma_{ee} & \sigma_{eu} & 0 \\ \sigma_{xu} & \sigma_{eu} & \sigma_{uu} & 0 \\ \sigma_{xW} & 0 & 0 & \sigma_{WW} \end{bmatrix}\right), \tag{1.4.9}$$

where $\mu_x = \pi_{12} + \pi_{22}\mu_W$ and we assume $\sigma_{xW} = \pi_{22}\sigma_{WW} \neq 0$. The model specifies two zero values in the mean vector and two zero covariances in the covariance matrix of (1.4.9), but the remaining parameters are unknown. In particular, we have not assumed σ_{eu} to be known or σ_{xu} to be zero. Under the assumptions, the observed vector (Y_t, X_t, W_t) is normally distributed with mean vector

$$(\mu_Y, \mu_X, \mu_W) = (\beta_0 + \beta_1\pi_{12} + \beta_1\pi_{22}\mu_W, \pi_{12} + \pi_{22}\mu_W, \mu_W)$$

and covariance matrix

$$\begin{bmatrix} \beta_1^2\sigma_{xx} + 2\beta_1\sigma_{xe} + \sigma_{ee} & \beta_1\sigma_{xx} + \sigma_{xe} + \beta_1\sigma_{xu} + \sigma_{eu} & \beta_1\pi_{22}\sigma_{WW} \\ \beta_1\sigma_{xx} + \sigma_{xe} + \beta_1\sigma_{xu} + \sigma_{eu} & \sigma_{xx} + 2\sigma_{xu} + \sigma_{uu} & \pi_{22}\sigma_{WW} \\ \beta_1\pi_{22}\sigma_{WW} & \pi_{22}\sigma_{WW} & \sigma_{WW} \end{bmatrix}. \tag{1.4.10}$$

The model (1.4.1), (1.4.9) contains 12 independent unknown parameters. One set of parameters is

$$(\mu_W, \pi_{12}, \pi_{22}, \beta_0, \beta_1, \sigma_{WW}, \sigma_{xx}, \sigma_{xe}, \sigma_{xu}, \sigma_{uu}, \sigma_{ee}, \sigma_{eu}).$$

The set of minimal sufficient statistics for a sample of n observations is the mean vector and the sample covariances of (Y_t, X_t, W_t). There are nine statistics in this set. Therefore, we cannot hope to estimate all 12 of the parameters of the model. However, we note that the ratio of covariances

$$\sigma_{XW}^{-1}\sigma_{YW} = (\pi_{22}\sigma_{WW})^{-1}\beta_1\pi_{22}\sigma_{WW} = \beta_1. \tag{1.4.11}$$

It follows that we can estimate β_1 and β_0 by

$$\hat{\beta}_1 = m_{XW}^{-1}m_{YW}, \tag{1.4.12}$$

$$\hat{\beta}_0 = \bar{Y} - \hat{\beta}_1\bar{X}, \tag{1.4.13}$$

where $(\bar{Y}, \bar{X}) = n^{-1}\sum_{t=1}^{n}(Y_t, X_t)$ and

$$(m_{YW}, m_{XW}) = (n-1)^{-1}\sum_{t=1}^{n}(Y_t - \bar{Y}, X_t - \bar{X})(W_t - \bar{W}).$$

Because the sample moments are consistent estimators of the population moments, the estimators of β_1 and β_0 will be consistent under the model assumptions. The assumption that $\sigma_{xW} \neq 0$ is critical. If $\sigma_{xW} = 0$, the denominator of $\hat{\beta}_1$ defined in (1.4.12) is estimating zero and $\hat{\beta}_1$ is not a consistent estimator of β_1. The limiting properties of the estimators are given in Theorem 1.4.1. The assumptions of the theorem are less restrictive than the assumption of trivariate normality used in introducing the estimator.

Theorem 1.4.1. Let model (1.4.1) hold. Let the vectors $(x_t, e_t, u_t, W_t)'$ be independently and identically distributed with mean $(\mu_x, 0, 0, \mu_W)'$ and finite fourth moments. Assume that the covariances σ_{We} and σ_{Wu} are zero and that $\sigma_{xW} \neq 0$. Also assume

$$\begin{aligned} E\{(W_t - \mu_W)^2 v_t^2\} &= \sigma_{WW}\sigma_{vv}, \\ E\{v_t^2(W_t - \mu_W)\} &= 0, \end{aligned} \tag{1.4.14}$$

where $v_t = e_t - \beta_1 u_t$. Let $(\hat{\beta}_0, \hat{\beta}_1)$ be defined by (1.4.13) and (1.4.12). Then

$$n^{1/2}\begin{bmatrix} \hat{\beta}_0 - \beta_0 \\ \hat{\beta}_1 - \beta_1 \end{bmatrix} \xrightarrow{L} N\left(\begin{bmatrix} 0 \\ 0 \end{bmatrix}, \begin{bmatrix} \sigma_{vv} + \mu_x^2 V_{22} & -\mu_x V_{22} \\ -\mu_x V_{22} & V_{22} \end{bmatrix}\right),$$

where $V_{22} = \sigma_{xW}^{-2}\sigma_{WW}\sigma_{vv}$.

Proof. The error in the sample moments is $O_p(n^{-1/2})$ because they are, approximately, the mean of n independent, identically distributed random variables with finite variance. Using the method of statistical differentials of Appendix 1.A, we have

$$\begin{aligned} \hat{\beta}_1 - \beta_1 &= m_{XW}^{-1}m_{Wv} = \sigma_{XW}^{-1}m_{Wv} + O_p(n^{-1}) \\ &= n^{-1}\sigma_{XW}^{-1}\sum_{t=1}^{n}(W_t - \mu_W)v_t + O_p(n^{-1}), \end{aligned}$$

where we have used $Y_t = \beta_0 + \beta_1 X_t + v_t$ and $m_{YW} = \beta_1 m_{XW} + m_{Wv}$. The variance of $n^{-1} \sum_{t=1}^{n} (W_t - \mu_W)v_t$ is $n^{-1}\sigma_{WW}\sigma_{vv}$ by assumption. The quantity $n^{-1/2}M_{Wv}$ converges in distribution to a normal random variable because the random variables $(W_t - \mu_W)v_t$ are independently and identically distributed with zero mean and finite variance. The limiting distribution of $n^{1/2}(\hat{\beta}_1 - \beta_1)$ then follows.

Writing

$$\hat{\beta}_0 - \beta_0 = \bar{v} - \mu_x(\hat{\beta}_1 - \beta_1) + O_p(n^{-1})$$

$$= n^{-1} \sum_{t=1}^{n} [v_t - \mu_x \sigma_{XW}^{-1}(W_t - \mu_W)v_t] + O_p(n^{-1})$$

and using $E\{v_t^2(W_t - \mu_W)\} = 0$, we obtain the result for $\hat{\beta}_0$. □

Theorem 1.4.1 covers a broad range of possibilities. In particular, it is possible for (e_t, u_t) to be correlated with x_t, and e_t may be correlated with u_t. The critical assumptions $\sigma_{We} = \sigma_{Wu} = 0$ and $\sigma_{xW} \neq 0$ permit the estimation of β_1 in the presence of measurement error that is correlated with the true values. The theorem was given for random (x_t, W_t), but the result holds for fixed x_t and for x_t that are the sums of fixed and random components, under mild conditions.

Because the sample moments are consistent estimators of the population moments, a consistent estimator of the covariance matrix of the approximate distribution of $(\hat{\beta}_0, \hat{\beta}_1)$ is

$$\hat{\mathbf{V}}\{(\hat{\beta}_0, \hat{\beta}_1)'\} = \begin{bmatrix} n^{-1}s_{vv} + \bar{X}^2\hat{V}\{\hat{\beta}_1\} & -\bar{X}\hat{V}\{\hat{\beta}_1\} \\ -\bar{X}\hat{V}\{\hat{\beta}_1\} & \hat{V}\{\hat{\beta}_1\} \end{bmatrix}, \quad (1.4.15)$$

where

$$\hat{V}\{\hat{\beta}_1\} = (n-1)^{-1} m_{XW}^{-2} m_{WW} s_{vv},$$

$$s_{vv} = (n-2)^{-1} \sum_{t=1}^{n} [Y_t - \bar{Y} - \hat{\beta}_1(X_t - \bar{X})]^2. \quad (1.4.16)$$

In most practical applications of the method of instrumental variables, one will wish to check the hypothesis that $\sigma_{WX} \neq 0$. A test that $\sigma_{WX} = 0$, under the assumption that $\sigma_{uW} = 0$, can be constructed by testing the hypothesis that the regression coefficient, π_{22}, in the regression of X_t on W_t is zero.

While the distribution of Theorem 1.4.1 is only approximate, one can construct an exact test of the hypothesis

$$H_0: \beta_1 = \beta_1^0$$

against the alternative, $H_A: \beta_1 \neq \beta_1^0$, under the stronger assumptions that (e_t, u_t) is normally distributed and independent of W_t. The random variable

$$h_t = Y_t - \beta_1^0 X_t = \beta_0 + (\beta_1 - \beta_1^0)x_t - \beta_1^0 u_t + e_t \quad (1.4.17)$$

is then independent of W_t if and only if $\beta_1 = \beta_1^0$. It follows that a test of the hypothesis that the population regression coefficient of h_t on W_t is zero is a test of the hypothesis that $\beta_1 = \beta_1^0$. Also, the set of values of β_1^0 for which the hypothesis is accepted at the η level constitutes a $1 - \eta$ confidence set for β_1. The confidence set is the set of β_1^0 for which

$$(n-2)^{-1} t_\eta^2 \geq (m_{hh} m_{WW} - m_{hW}^2)^{-1} m_{hW}^2$$

or, equivalently, the set of β_1^0 for which

$$t_\eta^2 \geq \frac{(n-2)(m_{YW} - \beta_1^0 m_{XW})^2}{[m_{YY} - 2\beta_1^0 m_{XY} + (\beta_1^0)^2 m_{XX}] m_{WW} - (m_{YW} - \beta_1^0 m_{XW})^2}, \quad (1.4.18)$$

where t_η is the η point of Student's t distribution with $n - 2$ degrees of freedom. The boundaries of the set are the solutions to the quadratic equation

$$\{m_{XX} m_{WW} - [1 + t_\eta^{-2}(n-2)] m_{XW}^2\} \beta_1^2 - 2\{m_{WW} m_{XY} - [1 + t_\eta^{-2}(n-2)] m_{YW} m_{XW}\} \beta_1 + m_{YY} m_{WW} - [1 + t_\eta^{-2}(n-2)] m_{YW}^2 = 0, \quad (1.4.19)$$

provided the solutions are real. The values of β_1 satisfying (1.4.18) are generally those in the closed interval with end points given by the solutions to (1.4.19). However, the set is sometimes the real line with the open interval deleted and, if the solutions to (1.4.19) are imaginary, the set is the entire real line. The condition (1.4.18) must be satisfied for the confidence set.

Given that one has an instrumental variable available, one might ask if the ordinary least squares estimator obtained by regressing Y_t on X_t is unbiased for β_1. The ordinary least squares estimator is unbiased when $\sigma_{XY} = \beta_1 \sigma_{XX}$, and under this condition, it follows from (1.4.10) that the population regression of Y_t on (X_t, W_t) is

$$\begin{bmatrix} \sigma_{XX} & \pi_{22}\sigma_{WW} \\ \pi_{22}\sigma_{WW} & \sigma_{WW} \end{bmatrix}^{-1} \begin{bmatrix} \beta_1 \sigma_{XX} \\ \pi_{22}\beta_1\sigma_{WW} \end{bmatrix} = \begin{bmatrix} \beta_1 \\ 0 \end{bmatrix}. \quad (1.4.20)$$

From (1.4.20) a test of the hypothesis that the ordinary least squares estimator of β_1 is unbiased is equivalent to a test of the hypothesis that the coefficient of W_t in the multiple regression of Y_t on X_t and W_t is zero. That is, we test the hypothesis that $\gamma = 0$ in the equation

$$Y_t = \beta_0 + \beta_1 X_t + \gamma W_t + v_t. \quad (1.4.21)$$

The analysis associated with (1.4.21) also explains why variables whose theoretical coefficients are zero are sometimes significant in an ordinary least squares regression. If theory specifies Y to be a function of x only, x is measured imperfectly by X, and W is correlated with x, then the coefficient for W in the multiple regression of Y on X and W is not zero.

Example 1.4.1. To illustrate the use of an instrumental variable we study reported magnitudes of Alaskan earthquakes for the period from 1969 to 1978. The data are from the National Oceanic and Atmospheric Administration's Hypocenter Data File (Meyers and von Hake, 1976). These data have been studied by Ganse, Amemiya, and Fuller (1983).

Three measures of earthquake magnitude are the logarithm of the seismogram amplitude of 20 second surface waves, denoted by Y_t, the logarithm of the seismogram amplitude of longitudinal body waves, denoted by X_t, and the logarithm of maximum seismogram trace amplitude at short distance, denoted by W_t. Table 1.4.1 contains the three reported magnitudes for 62 Alaskan earthquakes. These magnitudes are designed to be measures of earthquake "strength." Strength is a function of such things as rupture length and stress drop at the fault, both of which increase with strength. A model could be formulated to specify average rupture length and stress drop for a given strength. In addition to variations in fault length and stress drop from averages given by the strength model, there is a measurement error associated with the observations. The measurement error includes errors made in determining the amplitude of ground motion arising from such things as the orientation of a limited number of observation stations to the fault plane of the earthquake.

In this example the relationship between the amplitude of surface waves, Y_t, and the true value of body waves, x_t, is of interest. The proposed model is

$$Y_t = \beta_0 + \beta_1 x_t + e_t, \qquad X_t = x_t + u_t,$$

where x_t is the true earthquake strength in terms of body waves and (e_t, u_t) is the vector of measurement errors, it being understood that "measurement error" also includes the failure of the basic model to hold exactly for each earthquake. We assume (e_t, u_t) is uncorrelated with the measurement W_t. The vectors (Y_t, X_t, W_t) are assumed to satisfy the conditions of Theorem 1.4.1.

The sample mean vector for the data of Table 1.4.1 is

$$(\bar{Y}, \bar{X}, \bar{W}) = (5.0823, 5.2145, 5.2435)$$

and the sample covariance matrix is

$$\begin{bmatrix} m_{YY} & m_{YX} & m_{YW} \\ m_{XY} & m_{XX} & m_{XW} \\ m_{WY} & m_{WX} & m_{WW} \end{bmatrix} = \begin{bmatrix} 0.6198 & 0.2673 & 0.4060 \\ 0.2673 & 0.2121 & 0.2261 \\ 0.4060 & 0.2261 & 0.4051 \end{bmatrix}.$$

The test of the hypothesis that $\sigma_{xW} = 0$ is constructed by regressing X_t on W_t and testing the hypothesis that the coefficient of W_t is zero. The regression coefficient is 0.5581 and the t statistic is 9.39. Therefore, we are comfortable using W_t as an instrumental variable. By (1.4.12) and (1.4.13), the instrumental

TABLE 1.4.1. Three measures of strength for 62 Alaskan earthquakes

Observation t	Surface Wave Y_t	Body Wave X_t	Trace W_t	Observation t	Surface Wave Y_t	Body Wave X_t	Trace W_t
1	5.5	5.1	5.6	32	5.4	5.7	5.1
2	5.7	5.5	6.0	33	5.3	5.7	5.7
3	6.0	6.0	6.4	34	5.7	5.9	5.8
4	5.3	5.2	5.2	35	4.8	5.8	5.7
5	5.2	5.5	5.7	36	6.4	5.8	5.7
6	4.7	5.0	5.1	37	4.2	4.9	4.9
7	4.2	5.0	5.0	38	5.8	5.7	5.9
8	5.2	5.7	5.5	39	4.6	4.8	4.3
9	5.3	4.9	5.0	40	4.7	5.0	5.2
10	5.1	5.0	5.2	41	5.8	5.7	5.9
11	5.6	5.5	5.8	42	5.6	5.0	5.4
12	4.8	4.6	4.9	43	4.8	5.1	5.5
13	5.4	5.6	5.9	44	6.2	5.2	5.8
14	4.3	5.2	4.7	45	6.8	5.5	6.2
15	4.4	5.1	4.9	46	6.0	5.8	5.8
16	4.8	5.5	4.6	47	4.6	4.9	4.7
17	3.6	4.7	4.3	48	4.1	4.7	4.5
18	4.6	5.0	4.8	49	4.4	4.9	4.6
19	4.5	5.1	4.5	50	4.0	5.3	5.2
20	4.2	4.9	4.6	51	5.0	5.0	5.6
21	4.4	4.7	4.6	52	5.9	5.6	5.9
22	3.6	4.7	4.3	53	5.7	5.5	5.9
23	3.9	4.5	4.6	54	5.0	4.1	5.3
24	4.0	4.8	4.6	55	5.3	5.5	5.5
25	5.5	5.7	4.9	56	5.7	5.5	5.4
26	5.6	5.7	5.5	57	4.7	4.8	4.7
27	5.1	4.7	4.7	58	4.6	4.8	4.6
28	5.5	4.9	4.1	59	4.2	4.5	4.6
29	4.4	4.8	4.9	60	6.5	6.0	7.1
30	7.0	6.2	6.5	61	4.9	4.8	4.6
31	6.6	6.0	6.3	62	4.4	5.0	5.3

Source: Meyers and von Hake (1976).

variable estimator of (β_0, β_1) is

$$(\hat{\beta}_0, \hat{\beta}_1) = (-4.2829, 1.7960).$$

By (1.4.16) and (1.4.15), $s_{vv} = 0.3495$ and the estimated covariance matrix is

$$\hat{V}\{(\hat{\beta}_0, \hat{\beta}_1)'\} = \begin{bmatrix} 1.2404 & -0.2368 \\ -0.2368 & 0.0454 \end{bmatrix}.$$

Using $\hat{V}\{\hat{\beta}_1\}$, an approximate 95% confidence interval for β_1 is the interval (1.370, 2.222).

To illustrate the computation of an exact test for β_1, we test the hypothesis that $\beta_1 = 1.3$. We assume $(e_t, u_t)'$ to be normally distributed and regress the created variable

$$h_t = Y_t - 1.3X_t$$

on W_t. Student's t statistic for the hypothesis that the coefficient of W_t is zero is $t = 2.72$. If we calculate the t values for several h_t created with values of β_1 near 1.37 and for several h_t created with values of β_1 near 2.22, we find that we obtain a t value of -2.00 for $\beta_1 = 2.300$ and a t value of 2.00 for $\beta_1 = 1.418$. Note that the two-sided 5% point of Student's t distribution with 60 degrees of freedom is 2.00. Therefore, an exact 95% confidence interval for β_1 is (1.418, 2.300). Alternatively, one can construct the interval using (1.4.19) and (1.4.18).

To test the hypothesis that the ordinary least squares regression of Y_t on X_t would provide an unbiased estimator of β_1, we compute the multiple regression of Y_t on X_t and W_t. The estimated regression line is

$$\hat{Y}_t = \underset{(0.653)}{-1.258} + \underset{(0.195)}{0.474}X_t + \underset{(0.141)}{0.738}W_t,$$

where the numbers in parentheses are the estimated standard errors. The test of the hypothesis that the coefficient of W_t is zero is $t = 5.23$. Therefore, we reject the hypothesis that the ordinary least squares estimator of β_1 is unbiased.

See Exercise 1.42 for further analyses of these data. □ □

REFERENCES

Anderson (1976), Anderson and Rubin (1949, 1950), Basmann (1960), Durbin (1954), Fuller (1977), Halperin (1961), Johnston (1972), Sargan (1958), Sargan and Mikhail (1971).

EXERCISES

35. (Section 1.4)

(a) Let model (1.4.1) hold, let $\hat{\beta}_1$ be defined by (1.4.12), and let W_t be the instrumental variable. Show that

$$\sum_{t=1}^{n} W_t\hat{v}_t = \sum_{t=1}^{n} W_t[Y_t - \bar{Y} - \hat{\beta}_1(X_t - \bar{X})] = 0.$$

(b) Let model (1.4.1) hold with x_t fixed and let a fixed instrumental variable W_t be available. Let $m_{xW} = 0$. If $(e_t, u_t) \sim \text{NI}(\mathbf{0}, \mathbf{I}\sigma^2)$, what is the distribution of $\hat{\beta}_1$?

36. (Section 1.4) In Example 1.3.1 the estimated intercept was nonsignificant. Assuming the intercept to be zero, use the method of instrumental variables, with the constant function as the instrumental variable, to estimate β_1 of the model

$$Y_t = \beta_1 x_t + e_t, \qquad X_t = x_t + u_t.$$

Estimate the variance of the approximate distribution of the instrumental variable estimator of β_1.

37. (Section 1.4) Assume the model

$$Y_t = \beta_0 + \beta_1 x_t + e_t, \qquad X_t = x_t + u_t,$$
$$(x_t, e_t, u_t)' \sim \text{NI}[(\mu_x, \mathbf{0})', \text{diag}(\sigma_{xx}, \sigma_{ee}, \sigma_{uu})].$$

Let two independent identically distributed X determinations and a Y determination be made on a sample of n individuals. Consider the estimator

$$\tilde{\beta}_1 = \left[\sum_{t=1}^{n} (X_{t1} - \bar{X}_{.1})(X_{t2} - \bar{X}_{.2})\right]^{-1} \sum_{t=1}^{n} (X_{t1} - \bar{X}_{.1})(Y_t - \bar{Y}),$$

where X_{t1} and X_{t2} are the two determinations on X_t, and $\bar{X}_{.i} = n^{-1} \sum_{t=1}^{n} X_{ti}$. Give the variance of the limiting distribution of the estimator in terms of the parameters of the model.

38. (Section 1.4) Assume that only the value $W_{63} = 6.5$ is available for a new earthquake of the type studied in Example 1.4.1. Predict x_{63}. If the vector $(Y_{63}, X_{63}, W_{63}) = (6.5, 6.1, 6.5)$ is available, estimate x_{63} treating x_{63} as fixed. Compare the estimated variances of the two procedures under the added assumption that $\sigma_{eu} = 0$. In computing the variances, treat all estimates as if they were parameters. (Hint: Under the assumption $\sigma_{eu} = 0$, one can estimate σ_{ee} and σ_{uu}.)

39. (Section 1.4) Estimate the parameters of the model of Exercise 1.14 using condition, where condition is the value of i, as an instrumental variable. Estimate the parameters using $(i - 2)^2$ as an instrumental variable. Compare the estimated variance of $\hat{\beta}_1$ for the esimates constructed with the two different instrumental variables.

40. (Sections 1.1.2, 1.4) Let model (1.4.1) hold and let $W_t = x_t + b_t$, where (e_t, u_t, b_t) are independent identically distributed zero mean vectors with diagonal covariance matrix, finite fourth moments and $\sigma_{uu} = \sigma_{bb}$. Let (e_t, u_t, b_t) be independent of x_j for all t and j. Show that W_t satisfies the conditions for an instrumental variable. For x_t satisfying the conditions of Theorem 1.4.1 show that the correlation between X_t and W_t estimates κ_{xx}.

1.5. FACTOR ANALYSIS

The model of Section 1.4 contained a third variable and the assumption that the errors in the original (Y, X) pair were uncorrelated with the third variable. These assumptions enabled us to construct estimators of the parameters of the equation involving x_t but were not sufficient to permit estimation of all parameters. In this section we consider the model of Section 1.4 under added assumptions. Let

$$\begin{aligned} Y_{t1} &= \beta_{01} + \beta_{11}x_t + e_{t1}, \\ Y_{t2} &= \beta_{02} + \beta_{12}x_t + e_{t2}, \\ X_t &= x_t + u_t, \end{aligned} \tag{1.5.1}$$

where (Y_{t1}, Y_{t2}, X_t) can be observed. Assume that $\sigma_{xx} > 0$, $\beta_{11} \neq 0$, $\beta_{12} \neq 0$, and

$$(x_t, e_{t1}, e_{t2}, u_t)' \sim \mathrm{NI}[(\mu_x, 0, 0, 0)', \operatorname{diag}(\sigma_{xx}, \sigma_{ee11}, \sigma_{ee22}, \sigma_{uu})]. \tag{1.5.2}$$

It is the assumption that the covariance matrix is diagonal that will enable us to estimate all parameters of the model.

The model is the simplest form of the *factor* model used heavily in psychology and sociology. In the language of factor analysis, x_t is the *common factor* or *latent factor* in the three observed variables. The variables e_{t1}, e_{t2}, and u_t are sometimes called *unique factors*. Note that the first subscript of β_{ij} identifies the x variable (factor) and that the second subscript identifies the Y variable.

The two Y variables enter the model in a symmetric manner, and we choose to identify them with the same letter but different subscripts. This emphasizes the special assumptions of the factor model relative to the instrumental variable model. In fact, the three variables Y_{t1}, Y_{t2}, and X_t enter the model in a completely symmetric way, as we illustrate in Example 1.5.2.

Henceforth, we will often use the four subscript notation for population and sample moments. For example, the covariance matrix of $\mathbf{e}_t = (e_{t1}, e_{t2})$ is denoted by

$$\boldsymbol{\Sigma}_{ee} = \begin{bmatrix} \sigma_{ee11} & \sigma_{ee12} \\ \sigma_{ee21} & \sigma_{ee22} \end{bmatrix},$$

and the covariance between the first element of the vector $\mathbf{Y}_t = (Y_{t1}, Y_{t2})$ and the second element of $\mathbf{e}_t$ is denoted by $\sigma_{Ye12} = C\{Y_{t1}, e_{t2}\}$.

Under assumptions (1.5.1) and (1.5.2),

$$\begin{bmatrix} Y_{t1} \\ Y_{t2} \\ X_t \end{bmatrix} \sim \mathrm{NI}\left(\begin{bmatrix} \mu_{Y1} \\ \mu_{Y2} \\ \mu_x \end{bmatrix}, \begin{bmatrix} \beta_{11}^2\sigma_{xx} + \sigma_{ee11} & \beta_{11}\beta_{12}\sigma_{xx} & \beta_{11}\sigma_{xx} \\ \beta_{11}\beta_{12}\sigma_{xx} & \beta_{12}^2\sigma_{xx} + \sigma_{ee22} & \beta_{12}\sigma_{xx} \\ \beta_{11}\sigma_{xx} & \beta_{12}\sigma_{xx} & \sigma_{xx} + \sigma_{uu} \end{bmatrix} \right), \tag{1.5.3}$$

where $\mu_{Y1} = \beta_{01} + \beta_{11}\mu_x$ and $\mu_{Y2} = \beta_{02} + \beta_{12}\mu_x$.

The ratio of the variance associated with the common factor to the total variance of an observed variable is called the *communality* of the observed variable in factor analysis. Thus, the communality of Y_{t1} is

$$\kappa_{11}^2 = [\beta_{11}^2\sigma_{xx} + \sigma_{ee11}]^{-1}\beta_{11}^2\sigma_{xx} = 1 - \sigma_{YY11}^{-1}\sigma_{ee11}.$$

Note that κ_{11}^2 was called the reliability ratio in Section 1.1. The ratio $\sigma_{YY11}^{-1}\sigma_{ee11}$ is called the *uniqueness* of variable Y_{t1}.

Given a sample of n observations, the matrix of sample moments about the mean is

$$\mathbf{m}_{ZZ} = \begin{bmatrix} m_{YY11} & m_{YY12} & m_{YX11} \\ m_{YY21} & m_{YY22} & m_{YX21} \\ m_{XY11} & m_{XY12} & m_{XX11} \end{bmatrix},$$

where $\mathbf{Z}_t = (Y_{t1}, Y_{t2}, X_t)$, and, for example,

$$m_{YY12} = (n-1)^{-1} \sum_{t=1}^{n} (Y_{t1} - \bar{Y}_1)(Y_{t2} - \bar{Y}_2).$$

Since there is only one X variable, we can denote the sample variance of X by m_{XX} or by m_{XX11}.

The model contains nine independent unknown parameters. One set of parameters is $(\beta_{01}, \beta_{11}, \beta_{02}, \beta_{12}, \sigma_{ee11}, \sigma_{ee22}, \sigma_{uu}, \sigma_{xx}, \mu_x)$. Using (1.5.3), we can equate the sample moments to their expectations to obtain the estimators;

$$\begin{aligned} (\hat{\mu}_x, \hat{\beta}_{01}, \hat{\beta}_{02}) &= (\bar{X}, \bar{Y}_1 - \hat{\beta}_{11}\bar{X}, \bar{Y}_2 - \hat{\beta}_{12}\bar{X}), \\ (\hat{\beta}_{11}, \hat{\beta}_{12}) &= (m_{XY12}^{-1} m_{YY12}, m_{XY11}^{-1} m_{YY21}), \\ \hat{\sigma}_{eeii} &= m_{YYii} - \hat{\beta}_{1i}^2 \hat{\sigma}_{xx} \quad \text{for } i = 1, 2, \\ \hat{\sigma}_{uu} &= m_{XX} - \hat{\sigma}_{xx}, \\ \hat{\sigma}_{xx} &= m_{YY12}^{-1} m_{XY11} m_{XY12}. \end{aligned} \tag{1.5.4}$$

The estimator of β_{11} is the instrumental variable estimator for the relationship between Y_{t1} and x_t using Y_{t2} as the instrumental variable. In the same way, the estimator of β_{12} is the instrumental variable estimator of the relationship between Y_{t2} and x_t using Y_{t1} as the instrumental variable. For the model to be identified we must have $\sigma_{XY11} \neq 0$ and $\sigma_{XY12} \neq 0$. The conditions on the two covariances correspond to the conditions required of instrumental variables.

The estimators of the error variances are symmetric in the moments. For example, $m_{XY12}^{-1} m_{YY21} m_{XY11}$ is an estimator of that portion of the variance of Y_{t1} that is associated with the common factor. The expression for $\hat{\sigma}_{xx}$ is the analogous expression for X_t.

Alternative expressions for the estimators of the error variances are

$$\begin{aligned} \hat{\sigma}_{eeii} &= \hat{m}_{vvii} - \hat{\beta}_{1i}^2 \hat{\sigma}_{uu}, \qquad i = 1, 2, \\ \hat{\sigma}_{uu} &= (\hat{\beta}_{11}\hat{\beta}_{12})^{-1} \hat{m}_{vv12}, \end{aligned} \tag{1.5.5}$$

where $\hat{m}_{vvij} = (n-1)^{-1} \sum_{t=1}^{n} \hat{v}_{ti}\hat{v}_{tj}$, and $\hat{v}_{ti} = Y_{ti} - \bar{Y}_i - (X_t - \bar{X})\hat{\beta}_{1i}$. Because

$$\begin{aligned} E\{v_{t1}v_{t2}\} &= \beta_{11}\beta_{12}\sigma_{uu}, \\ E\{v_{ti}^2\} &= \sigma_{eeii} + \beta_{1i}^2 \sigma_{uu}, \qquad i = 1, 2, \end{aligned}$$

where $v_{ti} = e_{ti} - u_t\beta_{1i}$ for $i = 1, 2$, we see that the estimators of the error variances given in (1.5.5) are obtained by replacing (v_{t1}, v_{t2}) with $(\hat{v}_{t1}, \hat{v}_{t2})$, (β_{11}, β_{12}) with $(\hat{\beta}_{11}, \hat{\beta}_{12})$, and equating the resulting estimated sample moments of v_{ti} to the expectation of the true sample moments.

Under the normal distribution assumption, the estimators defined by (1.5.4) are maximum likelihood estimators adjusted for degrees of freedom, provided the solutions are in the parameter space. It is possible for some of the estimated variances of (1.5.4) to be negative. Maximum likelihood estimation for these types of samples is discussed in Section 4.3.

The estimators (1.5.4) are continuous differentiable functions of the sample moments and we can use the method of statistical differentials to express the estimators as approximate linear functions of moments. By the arguments of Theorem 1.4.1 we have

$$\begin{aligned} \hat{\beta}_{1i} - \beta_{1i} &= \sigma_{YXj1}^{-1} m_{Yvji} + O_p(n^{-1}), \\ \hat{\beta}_{0i} - \beta_{0i} &= \bar{v}_i - \bar{X}(\hat{\beta}_{1i} - \beta_{1i}) + O_p(n^{-1}), \end{aligned} \tag{1.5.6}$$

for $j \neq i$ and $i = 1, 2$. The vector of partial derivatives of $\hat{\sigma}_{uu}$ with respect to the vector $(m_{YX11}, m_{YY12}, m_{XX}, m_{XY12})$ is

$$(-m_{YY12}^{-1} m_{XY12},\ m_{YY12}^{-2} m_{XY11},\ 1,\ -m_{YY12}^{-1} m_{XY11}).$$

If we evaluate these derivatives at the expected values of the moments we have

$$\begin{aligned} \hat{\sigma}_{uu} - \sigma_{uu} &= (\beta_{11}\beta_{12})^{-1}[m_{vv12} - \sigma_{vv12}] + O_p(n^{-1}), \\ \hat{\sigma}_{eeii} - \sigma_{eeii} &= m_{vvii} - \beta_{1i}\beta_{1j}^{-1} m_{vvij} + O_p(n^{-1}), \end{aligned} \tag{1.5.7}$$

for $i \neq j$ and $i = 1, 2$. The expressions for the estimated error variances can also be obtained directly from (1.5.5). The approximate expressions for the estimators enable us to construct the covariance matrix of the limiting normal distribution of the estimators. Construction of the estimated covariance matrix is illustrated in Example 1.5.1.

The approximate expressions for the estimators are informative for several reasons. First, in the limit, the estimators of the error variances, $\sigma_{\varepsilon\varepsilon ii}$, are functions only of the moments of v_1 and v_2. That is, the nature of the distribution of the true x values has no influence on the distribution of the estimators of error variances. Second, given the independence of x_t and the (e_{t1}, e_{t2}, u_t), the covariance matrix of the limiting distribution of $(\hat{\beta}_{0i}, \hat{\beta}_{1i})$ can be given an explicit expression that depends only on the second moments of the original distribution. Finally, only the distribution of $\hat{\sigma}_{xx}$ depends on the higher moments of x_t. This means that expressions for the limiting distributions of the estimated coefficients and for the estimated error variances can be obtained under quite mild assumptions on the x_t.

An estimator of the uniqueness of variable Y_{t1} is

$$1 - \hat{\kappa}_{11} = m_{YY11}^{-1}\hat{\sigma}_{ee11}, \tag{1.5.8}$$

where $\hat{\kappa}_{11} = m_{YY11}^{-1}(\hat{\beta}_{11}^2\hat{\sigma}_{xx})$ is the estimated reliability ratio (communality) for Y_{t1}. An estimator of the variance of the approximate distribution of $\hat{\kappa}_{11}$ is

$$\hat{V}\{\hat{\kappa}_{11}\} = m_{YY11}^{-2}\hat{V}\{\hat{\sigma}_{ee11}\} + 2(n-1)^{-1}(1-\hat{\kappa}_{11})^2(2\hat{\kappa}_{11} - 1), \tag{1.5.9}$$

where $\hat{V}\{\hat{\sigma}_{ee11}\}$ is constructed in Example 1.5.1.

It is sometimes of interest to estimate the true x_t value that generated the vector (Y_{t1}, Y_{t2}, X_t). We can write model (1.5.1) as

$$\begin{bmatrix} Y_{t1} - \beta_{01} \\ Y_{t2} - \beta_{02} \\ X_t \end{bmatrix} = \begin{bmatrix} \beta_{11} \\ \beta_{12} \\ 1 \end{bmatrix} x_t + \begin{bmatrix} e_{t1} \\ e_{t2} \\ u_t \end{bmatrix}. \tag{1.5.10}$$

Therefore, if $(\beta_{01}, \beta_{02}, \beta_{11}, \beta_{12})$ and the covariance matrix of (e_{t1}, e_{t2}, u_t) are known, the generalized least squares estimator of x_t, treating x_t as fixed, is

$$\ddot{x}_t = \frac{\beta_{11}\sigma_{ee11}^{-1}(Y_{t1} - \beta_{01}) + \beta_{12}\sigma_{ee22}^{-1}(Y_{t2} - \beta_{02}) + \sigma_{uu}^{-1}X_t}{\beta_{11}^2\sigma_{ee11}^{-1} + \beta_{12}^2\sigma_{ee22}^{-1} + \sigma_{uu}^{-1}}. \tag{1.5.11}$$

See Exercise 1.43 and Section 1.2.3. The estimated value for x_t obtained by replacing the parameters with estimators is

$$\hat{x}_t = \frac{\hat{\beta}_{11}\hat{\sigma}_{ee11}^{-1}(Y_{t1} - \hat{\beta}_{01}) + \hat{\beta}_{12}\hat{\sigma}_{ee22}^{-1}(Y_{t2} - \hat{\beta}_{02}) + \hat{\sigma}_{uu}^{-1}X_t}{\hat{\beta}_{11}^2\hat{\sigma}_{ee11}^{-1} + \hat{\beta}_{12}^2\hat{\sigma}_{ee22}^{-1} + \hat{\sigma}_{uu}^{-1}}. \tag{1.5.12}$$

An alternative form for $\hat{x}_t$ is

$$\hat{x}_t = X_t - (\hat{v}_{t1}, \hat{v}_{t2})\hat{\mathbf{m}}_{vv}^{-1}\hat{\mathbf{m}}_{vu}, \tag{1.5.13}$$

where $\hat{\mathbf{m}}_{vv}$ and $(\hat{v}_{t1}, \hat{v}_{t2})$ are defined in (1.5.5) and

$$\hat{\mathbf{m}}_{vu} = (-\hat{\beta}_{11}\hat{\sigma}_{uu}, -\hat{\beta}_{12}\hat{\sigma}_{uu})'.$$

See Exercise 1.43 and Chapter 4. An estimator of the variance of $\hat{x}_t - x_t$ is

$$\hat{V}\{\hat{x}_t - x_t\} = (\hat{\beta}_{11}^2\hat{\sigma}_{ee11}^{-1} + \hat{\beta}_{12}^2\hat{\sigma}_{ee22}^{-1} + \hat{\sigma}_{uu}^{-1})^{-1}, \tag{1.5.14}$$

where the effect of estimating the parameters is ignored. The predictor of x_t, treating x_t as random, is given in Exercise 1.46 and is discussed in Section 4.3.

Example 1.5.1. In 1978 the U.S. Department of Agriculture conducted an experiment in which the area under specific crops was determined by three different methods. The three methods were digitized aerial photography, satellite imagery, and personal interview with the farm operator. We denote the hectares of corn determined for an area segment by the three methods by Y_{t1}, Y_{t2}, and X_t, respectively. An area segment is an area of the earth's surface of approximately 250 hectares. Observations for a sample of 37 area segments in north-central Iowa are given in Table 1.5.1. We begin by assuming the data satisfy model (1.5.1), (1.5.2). The sample mean vector is

$$(\bar{Y}_1, \bar{Y}_2, \bar{X}) = (123.28, 133.83, 120.32)$$

and the sample covariance matrix for (Y_{t1}, Y_{t2}, X_t) is

$$\mathbf{m}_{ZZ} = \begin{bmatrix} 1196.61 & 908.43 & 1108.74 \\ 908.43 & 1002.07 & 849.87 \\ 1108.74 & 849.87 & 1058.62 \end{bmatrix}.$$

By the formulas (1.5.4) we have

$$\begin{aligned} [\hat{\beta}_{01}, \hat{\beta}_{11}, \hat{\beta}_{02}, \hat{\beta}_{12}] &= [-5.340, 1.069\ \ 35.246, 0.819], \\ &\quad\ (4.646, 0.038, 11.738, 0.094) \\ [\hat{\sigma}_{ee11}, \hat{\sigma}_{ee22}, \hat{\sigma}_{uu}, \hat{\sigma}_{xx}] &= [11.47, 305.74,\ 21.36\ \ 1037.26]. \\ &\quad\ (23.10,\ \ 73.31,\ 20.70\ \ \ 250.27) \end{aligned}$$

The numbers in parentheses below the estimates are the estimated standard errors which will be obtained below.

The estimated covariances of (v_{t1}, v_{t2}) are

$$(\hat{m}_{vv11}, \hat{m}_{vv21}, \hat{m}_{vv22}) = (35.8776, 18.7056, 320.0740),$$

where $\hat{\mathbf{m}}_{vv} = (n-1)^{-1} \sum_{t=1}^{n} (\hat{v}_{t1}, \hat{v}_{t2})'(\hat{v}_{t1}, \hat{v}_{t2})$ and $\hat{v}_{ti} = Y_{ti} - \bar{Y}_1 - \hat{\beta}_{1i}(X_t - \bar{X})$. We note that

$$\hat{\sigma}_{uu} = 21.3557 = (\hat{\beta}_{11}\hat{\beta}_{12})^{-1}\hat{m}_{vv12}.$$

The $(\hat{v}_{t1}, \hat{v}_{t2})$ values are given in columns five and six of Table 1.5.1. The estimated values for m_{Yv21} and m_{Yv12} are zero by the construction of the estimators. Also see Exercise 1.35.

Under the normality assumption, the estimated covariance matrix of $(m_{Yv21}, m_{Yv12}, m_{vv11}, m_{vv12}, m_{vv22}, m_{XX})$ is

$$\begin{bmatrix} 998.663 & 569.442 & 0 & 304.701 & 317.726 & -1077.782 \\ 569.442 & 10638.993 & 11.921 & 101.989 & 0 & -1077.740 \\ 0 & 11.921 & 71.511 & 37.284 & 19.439 & 29.077 \\ 304.701 & 101.989 & 37.284 & 328.705 & 332.621 & 22.242 \\ 317.726 & 0 & 19.439 & 332.621 & 5691.520 & 17.014 \\ -1077.782 & -1077.740 & 29.077 & 22.242 & 17.014 & 62259.794 \end{bmatrix}.$$

TABLE 1.5.1. Hectares of corn determined by three methods

Segment	Photograph Y_{t1}	Satellite Y_{t2}	Interview X_t	$\hat{v}_{t1}$	$\hat{v}_{t2}$	$\hat{x}_t$
1	167.14	168.30	165.76	−4.702	−2.760	162.76
2	159.04	162.45	162.08	−8.869	−5.594	156.42
3	161.06	129.60	152.04	3.883	−30.218	153.93
4	163.49	166.05	161.75	−4.066	−1.724	159.17
5	97.12	94.05	96.32	−0.497	−20.115	95.65
6	123.02	140.85	114.12	6.376	12.101	118.34
7	111.29	110.70	100.60	9.098	−6.972	106.18
8	132.33	158.85	127.88	0.978	18.827	128.83
9	116.95	121.95	116.90	−2.665	−9.077	115.07
10	89.84	106.65	87.41	1.747	−0.215	88.50
11	84.17	99.00	88.59	−5.185	−8.831	85.18
12	88.22	153.00	88.59	−1.135	45.169	88.69
13	161.87	159.75	165.35	−9.534	−10.974	159.18
14	106.03	117.45	104.00	0.204	−3.007	104.07
15	87.01	84.15	88.63	−2.387	−23.714	86.71
16	159.85	157.50	153.70	0.899	−3.678	154.20
17	209.63	194.40	185.35	16.848	7.290	196.05
18	122.62	165.15	116.43	3.507	34.508	119.25
19	93.08	99.45	93.48	−1.502	−12.388	92.32
20	120.19	166.05	121.00	−3.808	31.664	119.18
21	115.74	154.35	109.91	3.596	29.050	112.69
22	125.45	153.90	122.66	−0.322	18.154	122.79
23	99.96	132.30	104.21	−6.091	11.671	100.60
24	99.55	92.70	92.88	5.610	−18.646	96.06
25	163.09	142.20	149.94	8.158	−15.898	154.77
26	60.30	65.25	64.75	−3.572	−23.048	62.10
27	101.98	113.40	99.96	0.472	−3.747	100.19
28	138.40	131.85	140.43	−6.367	−18.456	136.11
29	94.70	92.70	98.95	−5.729	−23.620	94.93
30	129.50	135.90	131.04	−5.230	−6.712	127.64
31	132.74	159.75	127.07	2.254	20.391	128.85
32	133.55	132.75	133.55	−3.863	−11.919	130.91
33	83.37	100.35	77.70	5.656	1.441	81.27
34	78.51	113.85	76.08	2.527	16.269	77.96
35	205.98	206.55	206.39	−9.292	2.201	200.60
36	110.07	130.50	108.33	−0.385	6.495	108.21
37	134.36	138.15	118.17	13.387	6.083	126.67

This matrix is calculated using the expressions of Appendix 1.B and the estimated moments, including $\hat{\mathbf{m}}_{vv}$.

It follows from expressions (1.5.6) and (1.5.7) that the estimated covariance matrix of the approximate distribution of $(10\hat{\beta}_{11}, 10\hat{\beta}_{12}, \hat{\sigma}_{ee11}, \hat{\sigma}_{ee22}, \hat{\sigma}_{uu}, 0.10\hat{\sigma}_{xx})$ is

$$\begin{bmatrix} 0.138 & 0.060 & -4.677 & 0.990 & 4.094 & -1.678 \\ 0.060 & 0.865 & -1.093 & -0.705 & 1.050 & -1.077 \\ -4.677 & -1.093 & 533.680 & -114.372 & -447.073 & 44.713 \\ 0.990 & -0.705 & -114.372 & 5374.730 & 92.102 & -9.214 \\ 4.094 & 1.050 & -447.073 & 92.102 & 428.550 & -40.315 \\ -1.678 & -1.077 & 44.713 & -9.214 & -40.315 & 626.376 \end{bmatrix}.$$

For example, the variance of the approximate distribution of $\hat{\sigma}_{ee11}$ is estimated with

$$\hat{V}\{m_{vv11}\} - 2\hat{\beta}_{11}\hat{\beta}_{12}^{-1}\hat{C}\{m_{vv11}, m_{vv12}\} + \hat{\beta}_{11}^2\hat{\beta}_{12}^{-2}\hat{V}\{m_{vv12}\} = 533.68,$$

where $\hat{V}\{m_{vv11}\} = 71.511$ and $\hat{V}\{m_{vv12}\} = 328.705$. We omit the portion of the covariance matrix associated with $(\hat{\beta}_{01}, \hat{\beta}_{02})$ because it is a simple function of the given covariance matrix and of the covariance matrix of $(\bar{v}_1, \bar{v}_2)$. See the equation for $(\hat{\beta}_{0i} - \beta_{0i})$ in (1.5.6).

The estimated communality and uniqueness for the three variables are given in Table 1.5.2. For example, the estimated uniqueness of variable Y_{t1} is

$$1 - \hat{\kappa}_{11} = m_{YY11}^{-1}\hat{\sigma}_{ee11} = 0.009586.$$

By (1.5.9), the estimated variance of the approximate distribution of $\hat{\kappa}_{11}$ is

$$\begin{aligned}\hat{V}\{\hat{\kappa}_{11}\} &= m_{YY11}^{-2}\hat{V}\{\hat{\sigma}_{ee11}\} + 2(n-1)^{-1}(1-\hat{\kappa}_{11})^2(2\hat{\kappa}_{11} - 1) \\ &= 0.0003777.\end{aligned}$$

The estimated values for x_t, treating x_t as fixed, are

$$\hat{x}_t = 2.7124 + 0.6270Y_{t1} + 0.0180Y_{t2} + 0.3151X_t,$$

TABLE 1.5.2. Estimated communality and uniqueness

Variable	Communality	Uniqueness	Standard Error
Y_{t1}	0.99041	0.00959	0.01944
Y_{t2}	0.69489	0.30511	0.08584
X_t	0.97983	0.02017	0.02008

where the coefficients are defined in (1.5.12). The estimated variance of the error in these estimators is given in (1.5.14) and is

$$\hat{V}\{\hat{x}_t - x_t\} = 6.7283.$$

The estimated true values are given in the last column of Table 1.5.1.

Figure 1.5.1 contains a plot of $\hat{v}_{t1}$ against $\hat{x}_t$. This plot is analogous to the plot of ordinary regression residuals against Y-hat. The variability of the $\hat{v}_{t1}$ seems to increase with $\hat{x}_t$. To test the hypothesis of homogeneous variance, we regress $\hat{v}_{t1}^2$ on $\hat{x}_t$. The t statistic for $\hat{x}_t$ in the regression containing an intercept is $t = 3.64$. Therefore, we are led to reject the original model that postulated identically distributed errors.

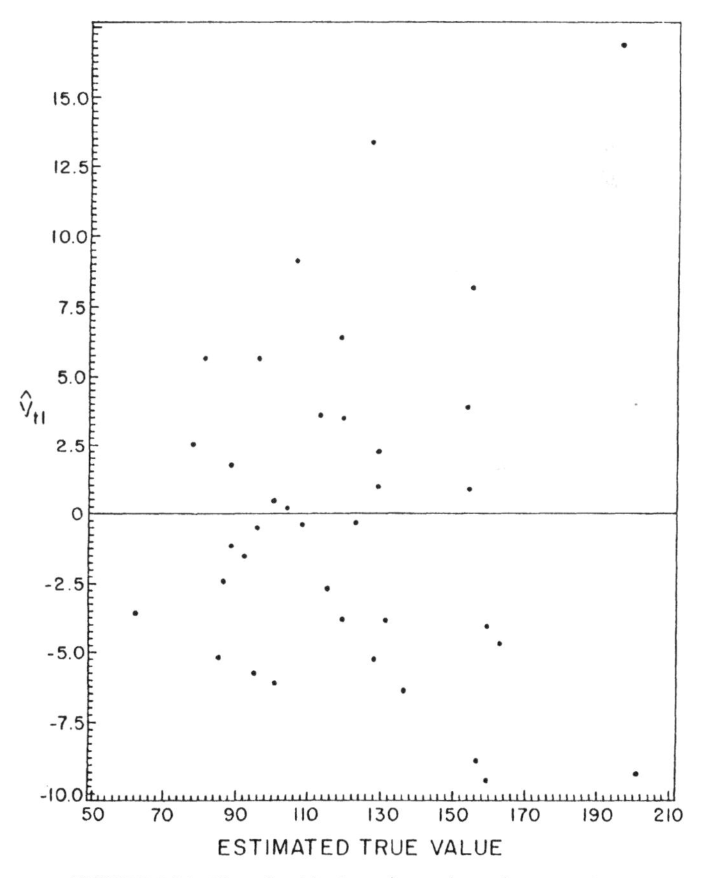

FIGURE 1.5.1. Plot of residuals against estimated true x values.

The estimated variances of the β-coefficients do not require normality (see Theorem 1.4.1), but the estimated variances of the error variances ($\hat{\sigma}_{ee11}$, $\hat{\sigma}_{ee22}$, $\hat{\sigma}_{uu}$) rest heavily on the normality assumption. The assumption of homoskedastic errors is required for all variance calculations. See Exercise 3.7 for alternative variance calculations for the instrumental variable estimator, Exercise 4.20 for alternative variance calculations for the vector of estimates of this example, and Exercise 1.44 for an alternative model. □ □

In the form (1.5.1) the systematic part of the variable X_t is called the factor. Assuming that $\beta_{11} \neq 0$ and $\beta_{12} \neq 0$, we could as easily have designated y_{t1} or y_{t2} as the factor. In factor analysis it is common to code the common factor by fixing its variance at one and its mean at zero. That is, the model is often written

$$Z_{ti} = \pi_{0i} + \pi_{1i} f_t + \varepsilon_{ti}, \qquad (1.5.15)$$
$$f_t \sim \text{NI}(0, 1),$$

for $i = 1, 2, 3$, where $\mathbf{Z}_t = (Z_{t1}, Z_{t2}, Z_{t3}) = (Y_{t1}, Y_{t2}, X_t)$, and $\boldsymbol{\varepsilon}_t = (e_{t1}, e_{t2}, u_t)$. The coefficients π_{11}, π_{12}, and π_{13} are called *factor loadings*. Under the model (1.5.15),

$$\mathbf{Z}_t' \sim \text{NI}(\boldsymbol{\pi}_0', \boldsymbol{\pi}_1 \boldsymbol{\pi}_1' + \boldsymbol{\Sigma}_{\varepsilon\varepsilon}) \qquad (1.5.16)$$

where $\boldsymbol{\Sigma}_{\varepsilon\varepsilon} = \text{diag}(\sigma_{\varepsilon\varepsilon 11}, \sigma_{\varepsilon\varepsilon 22}, \sigma_{\varepsilon\varepsilon 33})$ and $\boldsymbol{\pi}_j' = (\pi_{j1}, \pi_{j2}, \pi_{j3})$ for $j = 0, 1$. It follows that the parameters of model (1.5.15) can be estimated by

$$\hat{\sigma}_{\varepsilon\varepsilon ii} = m_{ZZii} - m_{ZZjk}^{-1} m_{ZZji} m_{ZZki},$$
$$\hat{\pi}_{1i}^2 = m_{ZZjk}^{-1} m_{ZZji} m_{ZZki}, \qquad (1.5.17)$$
$$\hat{\pi}_{0i} = \bar{Z}_i,$$

for $i = 1, 2, 3$ and $i \neq j \neq k$. The estimators of the error variances are identical to those of (1.5.4). There is a sign ambiguity in the definition of π_{ij} associated with model (1.5.15). If we choose the sign of π_{13} to be positive, then the sign of π_{11} is the sign of m_{ZZ13} and the sign of π_{12} is the sign of m_{ZZ23}. With this choice of standardization we see that

$$(\hat{\pi}_{11}, \hat{\pi}_{12}, \hat{\pi}_{13}) = \hat{\sigma}_{xx}^{1/2}(\hat{\beta}_{11}, \hat{\beta}_{12}, 1), \qquad (1.5.18)$$

where $\hat{\beta}_{11}$, $\hat{\beta}_{12}$, and $\hat{\sigma}_{xx}$ are defined in (1.5.4). Because the estimators $\hat{\pi}_{ij}$ are continuous differentiable functions of the moments, their sampling properties follow by the method of statistical differentials.

The covariance matrix of $(\hat{\pi}_{11}^2, \hat{\pi}_{12}^2, \hat{\pi}_{13}^2)$ is estimated by

$$\hat{\mathbf{G}}_{\pi\pi} \hat{\mathbf{V}}\{(m_{ZZ12}, m_{ZZ13}, m_{ZZ23})'\} \hat{\mathbf{G}}_{\pi\pi}', \qquad (1.5.19)$$

where

$$\hat{\mathbf{G}}_{\pi\pi} = \begin{bmatrix} \hat{A}_{12} & \hat{A}_{13} & -\hat{A}_{12}\hat{A}_{13} \\ \hat{A}_{21} & -\hat{A}_{21}\hat{A}_{23} & \hat{A}_{23} \\ -\hat{A}_{31}\hat{A}_{32} & \hat{A}_{31} & \hat{A}_{32} \end{bmatrix},$$

$\hat{A}_{ij} = m_{ZZjk}^{-1} m_{ZZki}$ for $i \neq j \neq k$, and $\hat{\mathbf{V}}\{(m_{ZZ12}, m_{ZZ13}, m_{ZZ23})'\}$ is the estimated covariance matrix of the vector of three moments. The covariance matrix of $\hat{\pi}_1$ is constructed by using

$$\hat{C}\{\hat{\pi}_{1i}, \hat{\pi}_{1j}\} \doteq (4\hat{\pi}_{1i}\hat{\pi}_{1j})^{-1}\hat{C}\{\hat{\pi}_{1i}^2, \hat{\pi}_{1j}^2\}.$$

Sometimes the observed variables are standardized and the factor parameters estimated from the sample correlation matrix. In this case the $\hat{\pi}_{1j}^2$ are the estimated communalities of the original variables and the estimated "error variance" for a standardized variable is the estimated uniqueness of that variable.

If we know the parameters of the model, the least squares estimator of f_t constructed on the basis of (1.5.15) is

$$\ddot{f}_t = \frac{\pi_{11}\sigma_{ee11}^{-1}(Y_{t1} - \pi_{01}) + \pi_{12}\sigma_{ee22}^{-1}(Y_{t2} - \pi_{02}) + \pi_{13}\sigma_{uu}^{-1}(X_t - \pi_{03})}{\pi_{11}^2\sigma_{ee11}^{-1} + \pi_{12}^2\sigma_{ee22}^{-1} + \pi_{13}^2\sigma_{uu}^{-1}} \tag{1.5.20}$$

The estimator is the generalized least squares estimator treating f_t as fixed. In practice, parameters on the right side of (1.5.20) can be replaced by sample estimators. The estimator $\hat{f}_t$ obtained in this manner is a linear function of the estimator defined in (1.5.12).

Example 1.5.2. Using the data of Example 1.5.1, the estimated coefficients of the standardized factor are

$$(\hat{\pi}_{11}, \hat{\pi}_{12}, \hat{\pi}_{13}) = (34.426, 26.387, 32.207),$$

where we have chosen $\hat{\pi}_{13}$ to be positive. Using (1.5.19), we find that the estimated covariance matrix of the approximate distribution of $(\hat{\pi}_{11}, \hat{\pi}_{12}, \hat{\pi}_{13})$, calculated under the assumption of trivariate normality, is

$$\hat{\mathbf{V}}\{\hat{\pi}_1\} = \begin{bmatrix} 16.8900 & 12.5856 & 15.2983 \\ 12.5856 & 18.2291 & 11.8308 \\ 15.2983 & 11.8308 & 15.0968 \end{bmatrix},$$

where the matrix $\hat{\mathbf{G}}_{\pi\pi}$ is constructed using

$$(\hat{A}_{12}, \hat{A}_{13}, \hat{A}_{23}) = (1.3046, 1.0689, 0.8193)$$

and the covariance matrix of $(m_{ZZ12}, m_{ZZ13}, m_{ZZ23})$ is constructed using the results of Appendix 1.B and the sample moments given in Example 1.5.1.

□ □

While the parameterization (1.5.15) is more common in factor analysis than the parameterization (1.5.1), we prefer the parameterization (1.5.1). The β parameters of (1.5.1) are analogous to regression coefficients and the π parameters of (1.5.15) are analogous to correlation coefficients. Generally, regression coefficients have proved to be much more portable from one situation to another. This is because regression coefficients are largely independent of the parameters of the distribution of x. The mean of a regression coefficient does not depend on the distribution of x and the variance of the coefficient depends only on the second moments of the x distribution. The same is true of the $\hat{\beta}_{1i}$ of the factor model. The distribution of a correlation coefficient is much more dependent on the form of the x distribution. For the factor model the mean of $\hat{\pi}_1$ is a function of the variance of x and the distribution of $\hat{\pi}_1$ depends on the distribution of x. Any screening of the data that changes the variance of x changes the π_1. If two different samples are to be compared, comparison of the π's typically requires the distribution of x to be the same in the two populations. On the other hand, a specification of equality for the β's of two samples can be made with the distribution of x in the two populations left unspecified.

In the models discussed to this point, the number of sample covariances has never exceeded the number of parameters to be estimated. In applications of factor analysis, it is often the case that the number of covariances is much larger than the number of parameters. Consider a model with $p > 3$ observed variables that are a function of a single latent factor

$$Z_{ti} = \pi_{0i} + \pi_{1i} f_t + e_{ti}, \qquad i = 1, 2, \ldots, p.$$

There will be p means, $\frac{1}{2}p(p + 1)$ sample covariances, and $3p$ parameters to be estimated from them. Inspection of a covariance matrix such as (1.5.3) will lead to several possible estimators for the parameters. In such a situation the parameters are said to be *overidentified*. The application of the method of maximum likelihood will lead to unique estimators of the parameters that are not equivalent to instrumental variable estimators. Estimation for models of higher dimension is discussed in Section 4.3.

REFERENCES

Barnett (1969), Harman (1976), Jöreskog (1978), Lawley (1940, 1941, 1943), Lawley and Maxwell (1971).

EXERCISES

41. (Section 1.5) As an example of a simple factor analysis model, consider some data studied by Grubbs (1948). The data given in the table are the time of burning for fuses on projectiles as recorded by three different observers. Assume the data satisfy model (1.5.1). The choice of observer to be identified with X_t is arbitrary because we have no reason to believe that one observer is superior to the others.
   (a) Fit model (1.5.1) to these data. Using the covariance matrix of the approximate distribution, test the hypothesis that the variance of the measurement error is the same for the three observers.
   (b) Write the solution in the parametric form

$$Y_{ti} = \pi_{0i} + \pi_{1i} f_t + e_{ti}, \qquad i = 1, 2, 3,$$

   where $\sigma_{ff} = 1$, Y_{ti} is the observation on the tth fuse made by the ith observer, and e_{ti} is the measurement error. Give the covariance matrix of $(\hat{\pi}_{01}, \hat{\pi}_{02}, \hat{\pi}_{03}, \hat{\pi}_{11}, \hat{\pi}_{12}, \hat{\pi}_{13})$.

Data for Exercise 41: Observed fuse burning times

Index t	Observer B Y_{t1}	Observer A Y_{t2}	Observer C X_t
1	10.07	10.10	10.07
2	9.90	9.98	9.90
3	9.85	9.89	9.86
4	9.71	9.79	9.70
5	9.65	9.67	9.65
6	9.83	9.89	9.83
7	9.75	9.82	9.79
8	9.56	9.59	9.59
9	9.68	9.76	9.72
10	9.89	9.93	9.92
11	9.61	9.62	9.64
12	10.23	10.24	10.24
13	9.83	9.84	9.86
14	9.58	9.62	9.63
15	9.60	9.60	9.65
16	9.73	9.74	9.74
17	10.32	10.32	10.34
18	9.86	9.86	9.86
19	9.64	9.65	9.65
20	9.49	9.50	9.50
21	9.56	9.56	9.55
22	9.53	9.54	9.54
23	9.89	9.89	9.88
24	9.52	9.53	9.51
25	9.52	9.52	9.53
26	9.43	9.44	9.45
27	9.67	9.67	9.67
28	9.76	9.77	9.78
29	9.84	9.86	9.86

Source: Grubbs (1948).

42. (Section 1.5) Assume that the data of Table 1.4.1 of Example 1.4.1 satisfy the factor model (1.5.1) with surface wave equal to Y_{t1}, trace amplitude equal to Y_{t2}, and body wave equal to X_t.
   (a) Estimate the parameters of the model.
   (b) Calculate $\hat{v}_{t1}$, $\hat{v}_{t2}$, and $\hat{x}_t$. Plot $\hat{v}_{t1}$ against $\hat{x}_t$. Plot $\hat{v}_{t2}$ against $\hat{x}_t$.
   (c) Estimate the parameters of the model with observation number 54 deleted. Plot $\hat{v}_{t1}$ and $\hat{v}_{t2}$ against $\hat{x}_t$ for the reduced data set.
   (d) Estimate the covariance matrix of the approximate distribution of the estimates obtained in part (c).
43. (Section 1.5)
   (a) Using the expressions given in (1.5.6) and (1.5.7), derive expression (1.5.9).
   (b) Using the method associated with (1.2.16), derive expression (1.5.11) for the estimated true value x_t, treating x_t as fixed.
   (c) Show that $\ddot{x}_t$ of (1.5.11) can also be written

$$\ddot{x}_t = X_t - \hat{u}_t,$$

   where $v_{ti} = e_{ti} - u_t\beta_{1i}$ and $\hat{u}_t = (v_{t1}, v_{t2})\Sigma_{vv}^{-1}(\sigma_{vu11}, \sigma_{vu21})'$.
44. (Section 1.5) Fit the one-factor model to the logarithms of the data of Example 1.5.1. Compute the covariance matrix of your estimates. Plot $\hat{v}_{t1}$ and $\hat{v}_{t2}$ against the estimated true values, all for the model in logarithms. Regress $|\hat{v}_{t1}|$ and $|\hat{v}_{t2}|$ on $\hat{x}_t$. What do you conclude?
45. (Section 1.5) Compute the ordinary least squares regression of X_t on $\hat{v}_{t1}$ and $\hat{v}_{t2}$ for the data of Table 1.5.1. Compare the calculated residuals for this regression with $\hat{x}_t$ of Table 1.5.1.
46. (Section 1.5) Assume that the parameters of the factor model (1.5.1) are known. Let an observation (Y_{t1}, Y_{t2}, X_t) be given. Treating x_t as random, obtain the best predictor of x_t. Show that this predictor is

$$\tilde{x}_t = \mu_x + [V\{\ddot{x}_t\}]^{-1}\sigma_{xx}(\ddot{x}_t - \mu_x),$$

where $\ddot{x}_t$ is defined in (1.5.11), $\gamma'_{uv} = \Sigma_{vv}^{-1}\Sigma_{vu}$, and

$$V\{\ddot{x}_t\} = E\{(\ddot{x}_t - \mu_x)^2\} = \sigma_{XX} - \gamma_{uv}\Sigma_{vu}.$$

Show that $V\{\tilde{x}_t - x_t\} = \sigma_{xx} - [V\{\ddot{x}_t\}]^{-1}\sigma_{xx}^2$.
47. (Section 1.5) Find a and b such that $\ddot{f}_t = a + b\ddot{x}_t$, where $\ddot{f}_t$ is defined in (1.5.20) and $\ddot{x}_t$ is defined in (1.5.11).
48. (Section 1.5) Prove that, at most, one of the variance estimates defined by Equations (1.5.4) can be negative.

1.6. OTHER METHODS AND MODELS

In the first five sections of this chapter we have discussed methods that have found considerable use in application. In this section we mention some other procedures that appear in the literature associated with errors in variables. We also describe two special situations in which ordinary least squares remains an appropriate procedure in the presence of measurement error.

1.6.1. Distributional Knowledge

In Sections 1.1–1.5 we often assumed both the errors and the true values to be normally distributed. Geary (1942, 1943) has demonstrated that the parameters of a model containing normal measurement error can be estimated if it is known that the distribution of x_t is not normal. Let

$$Y_t = \beta_0 + \beta_1 x_t + e_t, \qquad X_t = x_t + u_t, \qquad (1.6.1)$$
$$\boldsymbol{\varepsilon}_t' \sim \mathrm{NI}(\mathbf{0}, \boldsymbol{\Sigma}_{\varepsilon\varepsilon}),$$

where $\boldsymbol{\varepsilon}_t = (e_t, u_t)$ is independent of x_j for all t and j. Also, assume, for example, that it is known that

$$E\{(x_t - \mu_x)^3\} \neq 0. \qquad (1.6.2)$$

This knowledge can be interpreted as the availability of an instrumental variable. That is, if we set $W_t = (X_t - \bar{X})^2$, then

$$E\{(X_t - \mu_X)(W_t - \mu_W)\} = (1 - n^{-1})^2 E\{(x_t - \mu_x)^3\} \neq 0,$$

and, by the properties of the normal distribution,

$$E\{W_t u_t\} = E\{W_t e_t\} = 0.$$

Therefore, the methods of Section 1.4 can be used to construct the estimator

$$\hat{\beta}_1 = m_{XW}^{-1} m_{YW}, \qquad (1.6.3)$$

where

$$(m_{WY}, m_{WX}) = (n - 1)^{-1} \sum_{t=1}^{n} (X_t - \bar{X})^2 (Y_t - \bar{Y}, X_t - \bar{X}).$$

In this development, the assumption that the conditional mean of Y_t given x_t is linear is used in a very critical way. It is a part of the identifying information.

The distributional theory of Section 1.4 can be extended to cover estimator (1.6.3), but the estimated variance expression (1.4.15) is not appropriate for estimator (1.6.3) because the W_t of this section does not satisfy assumptions (1.4.14). See Exercise 3.7 for a variance estimator. Under model (1.6.1), (1.6.2), $(Y_t - \bar{Y})^2$ and $(X_t - \bar{X})(Y_t - \bar{Y})$ are also possible instrumental variables. See Section 2.4 for the use of multiple instrumental variables.

Reiersol (1950) showed that β_1 of (1.6.1) is identified for any model with nonnormal x. Estimation methods for β_1 using only the fact that the distribution of x is not normal are possible, but the procedures are complex and have been little used in practice. See Spiegelman (1979) and Bickel and Ritov (1985) for discussions of such estimation.

1.6.2. The Method of Grouping

Wald (1940) suggested an estimator of β_1 for model (1.6.1) constructed by dividing the observations into two groups. Let the observations (Y_1, X_1), $(Y_2, X_2), \ldots, (Y_r, X_r)$ constitute group one and the remaining observations group two. Then Wald's estimator is

$$\hat{\beta}_{1,W} = \frac{r^{-1} \sum_{t=1}^{r} Y_t - (n-r)^{-1} \sum_{t=r+1}^{n} Y_t}{r^{-1} \sum_{t=1}^{r} X_t - (n-r)^{-1} \sum_{t=r+1}^{n} X_t}. \tag{1.6.4}$$

Wald showed that the estimator is consistent for β_1 of model (1.6.1) if the grouping is independent of the errors and if

$$\lim_{n \to \infty} \inf |\bar{x}_{(1)} - \bar{x}_{(2)}| > 0, \tag{1.6.5}$$

where $\bar{x}_{(i)}$ is the mean of the true x values for the ith group. If the grouping is independent of the errors, this implies that there exists a variable,

$$W_t = \begin{cases} 1 & \text{if element is assigned to group one} \\ 0 & \text{otherwise,} \end{cases} \tag{1.6.6}$$

that is independent of the errors. Therefore, with grouping independent of the errors, Wald's method reduces to the method of instrumental variables.

Wald's method has often been interpreted incorrectly. For example, it has been suggested that the method can be applied by randomly assigning elements to two equal-sized groups. The random method of assigning elements is independent of the errors of measurement, but (1.6.5) is not satisfied because both sample means are converging to the true mean. It has also been suggested that the groups be formed by splitting the sample on the basis of the size of the observed X values. Splitting the sample on the basis of X will satisfy (1.6.5) but the group an element falls into will generally be a function of u_t. Therefore, the first condition of Wald's theorem will generally be violated when the observed X values are used to form the groups. See Exercises 1.50 and 2.19.

A method related to the method of grouping involves the use of ranks. It may be that the x values are so spaced, and the distribution of the errors is such that $X_t < X_s$ implies $x_t < x_s$. Given this situation, the rank of the observed X values is independent of the measurement error and highly correlated with x_t. Therefore, in the presence of these strong assumptions, the rank of the X can be used as an instrumental variable. Dorff and Gurland (1961a, b) and Ware (1972) have investigated this model.

1.6.3. Measurement Error and Prediction

One often hears the statement, "If the objective is prediction, it is not necessary to adjust for measurement error." As with all broad statements, this statement requires a considerable number of conditions to be correct. Let

$$\begin{gathered} Y_t = \beta_0 + \beta_1 x_t + e_t, \qquad X_t = x_t + u_t, \\ (x_t, \boldsymbol{\varepsilon}_t)' \sim \mathrm{NI}[(\mu_x, \mathbf{0})', \text{block diag}(\sigma_{xx}, \boldsymbol{\Sigma}_{\varepsilon\varepsilon})], \end{gathered} \tag{1.6.7}$$

where $(e_t, u_t) = \varepsilon_t$. Let a sample of n vectors (Y_t, X_t) be available and let X_{n+1} be observed. Because $(Y_t, X_t)'$ is distributed as a bivariate normal random variable, the best linear unbiased predictor of Y_{n+1} conditional on $(X_1, X_2, \ldots, X_{n+1})$ is

$$\hat{Y}_{n+1} = \hat{\gamma}_{0\ell} + \hat{\gamma}_{1\ell} X_{n+1}, \tag{1.6.8}$$

where

$$\hat{\gamma}_{1\ell} = \left[\sum_{t=1}^{n} (X_t - \bar{X})^2\right]^{-1} \sum_{t=1}^{n} (X_t - \bar{X})(Y_t - \bar{Y}),$$

$\hat{\gamma}_{0\ell} = \bar{Y} - \hat{\gamma}_{1\ell}\bar{X}$, and $\bar{X} = n^{-1} \sum_{t=1}^{n} X_t$.

The assumption of zero covariances between (e_t, u_t) and x_t is not required for the optimality of the predictor. Thus, if one chooses a random element from the same distribution, the simple regression of observed Y_t on observed X_t gives the optimal prediction of Y_{n+1} if (Y_t, X_t) is distributed as a bivariate normal random vector. If the joint distribution is not normal, the predictor is best in the class of linear predictors.

The introduction of normal measurement error will destroy the linearity of the relationship between X and Y when x is not normally distributed. If

$$Y_t = \beta_0 + \beta_1 x_t + e_t, \qquad \beta_1 \neq 0,$$

and $X_t = x_t + u_t$, where $(e_t, u_t)' \sim \mathrm{NI}(\mathbf{0}, \mathrm{diag}\{\sigma_{ee}, \sigma_{uu}\})$ and $\sigma_{uu} > 0$, then the expected value of Y given X is a linear function of X if and only if x is normally distributed. See Lindley (1947) and Kendall and Stuart (1979, p. 438).

It is worth emphasizing that the use of least squares for prediction requires the assumption that X_{n+1} be a random selection from the same distribution that generated the X_t of the estimation sample. For example, it would be improper to use the simple regression of Y on X of Example 1.2.1 to predict yield for a field in which the nitrogen was determined from twice as many locations within the field as the number used to obtain the data of Table 1.2.1.

Example 1.6.1. Assume that the fields in Example 1.2.1 were selected at random from the fields on Marshall soil. Assume further that a 12th field is selected at random from the population of fields. Assume that twice as many soil samples are selected in the field and twice as many laboratory determinations are made. Under these assumptions, the variance of the measurement error in X_{12} for the 12th field is $\frac{1}{2}\sigma_{uu} = 28.5$. Given that the observed soil nitrogen is 90, we wish to predict yield. We treat the estimates of Example

1.2.1 as if they were known parameters. Because the variance of the measurement error in X_{12} is $\frac{1}{2}\sigma_{uu}$, the covariance matrix for (Y_{12}, X_{12}) is

$$\begin{bmatrix} \beta_1^2\sigma_{xx} + \sigma_{ee} & \beta_1\sigma_{xx} \\ \beta_1\sigma_{xx} & \sigma_{xx} + \frac{1}{2}\sigma_{uu} \end{bmatrix} = \begin{bmatrix} 87.6727 & 104.8818 \\ 104.8818 & 276.3545 \end{bmatrix}.$$

It follows that, conditioning on the observed value of X_{12}, the predicted yield for the new randomly selected field is

$$\tilde{Y}_{12}(X_{12}) = 97.4546 + 0.3795(90 - 70.6364) = 104.80.$$

Under the assumption of known covariance structure, the expected value of the error in the predictions for fields with an observed nitrogen of 90 is zero. That is, the expected value of the prediction error conditional on X_{12} is zero. Note that the average of the predictions is for a fixed observed value, not for a fixed true value. The unobserved true value is a random variable.

The average squared prediction error for those samples with the observed X_{12} equal to 90 is

$$E\{[Y_{12} - \tilde{Y}(X_{12})]^2 \mid X_{12} = 90\} = \sigma_{YY} - 0.3795\sigma_{XY} = 47.87.$$

Because of the bivariate normality, the variance of the prediction error is the same for all observed values of X.

Let the true nitrogen value for the field be x_{12} and consider all possible predictions of yield that can be constructed for the field under consideration (or for all fields with true nitrogen equal to x_{12}). The conditionally unbiased predictor of yield, holding x_{12} fixed, is

$$\hat{Y}_{12}(x_{12} \text{ fixed}) = 97.4546 + 0.4232(90 - 70.6364) = 105.65.$$

Thus, if the true value of nitrogen is held fixed, the conditionally unbiased predictor of yield is obtained by using the structural equation as the prediction equation. The average squared prediction error for those fields with $x = x_{12}$ and the same number of soil determinations as used to obtain X_{12} is

$$\begin{aligned} E\{[Y_{12} - \hat{Y}(x_{12} \text{ fixed})]^2 \mid x = x_{12}\} &= \sigma_{ee} + (0.4232)^2(\tfrac{1}{2}\sigma_{uu}) \\ &= 53.50. \end{aligned}$$

The variance of the prediction error is a function of σ_{ee} and σ_{uu} only. The variance does not depend on the true, but unknown, x value.

Given that the variance of the prediction error for the prediction conditional on X is less than the variance conditional on x, why would one choose a prediction conditional on x? The model that permits one to use the prediction conditional on X assumes that x (the field) is selected at random from the population of fields. If this is not a reasonable assumption, the prediction constructed for fixed x should be used. As an example, assume that a farmer asks that his field be tested and corn yield be predicted for that field.

Is it reasonable to treat the field as randomly selected? Assume further that the farmer is known to sell a soil additive product "Magic Grow Sand" that is 100% silica. Is it reasonable to treat the field as randomly selected?

To investigate the random and fixed specification further, consider the problem of making statements about the true nitrogen x_t on the basis of the observed nitrogen X_t. If the sampled field is randomly chosen from the population of fields, it is reasonable to assume that the observed nitrogen and the true nitrogen are distributed as a bivariate normal. If we treat the estimates of Example 1.2.1 as the parameters, we have

$$\begin{bmatrix} X_t \\ x_t \end{bmatrix} \sim \mathrm{NI}\left(\begin{bmatrix} 70.6364 \\ 70.6364 \end{bmatrix}, \begin{bmatrix} 304.8545 & 247.8545 \\ 247.8545 & 247.8545 \end{bmatrix}\right).$$

Therefore, the best predictor of true nitrogen conditioning on the observed nitrogen value is

$$\tilde{x}_t(X_t) = 70.6364 + 0.81303(X_t - 70.6364).$$

Under bivariate normality, this predictor is conditionally unbiased,

$$E\{\tilde{x}_t(X_t) - x_t \mid X_t = \xi\} = 0 \quad \text{for all } \xi.$$

However, when we look at the conditional distribution of $\tilde{x}_t(X_t)$ holding x fixed, we have

$$E\{\tilde{x}_t(X_t) - x_t \mid x_t = \omega\} = -0.18697(\omega - 70.6364).$$

Also,

$$E\{[\tilde{x}_t(X_t) - x_t]^2 \mid x_t = \omega\} = 37.6780 + [0.18697(\omega - 70.6364)]^2.$$

It follows that the mean square error of $\tilde{x}_t(X_t)$ as an estimator for x_t is greater than the mean square error of X_t for $|x_t - 70.6364| > 23.5101$. Because the distribution of x_t has a standard deviation of 15.7434, the mean square error of $\tilde{x}_t(X_t)$ as an estimator of x_t is less than the mean square error of X_t for those fields whose true x is less than 1.493 standard deviations from the mean. See Exercise 1.51. □ □

In Example 1.6.1 we illustrated the construction of a predictor in a situation where the nature of the population of observables was changed by a change in the measuring procedure. In practice one may also find that the distribution of x in the prediction population differs from that of the estimation sample.

Example 1.6.2. To illustrate prediction for a second population, consider the earthquake data introduced in Example 1.4.1. In the National Oceanic and Atmospheric Administration's Hypocenter Data File, values for surface waves (Y) are generally available for large earthquakes but not for small ones, while values for longitudinal body waves (X) are generally available for small earthquakes but not for large ones. The group of earthquakes for which the triplet (Y_t, X_t, W_t) is available is a relatively small subset of the total. In the data file there are 5078 Alaskan earthquakes with only X values reported. It is of interest to construct Y values for these earthquakes. The sample mean and variance for the group of earthquakes with only X values is

$$(\bar{X}_{(2)}, \hat{\sigma}_{XX(2)}) = (4.5347, 0.2224),$$

while the corresponding vector for the data of Example 1.4.1 is

$$(\bar{X}_{(1)}, \hat{\sigma}_{XX(1)}) = (5.2145, 0.2121).$$

Clearly, the mean of X differs in the two populations from which the two samples were chosen. Let us assume that the model

$$\begin{aligned} Y_{ti} &= \beta_0 + \beta_1 x_{ti} + e_{ti}, \qquad X_{ti} = x_{ti} + u_{ti}, \\ W_{ti} &= \gamma_0 + \gamma_1 x_{ti} + c_{ti}, \\ (x_{ti}, \boldsymbol{\varepsilon}_{ti})' &\sim \mathrm{NI}[(\mu_{x(i)}, \mathbf{0})', \mathrm{diag}(\sigma_{xx(i)}, \sigma_{ee}, \sigma_{uu}, \sigma_{cc})], \end{aligned} \tag{1.6.9}$$

holds for $i = 1, 2$, where $\boldsymbol{\varepsilon}_{ti} = (e_{ti}, u_{ti}, c_{ti})$, $i = 1$ denotes the population from which the triplets were selected, and $i = 2$ denotes the population for which only X values are available. While the observations suggest that the X distribution is not exactly normal, it seems that normality remains a good working approximation. Model (1.6.9) is the factor model in one factor and contains stronger assumptions about the error covariance matrix than the instrumental variable model used in Example 1.4.1. Under the factor model, we can estimate the covariance matrix of (e_t, u_t, c_t). By Equations (1.5.6), the estimates are

$$(\hat{\sigma}_{ee}, \hat{\sigma}_{uu}, \hat{\sigma}_{cc}) = (0.1398, 0.0632, 0.0617).$$

The estimates of the variances of x in the two populations are

$$\begin{aligned} \hat{\sigma}_{xx(1)} &= \hat{\sigma}_{XX(1)} - \hat{\sigma}_{uu} = 0.1489, \\ \hat{\sigma}_{xx(2)} &= \hat{\sigma}_{XX(2)} - \hat{\sigma}_{uu} = 0.1592. \end{aligned}$$

From Example 1.4.1, the estimated structural equation is

$$\hat{Y}_t = -4.2829 + 1.7960x_t.$$

It follows that estimates of the remaining parameters of the (Y, X) distribution for population two are

$$\begin{aligned} \hat{\mu}_{Y(2)} &= -4.2829 + 1.7960\bar{X}_{(2)} = 3.8614, \\ \hat{\sigma}_{YY(2)} &= (1.7960)^2(0.1592) + 0.1398 = 0.6533, \\ \hat{\sigma}_{XY(2)} &= (1.7960)(0.1592) = 0.2859. \end{aligned}$$

Let us assume that we are asked to predict the Y value for an earthquake with an X value of 3.9 selected from population two. On the basis of our model

$$E\{Y_{(2)} | X_{(2)}\} = \mu_{Y(2)} + \sigma_{XX(2)}^{-1}\sigma_{XY(2)}(X_{(2)} - \mu_{X(2)}) \tag{1.6.10}$$

and

$$V\{Y_{(2)} | X_{(2)}\} = \sigma_{YY(2)} - \sigma_{XX(2)}^{-1}\sigma_{XY(2)}^{2}. \tag{1.6.11}$$

Our estimators of these quantities are

$$\hat{E}\{Y_{(2)} | X_{(2)}\} = 3.8614 + 1.2855(X_{(2)} - 4.5347),$$
$$\hat{V}\{Y_{(2)} | X_{(2)}\} = 0.2858.$$

Therefore, the predictor of Y for an observation in the second population with $X = 3.9$ is

$$\tilde{Y} = 3.8614 + 1.2855(-0.6347) = 3.0455,$$

and the estimated standard deviation of the prediction error is 0.5346. One could modify the estimated prediction standard deviation by recognizing the contribution to the error of estimating the parameters. See Ganse, Amemiya, and Fuller (1983).

Figure 1.6.1 illustrates the nature of prediction for the second population. The two clusters of points represent the two populations. The solid line represents the structural line common to the two populations. The two dashed lines are the least squares lines for the two populations. For our example the two clusters of points would overlap, but we have separated them in the figure for illustrative purposes. The point X_p represents the value of X for which we wish to predict Y in the second population. The prediction is denoted by $\hat{Y}_p$. The predictor computed by evaluating the least squares line for the first population at X_p is denoted by $\hat{Y}_\ell$. Also see Exercise 1.52. □ □

1.6.4. Fixed Observed X

We have discussed two forms of the model

$$Y_t = \beta_0 + \beta_1 x_t + e_t, \qquad X_t = x_t + u_t, \tag{1.6.12}$$
$$\varepsilon_t' \sim \mathrm{NI}(\mathbf{0}, \boldsymbol{\Sigma}_{\varepsilon\varepsilon}),$$

where $\varepsilon_t = (e_t, u_t)$. In the first form, the x_t are treated as random variables, often assumed to be normally distributed. In the second form of the model, the x_t are treated as fixed constants. There is a third experimental situation that leads to a model that appears very similar to (1.6.12), but for which the stochastic behavior is markedly different. Assume that we are conducting an experiment on the quality of cement being created in a continuous mixing operation. Assume that quality is a function of the amount of water used in

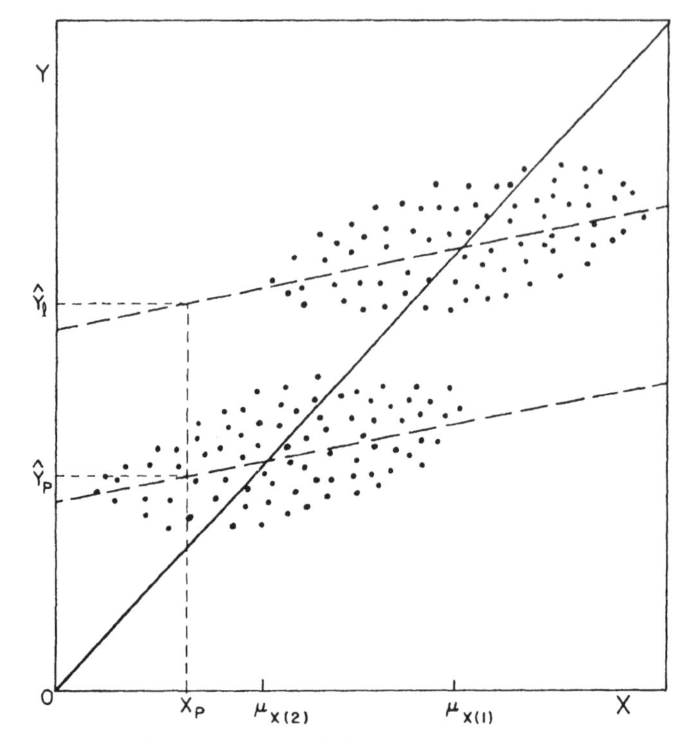

FIGURE 1.6.1. Prediction in a second population.

the mixture. We can set the reading on the dial of a water valve controlling water entering the mixture. However, because of random fluctuations in water pressure, the amount of water actually delivered per unit of time is not that set on the dial. The true amount of water x_t is equal to the amount set on the water dial X_t plus a random error u_t. If the dial has been calibrated properly, the average of the u_t is zero. Thus,

$$x_t = X_t - u_t, \tag{1.6.13}$$

where the u_t are $(0, \sigma_{uu})$ random variables. We used the negative sign on the right side of (1.6.13) so that the form of (1.6.12) is retained. It is important that in this experiment the observed X_t, the reading on the dial, is *fixed*. The observed value is *controlled* by the experimenter. Also, if it is assumed that variations in pressure are independent of the valve setting, the u_t are independent of X_t. In the model of earlier sections, x_t and u_t are independent, while X_t and u_t are correlated. In the cement experiment, x_t and u_t are correlated, while X_t and u_t are independent. Berkson (1950) observed that when X_t is controlled, ordinary least squares can be used to estimate the parameters of the line.

Theorem 1.6.1. Let

$$Y_t = \beta_0 + \beta_1 x_t + e_t, \qquad x_t = X_t - u_t,$$

where (e_t, u_t), $t = 1, 2, \ldots, n$, are independent vectors with zero mean and covariance matrix

$$E\{(e_t, u_t)'(e_t, u_t)\} = \text{diag}(\sigma_{ee}, \sigma_{uu}),$$

and $\mathbf{X}' = (X_1, X_2, \ldots, X_n)$ is a vector of fixed constants. Let

$$\hat{\beta}_{1\ell} = \left[\sum_{t=1}^{n} (X_t - \bar{X})^2\right]^{-1} \sum_{t=1}^{n} (X_t - \bar{X})(Y_t - \bar{Y}), \tag{1.6.14}$$

$$\hat{\beta}_{0\ell} = \bar{Y} - \hat{\beta}_{1\ell}\bar{X},$$

be the ordinary least squares estimators of β_0 and β_1. Then

$$E\{(\hat{\beta}_{0\ell}, \hat{\beta}_{1\ell})\} = (\beta_0, \beta_1)$$

and

$$\mathbf{V}\{(\hat{\beta}_{0\ell}, \hat{\beta}_{1\ell})'\} = \begin{bmatrix} n^{-1} + \bar{X}^2 A_{XX}^{-1} & -\bar{X} A_{XX}^{-1} \\ -\bar{X} A_{XX}^{-1} & A_{XX}^{-1} \end{bmatrix} \sigma_{vv}, \tag{1.6.15}$$

where $A_{XX} = (n-1)m_{XX}$, $v_t = e_t - u_t\beta_1$, and $\sigma_{vv} = \sigma_{ee} + \beta_1^2\sigma_{uu}$.

Proof. Substituting the definition of x_t into the equation for Y_t, we have

$$Y_t = \beta_0 + \beta_1 X_t + v_t, \tag{1.6.16}$$

where $v_t = e_t - u_t\beta_1$. By the assumptions, the vector (e_t, u_t) is independent of X_t. Therefore, v_t is independent of X_t. Substituting (1.6.16) into (1.6.14), we obtain

$$\hat{\beta}_{0\ell} - \beta_0 = \bar{v} - \bar{X}(\hat{\beta}_{1\ell} - \beta_1),$$

$$\hat{\beta}_{1\ell} - \beta_1 = \left[\sum_{t=1}^{n} (X_t - \bar{X})^2\right]^{-1} \sum_{t=1}^{n} (X_t - \bar{X})(v_t - \bar{v}).$$

The mean and variance results then follow because the X_t are fixed. □

Because (1.6.16) has the form of the classical fixed-X regression model, all the usual estimators of variance are appropriate. The residual mean square

$$s_{vv} = (n-2)^{-1} \sum_{t=1}^{n} [Y_t - \bar{Y} - (X_t - \bar{X})\hat{\beta}_{1\ell}]^2$$

is an unbiased estimator of σ_{vv} and replacing σ_{vv} of (1.6.15) with s_{vv} produces an unbiased estimator of $\mathbf{V}\{(\hat{\beta}_{0\ell}, \hat{\beta}_{1\ell})'\}$.

Theorem 1.6.1 and the discussion to this point suggest that errors of measurement associated with the controlled variable in an experiment can be ignored. This is not so. It is only estimation for the simplest model that remains unaffected. Berkson (1950) pointed out that tests in replicated experiments will be biased if the same error of measurement holds for several replicates. For example, assume that a fertilizer experiment is being conducted and consider the following two experimental procedures:

A. Fertilizer is applied with a spreader. A rate is set on the spreader and fertilizer is applied to every plot randomly selected to receive that rate. The spreader is then set for the next experimental rate and the process is repeated.
B. The fertilizer amounts are weighted separately for each plot. The amount shown by the scale is the same for all plots receiving the same treatment. The scale is calibrated with a standardized weight between each weighing. The quantity assigned to a particular plot is scattered evenly over the plot.

With procedure A, an error made in setting the spreader will be the same for all plots treated for that setting. We can write

$$x_{ti} = X_t - u_t - a_{ti},$$

where u_t is the error in the rate applied that is common for all plots treated with a single setting of the spreader and a_{ti} is the additional error arising from the fact that the material does not always feed through the spreader at the same rate. If the response Y is a linear function of the true application rate x_t, we have

$$\begin{aligned} Y_{ti} &= \beta_0 + \beta_1 x_t + e_{ti} \\ &= \beta_0 + \beta_1 X_t + e_{ti} - \beta_1 u_t - \beta_1 a_{ti}. \end{aligned}$$

Because a part of the measurement error, u_t, is common to the tth treatment, the treatment mean of the response variable is

$$\bar{Y}_{t\cdot} = \beta_0 + \beta_1 X_t + \bar{e}_{t.} - \beta_1 u_t - \beta_1 \bar{a}_{t.}$$

and the expected value of the within-treatment mean square is

$$E\left\{[n(r-1)]^{-1} \sum_{t=1}^{n} \sum_{i=1}^{r} (Y_{ti} - \bar{Y}_{t.})^2\right\} = \sigma_{ee} + \beta_1^2 \sigma_{aa},$$

where σ_{aa} is the variance of a_{ti}, we assume e_{ti}, u_t, and a_{ti} to be independent, and we assume each of n treatements is observed on r plots in a completely randomized design. The expected value of the residual mean square obtained in the regression of treatment means on X_t is

$$E\left\{(n-2)^{-1}\sum_{t=1}^{n} r(\bar{Y}_{t.}-\hat{\beta}_{0\ell}-X_t\hat{\beta}_{1\ell})^2\right\}=\sigma_{ee}+\beta_1^2\sigma_{aa}+r\beta_1^2\sigma_{uu}.$$

Therefore, with procedure A the estimator of β_1 is unbiased, but the usual within-treatment estimator of variance is a biased estimator of the true error variance.

With procedure B, the setting for a plot (the weighing operation) is repeated for each plot. If the calibration produces unbiased readings, we can write

$$X_{ti}=x_t+u_{ti},$$

where one can reasonably assume the u_{ti} to be uncorrelated. Then

$$Y_{ti}=\beta_0+\beta_1X_t+e_{ti}-u_{ti}\beta_1,$$

and the within-treatment mean square will be an unbiased estimator of $\sigma_{vv}=\sigma_{ee}+\beta_1^2\sigma_{uu}$.

Presence of measurement error in the experimental levels of a controlled experiment produces complications if the response is not linear. To investigate the nature of this problem, assume that

$$\begin{aligned} Y_t &= \beta_0+\beta_1x_t+\beta_2x_t^2+e_t, \qquad (1.6.17)\\ x_t &= X_t-u_t,\\ (e_t, u_t)' &\sim \text{NI}[\mathbf{0}, \text{diag}(\sigma_{ee}, \sigma_{uu})], \end{aligned}$$

where (e_t, u_t) is independent of X_j for all t and j. Then, using $x_t=X_t-u_t$, we obtain

$$E\{Y_t|X_t\}=(\beta_0+\beta_2\sigma_{uu})+\beta_1X_t+\beta_2X_t^2. \qquad (1.6.18)$$

If one estimates the quadratic function by ordinary least squares, one obtains unbiased estimators for β_1 and β_2, but the estimator of the intercept is an unbiased estimator of $\beta_0+\beta_2\sigma_{uu}$. Also, the conditional variance of Y_t given X_t is the variance of

$$e_t-\beta_1u_t-2\beta_2X_tu_t+\beta_2(u_t^2-\sigma_{uu}),$$

which is a function of X_t. Therefore, ordinary least squares will not be the most efficient estimation procedure and the ordinary estimates of the variance of the ordinary least squares estimators of $(\hat{\beta}_0, \hat{\beta}_1, \hat{\beta}_2)$ will be biased. The effects of measurement errors in the controlled variable of experiments with nonlinear response have been discussed by Box (1961).

REFERENCES

Section 1.6.1. Bickel and Ritov (1985), Geary (1942, 1943), Kendall and Stuart (1979), Madansky (1959), Malinvaud (1970), Neyman and Scott (1951), Reiersol (1950), Spiegelman (1979).

Section 1.6.2. Dorff and Gurland (1961a, 1961b), Pakes (1982), Wald (1940), Ware (1972).

Section 1.6.3. Ganse, Amemiya, and Fuller (1983), Lindley (1947).

Section 1.6.4. Berkson (1950), Box (1961), Draper and Beggs (1971).

EXERCISES

49. (Section 1.6.1) Assume that model (1.6.1) holds and that x_t is distributed as an exponential random variable. That is, the density of x_t is e^{-x}, $x \geqslant 0$. Give the variance of the approximate distribution of the estimator defined in (1.6.3).

50. (Section 1.6.2) Let the normal distribution model (1.2.1) of Section 1.2.1 hold. Let n be even and let $r = n/2$. Find the probability limit of Wald's estimator (1.6.4) if group one is composed of those observations with the r smallest X values.

51. (Section 1.6.3) Assume that a 13th field is randomly selected from the population of fields of Example 1.2.1. Assume that only one-half as many soil sites and chemical determinations are made for the field. It follows that the measurement error associated with the observed soil nitrogen for the 13th field is $2\sigma_u^2 = 114$. Assume that observed soil nitrogen is $X_{13} = 60$. Treating the estimates of Example 1.2.1 as parameters:

(a) Estimate true nitrogen treating x_{13} as fixed. Give the variance of your estimation error.

(b) Estimate true nitrogen conditioning on X_{13} and treating x_{13} as random. Give the variance of your prediction error.

(c) Predict observed yield conditioning on X_{13} and treating x_{13} as random. Give the variance of your prediction error.

52. (Section 1.6.3) Assume that the true parameters of a population of soil nitrogen values and yields such as that of Example 1.2.1 are $\sigma_{xx} = 200$, $\mu_x = 80$, $\beta_0 = 50$, $\beta_1 = 0.5$, and $\sigma_{ee} = 60$. Assume that a population of samples of X values with measurement error $\sigma_{uu} = 100$ is created by repeated sampling from a given field with true nitrogen equal to x_{13}.

(a) Find the mean and variance of the prediction error in the predicted value of true nitrogen constructed as

$$\tilde{x}(X_{13}) = \mu_x + \sigma_{XX}^{-1}\sigma_{xx}(X_{13} - \mu_x).$$

Compare the mean square error of this predictor to that of the unbiased estimator X_{13}. For what values of x_{13} does the predictor have smaller mean square error?

(b) Assuming that (Y_{13}, X_{13}) is observed, find the best predictor of x_{13} conditional on (Y_{13}, X_{13}). Find the best estimator of x_{13} treating x_{13} as fixed. For fixed x_{13} find the mean and variance of the error in the predictor of x_{13} that conditions on (Y_{13}, X_{13}).

53. (Section 1.6.3) Let (X, x) be distributed as a bivariate normal random vector with mean (μ_x, μ_x), $\sigma_{XX} = \sigma_{xx} + \sigma_{uu}$, and $\sigma_{Xx} = \sigma_{xx}$. Show that $\tilde{x}(X)$, defined in Example 1.6.1, has a smaller mean square error for x than that of X for $|x - \mu_x|^2 < a^2\sigma_{xx}$, where

$$a^2 = (\sigma_{xx}\sigma_{uu})^{-1}(\sigma_{XX}^2 - \sigma_{xx}^2).$$

APPENDIX 1.A. LARGE SAMPLE APPROXIMATIONS

In this appendix we state a theorem that is used in numerous places throughout the text.

Theorem 1.A.1. Let $g(\mathbf{a})$ be a real valued continuous function of the k-dimensional vector $\mathbf{a}$ for $\mathbf{a}$ an element of k-dimensional Euclidean space. Let $g(\mathbf{a})$ have continuous first derivatives at the point $\boldsymbol{\mu} = (\mu_1, \mu_2, \ldots, \mu_k)'$. Let the vector random variables $\mathbf{X}_t = (X_{t1}, X_{2t}, \ldots, X_{tk})'$, $t = 1, 2, \ldots$, be independently and identically distributed with mean $\boldsymbol{\mu}$ and covariance matrix Σ_{XX}. Let

$$\bar{\mathbf{X}} = n^{-1} \sum_{t=1}^{n} \mathbf{X}_t.$$

Then

$$n^{1/2}[g(\bar{\mathbf{X}}) - g(\boldsymbol{\mu})] \tag{1.A.1}$$

converges in distribution to a normal random variable with mean zero and variance

$$[g^{(1)}(\boldsymbol{\mu}), g^{(2)}(\boldsymbol{\mu}), \ldots, g^{(k)}(\boldsymbol{\mu})]\Sigma_{XX}[g^{(1)}(\boldsymbol{\mu}), g^{(2)}(\boldsymbol{\mu}), \ldots, g^{(k)}(\boldsymbol{\mu})]', \tag{1.A.2}$$

where $g^{(i)}(\boldsymbol{\mu})$ is the partial derivative of $g(\mathbf{a})$ with respect to a_i evaluated at $\mathbf{a} = \boldsymbol{\mu}$.

Proof. Because the derivatives are continuous at $\boldsymbol{\mu}$, there exists a closed ball B with $\boldsymbol{\mu}$ as an interior point such that the derivatives are continuous on B. The sample mean vector $\bar{\mathbf{X}}$ is converging to $\boldsymbol{\mu}$ in probability. Therefore, given $\varepsilon > 0$, there is an N such that

$$P\{\bar{\mathbf{X}} \in B\} > 1 - \tfrac{1}{2}\varepsilon \tag{1.A.3}$$

for $n > N$. For $\bar{\mathbf{X}} \in B$, by Taylor's theorem,

$$g(\bar{\mathbf{X}}) = g(\boldsymbol{\mu}) + \sum_{i=1}^{k} g^{(i)}(\mathbf{x}^*)(\bar{X}_i - \mu_i),$$

where $g^{(i)}(\mathbf{x}^*)$ is the derivative of $g(\mathbf{a})$ with respect to a_i evaluated at $\mathbf{x}^*$ and $\mathbf{x}^*$ is on the line segment joining $\bar{\mathbf{X}}$ and $\boldsymbol{\mu}$. Because $\bar{\mathbf{X}}$ is converging in probability to $\boldsymbol{\mu}$ and because $g^{(i)}(\mathbf{a})$ is continuous on B, $g^{(i)}(\mathbf{x}^*)$ is converging in probability to $g^{(i)}(\boldsymbol{\mu})$. It follows that, for $\bar{\mathbf{X}} \in B$,

$$\operatorname{plim} n^{1/2}\left\{\sum_{i=1}^{k} [g^{(i)}(\mathbf{x}^*) - g^{(i)}(\boldsymbol{\mu})](\bar{X}_i - \mu_i)\right\} = 0,$$

and that

$$\operatorname{plim} n^{1/2}\left\{g(\bar{\mathbf{X}}) - g(\boldsymbol{\mu}) - \sum_{i=1}^{k} g^{(i)}(\boldsymbol{\mu})(\bar{X}_i - \mu_i)\right\} = 0.$$

Therefore, the limiting distribution of $n^{1/2}[g(\bar{\mathbf{X}}) - g(\boldsymbol{\mu})]$ is the limiting distribution of

$$n^{1/2} \sum_{i=1}^{k} g^{(i)}(\boldsymbol{\mu})(\bar{X}_i - \mu_i) = n^{-1/2} \sum_{t=1}^{n} W_t,$$

where $W_t = \sum_{i=1}^{k} g^{(i)}(\boldsymbol{\mu})(X_{ti} - \mu_i)$. See, for example, Fuller (1976, p. 193). The W_t are independently and identically distributed with mean zero and variance given in (1.A.2). Therefore, the conclusion follows by the Lindeberg central limit theorem. □

The theorem extends immediately to vector valued functions.

Corollary 1.A.1. Let

$$\mathbf{g}(\mathbf{a}) = [g_1(\mathbf{a}), g_2(\mathbf{a}), \ldots, g_m(\mathbf{a})]'$$

be a vector valued function, where $g_i(\mathbf{a})$ are real valued functions satisfying the assumptions of Theorem 1.A.1. Let $\bar{\mathbf{X}}$ be as defined in Theorem 1.A.1. Then

$$n^{1/2}[\mathbf{g}(\bar{\mathbf{X}}) - \mathbf{g}(\boldsymbol{\mu})] \xrightarrow{L} N(\mathbf{0}, \mathbf{G}\boldsymbol{\Sigma}_{XX}\mathbf{G}),$$

where the ijth element of $\mathbf{G}$ is the derivative of $g_i(\mathbf{a})$ with respect to a_j evaluated at $\mathbf{a} = \boldsymbol{\mu}$.

Proof. Omitted. □

If the function has continuous second derivatives, it is possible to evaluate the order of the remainder in the approximation.

Corollary 1.A.2. Let the assumptions of Theorem 1.A.1 hold. In addition, assume that $g(\mathbf{a})$ has continuous second derivatives at the point $\boldsymbol{\mu}$. Then, given $\varepsilon > 0$, there is an N and an M_ε such that

$$P\left\{\left|g(\bar{\mathbf{X}}) - g(\boldsymbol{\mu}) - \sum_{i=1}^{k} g^{(i)}(\boldsymbol{\mu})(\bar{X}_i - \mu_i)\right| > n^{-1}M_\varepsilon\right\} < \varepsilon$$

for all $n > N$.

Proof. Let B be the compact ball defined in Theorem 1.A.1. Then for $\bar{\mathbf{X}} \in B$, the first-order Taylor expansion of $g(\bar{\mathbf{X}})$ about the point $\boldsymbol{\mu}$, with remainder, is

$$g(\bar{\mathbf{X}}) = g(\boldsymbol{\mu}) + \sum_{i=1}^{k} g^{(i)}(\boldsymbol{\mu})(\bar{X}_i - \mu_i) + \frac{1}{2}\sum_{i=1}^{k}\sum_{j=1}^{k} g^{(ij)}(\mathbf{x}^*)(\bar{X}_i - \mu_i)(\bar{X}_j - \mu_j), \tag{1.A.4}$$

where $g^{(ij)}(\mathbf{x}^*)$ is the second partial derivative of $g(\mathbf{a})$ with respect to a_i and a_j evaluated at $\mathbf{a} = \mathbf{x}^*$, and $\mathbf{x}^*$ is on the line segment joining $\boldsymbol{\mu}$ and $\bar{\mathbf{X}}$.

By the continuity of the derivatives on B, there is an M_1 such that $|g^{(ij)}(\mathbf{a})| < M_1$ for $\mathbf{a} \in B$. Furthermore, there exists an $M_{2\varepsilon}$ such that

$$P\{|(\bar{X}_i - \mu_i)(\bar{X}_j - \mu_j)| > n^{-1}M_{2\varepsilon}\} < \tfrac{1}{2}\varepsilon \tag{1.A.5}$$

for all n. Let $M_\varepsilon = M_1 M_{2\varepsilon}$. The conclusion follows from (1.A.3), (1.A.4), and (1.A.5). □

The application of Theorem 1.A.1 is sometimes called the method of statistical differentials or the delta method. These terms are appropriate because the mean and variance expressions are obtained by expanding the function $g(\mathbf{x})$ in a first-order Taylor series.

As an example of the use of Theorem 1.A.1, we consider the function

$$g(a) = a^{-1}$$

and let $\bar{X}$ be the mean of n normal independent (μ, σ^2) random variables, where $\mu \neq 0$. One method of expressing the fact that the distribution of $\bar{X}$ is centered about the true value with a standard error that is proportional to $n^{-1/2}$ is to write

$$\bar{X} - \mu = O_p(n^{-1/2}).$$

Formally, a sequence of random variables $\{W_n\}$ is $O_p(a_n)$ if, for every $\varepsilon > 0$, there exists a positive real number M_ε such that

$$P\{|W_n| > M_\varepsilon a_n\} \leqslant \varepsilon$$

for all n. See Fuller (1976, chap. 5). If we expand $g(\bar{X})$ in a first-order Taylor series about the point μ, we obtain

$$g(\bar{X}) = g(\mu) - \mu^{-2}(\bar{X} - \mu) + O_p(n^{-1}).$$

Ignoring the remainder, we have

$$g(\bar{X}) \doteq g(\mu) - \mu^{-2}(\bar{X} - \mu).$$

The mean and variance of the right side of this expression are $g(\mu)$ and $n^{-1}\mu^{-4}\sigma^2$, respectively, which agrees with the theorem. It is not formally correct to say that $\mu^{-4}\sigma^2$ is the limiting variance of $n^{1/2}[g(\bar{X}) - g(\mu)]$. One should say that $\mu^{-4}\sigma^2$ is the variance of the limiting distribution of $n^{1/2}[g(\bar{X}) - g(\mu)]$. Note that the theorem is not applicable if $\mu = 0$ because the function a^{-1} is not continuous at $a = 0$.

Very few restrictions are placed on the vector random variable $\mathbf{X}$ of Theorem 1.A.1. For example, X_2 might be a function of X_1. Consider the function

$$g(a_1, a_2) = a_2^{-1/2}a_1.$$

Let X_1 be a normal $(0, \sigma^2)$ random variable and let $X_2 = X_1^2$. Then we have

$$\begin{aligned} g(\bar{X}_1, \bar{X}_2) &= \left(n^{-1}\sum_{t=1}^{n} X_{1t}^2\right)^{-1/2}\bar{X}_1 = \bar{X}_2^{-1/2}\bar{X}_1 \\ &= \sigma^{-1}\bar{X}_1 + O_p(n^{-1}), \end{aligned}$$

because $g(\mu_1, \mu_2) = g^{(2)}(\mu_1, \mu_2) = 0$. We conclude that $n^{1/2}g(\bar{X}_1, \bar{X}_2)$ is approximately distributed as a $N(0, 1)$ random variable.

For a function with $g^{(i)}(\boldsymbol{\mu}) = 0$ for $i = 1, 2, \ldots, k$, the difference $n^{1/2}[g(\bar{\mathbf{X}}) - g(\boldsymbol{\mu})]$ converges in distribution to the constant zero. As an example, consider the function

$$g(a) = a^2$$

and let $\bar{X}$ be distributed as a normal $(0, n^{-1})$ random variable. Then

$$n^{1/2}(\bar{X}^2 - 0)$$

converges in probability and, hence, in distribution to the constant zero.

APPENDIX 1.B. MOMENTS OF THE NORMAL DISTRIBUTION

Let $\mathbf{Y}_t = (Y_{t1}, Y_{t2}, \ldots, Y_{tk})$ be distributed as a multivariate normal with mean zero and covariance matrix $\boldsymbol{\Sigma}$, where $\boldsymbol{\Sigma}$ has typical element σ_{ij}. Given a random sample of vectors $\mathbf{Y}_t$, $t = 1, 2, \ldots, n$, let

$$\bar{\mathbf{Y}} = n^{-1}\sum_{t=1}^{n} \mathbf{Y}_t,$$

$$s_{ij} = (n-1)^{-1}\sum_{t=1}^{n} (Y_{it} - \bar{Y}_i)(Y_{jt} - \bar{Y}_j), \qquad i = 1, 2, \ldots, k;\ j = 1, 2, \ldots, k.$$

It is well known that $\bar{\mathbf{Y}}$ is independent of all s_{ij} and that all odd moments of $\bar{\mathbf{Y}}$ are zero. See, for example, Anderson (1958). Also,

$$\begin{aligned}
E\{\bar{Y}_i\bar{Y}_j\} &= n^{-1}\sigma_{ij}, \\
E\{\bar{Y}_i\bar{Y}_j\bar{Y}_k\bar{Y}_m\} &= n^{-2}(\sigma_{ik}\sigma_{jm} + \sigma_{im}\sigma_{jk} + \sigma_{ij}\sigma_{km}), \\
E\{\bar{Y}_i\bar{Y}_j\bar{Y}_k\bar{Y}_m\bar{Y}_r\bar{Y}_t\} &= n^{-3}(\sigma_{ij}\sigma_{km}\sigma_{rt} + \sigma_{ij}\sigma_{kr}\sigma_{mt} + \sigma_{ij}\sigma_{kt}\sigma_{mr} \\
&\quad + \sigma_{ik}\sigma_{jm}\sigma_{rt} + \sigma_{ik}\sigma_{jr}\sigma_{mt} + \sigma_{ik}\sigma_{jt}\sigma_{mr} \\
&\quad + \sigma_{im}\sigma_{jk}\sigma_{rt} + \sigma_{im}\sigma_{jr}\sigma_{kt} + \sigma_{im}\sigma_{jt}\sigma_{kr} \\
&\quad + \sigma_{ir}\sigma_{jk}\sigma_{mt} + \sigma_{ir}\sigma_{jm}\sigma_{kt} + \sigma_{ir}\sigma_{jt}\sigma_{km} \\
&\quad + \sigma_{it}\sigma_{jk}\sigma_{mr} + \sigma_{it}\sigma_{jm}\sigma_{kr} + \sigma_{it}\sigma_{jr}\sigma_{km}) \\
C\{s_{ij}, s_{km}\} &= (n-1)^{-1}(\sigma_{ik}\sigma_{jm} + \sigma_{im}\sigma_{jk}), \\
E\{(s_{ij} - \sigma_{ij})(s_{km} - \sigma_{km})(s_{rt} - \sigma_{rt})\} &= (n-1)^{-2}[\sigma_{ik}\sigma_{jr}\sigma_{mt} + \sigma_{ik}\sigma_{jt}\sigma_{mr} \\
&\quad + \sigma_{im}\sigma_{jr}\sigma_{kt} + \sigma_{im}\sigma_{jt}\sigma_{kr} + \sigma_{ir}\sigma_{jk}\sigma_{mt} \\
&\quad + \sigma_{ir}\sigma_{jm}\sigma_{kt} + \sigma_{it}\sigma_{jk}\sigma_{mr} + \sigma_{it}\sigma_{jm}\sigma_{kr}],
\end{aligned}$$

for all i, j, k, m.

If we let $\mathbf{X}_t = (X_{t1}, X_{t2}, \ldots, X_{tk})$ be distributed as a multivariate normal with mean $\boldsymbol{\mu} = (\mu_1, \mu_2, \ldots, \mu_k)$ and covariance matrix $\boldsymbol{\Sigma}$, then

$$C\{X_{ti}X_{tj}, X_{tq}X_{tr}\} = \sigma_{ir}\sigma_{jq} + \sigma_{iq}\sigma_{jr} + \mu_i\mu_q\sigma_{jr} + \mu_i\mu_r\sigma_{jq} + \mu_j\mu_q\sigma_{ir} + \mu_j\mu_r\sigma_{iq}.$$

APPENDIX 1.C. CENTRAL LIMIT THEOREMS FOR SAMPLE MOMENTS

In this appendix we give the limiting distribution for the properly normalized sample mean and covariance matrix constructed from a random sample of n observations. Two situations are considered. In the first, the vector of observations is a random vector. In the second, the vector of observations is a sum of a fixed vector and a random vector. See Appendix 4.A for the definition of vech used in Theorem 1.C.1.

Theorem 1.C.1. Let $\{\mathbf{Z}_t\}$ be a sequence of independent identically distributed p-dimensional random vectors with mean $\boldsymbol{\mu}_Z$, covariance matrix $\boldsymbol{\Sigma}_{ZZ}$, and finite fourth moments. Let $\bar{\mathbf{Z}}$ be the sample mean, $\mathbf{m}_{ZZ}$ the sample covariance matrix,

$$\begin{aligned}
\operatorname{vech} \mathbf{m}_{ZZ} &= (m_{ZZ11}, m_{ZZ21}, \ldots, m_{ZZp1}, m_{ZZ22}, \ldots, m_{ZZp2}, \ldots, m_{ZZpp})', \\
\operatorname{vech} \boldsymbol{\Sigma}_{ZZ} &= (\sigma_{ZZ11}, \sigma_{ZZ21}, \ldots, \sigma_{ZZp1}, \sigma_{ZZ22}, \ldots, \sigma_{ZZp2}, \ldots, \sigma_{ZZpp})', \\
\mathbf{a}_t &= (\mathbf{Z}_t - \boldsymbol{\mu}_Z, [\operatorname{vech}\{(\mathbf{Z}_t - \boldsymbol{\mu}_Z)'(\mathbf{Z}_t - \boldsymbol{\mu}_Z) - \boldsymbol{\Sigma}_{ZZ}\}]'),
\end{aligned}$$

and $\boldsymbol{\Omega} = E\{\mathbf{a}_t\mathbf{a}_t'\}$. Then

$$n^{1/2}[(\bar{\mathbf{Z}} - \boldsymbol{\mu}_Z), (\operatorname{vech} \mathbf{m}_{ZZ} - \operatorname{vech} \boldsymbol{\Sigma}_{ZZ})']' \xrightarrow{L} N(\mathbf{0}, \boldsymbol{\Omega}).$$

Proof. We have

$$\mathbf{m}_{ZZ} = (n-1)^{-1}\left[\sum_{t=1}^{n} (\mathbf{Z}_t - \boldsymbol{\mu}_Z)'(\mathbf{Z}_t - \boldsymbol{\mu}_Z) + n(\bar{\mathbf{Z}} - \boldsymbol{\mu}_Z)'(\bar{\mathbf{Z}} - \boldsymbol{\mu}_Z)\right], \quad (1.\text{C}.1)$$

where $n^{1/2}(\bar{\mathbf{Z}} - \boldsymbol{\mu}_Z)'(\bar{\mathbf{Z}} - \boldsymbol{\mu}_Z)$ is converging to zero in probability. Therefore, the limiting distribution of $n^{1/2}(\text{vech } \mathbf{m}_{ZZ} - \text{vech } \boldsymbol{\Sigma}_{ZZ})$ is the same as that of

$$n^{1/2}(n-1)^{-1} \sum_{t=1}^{n} \text{vech}\{(\mathbf{Z}_t - \boldsymbol{\mu}_Z)'(\mathbf{Z}_t - \boldsymbol{\mu}_Z) - \boldsymbol{\Sigma}_{ZZ}\}.$$

For any arbitrary fixed row vector $\mathbf{d}$, where $\mathbf{dd}' \neq 0$, the sequence $\{\mathbf{da}_t\}$ is a sequence of independent identically distributed random variables with mean zero and variance $\mathbf{d\Omega d}'$. The conclusion follows by the Lindeberg central limit theorem. □

Corollary 1.C.1. Let the $\mathbf{Z}_t$ of Theorem 1.C.1 be normally distributed. Then

$$\boldsymbol{\Omega} = \text{block diag}(\boldsymbol{\Sigma}_{ZZ}, \boldsymbol{\Omega}_{22}),$$

where the element of $\boldsymbol{\Omega}_{22}$ associated with m_{ZZij} and $m_{ZZk\ell}$ is

$$\sigma_{ZZik}\sigma_{ZZj\ell} + \sigma_{ZZi\ell}\sigma_{ZZjk}.$$

Proof. For normal random variables $\bar{\mathbf{Z}}$ and $\mathbf{m}_{ZZ}$ are independent. See, for example, Kendall and Stuart (1977, Vol. 1, p. 384). The covariances of the sample moments from a normal distribution are given in Appendix 1.B. □

In Theorem 1.C.2 we give the limiting distribution of the vector composed of the means, mean squares, and mean products of variables that are the sum of fixed and random components. As a preliminary, we present two central limit theorems for weighted sums of random variables.

Lemma 1.C.1. Let $\mathbf{e}_t$ be a sequence of independently and identically distributed $(\mathbf{0}, \boldsymbol{\Sigma}_{ee})$ p-dimensional random row vectors, where $\boldsymbol{\Sigma}_{ee}$ is nonsingular. Let $\{\mathbf{c}_t\}$ be a sequence of fixed p-dimensional row vectors with $\mathbf{c}_1\mathbf{c}_1' \neq 0$ and

$$\lim_{n\to\infty} n^{-1} \sum_{t=1}^{n} \mathbf{c}_t\boldsymbol{\Sigma}_{ee}\mathbf{c}_t' = A > 0.$$

Then

$$V_n^{-1/2} \sum_{t=1}^{n} \mathbf{c}_t\mathbf{e}_t' \xrightarrow{\text{L}} N(0, 1),$$

where $V_n = \sum_{t=1}^{n} \mathbf{c}_t\boldsymbol{\Sigma}_{ee}\mathbf{c}_t'$.

Proof. The random variables $g_t = \mathbf{c}_t\mathbf{e}_t'$ are independent with zero means and variances,

$$E\{g_t^2\} = \mathbf{c}_t\mathbf{\Sigma}_{ee}\mathbf{c}_t'.$$

We have

$$V_n^{-1}\sum_{t=1}^{n}\int_{R_{1n}}(\mathbf{c}_t\mathbf{e}_t')^2 dF(\mathbf{e}) \leq V_n^{-1}\sum_{t=1}^{n}|\mathbf{c}_t|^2\int_{R_{2n}}|\mathbf{e}_t|^2 dF(\mathbf{e}),$$

where $F(\mathbf{e})$ is the distribution function of the vector $\mathbf{e}$,

$$R_{1n} = \{\mathbf{e}: (\mathbf{c}_t\mathbf{e}')^2 > \xi^2 V_n\},$$

$$R_{2n} = \left\{\mathbf{e}: |\mathbf{e}|^2 > \xi^2\left(V_n^{-1}\sup_{1\leq t\leq n}|\mathbf{c}_t|^2\right)^{-1}\right\},$$

and $|\mathbf{c}_t|^2 = \mathbf{c}_t\mathbf{c}_t'$. The ratio $V_n^{-1}\sum_{t=1}^{n}|\mathbf{c}_t|^2$ is bounded and, by the limit assumption,

$$\lim_{n\to\infty}\left(V_n^{-1}\sup_{1\leq t\leq n}|\mathbf{c}_t|^2\right)^{-1} = \infty.$$

Because the random variables $|\mathbf{e}_t|$ are independently and identically distributed with finite second moment,

$$\lim_{n\to\infty}\int_{R_{2n}}|\mathbf{e}|^2 dF(\mathbf{e}) = 0.$$

Therefore, the array $\{V_n^{-1/2}g_t\}$ satisfies the conditions of the Lindeberg central limit theorem. □

Lemma 1.C.2. Let $\mathbf{e}_t$ be a sequence of independently distributed $(\mathbf{0}, \mathbf{\Sigma}_{eett})$ p-dimensional random row vectors with uniformly bounded $2 + \delta$ $(\delta > 0)$ moments. Let $\{\mathbf{c}_t\}$ be a sequence of fixed p-dimensional row vectors with $\mathbf{c}_1\mathbf{c}_1' \neq 0$. Let $n^{-1}V_n$, where

$$V_n = \sum_{t=1}^{n}\mathbf{c}_t\mathbf{\Sigma}_{eett}\mathbf{c}_t',$$

be bounded above and below by positive real numbers for all n, assume

$$\lim_{n\to\infty} n^{-1}\sup_{1\leq t\leq n}|c_t|^2 = 0$$

and assume

$$n^{-1}\sum_{t=1}^{n}|c_t|^2$$

Proof. We have

$$V_n^{-1} \sum_{t=1}^{n} \int_{R_{1n}} (\mathbf{c}_t \mathbf{e}')^2 dF_t(\mathbf{e}) \leqslant V_n^{-1} \sum_{t=1}^{n} |\mathbf{c}_t|^2 \int_{R_{2n}} |\mathbf{e}|^2 dF_t(\mathbf{e})$$

$$\leqslant V_n^{-1} \sum_{t=1}^{n} |\mathbf{c}_t|^2 d_{n\xi}^{0.5\delta} \int_{R_{2n}} |\mathbf{e}|^{2+\delta} dF_t(\mathbf{e})$$

$$\leqslant V_n^{-1} \sum_{t=1}^{n} |\mathbf{c}_t|^2 d_{n\xi}^{0.5\delta} K,$$

where $d_{n\xi} = \xi^{-2} V_n^{-1} \sup_{1 \leqslant t \leqslant n} |\mathbf{c}_t|^2$, $R_{1n} = \{\mathbf{e}: (\mathbf{c}_t \mathbf{e}')^2 > \xi^2 V_n\}$, $R_{2n} = \{\mathbf{e}: |\mathbf{e}|^2 > d_{n\xi}^{-1}\}$, $F_t(\mathbf{e})$ is the distribution function of $\mathbf{e}_t$, and K is the uniform bound for $E\{|\mathbf{e}_t|^{2+\delta}\}$. By the assumptions

$$\lim_{n \to \infty} d_{n\xi}^{0.5\delta} V_n^{-1} \sum_{t=1}^{n} |\mathbf{c}_t|^2 = 0$$

and the conditions of the Lindeberg central limit theorem are satisfied. □

The limiting distribution for sample means, and mean products of random vectors with fixed components and identically distributed measurement errors, is given in Theorem 1.C.2.

Theorem 1.C.2. Let $\mathbf{Z}_t = \mathbf{z}_t + \boldsymbol{\varepsilon}_t$, where the $\boldsymbol{\varepsilon}_t$ are independent identically distributed p-dimensional random row vectors with zero mean vector, positive definite covariance matrix $\boldsymbol{\Sigma}_{\varepsilon\varepsilon}$, and finite fourth moments. Let $\{\mathbf{z}_t\}$ be a fixed sequence satisfying

$$\lim_{n \to \infty} \bar{\mathbf{z}} = \boldsymbol{\mu}_z,$$

$$\lim_{n \to \infty} \mathbf{m}_{zz} = \lim_{n \to \infty} (n-1)^{-1} \sum_{t=1}^{n} (\mathbf{z}_t - \bar{\mathbf{z}})'(\mathbf{z}_t - \bar{\mathbf{z}}) = \bar{\mathbf{m}}_{zz}.$$

Let $\hat{\boldsymbol{\theta}} = [\bar{\mathbf{Z}}, (\text{vech } \mathbf{m}_{ZZ})']'$ and $\boldsymbol{\theta}_n = [\bar{\mathbf{z}}, (\text{vech } \mathbf{m}_{zz} + \text{vech } \boldsymbol{\Sigma}_{\varepsilon\varepsilon})']'$, where vech is defined in Definition 4.A.2 of Appendix 4.A. Then

$$\mathbf{G}_n^{-1/2}(\hat{\boldsymbol{\theta}} - \boldsymbol{\theta}_n) \xrightarrow{L} N(\mathbf{0}, \mathbf{I}),$$

where the elements of $\mathbf{G}_n$ are the covariances of the elements of $\hat{\boldsymbol{\theta}}$,

$$C\{\bar{Z}_i, (\bar{Z}_j, m_{ZZjk})\} = n^{-1}(\sigma_{\varepsilon\varepsilon ij}, \tau_{ijk}),$$

$$C\{m_{ZZij}, m_{ZZk\ell}\} = (n-1)^{-1}(m_{zzik}\sigma_{\varepsilon\varepsilon j\ell} + m_{zzi\ell}\sigma_{\varepsilon\varepsilon jk} + m_{zzjk}\sigma_{\varepsilon\varepsilon i\ell} + m_{zzj\ell}\sigma_{\varepsilon\varepsilon ik} + \kappa_{ij,k\ell}) + O(n^{-2}),$$

$$\kappa_{ij,k\ell} = E\{(\varepsilon_i\varepsilon_j - \sigma_{\varepsilon\varepsilon ij})(\varepsilon_k\varepsilon_\ell - \sigma_{\varepsilon\varepsilon k\ell})\},$$

$\sigma_{\varepsilon\varepsilon ij} = E\{\varepsilon_i\varepsilon_j\}$, and $\tau_{ijk} = E\{\varepsilon_i\varepsilon_j\varepsilon_k\}$.

Proof. We first derive the elements of $\mathbf{G}_n$ associated with the higher moments. We have

$$\begin{aligned} C\{\bar{Z}_i, m_{ZZjk}\} &= E\{\bar{\varepsilon}_i(m_{z\varepsilon jk} + m_{z\varepsilon kj} + m_{\varepsilon\varepsilon jk})\} \\ &= n^{-1}\tau_{ijk}, \end{aligned}$$

where

$$\begin{aligned} E\{\bar{\varepsilon}_i m_{z\varepsilon jk}\} &= n^{-1}(n-1)^{-1}E\left\{\sum_{t=1}^{n}\sum_{r=1}^{n} \varepsilon_{ti}(z_{rj} - \bar{z}_j)\varepsilon_{rk}\right\} = 0, \\ E\{\bar{\varepsilon}_i m_{\varepsilon\varepsilon jk}\} &= n^{-1}(n-1)^{-1}E\left\{\sum_{t=1}^{n}\sum_{r=1}^{n} \varepsilon_{ti}\varepsilon_{rj}\varepsilon_{rk}\right\} \\ &\quad - n^{-2}(n-1)^{-1}E\left\{\sum_{t=1}^{n}\sum_{r=1}^{n}\sum_{s=1}^{n} \varepsilon_{ti}\varepsilon_{rj}\varepsilon_{sk}\right\} \\ &= n^{-1}\tau_{ijk}. \end{aligned}$$

Also $C\{m_{ZZij}, m_{ZZk\ell}\}$ is given by

$$C\{(m_{z\varepsilon ij} + m_{z\varepsilon ji} + m_{\varepsilon\varepsilon ij}), (m_{z\varepsilon k\ell} + m_{z\varepsilon\ell k} + m_{\varepsilon\varepsilon k\ell})\},$$

where

$$\begin{aligned} C\{m_{z\varepsilon ij}, m_{z\varepsilon k\ell}\} &= m_{zzik}\sigma_{\varepsilon\varepsilon j\ell}, \\ C\{m_{z\varepsilon ij}, m_{\varepsilon\varepsilon k\ell}\} &= (n-1)^{-1}E\left\{\sum_{t=1}^{n}\sum_{r=1}^{n} (z_{ti} - \bar{z}_i)\varepsilon_{tj}\varepsilon_{rk}\varepsilon_{r\ell}\right\} \\ &\quad - n^{-1}(n-1)^{-2}E\left\{\sum_{t=1}^{n}\sum_{r=1}^{n}\sum_{s=1}^{n} (z_{ti} - \bar{z}_i)\varepsilon_{tj}\varepsilon_{rk}\varepsilon_{s\ell}\right\} \\ &= 0, \\ C\{m_{\varepsilon\varepsilon ij}, m_{\varepsilon\varepsilon k\ell}\} &= (n-1)^{-2}E\left\{\sum_{t=1}^{n}\sum_{r=1}^{n} (\varepsilon_{ti}\varepsilon_{tj} - \sigma_{\varepsilon\varepsilon ij})(\varepsilon_{rk}\varepsilon_{r\ell} - \sigma_{\varepsilon\varepsilon k\ell})\right\} \\ &\quad - (n-1)^{-2}nE\left\{\bar{\varepsilon}_i\bar{\varepsilon}_j \sum_{r=1}^{n} (\varepsilon_{rk}\varepsilon_{r\ell} - \sigma_{\varepsilon\varepsilon k\ell})\right\} \\ &\quad - (n-1)^{-2}nE\left\{\bar{\varepsilon}_k\bar{\varepsilon}_\ell \sum_{t=1}^{n} (\varepsilon_{ti}\varepsilon_{tj} - \sigma_{\varepsilon\varepsilon ij})\right\} \\ &\quad + (n-1)^{-2}E\{(n\bar{\varepsilon}_i\bar{\varepsilon}_j - \sigma_{\varepsilon\varepsilon ij})(n\bar{\varepsilon}_j\bar{\varepsilon}_\ell - \sigma_{\varepsilon\varepsilon k\ell})\} \\ &= (n-1)^{-1}\kappa_{ij,k\ell} + O(n^{-2}). \end{aligned}$$

To establish the limiting normality of the vector of estimators, consider the linear combination

$$n^{1/2}\boldsymbol{\delta}'(\hat{\boldsymbol{\theta}} - \boldsymbol{\theta}_n) = n^{-1/2} \sum_{t=1}^{n} g_t + O_p(n^{-1}),$$

where

$$g_t = ([\mathbf{Z}_t - \bar{\mathbf{z}}], [\text{vech}\{(\mathbf{Z}_t - \bar{\mathbf{z}})'(\mathbf{Z}_t - \bar{\mathbf{z}}) - \mathbf{M}_{zz} - \boldsymbol{\Sigma}_{\varepsilon\varepsilon}\}]')\boldsymbol{\delta}$$

and $\boldsymbol{\delta}$ is an arbitrary vector such that $\boldsymbol{\delta}'\boldsymbol{\delta} \neq 0$. Using

$$\begin{aligned}(Z_{ti} - \bar{z}_i)(Z_{tj} - \bar{z}_j) - (z_{ti} - \bar{z}_i)(z_{tj} - \bar{z}_j) - \sigma_{\varepsilon\varepsilon ij} \\ = (z_{ti} - \bar{z}_i)\varepsilon_{tj} + (z_{tj} - \bar{z}_j)\varepsilon_{ti} + \varepsilon_{tj}\varepsilon_{ti} - \sigma_{\varepsilon\varepsilon ij},\end{aligned}$$

we can write

$$g_t = \sum_{i=1}^{p} b_{ti}\varepsilon_{ti} + \sum_{i=1}^{p} \sum_{j=i}^{p} d_{tij}(\varepsilon_{tj}\varepsilon_{ti} - \sigma_{\varepsilon\varepsilon ij}),$$

where b_{ti} is a linear function of $(z_{tr} - \bar{z}_r)$, $r = 1, 2, \ldots, p$, and d_{tij} is a linear function of $\boldsymbol{\delta}$. If we let

$$\begin{aligned}\mathbf{c}_t &= (b_{t1}, b_{t2}, \ldots, b_{tp}, d_{t11}, d_{t12}, \ldots, d_{tpp}),\\ \mathbf{e}_t &= (\varepsilon_{t1}, \varepsilon_{t2}, \ldots, \varepsilon_{tp}, \varepsilon_{t1}^2 - \sigma_{\varepsilon\varepsilon 11}, \varepsilon_{t1}\varepsilon_{t2} - \sigma_{\varepsilon\varepsilon 12}, \ldots, \varepsilon_{tp}^2 - \sigma_{\varepsilon\varepsilon pp}),\end{aligned}$$

the $\mathbf{c}_t$ and $\mathbf{e}_t$ satisfy the conditions of Lemma 1.C.1. The limiting normality is established because $\boldsymbol{\delta}$ is arbitrary. □

Corollary 1.C.2. Let the assumptions of Theorem 1.C.1 hold and, in addition, assume the $\boldsymbol{\varepsilon}_t$ to be normally distributed. Then

$$\mathbf{G}_n = \text{block diag}(\boldsymbol{\Sigma}_{\varepsilon\varepsilon}, \mathbf{G}_{22n}),$$

where the elements of $\mathbf{G}_{22n}$ are

$$\begin{aligned}C\{m_{ZZij}, m_{ZZk\ell}\} = (n-1)^{-1}(m_{zzik}\sigma_{\varepsilon\varepsilon j\ell} + m_{zzi\ell}\sigma_{\varepsilon\varepsilon jk} + m_{zzjk}\sigma_{\varepsilon\varepsilon i\ell} \\ + m_{zzj\ell}\sigma_{\varepsilon\varepsilon ik} + \sigma_{\varepsilon\varepsilon ik}\sigma_{\varepsilon\varepsilon j\ell} + \sigma_{\varepsilon\varepsilon i\ell}\sigma_{\varepsilon\varepsilon jk}).\end{aligned}$$

Proof. The result follows from the moment properties of the normal distribution. □

The limiting normal distribution for a vector of sample moments in which the measurement errors are not identically distributed, but possess finite $4 + \delta$ $(\delta > 0)$ moments, is given in Theorem 1.C.3.

Theorem 1.C.3. Let $\mathbf{Z}_t = \mathbf{z}_t + \boldsymbol{\varepsilon}_t$, where the $\boldsymbol{\varepsilon}_t$ are independent p-dimensional random row vectors with zero means, positive definite covariance matrices $\boldsymbol{\Sigma}_{\varepsilon\varepsilon tt}$, and bounded $4 + \delta$ $(\delta > 0)$ moments. Let $\mathbf{z}_t$ be a fixed sequence and let

$$\lim_{n\to\infty} \bar{\mathbf{z}} = \boldsymbol{\mu}_z, \qquad \lim_{n\to\infty} \mathbf{m}_{zz} = \bar{\mathbf{m}}_{zz},$$

$$\lim_{n\to\infty} n^{-1} \sum_{t=1}^{n} \boldsymbol{\Sigma}_{\varepsilon\varepsilon tt} = \bar{\boldsymbol{\Sigma}}_{\varepsilon\varepsilon},$$

where $\bar{\boldsymbol{\Sigma}}_{\varepsilon\varepsilon}$ is positive definite. Let $\hat{\boldsymbol{\theta}}$ and $\boldsymbol{\theta}_n$ be as defined in Theorem 1.C.2. Then

$$\mathbf{G}_n^{-1/2}(\hat{\boldsymbol{\theta}} - \boldsymbol{\theta}_n) \xrightarrow{L} N(\mathbf{0}, \mathbf{I}),$$

where $\mathbf{G}_n = \mathbf{V}\{\hat{\boldsymbol{\theta}}\}$.

Proof. The proof parallels that of Theorem 1.C.2, with Lemma 1.C.2 used to establish normality. □

APPENDIX 1.D. NOTES ON NOTATION

In this section we summarize, for reference, some of the notation that is used throughout the book. Because the treatment of measurement errors is a topic in many areas of statistics and in many areas of application, there is no standard notation at the present time. Therefore, any notation chosen for book length treatment will, at some point, be in conflict with the reader's previous experience.

We reserve capital Y_t and X_t to denote observable random variables. If these letters are boldface, $\mathbf{Y}_t$ and $\mathbf{X}_t$, they denote row vectors. Often $\mathbf{Y}_t$ and $\mathbf{X}_t$ are combined into a single row vector $\mathbf{Z}_t = (\mathbf{Y}_t, \mathbf{X}_t)$. Lowercase letters y_t, x_t, $\mathbf{y}_t$, $\mathbf{x}_t$, and $\mathbf{z}_t$ are reserved for the true values of Y_t, X_t, $\mathbf{Y}_t$, $\mathbf{X}_t$, and $\mathbf{Z}_t$, respectively. The variables denoted by lowercase letters may be either fixed or random, the nature of the variables being specified by the model. We desired a notation in which the observed and true values could be matched easily, and chose lowercase and capital letters as the simplest and most direct of the alternatives. The observed values, true values, and measurement errors are defined by

$$Y_t = y_t + e_t, \qquad \mathbf{X}_t = \mathbf{x}_t + \mathbf{u}_t,$$
$$\mathbf{Z}_t = (Y_t, \mathbf{X}_t) = \mathbf{z}_t + \boldsymbol{\varepsilon}_t,$$

where $\boldsymbol{\varepsilon}_t = (e_t, \mathbf{u}_t)$ is the row vector of measurement errors.

Measurement error models are related to the regression model and we chose a notation with close ties to regression. To permit us to write models in the regression form

$$Y_t = \mathbf{x}_t\boldsymbol{\beta} + e_t, \tag{1.D.1}$$

we define $\mathbf{x}_t$ to be a row vector and $\boldsymbol{\beta}$ to be a column vector. This causes our notation to differ from that often used for multivariate models. When the regression model with an error in the equation is being considered, the error in the equation is denoted by q_t and the pure measurement error in Y_t by w_t. In such models $e_t = w_t + q_t$ and

$$y_t = \mathbf{x}_t\boldsymbol{\beta} + q_t.$$

To be consistent with the linear model, we define the vector of parameters to be a column vector identified with a Greek symbol. For example,

$$\boldsymbol{\beta} = (\beta_0, \beta_1)'.$$

For models of the type (1.D.1) we reserve the letter v_t for the deviation

$$v_t = Y_t - \mathbf{X}_t\boldsymbol{\beta} = e_t - \mathbf{u}_t\boldsymbol{\beta}.$$

If $\mathbf{Y}_t$ is a row vector, there is a row vector of deviations

$$\mathbf{v}_t = \mathbf{Y}_t - \mathbf{X}_t\boldsymbol{\beta} = \mathbf{e}_t - \mathbf{u}_t\boldsymbol{\beta}.$$

We shall follow the common statistical practice and use the same notation for a random variable and for a realization of that random variable, with a few exceptions.

We use μ_Y to denote the mean of the random variable Y_t and $\boldsymbol{\mu}_Y$ to denote the mean of the random vector $\mathbf{Y}_t$. The three symbols σ_Y^2, σ_{YY}, and $V\{Y_t\}$ will be used to denote the variance of the random variable Y. The bold $\mathbf{V}\{\mathbf{X}'\}$ or $\boldsymbol{\Sigma}_{XX}$ will denote the covariance matrix of the column $\mathbf{X}'$. The ijth element of $\boldsymbol{\Sigma}_{XX}$ is σ_{XXij}. The covariance of the scalar random variables X and Y is denoted by σ_{XY} or by $C\{X, Y\}$. The matrix $\boldsymbol{\Sigma}_{XY} = E\{(\mathbf{X} - \boldsymbol{\mu}_X)'(\mathbf{Y} - \boldsymbol{\mu}_Y)\}$ is the covariance matrix of the column $\mathbf{X}'$ and the row $\mathbf{Y}$. The covariance between the ith element of $\mathbf{X}_t$ and the jth element of $\mathbf{Y}_t$ is σ_{XYij}.

Lowercase letter $\mathbf{m}$, appropriately subscripted, is used for the sample covariance matrix. For example,

$$\mathbf{m}_{ZZ} = (n-1)^{-1} \sum_{t=1}^{n} (\mathbf{Z}_t - \bar{\mathbf{Z}})'(\mathbf{Z}_t - \bar{\mathbf{Z}}),$$

where $\mathbf{Z}_t = (Z_{t1}, Z_{t2}, \ldots, Z_{tp})$, $\bar{\mathbf{Z}} = n^{-1} \sum_{t=1}^{n} \mathbf{Z}_t$, and the ijth element of $\mathbf{m}_{ZZ}$ is m_{ZZij}. Capital letter $\mathbf{M}$, appropriately subscripted, is used for the matrix of raw mean squares and products. Thus,

$$\mathbf{M}_{XY} = n^{-1} \sum_{t=1}^{n} \mathbf{X}_t'\mathbf{Y}_t.$$

We use the letter S, appropriately subscripted, for an estimator of the covariance matrix of the error vector. Thus, $\mathbf{S}_{\varepsilon\varepsilon}$ is an estimator of the covariance matrix of ε_t.

We shall generally use column vectors when discussing the distribution of a vector random variable. Thus, we write

$$\mathbf{Z}_t' \sim \mathrm{NI}(\mathbf{0}, \mathbf{\Sigma}_{ZZ}),$$

where $\sim \mathrm{NI}(\mathbf{0}, \mathbf{\Sigma}_{ZZ})$ signifies that the random column vectors $\mathbf{Z}_t'$, $t = 1, 2, \ldots,$ are distributed normally and independently with zero mean vector and covariance matrix $\mathbf{\Sigma}_{ZZ}$.

Special notation for mapping a matrix into a vector is defined in Appendix 4.A.

REFERENCES

Appendix 1.A. Cramér (1946), Fuller (1976), Mann and Wald (1943).
Appendix 1.B. Anderson (1958).
Appendix 1.C. Kendall and Stuart (1977), Fuller (1976, 1980).

EXERCISES

54. (Appendix 1.A) Using the method of statistical differentials, find the mean and variance of the limiting distribution of $n^{1/2}(\hat{\sigma}_X - \sigma_X)$, where

$$\hat{\sigma}_X^2 = n^{-1} \sum_{t=1}^{n} (X_t - \bar{X})^2$$

and the X_t are normal independent (μ, σ_X^2) random variables.

55. (Appendix 1.A) Using the method of statistical differentials, obtain the mean and variance of the limiting distribution of $\hat{R} = \bar{Y}\bar{X}^{-1}$, where (Y_t, X_t) are normal independent vectors with $\mu_X \neq 0$ and $\mu_Y \mu_X^{-1} = R$.

56. (Appendix 1.A) Let $X_t \sim (\mu_X, \sigma_{XX})$, where $\mu_X > 0$. Let $C_{XX} = \mu_X^{-2}\sigma_{XX}$ be the squared coefficient of variation of X.

   (a) Show that the coefficient of variation of the approximate distribution of $\bar{X}^{-1}$ is equal to the coefficient of variation of $\bar{X}$.

   (b) Show that, for $\mu_Y \neq 0$ and $\mu_X \neq 0$, the squared coefficient of variation of the approximate distribution of $\hat{R}$ obtained in Exercise 55 can be written

$$C_{\hat{R}\hat{R}} = n^{-1}(C_{YY} - 2C_{XY} + C_{XX}),$$

   where $C_{XY} = \mu_Y^{-1}\mu_X^{-1}\sigma_{XY}$.

57. (Appendixes 1.A, 1.C, Section 1.1) In Section 1.1 it was shown that $E\{\hat{\gamma}_{1\ell}\} = \beta_1 \sigma_{XX}^{-1}\sigma_{xx}$ under the assumption that the (x_t, e_t, u_t) are normally and independently distributed with diagonal covariance matrix.

(a) Show that, for $n > 3$,

$$V\{\hat{\gamma}_{1\ell}\} = (n-3)^{-1}\sigma_{XX}^{-2}(\sigma_{XX}\sigma_{ee} + \beta_1^2\sigma_{xx}\sigma_{uu})$$

under the assumption that the (x_t, e_t, u_t) are normally and independently distributed with diagonal covariance matrix.

(b) Assume model (1.1.1), (1.1.2) and assume $\{x_t\}$ is a fixed sequence with

$$\lim_{n\to\infty} n^{-1} \sum_{t=1}^{n} [x_t, (x_t - \bar{x})^2] = (0, \sigma_{xx}).$$

Assume $(e_t, u_t)' \sim \mathrm{NI}[\mathbf{0}, \mathrm{diag}(\sigma_{ee}, \sigma_{uu})]$ and show that

$$n^{1/2}(\hat{\gamma}_{1\ell} - \gamma_{1n}) \xrightarrow{L} N(0, V),$$

where $\gamma_{1n} = (m_{xx} + \sigma_{uu})^{-1}\beta_1 m_{xx}$ and

$$V = (\sigma_{xx} + \sigma_{uu})^{-1}\sigma_{ee} + (\sigma_{xx} + \sigma_{uu})^{-4}\beta_1^2\sigma_{xx}\sigma_{uu}(\sigma_{xx}^2 + \sigma_{uu}^2).$$

(c) Show that, under the assumptions of part (a),

$$V\{E[\hat{\gamma}_{1\ell} | m_{xx}]\} \doteq 2n^{-1}\beta_1^2(m_{xx} + \sigma_{uu})^{-4}(m_{xx}\sigma_{uu})^2,$$

where the approximation is understood to mean that only the expectation of the leading term of the Taylor series is evaluated.

58. (Appendix 1.A, Section 1.2) Using the method of statistical differentials, find the limiting distribution of $n^{1/2}(s_{vv} - \sigma_{vv})$, where s_{vv} is defined in (1.2.7).

59. (Appendix 1.A, Section 1.2) Let $X_t \sim \mathrm{NI}(\mu_x, \sigma_{XX})$ and let $m_{XX} = (n-1)^{-1}\sum_{t=1}^{n}(X_t - \bar{X})^2$.

(a) Obtain the exact mean and variance of m_{XX}^{-1} for $n > 5$. Compare this expression with the mean and variance of the approximate distribution derived by the methods of Appendix 1.A.

(b) Obtain the coefficient of variation of the approximate distribution of $D = (1 - m_{XX}^{-1}\sigma_{uu})$.

(c) Let $\hat{\beta}_1 = \hat{\gamma}(1 - m_{XX}^{-1}\sigma_{uu})^{-1}$, where $\hat{\gamma} = m_{XX}^{-1}m_{XY}$ and (Y_t, X_t) are bivariate normal and satisfy model (1.2.1). Assume $\beta_1 \neq 0$. Show that the squared coefficient of variation of the approximate distribution of $\hat{\beta}_1$ is $C_{\beta\beta} = C_{\gamma\gamma} + C_{DD}$, where $C_{\gamma\gamma}$ and C_{DD} are the squared coefficients of variation of the approximate distributions of $\hat{\gamma}$ and D, respectively.

(d) On the basis of higher-order approximations, Cochran (1977, p. 162) suggests that the approximate squared coefficient of variation of a ratio of uncorrelated approximately normally distributed random variables will be within $9\delta\%$ of the true squared coefficient of variation if the squared coefficient of variation of the denominator is less than $\delta\%$. For what values of $\sigma_{XX}^{-1}\sigma_{uu}$ and n will the coefficient of variation of the approximate distribution of the denominator D be less than $\delta\%$?

60. (Appendix 1.A, Section 1.2) Let $\{x_t\}_1^{\infty}$ be a fixed sequence and let $X_t = x_t + u_t$.

(a) Show that for $u_t \sim \mathrm{NI}(0, \sigma_{uu})$,

$$V\{m_{XX} - m_{xx}\} = 4(n-1)^{-2}\sum_{t=1}^{n}(x_t - \bar{x})^2\sigma_{uu} + 2(n-1)^{-1}\sigma_{uu}^2.$$

(b) Using result (a), show that the coefficient of variation of $\hat{m}_{xx} = m_{XX} - \sigma_{uu}$ is less than 0.2 if

$$m_{xx}^{-1}\sigma_{uu} < [0.02(n-1) + 1]^{1/2} - 1.$$

Show that the coefficient of variation of $\hat{m}_{xx}$ is less than 0.2 if

$$[(n-1)m_{xx}^2]^{-1}\sigma_{uu}(\sigma_{uu} + m_{xx}) < 0.01.$$

(c) Using Exercise 56, show that the squared coefficient of variation of the approximate distribution of $(m_{XX} - \sigma_{uu})^{-1}m_{XX}$ is

$$\frac{2[(m_{xx} + \sigma_{uu})^2 - m_{xx}^2]\sigma_{uu}^2}{(n-1)(m_{xx} + \sigma_{uu})^2 m_{xx}^2} = \frac{2[2+\xi]\xi^3}{(1+\xi)^2(n-1)}$$
$$\leqslant 2(n-1)^{-1}m_{xx}^{-2}\sigma_{uu}^2,$$

where $\xi = m_{xx}^{-1}\sigma_{uu}$.

61. (Appendix 1.A, Section 1.3) Let $\hat{\beta}_1$, the estimator of β_1 of model (1.3.1), be defined by (1.3.7).

(a) Let $\theta = \arctan \beta_1$ and $\hat{\theta} = \arctan \hat{\beta}_1$. Show that, for $\sigma_{uu} = \sigma_{ee}$,

$$\tfrac{1}{2}\tan 2\hat{\theta} = [m_{XX} - m_{YY}]^{-1}m_{XY}.$$

Obtain the limiting distribution of $\hat{\theta}$.

(b) Show that the variance of the limiting distribution of $n^{1/2}(s_{vv} - \sigma_{vv})$ is $2\sigma_{vv}^2 + \sigma_{uv}^2\Gamma_{22}$, where s_{vv} is defined in (1.3.12) and Γ_{22} is defined in Theorem 1.3.1.

62. (Appendix 1.A, Section 1.5) The estimates of σ_{ee11} and σ_{uu} of Example 1.5.1 are quite similar. Using the estimates and the estimated covariance matrix of Example 1.5.1, construct estimates of the parameters subject to the restriction $\sigma_{ee11} = \sigma_{uu}$. That is, use generalized least squares to construct restricted estimates under the assumption that the estimators of Example 1.5.1 are normally distributed with the estimated covariance matrix constructed in Example 1.5.1. Do not use the original data in your calculations. Estimate the covariance matrix of your estimators.

63. (Appendix 1.A) Let (Y_t, X_t) be distributed as a bivariate normal random vector. Using the methods of Appendix 1.A, obtain the approximate distribution of $\hat{\gamma}_{XY}^{-1}$, where $\hat{\gamma}_{XY} = m_{YY}^{-1}m_{XY}$.

64. (Appendix 1.B) Let $Y_t \sim \text{NI}(\mu_Y, \sigma_{YY})$. Prove that

$$\hat{V}\{m_{YY}\} = 2(n+1)^{-1}m_{YY}^2$$

is an unbiased estimator of $V\{m_{YY}\} = 2(n-1)^{-1}\sigma_{YY}^2$. Extend this result to the estimation of $C\{m_{YY}, m_{XY}\}$.

65. (Appendix 1.B) Prove the following lemma (Jöreskog, 1973).

Lemma. Let $\mathbf{Z}_t' \sim \text{NI}(\mathbf{0}, \boldsymbol{\Sigma})$, $t = 1, 2, \ldots,$ and let

$$\mathbf{W} = \boldsymbol{\Sigma}^{-1}(\mathbf{m} - \boldsymbol{\Sigma})\boldsymbol{\Sigma}^{-1}, \qquad \mathbf{m} = (n-1)^{-1}\sum_{t=1}^{n}(\mathbf{Z}_t - \bar{\mathbf{Z}})'(\mathbf{Z}_t - \bar{\mathbf{Z}}),$$

and $\bar{\mathbf{Z}} = n^{-1}\sum_{t=1}^{n}\mathbf{Z}_t$. Then

$$C\{w_{ij}, w_{rt}\} = (n-1)^{-1}(\sigma^{ir}\sigma^{jt} + \sigma^{it}\sigma^{jr}),$$

where σ^{ij} is the ijth element of $\boldsymbol{\Sigma}^{-1}$ and w_{ij} is the ijth element of $\mathbf{W}$.

66. (Appendix 1.B) Let $\mathbf{Z}_t' = (Z_{1t}, Z_{2t})' \sim \text{NI}(\mathbf{0}, \boldsymbol{\Sigma})$, where $\boldsymbol{\Sigma}$ is a 2×2 matrix with ijth element equal to σ_{ij}. Prove that the conditional variance of Z_{1t} given Z_{2t} is $(\sigma^{11})^{-1}$, where σ^{ij} is the ijth element of $\boldsymbol{\Sigma}^{-1}$. Generalize this result to higher dimensions.

CHAPTER 2

Vector Explanatory Variables

In this chapter we study models with more than one x variable. Maximum likelihood estimation is investigated and certain problems associated with the likelihood function are identified. Estimators of parameters that are extensions of the estimators of Chapter 1 are presented.

The first four sections of this chapter parallel the first four sections of Chapter 1. Section 2.5 is devoted to the small sample properties of estimators and to modifications designed to improve the small sample behavior. The calibration problem is also considered in Section 2.5.

2.1. BOUNDS FOR COEFFICIENTS

In Section 1.1.3 we demonstrated that it is possible to establish bounds for the population regression coefficient in a simple regression with the independent variable subject to measurement error, provided the measurement error is independent of the error in Y_t. Given a diagonal error covariance matrix, such bounds can be constructed for the model with a vector of explanatory variables.

Let

$$y_t = \beta_0 + \mathbf{x}_t\boldsymbol{\beta}, \qquad \mathbf{Z}_t = \mathbf{z}_t + \boldsymbol{\varepsilon}_t, \tag{2.1.1}$$
$$(\mathbf{x}_t, \boldsymbol{\varepsilon}_t)' \sim \text{NI}[(\boldsymbol{\mu}_x, \mathbf{0})', \text{block diag}(\boldsymbol{\Sigma}_{xx}, \boldsymbol{\Sigma}_{\varepsilon\varepsilon})],$$

where $\mathbf{Z}_t = (Y_t, \mathbf{X}_t)$, $\mathbf{x}_t$ is a k-dimensional row vector, $\boldsymbol{\beta}$ is a k-dimensional column vector, and the covariance matrix of $\mathbf{Z}_t$, denoted by $\boldsymbol{\Sigma}_{ZZ}$, is positive definite. Under this model, the unknown parameter vector must satisfy

$$(\boldsymbol{\Sigma}_{ZZ} - \boldsymbol{\Sigma}_{\varepsilon\varepsilon})(1, -\boldsymbol{\beta}')' = \mathbf{0}.$$

Thus, if $\ddot{\Sigma}_{\varepsilon\varepsilon}$ is any positive semidefinite matrix, the vector $\ddot{\boldsymbol{\beta}}$ that satisfies

$$(\Sigma_{ZZ} - \hat{\lambda}\ddot{\Sigma}_{\varepsilon\varepsilon})(1, -\ddot{\boldsymbol{\beta}}')' = \mathbf{0},$$

where $\hat{\lambda}$ is the smallest root of

$$|\Sigma_{ZZ} - \hat{\lambda}\ddot{\Sigma}_{\varepsilon\varepsilon}| = 0,$$

is an acceptable parameter vector. We now assume that $\Sigma_{\varepsilon\varepsilon}$ is a positive semidefinite diagonal covariance matrix. This permits us to establish upper bounds for the error variances. If we solve the equation

$$|\Sigma_{ZZ} - \lambda_i \mathbf{D}_{ii}| = 0$$

for λ_i, where $\mathbf{D}_{ii}$ is a diagonal matrix with the ith diagonal element equal to one and all other entries equal to zero, we obtain the upper bound for the error variance of the ith component of $\mathbf{Z}_t$. The coefficient λ_i is the population regression residual mean square obtained in the population regression of the ith element of $\mathbf{Z}_t$ on the remaining k elements of $\mathbf{Z}_t$.

Setting bounds for the elements of $\boldsymbol{\beta}$ is more difficult. We illustrate by considering a model with two explanatory variables. Let

$$Y_t = \beta_0 + \beta_1 x_{t1} + \beta_2 x_{t2} + e_t$$

and retain the assumptions of model (2.1.1). Then we can write the model as

$$W_{t0} = \beta_0^* + \beta_2 w_{t2} + e_t,$$

where w_{t2} is the portion of x_{t2} that is orthogonal to x_{t1}, W_{t0} is the portion of Y_t that is orthogonal to x_{t1}, and $(W_{t0}, W_{t2}) = (w_{t0}, w_{t2}) + (e_t, u_{t2})$. If we fix a value for σ_{uu11}, say $\ddot{\sigma}_{uu11}$, in the permissible range for σ_{uu11}, then one can define the new variables

$$(W_{t0}, w_{t2}) = (Y_t - C_{01}x_{t1}, x_{t2} - C_{21}x_{t1}),$$

where

$$(C_{01}, C_{21}) = \ddot{\sigma}_{xx11}^{-1}(\sigma_{XY11}, \sigma_{XX12}),$$

and $\ddot{\sigma}_{xx11} = \sigma_{XX11} - \ddot{\sigma}_{uu11}$. It follows that the bounds of Section 1.1.3 apply to the reduced model and, for $\sigma_{uu11} = \ddot{\sigma}_{uu11}$ and $\sigma_{WW02} > 0$,

$$\sigma_{WW22}^{-1}\sigma_{WW02} \leqslant \beta_2 \leqslant \sigma_{WW02}^{-1}\sigma_{WW00}.$$

If σ_{WW02} has the same sign for all $\ddot{\sigma}_{uu11}$ in the permissible range, then the derivatives of the bounds with respect to $\ddot{\sigma}_{uu11}$ also never change sign. It follows that the unconditional bounds for β_1 will be in the set of conditional bounds associated with the minimum and maximum values of $\ddot{\sigma}_{uu11}$. If $\ddot{\sigma}_{uu11} = 0$, the lower bound for β_2 is the regression coefficient of X_2 obtained in the regression of Y on X_1 and X_2, and the upper bound for β_2 is the inverse of the regression coefficient for Y in the regression of X_2 on Y and X_1. At the maximum value for $\ddot{\sigma}_{uu11}$, the conditional upper and lower bounds for β_2 coincide. The value of β_2 is the negative of the ratio of the coefficient for X_2 to the coefficient for Y in the regression of X_1 on Y and X_2.

If the conditional covariance between W_{t0} and W_{t2} has a different sign for two values of $\ddot{\sigma}_{uu11}$ in the permissible range, then there is a $\ddot{\sigma}_{uu11}$ for which $\sigma_{WW02} = 0$ and the range of possible values for the β's is infinite. If the conditional covariances do not change sign, then the range of possible values is finite.

For the general problem, consider the set of $k + 1$ values of $\boldsymbol{\beta}$ constructed by using each of the $k + 1$ variables in $\mathbf{Z}_t$ as the dependent variable in an ordinary least squares regression. In all cases the least squares solution is standardized so that the equation is written in the form $Y = \mathbf{X}\boldsymbol{\beta}$. The bounds for $\boldsymbol{\beta}$ will be finite if each coefficient vector in the set of $k + 1$ vectors obtained from the $k + 1$ regressions has the same sign structure. If the $k + 1$ vectors are all in the same orthant of the k-dimensional space, the set of possible values for $\boldsymbol{\beta}$ is composed of the convex linear combinations of the $k + 1$ coefficients obtained from the $k + 1$ regressions. If the set of $k + 1$ population vectors are not all in the same orthant, the set of possible values for $\boldsymbol{\beta}$ is unbounded. See Klepper and Leamer (1984) and Patefield (1981).

For the three-variable case, the possible set of values for (β_1, β_2) is a triangle in (β_1, β_2) space with vertices corresponding to the values obtained from the three ordinary least squares regressions using each of the three variables as the dependent variable, provided all three solutions are in the same quadrant.

REFERENCES

Bekker, Wansbeek, and Kapteyn (1985), Klepper and Leamer (1984), Koopmans (1937), Patefield (1981).

EXERCISES

1. (Section 2.1) Let model (2.1.1) hold with $\boldsymbol{\Sigma}_{ue} = \mathbf{0}$. For one explanatory variable and simple uncorrelated measurement error it was shown in Section 1.1.1 that the expected value of the ordinary least squares estimator of the slope was closer to zero than the true value.

(a) Show that

$$\boldsymbol{\beta}'\boldsymbol{\Sigma}_{XX}\boldsymbol{\beta} \geqslant \boldsymbol{\gamma}'\boldsymbol{\Sigma}_{XX}\boldsymbol{\gamma} \quad \text{and} \quad \boldsymbol{\beta}'\boldsymbol{\Sigma}_{xx}\boldsymbol{\beta} \geqslant \boldsymbol{\gamma}'\boldsymbol{\Sigma}_{xx}\boldsymbol{\gamma},$$

where $\boldsymbol{\gamma} = \boldsymbol{\Sigma}_{XX}^{-1}\boldsymbol{\Sigma}_{XY}$. (Hint: See (4.A.20) of Appendix 4.A.)

(b) Let the two squared multiple correlations be defined by

$$R^2_{YX} = \sigma_{YY}^{-1}\Sigma_{YX}\Sigma_{XX}^{-1}\Sigma_{XY} \quad \text{and} \quad R^2_{Yx} = \sigma_{YY}^{-1}\Sigma_{Yx}\Sigma_{xx}^{-1}\Sigma_{xY}.$$

Show that $R^2_{YX} \leqslant R^2_{Yx}$. Show that the inequality is strict if at least one element of Σ_{uu} is positive.

2. (Section 2.1) Let model (2.1.1) hold and let $\mathbf{X}_t = (\mathbf{x}_{t1}, \mathbf{X}_{t2})$ be observed, where $\boldsymbol{\beta}' = (\boldsymbol{\beta}'_1, \boldsymbol{\beta}'_2)$, $\mathbf{x}_{t1}$ is measured without error, and $\mathbf{X}_{t2} = \mathbf{x}_{t2} + u_{t2}$. Show that, typically, the least squares coefficients obtained in the regression of Y_t on $\mathbf{X}_t$ are biased for both $\boldsymbol{\beta}_1$ and $\boldsymbol{\beta}_2$. Show that the least squares estimator of $\boldsymbol{\beta}_1$ is unbiased for $\boldsymbol{\beta}_1$ if $E\{\mathbf{x}'_{t1}\mathbf{x}_{t2}\} = \mathbf{0}$. See Carroll, Gallo, and Gleser (1985).

2.2. THE MODEL WITH AN ERROR IN THE EQUATION

2.2.1. Estimation of Slope Parameters

The generalization of model (1.2.1) of Section 1.2 to a model containing a vector of x variables is

$$Y_t = \mathbf{x}_t\boldsymbol{\beta} + e_t, \qquad \mathbf{X}_t = \mathbf{x}_t + \mathbf{u}_t, \tag{2.2.1}$$

for $t = 1, 2, \ldots, n$, where $\mathbf{x}_t$ is a k-dimensional row vector, $\boldsymbol{\beta}$ is a k-dimensional column vector, and the $(k + 1)$-dimensional vectors $\boldsymbol{\varepsilon}'_t = (e_t, \mathbf{u}_t)'$ are independent normal $(\mathbf{0}, \Sigma_{\varepsilon\varepsilon})$ random vectors. It is assumed that the vector of covariances between e_t and $\mathbf{u}_t$, Σ_{eu}, and the covariance matrix of $\mathbf{u}_t$, Σ_{uu}, are known. The variance of e_t, σ_{ee}, is unknown. Instead of assuming the x variables to be random variables, as we did in Section 1.2, we initially assume that $\{\mathbf{x}_t\}$ is a sequence of fixed k-dimensional row vectors.

We examine the likelihood function assuming that $\Sigma_{\varepsilon\varepsilon}$ is nonsingular and, with no loss of generality, that

$$\Sigma_{uu} = \mathbf{I} \quad \text{and} \quad \Sigma_{ue} = \mathbf{0}. \tag{2.2.2}$$

With these assumptions, the density of $(e_t, \mathbf{u}_t)'$ is

$$(2\pi)^{-(k+1)/2}\sigma_{ee}^{-1/2}\exp\{-\tfrac{1}{2}[\sigma_{ee}^{-1}e_t^2 + \mathbf{u}_t\mathbf{u}'_t]\}. \tag{2.2.3}$$

Because $\mathbf{x}_t$ is fixed, the Jacobian of the transformation of $(e_t, \mathbf{u}_t)$ into $(Y_t, \mathbf{X}_t)$ is one, and the logarithm of the likelihood for a sample of n observations is

$$\begin{aligned} \log L = {} & -\tfrac{1}{2}n[(k+1)\log 2\pi + \log \sigma_{ee}] \\ & -\tfrac{1}{2}\left\{\sum_{t=1}^{n}\sigma_{ee}^{-1}(Y_t - \mathbf{x}_t\boldsymbol{\beta})^2 + \sum_{t=1}^{n}(\mathbf{X}_t - \mathbf{x}_t)(\mathbf{X}_t - \mathbf{x}_t)'\right\}, \end{aligned} \tag{2.2.4}$$

To obtain the maximum likelihood estimators, the likelihood is to be maximized with respect to $\boldsymbol{\beta}$, σ_{ee}, and $\mathbf{x}_t$, $t = 1, 2, \ldots, n$. Differentiating (2.2.4), we obtain

$$\frac{\partial \log L}{\partial \boldsymbol{\beta}} = \sigma_{ee}^{-1} \sum_{t=1}^{n} \mathbf{x}_t'(Y_t - \mathbf{x}_t\boldsymbol{\beta}),$$

$$\frac{\partial \log L}{\partial \sigma_{ee}} = -\tfrac{1}{2}n\sigma_{ee}^{-1} + \tfrac{1}{2}\sigma_{ee}^{-2} \sum_{t=1}^{n} (Y_t - \mathbf{x}_t\boldsymbol{\beta})^2, \tag{2.2.5}$$

$$\frac{\partial \log L}{\partial \mathbf{x}_t} = \sigma_{ee}^{-1}(Y_t - \mathbf{x}_t\boldsymbol{\beta})\boldsymbol{\beta}' + (\mathbf{X}_t - \mathbf{x}_t), \qquad t = 1, 2, \ldots, n.$$

Equating the partial derivative with respect to $\mathbf{x}_t$ to zero and using the hat ($\hat{\ }$) to denote estimators, we obtain

$$\hat{\mathbf{x}}_t' = [\hat{\sigma}_{ee}^{-1}\hat{\boldsymbol{\beta}}\hat{\boldsymbol{\beta}}' + \mathbf{I}]^{-1}[\hat{\sigma}_{ee}^{-1}Y_t\hat{\boldsymbol{\beta}} + \mathbf{X}_t'], \qquad t = 1, 2, \ldots, n. \tag{2.2.6}$$

Noting that

$$[\hat{\sigma}_{ee}^{-1}\hat{\boldsymbol{\beta}}\hat{\boldsymbol{\beta}}' + \mathbf{I}]^{-1} = \mathbf{I} - [1 + \hat{\sigma}_{ee}^{-1}\hat{\boldsymbol{\beta}}'\hat{\boldsymbol{\beta}}]^{-1}\hat{\boldsymbol{\beta}}\hat{\boldsymbol{\beta}}'\hat{\sigma}_{ee}^{-1},$$

we obtain

$$Y_t - \hat{\mathbf{x}}_t\hat{\boldsymbol{\beta}} = (\hat{\sigma}_{ee} + \hat{\boldsymbol{\beta}}'\hat{\boldsymbol{\beta}})^{-1}(Y_t - \mathbf{X}_t\hat{\boldsymbol{\beta}})\hat{\sigma}_{ee}, \tag{2.2.7}$$

$$\mathbf{X}_t - \hat{\mathbf{x}}_t = (\hat{\sigma}_{ee} + \hat{\boldsymbol{\beta}}'\hat{\boldsymbol{\beta}})^{-1}(Y_t - \mathbf{X}_t\hat{\boldsymbol{\beta}})\hat{\boldsymbol{\beta}}'. \tag{2.2.8}$$

If we substitute (2.2.7) into the second equation of (2.2.5) and set the derivative equal to zero, we obtain

$$\hat{\sigma}_{ee}^{-1}(\hat{\sigma}_{ee} + \hat{\boldsymbol{\beta}}'\hat{\boldsymbol{\beta}}) = (\hat{\sigma}_{ee} + \hat{\boldsymbol{\beta}}'\hat{\boldsymbol{\beta}})^{-1}\left[n^{-1}\sum_{t=1}^{n}(Y_t - \mathbf{X}_t\hat{\boldsymbol{\beta}})^2\right]. \tag{2.2.9}$$

Equation (2.2.9) signals a problem. We know that the variance of

$$Y_t - \mathbf{X}_t\boldsymbol{\beta} = e_t - \mathbf{u}_t\boldsymbol{\beta}$$

is $\sigma_{ee} + \boldsymbol{\beta}'\boldsymbol{\beta}$. Therefore, the quantity on the right side of Equation (2.2.9) should be estimating one, but the quantity on the left side of the equation is clearly not estimating one.

We conclude that the method of maximum likelihood has failed to yield consistent estimators for all parameters of the model. This is a common occurrence when the likelihood method is applied to a model in which the number of parameters increases with n. In our model the parameters are $(\sigma_{ee}, \boldsymbol{\beta}, \mathbf{x}_1, \mathbf{x}_2, \ldots, \mathbf{x}_n)$. We should not be surprised that we cannot obtain consistent estimators of $n + 2$ parameters from n observations. Estimation in the presence of an increasing number of parameters has been discussed by Neyman and Scott (1948), Kiefer and Wolfowitz (1956), Kalbfleisch and Sprott (1970), and Morton (1981a). Anderson and Rubin (1956) showed that the likelihood function (2.2.4) is unbounded.

Let us now examine maximum likelihood estimation for the model with random $\mathbf{x}$ values. We shall see that the application of likelihood methods to the normal structural model produces estimators of the unknown parameters that are consistent and asymptotically normal under a wide range of assumptions.

Let

$$Y_t = \beta_0 + \mathbf{x}_t\boldsymbol{\beta}_1 + e_t, \qquad \mathbf{X}_t = \mathbf{x}_t + \mathbf{u}_t, \tag{2.2.10}$$

where

$$\begin{bmatrix} \mathbf{x}_t' \\ e_t \\ \mathbf{u}_t' \end{bmatrix} \sim \mathrm{NI}\left(\begin{bmatrix} \boldsymbol{\mu}_x' \\ 0 \\ \mathbf{0} \end{bmatrix}, \begin{bmatrix} \boldsymbol{\Sigma}_{xx} & \mathbf{0} & \mathbf{0} \\ \mathbf{0} & \sigma_{ee} & \boldsymbol{\Sigma}_{eu} \\ \mathbf{0} & \boldsymbol{\Sigma}_{ue} & \boldsymbol{\Sigma}_{uu} \end{bmatrix} \right).$$

Assume that $\boldsymbol{\Sigma}_{uu}$ and $\boldsymbol{\Sigma}_{ue}$ are known. To be conformable with the dimensions of (2.2.1), we let $\mathbf{X}_t$ be a $(k-1)$-dimensional vector. Let a sample of n vectors $\mathbf{Z}_t = (Y_t, \mathbf{X}_t)$ be available. Then twice the logarithm of the likelihood adjusted for degrees of freedom is

$$\begin{aligned} 2 \log L_c(\boldsymbol{\theta}) = &-kn \log 2\pi - (n-1) \log|\boldsymbol{\Sigma}_{ZZ}| - (n-1) \operatorname{tr}\{\mathbf{m}_{ZZ}\boldsymbol{\Sigma}_{ZZ}^{-1}\} \\ &-n(\bar{\mathbf{Z}} - \boldsymbol{\mu}_Z)\boldsymbol{\Sigma}_{ZZ}^{-1}(\bar{\mathbf{Z}} - \boldsymbol{\mu}_Z)', \end{aligned} \tag{2.2.11}$$

where

$$\boldsymbol{\theta}' = [\boldsymbol{\mu}_x, \beta_0, \boldsymbol{\beta}_1', \sigma_{ee}, (\operatorname{vech} \boldsymbol{\Sigma}_{xx})']$$

and

$$\boldsymbol{\Sigma}_{ZZ} = (\boldsymbol{\beta}_1, \mathbf{I})'\boldsymbol{\Sigma}_{xx}(\boldsymbol{\beta}_1, \mathbf{I}) + \boldsymbol{\Sigma}_{\varepsilon\varepsilon}.$$

(See Appendix 4.A for the definition of vech $\boldsymbol{\Sigma}_{xx}$.) The number of unknown parameters in $\boldsymbol{\theta}$ is equal to the number of elements in the vector of statistics $[\bar{\mathbf{Z}}, (\operatorname{vech} \mathbf{m}_{zz})']$. It is well known that $\hat{\boldsymbol{\mu}}_Z = \bar{\mathbf{Z}}$ and $\hat{\boldsymbol{\Sigma}}_{ZZ} = \mathbf{m}_{ZZ}$ maximize the likelihood (2.2.11) with respect to $\boldsymbol{\mu}_Z$ and $\boldsymbol{\Sigma}_{ZZ}$ when there are no restrictions on $\boldsymbol{\mu}_Z$ and $\boldsymbol{\Sigma}_{ZZ}$. Therefore, due to the functional invariance property of the method of maximum likelihood, the maximum likelihood estimators adjusted for degrees of freedom are

$$\begin{aligned} (\hat{\boldsymbol{\mu}}_x, \hat{\beta}_0) &= (\bar{\mathbf{X}}, \bar{Y} - \bar{\mathbf{X}}\hat{\boldsymbol{\beta}}_1), \\ \hat{\boldsymbol{\beta}}_1 &= (\mathbf{m}_{XX} - \boldsymbol{\Sigma}_{uu})^{-1}(\mathbf{m}_{XY} - \boldsymbol{\Sigma}_{ue}), \\ \hat{\sigma}_{ee} &= m_{YY} - 2\mathbf{m}_{YX}\hat{\boldsymbol{\beta}}_1 + \hat{\boldsymbol{\beta}}_1'\mathbf{m}_{XX}\hat{\boldsymbol{\beta}}_1 + 2\boldsymbol{\Sigma}_{eu}\hat{\boldsymbol{\beta}}_1 - \hat{\boldsymbol{\beta}}_1'\boldsymbol{\Sigma}_{uu}\hat{\boldsymbol{\beta}}_1, \end{aligned} \tag{2.2.12}$$

and $\hat{\boldsymbol{\Sigma}}_{xx} = \mathbf{m}_{XX} - \boldsymbol{\Sigma}_{uu}$, provided $\hat{\boldsymbol{\Sigma}}_{xx}$ is positive definite and $\hat{\sigma}_{ee} \geq \boldsymbol{\Sigma}_{eu}\boldsymbol{\Sigma}_{uu}^{\dagger}\boldsymbol{\Sigma}_{ue}$, where $\boldsymbol{\Sigma}_{uu}^{\dagger}$ is the Moore–Penrose generalized inverse of $\boldsymbol{\Sigma}_{uu}$. If either of these conditions is violated, the estimators fall on the boundary of the parameter space. Let $\hat{\lambda}_{\ell+1}^{-1} \leq \hat{\lambda}_{\ell+2}^{-1} \leq \cdots \leq \hat{\lambda}_{k-1}^{-1} \leq \hat{\lambda}_k^{-1}$ be the positive values of λ^{-1} that satisfy

$$|\ddot{\boldsymbol{\Sigma}}_{aa} - \lambda^{-1}\mathbf{m}_{ZZ}| = 0, \tag{2.2.13}$$

where $\ddot{\sigma}_{aa11} = \boldsymbol{\Sigma}_{eu}\boldsymbol{\Sigma}_{uu}^{\dagger}\boldsymbol{\Sigma}_{ue}$ and

$$\ddot{\boldsymbol{\Sigma}}_{aa} = \begin{bmatrix} \ddot{\sigma}_{aa11} & \boldsymbol{\Sigma}_{eu} \\ \boldsymbol{\Sigma}_{ue} & \boldsymbol{\Sigma}_{uu} \end{bmatrix}. \tag{2.2.14}$$

If $\hat{\lambda}_k < 1$ and $\hat{\lambda}_{k-1} > 1$, the maximum likelihood estimators are

$$\hat{\boldsymbol{\beta}}_1 = (\mathbf{m}_{XX} - \hat{\lambda}_k\boldsymbol{\Sigma}_{uu})^{-1}(\mathbf{m}_{XY} - \hat{\lambda}_k\boldsymbol{\Sigma}_{ue}), \tag{2.2.15}$$

$\hat{\boldsymbol{\Sigma}}_{xx} = \mathbf{m}_{XX} - \hat{\lambda}_k\boldsymbol{\Sigma}_{uu}$, and $\hat{\sigma}_{ee} = \boldsymbol{\Sigma}_{eu}\boldsymbol{\Sigma}_{uu}^{\dagger}\boldsymbol{\Sigma}_{ue}$. If $\hat{\lambda}_{k-1} \leqslant 1$, the maximum likelihood estimator of $\boldsymbol{\Sigma}_{xx}$ is singular and the estimator of $\boldsymbol{\beta}_1$ is indeterminate. Estimator (2.2.15) is derived in the proof of Theorem 4.C.1. The estimator of $\boldsymbol{\beta}_1$ given in (2.2.12) is the vector analogue of the estimator given in (1.2.3).

On the basis of degrees-of-freedom arguments, we suggest that the estimator of σ_{ee} in (2.2.12) be replaced by

$$\hat{\sigma}_{ee} = s_{vv} - \hat{\boldsymbol{\beta}}_1'\boldsymbol{\Sigma}_{uu}\hat{\boldsymbol{\beta}}_1 + 2\boldsymbol{\Sigma}_{eu}\hat{\boldsymbol{\beta}}_1, \tag{2.2.16}$$

where $s_{vv} = (n-k)^{-1}\sum_{t=1}^{n}[Y_t - \bar{Y} - (\mathbf{X}_t - \bar{\mathbf{X}})\hat{\boldsymbol{\beta}}_1]^2$.

In many practical situations the matrices $\boldsymbol{\Sigma}_{uu}$ and $\boldsymbol{\Sigma}_{ue}$ are not known but are estimated. For example, it is common for laboratories to make repeated determinations on the same material to establish the magnitude of the measurement error.

If the covariance matrices for measurement error are estimated, a more detailed specification of the model than that given in (2.2.1) is required. This is because the random variable e_t entering Equation (2.2.1) may be composed of two parts. The true values y_t and $\mathbf{x}_t$ will not be perfectly related if factors other than $\mathbf{x}_t$ are responsible for variation in y_t. Thus, one might specify

$$y_t = \mathbf{x}_t\boldsymbol{\beta} + q_t, \tag{2.2.17}$$

where the q_t are independent $(0, \sigma_{qq})$ random variables, and q_t is independent of $\mathbf{x}_j$ for all t and j. The random variable q_t is called the *error in the equation.* We observe

$$(Y_t, \mathbf{X}_t) = (y_t, \mathbf{x}_t) + (w_t, \mathbf{u}_t), \tag{2.2.18}$$

where $(w_t, \mathbf{u}_t) = \mathbf{a}_t$ is a vector of measurement errors,

$$\mathbf{a}_t' \sim \mathrm{NI}(\mathbf{0}, \boldsymbol{\Sigma}_{aa}), \tag{2.2.19}$$

and $\mathbf{a}_t$ is independent of $(q_j, \mathbf{x}_j)$ for all t and j. In terms of the original model (2.2.1), $e_t = w_t + q_t$ is the sum of an error in the equation and an error made in measuring y_t. Typically, the variance of q_t is unknown, but it is possible to conduct experiments to estimate the covariance matrix of $\mathbf{a}_t = (w_t, \mathbf{u}_t)$. Because w_t and q_t are assumed to be independent, the covariance between $\mathbf{u}_t$ and w_t is equal to the covariance between $\mathbf{u}_t$ and e_t.

Let $\mathbf{S}_{aa}$ denote an unbiased estimator of $\boldsymbol{\Sigma}_{aa}$. Then the estimator of $\boldsymbol{\beta}$ analogous to (2.2.12) is

$$\tilde{\boldsymbol{\beta}} = (\mathbf{M}_{XX} - \mathbf{S}_{uu})^{-1}(\mathbf{M}_{XY} - \mathbf{S}_{uw}). \tag{2.2.20}$$

We use matrices of uncorrected sums of squares and products to define the estimator so that the estimator is conformable with model (2.2.17). The definition of the estimators permits the matrix $\mathbf{S}_{uu}$ to be singular. For example, if the model contains an intercept term, one of the elements of $\mathbf{X}_t$ is always one and the corresponding row and column of $\mathbf{S}_{uu}$ are zero vectors.

Given the estimated error covariance matrix $\mathbf{S}_{aa}$, a consistent estimator of σ_{qq} is

$$\tilde{\sigma}_{qq} = s_{vv} - (S_{ww} - 2\tilde{\boldsymbol{\beta}}'\mathbf{S}_{uw} + \tilde{\boldsymbol{\beta}}'\mathbf{S}_{uu}\tilde{\boldsymbol{\beta}}), \tag{2.2.21}$$

where $s_{vv} = (n - k)^{-1} \sum_{t=1}^{n} (Y_t - \mathbf{X}_t\tilde{\boldsymbol{\beta}})^2$.

In Theorem 2.2.1 we give the limiting distribution of the standardized estimators of $\boldsymbol{\beta}$ and σ_{qq} under less restrictive conditions than those used to obtain the maximum likelihood estimator. The $\mathbf{x}_t$ are assumed to be distributed with a mean $\boldsymbol{\mu}_{xt}$ and covariance matrix $\boldsymbol{\Sigma}_{xx}$, where the covariance matrix $\boldsymbol{\Sigma}_{xx}$ can be singular. This specification contains both the functional and structural models as special cases. Under the fixed model, $\mathbf{x}_t \equiv \boldsymbol{\mu}_{xt}$ are fixed vectors. Under the random model $\boldsymbol{\mu}_{xt} \equiv (1, \boldsymbol{\mu}_x)$, where $\boldsymbol{\mu}_x$ is fixed over t, and the lower right $(k - 1) \times (k - 1)$ portion of $\boldsymbol{\Sigma}_{xx}$ is positive definite. Dolby (1976) suggested that the model with random true values whose means are a function of t be called the *ultrastructural* model. To obtain a limiting normal distribution, some conditions must be imposed on the sequence of means of the true values. We assume that

$$\lim_{n \to \infty} n^{-1} \sum_{t=1}^{n} \boldsymbol{\mu}_{xt}'\boldsymbol{\mu}_{xt} = \mathbf{M}_{\mu\mu} \tag{2.2.22}$$

and let $\bar{\mathbf{M}}_{xx} = \mathbf{M}_{\mu\mu} + \boldsymbol{\Sigma}_{xx}$, where $\bar{\mathbf{M}}_{xx}$ is positive definite.

Commonly, the error covariance matrix will be estimated from a source independent of the $\mathbf{M}_{ZZ}$ matrix. To assure that the error made in estimating the error covariance matrix enters the covariance matrix of the limiting distribution, we assume that the error in the estimator of $\boldsymbol{\Sigma}_{aa}$ is of the same order as the error in $\mathbf{M}_{ZZ}$. This is accomplished by assuming that the degrees of freedom for $\mathbf{S}_{aa}$ is nearly proportional to n. Thus, we assume that

$$\lim_{n \to \infty} d_f^{-1} n = \nu, \tag{2.2.23}$$

where d_f is the degrees of freedom for $\mathbf{S}_{aa}$ and ν is a fixed number. If the covariance matrix of the measurement error is known, then $\nu = 0$.

Theorem 2.2.1. Let model (2.2.17)–(2.2.19) hold and let $(\mathbf{x}_t - \boldsymbol{\mu}_{xt}, q_t)$ be independently and identically distributed with zero mean vector and covariance matrix,

$$E\{(\mathbf{x}_t - \boldsymbol{\mu}_{xt}, q_t)'(\mathbf{x}_t - \boldsymbol{\mu}_{xt}, q_t)\} = \text{block diag}(\boldsymbol{\Sigma}_{xx}, \sigma_{qq}),$$

where $\{\boldsymbol{\mu}_{xt}\}$ is a sequence of fixed k-dimensional vectors, q_t has finite fourth moments, and q_t is independent of $\mathbf{x}_t$. Let $\mathbf{S}_{aa}$ be an unbiased estimator of $\boldsymbol{\Sigma}_{aa}$ that is distributed as a multiple of a Wishart matrix with d_f degrees of freedom, independent of $(Y_t, \mathbf{X}_t)$ for all t. Let (2.2.22) and (2.2.23) hold. Let $\boldsymbol{\theta} = (\boldsymbol{\beta}', \sigma_{qq})'$ and let $\tilde{\boldsymbol{\theta}} = (\tilde{\boldsymbol{\beta}}', \tilde{\sigma}_{qq})'$, where $\tilde{\boldsymbol{\beta}}$ is defined in (2.2.20) and $\tilde{\sigma}_{qq}$ is defined in (2.2.21). Then

$$n^{1/2}(\tilde{\boldsymbol{\theta}} - \boldsymbol{\theta}) \xrightarrow{L} N(\mathbf{0}, \boldsymbol{\Gamma}),$$

where the submatrices of $\boldsymbol{\Gamma}$ are

$$\begin{aligned}
\boldsymbol{\Gamma}_{\beta\beta} &= \bar{\mathbf{M}}_{xx}^{-1}\sigma_{vv} + \bar{\mathbf{M}}_{xx}^{-1}[\boldsymbol{\Sigma}_{uu}\sigma_{vv} + \boldsymbol{\Sigma}_{uv}\boldsymbol{\Sigma}_{vu}]\bar{\mathbf{M}}_{xx}^{-1} \\
&\quad + \nu\bar{\mathbf{M}}_{xx}^{-1}[\boldsymbol{\Sigma}_{uu}\sigma_{rr} + \boldsymbol{\Sigma}_{uv}\boldsymbol{\Sigma}_{vu}]\bar{\mathbf{M}}_{xx}^{-1}, \\
\Gamma_{qq} &= V\{v_t^2\} + 2\nu\sigma_{rr}^2, \\
\boldsymbol{\Gamma}_{\beta q} &= 2\bar{\mathbf{M}}_{xx}^{-1}\boldsymbol{\Sigma}_{uv}(\sigma_{vv} + \nu\sigma_{rr}),
\end{aligned}$$

and $\sigma_{rr} = \sigma_{ww} - 2\boldsymbol{\Sigma}_{wu}\boldsymbol{\beta} + \boldsymbol{\beta}'\boldsymbol{\Sigma}_{uu}\boldsymbol{\beta}$.

Proof. Using the fact that $\mathbf{X}_t = \mathbf{x}_t + \mathbf{u}_t$, we may write

$$Y_t = \mathbf{X}_t\boldsymbol{\beta} + v_t,$$

where $v_t = e_t - \mathbf{u}_t\boldsymbol{\beta}$. It follows that

$$\begin{aligned}
\tilde{\boldsymbol{\beta}} &= (\mathbf{M}_{XX} - \mathbf{S}_{uu})^{-1}(\mathbf{M}_{XX}\boldsymbol{\beta} + \mathbf{M}_{Xv} - \mathbf{S}_{uw} + \mathbf{S}_{uu}\boldsymbol{\beta} - \mathbf{S}_{uu}\boldsymbol{\beta}) \\
&= \boldsymbol{\beta} + (\mathbf{M}_{XX} - \mathbf{S}_{uu})^{-1}[(\mathbf{M}_{Xv} - \boldsymbol{\Sigma}_{ur}) - (\mathbf{S}_{ur} - \boldsymbol{\Sigma}_{ur})],
\end{aligned}$$

where $\mathbf{M}_{Xv} = n^{-1}\sum_{t=1}^{n}\mathbf{X}_t v_t$, $r_t = w_t - \mathbf{u}_t\boldsymbol{\beta}$, $\mathbf{S}_{ur} = \mathbf{S}_{uw} - \mathbf{S}_{uu}\boldsymbol{\beta}$, and $\boldsymbol{\Sigma}_{ur} = \boldsymbol{\Sigma}_{uw} - \boldsymbol{\Sigma}_{uu}\boldsymbol{\beta} = \boldsymbol{\Sigma}_{uv}$. Now

$$\mathbf{M}_{XX} = n^{-1}\sum_{t=1}^{n}(\mathbf{x}_t'\mathbf{x}_t + \mathbf{x}_t'\mathbf{u}_t + \mathbf{u}_t'\mathbf{x}_t + \mathbf{u}_t'\mathbf{u}_t)$$

and, by the weak law of large numbers,

$$\operatorname{plim} n^{-1}\sum_{t=1}^{n}(\mathbf{x}_t - \boldsymbol{\mu}_{xt})'(\mathbf{x}_t - \boldsymbol{\mu}_{xt}) = \boldsymbol{\Sigma}_{xx},$$

$$\operatorname{plim} n^{-1}\sum_{t=1}^{n}(\mathbf{x}_t - \boldsymbol{\mu}_{xt})'\mathbf{u}_t = \mathbf{0}.$$

Also, $\mathbf{M}_{XX} - \mathbf{S}_{uu} = \mathbf{M}_{xx} + O_p(n^{-1/2})$ and

$$\tilde{\boldsymbol{\beta}} - \boldsymbol{\beta} = \mathbf{M}_{xx}^{-1}[(\mathbf{M}_{Xv} - \boldsymbol{\Sigma}_{uv}) - (\mathbf{S}_{ur} - \boldsymbol{\Sigma}_{ur})] + O_p(n^{-1}).$$

Consider the linear combination

$$n^{1/2}\boldsymbol{\delta}'(\mathbf{M}_{Xv} - \boldsymbol{\Sigma}_{uv}) = n^{-1/2} \sum_{t=1}^{n} \sum_{i=1}^{k} \delta_i[x_{ti}v_t + u_{ti}v_t - \sigma_{uvi1}]$$

$$= n^{-1/2} \sum_{t=1}^{n} g_t \tag{2.2.24}$$

where σ_{uvi1} is the covariance between u_{it} and v_t, $\boldsymbol{\delta} = (\delta_1, \delta_2, \ldots, \delta_k)'$ is an arbitrary vector such that $\boldsymbol{\delta}'\boldsymbol{\delta} \neq 0$, and the random variables,

$$g_t = \sum_{i=1}^{k} \delta_i[x_{ti}v_t + u_{ti}v_t - \sigma_{uvi1}],$$

are linear functions of independent and identically distributed random vectors. The mean of g_t is zero and the variance is

$$E\{g_t^2\} = \boldsymbol{\delta}'[(\boldsymbol{\mu}_{xt}'\boldsymbol{\mu}_{xt} + \boldsymbol{\Sigma}_{xx} + \boldsymbol{\Sigma}_{uu})\sigma_{vv} + \boldsymbol{\Sigma}_{uv}\boldsymbol{\Sigma}_{vu}]\boldsymbol{\delta}.$$

Therefore, by Lemma 1.C.1,

$$n^{-1/2} \sum_{t=1}^{n} g_t \xrightarrow{L} N\{0, \boldsymbol{\delta}'\bar{\mathbf{M}}_{xx}\boldsymbol{\delta}\sigma_{vv} + \boldsymbol{\delta}'(\boldsymbol{\Sigma}_{uu}\sigma_{vv} + \boldsymbol{\Sigma}_{uv}\boldsymbol{\Sigma}_{vu})\boldsymbol{\delta}\}.$$

Because the nonzero $\boldsymbol{\delta}$ was arbitrary,

$$n^{1/2}(\mathbf{M}_{Xv} - \boldsymbol{\Sigma}_{uv}) \xrightarrow{L} N(\mathbf{0}, \bar{\mathbf{M}}_{xx}\sigma_{vv} + \boldsymbol{\Sigma}_{uu}\sigma_{vv} + \boldsymbol{\Sigma}_{uv}\boldsymbol{\Sigma}_{vu}).$$

The limiting distribution for $n^{1/2}(\tilde{\boldsymbol{\beta}} - \boldsymbol{\beta})$ follows because $n^{1/2}(\mathbf{S}_{ur} - \boldsymbol{\Sigma}_{ur})$ converges in distribution to a multivariate normal random vector with mean zero and covariance matrix

$$v(\boldsymbol{\Sigma}_{uu}\sigma_{rr} + \boldsymbol{\Sigma}_{ur}\boldsymbol{\Sigma}_{ru})$$

and because $\mathbf{S}_{ur}$ is independent of $\mathbf{M}_{Xv}$.

We have shown that $\tilde{\boldsymbol{\beta}} - \boldsymbol{\beta} = O_p(n^{-1/2})$, and it follows that

$$s_{vv} = (n-k)^{-1} \sum_{t=1}^{n} v_t^2 - 2\boldsymbol{\Sigma}_{vu}(\tilde{\boldsymbol{\beta}} - \boldsymbol{\beta}) + O_p(n^{-1}).$$

Also

$$S_{ww} - 2\mathbf{S}_{wu}\tilde{\boldsymbol{\beta}} + \tilde{\boldsymbol{\beta}}'\mathbf{S}_{uu}\tilde{\boldsymbol{\beta}} = (1, -\boldsymbol{\beta}')\mathbf{S}_{aa}(1, -\boldsymbol{\beta}')' - 2\boldsymbol{\Sigma}_{vu}(\tilde{\boldsymbol{\beta}} - \boldsymbol{\beta}) + O_p(n^{-1}),$$

and, from (2.2.21),

$$\tilde{\sigma}_{qq} = (n-k)^{-1} \sum_{t=1}^{n} v_t^2 - (1, -\boldsymbol{\beta}')\mathbf{S}_{aa}(1, -\boldsymbol{\beta}')' + O_p(n^{-1}).$$

Because $\mathbf{S}_{aa}$ is independent of $\mathbf{M}_{Xv}$ and m_{vv}, the conclusion follows. □

The variance of the approximate distribution of $\tilde{\boldsymbol{\beta}}$ can be estimated by

$$\hat{\mathbf{V}}\{\tilde{\boldsymbol{\beta}}\} = n^{-1}[\tilde{\mathbf{M}}_{xx}^{-1}s_{vv} + \tilde{\mathbf{M}}_{xx}^{-1}(\mathbf{S}_{uu}s_{vv} + \tilde{\mathbf{S}}_{uv}\tilde{\mathbf{S}}_{vu})\tilde{\mathbf{M}}_{xx}^{-1}] + d_f^{-1}\tilde{\mathbf{M}}_{xx}^{-1}[\mathbf{S}_{uu}s_{rr} + \tilde{\mathbf{S}}_{uv}\tilde{\mathbf{S}}_{vu}]\tilde{\mathbf{M}}_{xx}^{-1}, \tag{2.2.25}$$

where $\tilde{\mathbf{M}}_{xx} = \mathbf{M}_{XX} - \mathbf{S}_{uu}$, $\tilde{\mathbf{S}}_{uv} = \mathbf{S}_{uw} - \mathbf{S}_{uu}\tilde{\boldsymbol{\beta}}$,

$$s_{rr} = (1, -\tilde{\boldsymbol{\beta}}')\mathbf{S}_{aa}(1, -\tilde{\boldsymbol{\beta}}')',$$

and s_{vv} is defined in (2.2.21). If $\boldsymbol{\Sigma}_{uu}$ and $\boldsymbol{\Sigma}_{ue}$ are known, the estimated covariance matrix is given by (2.2.25) with $d_f^{-1} = 0$. The normality of $\mathbf{u}_t$ and the assumption that $\mathbf{S}_{aa}$ is a Wishart matrix were used in Theorem 2.2.1 to obtain an explicit expression for the covariance matrix of the limiting distribution. If $\mathbf{M}_{xx}^{-1}\mathbf{S}_{uu}$ is not large, the variance estimator (2.2.25) will be satisfactory for modest departures from normality.

Occasionally, some of the variances and (or) covariances in the matrix $\boldsymbol{\Sigma}_{aa}$ are known to be zero, while others must be estimated. A common model in sociology and psychology postulates a diagonal covariance matrix for the measurement error. In most such cases the estimators of the individual error variances will be independent. An estimator of $\boldsymbol{\beta}$ for the model with diagonal covariance matrix is

$$\tilde{\boldsymbol{\beta}} = (\mathbf{M}_{XX} - \ddot{\mathbf{S}}_{uu})^{-1}\mathbf{M}_{XY}, \tag{2.2.26}$$

where $\ddot{\mathbf{S}}_{uu} = \text{diag}(S_{uu11}, S_{uu22}, \ldots, S_{uukk})$ and the elements of $(S_{uu11}, S_{uu22}, \ldots, S_{uukk})$ are independent unbiased estimators of $(\sigma_{uu11}, \sigma_{uu22}, \ldots, \sigma_{uukk})$ distributed as multiples of chi-square random variables with $d_{f1}, d_{f2}, \ldots, d_{fk}$ degrees of freedom, independent of $(Y_t, \mathbf{X}_t)$ for all t. The estimator of the covariance matrix of the approximate distribution of $\tilde{\boldsymbol{\beta}}$ is

$$\tilde{\mathbf{V}}\{\tilde{\boldsymbol{\beta}}\} = \tilde{\mathbf{M}}_{xx}^{-1}[n^{-1}(\mathbf{M}_{XX}s_{vv} + \tilde{\mathbf{S}}_{uv}\tilde{\mathbf{S}}_{vu}) + 2\tilde{\mathbf{R}}_{uu}]\tilde{\mathbf{M}}_{xx}^{-1}, \tag{2.2.27}$$

where $\tilde{\mathbf{M}}_{xx} = \mathbf{M}_{XX} - \ddot{\mathbf{S}}_{uu}$, $\tilde{\mathbf{S}}_{uv} = -\ddot{\mathbf{S}}_{uu}\tilde{\boldsymbol{\beta}}$,

$$\tilde{\mathbf{R}}_{uu} = \text{diag}\{d_{f1}^{-1}\tilde{\beta}_1^2 S_{uu11}^2, d_{f2}^{-1}\tilde{\beta}_2^2 S_{uu22}^2, \ldots, d_{fk}^{-1}\tilde{\beta}_k^2 S_{uukk}^2\},$$

and s_{vv} is defined in (2.2.21). Because the knowledge that some covariances are zero is used in constructing the estimator, the variance estimated by (2.2.27) is smaller than the variance estimated by (2.2.25), for comparable degrees of freedom.

Example 2.2.1. In this example we consider some data studied by Warren, White, and Fuller (1974). In the original study the responses of 98 managers of Iowa farmer cooperatives were analyzed. We use a subsample of the original data containing 55 observations. The data are given in Table

TABLE 2.2.1. Data from role performance study

Observation	Knowledge	Value Orientation	Role Satisfaction	Past Training	Role Performance
1	1.193	2.656	2.333	2.000	−0.054
2	1.654	3.300	2.320	2.000	0.376
3	1.193	2.489	2.737	2.000	0.072
4	1.077	2.478	2.203	2.000	−0.150
5	1.539	2.822	2.840	2.000	0.171
6	1.385	3.000	2.373	2.000	0.042
7	1.462	3.111	2.497	2.667	0.188
8	1.385	2.545	2.617	2.167	−0.052
9	1.539	2.556	2.997	2.000	0.310
10	1.654	2.945	2.150	2.167	0.147
11	1.462	2.778	2.227	1.833	0.005
12	1.154	2.545	2.017	2.167	−0.088
13	1.424	3.611	2.303	2.333	0.044
14	1.116	2.956	2.517	2.333	−0.073
15	1.270	2.856	1.770	2.333	0.224
16	1.347	2.956	2.430	2.000	0.103
17	1.116	2.545	2.043	2.000	−0.108
18	1.077	3.356	2.410	2.000	−0.019
19	1.423	3.211	2.150	2.000	−0.062
20	0.923	2.556	2.180	2.000	−0.239
21	1.385	2.589	2.490	2.000	−0.159
22	1.270	2.900	1.920	2.333	0.069
23	1.116	2.167	2.663	1.333	−0.118
24	1.346	2.922	2.520	2.000	0.083
25	0.846	1.711	3.150	1.500	−0.255
26	1.077	2.556	2.297	2.000	−0.159
27	1.231	3.567	2.307	2.167	0.014
28	0.962	2.689	2.830	1.333	0.102
29	1.500	2.978	2.737	2.000	0.109
30	1.577	2.945	3.117	2.167	0.006
31	1.885	3.256	2.647	2.667	0.334
32	1.231	2.956	2.217	1.667	−0.076
33	1.808	2.811	2.327	2.000	−0.043
34	1.039	2.733	2.447	2.000	−0.126
35	1.385	2.400	2.347	2.167	−0.056
36	1.846	2.944	2.410	2.000	0.203
37	1.731	3.200	2.277	2.000	0.023
38	1.500	2.911	2.577	2.000	0.047
39	1.231	3.167	2.507	2.333	0.011
40	1.346	3.322	2.653	2.500	0.153
41	1.347	2.833	2.587	2.667	0.100

TABLE 2.2.1. (Continued)

Observation	Knowledge	Value Orientation	Role Satisfaction	Past Training	Role Performance
42	1.154	2.967	3.140	2.000	−0.089
43	0.923	2.700	2.557	1.833	0.007
44	1.731	3.033	2.423	2.000	0.089
45	1.808	2.911	2.793	2.000	0.182
46	1.193	3.311	2.283	2.333	0.259
47	1.308	2.245	2.210	2.000	0.007
48	1.424	2.422	2.350	2.000	−0.015
49	1.385	2.744	2.330	2.000	−0.023
50	1.385	2.956	2.130	2.000	−0.150
51	1.347	2.933	2.837	2.167	0.152
52	1.539	3.411	2.600	2.167	0.377
53	1.385	1.856	2.790	2.000	0.043
54	1.654	3.089	2.500	2.000	0.184
55	1.308	2.967	2.813	2.667	0.127

2.2.1. The postulated model is

$$y_t = \beta_0 + \sum_{i=1}^{4} x_{ti}\beta_i + q_t, \tag{2.2.28}$$

where y_t is the role performance of the tth manager, x_1 is knowledge of the economic phases of management, x_2 is value orientation, x_3 is role satisfaction, and x_4 is past training. The random variable q_t is assumed to be a normal $(0, \sigma_{qq})$ random variable. Value orientation is the tendency to rationally evaluate means to an economic end, role satisfaction is the gratification obtained from the managerial role, and past training is the amount of formal education. The amount of past training is the total years of formal schooling divided by six and is assumed to be measured without error. Role performance, knowledge, value orientation, and role satisfaction were measured on the basis of responses to several questions for each item. Using replicated determinations on the same individuals, measurement error variances were estimated to be 0.0037, 0.0203, 0.0438, and 0.0180 for role performance, knowledge, value orientation, and role satisfaction, respectively. Each error variance estimate is based on 97 degrees of freedom and it is assumed that $\boldsymbol{\Sigma}_{uu}$ is diagonal. This is the model for which estimator (2.2.26) is appropriate. The matrix of mean squares and products corrected for the mean is

$$\mathbf{m}_{XX} = \begin{bmatrix} 0.0598 & 0.0345 & 0.0026 & 0.0188 \\ 0.0345 & 0.1414 & -0.0186 & 0.0474 \\ 0.0026 & -0.0186 & 0.0887 & -0.0099 \\ 0.0188 & 0.0474 & -0.0099 & 0.0728 \end{bmatrix}$$

and

$$\mathbf{m}'_{XY} = (0.0210, 0.0271, 0.0063, 0.0155).$$

We compute the estimate of $\boldsymbol{\beta}$ using the program SUPER CARP. For reasons that will be explained in Section 2.5, the estimator is computed by subtracting a multiple, slightly less than one, of $\ddot{\mathbf{S}}_{uu}$ from the moment matrix. The estimator of $\boldsymbol{\beta}$ expressed in terms of uncorrected moments is

$$\begin{aligned} \tilde{\boldsymbol{\beta}} &= [\mathbf{M}_{XX} - n^{-1}(n-6)\ddot{\mathbf{S}}_{uu}]^{-1}\mathbf{M}_{XY} \\ &= [-1.24,\ 0.36,\ 0.149,\ 0.117,\ 0.040]', \\ &\quad\ (0.27)\ (0.13)\ (0.094)\ (0.069)\ (0.075) \end{aligned} \tag{2.2.29}$$

where the first coefficient is the intercept and the numbers in parentheses are the estimated standard errors of the coefficients. The sample size is $n = 55$ and 5 is the number of parameters estimated.

The standard errors are obtained from the covariance matrix of expression (2.2.27). In this example the increase in the variance associated with the estimation of the error variances is small. For example, the estimated variance of the coefficient for knowledge, assuming the error variance to be known, is 0.0157. The estimated variance of the knowledge coefficient, under the assumption that the estimated error variance is based on 97 degrees of freedom, is 0.0173. By Equation (2.2.21), an estimator of σ_{qq} is

$$\tilde{\sigma}_{qq} = 0.0129 - 0.0075 = 0.0054,$$

where $s_{vv} = 0.0129$ and $S_{ww} + \sum_{i=1}^{4} S_{uuii}\tilde{\beta}_i^2 = 0.0075$. By the arguments used in the proof of Theorem 2.2.1,

$$\tilde{\sigma}_{qq} = m_{vv} - \left(S_{ww} + \sum_{i=1}^{4} S_{uuii}\beta_i^2\right) + O_p(n^{-1}) \tag{2.2.30}$$

and an estimator of the variance of the approximate distribution of $\tilde{\sigma}_{qq}$ is

$$\begin{aligned} \hat{V}\{\tilde{\sigma}_{qq}\} &= 2\left[(n-k)^{-1}s_{vv}^2 + d_{f0}^{-1}S_{ww}^2 + \sum_{i=1}^{4} d_{fi}^{-1}S_{uuii}^2\tilde{\beta}_i^4\right] \\ &= (0.00266)^2. \end{aligned} \tag{2.2.31}$$

The estimated standard error for $\tilde{\sigma}_{qq}$ is about one-half of the estimate. An alternative test of the hypothesis that $\sigma_{qq} = 0$ is developed in Section 2.4 and illustrated in Example 2.4.2. □ □

2.2.2. Estimation of True Values

In Section 1.2.3 we obtained, for fixed x_t, the best estimators of the x_t using the model information. We now extend those estimators to the model of this section. We assume model (2.2.17)–(2.2.19) and, as usual, we let $v_t = e_t - \mathbf{u}_t\boldsymbol{\beta}$

and $\mathbf{a}_t = (w_t, \mathbf{u}_t)$. In the model (2.2.17) the true y_t differs from $\mathbf{x}_t\boldsymbol{\beta}$ by q_t. Hence, it is of interest to estimate both y_t and $\mathbf{x}_t$.

To estimate $(y_t, \mathbf{x}_t)$, we predict $(e_t, \mathbf{u}_t)$ and subtract the predictor of $(e_t, \mathbf{u}_t)$ from $(Y_t, \mathbf{X}_t)$. Under normality, the best predictor of $(e_t, \mathbf{u}_t)$, given v_t, is

$$(\ddot{e}_t, \ddot{\mathbf{u}}_t) = v_t\boldsymbol{\delta}', \tag{2.2.32}$$

where $\boldsymbol{\delta}' = \sigma_{vv}^{-1}\boldsymbol{\Sigma}_{va}$. It follows that the best estimator of $(y_t, \mathbf{x}_t)$, treating $\mathbf{x}_t$ as fixed, is

$$(\ddot{y}_t, \ddot{\mathbf{x}}_t) = (Y_t, \mathbf{X}_t) - v_t\boldsymbol{\delta}'. \tag{2.2.33}$$

For the estimator of $\mathbf{z}_t$ constructed with estimated model parameters, we let

$$\hat{\mathbf{z}}_t = (\hat{y}_t, \hat{\mathbf{x}}_t) = (Y_t, \mathbf{X}_t) - \hat{v}_t\hat{\boldsymbol{\delta}}', \tag{2.2.34}$$

where $\hat{\boldsymbol{\delta}} = \hat{\sigma}_{vv}^{-1}\hat{\boldsymbol{\Sigma}}_{va}$, $\hat{\boldsymbol{\Sigma}}_{va} = (1, -\tilde{\boldsymbol{\beta}}')\mathbf{S}_{aa}$, and $\hat{\sigma}_{vv} = (1, -\tilde{\boldsymbol{\beta}}')\mathbf{M}_{ZZ}(1, -\tilde{\boldsymbol{\beta}}')'$. An estimator of the covariance matrix of the error in the approximate distribution of $\hat{\mathbf{z}}_t$ is

$$\hat{\mathbf{V}}\{\hat{\mathbf{z}}_t' \mid \mathbf{x}_t\} = \mathbf{S}_{aa} - \hat{\sigma}_{vv}^{-1}\hat{\boldsymbol{\Sigma}}_{av}\hat{\boldsymbol{\Sigma}}_{va}. \tag{2.2.35}$$

The estimator (2.2.35) ignores the contribution of estimation error in $\tilde{\boldsymbol{\beta}}$ and $\hat{\sigma}_{ee}$ to the variance. An estimator of the variance of $(\hat{y}_t, \hat{\mathbf{x}}_t)$ that contains order n^{-1} terms associated with parameter estimation is given in (2.2.41) of Section 2.2.3.

Example 2.2.2. We construct estimates of the individual y and x values for Example 2.2.1, treating $\mathbf{x}_t$ as fixed. Let

$$\mathbf{x}_t = (1, x_{t1}, x_{t2}, x_{t3}, x_{t4}),$$

where x_1 is knowledge, x_2 is value orientation, x_3 is role satisfaction, and x_4 is past training. Then the estimated covariance matrix of the measurement error is

$$\hat{\mathbf{V}}\{\mathbf{a}_t\} = \text{diag}(0.0037, 0.0, 0.0203, 0.0438, 0.0180, 0.0).$$

The matrix $\mathbf{I} - (1, -\tilde{\boldsymbol{\beta}}')'\sigma_{vv}^{-1}(\hat{\sigma}_{vw}, \hat{\boldsymbol{\Sigma}}_{vu})$ is

$$\begin{bmatrix} 0.712 & 0.0 & 0.564 & 0.507 & 0.164 & 0.0 \\ -0.356 & 1.0 & 0.698 & 0.627 & 0.204 & 0.0 \\ 0.103 & 0.0 & 0.799 & -0.181 & -0.059 & 0.0 \\ 0.043 & 0.0 & -0.084 & 0.925 & -0.024 & 0.0 \\ 0.034 & 0.0 & -0.066 & -0.059 & 0.981 & 0.0 \\ 0.011 & 0.0 & -0.022 & -0.020 & -0.007 & 1.0 \end{bmatrix},$$

TABLE 2.2.2. Estimated true values for role performance data

		$\hat{z}$			
Observation	$\hat{v}$	Knowledge	Value Orientation	Role Satisfaction	Role Performance
1	0.010	1.198	2.661	2.335	−0.057
2	0.182	1.756	3.392	2.350	0.324
3	0.114	1.257	2.546	2.755	0.039
4	−0.003	1.075	2.476	2.203	−0.149
5	0.028	1.554	2.836	2.845	0.163
6	−0.018	1.374	2.991	2.370	0.047
7	0.043	1.486	3.133	2.504	0.176
8	−0.080	1.340	2.504	2.604	−0.029
9	0.188	1.644	2.651	3.028	0.256
10	0.019	1.664	2.954	2.153	0.142
11	−0.026	1.447	2.765	2.222	0.012
12	0.037	1.175	2.563	2.023	−0.099
13	−0.126	1.353	3.547	2.283	0.080
14	−0.061	1.081	2.925	2.507	−0.056
15	0.284	1.430	2.999	1.817	0.142
16	0.056	1.378	2.984	2.439	0.087
17	0.034	1.135	2.561	2.049	−0.118
18	−0.027	1.062	3.342	2.406	−0.011
19	−0.141	1.344	3.140	2.127	−0.021
20	−0.046	0.897	2.533	2.173	−0.226
21	−0.172	1.288	2.502	2.462	−0.110
22	0.105	1.329	2.953	1.937	0.039
23	0.034	1.135	2.184	2.669	−0.128
24	0.031	1.363	2.938	2.525	0.074
25	−0.003	0.845	1.710	3.150	−0.254
26	−0.034	1.058	2.538	2.291	−0.149
27	−0.074	1.189	3.529	2.295	0.035
28	0.212	1.081	2.796	2.865	0.041
29	−0.032	1.482	2.962	2.732	0.118
30	−0.208	1.460	2.839	3.082	0.066

where $s_{vv} = 0.0129$,

$$(1, -\tilde{\boldsymbol{\beta}}') = (1,\ 1.24,\ -0.36,\ -0.149,\ -0.117,\ -0.040),$$

$$(\hat{\sigma}_{vw}, \hat{\boldsymbol{\Sigma}}_{vu}) = (0.0037,\ 0.0,\ -0.00731,\ -0.00653,\ -0.00211,\ 0.0).$$

The estimated y and x values for the first 30 observations are given in Table 2.2.2. Because $x_0 = X_0 \equiv 1$ and because past training is assumed to be measured without error, estimates for x_0 and x_4 are not given in Table 2.2.2. The estimated covariance matrix of the estimator of $\mathbf{z}_t$ defined in (2.2.35) is

$$\begin{bmatrix} 0.0026 & 0.0 & 0.0021 & 0.0019 & 0.0006 & 0.0 \\ 0.0 & 0.0 & 0.0 & 0.0 & 0.0 & 0.0 \\ 0.0021 & 0.0 & 0.0162 & -0.0037 & -0.0012 & 0.0 \\ 0.0019 & 0.0 & -0.0037 & 0.0405 & -0.0011 & 0.0 \\ 0.0006 & 0.0 & -0.0012 & -0.0011 & 0.0177 & 0.0 \\ 0.0 & 0.0 & 0.0 & 0.0 & 0.0 & 0.0 \end{bmatrix}.$$

The quantity $\hat{v}_t = Y_t - \sum_{i=0}^{4} \tilde{\beta}_i X_{ti}$ is also given in Table 2.2.2. Under the model, v_t and $\ddot{\mathbf{x}}_t$ of (2.2.33) are uncorrelated. Therefore, as with the simple

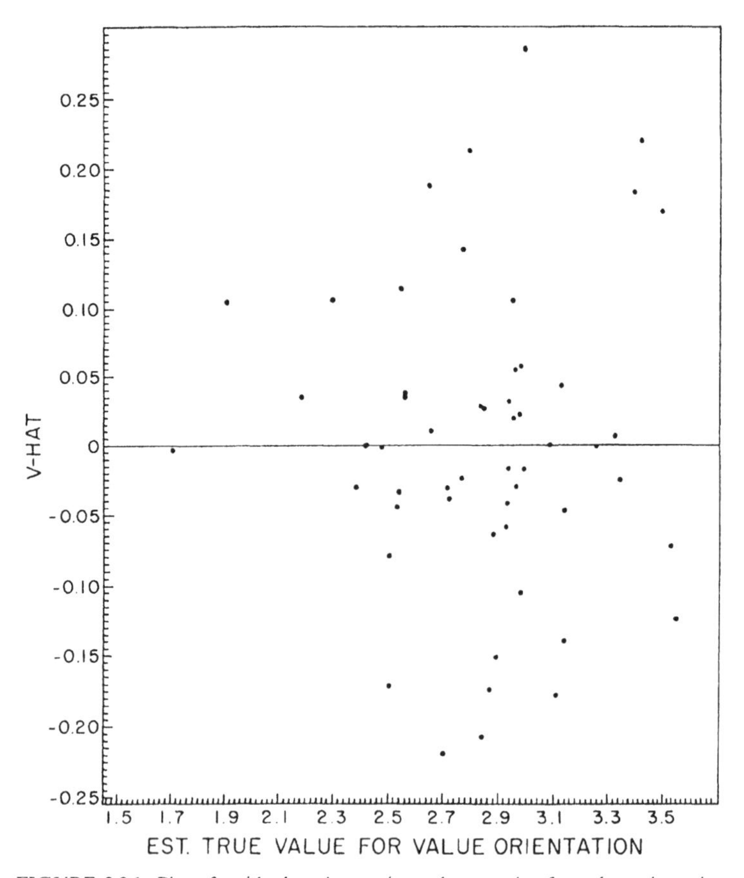

FIGURE 2.2.1. Plot of residual against estimated true value for value orientation.

model of Section 1.2, plots of $\hat{v}_t$ against the elements of $\hat{\mathbf{x}}_t$ should be constructed as model checks. Figure 2.2.1 contains a plot of $\hat{v}_t$ against the estimated true value for value orientation, $\hat{x}_{t2}$. This plot gives no reason to reject the assumptions of linearity and normality. The plots of $\hat{v}_t$ against the other estimated true values also gave no reason to reject the original model.

Figure 2.2.2 contains a plot of the ordered $\hat{v}_t$ against the expected value of the normal order statistics for a sample of size 55. The Kolmogorov–Smirnov statistic for the test of normality has a significance level of 0.084 when computed under the assumption that the $\hat{v}_t$ are a simple random sample of size 55. Given that the $\hat{v}_t$ are based on estimated parameters it

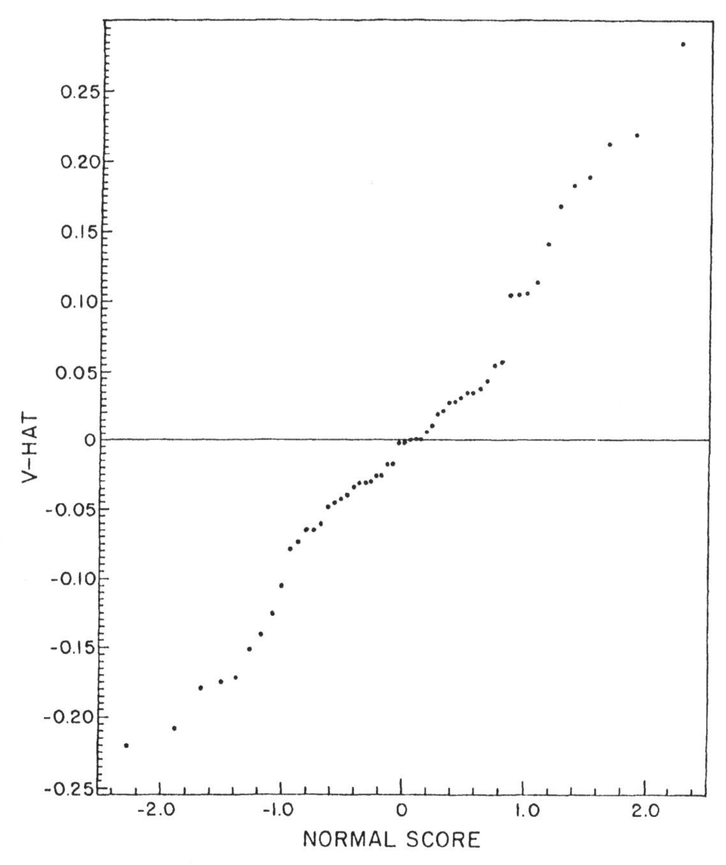

FIGURE 2.2.2. Normal probability plot for residuals.

seems that one would be willing to use calculations based on the normal model. □ □

2.2.3. Higher-Order Approximations for Residuals and True Values

The estimator of the variance for the estimated $\mathbf{x}_t$ given in (2.2.35) ignored the increase in variance of the estimator that comes from replacing parameters with estimators. In this section, we develop an estimator for the variance of the approximate distribution of $\hat{\mathbf{x}}_t$ that contains an estimator of the contribution to the variance arising from the estimation of the parameters.

We retain model (2.2.17)–(2.2.19) with the assumption of normal q_t and the assumption that the covariance matrix $\boldsymbol{\Sigma}_{aa}$ is known. Assume that we have completed estimation for a sample of size n so that we have estimators of $\boldsymbol{\beta}$, σ_{qq}, and σ_{vv} and an estimator of the covariance matrix of $\hat{\boldsymbol{\beta}}$. We wish to estimate the vector of true values $(y_t, \mathbf{x}_t)$ and the covariance matrix of the estimation error, treating $\mathbf{x}_t$ as fixed. Let

$$\hat{\boldsymbol{\alpha}}' = (1, -\hat{\boldsymbol{\beta}}') \quad \text{and} \quad \boldsymbol{\alpha}' = (1, -\boldsymbol{\beta}'),$$
$$\hat{\boldsymbol{\delta}}' = \hat{\sigma}_{vv}^{-1}\hat{\boldsymbol{\Sigma}}_{va} \quad \text{and} \quad \boldsymbol{\delta}' = \sigma_{vv}^{-1}\boldsymbol{\Sigma}_{va},$$
$$\hat{v}_t = \mathbf{Z}_t\hat{\boldsymbol{\alpha}} \quad \text{and} \quad v_t = \mathbf{Z}_t\boldsymbol{\alpha} = \boldsymbol{\varepsilon}_t\boldsymbol{\alpha},$$

where $\mathbf{Z}_t = (Y_t, \mathbf{X}_t)$, $\mathbf{a}_t = (w_t, \mathbf{u}_t)$, $\boldsymbol{\varepsilon}_t = (w_t + q_t, \mathbf{u}_t) = (e_t, \mathbf{u}_t)$, $\boldsymbol{\Sigma}_{va} = \boldsymbol{\alpha}'\boldsymbol{\Sigma}_{aa}$, $\hat{\boldsymbol{\Sigma}}_{va} = \hat{\boldsymbol{\alpha}}'\boldsymbol{\Sigma}_{aa}$, $\hat{\sigma}_{vv} = n^{-1}\sum_{t=1}^{n}\hat{v}_t^2$, and $\hat{\boldsymbol{\beta}}$ is defined by (2.2.20) with $\boldsymbol{\Sigma}_{aa}$ replacing $\mathbf{S}_{aa}$. The estimator of $\mathbf{z}_t'$ is

$$\hat{\mathbf{z}}_t' = \mathbf{Z}_t' - \hat{\boldsymbol{\delta}}\hat{v}_t \tag{2.2.36}$$
$$= \mathbf{Z}_t' - \boldsymbol{\delta}v_t - (\hat{\boldsymbol{\delta}} - \boldsymbol{\delta})v_t - \boldsymbol{\delta}\mathbf{Z}_t(\hat{\boldsymbol{\alpha}} - \boldsymbol{\alpha}) + O_p(n^{-1}), \tag{2.2.37}$$

where

$$\hat{\boldsymbol{\delta}} - \boldsymbol{\delta} = \sigma_{vv}^{-1}[(\boldsymbol{\Sigma}_{aa} - 2\boldsymbol{\delta}\boldsymbol{\Sigma}_{va})(\hat{\boldsymbol{\alpha}} - \boldsymbol{\alpha}) - \boldsymbol{\delta}(m_{vv} - \sigma_{vv})] + O_p(n^{-1}).$$

Therefore,

$$\begin{aligned}\hat{\mathbf{z}}_t' - \mathbf{z}_t' &= \ddot{\mathbf{z}}_t' - \mathbf{z}_t' - [\boldsymbol{\delta}\ddot{\mathbf{z}}_t + v_t\sigma_{vv}^{-1}(\boldsymbol{\Sigma}_{aa} - \boldsymbol{\delta}\boldsymbol{\Sigma}_{va})](\hat{\boldsymbol{\alpha}} - \boldsymbol{\alpha}) \\ &\quad + v_t\sigma_{vv}^{-1}\boldsymbol{\delta}(m_{vv} - \sigma_{vv}) + O_p(n^{-1}) \\ &= \ddot{\mathbf{z}}_t' - \mathbf{z}_t' + [\boldsymbol{\delta}\ddot{\mathbf{z}}_t + v_t\sigma_{vv}^{-1}(\boldsymbol{\Sigma}_{aa} - \boldsymbol{\delta}\boldsymbol{\Sigma}_{va})][0, (\mathbf{M}_{vX} - \boldsymbol{\Sigma}_{vu})\mathbf{M}_{xx}^{-1}]' \\ &\quad + v_t\sigma_{vv}^{-1}\boldsymbol{\delta}(m_{vv} - \sigma_{vv}) + O_p(n^{-1}),\end{aligned} \tag{2.2.38}$$

where $\ddot{\mathbf{z}}_t = \mathbf{Z}_t - v_t\boldsymbol{\delta}'$.

We drop the term of $O_p(n^{-1})$ in (2.2.38) and evaluate the covariance matrix of the leading terms. The random vector

$$\ddot{\mathbf{z}}_t - \mathbf{z}_t = \boldsymbol{\varepsilon}_t - v_t\boldsymbol{\delta}' \tag{2.2.39}$$

is independent of v_t and it follows that the covariance between $\boldsymbol{\varepsilon}_t - v_t\boldsymbol{\delta}'$ and the remaining $O_p(n^{-1/2})$ terms of (2.2.38) is zero. The only cross product term with an $O(n^{-1})$ expectation is

$$\begin{aligned}
E\{v_t^2\sigma_{vv}^{-2}[(\Sigma_{aa} - \delta\Sigma_{va})(\hat{\alpha} - \alpha)(m_{vv} - \sigma_{vv})\delta' \\
+ \delta(m_{vv} - \sigma_{vv})(\hat{\alpha} - \alpha)'(\Sigma_{aa} - \delta\Sigma_{va})]\} \\
= 2n^{-1}\sigma_{vv}^{-1}[(\Sigma_{aa} - \delta\Sigma_{va})(0, \Sigma_{vu}\mathbf{M}_{xx}^{-1})'\Sigma_{va} \\
+ \Sigma_{av}(0, \Sigma_{vu}\mathbf{M}_{xx}^{-1})(\Sigma_{aa} - \delta\Sigma_{va})] + O(n^{-2}).
\end{aligned}$$

Therefore, the covariance matrix of the approximate distribution of $\hat{\mathbf{z}}_t - \mathbf{z}_t$ is

$$\begin{aligned}
\Sigma_{aa} - \delta\Sigma_{av} + \delta\delta' E\{\ddot{\mathbf{z}}_t \mathbf{V}_{\alpha\alpha}\ddot{\mathbf{z}}_t'\} + \sigma_{vv}^{-1}(\Sigma_{aa} - \delta\Sigma_{va})\mathbf{V}_{\alpha\alpha}(\Sigma_{aa} - \delta\Sigma_{va}) \\
+ 2(n-k)^{-1}\delta\delta'\sigma_{vv} + 2n^{-1}\sigma_{vv}^{-1}[(\Sigma_{aa} - \delta\Sigma_{va})(0, \Sigma_{vu}\mathbf{M}_{xx}^{-1})'\Sigma_{va} \\
+ \Sigma_{av}(0, \Sigma_{vu}\mathbf{M}_{xx}^{-1})(\Sigma_{aa} - \delta\Sigma_{va})],
\end{aligned} \tag{2.2.40}$$

where $\mathbf{V}_{\alpha\alpha}$ is the covariance matrix of the approximate distribution of $\hat{\alpha}$ and $\mathbf{M}_{xx}$ is the matrix of mean squares and products of the true $\mathbf{x}_t$. Because the first row and column of $\mathbf{V}_{\alpha\alpha}$ contain only zeros,

$$\ddot{\mathbf{z}}_t \mathbf{V}_{\alpha\alpha}\ddot{\mathbf{z}}_t' = \ddot{\mathbf{x}}_t \mathbf{V}_{\beta\beta}\ddot{\mathbf{x}}_t',$$

where $\mathbf{V}_{\beta\beta}$ is the covariance matrix of the approximate distribution of $\hat{\boldsymbol{\beta}}$.

All terms in expression (2.2.40) have sample analogues. Therefore, an estimator of the variance of the approximate distribution of $\hat{\mathbf{z}}_t - \mathbf{z}_t$ is

$$\begin{aligned}
\tilde{\mathbf{V}}\{\hat{\mathbf{z}}_t - \mathbf{z}_t\} = \Sigma_{aa} - \hat{\delta}\hat{\Sigma}_{va} + \hat{\delta}\hat{\delta}'[\hat{\mathbf{x}}_t\hat{\mathbf{V}}_{\beta\beta}\hat{\mathbf{x}}_t' + 2(n-k)^{-1}\hat{\sigma}_{vv}] \\
+ \hat{\sigma}_{vv}^{-1}(\Sigma_{aa} - \hat{\delta}\hat{\Sigma}_{va})\hat{\mathbf{V}}_{\alpha\alpha}(\Sigma_{aa} - \hat{\delta}\hat{\Sigma}_{va}) \\
+ 2n^{-1}[(\Sigma_{aa} - \hat{\delta}\hat{\Sigma}_{va})(0, \hat{\Sigma}_{vu}\hat{\mathbf{M}}_{xx}^{-1})'\hat{\delta}' \\
+ \hat{\delta}(0, \hat{\Sigma}_{vu}\hat{\mathbf{M}}_{xx}^{-1})(\Sigma_{aa} - \hat{\delta}\hat{\Sigma}_{va})],
\end{aligned} \tag{2.2.41}$$

where $\hat{\mathbf{V}}_{\alpha\alpha} = \text{block diag}(0, \hat{\mathbf{V}}_{\beta\beta})$ and $\hat{\mathbf{V}}_{\beta\beta}$ is the estimator of the covariance matrix of the approximate distribution of $\hat{\boldsymbol{\beta}}$.

The last three terms of (2.2.41) are products of three matrices, where two of the matrices are matrices of error variances and the third is a function of the inverse of an estimator of $\mathbf{M}_{xx}$. Therefore, the last three terms of (2.2.41) will be small relative to the other terms when the error variance is small relative to the variation in $\mathbf{x}_t$. An approximation to the first three terms of (2.2.41) is

$$\dot{\mathbf{V}}\{\hat{\mathbf{z}}_t - \mathbf{z}_t\} = \Sigma_{aa} - n^{-1}(n-k-2)\hat{\delta}\hat{\Sigma}_{va}. \tag{2.2.42}$$

To develop higher-order approximations for the predictor constructed under the structural model, we let $\mathbf{z}_{t1} = (y_t, \mathbf{x}_{t1})$, where $\mathbf{x}_t = (1, \mathbf{x}_{t1})$ and $\mathbf{x}_{t1}$ is the $(k-1)$-dimensional vector that is the random portion of $\mathbf{x}_t$. Then a predictor of $\mathbf{z}_{t1}$ is

$$\tilde{\mathbf{z}}_{t1} = \bar{\mathbf{Z}}_1 + (\mathbf{Z}_{t1} - \bar{\mathbf{Z}}_1)(\mathbf{I} - c\mathbf{m}_{ZZ11}^{-1}\Sigma_{aa11}), \tag{2.2.43}$$

where $c = (n-1)^{-1}(n-k-2)$, $\mathbf{m}_{ZZ11}$ is the sample moment matrix of $\mathbf{Z}_{t1}$, $\boldsymbol{\Sigma}_{aa11}$ is the covariance matrix of $\mathbf{a}_{t1}$, and $\mathbf{a}_{t1}$ is the error associated with $\mathbf{z}_{t1}$. The predictor is an estimator of the conditional expectation of $\mathbf{z}_{t1}$ given $\mathbf{Z}_{t1}$. Under the normal distribution assumption, an estimator of the variance of $\tilde{\mathbf{z}}_{t1} - \mathbf{z}_{t1}$ is

$$\tilde{\mathbf{V}}\{\hat{\mathbf{z}}_{t1} - \mathbf{z}_{t1}\} = \boldsymbol{\Sigma}_{aa11} - c^2\boldsymbol{\Sigma}_{aa11}\mathbf{m}_{ZZ11}^{-1}\boldsymbol{\Sigma}_{aa11}. \tag{2.2.44}$$

The multiplier c appearing in expressions (2.2.43) and (2.2.44) is based on the work of Fuller and Harter (1987).

One can also use Taylor series to derive the covariance matrix of the approximate distribution of the residuals $\hat{v}_t$. Let $\hat{\mathbf{v}}$ denote the n-dimensional column vector

$$\hat{\mathbf{v}} = (\hat{v}_1, \hat{v}_2, \ldots, \hat{v}_n)' = \mathbf{Y} - \mathbf{X}\hat{\boldsymbol{\beta}},$$

where $\mathbf{Y} = (Y_1, Y_2, \ldots, Y_n)'$, and $\mathbf{X}' = (\mathbf{X}_1', \mathbf{X}_2', \ldots, \mathbf{X}_n')$. By our usual expansions,

$$\begin{aligned}\hat{\mathbf{v}} &= \mathbf{v} - \mathbf{X}(\hat{\boldsymbol{\beta}} - \boldsymbol{\beta}) \\ &= \mathbf{v} - \mathbf{X}\mathbf{M}_{xx}^{-1}(\mathbf{M}_{Xv} - \boldsymbol{\Sigma}_{uv}) + O_p(n^{-1}).\end{aligned} \tag{2.2.45}$$

Now, for fixed $\mathbf{x}_t$,

$$E\{v_j\mathbf{X}_t\mathbf{M}_{xx}^{-1}(\mathbf{M}_{Xv} - \boldsymbol{\Sigma}_{uv})\} = n^{-1}\operatorname{tr}\{[\mathbf{x}_j'\mathbf{x}_t\sigma_{vv} + (\boldsymbol{\Sigma}_{uu}\sigma_{vv} + \boldsymbol{\Sigma}_{uv}\boldsymbol{\Sigma}_{vu})\delta_{jt}]\mathbf{M}_{xx}^{-1}\},$$

where δ_{jt} is Kronecker's delta. It follows that

$$\begin{aligned}E\{[\mathbf{v} - n^{-1}\mathbf{X}\mathbf{M}_{xx}^{-1}(\mathbf{M}_{Xv} - \boldsymbol{\Sigma}_{uv})][\mathbf{v} - n^{-1}\mathbf{X}\mathbf{M}_{xx}^{-1}(\mathbf{M}_{xv} - \boldsymbol{\Sigma}_{uv})]'\} \\ = \mathbf{I}_n[\sigma_{vv} - 2n^{-1}\operatorname{tr}\{(\boldsymbol{\Sigma}_{uu}\sigma_{vv} + \boldsymbol{\Sigma}_{uv}\boldsymbol{\Sigma}_{vu})\mathbf{M}_{xx}^{-1}\} + \operatorname{tr}\{\boldsymbol{\Sigma}_{uu}\mathbf{V}_{\beta\beta}\}] \\ - 2n^{-1}\mathbf{x}\mathbf{M}_{xx}^{-1}\mathbf{x}'\sigma_{vv} + n^{-1}\mathbf{x}\mathbf{V}_{\beta\beta}\mathbf{x}' + O(n^{-2}).\end{aligned}$$

The quantity of most interest for model checking is the vector of standardized residuals $s_{vv}^{-1/2}\hat{\mathbf{v}}$. Using

$$\begin{aligned}s_{vv}^{-1/2} &= A_{vv}^{-1/2} - \tfrac{1}{2}A_{vv}^{-3/2}[-2A_{vv}\sigma_{vv}^{-1}\boldsymbol{\Sigma}_{vu}(\hat{\boldsymbol{\beta}} - \boldsymbol{\beta})] + O_p(n^{-1}) \\ &= A_{vv}^{-1/2}[1 - \sigma_{vv}^{-1}\boldsymbol{\Sigma}_{uv}(\hat{\boldsymbol{\beta}} - \boldsymbol{\beta})] + O_p(n^{-1}),\end{aligned}$$

we have

$$\begin{aligned}s_{vv}^{-1/2}\hat{\mathbf{v}} &= A_{vv}^{-1/2}[\mathbf{v} - \mathbf{X}(\hat{\boldsymbol{\beta}} - \boldsymbol{\beta})][1 - \sigma_{vv}^{-1}\boldsymbol{\Sigma}_{vu}(\hat{\boldsymbol{\beta}} - \boldsymbol{\beta})] + O_p(n^{-1}) \\ &= A_{vv}^{-1/2}[\mathbf{I}_n - n^{-1}\ddot{\mathbf{x}}\mathbf{M}_{xx}^{-1}\ddot{\mathbf{x}}']\mathbf{v} + O_p(n^{-1}),\end{aligned} \tag{2.2.46}$$

where $\ddot{\mathbf{x}} = \mathbf{X} - \mathbf{v}\sigma_{vv}^{-1}\boldsymbol{\Sigma}_{vu}$, and

$$A_{vv} = (n-k)^{-1}\mathbf{v}'(\mathbf{I}_n - n^{-1}\ddot{\mathbf{x}}'\mathbf{M}_{xx}^{-1}\ddot{\mathbf{x}}')^2\mathbf{v}.$$

In ordinary least squares estimation, with $\mathbf{x}$ observed, it is common practice to estimate the covariance matrix of the residuals with

$$[\mathbf{I}_n - \mathbf{x}(\mathbf{x}'\mathbf{x})^{-1}\mathbf{x}']s^2,$$

where $\mathbf{I}_n$ is the n-dimensional identity matrix and s^2 is the residual mean square. By (2.2.46), the analogous estimator for the measurement error problem is

$$[\mathbf{I}_n + \hat{\mathbf{x}}(s_{vv}^{-1}\hat{\mathbf{V}}_{\beta\beta} - 2n^{-1}\hat{\mathbf{M}}_{xx}^{-1})\hat{\mathbf{x}}']s_{vv}, \tag{2.2.47}$$

where $\hat{\mathbf{x}}$ is the $n \times k$ matrix of estimated true values and $n\hat{\mathbf{M}}_{xx}$ is the estimator of $\sum_{t=1}^{n} \mathbf{x}_t'\mathbf{x}_t$.

Expression (2.2.47) for the approximate covariance matrix of the $\hat{v}_t$ contains terms that are not present in the covariance matrix of ordinary least squares residuals. Nonetheless, the covariance matrix

$$[\mathbf{I}_n - \hat{\mathbf{x}}(\hat{\mathbf{x}}'\hat{\mathbf{x}})^{-1}\hat{\mathbf{x}}']s_{vv}$$

furnishes a useful approximation in many applications. For example, if one has available a program for diagnostic checking for ordinary least squares, the $\hat{v}_t$ and $\hat{\mathbf{x}}_t$ can be read into the program as the "dependent variable" and the vector of "independent variables," respectively. The resulting diagnostics can be treated, approximately, in the same manner as one would treat the statistics for ordinary least squares.

Example 2.2.3. We use the data and model of Example 1.2.1 to compute estimates of the variances of $\hat{x}_t$ and $\hat{y}_t$. From Examples 1.2.1 and 1.2.2, we have $n = 11$, $\hat{\beta}_1 = 0.4232$, $\sigma_{uu} = 57$, $\hat{\sigma}_{vv} = 53.4996$, $\hat{\sigma}_{xx} = 247.8545$, and $\hat{\sigma}_{uv} = -24.1224$.

We assume that the measurement error made in determining yield for a sample of plots within the field is $\sigma_{ww} = 25$ and we also assume that $\sigma_{uw} = 0$. It follows that

$$\hat{\boldsymbol{\delta}}' = \hat{\sigma}_{vv}^{-1}\hat{\boldsymbol{\Sigma}}_{va} = \hat{\sigma}_{vv}^{-1}(\sigma_{vw}, 0, \hat{\sigma}_{vu}) = (0.4673, 0, -0.4509),$$

where the zero element in the vector is for the intercept. The first-order approximation to the covariance matrix of $(\hat{y}_t - y_t, 1, \hat{x}_t - x_t)$ is

$$\boldsymbol{\Sigma}_{aa} - \hat{\boldsymbol{\delta}}\hat{\boldsymbol{\Sigma}}_{va} = \begin{bmatrix} 13.3177 & 0 & 11.2722 \\ 0 & 0 & 0 \\ 11.2722 & 0 & 46.1235 \end{bmatrix}.$$

For a model with an intercept, the computations are somewhat simplified and are less subject to rounding error if all independent variables, except the variable that is identically equal to one, are coded as deviations from the mean. If we let

$$\mathbf{Z}_t = (Y_t, 1, X_t - 70.6364),$$

then $\hat{\boldsymbol{\alpha}}' = (1, -97.4545, -0.4232)$, $n\hat{\mathbf{M}}_{xx} = \text{diag}(11, 2478.545)$, and

$$\hat{\mathbf{V}}_{\alpha\alpha} = \text{diag}(0, 4.8636, 0.0304).$$

TABLE 2.2.3. Order n^{-1} approximations to the variance of $(\hat{x}_t, \hat{y}_t, \hat{v}_t)$ for corn yield and soil nitrogen

Site	$\hat{x}_t - \bar{X}$	$\tilde{V}\{\hat{x}_t - x_t\}$	$\tilde{V}\{\hat{y}_t - y_t\}$	$\tilde{V}\{\hat{v}_t\}$	$[\tilde{V}\{\hat{v}_t\}]^{-1/2}\hat{v}_t$
1	−5.68	51.5	16.8	54.2	−1.52
2	29.24	56.6	22.2	43.7	0.97
3	−17.64	53.2	18.6	50.7	0.00
4	−10.54	52.0	17.3	53.2	−1.19
5	25.37	55.3	20.9	46.4	0.33
6	−8.28	51.7	17.0	53.8	−0.50
7	−16.00	52.9	18.3	51.4	1.43
8	−1.17	51.3	16.6	54.6	−0.16
9	19.60	53.7	19.1	49.7	−1.18
10	1.62	51.3	16.6	55.2	0.97
11	−16.55	53.0	18.4	51.7	0.95

From (2.2.41), the sum of the terms in the covariance matrix of the approximate distribution of $\hat{\mathbf{z}}_t - \mathbf{z}_t$ that are constant over t is

$$\begin{bmatrix} 15.5213 & 0 & 8.9926 \\ 0 & 0 & 0 \\ 8.9926 & 0 & 50.3172 \end{bmatrix} \tag{2.2.48}$$

The estimator (2.2.41) is the matrix (2.2.48) plus $\hat{\delta}\hat{\delta}'\hat{\mathbf{x}}_t\hat{\mathbf{V}}_{\beta\beta}\hat{\mathbf{x}}_t'$. Because the row and column of $\tilde{\mathbf{V}}\{\hat{\mathbf{z}}_t - \mathbf{z}_t\}$ associated with the intercept will always be zero, one can easily restrict the computations to 2×2 matrices. Of course, $\hat{\mathbf{x}}_t\hat{\mathbf{V}}_{\beta\beta}\hat{\mathbf{x}}_t'$ must include the terms for the intercept. We have carried out the computations for the full 3×3 matrix for illustrative purposes. The estimated variances of $\hat{x}_t - x_t$ and $\hat{y}_t - y_t$ are given in Table 2.2.3. The average increase in the estimated variance of $\hat{x}_t - x_t$ due to estimating the parameters is 6.8, an increase of about 15%. The average increase in the estimated variance of $\hat{y}_t - y_t$ due to estimating the parameters is 5.0, an increase of about 38%. The largest percentage increase in the estimated variance of $\hat{y}_t - y_t$ is about 67% and occurs for the second observation, the observation with the largest absolute value of $\hat{x}_t - \bar{X}$. One should remember that this is a small sample and, hence, the estimation effect is relatively large.

Expression (2.2.47) for the estimated variance of $\hat{\mathbf{v}}$ can be written as a function of the original X values or as a function of deviations from the mean. Using deviations from the mean, we have

$$\tilde{\mathbf{V}}\{\hat{\mathbf{v}}\} = (59.4440)\mathbf{I}_{11} + \hat{\mathbf{x}}(\hat{\mathbf{V}}_{\beta\beta} - 2n^{-1}\hat{\mathbf{M}}_{xx}^{-1}\hat{\sigma}_{vv})\hat{\mathbf{x}}'$$

where the tth row of $\hat{\mathbf{x}}$ is $(1, \hat{x}_t - 70.6364)$ and $s_{vv} = 59.4440$. The $\hat{x}_t - \bar{X}$ and the diagonal elements of $\tilde{\mathbf{V}}\{\hat{\mathbf{v}}\}$ are given in Table 2.2.3. The last column of Table 2.2.3 contains $\hat{v}_t$ divided by the estimated standard error of v_t. These quantities can be used in place of $\hat{v}_t$ for model checking. □ □

REFERENCES

Anderson and Rubin (1956), Fuller (1975, 1980), Kalbfleisch and Sprott (1970), Kiefer and Wolfowitz (1956), Morton (1981a).

EXERCISES

3. (Sections 2.2.1, 2.2.2) Using the data of Table 2.2.1 and the associated error covariance matrix, estimate the parameters of the model

$$y_t = \beta_0 + \sum_{i=1}^{3} x_{ti}\beta_i + q_t.$$

Estimate $\mathbf{x}_t = (x_{t1}, x_{t2}, x_{t3})$ for $t = 1, 2, \ldots, 20$ treating $\mathbf{x}_{t1}$ as fixed. Give the estimated covariance matrix for your estimates of $\mathbf{x}_{t1}$. Predict the true values treating $\mathbf{x}_{t1}$ as random. Give the estimated covariance matrix of the prediction error defined in (2.2.44).

4. (Section 2.2.1) Show that the likelihood in (2.2.4) increases without bound as $\sigma_{ee} \to 0$. (Hint: Fix σ_{ee} and maximize (2.2.4).)

5. (Section 2.2.3) Using expression (2.2.45), show that the covariance matrix of the approximate distribution of $(\hat{v}_t, u_{tj})$ is

$$\sigma_{uvj1}(1 - n^{-1}\mathbf{x}_t\mathbf{M}_{xx}^{-1}\mathbf{x}_t') - n^{-1}\operatorname{tr}[(\boldsymbol{\Sigma}_{uu}\sigma_{uvj1} + E\{u_{tj}\mathbf{u}_t'\}\boldsymbol{\Sigma}_{vu})\mathbf{M}_{xx}^{-1}].$$

Using this expression for $\operatorname{cov}\{\hat{v}_t, u_{tj}\}$ and expression (2.2.46), construct an alternative estimator for $\mathbf{x}_t$.

6. (Section 2.2.1) A random sample of 40 of the plots used in the soil moisture experiment described in Example 2.3.2 gives the sample moments

$$(m_{YY}, m_{XY}, m_{XX}) = (0.5087, 0.1838, 0.6006).$$

Assume that these data satisfy the model

$$Y_t = \beta_0 + \beta_1 x_t + e_t, \qquad X_t = x_t + u_t,$$
$$(x_t, e_t, u_t)' \sim \operatorname{NI}[(\mu_x, 0, 0)', \operatorname{diag}(\sigma_{xx}, \sigma_{ee}, 0.64)].$$

Estimate σ_{ee}, σ_{xx}, and β_1. Give the estimated covariance matrix for the approximate distribution of the estimators.

7. (Section 2.2.1)
   (a) Prove (2.2.7) from formulas (2.2.5) and (2.2.6).
   (b) Verify the formula for $E\{g_t^2\}$ that follows (2.2.24). (Hint: First show that x_t and v_t are uncorrelated under the conditions of Theorem 2.2.1.)

8. (Section 2.2.3) Verify the formula for $\hat{\delta} - \hat{\delta}$ that follows (2.2.37). (Hint: First show that $\sigma_{vv} = \boldsymbol{\alpha}'\boldsymbol{\Sigma}_{ZZ}\boldsymbol{\alpha} = \boldsymbol{\alpha}'\boldsymbol{\Sigma}_{\epsilon\epsilon}\boldsymbol{\alpha}$ under the assumptions of the model.)

9. (Section 2.2.1) The estimator of $\boldsymbol{\beta}$ given in (2.2.12) was obtained by maximizing the likelihood for normal $\mathbf{x}_t$ and is a method of moments estimator for any random $\mathbf{x}_t$. Let $\boldsymbol{\Sigma}_{eu} = \mathbf{0}$ for model (2.2.1).

(a) Show that $\hat{\boldsymbol{\beta}}$ of (2.2.12) (or $\tilde{\boldsymbol{\beta}}$ of (2.2.20) with $\mathbf{S}_{uu} = \boldsymbol{\Sigma}_{uu}$ and $\mathbf{S}_{uw} = \mathbf{0}$) is the value of $\boldsymbol{\beta}$ that minimizes

$$Q(\boldsymbol{\beta}) = \sum_{t=1}^{n} (Y_t - \mathbf{X}_t\boldsymbol{\beta})^2 - n\boldsymbol{\beta}'\boldsymbol{\Sigma}_{uu}\boldsymbol{\beta}.$$

Does the function have a minimum for all values of $\mathbf{X}_t$ and $\boldsymbol{\Sigma}_{uu}$? Explain.

(b) Show that the minimum over $\boldsymbol{\theta}$ of

$$E\left\{\sum_{t=1}^{n} [(Y_t - \mathbf{X}_t\boldsymbol{\theta})^2 - \boldsymbol{\theta}'\boldsymbol{\Sigma}_{uu}\boldsymbol{\theta}]\right\}$$

occurs with $\boldsymbol{\theta} = \boldsymbol{\beta}^0$, where $\boldsymbol{\beta}^0$ is the population parameter, and that the minimum of the expectation is $n\sigma_{ee}$.

2.3. THE MODEL WITH NO ERROR IN THE EQUATION

In this section we consider maximum likelihood estimation of the errors-in-variables model when the entire error covariance structure, including σ_{ee}, is known or is known up to a scalar multiple. The model is the extension of the model of Section 1.3 to a model with a vector of explanatory variables.

2.3.1. The Functional Model

We first derive the maximum likelihood estimator treating the vector of true values $\mathbf{x}_t$ as fixed in repeated sampling. The model is

$$y_t = \mathbf{x}_t\boldsymbol{\beta}, \qquad (2.3.1)$$
$$(Y_t, \mathbf{X}_t) = (y_t, \mathbf{x}_t) + (e_t, \mathbf{u}_t),$$

for $t = 1, 2, \ldots, n$, where $\{\mathbf{x}_t\}$ is a sequence of fixed k-dimensional row vectors and $\boldsymbol{\varepsilon}_t = (e_t, \mathbf{u}_t)'$ is the vector of measurement errors. We can also write the defining equations (2.3.1) as

$$\mathbf{z}_t\boldsymbol{\alpha} = 0, \qquad \mathbf{Z}_t = \mathbf{z}_t + \boldsymbol{\varepsilon}_t, \qquad (2.3.2)$$

where $\mathbf{z}_t = (y_t, \mathbf{x}_t)$, $\mathbf{Z}_t = (Y_t, \mathbf{X}_t)$, and

$$\boldsymbol{\alpha}' = (\alpha_1, \alpha_2, \ldots, \alpha_{k+1}) = (1, -\boldsymbol{\beta}').$$

Theorem 2.3.1. Let model (2.3.1) and the associated assumptions hold. Let $\boldsymbol{\varepsilon}_t \sim \text{NI}(\mathbf{0}, \boldsymbol{\Sigma}_{\varepsilon\varepsilon})$, where $\boldsymbol{\Sigma}_{\varepsilon\varepsilon} = \boldsymbol{\Upsilon}_{\varepsilon\varepsilon}\sigma^2$ and $\boldsymbol{\Upsilon}_{\varepsilon\varepsilon}$ is known. Then the maximum likelihood estimators of $\boldsymbol{\beta}$ and σ^2 are

$$\hat{\boldsymbol{\beta}} = (\mathbf{M}_{XX} - \hat{\lambda}\boldsymbol{\Upsilon}_{uu})^{-1}(\mathbf{M}_{XY} - \hat{\lambda}\boldsymbol{\Upsilon}_{ue}),$$
$$\hat{\sigma}_m^2 = (k + 1)^{-1}\hat{\lambda},$$

where $\mathbf{M}_{ZZ} - n^{-1} \Sigma_{t=1}^{n} \mathbf{Z}_t'\mathbf{Z}_t$ and $\hat{\lambda}$ is the smallest root of

$$|\mathbf{M}_{ZZ} - \lambda \boldsymbol{\Upsilon}_{\varepsilon\varepsilon}| = 0. \tag{2.3.3}$$

The maximum likelihood estimator of $\mathbf{z}_t$, $t = 1, 2, \ldots, n$, is

$$\hat{\mathbf{z}}_t = \mathbf{Z}_t - (\mathbf{Y}_t - \mathbf{X}_t\hat{\boldsymbol{\beta}})[(1, -\hat{\boldsymbol{\beta}}')\boldsymbol{\Upsilon}_{\varepsilon\varepsilon}(1, -\hat{\boldsymbol{\beta}}')']^{-1}(1, -\hat{\boldsymbol{\beta}}')\boldsymbol{\Upsilon}_{\varepsilon\varepsilon}. \tag{2.3.4}$$

Proof. We derive the estimator under the assumption that $\boldsymbol{\Upsilon}_{\varepsilon\varepsilon}$ is nonsingular. The density for $\boldsymbol{\varepsilon}_t$ is

$$|2\pi\boldsymbol{\Upsilon}_{\varepsilon\varepsilon}\sigma^2|^{-1/2} \exp\{-(2\sigma^2)^{-1}(\boldsymbol{\varepsilon}_t\boldsymbol{\Upsilon}_{\varepsilon\varepsilon}^{-1}\boldsymbol{\varepsilon}_t')\}. \tag{2.3.5}$$

Because the $\mathbf{z}_t$ are fixed, the logarithm of the likelihood for a sample of n observations is

$$\log L = -2^{-1}n \log|2\pi\boldsymbol{\Upsilon}_{\varepsilon\varepsilon}\sigma^2| - (2\sigma^2)^{-1} \sum_{t=1}^{n} (\mathbf{Z}_t - \mathbf{z}_t)\boldsymbol{\Upsilon}_{\varepsilon\varepsilon}^{-1}(\mathbf{Z}_t - \mathbf{z}_t)'. \tag{2.3.6}$$

We first maximize (2.3.5) with respect to $\mathbf{z}_t$, $t = 1, 2, \ldots, n$, for a given $\boldsymbol{\beta}$. This reduces to the least squares problem of minimizing

$$(Y_t - \mathbf{x}_t\boldsymbol{\beta}, \mathbf{X}_t - \mathbf{x}_t)\boldsymbol{\Upsilon}_{\varepsilon\varepsilon}^{-1}(Y_t - \mathbf{x}_t\boldsymbol{\beta}, \mathbf{X}_t - \mathbf{x}_t)' \tag{2.3.7}$$

with respect to $\mathbf{x}_t$. It follows that the least squares estimator of $\mathbf{z}_t'$ is

$$\ddot{\mathbf{z}}_t' = (\boldsymbol{\beta}, \mathbf{I}_k)'[(\boldsymbol{\beta}, \mathbf{I}_k)\boldsymbol{\Upsilon}_{\varepsilon\varepsilon}^{-1}(\boldsymbol{\beta}, \mathbf{I}_k)']^{-1}(\boldsymbol{\beta}, \mathbf{I}_k)\boldsymbol{\Upsilon}_{\varepsilon\varepsilon}^{-1}\mathbf{Z}_t' \tag{2.3.8}$$

$$= \mathbf{Z}_t' - \boldsymbol{\Upsilon}_{\varepsilon\varepsilon}\boldsymbol{\alpha}(\boldsymbol{\alpha}'\boldsymbol{\Upsilon}_{\varepsilon\varepsilon}\boldsymbol{\alpha})^{-1}\boldsymbol{\alpha}'\mathbf{Z}_t'. \tag{2.3.9}$$

Substituting expression (2.3.8) into (2.3.5), we have

$$\log L = -2^{-1}n \log|2\pi\boldsymbol{\Upsilon}_{\varepsilon\varepsilon}\sigma^2| - (2\sigma^2)^{-1} \sum_{t=1}^{n} \mathbf{Z}_t\boldsymbol{\alpha}(\boldsymbol{\alpha}'\boldsymbol{\Upsilon}_{\varepsilon\varepsilon}\boldsymbol{\alpha})^{-1}\boldsymbol{\alpha}'\mathbf{Z}_t'. \tag{2.3.10}$$

Therefore, the $\boldsymbol{\alpha}$ that maximizes (2.3.6) is the vector that minimizes

$$n^{-1} \sum_{t=1}^{n} (\boldsymbol{\alpha}'\boldsymbol{\Upsilon}_{\varepsilon\varepsilon}\boldsymbol{\alpha})^{-1}\boldsymbol{\alpha}'\mathbf{Z}_t'\mathbf{Z}_t\boldsymbol{\alpha} = (\boldsymbol{\alpha}'\boldsymbol{\Upsilon}_{\varepsilon\varepsilon}\boldsymbol{\alpha})^{-1}\boldsymbol{\alpha}'\mathbf{M}_{ZZ}\boldsymbol{\alpha}, \tag{2.3.11}$$

where we have introduced the factor n^{-1} for notational convenience.

By Corollary 4.A.10 of Appendix 4.A, the $\boldsymbol{\alpha}$ minimizing (2.3.11) is given by

$$(\mathbf{M}_{ZZ} - \hat{\lambda}\boldsymbol{\Upsilon}_{\varepsilon\varepsilon})\hat{\boldsymbol{\alpha}} = \mathbf{0}, \tag{2.3.12}$$

where $\hat{\lambda}$ is the smallest root of (2.3.3). The smallest root $\hat{\lambda}$ is also the minimum value for the ratio (2.3.11)

$$\hat{\lambda} = (\hat{\boldsymbol{\alpha}}'\boldsymbol{\Upsilon}_{\varepsilon\varepsilon}\hat{\boldsymbol{\alpha}})^{-1}\hat{\boldsymbol{\alpha}}'\mathbf{M}_{ZZ}\hat{\boldsymbol{\alpha}}. \tag{2.3.13}$$

With probability one, the rank of $\mathbf{M}_{ZZ} - \hat{\lambda}\boldsymbol{\Upsilon}_{\varepsilon\varepsilon}$ is k and $\hat{\boldsymbol{\alpha}}$ is determined up to a multiple. Using $\boldsymbol{\alpha}' = (1, -\boldsymbol{\beta}')$, the estimator of $\boldsymbol{\beta}$ is

$$\hat{\boldsymbol{\beta}} = (\mathbf{M}_{XX} - \hat{\lambda}\boldsymbol{\Upsilon}_{uu})^{-1}(\mathbf{M}_{XY} - \hat{\lambda}\boldsymbol{\Upsilon}_{ue}), \tag{2.3.14}$$

where $\boldsymbol{\Upsilon}_{ee}$, $\boldsymbol{\Upsilon}_{eu}$, and $\boldsymbol{\Upsilon}_{uu}$ are the submatrices of $\boldsymbol{\Upsilon}_{\varepsilon\varepsilon}$, and $\mathbf{M}_{XX} - \hat{\lambda}\boldsymbol{\Upsilon}_{uu}$ is nonsingular with probability one. See the discussion in the text below for singular $\boldsymbol{\Upsilon}_{\varepsilon\varepsilon}$.

By (2.3.9), the maximum likelihood estimator of $\mathbf{z}_t$ is

$$\hat{\mathbf{z}}_t = \mathbf{Z}_t[\mathbf{I} - \hat{\boldsymbol{\alpha}}(\hat{\boldsymbol{\alpha}}'\boldsymbol{\Upsilon}_{\varepsilon\varepsilon}\hat{\boldsymbol{\alpha}})^{-1}\hat{\boldsymbol{\alpha}}'\boldsymbol{\Upsilon}_{\varepsilon\varepsilon}]. \tag{2.3.15}$$

Differentiating (2.3.5) with respect to σ^2, we obtain

$$\frac{\partial \log L}{\partial \sigma^2} = -2^{-1}(k+1)n\sigma^{-2} + 2^{-1}\sigma^{-4} \sum_{t=1}^{n} (\mathbf{Z}_t - \mathbf{z}_t)\boldsymbol{\Upsilon}_{\varepsilon\varepsilon}^{-1}(\mathbf{Z}_t - \mathbf{z}_t)'$$

and the maximum likelihood estimator of σ^2 is

$$\begin{aligned}
\hat{\sigma}_m^2 &= [(k+1)n]^{-1} \sum_{t=1}^{n} (\mathbf{Z}_t - \hat{\mathbf{z}}_t)\boldsymbol{\Upsilon}_{\varepsilon\varepsilon}^{-1}(\mathbf{Z}_t - \hat{\mathbf{z}}_t)' \\
&= [(k+1)n]^{-1} \sum_{t=1}^{n} (\hat{\boldsymbol{\alpha}}'\boldsymbol{\Upsilon}_{\varepsilon\varepsilon}\hat{\boldsymbol{\alpha}})^{-1}\hat{\boldsymbol{\alpha}}'\mathbf{Z}_t'\mathbf{Z}_t\hat{\boldsymbol{\alpha}} \\
&= (k+1)^{-1}\hat{\lambda},
\end{aligned} \tag{2.3.16}$$

where $\hat{\lambda}$ is defined in (2.3.13) and (2.3.3). □

Observe that the maximum likelihood estimators of $\boldsymbol{\alpha}$ and $\mathbf{z}_t$ given in Theorem 2.3.1 would be the same if σ^2 were known. That is, the estimators of $\boldsymbol{\alpha}$ and $\mathbf{z}_t$ are the same whether $\boldsymbol{\Sigma}_{\varepsilon\varepsilon}$ or $\boldsymbol{\Upsilon}_{\varepsilon\varepsilon}$ is known.

In deriving the maximum likelihood estimators we assumed $\boldsymbol{\Sigma}_{\varepsilon\varepsilon}$ to be nonsingular. This was an assumption of convenience. The definition of $\hat{\boldsymbol{\beta}}$ given by Equation (2.3.14) does not require $\boldsymbol{\Upsilon}_{\varepsilon\varepsilon}$ to be nonsingular, provided $\mathbf{m}_{ZZ}$ is nonsingular. Likewise, the estimator of $\mathbf{z}_t$ defined in (2.3.15) is the maximum likelihood estimator for singular $\boldsymbol{\Upsilon}_{\varepsilon\varepsilon}$.

If $\hat{\boldsymbol{\alpha}}$ is a consistent estimator for $\boldsymbol{\alpha}$, then the maximum likelihood estimator of σ^2 will be a consistent estimator for $(k+1)^{-1}\sigma^2$. As in Section 2.2, maximum likelihood fails to produce consistent estimators of all parameters for the functional model. The problem is essentially a "degrees of freedom" problem. We estimate $(nk + k)$ parameters, but there is no adjustment in the maximum likelihood estimator for this fact. The estimator (2.3.14) of $\boldsymbol{\beta}$ has considerable appeal and we shall demonstrate that it possesses desirable large sample properties. In line with our earlier results, a reasonable consistent estimator of σ^2 is

$$\hat{\sigma}^2 = (n-k)^{-1}n\hat{\lambda}. \tag{2.3.17}$$

In Theorem 2.3.2 we demonstrate that the estimator of $\boldsymbol{\beta}$ is approximately normally distributed if the error variances are small relative to the variation in $\mathbf{x}_t$ or (and) if the sample size is large. To accommodate the two types of limits, the sequence of estimators is indexed by v, where $v = n_v T_v$, n_v is the sample size, and the error covariance matrix for a particular v is a fixed matrix multiplied by T_v^{-1}. It is assumed that the sequences $\{n_v\}$ and $\{T_v\}$ are nondecreasing in v. Henceforth, we shall suppress the subscript v on n and T.

In practice, we often have an estimator of $\boldsymbol{\Sigma}_{\varepsilon\varepsilon}$, rather than knowing the matrix $\boldsymbol{\Upsilon}_{\varepsilon\varepsilon}$. We give the limiting distribution of the estimators for such a situation, and the estimation of the error covariance matrix contributes a term to the covariance matrix of the approximate distribution of $\tilde{\boldsymbol{\beta}}$.

Theorem 2.3.2. Let

$$Y_t = \mathbf{x}_t\boldsymbol{\beta} + e_t, \qquad \mathbf{X}_t = \mathbf{x}_t + \mathbf{u}_t,$$
$$(e_t, \mathbf{u}_t)' \sim \mathrm{NI}(\mathbf{0}, T^{-1}\boldsymbol{\Omega}),$$

for $t = 1, 2, \ldots, n$, where $\boldsymbol{\Omega}$ is a fixed positive semidefinite matrix and $\{\mathbf{x}_t\}$ is a fixed sequence of k-dimensional vectors. Let $n \geqslant k + 1$, $T \geqslant 1$, $v = nT$, and $|\mathbf{M}_{xx}| > 0$ and let

$$\lim_{v \to \infty} \bar{\mathbf{x}} = \boldsymbol{\mu}_x, \qquad \lim_{v \to \infty} \mathbf{M}_{xx} = \bar{\mathbf{M}}_{xx},$$

where $\bar{\mathbf{M}}_{xx}$ is a positive definite symmetric matrix. Let $\mathbf{S}_{\varepsilon\varepsilon}$ be an unbiased estimator of $\boldsymbol{\Omega}$, where $\mathbf{S}_{uu}$ is the lower right $k \times k$ portion of $\mathbf{S}_{\varepsilon\varepsilon}$. Let $\mathbf{S}_{\varepsilon\varepsilon}$ be distributed as a multiple of a Wishart matrix with d_f degrees of freedom, independent of $(Y_t, \mathbf{X}_t)$ for all t. Let $d_f \geqslant n\xi_1$, where ξ_1 is a fixed positive number. Let

$$\tilde{\boldsymbol{\beta}} = (\mathbf{M}_{XX} - \tilde{\lambda}T^{-1}\mathbf{S}_{uu})^{-1}(\mathbf{M}_{XY} - \tilde{\lambda}T^{-1}\mathbf{S}_{ue}),$$

where $\tilde{\lambda}$ is the smallest root of

$$|\mathbf{M}_{ZZ} - \lambda T^{-1}\mathbf{S}_{\varepsilon\varepsilon}| = 0. \tag{2.3.18}$$

Then, for $n > k + 1$,

$$n(n-k)^{-1}\tilde{\lambda} = F + O_p(n^{-1/2}v^{-1/2}), \tag{2.3.19}$$

where F is a random variable distributed as Snedecor's F with $n - k$ and d_f degrees of freedom. Furthermore,

$$\boldsymbol{\Gamma}_v^{-1/2}(\tilde{\boldsymbol{\beta}} - \boldsymbol{\beta}) \xrightarrow{L} N(\mathbf{0}, \mathbf{I})$$

as $v \to \infty$, where

$$\boldsymbol{\Gamma}_v = n^{-1}\mathbf{M}_{xx}^{-1}\sigma_{vv} + (n^{-1} + d_f^{-1})\mathbf{M}_{xx}^{-1}(\boldsymbol{\Sigma}_{uu}\sigma_{vv} - \boldsymbol{\Sigma}_{uv}\boldsymbol{\Sigma}_{vu})\mathbf{M}_{xx}^{-1},$$

and, for example, $\sigma_{vv} = T^{-1}(1, -\boldsymbol{\beta}')\boldsymbol{\Omega}(1, -\boldsymbol{\beta}')'$.

Proof. Now, for example, $M_{XXij} = M_{xxij} + M_{xuij} + M_{xuji} + M_{uuij}$, and

$$V\{M_{xuij}\} = n^{-1}M_{xxii}\sigma_{uujj} = (nT)^{-1}M_{xxii}\omega_{uuij} = O(\nu^{-1}),$$
$$V\{M_{uuij}\} = n^{-1}T^{-2}(\omega_{uuii}\omega_{uujj} + \omega^2_{uuij}) = O(\nu^{-1}T^{-1}),$$

where ω_{uuij} is the ijth element of $\mathbf{\Omega}_{uu}$ and $\mathbf{\Omega}_{uu}$ is the lower right $k \times k$ portion of $\mathbf{\Omega}$. It follows that

$$\mathbf{M}_{ZZ} = \mathbf{M}_{zz} + \mathbf{\Sigma}_{\varepsilon\varepsilon} + O_p(\nu^{-1/2}),$$

where $\mathbf{\Sigma}_{\varepsilon\varepsilon} = T^{-1}\mathbf{\Omega}$. Because $\tilde{\lambda}$ is the minimum root of (2.3.18), we have

$$\begin{aligned}\tilde{\lambda} &= T[(1, -\tilde{\boldsymbol{\beta}}')\mathbf{S}_{\varepsilon\varepsilon}(1, -\tilde{\boldsymbol{\beta}}')']^{-1}(1, -\tilde{\boldsymbol{\beta}}')\mathbf{M}_{ZZ}(1, -\tilde{\boldsymbol{\beta}}')' \\ &\leqslant TM_{vv}[(1, -\boldsymbol{\beta}')S_{\varepsilon\varepsilon}(1, -\boldsymbol{\beta}')']^{-1}.\end{aligned} \tag{2.3.20}$$

The ratio on the right side of the inequality in (2.3.20) is distributed as an F random variable with n and d_f degrees of freedom. Hence, $\tilde{\lambda}$ is bounded in probability. The root $\tilde{\lambda}$ is a continuous function of the elements of $\hat{\boldsymbol{\theta}} = [(\mathbf{M}_{ZZ}, \mathbf{S}_{\varepsilon\varepsilon}) - (\mathbf{M}_{zz} + \mathbf{\Sigma}_{\varepsilon\varepsilon}, \mathbf{\Omega})]$ with continuous first derivatives in a region about $\boldsymbol{\theta} = \mathbf{0}$. Because $\hat{\boldsymbol{\theta}} = O_p(n^{-1/2})$, it follows that

$$\tilde{\lambda} - 1 = O_p(n^{-1/2}), \tag{2.3.21}$$

where the result holds for fixed n because $\tilde{\lambda}$ is bounded in probability. From the definition of $\tilde{\boldsymbol{\beta}}$,

$$\begin{aligned}\tilde{\boldsymbol{\beta}} - \boldsymbol{\beta} &= (\mathbf{M}_{XX} - \tilde{\lambda}T^{-1}\mathbf{S}_{uu})^{-1}[\mathbf{M}_{Xv} - \tilde{\lambda}T^{-1}(\mathbf{S}_{eu} - \mathbf{S}_{uu}\boldsymbol{\beta})] \\ &= \mathbf{M}_{xx}^{-1}[\mathbf{M}_{xv} + \mathbf{M}_{uv} - T^{-1}(\mathbf{S}_{eu} - \mathbf{S}_{uu}\boldsymbol{\beta}) - (\tilde{\lambda} - 1)\mathbf{\Sigma}_{uv}] + O_p(\nu^{-1}) \\ &= O_p(\nu^{-1/2}),\end{aligned} \tag{2.3.22}$$

where we have used

$$\begin{aligned}(\tilde{\lambda} - 1)T^{-1}\mathbf{S}_{\varepsilon\varepsilon} &= O_p(n^{-1/2}T^{-1}), \\ \mathbf{M}_{XX} - \tilde{\lambda}T^{-1}\mathbf{S}_{uu} &= \mathbf{M}_{xx} + O_p(\nu^{-1/2}), \\ \mathbf{M}_{Xv} - (\tilde{\lambda} - 1)T^{-1}(\mathbf{S}_{ue} - \mathbf{S}_{uu}\boldsymbol{\beta}) &= \mathbf{M}_{Xv} - (\tilde{\lambda} - 1)\mathbf{\Sigma}_{uv} + O_p(\nu^{-1}).\end{aligned}$$

By the definitions of $\tilde{\lambda}$ and $\tilde{\boldsymbol{\beta}}$,

$$\begin{aligned}\tilde{\lambda} - 1 &= \frac{T(M_{YY} - 2\tilde{\boldsymbol{\beta}}'\mathbf{M}_{XY} + \tilde{\boldsymbol{\beta}}'\mathbf{M}_{XX}\tilde{\boldsymbol{\beta}})}{s_{ee} - 2\tilde{\boldsymbol{\beta}}'\mathbf{S}_{ue} + \tilde{\boldsymbol{\beta}}'\mathbf{S}_{uu}\tilde{\boldsymbol{\beta}}} - 1 \\ &= \frac{T[M_{vv} - \tilde{s}_{vv} - 2(\tilde{\boldsymbol{\beta}} - \boldsymbol{\beta})'(\mathbf{M}_{Xv} - \tilde{\mathbf{S}}_{uv}) + (\tilde{\boldsymbol{\beta}} - \boldsymbol{\beta})'(\mathbf{M}_{XX} - \tilde{\mathbf{S}}_{uu})(\tilde{\boldsymbol{\beta}} - \boldsymbol{\beta})]}{T[\tilde{s}_{vv} - 2(\tilde{\boldsymbol{\beta}} - \boldsymbol{\beta})'\tilde{\mathbf{S}}_{uv} + (\tilde{\boldsymbol{\beta}} - \boldsymbol{\beta})'\tilde{\mathbf{S}}_{uu}(\tilde{\boldsymbol{\beta}} - \boldsymbol{\beta})]} \\ &= \tilde{s}_{vv}^{-1}[M_{vv} - \tilde{s}_{vv} - 2(\tilde{\boldsymbol{\beta}} - \boldsymbol{\beta})'(\mathbf{M}_{Xv} - \tilde{\mathbf{S}}_{uv}) \\ &\quad + (\tilde{\boldsymbol{\beta}} - \boldsymbol{\beta})'(\mathbf{M}_{XX} - \tilde{\mathbf{S}}_{uu})(\tilde{\boldsymbol{\beta}} - \boldsymbol{\beta})] + O_p(\nu^{-1/2}n^{-1/2}),\end{aligned}$$

where $\tilde{\mathbf{S}}_{uu} = T^{-1}\mathbf{S}_{uu}$, $\tilde{\mathbf{S}}_{uv} = T^{-1}(\mathbf{S}_{eu} - \mathbf{S}_{uu}\boldsymbol{\beta})$, and

$$\tilde{s}_{vv} = T^{-1}(1, -\boldsymbol{\beta}')\mathbf{S}_{\varepsilon\varepsilon}(1, -\boldsymbol{\beta}')'.$$

By the second expression of (2.3.22) and the properties of the sample moments,

$$\tilde{\boldsymbol{\beta}} - \boldsymbol{\beta} = \mathbf{M}_{xx}^{-1}\mathbf{M}_{xv} + O_p(v^{-1/2}T^{-1/2}),$$
$$\mathbf{M}_{Xv} - \tilde{\mathbf{S}}_{uv} = \mathbf{M}_{xv} + O_p(v^{-1/2}T^{-1/2}),$$
$$(\mathbf{M}_{XX} - \tilde{\lambda}\tilde{\mathbf{S}}_{uu})^{-1}(\mathbf{M}_{XX} - \tilde{\mathbf{S}}_{uu}) = \mathbf{I} + O_p(v^{-1/2}T^{-1/2}).$$

It follows that

$$(\tilde{\boldsymbol{\beta}} - \boldsymbol{\beta})'(\mathbf{M}_{XX} - \tilde{\mathbf{S}}_{uu})(\tilde{\boldsymbol{\beta}} - \boldsymbol{\beta}) = \mathbf{M}_{vx}\mathbf{M}_{xx}^{-1}\mathbf{M}_{xv} + O_p(v^{-1}T^{-1/2})$$

and

$$\tilde{\lambda} = \tilde{s}_{vv}^{-1}(M_{vv} - \mathbf{M}_{vx}\mathbf{M}_{xx}^{-1}\mathbf{M}_{xv}) + O_p(v^{-1/2}n^{-1/2}). \tag{2.3.23}$$

By assumption, the numerator and denominator of (2.3.23) are independent, and $d_f\sigma_{vv}^{-1}\tilde{s}_{vv}$ is distributed as a chi-square random variable with d_f degrees of freedom. The limiting distribution of $n(n-k)^{-1}\tilde{\lambda}$ follows because $(n-1)\sigma_{vv}^{-1}(M_{vv} - \mathbf{M}_{vx}\mathbf{M}_{xx}^{-1}\mathbf{M}_{xv})$ is distributed as a chi-square random variable with $n-k$ degrees of freedom.

If T increases without bound, then

$$\tilde{\boldsymbol{\beta}} - \boldsymbol{\beta} = \mathbf{M}_{xx}^{-1}\mathbf{M}_{xv} + o_p(v^{-1/2}).$$

Because $\mathbf{x}_t$ is fixed, $[\mathbf{V}\{\mathbf{M}_{xv}\}]^{-1/2}\mathbf{M}_{xv}$ is a $N(\mathbf{0}, \mathbf{I})$ random vector for all v. Now

$$\mathbf{V}\{\mathbf{M}_{xv}\} = T^{-1}n^{-1}\mathbf{M}_{xx}(1, -\boldsymbol{\beta}')\boldsymbol{\Omega}(1, -\boldsymbol{\beta}')'$$

and $\boldsymbol{\Gamma}_v = \mathbf{V}\{\mathbf{M}_{xx}^{-1}\mathbf{M}_{xv}\} + O_p(v^{-1}T^{-1})$. Therefore, the limiting distribution of $\boldsymbol{\Gamma}_v^{-1/2}(\tilde{\boldsymbol{\beta}} - \boldsymbol{\beta})$ is established for increasing T.

If $n^{-1} = O(v^{-1})$, we write

$$n^{1/2}(\tilde{\boldsymbol{\beta}} - \boldsymbol{\beta}) = n^{-1/2}\mathbf{M}_{xx}^{-1}\sum_{t=1}^{n}\mathbf{g}_t - n^{1/2}\mathbf{M}_{xx}^{-1}(\mathbf{S}_{uv} - \sigma_{vv}s_{vv}\boldsymbol{\Sigma}_{uv}) + O_p(n^{-1/2}),$$

where $\mathbf{g}_t = \mathbf{X}_t'v_t - \boldsymbol{\Sigma}_{uv} - \sigma_{vv}^{-1}(v_t^2 - \sigma_{vv})\boldsymbol{\Sigma}_{uv}$ and $\mathbf{g}_t$ is independent of $\mathbf{S}_{uv}$. The limiting normality of $\boldsymbol{\Gamma}_v^{-1/2}(\tilde{\boldsymbol{\beta}} - \boldsymbol{\beta})$ then follows by the arguments of Theorem 2.2.1. □

If $\boldsymbol{\Omega}$ is known or known up to a multiple, Theorem 2.3.2 holds with $d_f^{-1} = 0$. If $\boldsymbol{\Omega}$ is known up to a multiple, it follows from Theorem 2.3.2 that the estimator of σ^2 given in (2.3.17) is approximately distributed as a multiple of a chi-square random variable with $n-k$ degrees of freedom.

If the error variances become small as $v \to \infty$, then one could normalize $\tilde{\boldsymbol{\beta}} - \boldsymbol{\beta}$ with $\mathbf{V}\{\mathbf{M}_{xx}^{-1}\mathbf{M}_{xv}\}$ to obtain the limiting distribution. From (2.3.23),

$$\tilde{\lambda} - 1 = \sigma_{vv}^{-1}[M_{vv} - \mathbf{M}_{vx}\mathbf{M}_{xx}^{-1}\mathbf{M}_{xv} - \tilde{s}_{vv}] + O_p(n^{-1})$$

and substituting this expression into (2.3.22), we obtain

$$\tilde{\boldsymbol{\beta}} - \boldsymbol{\beta} = \mathbf{M}_{xx}^{-1}[\mathbf{M}_{xv} - \tilde{\mathbf{S}}_{uv} - \boldsymbol{\Sigma}_{uv}\sigma_{vv}^{-1}(M_{vv} - \tilde{s}_{vv})] + O_p(n^{-1}).$$

It is because $\boldsymbol{\Gamma}_v$ is the variance of the leading term on the right side of the equality that we prefer $\boldsymbol{\Gamma}_v$ to $\mathbf{V}\{\mathbf{M}_{xx}^{-1}\mathbf{M}_{xv}\}$ as the variance of the approximate distribution of $\tilde{\boldsymbol{\beta}}$, even for small error variances.

On the basis of Theorem 2.3.2 we may use the variance expression of the approximate distribution for samples in which the variances of the elements of the difference

$$\mathbf{M}_{XX} - (\mathbf{M}_{xx} + \boldsymbol{\Sigma}_{uu})$$

are small relative to the diagonal elements of $\mathbf{M}_{xx}$. This will be true if n is large so that the variances of the elements of $\mathbf{M}_{xu}$ and $\mathbf{M}_{uu}$ are small. Likewise, the variances of the elements of $\mathbf{M}_{xu}$ and $\mathbf{M}_{uu}$ will be small if $\boldsymbol{\Sigma}_{uu}$ is small (T large).

Theorem 2.3.2 gives the distribution of $\tilde{\lambda}$ only for the null model. From expressions (2.3.10) and the fact that the sample moments converge in probability, we see that a test based on $\tilde{\lambda}$ will have power against models in which the relationship between y_t and $\mathbf{x}_t$ is not linear.

An estimator of the covariance matrix of the approximate distribution of $\tilde{\boldsymbol{\beta}}$ is

$$\hat{\mathbf{V}}\{\tilde{\boldsymbol{\beta}}\} = n^{-1}\hat{\mathbf{M}}_{xx}^{-1}\hat{\sigma}_{vv} + (n^{-1} + d_f^{-1})\hat{\mathbf{M}}_{xx}^{-1}(\hat{\boldsymbol{\Sigma}}_{uu}\hat{\sigma}_{vv} - \hat{\boldsymbol{\Sigma}}_{uv}\hat{\boldsymbol{\Sigma}}_{vu})\hat{\mathbf{M}}_{xx}^{-1}, \quad (2.3.24)$$

where

$$\begin{aligned}
\hat{\boldsymbol{\Sigma}}_{\varepsilon\varepsilon} &= (n - k + d_f)^{-1}[n(\mathbf{M}_{ZZ} - \hat{\mathbf{M}}_{zz}) + d_f\mathbf{S}_{\varepsilon\varepsilon}],\\
\hat{\mathbf{M}}_{zz} &= (\tilde{\boldsymbol{\beta}}, \mathbf{I})'\hat{\mathbf{M}}_{xx}(\tilde{\boldsymbol{\beta}}, \mathbf{I}),\\
\hat{\mathbf{M}}_{xx} &= \hat{\mathbf{H}}_2'(\mathbf{M}_{ZZ} - \mathbf{S}_{\varepsilon\varepsilon})\hat{\mathbf{H}}_2,\\
\hat{\mathbf{H}}_2 &= (\mathbf{0}, \mathbf{I}_k)' - (1, -\tilde{\boldsymbol{\beta}}')'[(1, -\tilde{\boldsymbol{\beta}}')\mathbf{S}_{\varepsilon\varepsilon}(1, -\tilde{\boldsymbol{\beta}}')']^{-1}(1, -\tilde{\boldsymbol{\beta}}')\mathbf{S}_{\varepsilon u},\\
\hat{\sigma}_{vv} &= (n - k + d_f)^{-1}\left[\sum_{t=1}^{n}(Y_t - \mathbf{X}_t\tilde{\boldsymbol{\beta}})^2 + d_f(1, -\tilde{\boldsymbol{\beta}}')\mathbf{S}_{\varepsilon\varepsilon}(1, -\tilde{\boldsymbol{\beta}}')'\right],
\end{aligned}$$

and $\hat{\boldsymbol{\Sigma}}_{uv} = \hat{\boldsymbol{\Sigma}}_{ue} - \hat{\boldsymbol{\Sigma}}_{uu}\tilde{\boldsymbol{\beta}}$. The theoretical basis for the estimator $\hat{\mathbf{V}}\{\tilde{\boldsymbol{\beta}}\}$ is given in Theorem 4.1.5. The estimator of $\boldsymbol{\Sigma}_{\varepsilon\varepsilon}$ is derived in Theorem 4.1.2.

Example 2.3.1. The data in Table 2.3.1 are the means of six trees for log apple crop, log growth in wood, and log growth in girth of Cox's Orange Pippin apple trees for 5 years. Different trees were observed in each year. The data were collected at East Malling Research Station and are taken from Sprent (1969, p. 121ff). Sprent (1966, 1969) suggested that log crop (y), log extension wood growth (x_1), and log girth increment (x_2) would satisfy

TABLE 2.3.1. Crop, wood growth, and girth increment of apple trees

Year	Log Crop (Y)	Log Wood (X_1)	Log Girth (X_2)
1954	1.015	3.442	0.477
1955	1.120	3.180	0.610
1956	1.937	3.943	0.505
1957	1.743	3.982	0.415
1958	2.173	4.068	0.610

Source: Sprent (1969, p. 122).

the model

$$y_t = \beta_0 + \beta_1 x_{t1} + \beta_2 x_{t2}. \tag{2.3.25}$$

The model was put forward on the basis that reproductive and vegetative growth tend to balance each other. We assume that

$$(Y_t, 1, X_{t1}, X_{t2}) = (y_t, x_{t0}, x_{t1}, x_{t2}) + (e_t, u_{t0}, u_{t1}, u_{t2}),$$
$$\boldsymbol{\varepsilon}_t' = (e_t, u_{t0}, u_{t1}, u_{t2})' \sim \text{NI}(\mathbf{0}, \boldsymbol{\Sigma}_{\varepsilon\varepsilon}).$$

The trees are relatively young and the crop is increasing. It seems most reasonable to assume the x_{ti} to be fixed.

The estimator of $\boldsymbol{\Sigma}_{\varepsilon\varepsilon}$ is the matrix that is one-sixth of the sample covariance matrix obtained by pooling the five covariance matrices for the variation among trees within years. This estimator is

$$\mathbf{S}_{\varepsilon\varepsilon} = \begin{bmatrix} 0.5546 & 0 & -0.1079 & -0.0691 \\ 0 & 0 & 0 & 0 \\ -0.1079 & 0 & 0.2756 & 0.1247 \\ -0.0691 & 0 & 0.1247 & 0.0878 \end{bmatrix} 10^{-2}.$$

There is evidence that the within-year covariance matrices are not equal, but for purposes of this example we assume that they are equal and that $\mathbf{S}_{\varepsilon\varepsilon}$ is distributed as a multiple of a Wishart matrix. The moment matrix for the yearly means of Table 2.3.1 is

$$\mathbf{M}_{ZZ} = \begin{bmatrix} 2.75958 & 1.59767 & 6.09486 & 0.83885 \\ 1.59767 & 1.00000 & 3.72293 & 0.52333 \\ 6.09486 & 3.72293 & 13.98190 & 1.94112 \\ 0.83885 & 0.52333 & 1.94112 & 0.27973 \end{bmatrix}.$$

The smallest root of the determinantal equation

$$|\mathbf{M}_{ZZ} - \lambda \mathbf{S}_{\varepsilon\varepsilon}| = 0$$

is $\tilde{\lambda} = 0.2089$ and the F statistic of (2.3.19) is $F = 2.5\tilde{\lambda} = 0.52$. Under the null, the approximate distribution of F is that of Snedecor's F with 2 and 25 degrees of freedom. One easily accepts the hypothesis that the matrix of mean squares and products for the true vector $\mathbf{z}_t$ is singular. The test of the hypothesis that the matrix of mean squares and products for $\mathbf{x}_t$ is singular is given by

$$F = 3^{-1}(4)\tilde{\lambda}_x = 8.08,$$

where $\tilde{\lambda}_x = 6.0605$ is the smallest root of

$$\left|\begin{bmatrix} 15.2168 & -0.8967 \\ -0.8967 & 0.7334 \end{bmatrix} - \lambda_x \begin{bmatrix} 0.2756 & 0.1247 \\ 0.1247 & 0.0878 \end{bmatrix}\right| = 0.$$

Therefore, one is reasonably comfortable with the assumption that $\sum_{t=1}^{n} (\mathbf{x}_t - \bar{\mathbf{x}}_t)'(\mathbf{x}_t - \bar{\mathbf{x}})$ is nonsingular. We calculated the test for singularity using the matrix of corrected mean squares for (X_{t1}, X_{t2}), but the test is equivalent to that computed using the 3×3 matrix $\mathbf{M}_{XX}$. The maximum likelihood estimator of $\boldsymbol{\beta}$ is

$$\begin{aligned} \tilde{\boldsymbol{\beta}} &= (\mathbf{M}_{XX} - \tilde{\lambda}\mathbf{S}_{uu})^{-1}(\mathbf{M}_{XY} - \tilde{\lambda}\mathbf{S}_{ue}) \\ &= (-4.6518, 1.3560, 2.2951)'. \end{aligned}$$

Using (2.3.24), the estimated covariance matrix for $\tilde{\boldsymbol{\beta}}$ is

$$\hat{\mathbf{V}}\{\tilde{\boldsymbol{\beta}}\} = \begin{bmatrix} 1.3417 & -0.2322 & -0.9009 \\ -0.2322 & 0.0517 & 0.0757 \\ -0.9009 & 0.0757 & 1.1828 \end{bmatrix},$$

where $\hat{\sigma}_{vv} = 0.02803$, $\hat{\mathbf{\Sigma}}_{vu} = (0, -0.7406, -0.4241)10^{-2}$,

$$\hat{\mathbf{\Sigma}}_{\varepsilon\varepsilon} = \begin{bmatrix} 0.5773 & 0 & -0.0860 & -0.0574 \\ 0 & 0 & 0 & 0 \\ -0.0860 & 0 & 0.2768 & 0.1217 \\ -0.0574 & 0 & 0.1217 & 0.0879 \end{bmatrix} 10^{-2},$$

$$\hat{\mathbf{M}}_{xx} = \begin{bmatrix} 1.0000 & 3.7229 & 0.5233 \\ 3.7229 & 13.9807 & 1.9408 \\ 0.5233 & 1.9408 & 0.2794 \end{bmatrix},$$

$$\hat{\mathbf{H}}_2 = \begin{bmatrix} 0 & 0.2639 & 0.1511 \\ 1 & 1.2276 & 0.7029 \\ 0 & 0.6422 & -0.2049 \\ 0 & -0.6057 & 0.6532 \end{bmatrix}.$$

This example provides a simple illustration of the computations because of the small number of observations. Also, the large sample theory is adequate because the error variances are small relative to the variation in $\mathbf{x}_t$. In situations where the error variance is small relative to the variation in $\mathbf{x}_t$, the estimated variance of the estimator will be dominated by the term $n^{-1}\hat{\mathbf{M}}_{xx}^{-1}\hat{\sigma}_{vv}$. The fact that $\mathbf{\Sigma}_{\varepsilon\varepsilon}$ is estimated results in a very modest contribution to the estimated error variance in such situations. In this sample the contribution is

$$(25)^{-1}\hat{\mathbf{M}}_{xx}^{-1}(\hat{\mathbf{\Sigma}}_{uu}\hat{\sigma}_{vv} - \hat{\mathbf{\Sigma}}_{uv}\hat{\mathbf{\Sigma}}_{vu})\hat{\mathbf{M}}_{xx}^{-1} = \begin{bmatrix} 0.882 & -0.103 & -0.949 \\ -0.103 & 0.014 & 0.097 \\ -0.949 & 0.097 & 1.121 \end{bmatrix} 10^{-2}.$$

The estimated values for (x_{t1}, x_{t2}) are given in Table 2.3.2. These values were computed by the formula

$$\hat{\mathbf{x}}_t = \bar{\mathbf{X}} + (\mathbf{Z}_t - \bar{\mathbf{Z}})\hat{\mathbf{H}}_2,$$

which is equivalent to (2.3.15). The first entry in the vector $\hat{\mathbf{x}}_t$ is always one in this example and is not included in the table. The estimated covariance matrix for the error in the estimated true values $(\hat{x}_{t1}, \hat{x}_{t2})$ is the lower right 2×2 portion of $\hat{\mathbf{H}}_2'\hat{\mathbf{\Sigma}}_{\varepsilon\varepsilon}\hat{\mathbf{H}}_2$ and is

$$\begin{bmatrix} 0.0811 & 0.0097 \\ 0.0097 & 0.0237 \end{bmatrix}.$$

□ □

Example 2.3.1 illustrates an experimental situation in which measurement error played a modest role. Because the variance of the measurement error is small relative to the variation in the true values, the maximum likelihood estimate of the structural equation is not greatly different from the ordinary least squares estimate. The following example presents the analysis of a large experiment in which measurement error was very important.

TABLE 2.3.2. Estimated true values for apple trees

Year	Log Wood Growth $\hat{x}_1$	Log Girth Increment $\hat{x}_2$	$\hat{v}$
1954	3.417	0.462	−0.094
1955	3.196	0.619	0.060
1956	3.965	0.517	0.082
1957	3.993	0.421	0.043
1958	4.044	0.596	−0.091

Example 2.3.2. We discuss an experiment conducted by the Iowa Agriculture Experiment station at the Doon Experimental Farm in northwest Iowa. The experiment is described by Mowers (1981) and in Mowers, Fuller, and Shrader (1981). The experiment consisted of growing crops in the sequence corn–oats–meadow–meadow with meadow–kill treatments applied to the second-year meadow at various times of the growing season. In the control treatment, the second-year meadow was harvested two or three times. Treatment two was a "short-fallow" treatment, in which second-year meadow was killed with herbicides in the early fall after the second cutting of hay. The third treatment was a longer fallow treatment, with meadow killed in midsummer after the first cutting of hay. All plots were plowed in the spring before corn was planted. Killing the meadow crop increased the amount of soil moisture available to the following corn crop. The longer the fallow period, the more soil moisture was increased relative to treatment one.

Table 2.3.3 contains a portion of the data on corn yield and soil moisture collected in the experiment. The response to moisture is not linear over the full range of observations. The data in Table 2.3.3 are a subset of the data for which linearity is a reasonable approximation. The yield is the yield of corn grain in tens of bushels per acre. The soil moisture is the inches of available moisture in the soil at corn planting time. Table 2.3.4 contains the

TABLE 2.3.3. Treatment means for yield–soil moisture experiment

	Treatment One		Treatment Two		Treatment Three	
Year	Yield	Soil Moisture	Yield	Soil Moisture	Yield	Soil Moisture
1958	4.83	4.58	6.23	6.14	7.48	7.50
1959	4.49	2.04	6.07	2.24	5.13	2.12
1960	10.55	7.67	9.89	8.69	10.02	9.47
1961	7.24	3.37	9.38	4.11	9.95	5.67
1962	10.50	4.40	9.05	3.73	10.08	4.63
1963	3.17	1.16	5.66	4.06	6.00	4.30
1964	9.71	2.94	9.42	2.97	9.94	3.66
1965	10.84	3.64	11.42	4.03	10.80	3.32
1966	2.20	4.42	5.83	5.88	7.69	6.83
1967	0.85	2.36	2.73	4.09	2.37	4.20
1969	14.23	6.01	14.88	5.81	15.46	6.32
1971	3.39	1.31	4.60	1.97	5.07	1.83
1975	5.80	3.63	6.61	3.39	6.25	3.67
1977	7.46	0.87	7.33	0.64	8.55	1.44
Mean	6.80	3.46	7.79	4.12	8.20	4.64

Source: Mowers, Fuller, and Shrader (1981).

TABLE 2.3.4. Analysis of variance and covariance for yield (Y) and soil moisture (X)

Source	Degrees of Freedom	Mean Squares and Products		
		YY	XY	XX
Total	125	11.7907	3.2278	4.8560
Years	13	101.2508	23.8933	36.3441
Treatments	2	21.5797	17.6327	14.7826
Years × treatments	26	2.5389	1.2677	1.3609
Error	84	0.5762	0.2932	0.8284

mean squares and products for yield and soil moisture. Clearly, the treatment of killing the meadow had an effect on soil moisture the following spring and on the yield of the corn crop planted that spring.

Two questions are of interest. First, did the treatments have any effect on yield other than the indirect effect through soil moisture? Second, how much is yield increased by an additional inch of soil moisture?

We analyze the data recognizing that soil moisture is measured with error. We postulate the model

$$y = \beta_0 + \beta_1 x, \qquad (Y, X) = (y, x) + (e, u),$$

where $\boldsymbol{\varepsilon}' \sim \mathrm{NI}(\mathbf{0}, \boldsymbol{\Sigma}_{\varepsilon\varepsilon})$ and $\boldsymbol{\varepsilon} = (e, u)$ is independent of x. We have purposely omitted subscripts. We treat the components of the error line in the analysis of variance as the elements of the matrix $\mathbf{S}_{\varepsilon\varepsilon}$, where $\mathbf{S}_{\varepsilon\varepsilon}$ is an unbiased estimator of $\boldsymbol{\Sigma}_{\varepsilon\varepsilon}$. The means (or multiples of the means) associated with different lines of the analysis of variance will serve as the Y and X observations in the analysis.

In this example we analyze deviations from the mean. All results presented in the text for $\mathbf{M}_{ZZ}$ are applicable for $\mathbf{m}_{ZZ}$ with proper degrees-of-freedom modifications. Under the model, a multiple of the smallest root of the determinantal equation

$$|\mathbf{m}_{ZZ} - \lambda \mathbf{S}_{\varepsilon\varepsilon}| = 0$$

is approximately distributed as an F random variable. If we let the y and x values be the true values of year-by-treatment effects, we compute the smallest root of

$$\left| \begin{bmatrix} 2.5389 & 1.2677 \\ 1.2677 & 1.3609 \end{bmatrix} - \lambda \begin{bmatrix} 0.5762 & 0.2932 \\ 0.2932 & 0.8284 \end{bmatrix} \right| = 0.$$

This root is $\tilde{\lambda} = 1.072$ and $F_{84}^{25} = (26)(25)^{-1}\tilde{\lambda} = 1.11$. We conclude that the observed year-by-treatment effects are consistent with the model.

If we let the y and x values be the true values of treatment means, we compute the smallest root of

$$\left|\begin{bmatrix} 21.5797 & 17.6327 \\ 17.6327 & 14.7826 \end{bmatrix} - \lambda\begin{bmatrix} 0.5762 & 0.2932 \\ 0.2932 & 0.8284 \end{bmatrix}\right| = 0.$$

The smallest root is $\tilde{\lambda} = 0.510$ and $F_{84}^{1} = 2\tilde{\lambda} = 1.02$. We conclude that the treatment means are consistent with the model.

If we let the y and x values be the true values of the year means, we compute the smallest root of

$$\left|\begin{bmatrix} 101.2508 & 23.8933 \\ 23.8933 & 36.3441 \end{bmatrix} - \lambda\begin{bmatrix} 0.5762 & 0.2932 \\ 0.2932 & 0.8284 \end{bmatrix}\right| = 0.$$

The smallest root is $\tilde{\lambda} = 41.749$, $F_{84}^{12} = 13(12)^{-1}\tilde{\lambda} = 45.23$, and the model is rejected for year means. This is not surprising because there are many factors other than soil moisture that change the yearly environment for corn.

To summarize, one can conclude that treatment effects and treatment-by-year effects are due to soil moisture effects created by the treatments. We can accept the hypothesis that treatments had no effect on yield beyond that due to moisture. However, there is variation in the year means that is not associated with variation in soil moisture.

On the basis of these conclusions, we pool the treatment and treatment-by-years sums of squares to estimate the slope of the response of corn yield to moisture. The determinantal equation for the pooled data is

$$\left|\begin{bmatrix} 3.8990 & 2.4366 \\ 2.4366 & 2.3196 \end{bmatrix} - \lambda\begin{bmatrix} 0.5762 & 0.2932 \\ 0.2932 & 0.8284 \end{bmatrix}\right| = 0.$$

The smallest root is $\tilde{\lambda} = 1.1573$ and the associated F statistic is $F_{84}^{27} = 1.2002$. The estimator of β_1 is

$$\tilde{\beta}_1 = (2.3196 - 0.9587)^{-1}(2.4366 - 0.3393) = 1.5411$$

and the estimated variance of the approximate distribution of $\tilde{\beta}_1$, constructed with moments corrected for the mean, is

$$\hat{V}\{\tilde{\beta}_1\} = 0.06083,$$

where $n - 1 = 28$, $d_f = 84$, $\hat{m}_{xx} = \hat{\mathbf{H}}_2'(\mathbf{m}_{ZZ} - \mathbf{S}_{\varepsilon\varepsilon})\hat{\mathbf{H}}_2 = 1.4385$, $\hat{\sigma}_{vv} = 1.7027$,

$$\hat{\mathbf{\Sigma}}_{\varepsilon\varepsilon} = (n + d_f - 2)^{-1}[d_f\mathbf{S}_{\varepsilon\varepsilon} + (n - 1)(\mathbf{m}_{ZZ} - \hat{\mathbf{m}}_{zz})] = \begin{bmatrix} 0.5534 & 0.2753 \\ 0.2753 & 0.8412 \end{bmatrix},$$

$\hat{\mathbf{m}}_{zz} = (\tilde{\beta}_1, 1)'\hat{m}_{xx}(\tilde{\beta}_1, 1)$, $\hat{\sigma}_{uv} = \hat{\sigma}_{ue} - \tilde{\beta}_1\hat{\sigma}_{uu} = -1.0211$, and $\hat{\mathbf{H}}_2' = (0.5997, 0.0758)$.

The error line in the analysis of variance table is not for "pure" measurement error. Because of variations in the topography and in the soil, there are differences in the true available soil moisture of different plots. It follows that the three estimates in the error line of the analysis of variance are the sums of two components, one that arises from variation in true values and one that is due to measurement errors. This fact does not impair the validity of our original measurement error model.

For this experiment it seems reasonable to assume the measurement error associated with the moisture determination to be independent of the plot-to-plot variability and of any measurement error made in determining yield. Under this assumption, the mean cross product on the error line is an estimator of $\beta_1 \sigma_{pp}$, where σ_{pp} is the plot variability in true soil moisture. Therefore, using the estimate from $\hat{\Sigma}_{\varepsilon\varepsilon}$, we have

$$\hat{\sigma}_{pp} = (1.5411)^{-1} 0.2753 = 0.1786.$$

The error mean square for soil moisture is estimating $\sigma_{pp} + \sigma_{mm}$, where σ_{mm} is the variance of pure measurement error. Hence,

$$\hat{\sigma}_{mm} = 0.8412 - 0.1786 = 0.6626.$$

This estimate of measurement error variance is similar to direct estimates obtained in studies such as that of Shaw, Nielsen, and Runkles (1959). (See Exercise 4.6 of Chapter 4.)

For the Doon experiment the ordinary least squares estimate of the effect of an inch of soil moisture computed from the error line of the analysis of variance is 0.354 with a standard error of 0.083. The analysis that recognized the measurement error in soil moisture produced an estimate of 1.541 with a standard error of 0.25. The analysis recognizing measurement error produced an estimate more than four times that of the (incorrect) ordinary least squares procedure. At a price of \$3.00 per bushel this is a difference of \$35.61 in the estimated marginal value of one acre inch of soil moisture. □ □

When the error variances are estimated, the knowledge that some error covariances are zero changes the covariance matrix of the limiting distribution. One model that arises frequently in practice is the model of (2.3.1) with the additional specification that the error covariance matrix is diagonal. Let the diagonal covariance matrix be estimated by

$$\ddot{\mathbf{S}}_{\varepsilon\varepsilon} = \operatorname{diag}(S_{ee}, S_{uu11}, S_{uu22}, \ldots, S_{uukk}),$$

where $E\{\ddot{\mathbf{S}}_{\varepsilon\varepsilon}\} = \boldsymbol{\Sigma}_{\varepsilon\varepsilon}$ and the elements of $\ddot{\mathbf{S}}_{\varepsilon\varepsilon}$ are independently distributed as multiples of chi-square random variables with d_{fi} degrees of freedom, $i = 0, 1, \ldots, k$. Let the estimator $\hat{\boldsymbol{\beta}}$ be defined by (2.3.14) with $\ddot{\mathbf{S}}_{\varepsilon\varepsilon}$ replacing $\boldsymbol{\Upsilon}_{\varepsilon\varepsilon}$ and let $\hat{\lambda}$ be the smallest root of (2.3.12) with $\ddot{\mathbf{S}}_{\varepsilon\varepsilon}$ replacing $\boldsymbol{\Upsilon}_{\varepsilon\varepsilon}$. Given regularity conditions, the estimator of $\boldsymbol{\beta}$ constructed with $\ddot{\mathbf{S}}_{\varepsilon\varepsilon}$ is approximately normally distributed with covariance matrix

$$n^{-1}[\bar{\mathbf{M}}_{xx}^{-1}\sigma_{vv} + \bar{\mathbf{M}}_{xx}^{-1}(\boldsymbol{\Sigma}_{uu}\sigma_{vv} - \boldsymbol{\Sigma}_{uv}\boldsymbol{\Sigma}_{vu})\bar{\mathbf{M}}_{xx}^{-1}]$$
$$+ 2\bar{\mathbf{M}}_{xx}^{-1}\left\{\mathbf{R} + \boldsymbol{\Sigma}_{uv}\boldsymbol{\Sigma}_{vu}\sigma_{vv}^{-2}\left(d_{f0}^{-1}\sigma_{ee}^2 + \sum_{i=1}^{k} d_{fi}^{-1}\beta_i^4\sigma_{uuii}^2\right)\right.$$
$$\left. + \sigma_{vv}^{-1}(\mathbf{R}\boldsymbol{\beta}\boldsymbol{\Sigma}_{vu} + \boldsymbol{\Sigma}_{uv}\boldsymbol{\beta}'\mathbf{R})\right\}\bar{\mathbf{M}}_{xx}^{-1}, \tag{2.3.26}$$

where $\mathbf{R} = \text{diag}(d_{f1}^{-1}\beta_1^2\sigma_{uu11}^2, d_{f2}^{-1}\beta_2^2\sigma_{uu22}^2, \ldots, d_{fk}^{-1}\beta_k^2\sigma_{uukk}^2)$.

In Theorem 2.3.2 the approximate distribution of the smallest root of (2.3.18) was shown to be that of an F random variable with numerator degrees of freedom equal to $n - k$ and denominator degrees of freedom equal to d_f. Therefore, a reasonable approximation for the distribution of $\hat{\lambda}$ based on $\ddot{\mathbf{S}}_{\varepsilon\varepsilon}$ is the F distribution with numerator degrees of freedom equal to $n - k$ and denominator degrees of freedom determined by the variance introduced by the estimation of $\boldsymbol{\Sigma}_{\varepsilon\varepsilon}$ with $\ddot{\mathbf{S}}_{\varepsilon\varepsilon}$. Recall that the variance of the F distribution is

$$[\nu_1(\nu_2 - 2)^2(\nu_2 - 4)]^{-1}2\nu_2^2(\nu_1 + \nu_2 - 2), \qquad \nu_2 > 4,$$

where ν_1 is the numerator degrees of freedom and ν_2 is the denominator degrees of freedom. See Kendall and Stuart (1977, Vol. I, p. 406). For large ν_2 the variance is approximately $2\nu_1^{-1} + 2\nu_2^{-1}$.

It follows that an approximation for the distribution of $n(n - k)^{-1}\hat{\lambda}$ based on $\ddot{\mathbf{S}}_{\varepsilon\varepsilon}$ is that of the F distribution with $n - k$ and ν_2 degrees of freedom where ν_2 is estimated by

$$\hat{\nu}_2 = \left(d_{f0}^{-1}S_{ee}^2 + \sum_{i=1}^{k} d_{fi}^{-1}\hat{\beta}_i^4 S_{uuii}^2\right)^{-1}\left(S_{ee} + \sum_{i=1}^{k}\hat{\beta}_i^2 S_{uuii}\right)^2. \tag{2.3.27}$$

Example 2.3.3. We study further the data of Example 2.2.1. We are now in a position to construct a second test of the hypothesis that $\sigma_{qq} = 0$. Let $\hat{\lambda}$ be the smallest root of

$$|\mathbf{M}_{ZZ} - \lambda\ddot{\mathbf{S}}_{\varepsilon\varepsilon}| = 0, \tag{2.3.28}$$

where

$$\ddot{\mathbf{S}}_{\varepsilon\varepsilon} = \text{diag}(0.0037, 0.0, 0.0203, 0.0438, 0.0180, 0.0)$$

and the elements of $\mathbf{Z}_t$ are the observations on role performance, $X_{t1} \equiv 1$, knowledge, value orientation, role satisfaction, and past training, respectively. If $\sigma_{qq} = 0$, $\hat{\lambda}$ is approximately distributed as an F random variable with 50 and ν_2 degrees of freedom, where $\hat{\nu}_2$ is given in (2.3.27). For the data of

Example 2.2.1, the smallest root of (2.3.28) is $\hat{\lambda} = 1.46$ and $F = 1.61$, where the numerator degrees of freedom is $n - k = 50$. Using (2.3.27) we have $\hat{\nu}_2 = 243$, where

$$\hat{\boldsymbol{\beta}}' = (-1.396, 0.506, 0.157, 0.126, -0.0028),$$

is the vector of estimates computed under the assumption $q_t \equiv 0$. Because 1.61 is close to the 1% tabular point of the F distribution with 50 and 243 degrees of freedom, we reject the hypothesis that $\sigma_{qq} = 0$ and accept $\tilde{\sigma}_{qq} = 0.0054$ as an estimate of σ_{qq}.

Because we reject the hypothesis that the covariance matrix of $(y_t, \mathbf{x}_t)$ is singular, it is clear that we also reject the hypothesis that the covariance matrix of $\mathbf{x}_t$ is singular. Nevertheless, we illustrate the computations by computing a test of the hypothesis that the covariance matrix of $\mathbf{x}_t$ is singular. The smallest root of

$$|\mathbf{M}_{XX} - \lambda \ddot{\mathbf{S}}_{uu}| = 0$$

is $\hat{\lambda} = 1.75$ and the F statistic is $55(51)^{-1}\hat{\lambda} = 1.89$. The vector $\hat{\boldsymbol{\theta}}$ satisfying

$$(\mathbf{M}_{XX} - \hat{\lambda}\ddot{\mathbf{S}}_{uu})\hat{\boldsymbol{\theta}} = \mathbf{0}$$

is $\hat{\boldsymbol{\theta}}' = (-0.371, -1.0, 1.935, 0.595, -0.920)$, where the first entry in $\mathbf{X}_t$ is always one and the other entries are for knowledge, value orientation, role satisfaction, and past training. From (2.3.27), the estimated degrees of freedom for the denominator of the F statistic is $\hat{\nu}_2 = 129$. The 1% tabular value of F with 50 and 129 degrees of freedom is 1.68 and the hypothesis that the covariance matrix of x_t is singular is rejected at that level. □ □

2.3.2. The Structural Model

The majority of the results developed for the estimator of $\boldsymbol{\beta}$ for the model with fixed $\mathbf{x}_t$ are also appropriate for the model with random $\mathbf{x}_t$. If

$$\boldsymbol{\Sigma}_{\varepsilon\varepsilon} = \boldsymbol{\Upsilon}_{\varepsilon\varepsilon}\sigma^2,$$

where $\boldsymbol{\Upsilon}_{\varepsilon\varepsilon}$ is known, the maximum likelihood estimator of $\boldsymbol{\beta}$ for the structural model is that given in Theorem 2.3.1 for the functional model. See Theorem 4.1.1 and Exercise 4.11. The maximum likelihood estimator for the structural model with $\boldsymbol{\Sigma}_{\varepsilon\varepsilon}$ estimated by $\mathbf{S}_{\varepsilon\varepsilon}$ is given in Theorem 4.1.2.

The sample moment matrix $\mathbf{m}_{xx}$ will converge to $\boldsymbol{\Sigma}_{xx}$ almost surely as $n \to \infty$, for any x distribution with finite second moments. Therefore, as $n \to \infty$, the limiting distribution of $\boldsymbol{\Gamma}_v^{-1/2}(\tilde{\boldsymbol{\beta}} - \boldsymbol{\beta})$ given in Theorem 2.3.2 holds for any such x distribution. Also, the variance of the approximate distribution of $\tilde{\boldsymbol{\beta}}$ can be estimated by the expression given in (2.3.24) for such models.

2.3.3. Higher-Order Approximations for Residuals and True Values

In this section we present an estimator for the variance of the estimated true values, $\hat{\mathbf{x}}_t$, that recognizes the fact that an estimator of $\boldsymbol{\beta}$ is used to construct $\hat{\mathbf{x}}_t$. We asssume

$$y_t = \mathbf{x}_t\boldsymbol{\beta}, \qquad \mathbf{Z}_t = \mathbf{z}_t + \boldsymbol{\varepsilon}_t, \qquad (2.3.29)$$
$$\boldsymbol{\varepsilon}_t' \sim \mathrm{NI}(\mathbf{0}, \boldsymbol{\Sigma}_{\varepsilon\varepsilon}),$$

where $\mathbf{Z}_t = (Y_t, \mathbf{X}_t)$, $\boldsymbol{\Sigma}_{\varepsilon\varepsilon}$ is known, and the $\mathbf{x}_t$, $t = 1, 2, \ldots$, are fixed. The estimator of the vector of true values is

$$\hat{\mathbf{z}}_t = \mathbf{Z}_t - \hat{v}_t\hat{\boldsymbol{\delta}}', \qquad (2.3.30)$$

where $\hat{v}_t = \mathbf{Z}_t\hat{\boldsymbol{\alpha}}$, $\hat{\boldsymbol{\alpha}}' = (1, -\hat{\boldsymbol{\beta}}')$, $\hat{\boldsymbol{\delta}}' = \hat{\sigma}_{vv}^{-1}\hat{\boldsymbol{\Sigma}}_{v\varepsilon}$, $\hat{\sigma}_{vv} = \hat{\boldsymbol{\alpha}}'\boldsymbol{\Sigma}_{\varepsilon\varepsilon}\hat{\boldsymbol{\alpha}}$, $\hat{\boldsymbol{\Sigma}}_{v\varepsilon} = \hat{\boldsymbol{\alpha}}'\boldsymbol{\Sigma}_{\varepsilon\varepsilon}$, and $\hat{\boldsymbol{\beta}}$ is defined by (2.3.14) with $\boldsymbol{\Sigma}_{\varepsilon\varepsilon}$ replacing $\boldsymbol{\Upsilon}_{\varepsilon\varepsilon}$. Using (2.3.30), $\hat{\boldsymbol{\alpha}} - \boldsymbol{\alpha} = O_p(n^{-1/2})$, and $\hat{\boldsymbol{\delta}} - \boldsymbol{\delta} = O_p(n^{-1/2})$, we have

$$\hat{\mathbf{z}}_t' - \mathbf{z}_t' = \ddot{\mathbf{z}}_t' - \mathbf{z}_t' - (\hat{\boldsymbol{\delta}} - \boldsymbol{\delta})v_t - (\hat{v}_t - v_t)\boldsymbol{\delta} + O_p(n^{-1}),$$

where $\ddot{\mathbf{z}}_t = \mathbf{Z}_t - v_t\boldsymbol{\delta}'$, $\boldsymbol{\delta}' = \sigma_{vv}^{-1}\boldsymbol{\Sigma}_{v\varepsilon}$,

$$\hat{\boldsymbol{\delta}} - \boldsymbol{\delta} = (\sigma_{vv}^{-1}\boldsymbol{\Sigma}_{\varepsilon\varepsilon} - 2\boldsymbol{\delta}\boldsymbol{\delta}')(\hat{\boldsymbol{\alpha}} - \boldsymbol{\alpha}) + O_p(n^{-1}),$$

and $\hat{v}_t - v_t = \mathbf{Z}_t(\hat{\boldsymbol{\alpha}} - \boldsymbol{\alpha})$. Using the definition of $\ddot{\mathbf{z}}_t$ we obtain

$$\hat{\mathbf{z}}_t' = \ddot{\mathbf{z}}_t' - [\boldsymbol{\delta}\ddot{\mathbf{z}}_t + \sigma_{vv}^{-1}v_t(\boldsymbol{\Sigma}_{\varepsilon\varepsilon} - \boldsymbol{\delta}\boldsymbol{\Sigma}_{v\varepsilon})](\hat{\boldsymbol{\alpha}} - \boldsymbol{\alpha}) + O_p(n^{-1}).$$

It follows that the covariance matrix of the approximate distribution of $\hat{\mathbf{z}}_t - \mathbf{z}_t$ is

$$\begin{aligned} \mathbf{V}\{\hat{\mathbf{z}}_t - \mathbf{z}_t\} &= \boldsymbol{\Sigma}_{\varepsilon\varepsilon} - \boldsymbol{\delta}\boldsymbol{\Sigma}_{v\varepsilon} + \boldsymbol{\delta}\boldsymbol{\delta}'E\{\ddot{\mathbf{z}}_t\mathbf{V}_{\alpha\alpha}\ddot{\mathbf{z}}_t'\} \\ &\quad + \sigma_{vv}^{-1}(\boldsymbol{\Sigma}_{\varepsilon\varepsilon} - \boldsymbol{\delta}\boldsymbol{\Sigma}_{v\varepsilon})\mathbf{V}_{\alpha\alpha}(\boldsymbol{\Sigma}_{\varepsilon\varepsilon} - \boldsymbol{\delta}\boldsymbol{\Sigma}_{v\varepsilon}), \end{aligned} \qquad (2.3.31)$$

where $\mathbf{V}_{\alpha\alpha}$ is the covariance matrix of the approximate distribution of $\hat{\boldsymbol{\alpha}}$ and we used the independence of v_t and $\ddot{\mathbf{z}}_t$ in deriving the result. Replacing parameters in (2.3.31) with their sample analogues, an estimator of the variance of the approximate distribution of $\hat{\mathbf{z}}_t - \mathbf{z}_t$ is

$$\begin{aligned} \tilde{\mathbf{V}}\{\hat{\mathbf{z}}_t - \mathbf{z}_t\} &= \boldsymbol{\Sigma}_{\varepsilon\varepsilon} - \hat{\boldsymbol{\delta}}\hat{\boldsymbol{\Sigma}}_{v\varepsilon} + \hat{\boldsymbol{\delta}}\hat{\boldsymbol{\delta}}'\hat{\mathbf{x}}_t\hat{\mathbf{V}}_{\beta\beta}\hat{\mathbf{x}}_t' \\ &\quad + \hat{\sigma}_{vv}^{-1}(\boldsymbol{\Sigma}_{\varepsilon\varepsilon} - \hat{\boldsymbol{\delta}}\hat{\boldsymbol{\Sigma}}_{v\varepsilon})\hat{\mathbf{V}}_{\alpha\alpha}(\boldsymbol{\Sigma}_{\varepsilon\varepsilon} - \hat{\boldsymbol{\delta}}\hat{\boldsymbol{\Sigma}}_{v\varepsilon}). \end{aligned} \qquad (2.3.32)$$

To develop an approximation for the covariance matrix of the limiting distribution of the estimated v_t, let

$$\hat{\mathbf{v}} = (\hat{v}_1, \hat{v}_2, \ldots, \hat{v}_n)' \quad \text{and} \quad \mathbf{Z} = (\mathbf{Z}_1', \mathbf{Z}_2', \ldots, \mathbf{Z}_n')'.$$

Then

$$\begin{aligned}\hat{\mathbf{v}} &= \mathbf{v} - \mathbf{X}(\hat{\boldsymbol{\beta}} - \boldsymbol{\beta}) \\ &= \mathbf{v} - \mathbf{X}\mathbf{M}_{xx}^{-1}\left[n^{-1}\sum_{t=1}^{n}(\mathbf{X}_t'v_t - v_t^2\boldsymbol{\delta}_1)\right] + O_p(n^{-1}) \\ &= \mathbf{v} - n^{-1}(\ddot{\mathbf{x}} + \boldsymbol{\xi})\mathbf{M}_{xx}^{-1}\ddot{\mathbf{x}}'\mathbf{v} + O_p(n^{-1}),\end{aligned}$$

where $\boldsymbol{\delta}' = (\sigma_{vv}^{-1}\sigma_{ve}, \boldsymbol{\delta}_1')$, $\ddot{\mathbf{x}} = \mathbf{X} - \boldsymbol{\xi}$, and $\boldsymbol{\xi}_t = v_t\boldsymbol{\delta}_1'$. Using the fact that $\ddot{\mathbf{x}}$ and $\mathbf{v}$ are independent, we have

$$\begin{aligned}&E\{[\mathbf{v} - n^{-1}(\ddot{\mathbf{x}} + \boldsymbol{\xi})\mathbf{M}_{xx}^{-1}\ddot{\mathbf{x}}'\mathbf{v}][\mathbf{v} - n^{-1}(\ddot{\mathbf{x}} + \boldsymbol{\xi})\mathbf{M}_{xx}^{-1}\ddot{\mathbf{x}}'\mathbf{v}]'\} \\ &\qquad = \mathbf{I}_n\sigma_{vv} - 2n^{-1}E\{\ddot{\mathbf{x}}\mathbf{M}_{xx}^{-1}\ddot{\mathbf{x}}'\}\sigma_{vv} + E\{\ddot{\mathbf{x}}\mathbf{V}_{\beta\beta}\ddot{\mathbf{x}}'\} + \mathbf{I}_n\operatorname{tr}\{\mathbf{V}_{\beta\beta}\mathbf{V}_{\xi\xi}\},\end{aligned}$$

where $\mathbf{V}_{\xi\xi} = E\{\boldsymbol{\xi}_t'\boldsymbol{\xi}_t\} = \boldsymbol{\Sigma}_{uv}\boldsymbol{\delta}_1'$ and $\mathbf{V}_{\beta\beta}$ is the variance of the approximate distribution of $\hat{\boldsymbol{\beta}}$. It follows that the covariance matrix of the approximate distribution of $\hat{\mathbf{v}}$ is

$$\mathbf{I}_n(\sigma_{vv} + \boldsymbol{\delta}_1'\mathbf{V}_{\beta\beta}\boldsymbol{\Sigma}_{uv}) + E\{\ddot{\mathbf{x}}(\mathbf{V}_{\beta\beta} - 2n^{-1}\mathbf{M}_{xx}^{-1}\sigma_{vv})\ddot{\mathbf{x}}'\}.$$

The residuals divided by an estimate of the standard error of the residual are of interest for diagnostic checking. We have

$$\begin{aligned}s_{vv}^{-1/2} &= A_{vv}^{-1/2} - \tfrac{1}{2}A_{vv}^{-3/2}[-2\mathbf{v}'\boldsymbol{\xi}(\hat{\boldsymbol{\beta}} - \boldsymbol{\beta})] + O_p(n^{-1}) \\ &= A_{vv}^{-1/2}[1 + \boldsymbol{\delta}_1'(\hat{\boldsymbol{\beta}} - \boldsymbol{\beta})] + O_p(n^{-1}),\end{aligned}$$

where

$$\begin{aligned}A_{vv} &= (n-k)^{-1}[\mathbf{v} - \ddot{\mathbf{x}}(\hat{\boldsymbol{\beta}} - \boldsymbol{\beta})]'[\mathbf{v} - \ddot{\mathbf{x}}(\hat{\boldsymbol{\beta}} - \boldsymbol{\beta})] \\ &= (n-k)^{-1}\mathbf{v}'[\mathbf{I}_n - \ddot{\mathbf{x}}(\mathbf{x}'\mathbf{x})^{-1}\ddot{\mathbf{x}}']^2\mathbf{v} + O_p(n^{-1}).\end{aligned}$$

It follows that

$$\begin{aligned}s_{vv}^{-1/2}\hat{\mathbf{v}} &= A_{vv}^{-1/2}[\mathbf{v} - \mathbf{X}(\hat{\boldsymbol{\beta}} - \boldsymbol{\beta})][1 + \boldsymbol{\delta}_1'(\hat{\boldsymbol{\beta}} - \boldsymbol{\beta})] + O_p(n^{-1}) \\ &= A_{vv}^{-1/2}[\mathbf{v} - \ddot{\mathbf{x}}(\hat{\boldsymbol{\beta}} - \boldsymbol{\beta})] + O_p(n^{-1}) \\ &= A_{vv}^{-1/2}[\mathbf{I}_n - \ddot{\mathbf{x}}(\mathbf{x}'\mathbf{x})^{-1}\ddot{\mathbf{x}}']\mathbf{v} + O_p(n^{-1}).\end{aligned} \tag{2.3.33}$$

If $\boldsymbol{\Sigma}_{\varepsilon\varepsilon}$ is known and σ_{vv} is estimated with $\hat{\sigma}_{vv} = (1, -\hat{\boldsymbol{\beta}}')\boldsymbol{\Sigma}_{\varepsilon\varepsilon}(1, -\hat{\boldsymbol{\beta}}')'$, then

$$\begin{aligned}\hat{\sigma}_{vv}^{-1/2}\hat{\mathbf{v}} &= \sigma_{vv}^{-1/2}[\mathbf{v} - \mathbf{X}(\hat{\boldsymbol{\beta}} - \boldsymbol{\beta})][1 + \boldsymbol{\delta}_1'(\hat{\boldsymbol{\beta}} - \boldsymbol{\beta})] + O_p(n^{-1}) \\ &= \sigma_{vv}^{-1/2}[\mathbf{I}_n - \ddot{\mathbf{x}}(\mathbf{x}'\mathbf{x})^{-1}\ddot{\mathbf{x}}']\mathbf{v} + O_p(n^{-1}).\end{aligned} \tag{2.3.34}$$

The leading term of expression (2.3.33) differs from the corresponding expression for the ordinary least squares residuals associated with the regression of v_t on $\ddot{\mathbf{x}}_t$ in that $(\mathbf{x}'\mathbf{x})^{-1}$ replaces $(\ddot{\mathbf{x}}'\ddot{\mathbf{x}})^{-1}$. An estimator of $[\mathbf{I}_n - \ddot{\mathbf{x}}(\mathbf{x}'\mathbf{x})^{-1}\ddot{\mathbf{x}}']^2$

is

$$\mathbf{I}_n + \hat{\mathbf{x}}(\hat{\mathbf{V}}_{\beta\beta}s_{vv}^{-1} - 2n^{-1}\hat{\mathbf{M}}_{xx}^{-1})\hat{\mathbf{x}}'.$$

Consequently, the standardized residuals $[\text{diag } \hat{\mathbf{V}}\{\hat{\mathbf{v}}\}]^{-1/2}\hat{\mathbf{v}}$, where

$$\hat{\mathbf{V}}\{\hat{\mathbf{v}}\} = s_{vv}\mathbf{I}_n + \hat{\mathbf{x}}(\hat{\mathbf{V}}_{\beta\beta} - 2n^{-1}\hat{\mathbf{M}}_{xx}^{-1}s_{vv})\hat{\mathbf{x}}' \tag{2.3.35}$$

and diag $\hat{\mathbf{V}}\{\hat{\mathbf{v}}\}$ is the diagonal matrix composed of the diagonal elements of $\hat{\mathbf{V}}\{\hat{\mathbf{v}}\}$, will have a distribution approximated by that of the standardized ordinary least squares residuals. See Miller (1986).

As noted in Section 2.2.3, the diagnostic statistics computed by replacing the dependent variable and vector of independent variables by $\hat{v}_t$ and $\hat{\mathbf{x}}_t$, respectively, in the ordinary least squares forms will behave, approximately, as the ordinary least squares statistics. The smaller the measurement variance relative to the variation in $\mathbf{x}_t$, the better the approximation.

Example 2.3.4. We use the results of this section to compute the approximate variances for the $\hat{x}_t$ values of Example 1.3.2. If the number of cells forming rosettes is expressed as a deviation from the mean, the estimated model becomes

$$\hat{y}_t = 3.5748 + 0.6079(x_t - 12.7374).$$

In this parameterization $\mathbf{Z}_t = (Y_t, 1, X_t - 12.7374)$,

$$\hat{\alpha}' = (1, -3.5748, -0.6079), \qquad \hat{\delta}' = (0.3424)^{-1}(0.25, 0, -0.1520),$$
$$\hat{\mathbf{V}}_{\beta\beta} = \hat{\mathbf{V}}\{(\bar{v}, \hat{\beta}_1)'\} = \text{diag}\{0.06848, 0.00742\},$$
$$n\hat{\mathbf{M}}_{xx} = \text{diag}\{5, 46.8768\}, \quad \text{and} \quad \boldsymbol{\Sigma}_{\varepsilon\varepsilon} = \text{diag}\{0.25, 0, 0.25\}.$$

Then the first-order approximation to the covariance matrix of $\hat{\mathbf{z}}_t$ is

$$\hat{\boldsymbol{\Sigma}}_{\varepsilon\varepsilon} - \hat{\delta}\hat{\boldsymbol{\Sigma}}_{v\varepsilon} = \begin{bmatrix} 0.0675 & 0 & 0.1110 \\ 0 & 0 & 0 \\ 0.1110 & 0 & 0.1825 \end{bmatrix}.$$

The $(\hat{y}_t, \hat{x}_t)$ portion of this matrix is singular, because, to the first-order of approximation, $\hat{y}_t$ is a linear function of $\hat{x}_t$. The second-order approximation to the variance of $\hat{\mathbf{z}}_t$ is

$$\begin{bmatrix} 0.0677 & 0 & 0.1114 \\ 0 & 0 & 0 \\ 0.1114 & 0 & 0.1833 \end{bmatrix} + \begin{bmatrix} 0.5330 & 0 & -0.3241 \\ 0 & 0 & 0 \\ -0.3241 & 0 & 0.1970 \end{bmatrix} \hat{\mathbf{x}}_t\hat{\mathbf{V}}_{\beta\beta}\hat{\mathbf{x}}_t',$$

where the first matrix is

$$\boldsymbol{\Sigma}_{\varepsilon\varepsilon} - \hat{\delta}\hat{\boldsymbol{\Sigma}}_{v\varepsilon} + \hat{\sigma}_{vv}^{-1}(\boldsymbol{\Sigma}_{\varepsilon\varepsilon} - \hat{\delta}\hat{\boldsymbol{\Sigma}}_{v\varepsilon})\hat{\mathbf{V}}_{\alpha\alpha}(\boldsymbol{\Sigma}_{\varepsilon\varepsilon} - \hat{\delta}\hat{\boldsymbol{\Sigma}}_{v\varepsilon}),$$

TABLE 2.3.5. Estimated variances of $(\hat{y}_t, \hat{x}_t)$ and $\hat{v}_t$ for the cell data

t	$\hat{x}_t - \bar{X}$	$\tilde{V}\{\hat{x}_t\}$	$\tilde{V}\{\hat{y}_t\}$	$\tilde{V}\{\hat{v}_t\}$	$[\tilde{V}\{\hat{v}_t\}]^{-1/2}\hat{v}_t$
1	5.7189	0.2446	0.2336	0.0388	1.11
2	−1.1296	0.1986	0.1093	0.2647	−1.17
3	0.4886	0.1971	0.1052	0.2721	−0.34
4	−2.0303	0.2028	0.1205	0.2442	−0.29
5	−3.0445	0.2103	0.1409	0.2072	1.54

the second matrix is $\hat{\boldsymbol{\delta}}\hat{\boldsymbol{\delta}}'$, and $\hat{\mathbf{x}}_t = (1, \hat{x}_t - 12.7374)$. The estimated variances are given in Table 2.3.5. In this example, the variance of u_1 is 0.25 and the estimated variance of $\hat{x}_1$ is close to that value. Figure 1.3.2 provides the explanation. The first observation is separated from the other observations by a considerable distance. Therefore, the least squares procedure will always produce a line that is close to the observation at $t = 1$. Because the X_1 and $\hat{x}_1$ values will be close together, the variance of $\hat{x}_1$ will be close to that of X_1. Observations two through five are closer to the sample mean than observation one, and the first-order and second-order approximations to the variance of $\hat{x}_t$ are in better agreement for these observations than are the two variance approximations for $\hat{x}_1$.

The estimator of the covariance matrix of the approximate distribution of $\mathbf{v}$ is

$$0.3424\mathbf{I} + \hat{\mathbf{x}}(\hat{\mathbf{V}}_{\beta\beta} - 2n^{-1}\hat{\mathbf{M}}_{xx}^{-1}\hat{\sigma}_{vv})\hat{\mathbf{x}}',$$

where $\hat{\sigma}_{vv} = 0.3424$,

$$\hat{\mathbf{V}}_{\beta\beta} - 2n^{-1}\hat{\mathbf{M}}_{xx}^{-1}\hat{\sigma}_{vv} = \text{diag}(-0.06848, \ -0.00718),$$

and the $\hat{x}$ values are expressed as deviations from the mean. The estimated variances of the $\hat{v}_t$ are given in Table 2.3.5. The estimated variance for $\hat{v}_1$ is small for the same reason that the estimated variance of $\hat{x}_1$ is large. The sum of squares of the standardized residuals of the last column of Table 2.3.5 is approximately equal to n, as it should be. □ □

REFERENCES

Anderson (1951b), Cox (1976), Fuller (1977, 1980, 1985), Koopmans (1937), Miller (1986), Nussbaum (1977, 1978), Sprent (1966, 1969), Takemura, Momma, and Takeuchi (1985), Tintner (1945, 1952), Tukey (1951).

EXERCISES

10. (Section 2.3.1)

(a) Verify that the estimator (2.3.15) is the least squares estimator of $\mathbf{z}_t$ obtained from the observation $\mathbf{Z}_t$ under the assumptions that

$$\mathbf{z}_t\hat{\boldsymbol{\alpha}} = 0, \qquad \mathbf{Z}_t = \mathbf{z}_t + \boldsymbol{\varepsilon}_t,$$

where $E\{\boldsymbol{\varepsilon}_t\} = 0$ and $E\{\boldsymbol{\varepsilon}_t'\boldsymbol{\varepsilon}_t\} = \boldsymbol{\Upsilon}_{\varepsilon\varepsilon}\sigma^2$. First, assume that $\boldsymbol{\Upsilon}_{\varepsilon\varepsilon}$ is nonsingular. Next, extend the estimator to singular $\boldsymbol{\Upsilon}_{\varepsilon\varepsilon}$.

(b) Show that the estimator $\hat{\mathbf{V}}\{\hat{\boldsymbol{\beta}}\}$ given in (2.3.24) for $d_f^{-1} = 0$ and $\mathbf{S}_{\varepsilon\varepsilon} = \boldsymbol{\Sigma}_{\varepsilon\varepsilon}$ can be written

$$\hat{\mathbf{V}}\{\hat{\boldsymbol{\beta}}\} = \mathbf{M}_{xx}^{-1}\hat{\boldsymbol{\Sigma}}_{\eta\eta}\mathbf{M}_{xx}^{-1}s_{vv},$$

where $\hat{\mathbf{x}}_t = \mathbf{X}_t - \hat{v}_t\hat{\boldsymbol{\delta}}_x'$, $\hat{\boldsymbol{\delta}}_x' = [(1, -\hat{\boldsymbol{\beta}}')\hat{\boldsymbol{\Sigma}}_{\varepsilon\varepsilon}(1 - \hat{\boldsymbol{\beta}}')']^{-1}(1, -\hat{\boldsymbol{\beta}}')\hat{\boldsymbol{\Sigma}}_{\varepsilon u}$, $\hat{\boldsymbol{\Sigma}}_{\eta\eta} = n^{-1}\sum_{t=1}^{n}\hat{\mathbf{x}}_t'\hat{\mathbf{x}}_t$, and $\hat{\mathbf{M}}_{xx}$ and $\hat{\boldsymbol{\Sigma}}_{\varepsilon\varepsilon}$ are defined in (2.3.24).

11. (Section 2.3.1) Show that $\mathbf{I} - (\hat{\boldsymbol{\alpha}}'\boldsymbol{\Upsilon}_{\varepsilon\varepsilon}\hat{\boldsymbol{\alpha}})^{-1}\hat{\boldsymbol{\alpha}}\hat{\boldsymbol{\alpha}}'\boldsymbol{\Upsilon}_{\varepsilon\varepsilon}$ of (2.3.15) is idempotent. What is the rank of $\mathbf{I} - (\hat{\boldsymbol{\alpha}}'\boldsymbol{\Upsilon}_{\varepsilon\varepsilon}\hat{\boldsymbol{\alpha}})^{-1}\hat{\boldsymbol{\alpha}}\hat{\boldsymbol{\alpha}}'\boldsymbol{\Upsilon}_{\varepsilon\varepsilon}$?

12. (Section 2.3.1) The data in the table were generated by the model

$$y_t = \beta_0 + \beta_1 x_{t1} + \beta_2 x_{t2},$$
$$(Y_t, X_{t1}, X_{t2}) = (y_t, x_{t1}, x_{t2}) + (e_t, u_{t1}, u_{t2}),$$

Observation	Y_t	X_{t1}	X_{t2}
1	0.6	7.8	6.5
2	4.0	8.4	7.3
3	14.5	8.3	12.9
4	15.6	12.1	9.8
5	13.8	8.4	10.9
6	6.4	8.9	10.4
7	3.7	7.3	9.8
8	5.9	9.1	8.6
9	7.2	8.4	9.6
10	16.1	9.7	11.2
11	9.6	11.5	9.6
12	1.5	8.7	7.1
13	13.0	11.8	11.7
14	8.3	10.9	9.0
15	11.1	12.6	12.9
16	8.3	8.0	10.1
17	13.3	10.8	8.7
18	11.3	14.5	16.0
19	2.0	8.2	8.6
20	12.3	10.7	10.9

where $(e_t, u_{t1}, u_{t2})' \sim \text{NI}(\mathbf{0}, \mathbf{I}\sigma^2)$. Estimate the parameters of the model. Test the hypothesis that the matrix of sums of squares and products of the true values is of full rank. Test the hypothesis that $\sigma^2 = 1$. Estimate the variance of $\hat{\beta}_1 + \hat{\beta}_2$. Plot $\hat{v}_t$ against $\hat{x}_{t1}$ and against $\hat{x}_{t2}$.

13. (Sections 2.3.1, 1.3.4) Extend the method of Section 1.3.4 to vector $\mathbf{x}_t$ by using the fact that v_t is independent of

$$\ddot{\mathbf{x}}_t = \mathbf{X}_t - v_t\sigma_{vv}^{-1}\boldsymbol{\Sigma}_{uv}.$$

Hence, the F statistic calculated for the null hypothesis that the coefficient vector is zero in the regression of v_t on $\ddot{\mathbf{x}}_t$ is distributed as Snedecor's F. If the covariance matrix $\boldsymbol{\Sigma}_{\varepsilon\varepsilon}$ is known, the test statistic can be calculated as a chi-square. Use this result to test the hypothesis that $(\beta_1, \beta_2) = (2, 3)$ against the alternative that both coefficients are positive for the model of Example 2.3.1.

14. (Section 2.3.1) Sprent (1969) suggested that the vector (β_1, β_2) for the model of Example 2.3.1 might be equal to $(1, 2)$.
   (a) Using the estimated covariance matrix of the approximate distribution of $(\hat{\beta}_1, \hat{\beta}_2)$ of Example 2.3.1, test the hypothesis $(\beta_1, \beta_2) = (1, 2)$.
   (b) Construct the likelihood ratio test of the hypothesis $(\beta_1, \beta_2) = (1, 2)$.
   (c) Assume that $\beta_2 = 2\beta_1$ and estimate the equation subject to this restriction. (Hint: Transform the $\mathbf{X}_t$ vector.)
   (d) Give the estimated covariance matrix of the approximate distribution for your restricted estimator of part (c).

15. (Section 2.3.1) Assume the model

$$Y_t = \beta_0 + \beta_1 x_t + e_t, \qquad X_t = x_t + u_t,$$
$$(e_t, u_t)' \sim \mathrm{NI}(\mathbf{0}, \mathbf{I}\sigma_{uu}).$$

Assume that for a sample of 100 observations we observe

$$(m_{YY}, m_{XY}, m_{XX}) = (5.00, 0.01, 0.98).$$

What do you conclude about the unknown parameters if it is known that $\sigma_{uu} = 1$? Does your conclusion change if σ_{uu} is unknown?

16. (Section 2.3) Construct an analysis of variance table for the Reilly–Pantino-Leal data of Exercise 1.14. Assume that the data satisfy the model

$$Y_{ij} = \beta_0 + \beta_1 x_i + e_{ij}, \qquad X_{ij} = x_i + u_{ij},$$

where $(e_{ij}, u_{ij})' \sim \mathrm{NI}(\mathbf{0}, \boldsymbol{\Sigma}_{\varepsilon\varepsilon})$, (x_1, x_2, x_3) is fixed, the covariance matrix of (e_{ij}, u_{ij}) is unknown, i denotes condition, and j denotes replicate within condition. Estimate $(\beta_0, \beta_1, \sigma_{ee}, \sigma_{eu}, \sigma_{uu})$. Compute the estimate of the covariance matrix of the approximate distribution of $(\hat{\beta}_0, \hat{\beta}_1)'$. Compute the estimated covariance matrix of the approximate distribution of $(\hat{\sigma}_{ee}, \hat{\sigma}_{eu}, \hat{\sigma}_{uu})$. Do you feel the large sample approximation will perform well? Estimate (x_1, x_2, x_3) treating these quantities as fixed. Hint: Note that the means $(\bar{Y}_{i.}, \bar{X}_{i.})$ satisfy the model

$$r_i^{1/2}\bar{Y}_{i.} = r_i^{1/2}\beta_0 + r_i^{1/2}\bar{x}_{i.}\beta_1 + r_i^{1/2}\bar{e}_{i.},$$
$$r_i^{1/2}\bar{X}_{i.} = r_i^{1/2}\bar{x}_{i.} + r_i^{1/2}\bar{u}_{i.},$$

$r_i^{1/2}(\bar{e}_{i.}, \bar{u}_{i.})' \sim \mathrm{NI}(\mathbf{0}, \boldsymbol{\Sigma}_{\varepsilon\varepsilon})$, where r_i is the number of replicates.

17. (Section 2.3.1)
   (a) Using the data of Table 2.3.3, construct a classical analysis of covariance using soil moisture as the covariate and ignoring measurement error. Test for the effect of treatments after adjusting for soil moisture. Compare the conclusions reached on the basis of such a naive analysis with the conclusions of Example 2.3.2.
   (b) Let the observations of Table 2.3.3 be denoted by Y_{tj}. Let the model in soil moisture be expressed as

$$Y_{tj} = \sum_{i=1}^{14} \gamma_i w_{ij} + \beta x_{tj} + e_{tj}, \qquad X_{tj} = x_{tj} + u_{tj},$$

$$(e_{tj}, u_{tj}) \sim \mathrm{NI}(\mathbf{0}, \boldsymbol{\Sigma}_{\varepsilon\varepsilon}),$$

where w_{ij} are year indicator variables with $w_{ij} = 1$ for $i = t$ and zero otherwise, and the γ_i are the year effects. Using the general formulas for the estimators, estimate all parameters of the model and estimate the covariance matrix of the estimators. Plot the residuals $\hat{v}_{tj}$ against the estimated true values $\hat{x}_{tj}$. What do you conclude?

18. (Section 2.3.3) Verify the formula for $\hat{\boldsymbol{\delta}} - \boldsymbol{\delta}$ given after (2.3.30).

19. (Sections 2.3.1, 1.6.2) Assume the existence of a sequence of experiments indexed by n. At the nth experiment, we observe (Y_{nt}, X_{nt}) satisfying

$$Y_{nt} = \beta_0 + \beta_1 x_{nt} + e_{nt}, \qquad X_{nt} = x_{nt} + u_{nt},$$
$$(e_{nt}, u_{nt})' \sim \mathrm{NI}(\mathbf{0}, \mathrm{diag}[\sigma_{ee}, a_n^{-1}\sigma_{ww}]),$$

for $t = 1, 2, \ldots, b_n$, where $\{b_n\}_{n=1}^{\infty}$ is a sequence of even integers and the x_{nt} are fixed. Let d_n be the distance between the $(b_n/2)$th smallest x_{nt} and the $(b_n/2 + 1)$st smallest x_{nt}. Assume that $d_n^{-1} = O(b_n)$ and $a_n^{-1/3} = o(b_n^{-1})$. Let $\hat{\beta}_w$ be the estimator (1.6.4), where the first group is composed of the $b_n/2$ smallest X_{nt}. That is, the groups are formed on the basis of the observed X_{nt}. Assume

$$\lim_{n \to \infty} (\bar{x}_{n(2)} - \bar{x}_{n(1)}) = C,$$

where $C > 0$, $\bar{x}_{n(1)}$ is the mean of the $b_n/2$ smallest x_{nt}, and $\bar{x}_{n(2)}$ is the mean of the $b_n/2$ largest x_{nt}. Show that, as $n \to \infty$, $\hat{\beta}_w \xrightarrow{P} \beta_1$,

$$\hat{\beta}_w - (\bar{x}_{(1)} - \bar{x}_{(2)})^{-1}(\bar{Y}_{(1)} - \bar{Y}_{(2)}) = O_p(a_n^{-1/2} b_n^{-1/2}),$$
$$b_n^{1/2}(\hat{\beta}_w - \beta_1) \xrightarrow{L} N(0, C^{-2}\sigma_{ee}).$$

20. (Sections 2.3.1, 2.2.1) Assume the model

$$Y_t = \beta_0 + \beta_1 x_t + e_t, \qquad X_t = x_t + u_t,$$

for $t = 1, 2, \ldots, n$, where $(x_t, e_t, u_t)' \sim \mathrm{NI}[(\mu_x, 0, 0)', \mathrm{diag}\,(\sigma_{xx}, \sigma_{ee}, \sigma_{uu})]$.

(a) Assume σ_{ee} known and σ_{uu} unknown.

(i) Derive the maximum likelihood estimator adjusted for degrees of freedom of $(\sigma_{xx}, \sigma_{uu}, \beta_0, \beta_1)$ for samples with

$$(m_{YY} - \sigma_{ee})m_{XX} - m_{XY}^2 > 0.$$

(ii) Derive the maximum likelihood estimator adjusted for degrees of freedom for all samples.

(b) Assume σ_{ee} and σ_{uu} unknown and σ_{xx} known.

(i) Derive the maximum likelihood estimator adjusted for degrees of freedom of $(\sigma_{ee}, \sigma_{uu}, \beta_0, \beta_1)$ for samples with $m_{XX} - \sigma_{xx} > 0$ and $m_{YY}\sigma_{xx} - m_{XY}^2 > 0$.

(ii) Derive the maximum likelihood estimator adjusted for degrees of freedom for all samples.

21. (Sections 2.3.1, 2.2.1) Prove the following.

Theorem. Let

$$Y_t = \beta_0 + \beta_1 x_t + e_t, \qquad X_t = x_t + u_t,$$
$$(e_t, u_t)' \sim \mathrm{NI}(\mathbf{0}, \mathrm{diag}\{\sigma_{ee}, T^{-1}\omega_{22}\}),$$

where $\{x_t\}$ is a fixed sequence satisfying

$$\lim_{n \to \infty} \bar{x} = \mu_x, \qquad \lim_{n \to \infty} m_{xx} = \bar{m}_{xx}.$$

Let s_{22} be an unbiased estimator of ω_{22} distributed as a multiple of a chi-square random variable with d_f degrees of freedom independent of (Y_t, X_t) for all t, where $d_f^{-1} = O(n^{-1})$. Let

$$\hat{\beta}_1 = (m_{xx} - T^{-1}s_{22})^{-1}m_{XY}, \qquad \hat{\beta}_0 = \bar{Y} - \hat{\beta}_1\bar{X}.$$

Then

$$[\hat{V}\{\hat{\beta}_1\}]^{-1/2}(\hat{\beta}_1 - \beta_1) = t_{n-2} + o_p(1)$$

as $v \to \infty$, where $v = nT$,

$$\hat{V}\{\hat{\beta}_1\} = (n-1)^{-1}[\hat{m}_{xx}^{-1}s_{vv} + \hat{m}_{xx}^{-2}(T^{-1}s_{22}s_{vv} + \hat{\beta}_1^2T^{-2}s_{22}^2)] + \hat{m}_{xx}^{-2}d_f^{-1}\hat{\beta}_1^2T^{-2}s_{22}^2,$$

$$s_{vv} = (n-2)^{-1}\sum_{t=1}^{n}(Y_t - \hat{\beta}_0 - \hat{\beta}_1X_t)^2,$$

$\hat{m}_{xx} = m_{XX} - T^{-1}s_{22}$, and t_{n-2} is Student's t with $n-2$ degrees of freedom.

22. (Sections 2.3, 2.2) Beaton, Rubin, and Barone (1976) used the data of Longley (1967) in a discussion of the effect of measurement error on regression coefficients. Longley (1967) originally used the data to test the computational accuracy of regression programs. Therefore, one should not be overly concerned with the economic content of the model. The model of Beaton, Rubin, and Barone might be written

$$y_t = \beta_0 + \sum_{i=1}^{6}\beta_i x_{ti} + q_t, \qquad (Y_t, \mathbf{X}_t) = (y_t, \mathbf{x}_t) + \mathbf{a}_t,$$

Longley Data

Total Employment Y_t	GNP Price Deflator X_{t1}	GNP X_{t2}	Total Unemployed X_{t3}	Size of Armed Forces X_{t4}	Population > 14 Years X_{t5}	Year X_{t6}
60,323	83.0	234,289	2,356	1,590	107,608	1947
61,122	88.5	259,426	2,325	1,456	108,632	1948
60,171	88.2	258,054	3,682	1,616	109,773	1949
61,187	89.5	284,599	3,351	1,650	110,929	1950
63,221	96.2	328,975	2,099	3,099	112,075	1951
63,639	98.1	346,999	1,932	3,594	113,270	1952
64,989	99.0	365,385	1,870	3,547	115,094	1953
63,761	100.0	363,112	3,578	3,350	116,219	1954
66,019	101.2	397,469	2,904	3,048	117,388	1955
67,857	104.6	419,180	2,822	2,857	118,734	1956
68,169	108.4	442,769	2,936	2,798	120,445	1957
66,513	110.8	444,546	4,681	2,637	121,950	1958
68,655	112.6	482,704	3,813	2,552	123,366	1959
69,564	114.2	502,601	3,931	2,514	125,368	1960
69,331	115.7	518,173	4,806	2,572	127,852	1961
70,551	116.9	554,894	4,007	2,827	130,081	1962

Source: Longley (1967).

where $\mathbf{a}_t = (w_t, \mathbf{u}_t)$. Beaton, Rubin, and Barone argued that rounding error furnishes a lower bound for the measurement error. The covariance matrix

$$\Sigma_{aa} = (12)^{-1} \operatorname{diag}\{1, 10^{-2}, 1, 1, 1, 1, 1\}$$

is a possible covariance matrix for rounding error.

(a) Compute the smallest root of

$$|\mathbf{m}_{ZZ} - \lambda \Sigma_{aa}| = 0,$$

where $\mathbf{Z}_t = (Y_t, \mathbf{X}_t)$. Do you feel that $\mathbf{m}_{zz}$ is singular? Do you feel that the measurement error model is appropriate and supported by this test?

(b) Assume that the measurement error for year is zero so that

$$\Sigma_{aa} = (12)^{-1} \operatorname{diag}\{1, 10^{-2}, 1, 1, 1, 1, 0\}.$$

Compute the smallest root of $|\mathbf{m}_{ZZ} - \lambda \Sigma_{aa}| = 0$. What do you conclude?

(c) If the reliability ratios for the variables of $Z_t = (Y_t, X_{t1}, X_{t2}, X_{t3}, X_{t4}, X_{t5})$ are all the same, if the reliability ratio for x_{t6} is one, and if the rank of $\mathbf{m}_{zz}$ is five, what is your estimate of the common reliability ratio?

(d) Using the error covariance matrix of part (b), estimate the parameters of the model and estimate the covariance matrix of the approximate distribution of your estimator.

23. (Sections 2.2, 2.3) The yearly means of the data of Table 2.3.3 of Example 2.3.2 are

$$\bar{\mathbf{Y}}_{t.} = (6.180, 5.230, \ldots, 7.780),$$
$$\bar{\mathbf{X}}_{t.} = (6.073, 2.133, \ldots, 0.983).$$

Assume that these means satisfy the model

$$\bar{Y}_{t.} = \beta_0 + \beta_1 \bar{x}_{t.} + b_t, \qquad \bar{X}_{t.} = \bar{x}_{t.} + \bar{u}_{t.},$$

where $(b_t, \bar{u}_{t.})' \sim \mathrm{NI}[\mathbf{0}, \operatorname{diag}(\sigma_{bb}, 9^{-1}\sigma_{uu})]$. An estimator of σ_{uu} based on 84 degrees of freedom is given in Table 2.3.4 as 0.8284.

(a) Using the yearly means, estimate $(\beta_0, \beta_1, \sigma_{bb})$. Estimate the covariance matrix of the approximate distribution of the estimators.

(b) Combine the estimators of (β_0, β_1) of part (a) with those of Example 2.3.2 to obtain improved estimators. Test the hypothesis that the two sets of estimators are estimating the same quantity. Give an estimated covariance matrix for the approximate distribution of your combined estimator of (β_0, β_1).

2.4. INSTRUMENTAL VARIABLE ESTIMATION

The method of instrumental variables introduced in Section 1.4 extends to equations with several explanatory variables, some of which are measured with error, provided the number of instrumental variables equals or exceeds the number of variables measured with error. Let

$$Y_t = \mathbf{x}_t \boldsymbol{\beta} + e_t, \qquad \mathbf{X}_t = \mathbf{x}_t + \mathbf{u}_t, \tag{2.4.1}$$

for $t = 1, 2, \ldots, n$, where $\mathbf{x}_t$ is a k-dimensional row vector of explanatory variables, $(Y_t, \mathbf{X}_t)$ is observed, and $\mathbf{u}_t$ is the vector of measurement errors. Let a q-dimensional vector of instrumental variables, denoted by $\mathbf{W}_t$, be available and assume that $n > q \geqslant k$. Let n and $\mathbf{W}_t$ be such that $\sum_{t=1}^n \mathbf{W}_t'\mathbf{W}_t$ is nonsingular with probability one,

$$E\{\mathbf{W}_t'(e_t, \mathbf{u}_t)\} = (\mathbf{0}, \mathbf{0}),$$

and the rank of $(\sum_{t=1}^n \mathbf{W}_t'\mathbf{W}_t)^{-1} \sum_{t=1}^n \mathbf{W}_t'\mathbf{X}_t$ is k with probability one.

This specification permits some of the elements of $\mathbf{X}_t$ to have zero error variance. An x_{ti} measured without error must be a linear function of $\mathbf{W}_t$. Thus, a variable measured without error can serve as the instrumental variable for itself.

Assume that n observations are available and write the model in matrix notation as

$$\mathbf{Y} = \mathbf{x}\boldsymbol{\beta} + \mathbf{e}, \qquad \mathbf{X} = \mathbf{x} + \mathbf{u}, \tag{2.4.2}$$

where $\mathbf{Y}$ is the n-dimensional column vector of observations on Y_t, and $\mathbf{X} = (\mathbf{X}_1', \mathbf{X}_2', \ldots, \mathbf{X}_n')'$ is the $n \times k$ matrix of observations on $\mathbf{X}_t$. Following the approach of Section 1.4, we express $\mathbf{X}$ as a function of $\mathbf{W}$ and an error, denoted by $\mathbf{a}_2$, by using the population regression of $\mathbf{X}_t$ on $\mathbf{W}_t$. Thus, we write

$$\mathbf{X} = \mathbf{W}\boldsymbol{\pi}_2 + \mathbf{a}_2, \tag{2.4.3}$$

where $\boldsymbol{\pi}_2 = [E\{\mathbf{W}'\mathbf{W}\}]^{-1}E\{\mathbf{W}'\mathbf{X}\}$ is a $q \times k$ matrix of population regression coefficients. Substituting the regression expression for $\mathbf{X}$ into the system (2.4.2) we obtain the system

$$\mathbf{Y} = \mathbf{W}\boldsymbol{\pi}_1 + \mathbf{a}_1, \tag{2.4.4}$$

$$\mathbf{X} = \mathbf{W}\boldsymbol{\pi}_2 + \mathbf{a}_2, \tag{2.4.5}$$

where $\boldsymbol{\pi}_1 = \boldsymbol{\pi}_2\boldsymbol{\beta}$, $\mathbf{a}_1 = \mathbf{v} + \mathbf{a}_2\boldsymbol{\beta}$, $\mathbf{v} = (v_1, v_2, \ldots, v_n)'$, and $v_t = e_t - \mathbf{u}_t\boldsymbol{\beta}$. The set of equations (2.4.4) and (2.4.5) is sometimes called the *reduced form*. For the model to be identified, we assume

$$|\boldsymbol{\pi}_2'\mathbf{M}_{WW}\boldsymbol{\pi}_2| \neq 0. \tag{2.4.6}$$

To motivate the estimator we make two assumptions, which we will later relax. First, and with no loss of generality, we assume

$$\mathbf{W}'\mathbf{W} = n\mathbf{I}_q. \tag{2.4.7}$$

Next, let $\mathbf{a}_t = (a_{t1}, \mathbf{a}_{t2})$ be the tth row of $\mathbf{a} = (\mathbf{a}_1, \mathbf{a}_2)$, and assume the $\mathbf{a}_t$ to be independent $(\mathbf{0}, \boldsymbol{\Sigma}_{aa})$ random vectors. Then the least squares estimators

$$(\hat{\boldsymbol{\pi}}_1, \hat{\boldsymbol{\pi}}_2) = (\mathbf{W}'\mathbf{W})^{-1}(\mathbf{W}'\mathbf{Y}, \mathbf{W}'\mathbf{X}) \tag{2.4.8}$$

satisfy

$$\hat{\boldsymbol{\pi}}_1 = \boldsymbol{\pi}_2\boldsymbol{\beta} + \boldsymbol{\zeta}_1, \qquad \hat{\boldsymbol{\pi}}_2 = \boldsymbol{\pi}_2 + \boldsymbol{\zeta}_2, \tag{2.4.9}$$

where the q rows of $\boldsymbol{\zeta} = (\boldsymbol{\zeta}_1, \boldsymbol{\zeta}_2)$ are uncorrelated with zero mean and covariance matrix

$$\boldsymbol{\Sigma}_{\zeta\zeta} = n^{-1}\boldsymbol{\Sigma}_{aa}.$$

The covariance result for $\boldsymbol{\zeta}$ follows from the fact that, for example,

$$V\{(\hat{\boldsymbol{\pi}}'_1, \hat{\boldsymbol{\pi}}'_{2i})' \mid \mathbf{W}\} = \begin{bmatrix} (\mathbf{W}'\mathbf{W})^{-1}\sigma_{aa11} & (\mathbf{W}'\mathbf{W})^{-1}\sigma_{aa1,2i} \\ (\mathbf{W}'\mathbf{W})^{-1}\sigma_{aa1,2i} & (\mathbf{W}'\mathbf{W})^{-1}\sigma_{aa,2i,2i} \end{bmatrix},$$

where $\mathbf{W}'\mathbf{W} = \mathbf{I}$, $\hat{\boldsymbol{\pi}}_{2i}$ is the ith column of $\hat{\boldsymbol{\pi}}_2$, σ_{aa11} is the variance of a_{t1}, $\sigma_{aa,2i,2i}$ is the variance of the ith element of $\mathbf{a}_{t2}$, and $\sigma_{aa1,2i}$ is the covariance between a_{t1} and the ith element of $\mathbf{a}_{t2}$. Thus, the ℓth element of $\boldsymbol{\zeta}_1$ is correlated only with the ℓth row of $\boldsymbol{\zeta}_2$. Furthermore, the covariance between the ℓth element of $\boldsymbol{\zeta}_1$ and the ℓth row of $\boldsymbol{\zeta}_2$ is the covariance between a_{t1} and $\mathbf{a}_{t2}$ multiplied by n^{-1}.

If $q = k$, one can construct an estimator for $\boldsymbol{\beta}$ by setting $\hat{\boldsymbol{\pi}}_1$ and $\hat{\boldsymbol{\pi}}_2$ equal to their expectations and solving. Thus,

$$\hat{\boldsymbol{\beta}} = \hat{\boldsymbol{\pi}}_2^{-1}\hat{\boldsymbol{\pi}}_1 \tag{2.4.10}$$

when $q = k$. In the econometric literature a model with $q = k$ is said to be *just identified.* Note that (2.4.10) can be written

$$\hat{\boldsymbol{\beta}} = (\mathbf{W}'\mathbf{X})^{-1}\mathbf{W}'\mathbf{Y},$$

which is the matrix generalization of expression (1.4.12).

Under the assumption that the $\mathbf{a}_t$ are random vectors with zero mean and common covariance matrix, an unbiased estimator of $\boldsymbol{\Sigma}_{aa}$ is

$$\mathbf{S}_{aa} = (n - q)^{-1}[(\mathbf{Y}, \mathbf{X})'(\mathbf{Y}, \mathbf{X}) - (\mathbf{Y}, \mathbf{X})'\mathbf{W}(\mathbf{W}'\mathbf{W})^{-1}\mathbf{W}'(\mathbf{Y}, \mathbf{X})]$$

and an unbiased estimator of $\boldsymbol{\Sigma}_{\zeta\zeta}$ is $n^{-1}\mathbf{S}_{aa}$. If any of the x-variables are measured without error, the row and column of $\mathbf{S}_{aa}$ associated with that variable are vectors of zeros. The estimator $(\hat{\boldsymbol{\pi}}_1, \hat{\boldsymbol{\pi}}_2)$ satisfies the system (2.4.9) in the unknown parameters $\boldsymbol{\pi}_2$ and $\boldsymbol{\beta}$. We recognize the model (2.4.9) as the measurement error model (2.3.1) with $\hat{\boldsymbol{\pi}}_1$ replacing $(Y_1, Y_2, \ldots, Y_n)'$, $\hat{\boldsymbol{\pi}}_2$ replacing $(\mathbf{X}'_1, \mathbf{X}'_2, \ldots, \mathbf{X}'_n)$, $\boldsymbol{\zeta}$ replacing $(\boldsymbol{\varepsilon}'_1, \boldsymbol{\varepsilon}'_2, \ldots \boldsymbol{\varepsilon}'_n)$, and the number of observations equal to q rather than n. It follows that the estimator of $\boldsymbol{\beta}$ for model (2.4.9) defined in Theorem 2.3.2 is

$$\hat{\boldsymbol{\beta}} = [q^{-1}\hat{\boldsymbol{\pi}}'_2\hat{\boldsymbol{\pi}}_2 - \tilde{\lambda}n^{-1}\mathbf{S}_{aa22}]^{-1}[q^{-1}\hat{\boldsymbol{\pi}}'_2\hat{\boldsymbol{\pi}}_1 - \tilde{\lambda}n^{-1}\mathbf{S}_{aa21}], \tag{2.4.11}$$

where S_{aa11}, $\mathbf{S}_{aa12}$, and $\mathbf{S}_{aa22}$ are the submatrices of $\mathbf{S}_{aa}$, $\tilde{\lambda}$ is the smallest root of

$$|q^{-1}\hat{\boldsymbol{\pi}}'\hat{\boldsymbol{\pi}} - \lambda n^{-1}\mathbf{S}_{aa}| = 0, \tag{2.4.12}$$

and $\hat{\boldsymbol{\pi}} = (\hat{\boldsymbol{\pi}}_1, \hat{\boldsymbol{\pi}}_2)$. If $q = k$, the smallest root of (2.4.12) is zero and estimator (2.4.11) reduces to estimator (2.4.10).

The estimator of $\boldsymbol{\beta}$ usually is calculated by noting that $n\hat{\boldsymbol{\pi}}'\hat{\boldsymbol{\pi}}$ is the sum of squares and products of the estimated values for $(\mathbf{Y}, \mathbf{X})$,

$$n\hat{\boldsymbol{\pi}}'\hat{\boldsymbol{\pi}} = (\hat{\mathbf{Y}}, \hat{\mathbf{X}})'(\hat{\mathbf{Y}}, \hat{\mathbf{X}}) = (\mathbf{Y}, \mathbf{X})'\mathbf{W}(\mathbf{W}'\mathbf{W})^{-1}\mathbf{W}'(\mathbf{Y}, \mathbf{X}),$$

where $(\hat{\mathbf{Y}}, \hat{\mathbf{X}}) = \mathbf{W}(\mathbf{W}'\mathbf{W})^{-1}\mathbf{W}'(\mathbf{Y}, \mathbf{X})$. Also, the smallest root $\tilde{v}$ of

$$|(\hat{\mathbf{Y}}, \hat{\mathbf{X}})'(\hat{\mathbf{Y}}, \hat{\mathbf{X}}) - v\mathbf{S}_{aa}| = 0 \tag{2.4.13}$$

is the smallest root of (2.4.12) multiplied by q. That is, $\tilde{v} = q\tilde{\lambda}$. Therefore, the estimator of $\boldsymbol{\beta}$ given in (2.4.11) can be expressed as

$$\hat{\boldsymbol{\beta}} = (\hat{\mathbf{X}}'\hat{\mathbf{X}} - \tilde{v}\mathbf{S}_{aa22})^{-1}(\hat{\mathbf{X}}'\hat{\mathbf{Y}} - \tilde{v}\mathbf{S}_{aa21}) \tag{2.4.14}$$

or as

$$\hat{\boldsymbol{\beta}} = (\mathbf{X}'\mathbf{X} - \tilde{\gamma}\mathbf{S}_{aa22})^{-1}(\mathbf{X}'\mathbf{Y} - \tilde{\gamma}\mathbf{S}_{aa21}), \tag{2.4.15}$$

where $\tilde{\gamma}$ is the smallest root of

$$|(\mathbf{Y}, \mathbf{X})'(\mathbf{Y}, \mathbf{X}) - \gamma\mathbf{S}_{aa}| = 0. \tag{2.4.16}$$

Expressions (2.4.14) and (2.4.15) do not require $\mathbf{W}'\mathbf{W}$ to be a multiple of the identity matrix.

In Theorem 2.4.1 we give the large sample properties of the instrumental variable estimator under much weaker assumptions than we used to motivate the estimator.

Theorem 2.4.1. Let

$$Y_t = \mathbf{x}_t\boldsymbol{\beta} + e_t, \qquad \mathbf{X}_t = \mathbf{x}_t + \mathbf{u}_t,$$

where $\mathbf{x}_t$ are k-dimensional vectors. Let $\mathbf{W}_t$ be a q-dimensional vector, where $n > q \geqslant k$, and let $(\boldsymbol{\varepsilon}_t, \mathbf{x}_t - \boldsymbol{\mu}_{xt}, \mathbf{W}_t - \boldsymbol{\mu}_{Wt})$ be independently and identically distributed with mean zero and finite fourth moments, where $\boldsymbol{\varepsilon}_t = (e_t, \mathbf{u}_t)$. Let $\boldsymbol{\mu}_t = (\boldsymbol{\mu}_{xt}, \boldsymbol{\mu}_{Wt})$ be a sequence of fixed $(k + q)$-dimensional row vectors satisfying

$$\lim_{n\to\infty} n^{-1} \sum_{t=1}^{n} \boldsymbol{\mu}_t'\boldsymbol{\mu}_t = \bar{\mathbf{M}}_{\mu\mu}.$$

Assume

$$E\{\boldsymbol{\varepsilon}_t | \mathbf{W}_t\} = \mathbf{0}, \qquad E\{\boldsymbol{\varepsilon}_t'\boldsymbol{\varepsilon}_t | \mathbf{W}_t\} = \boldsymbol{\Sigma}_{\varepsilon\varepsilon}, \tag{2.4.17}$$

and let $\boldsymbol{\pi} = (\boldsymbol{\pi}_1, \boldsymbol{\pi}_2) = \bar{\mathbf{M}}_{WW}^{-1}(\bar{\mathbf{M}}_{WY}, \bar{\mathbf{M}}_{WX})$, where

$$\text{plim}\, n^{-1} \sum_{t=1}^{n} \mathbf{W}_t'(\mathbf{W}_t, Y_t, \mathbf{X}_t) = (\bar{\mathbf{M}}_{WW}, \bar{\mathbf{M}}_{WY}, \bar{\mathbf{M}}_{WX}),$$

and $\bar{\mathbf{M}}_{WW}$ and $\boldsymbol{\pi}_2'\bar{\mathbf{M}}_{WW}\boldsymbol{\pi}_2$ are nonsingular. Let $\hat{\boldsymbol{\beta}}$ be defined by (2.4.14) and let $\tilde{\nu}$ be the smallest root of (2.4.13). Then

$$n^{1/2}(\hat{\boldsymbol{\beta}} - \boldsymbol{\beta}) \xrightarrow{L} N[\mathbf{0}, (\boldsymbol{\pi}_2'\bar{\mathbf{M}}_{WW}\boldsymbol{\pi}_2)^{-1}\sigma_{vv}],$$
$$\tilde{\nu} \xrightarrow{L} \chi^2_{q-k},$$

where $\sigma_{vv} = (1, -\boldsymbol{\beta}')\boldsymbol{\Sigma}_{\varepsilon\varepsilon}(1, -\boldsymbol{\beta}')'$ and χ^2_{q-k} is a chi-square random variable with $q - k$ degrees of freedom for $q > k$. Furthermore, $n^{1/2}(\hat{\boldsymbol{\beta}} - \boldsymbol{\beta})$ and $\tilde{\nu}$ are independent in the limit.

Proof. If the assumptions of the original model specification (2.4.1) are retained, $\hat{\boldsymbol{\beta}}$ is defined with probability one. If only the weaker assumptions of the theorem statement hold, given $\varepsilon > 0$ there is some N such that $\hat{\boldsymbol{\beta}}$ is defined with probability greater than $1 - \varepsilon$ for all $n > N$. We complete the proof assuming $\hat{\boldsymbol{\beta}}$ is defined with probability one.

Because $\hat{\boldsymbol{\beta}}$ is the value of $\boldsymbol{\beta}$ that minimizes the ratio

$$nq^{-1}[(1, -\boldsymbol{\beta}')\mathbf{S}_{aa}(1, -\boldsymbol{\beta}')']^{-1}[(1, -\boldsymbol{\beta}')\hat{\boldsymbol{\pi}}'\mathbf{W}'\mathbf{W}\hat{\boldsymbol{\pi}}(1, -\boldsymbol{\beta}')']$$

and $\tilde{\lambda}$ is the minimum value of the ratio, it follows that $\tilde{\lambda}$ is less than the ratio evaluated at the true $\boldsymbol{\beta}$. The ratio evaluated at the true $\boldsymbol{\beta}$ is

$$\begin{aligned} nq^{-1}[(1, -\boldsymbol{\beta}')\mathbf{S}_{aa}(1, -\boldsymbol{\beta}')']^{-1}(1, -\boldsymbol{\beta}')\mathbf{Z}'\mathbf{W}(\mathbf{W}'\mathbf{W})^{-1}\mathbf{W}'\mathbf{Z}(1, -\boldsymbol{\beta}')' \\ = nq^{-1}[(1, -\boldsymbol{\beta}')\mathbf{S}_{aa}(1, -\boldsymbol{\beta}')']^{-1}\mathbf{v}'\mathbf{W}(\mathbf{W}'\mathbf{W})^{-1}\mathbf{W}'\mathbf{v}, \end{aligned} \tag{2.4.18}$$

where $\mathbf{Z} = (\mathbf{Y}, \mathbf{X})$, $\mathbf{v} = \mathbf{e} - \mathbf{u}\boldsymbol{\beta}$, and the expressions hold for general $\mathbf{W}$ of rank q. Now

$$\begin{aligned} (1, -\boldsymbol{\beta}')\mathbf{S}_{aa}(1, -\boldsymbol{\beta}')' &= (n - q)^{-1}(1, -\boldsymbol{\beta}')\mathbf{Z}'\mathbf{R}\mathbf{Z}(1, -\boldsymbol{\beta}')' \\ &= (n - q)^{-1}\mathbf{v}'\mathbf{R}\mathbf{v}, \end{aligned}$$

where $\mathbf{R} = \mathbf{I} - \mathbf{W}(\mathbf{W}'\mathbf{W})^{-1}\mathbf{W}'$. Under our assumptions

$$(n - q)^{-1}\mathbf{v}'\mathbf{R}\mathbf{v} \xrightarrow{P} \sigma_{vv}$$

and

$$\left(\sum_{t=1}^{n} \mathbf{W}_t'\mathbf{W}_t\right)^{-1/2} \sum_{t=1}^{n} \mathbf{W}_t'v_t \xrightarrow{L} N(\mathbf{0}, \mathbf{I}_q\sigma_{vv}). \tag{2.4.19}$$

The normality result of (2.4.19) follows because the random variables

$$\mathbf{W}_t'v_t = [\boldsymbol{\mu}_{Wt} + (\mathbf{W}_t - \boldsymbol{\mu}_{Wt})]'v_t$$

are independently distributed with finite second moments. See the proof of Theorem 2.2.1 and Theorem 1.C.2. The zero mean for $\mathbf{W}_t'v_t$ and the form of the covariance matrix in (2.4.19) follows from assumptions (2.4.17). By (2.4.18) and (2.4.19), $q\tilde{\lambda} = \tilde{v}$ is bounded by a random variable whose limiting distribution is that of a chi-square random variable with q degrees of freedom. Hence, $\tilde{\lambda} = O_p(1)$. From the definition of $\hat{\boldsymbol{\beta}}$ we have

$$\begin{aligned}\hat{\boldsymbol{\beta}} - \boldsymbol{\beta} &= [q^{-1}\hat{\boldsymbol{\pi}}_2'\mathbf{M}_{WW}\hat{\boldsymbol{\pi}}_2 - \tilde{\lambda}n^{-1}\mathbf{S}_{aa22}]^{-1}[q^{-1}\hat{\boldsymbol{\pi}}_2'\mathbf{M}_{WW}\hat{\boldsymbol{\pi}}_1 - \tilde{\lambda}n^{-1}\mathbf{S}_{aa21}] - \boldsymbol{\beta} \\ &= [\boldsymbol{\pi}_2'\bar{\mathbf{M}}_{WW}\boldsymbol{\pi}_2]^{-1}\boldsymbol{\pi}_2'\mathbf{M}_{WW}\boldsymbol{\omega} + o_p(n^{-1/2}),\end{aligned} \tag{2.4.20}$$

where $\boldsymbol{\omega} = \hat{\boldsymbol{\pi}}_1 - \hat{\boldsymbol{\pi}}_2\boldsymbol{\beta} = \mathbf{M}_{WW}^{-1}\mathbf{M}_{Wv}$. The vector $\boldsymbol{\omega}$ is the vector of regression coefficients obtained by regressing $v_t = Y_t - \mathbf{X}_t\boldsymbol{\beta}$ on $\mathbf{W}_t$, and by (2.4.19),

$$n^{1/2}\boldsymbol{\omega} \xrightarrow{L} N(\mathbf{0}, \bar{\mathbf{M}}_{WW}^{-1}\sigma_{vv}).$$

The covariance matrix of the limiting distribution of $n^{1/2}(\hat{\boldsymbol{\beta}} - \boldsymbol{\beta})$ follows from (2.4.20).

Now

$$\begin{aligned}\tilde{v} &= n[(1, -\hat{\boldsymbol{\beta}}')\mathbf{S}_{aa}(1, -\hat{\boldsymbol{\beta}}')']^{-1}(1, -\hat{\boldsymbol{\beta}}')\hat{\boldsymbol{\pi}}'\mathbf{M}_{WW}\hat{\boldsymbol{\pi}}(1, -\hat{\boldsymbol{\beta}}')' \\ &= n[(1, -\boldsymbol{\beta}')\mathbf{S}_{aa}(1, -\boldsymbol{\beta}')']^{-1}\boldsymbol{\omega}'\mathbf{K}\boldsymbol{\omega} + O_p(n^{-1/2}) \qquad (2.4.21) \\ &= \sigma_{vv}^{-1}n\boldsymbol{\omega}'\mathbf{K}\boldsymbol{\omega} + O_p(n^{-1/2}), \qquad (2.4.22)\end{aligned}$$

where

$$\mathbf{K} = \mathbf{M}_{WW} - \mathbf{M}_{WW}\boldsymbol{\pi}_2(\boldsymbol{\pi}_2'\mathbf{M}_{WW}\boldsymbol{\pi}_2)^{-1}\boldsymbol{\pi}_2'\mathbf{M}_{WW}.$$

Because $n^{1/2}\mathbf{M}_{WW}^{1/2}\sigma_{vv}^{-1/2}\boldsymbol{\omega}$ is converging in distribution to a $N(\mathbf{0}, \mathbf{I}_k)$ random vector, the leading term of (2.4.22) is converging in distribution to a chi-square random variable with $q - k$ degrees of freedom. The independence of $\boldsymbol{\omega}'\mathbf{K}\boldsymbol{\omega}$ and $\boldsymbol{\pi}_2'\mathbf{M}_{Wv}$ in the limit follows from standard least squares results. □

The limiting distribution of $\tilde{v}$ is that of a chi-square random variable with $q - k$ degrees of freedom. For normally distributed ε_t, expression (2.4.21) demonstrates that $(q - k)^{-1}\tilde{v}$ is equal to an F random variable plus a remainder that is $O_p(n^{-1/2})$. Therefore, it is suggested that the distribution of $(q - k)^{-1}\tilde{v}$ be approximated by the central F distribution with $q - k$ and $n - q$ degrees of freedom. A test of model specification can be performed by comparing $(q - k)^{-1}\tilde{v}$ with the tabulated F distribution.

The covariance matrix of the approximate distribution of $\hat{\boldsymbol{\beta}}$ can be estimated with

$$\hat{\mathbf{V}}\{\hat{\boldsymbol{\beta}}\} = (\hat{\mathbf{X}}'\hat{\mathbf{X}} - \tilde{v}\mathbf{S}_{aa22})^{-1}s_{vv}, \tag{2.4.23}$$

where

$$s_{vv} = (n - k)^{-1}\sum_{t=1}^{n}(Y_t - \mathbf{X}_t\hat{\boldsymbol{\beta}})^2. \tag{2.4.24}$$

The matrix $\boldsymbol{\pi}_2'\mathbf{M}_{WW}\boldsymbol{\pi}_2$ must be nonsingular in order for the model to be identified. Therefore, one should check this assumption by computing the smallest root of

$$|\hat{\boldsymbol{\pi}}_2'\mathbf{M}_{WW}\hat{\boldsymbol{\pi}}_2 - \theta n^{-1}\mathbf{S}_{aa22}| = 0. \tag{2.4.25}$$

If the rank of $\boldsymbol{\pi}_2'\mathbf{M}_{WW}\boldsymbol{\pi}_2$ is $k - 1$, the smallest root $\hat{\theta}$ divided by $(q - k + 1)$ is approximately distributed as a central F random variable with $q - k + 1$ and $n - q$ degrees of freedom. A large F is desired because a large F indicates that the model is identified. If F is small it may be possible to identify the model by adding instrumental variables.

It is suggested that the modified estimator given in Theorem 2.5.3 of Section 2.5 be used in practice. The estimator (2.5.13) can be written

$$\hat{\boldsymbol{\beta}} = [\hat{\mathbf{X}}'\hat{\mathbf{X}} - (\tilde{\nu} - \alpha)\mathbf{S}_{aa22}]^{-1}[\hat{\mathbf{X}}'\hat{\mathbf{Y}} - (\tilde{\nu} - \alpha)\mathbf{S}_{aa21}] \tag{2.4.26}$$

in the notation of the instrumental variable model of this section. Note that the n of Theorem 2.5.3 is equal to the q of this section. The theory of Theorem 2.5.3 is not directly applicable to the instrumental variable problem, but an analogous theorem can be proved for the instrumental variable problem. Setting α equal to one in (2.4.26) produces an estimator of $\boldsymbol{\beta}$ that is nearly unbiased.

For the model with $(\mathbf{a}_t, v_t) \sim \text{NI}(\mathbf{0}, \boldsymbol{\Sigma})$, Anderson and Rubin (1949) demonstrated that the estimator (2.4.11) is a type of maximum likelihood estimator for the simultaneous equation model. For the simultaneous equation model, estimator (2.4.11) is called the *limited information maximum likelihood estimator*. Sargan (1958) obtained the estimator in the general instrumental variable setting. Anderson (1976, 1984) discusses the relationships between the limited information estimator and the errors-in-variables estimator. Fuller (1977) used the analogy to errors in variables associated with (2.4.9).

Example 2.4.1. In November 1983, the Department of English at Iowa State University conducted a study in which members of the general university faculty were asked to evaluate two essays. The essays were presented to the faculty as essays prepared as part of a placement examination by two foreign graduate students who were nonnative speakers of English. Three pairs of essays were used in the study: a pair containing errors in the use of articles; a pair containing errors in spelling; and a pair containing errors in verb tense. The faculty members were asked to read the essays and to score them using a five point scale for 11 items. The study is described in Vann and Lorenz (1984).

We analyze the responses for eight items. The eight items are divided into three groups; three items pertaining to the essay, three items pertaining to the language used, and two items pertaining to the writer. The low and high points of the scale for the eight items were described as follows:

A. The essay is

Z_1 poorly developed—well developed
Z_2 difficult to understand—easy to understand
Z_3 illogical—logical

B. The writer uses language that is

Z_4 inappropriate—appropriate
Z_5 unacceptable—acceptable
Z_6 irritating—not irritating

C. The writer seems

Z_7 careless—careful
Z_8 unintelligent—intelligent

Table 2.A.1 of Appendix 2.A contains the data for 100 respondents. This is a subsample of the 219 faculty members who participated in the study. The score on each item is the sum of the scores on that item for the two essays scored by the faculty member. The sample covariance matrix is given in Table 2.4.1. A possible model for these data is the factor model, where

$$Z_{ti} = \beta_{i0} + \beta_{i3}z_{t3} + \beta_{i6}z_{t6} + \beta_{i8}z_{t8} + \varepsilon_{ti}, \quad \text{for } i = 1, 2, 4, 5, 7,$$
$$Z_{ti} = z_{ti} + \varepsilon_{ti}, \quad \text{for } i = 3, 6, 8,$$

and the $\boldsymbol{\varepsilon}_t$ are independent $[\mathbf{0}, \operatorname{diag}(\sigma_{\varepsilon\varepsilon 11}, \sigma_{\varepsilon\varepsilon 22}, \ldots, \sigma_{\varepsilon\varepsilon 88})]$ random vectors. In this model each observation is expressed as a linear function of three of the unknown true values, where one of the unknown true values is chosen from each of the three groups of equations. We shall estimate the equation for Z_{t1}. Because the measurement errors in different variables are assumed to be uncorrelated, the observed variables not entering the equation for Z_{t1} can be used as instrumental variables for that equation. Thus, we estimate the equation for Z_{t1} using Z_{t2}, Z_{t4}, Z_{t5}, and Z_{t7} as instrumental variables. In the notation of this section,

$$(Z_{t1}, Z_{t3}, Z_{t6}, Z_{t8}, Z_{t2}, Z_{t4}, Z_{t5}, Z_{t7}) = (Y_t, X_{t1}, X_{t2}, X_{t3}, W_{t1}, W_{t2}, W_{t3}, W_{t4}).$$

We have not included the constant function in either the set of X variables or in the set of W variables. This will permit us to use matrices of corrected sums of squares and products in the calculations. To be strictly conformable

TABLE 2.4.1. Sample moments for 100 observations on language evaluation

	Developed Y	Logical X_1	Irritating X_2	Intelligent X_3	Understand W_1	Appropriate W_2	Acceptable W_3	Careful W_4
Y	2.9798	2.1364	1.2929	1.3182	1.7727	1.5758	1.5960	1.3131
X_1	2.1364	2.4092	1.1483	1.1839	1.8181	1.4675	1.3895	1.1271
X_2	1.2929	1.1483	2.3055	1.0533	1.3871	1.6166	1.8392	1.1459
X_3	1.3182	1.1839	1.0533	1.7264	1.2151	1.2210	1.2804	1.4766
W_1	1.7727	1.8181	1.3871	1.2151	2.6516	1.6133	1.5802	1.4992
W_2	1.5758	1.4675	1.6166	1.2210	1.6133	2.9681	2.1685	1.2550
W_3	1.5960	1.3895	1.8392	1.2804	1.5802	2.1685	3.2663	1.3685
W_4	1.3131	1.1271	1.1459	1.4766	1.4992	1.2550	1.3685	2.6428
Mean	6.50	7.07	7.76	7.97	7.43	6.96	6.92	7.06

with (2.4.1), (2.4.4), and (2.4.5), the X vector and W vector would be

$$\mathbf{X}_t = (1, Z_{t3}, Z_{t6}, Z_{t8}) \quad \text{and} \quad \mathbf{W}_t = (1, Z_{t2}, Z_{t4}, Z_{t5}, Z_{t7}).$$

The ordinary least squares estimates of the reduced form are

$$\begin{aligned}
\hat{Y}_t &= \underset{(0.696)}{0.523} + \underset{(0.109)}{0.436}W_{t1} + \underset{(0.112)}{0.151}W_{t2} + \underset{(0.105)}{0.131}W_{t3} + \underset{(0.100)}{0.110}W_{t4}, \\
\hat{X}_{t1} &= \underset{(0.572)}{1.447} + \underset{(0.090)}{0.557}W_{t1} + \underset{(0.092)}{0.148}W_{t2} + \underset{(0.086)}{0.052}W_{t3} + \underset{(0.082)}{0.013}W_{t4}, \\
\hat{X}_{t2} &= \underset{(0.570)}{2.396} + \underset{(0.089)}{0.175}W_{t1} + \underset{(0.092)}{0.176}W_{t2} + \underset{(0.086)}{0.328}W_{t3} + \underset{(0.082)}{0.081}W_{t4}, \\
\hat{X}_{t3} &= \underset{(0.474)}{2.918} + \underset{(0.074)}{0.108}W_{t1} + \underset{(0.076)}{0.114}W_{t2} + \underset{(0.071)}{0.100}W_{t3} + \underset{(0.068)}{0.392}W_{t4}.
\end{aligned}$$

The matrix $\mathbf{S}_{aa}$ that is the estimator of the covariance matrix of the reduced form errors is

$$\begin{bmatrix} 1.6825 & 0.8494 & 0.0794 & 0.2849 \\ 0.8494 & 1.1379 & 0.0264 & 0.2503 \\ 0.0794 & 0.0264 & 1.1284 & 0.0903 \\ 0.2849 & 0.2503 & 0.0903 & 0.7813 \end{bmatrix}.$$

The smallest root of Equation (2.4.13) is $\tilde{\nu} = 0.010$. Because $q - k = 1$, the F statistic is equal to 0.010. Under the null, the distribution of the F statistic is approximately that of a central F with 1 and 95 degrees of freedom. The small, but not overly small, value for F gives us no reason to question the model. To check the identification status of the model, we compute the smallest root of Equation (2.4.25). The smallest root is $\hat{\theta} = 10.71$ and the associated F statistic is $F = 5.35$. If the rank of $\boldsymbol{\pi}'_2\mathbf{m}_{WW}\boldsymbol{\pi}_2$ is two, the distribution of the statistic is approximately that of a central F with 2 and 95 degrees of freedom. Because the F statistic is large, we are comfortable with the assumption that the rank of $\boldsymbol{\pi}'_2\mathbf{m}_{WW}\boldsymbol{\pi}_2$ is three and we proceed with estimation of the equation for developed. The estimators we present were computed using SYSREG of SAS. The limited information maximum likelihood option is used in which Y, X_1, X_2, X_3 are called endogenous variables and W_1, W_2, W_3, W_4 are called exogenous variables in the block statement. We set ALPHA of the program equal to one, which results in the calculation of estimator (2.4.26) with $\alpha = 1$. The estimated equation is

$$\hat{Y}_t = \underset{(0.843)}{-1.566} + \underset{(0.193)}{0.677}x_{t1} + \underset{(0.235)}{0.202}x_{t2} + \underset{(0.225)}{0.215}x_{t3},$$

where the estimated standard errors are those output by the program and $s_{vv} = 1.058$. The program calculates the standard errors as the square roots of the diagonal elements of

$$[\hat{\mathbf{X}}'\hat{\mathbf{X}} - (\tilde{\nu} - \alpha)\mathbf{S}_{aa22}]^{-1}s_{vv}. \qquad \square\ \square$$

Example 2.4.2. This example is taken from Miller and Modigliani (1966). The authors developed a model for the value of a firm and applied it to a sample of 63 large electric utilities. We write their model as

$$Y_t = \beta_0 + \beta_1 x_t + e_t,$$

where Y_t is the current market value of the firm multiplied by 10 and divided by book value of assets, x_t is the market's expectation of the long run, future tax adjusted earning power of the firm's assets multiplied by 100 and divided by book value of assets, and e_t is the error in the equation. The coefficient β_1 is the capitalization rate for the expected earning power for utility firms. Because x_t is an expectation, it cannot be observed directly. Each year the firms report their earnings and we let X_t be the measured tax adjusted earnings multiplied by 100 and divided by book value of assets, where $X_t = x_t + u_t$ and u_t is the difference between current reported earnings and long run expected earnings. Miller and Modigliani suggest three instrumental variables:

- $W_{t2}^{\dagger}$ Current dividends paid multiplied by 100 and divided by book value of assets
- $W_{t3}^{\dagger}$ Market value of debt multiplied by 10 and divided by book value of assets
- $W_{t4}^{\dagger}$ Market value of preferred stock multiplied by 10 and divided by book value of assets

Because the firms practice dividend stabilization, dividends paid should reflect management's expectation of long terms profits. In turn, management's expectations should be correlated with the market's expectations. Miller and Modigliani explain the more subtle reasons for including the other two variables in the set of instrumental variables. Miller and Modigliani (1966, p. 355) discuss possible correlations between the $W_t^{\dagger}$ and u_t and state: "But while complete independence is not to be expected, we would doubt that such correlation as does exist is so large as to dash all hopes for substantially improving the estimates by the instrumental variable procedure." Therefore, our operating assumption is that

$$(e_t, u_t)' \sim \mathrm{NI}(\mathbf{0}, \boldsymbol{\Sigma}_{\varepsilon\varepsilon})$$

independent of $(W_{t2}^{\dagger}, W_{t3}^{\dagger}, W_{t4}^{\dagger})$.

TABLE 2.4.2. Matrix of corrected mean squares and products

Y	X	$W_2^\dagger$	$W_3^\dagger$	$W_4^\dagger$
0.8475	0.4127	0.2336	−0.0005	−0.0466
0.4127	0.3020	0.1305	0.0092	0.0106
0.2336	0.1305	0.1743	−0.0065	−0.0786
−0.0005	0.0092	−0.0065	0.1780	−0.0792
0.0466	0.0106	−0.0786	−0.0792	0.3900

Source: Miller and Modigliani (1966).

Miller and Modigliani studied data for 3 years. We use the data for 1954. The sample mean vector is

$$(\bar{Y}, \bar{X}, \bar{W}_2^\dagger, \bar{W}_3^\dagger, \bar{W}_4^\dagger) = (8.9189, 4.5232, 2.5925, 4.9306, 1.0974)$$

and the matrix of mean squares and products is given in Table 2.4.2. The matrix of mean squares and products has been adjusted for two variables, the reciprocal of assets and a measure of growth. Therefore, the degrees of freedom is 60 for the mean squares and products presented in the table.

We define the standardized vector $\mathbf{W}_t$ by

$$\begin{aligned}\mathbf{W}_t &= [1, (W_{t2}^\dagger - \bar{W}_2^\dagger, W_{t3}^\dagger - \bar{W}_3^\dagger, W_{t4}^\dagger - \bar{W}_4^\dagger)\mathbf{T}'] \\ &= [1, \mathbf{\Phi}_t],\end{aligned} \tag{2.4.27}$$

where $\mathbf{T}$ is the lower triangular (Gram–Schmidt) transformation

$$\mathbf{T} = \begin{bmatrix} 0.3953 & 0 & 0 \\ 0.0879 & 2.3719 & 0 \\ 0.8318 & 0.8206 & 1.7771 \end{bmatrix}.$$

Then the matrix of sample raw mean squares and products of $\mathbf{W}_t$ is the identity matrix. Because we shall deal with moments about the mean, we have chosen to define the subvector $\mathbf{\Phi}_t$. The matrix of corrected mean squares and products for $(Y_t, X_t, \mathbf{\Phi}_t)$ is

$$\begin{bmatrix} 0.8475 & 0.4127 & 0.5594 & 0.0193 & 0.2767 \\ 0.4127 & 0.3020 & 0.3126 & 0.0334 & 0.1350 \\ 0.5594 & 0.3126 & 1.0000 & 0 & 0 \\ 0.0193 & 0.0334 & 0 & 1.0000 & 0 \\ 0.2767 & 0.1350 & 0 & 0 & 1.0000 \end{bmatrix}.$$

The model in the reduced form (2.4.4), (2.4.5) is

$$\begin{aligned} Y_t &= \pi_{11} + \pi_{21} W_{t2} + \pi_{31} W_{t3} + \pi_{41} W_{t4} + a_{t1}, \\ X_t &= \pi_{12} + \pi_{22} W_{t2} + \pi_{32} W_{t3} + \pi_{42} W_{t4} + a_{t2}, \end{aligned}$$

where $\pi_{i1} = \pi_{i2}\beta_1$ for $i = 2, 3, 4$. The reduced form estimated equations are

$$\hat{Y}_t = \underset{(0.0896)}{8.9189} + \underset{(0.0896)}{0.5594}W_{t2} + \underset{(0.0896)}{0.0193}W_{t3} + \underset{(0.0896)}{0.2767}W_{t4},$$

$$\hat{X}_t = \underset{(0.0570)}{4.5232} + \underset{(0.0570)}{0.3126}W_{t2} + \underset{(0.0570)}{0.0334}W_{t3} + \underset{(0.0570)}{0.1350}W_{t4},$$

where the numbers in parentheses are the estimated standard errors of the regression coefficients. The matrix

$$(n-4)(n-1)^{-1}\mathbf{S}_{aa} = \mathbf{m}_{ZZ} - \mathbf{m}_{Z\Phi}\mathbf{m}_{\Phi\Phi}^{-1}\mathbf{m}_{\Phi Z} = \begin{bmatrix} 0.4576 & 0.1998 \\ 0.1998 & 0.1849 \end{bmatrix}$$

and $\mathbf{S}_{\zeta\zeta} = (60)^{-1}\mathbf{S}_{aa}$, where $n - 1 = 60$. The determinantal equation (2.4.12) is

$$\left| \begin{bmatrix} 0.1300 & 0.0710 \\ 0.0710 & 0.0390 \end{bmatrix} - \lambda \begin{bmatrix} 0.0080 & 0.0035 \\ 0.0035 & 0.0032 \end{bmatrix} \right| = 0,$$

and the smallest root is $\hat{\lambda} = 0.153$. The corresponding F statistic is $F = 1.5(0.153) = 0.23$, which is approximately distributed with 2 and 57 degrees of freedom for a correctly specified model. The F test supports the model specification. If this F test were large, it would call into question the model specification. The statistic is small because the ratios of the three pairs of regression coefficients are roughly the same. The coefficients of W_{t3} are small relative to the standard errors so that a value of 1.8 is totally acceptable for the ratio of the two coefficients.

In this example there is little doubt that the model is identified. The test statistic for the hypothesis that $\sum_{i=2}^{4} \pi_{i2}^2 = 0$ is

$$F = (0.00324)^{-1}(0.03902) = 12.04.$$

If $\sum_{i=2}^{4} \pi_{i2}^2 = 0$, the distribution of the test statistic is approximately that of Snedecor's F with 3 and 57 degrees of freedom. One easily rejects the hypothesis that $\sum_{i=2}^{4} \pi_{i2}^2 = 0$ and accepts the hypothesis that the model is identified.

The estimator of β_1 is

$$\hat{\beta}_1 = (0.0390 - 0.0005)^{-1}(0.0710 - 0.0005) = 1.8281$$

and the estimator of β_0 is $\hat{\beta}_0 = 0.6500$. By (2.4.23), the estimated covariance matrix for the approximate distribution of $(\hat{\beta}_0, \hat{\beta}_1)$ is

$$\hat{\mathbf{V}}\{(\hat{\beta}_0, \hat{\beta}_1)'\} = \begin{bmatrix} 1.0488 & -0.2306 \\ -0.2306 & 0.0510 \end{bmatrix},$$

where $s_{vv} = 0.3535$.

The tests we have performed on the model indicate that the model is identified, but the sample size and error variances are such that the approximations based on the normal distribution of Theorem 2.4.1 should be used with care.

In this example we carried out all numerical calculations parallel to the theoretical development. As illustrated in Example 2.4.1, the calculations are performed most easily in practice by using the limited information single equation option of a statistical package such as SAS (Barr et al, 1979).

□ □

REFERENCES

Anderson (1951b, 1976), Anderson and Rubin (1949, 1950), Fuller (1976, 1977), Miller and Modigliani (1966), Sargan (1958), Sargan and Mikhail (1971).

EXERCISES

24. (Section 2.4) The full model for the Miller–Modigliani study of Example 2.4.2 is

$$Y_t = \beta_0 + \beta_1 x_t + \beta_2 z_{t2} + \beta_3 z_{t3} + e_t,$$

where z_{t2} is the reciprocal of book value of assets multiplied by 10^7, z_{t3} is the average growth in assets for 5 years divided by assets multiplied by 10, and the remaining variables are defined in Example 2.4.2. It is assumed that z_{t2} and z_{t3} are measured without error. The sample correlation matrix for 1956 is

$$\begin{array}{ccccccc} Y_t & X_t & z_{t2} & z_{t3} & W_{t1} & W_{t2} & W_{t3} \end{array}$$
$$\begin{bmatrix} 1.000 & 0.833 & -0.162 & 0.179 & 0.432 & 0.014 & -0.022 \\ 0.833 & 1.000 & -0.040 & -0.083 & 0.630 & -0.017 & -0.077 \\ -0.162 & -0.040 & 1.000 & -0.106 & 0.104 & 0.249 & -0.078 \\ 0.179 & -0.083 & -0.106 & 1.000 & -0.384 & 0.344 & 0.170 \\ 0.432 & 0.630 & 0.104 & -0.384 & 1.000 & -0.199 & -0.442 \\ 0.014 & -0.017 & 0.249 & 0.344 & -0.199 & 1.000 & -0.288 \\ -0.022 & -0.077 & -0.078 & 0.170 & -0.442 & -0.288 & 1.000 \end{bmatrix}.$$

The means of the variables are (8.69, 4.84, 0.739, 0.799, 2.49, 4.88, 1.09) and the standard deviations are (1.1, 0.55, 0.98, 0.80, 0.42, 0.49, 0.52). Estimate the $\boldsymbol{\beta}$ vector by the method of instrumental variables. Estimate the covariance matrix of the approximate distribution of the estimators.

25. (Section 2.4) Assume that the model of Example 2.4.2 and Exercise 2.24 is replaced by the model with $\beta_0 \equiv 0$. Miller and Modigliani argue that it is reasonable for the model to have a zero intercept. Noting that the column of ones is an instrumental variable, estimate $(\beta_1, \beta_2, \beta_3)$. Test the model specification. Estimate the variance of the approximate distribution of your estimator.

26. (Sections 2.4, 2.3) Let the model and assumptions of Theorem 2.4.1 hold. Let $\hat{\boldsymbol{\beta}}$ be defined by (2.4.20) or, equivalently, by

$$\hat{\boldsymbol{\beta}} = (\mathbf{M}_{22} - \tilde{\lambda}\mathbf{S}_{\varepsilon\varepsilon 22})^{-1}(\mathbf{M}_{21} - \tilde{\lambda}\mathbf{S}_{\varepsilon\varepsilon 21}),$$

where

$$\mathbf{M} = \begin{bmatrix} \mathbf{M}_{11} & \mathbf{M}_{12} \\ \mathbf{M}_{21} & \mathbf{M}_{22} \end{bmatrix} = q^{-1}\hat{\boldsymbol{\pi}}'\hat{\boldsymbol{\pi}},$$

$\mathbf{S}_{\varepsilon\varepsilon} = n^{-1}\mathbf{S}_{aa}$, $\hat{\boldsymbol{\pi}} = \mathbf{M}_{WW}^{-1/2}\mathbf{M}_{WZ}$, and $\tilde{\lambda}$ is the smallest root of $|\mathbf{M} - \lambda\mathbf{S}_{\varepsilon\varepsilon}| = 0$. If we use (2.3.24), the covariance matrix of the approximate distribution of $\hat{\boldsymbol{\beta}}$ is estimated by

$$\hat{\mathbf{V}}\{\hat{\boldsymbol{\beta}}\} = q^{-1}\hat{\mathbf{M}}_{22}^{-1}\hat{\sigma}_{vv} + [q^{-1} + (n-q)^{-1}]\hat{\mathbf{M}}_{22}^{-1}\hat{\boldsymbol{\Sigma}}_{pp}\hat{\mathbf{M}}_{22}^{-1}\hat{\sigma}_{vv},$$

where $\hat{\boldsymbol{\Sigma}}_{pp} = n(n-k)^{-1}\hat{\mathbf{H}}_2'\mathbf{S}_{\varepsilon\varepsilon}\hat{\mathbf{H}}_2$, $\hat{\mathbf{M}}_{22} = \hat{\mathbf{H}}_2'(\mathbf{M} - \mathbf{S}_{\varepsilon\varepsilon})\hat{\mathbf{H}}_2$,

$$\hat{\mathbf{H}}_2 = (\mathbf{0}, \mathbf{I}_k)' - (1, -\hat{\boldsymbol{\beta}}')'[(1, -\hat{\boldsymbol{\beta}}')\mathbf{S}_{\varepsilon\varepsilon}(1, -\hat{\boldsymbol{\beta}}')']^{-1}(1, -\hat{\boldsymbol{\beta}}')\mathbf{S}_{\varepsilon\varepsilon},$$

$$\hat{\sigma}_{vv} = (n-k)^{-1}[q(1, -\hat{\boldsymbol{\beta}}')\mathbf{M}(1, -\hat{\boldsymbol{\beta}}')' + (n-q)(1, -\hat{\boldsymbol{\beta}}')\mathbf{S}_{\varepsilon\varepsilon}(1, -\hat{\boldsymbol{\beta}}')'].$$

Note that $(n, n-k, d_f, \mathbf{Z}_t)$ of (2.3.24) corresponds to $(q, q-k, n-q, \hat{\boldsymbol{\pi}}_i)$ of the instrumental variable model. Prove that

$$[\hat{\mathbf{V}}\{\hat{\boldsymbol{\beta}}\}]^{-1/2}(\hat{\boldsymbol{\beta}} - \boldsymbol{\beta}) \xrightarrow{L} N(\mathbf{0}, \mathbf{I}_k), \quad \text{as } n \to \infty.$$

27. (Sections 2.4, 2.3, 1.3) Let

$$Y_t = \beta_1 x_t + e_t, \qquad X_t = x_t + u_t,$$
$$(x_t, e_t, u_t)' \sim \mathrm{NI}[(\mu_x, 0, 0)', \mathrm{diag}(\sigma_{xx}, \sigma_{ee}, \sigma_{uu})],$$

where $\mu_x \neq 0$.

(a) Find the maximum likelihood estimator of $(\beta_1, \mu_x, \sigma_{xx}, \sigma_{ee}, \sigma_{uu})$ for a sample of n vectors (Y_t, X_t).

(b) Give the covariance matrix of the limiting distribution of the maximum likelihood estimator of part (a).

(c) Find the covariance matrix of the limiting distribution of the maximum likelihood estimator of $(\beta_1, \mu_x, \sigma_{xx}, \sigma_{uu})$ under the assumption that it is known that $\sigma_{ee} = \sigma_{uu}$. Show that the variance of the limiting distributtion of the estimator of β_1 is that of the optimal linear combination of $\bar{X}^{-1}\bar{Y}$ and estimator (1.3.7).

28. (Section 2.4)

(a) Estimate the equation

$$Z_{t1} = \beta_{10} + \beta_{13}z_{t3} + \varepsilon_{t3}$$

for the data of Example 2.4.1, using $(Z_{t2}, Z_{t4}, Z_{t5}, Z_{t6}, Z_{t7}, Z_{t8})$ as instrumental variables. Is the model acceptable?

(b) Estimate the equation

$$Z_{t7} = \beta_{70} + \beta_{73}z_{t3} + \beta_{76}z_{t6} + \beta_{78}z_{t8} + \varepsilon_{t7}$$

for the data of Example 2.4.1, using $(Z_{t1}, Z_{t2}, Z_{t4}, Z_{t5})$ as instrumental variables. What do you conclude?

29. (Section 2.4) Show that when $q > k$ the estimated error variance s_{vv} given in (2.4.24) is the weighted sum of two independent estimates, one that is a function of $\mathbf{S}_{aa}$ and one that is a function of $\hat{\boldsymbol{\pi}}$.

30. (Section 2.4) Verify, through the following steps, that (2.4.14) and (2.4.15) give the same estimator of $\boldsymbol{\beta}$:

(i) Define $\mathbf{P}_w = \mathbf{W}(\mathbf{W}'\mathbf{W})^{-1}\mathbf{W}'$. In linear model theory, $\mathbf{P}_w$ is called the orthogonal projection onto the column space of $\mathbf{W}$. Show that

$$\mathbf{S}_{aa} = (n-q)^{-1}(\mathbf{Y}, \mathbf{X})'[\mathbf{I}_q - \mathbf{P}_w](\mathbf{Y}, \mathbf{X}).$$

(ii) Show that $(\hat{\mathbf{Y}}, \hat{\mathbf{X}})'(\hat{\mathbf{Y}}, \hat{\mathbf{X}}) - v\mathbf{S}_{aa} = (\mathbf{Y}, \mathbf{X})'(\mathbf{Y}, \mathbf{X}) - [v + (n-q)]\mathbf{S}_{aa}$ and, hence, that $\tilde{v}$ defined in (2.4.13) and $\tilde{\gamma}$ defined in (2.4.16) satisfy the relationship

$$\tilde{\gamma} = [\tilde{v} + (n-q)].$$

(iii) Show that, for $\tilde{v}$ and $\tilde{\gamma}$ as defined in (ii),

$$\hat{\mathbf{X}}'\hat{\mathbf{X}} - \tilde{v}\mathbf{S}_{aa22} = \mathbf{X}'\mathbf{X} - \tilde{\gamma}\mathbf{S}_{aa22},$$
$$\hat{\mathbf{X}}'\hat{\mathbf{Y}} - \tilde{v}\mathbf{S}_{aa21} = \mathbf{X}'\mathbf{Y} - \tilde{\gamma}\mathbf{S}_{aa21}.$$

31. (Section 2.4) Verify formula (2.4.21). (Hint: First derive an expression for $\hat{\boldsymbol{\beta}} - \boldsymbol{\beta}$ similar to (2.4.20), but with $\bar{\mathbf{M}}_{WW}$ replaced by $\mathbf{M}_{WW}$.)

32. (Section 2.4) Let q instrumental variables $(W_{t1}, W_{t2}, \ldots, W_{tq})$ be available for the single-x model (1.4.1). Let $\hat{\beta}_1$ be given by (2.4.14) and let

$$\hat{v}_t = Y_t - \bar{Y} - \hat{\beta}_1(X_t - \bar{X}).$$

Is it true that $\Sigma_{t=1}^{n} W_{ti}\hat{v}_t = 0$ for $i = 1, 2, \ldots, q$?

33. (Sections 2.4, 2.2) The $(\hat{\mathbf{Y}}, \hat{\mathbf{X}})$ used in the instrumental variable estimator (2.4.14) can be viewed as the best predictor of $(\mathbf{y}, \mathbf{x})$ given $\mathbf{W}$, where $\mathbf{y} = \mathbf{x}\boldsymbol{\beta}$. Recall that

$$\tilde{\mathbf{x}}_t = \bar{\mathbf{X}} + (\mathbf{X}_t - \bar{\mathbf{X}})\mathbf{m}_{XX}^{-1}(\mathbf{m}_{XX} - \boldsymbol{\Sigma}_{uu})$$

is an estimator of the best predictor of $\mathbf{x}_t$ given $\mathbf{X}_t$ when $\boldsymbol{\Sigma}_{uu}$ is known and

$$\tilde{y}_t = \bar{Y} + (\mathbf{X}_t - \bar{\mathbf{X}})\mathbf{m}_{XX}^{-1}\mathbf{m}_{XY}$$

is an estimator of the best predictor of y_t given $\mathbf{X}_t$. Let

$$\mathbf{S}_{aa} = (n-k)^{-1}\sum_{t=1}^{n}(Y_t - \tilde{y}_t, \mathbf{X}_t - \tilde{\mathbf{x}}_t)'(Y_t - \tilde{y}_t, \mathbf{X}_t - \tilde{\mathbf{x}}_t)$$

and assume $\boldsymbol{\Sigma}_{ue} = \mathbf{0}$.

Show that the estimator (2.4.14) with the rows of $(\hat{\mathbf{Y}}, \hat{\mathbf{X}})$ equal to $(\tilde{y}_t, \tilde{\mathbf{x}}_t)$ is the estimator of $\boldsymbol{\beta}$ defined in (2.2.12).

2.5. MODIFICATIONS TO IMPROVE MOMENT PROPERTIES

The estimators that we have been studying have limiting distributions, but the estimators do not necessarily have finite means and variances in small samples. In this section we introduce modified estimators that possess finite moments and that have small sample properties that are superior to those of the maximum likelihood estimators.

2.5.1. An Error in the Equation

The model of Section 1.2 is

$$Y_t = \beta_0 + x_t\beta_1 + e_t, \qquad X_t = x_t + u_t, \tag{2.5.1}$$
$$(x_t, e_t, u_t)' \sim \text{NI}[(\mu_x, 0, 0)', \text{diag}(\sigma_{xx}, \sigma_{ee}, \sigma_{uu})],$$

where σ_{uu} is known and $\sigma_{xx} > 0$. The estimator

$$\hat{\beta}_1 = (m_{XX} - \sigma_{uu})^{-1} m_{XY}, \tag{2.5.2}$$

introduced in Section 1.2, is a ratio of two random variables. Such ratios are typically biased estimators of the ratio of the expectations. In fact, the expectation of the estimator (2.5.2) is not defined. See Exercise 1.13. We shall demonstrate that it is possible to modify the estimator (2.5.2) to produce an estimator that is nearly unbiased for β_1. To this end we define the alternative estimator,

$$\tilde{\beta}_1 = [\hat{H}_{xx} + \alpha(n-1)^{-1}\sigma_{uu}]^{-1} m_{XY}, \tag{2.5.3}$$

where $\alpha > 0$ is a fixed number to be determined,

$$\hat{H}_{xx} = \begin{cases} m_{XX} - \sigma_{uu} & \text{if } \hat{\lambda} \geqslant 1 + (n-1)^{-1} \\ m_{XX} - [\hat{\lambda} - (n-1)^{-1}]\sigma_{uu} & \text{if } \hat{\lambda} < 1 + (n-1)^{-1}, \end{cases} \tag{2.5.4}$$

and $\hat{\lambda}$ is the root of

$$|\mathbf{m}_{ZZ} - \lambda \, \text{diag}(0, \sigma_{uu})| = 0. \tag{2.5.5}$$

The estimator $\tilde{\beta}_1$ is a modification of the maximum likelihood estimator. Note that the estimator of σ_{xx}, denoted by $\hat{H}_{xx}$, is never less than $(n-1)^{-1}\sigma_{uu}$. The additional modification associated with α produces a denominator in the estimator of β_1 that is never less than $(n-1)^{-1}(1+\alpha)\sigma_{uu}$. Because the denominator in the estimator of β_1 is bounded below by a positive number, the estimator has moments. The approximate mean of the estimator is given in Theorem 2.5.1.

Theorem 2.5.1. Let model (2.5.1) hold with $\sigma_{xx} > 0$, $\sigma_{uu} > 0$, and $\sigma_{ee} > 0$. Then

$$E\{\tilde{\beta}_1 - \beta_1\} = (n-1)^{-1}\sigma_{xx}^{-1}\{\alpha - 2 - 2\sigma_{xx}^{-1}\sigma_{uu}\}\sigma_{uv} + O(n^{-2}), \tag{2.5.6}$$

where $v_t = e_t - u_t\beta_1$. Furthermore, the mean square error of $\tilde{\beta}_1$ through terms of $O(n^{-2})$ is smaller for $\alpha = 5$ than for any smaller α, uniformly in the parameters.

Proof. The root $\hat{\lambda}$ is given by

$$\hat{\lambda} = \sigma_{uu}^{-1}[m_{XX} - m_{YY}^{-1} m_{YX}^2].$$

The quantity in the square brackets is the residual sum of squares obtained in the regression of X on Y multiplied by $(n-1)^{-1}$. The residual sum of squares divided by $\sigma_{uu} + \sigma_{YY}^{-1}\sigma_{ee}\sigma_{xx}$ is distributed as a chi-square random variable with $n-2$ degrees of freedom. Therefore,

$$E\{\hat{\lambda}\} = (n-1)^{-1}(n-2)[1 + (\sigma_{uu}\sigma_{YY})^{-1}\sigma_{ee}\sigma_{xx}]$$

and, for n sufficiently large,

$$\begin{aligned} P\{\hat{\lambda} < 1 + n^{-1}\} &= P\{\hat{\lambda} - E(\hat{\lambda}) < 1 + n^{-1} - E(\hat{\lambda})\} \\ &\leqslant P\{|\hat{\lambda} - E(\hat{\lambda})| > |1 + n^{-1} - E(\hat{\lambda})|\} \\ &\leqslant [1 + n^{-1} - E(\hat{\lambda})]^{-4} E\{[\hat{\lambda} - E(\hat{\lambda})]^4\} \\ &= O(n^{-2}), \end{aligned}$$

where we have used Chebyshev's inequality and the moment properties of the chi-square distribution. Therefore, it is sufficient to consider the distribution of the estimator when $\hat{\lambda} > 1 + (n-1)^{-1}$ and we write

$$\begin{aligned} \tilde{\beta}_1 - \beta_1 = {} & [m_{XX} - (n-1)^{-1}(n-1-\alpha)\sigma_{uu}]^{-1} \\ & \times [m_{Xv} + (n-1)^{-1}(n-1-\alpha)\sigma_{uu}\beta_1] + O_p(n^{-2}). \end{aligned}$$

Expanding $\tilde{\beta}_1 - \beta_1$ in a Taylor series we have

$$\begin{aligned} \tilde{\beta}_1 - \beta_1 = {} & \sigma_{xx}^{-1}[m_{xv} + m_{uv} - \sigma_{uv} + \alpha(n-1)^{-1}\sigma_{uv}] \\ & - \sigma_{xx}^{-2}[m_{xv} + m_{uv} - \sigma_{uv} + \alpha(n-1)^{-1}\sigma_{uv}] \\ & \times [2m_{xu} + m_{xx} - \sigma_{xx} + m_{uu} - \sigma_{uu} + \alpha(n-1)^{-1}\sigma_{uu}] + O_p(n^{-3/2}), \end{aligned}$$

where, for example, $m_{xv} = (n-1)^{-1}\sum_{t=1}^{n}(x_t - \bar{x})(v_t - \bar{v})$. It can be verified that the conditions of Theorems 5.4.3 and 5.4.4 of Fuller (1976) are satisfied. Therefore, we may take the expectations of the first two terms in the Taylor series to obtain the bias result. (Recall the moment properties of the normal distribution.)

The terms in the square $(\tilde{\beta}_1 - \beta_1)^2$ that contain α and whose expectations are $O(n^{-2})$ are

$$\begin{aligned} & (\alpha n^{-1}\sigma_{uv})^2 - 4\sigma_{xx}^{-3}(\alpha n^{-1}\sigma_{uv})(m_{xv} + m_{uv} - \sigma_{uv})(2m_{xu} + m_{uu} - \sigma_{uu}) \\ & - 2\sigma_{xx}^{-3}(\alpha n^{-1}\sigma_{uu})(m_{xv} + m_{uv} - \sigma_{uv})^2. \end{aligned}$$

The expected value of this expression is

$$n^{-2}\sigma_{xx}^{-2}[\sigma_{uv}^2(\alpha^2 - 8\alpha) - 2\alpha\sigma_{uu}\sigma_{vv} - 2\alpha\sigma_{xx}^{-1}\sigma_{uu}(\sigma_{uu}\sigma_{vv} + 3\sigma_{uv}^2)] + O(n^{-3}).$$

Because $\sigma_{uu}\sigma_{vv} > \sigma_{uv}^2$, the expression is smaller for $\alpha = 5$ than for $\alpha < 5$.

□

Monte Carlo methods can be used to demonstrate that the theoretical superiority of the modified estimator is realized in small samples. Table 2.5.1

TABLE 2.5.1. The maximum likelihood estimator adjusted for degrees of freedom ($\hat{\beta}_1$) and modified estimator ($\tilde{\beta}_1$) for 25 samples of size 21

Sample Number	m_{XX}	m_{XY}	m_{YY}	$\hat{\lambda}$	$\hat{\beta}_1$	$\tilde{\beta}_1$ $\alpha = 2$	$\tilde{\beta}_1$ $\alpha = 4 - 2\hat{\kappa}_{xx}$
1	0.89	0.12	1.46	0.88	11.67	0.78	0.45
2	1.14	0.90	1.79	0.69	1.99	1.49	1.30
3	1.27	0.23	1.83	1.24	0.84	0.61	0.51
4	1.28	0.83	2.15	0.96	2.58	1.76	1.51
5	1.56	0.65	1.64	1.31	1.15	0.98	0.89
6	1.61	1.11	2.22	1.06	1.81	1.55	1.43
7	1.69	0.41	1.52	1.58	0.60	0.52	0.48
8	1.71	1.18	1.87	0.97	1.58	1.32	1.24
9	1.74	0.56	1.55	1.53	0.76	0.67	0.63
10	1.74	1.53	3.05	0.97	1.99	1.67	1.57
11	1.86	0.88	1.21	1.23	1.02	0.91	0.86
12	1.93	1.37	3.07	1.31	1.48	1.34	1.27
13	1.94	0.76	1.66	1.59	0.81	0.73	0.70
14	1.95	1.34	2.37	1.19	1.41	1.28	1.22
15	2.01	0.56	2.19	1.87	0.56	0.51	0.49
16	2.02	1.59	1.74	0.58	1.10	0.99	0.96
17	2.09	0.74	2.01	1.82	0.68	0.62	0.59
18	2.11	1.65	3.85	1.41	1.48	1.36	1.31
19	2.17	1.05	1.98	1.62	0.89	0.82	0.80
20	2.29	1.44	2.52	1.47	1.11	1.03	1.00
21	2.34	0.66	1.10	1.95	0.49	0.46	0.44
22	2.59	1.45	1.92	1.50	0.91	0.86	0.84
23	2.95	1.20	2.40	2.34	0.62	0.59	0.58
24	3.38	2.19	2.24	1.24	0.92	0.88	0.87
25	3.70	1.41	2.19	2.80	0.52	0.50	0.50

contains the sample moments and estimators for 25 samples of 21 observations generated by the model

$$Y_t = x_t + e_t, \qquad X_t = x_t + u_t, \tag{2.5.7}$$
$$(x_t, e_t, u_t)' \sim \mathrm{NI}(\mathbf{0}, \mathbf{I}).$$

The samples were constructed so that the mean of m_{XX} over the samples is equal to the population σ_{XX}. Therefore, this set of samples contains more information about the population of samples than would a simple random sample of 25 samples.

Three estimators of β_1 are given in the table. The first estimator is the maximum likelihood estimator adjusted for degrees of freedom and is defined by

$$\hat{\beta}_1 = \begin{cases} (m_{XX} - \sigma_{uu})^{-1} m_{XY} & \text{if } \hat{\lambda} > 1 \\ m_{XY}^{-1} m_{YY} & \text{if } \hat{\lambda} \leqslant 1, \end{cases} \tag{2.5.8}$$

where $\hat{\lambda}$ is the root of (2.5.5). The root $\hat{\lambda}$ is given in the fifth column of the table. The second estimator is the modified estimator (2.5.3) with $\alpha = 2$.

The third estimator is the modified estimator with

$$\alpha = 2 + 2m_{XX}^{-1}\sigma_{uu} = 4 - 2\hat{\kappa}_{xx},$$

where $\hat{\kappa}_{xx} = m_{XX}^{-1}(m_{XX} - \sigma_{uu})$. The bias expression of Theorem 2.5.1 furnishes the motivation for the third estimator. The quantity $m_{XX}^{-1}\sigma_{uu}$ is a biased estimator of $\sigma_{xx}^{-1}\sigma_{uu}$, but it is preferred to other estimators of the ratio because of its smaller variance.

The samples of Table 2.5.1 are ordered on m_{XX}. This makes it clear that the α modification produces large improvements in the estimator for small m_{XX}, while producing modest losses for large m_{XX}. Sample number one is an example of the type of sample that produces estimates of large absolute value. It is because of such samples that the maximum likelihood estimator does not have moments and it is for samples of this type that the modification produces large improvements.

Figure 2.5.1 contains a histogram for the maximum likelihood estimator adjusted for degrees of freedom constructed for 2000 samples generated by model (2.5.7). Figure 2.5.2 is the corresponding histogram for the modified estimator with $\alpha = 2$. The superiority of the modified estimator is clear in these figures. The empirical distribution of the maximum likelihood estimator has a very heavy tail with a few observations well beyond the range of the figure. The spike to the right of the break in the scale of the abscissa of Figure 2.5.1 indicates that about 3% of the maximum likelihood estimates are greater than 3.5. On the other hand, the largest value for $\tilde{\beta}_1$ in the 2000 samples is 2.36. About 1% of the samples gave negative values for both estimators.

The percentiles of the empirical distributions of $\hat{\beta}_1$ and $\tilde{\beta}_1$ for 2000 samples are given in Table 2.5.2. Percentiles are given for the population with $\sigma_{xx} = 1$ and for the population with $\sigma_{xx} = 4$. The entire distribution of the modified estimator is shifted to the left relative to the distribution of the maximum likelihood estimator. The shifts of the upper percentiles toward one are much larger than the shifts in the lower percentiles away from one. The empirical mean for the modified estimator for the population with $\sigma_{xx} = 1$ is 0.963 and the empirical variance is 0.190. The variance of the approximate distribution

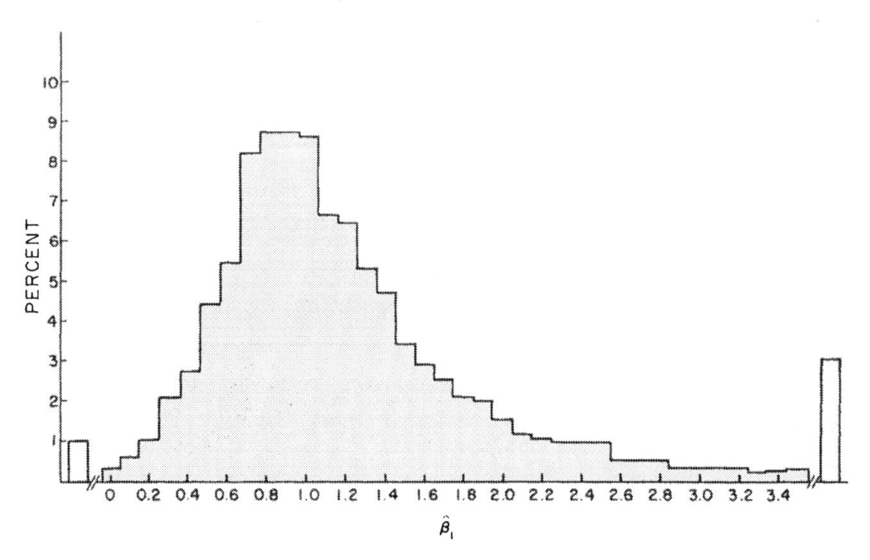

FIGURE 2.5.1. Histogram for 2000 maximum likelihood estimates.

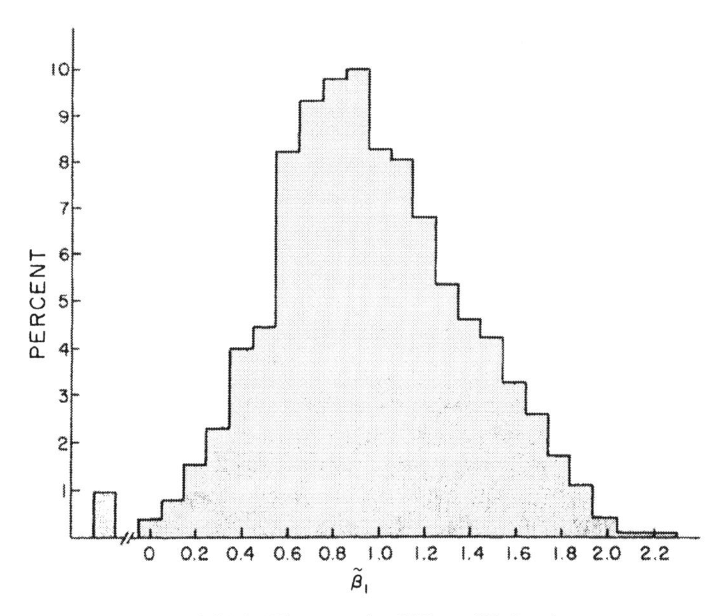

FIGURE 2.5.2. Histogram for 2000 modified estimates.

TABLE 2.5.2. Monte Carlo percentiles for maximum likelihood estimator adjusted for degrees of freedom ($\hat{\beta}_1$) and modified estimator ($\tilde{\beta}_1$ with $\alpha = 2$), $\sigma_{ee} = 1$, $\sigma_{uu} = 1$, $\beta_1 = 1$, and $n = 21$

	$\sigma_{xx} = 1$		$\sigma_{xx} = 4$	
Percentile	$\hat{\beta}_1$	$\tilde{\beta}_1$	$\hat{\beta}_1$	$\tilde{\beta}_1$
0.01	−0.07	−0.04	0.63	0.61
0.05	0.35	0.30	0.72	0.71
0.10	0.51	0.45	0.78	0.76
0.25	0.74	0.68	0.89	0.87
0.50	1.03	0.93	1.01	0.99
0.75	1.46	1.25	1.15	1.11
0.90	2.16	1.54	1.31	1.25
0.95	2.82	1.71	1.42	1.34
0.99	5.94	1.96	1.67	1.55

of the estimator for the population with $\sigma_{xx} = 1$ is

$$V\{\hat{\beta}_1\} = V\{\tilde{\beta}_1\} = (n-1)^{-1}\sigma_{xx}^{-2}(\sigma_{xx}\sigma_{vv} + \sigma_{uv}^2)$$
$$= (20)^{-1}[2(2) + 1] = 0.25,$$

which is larger than the empirical variance. There are at least two explanations for this. First, the distributional approximation makes no allowance for the $\hat{\lambda}$ adjustment in the estimator. For the parametric configuration of our example this adjustment takes place about 19% of the time. Second, the modification has a definite effect on the variance of the estimator for samples of this size. The variance effect is evident in Figure 2.5.1.

The Monte Carlo mean and variance of $\tilde{\beta}_1$ for the population with $\sigma_{xx} = 4$ are 1.000 and 0.0382, respectively. The Monte Carlo variance is larger than 0.0344 which is the variance of the approximate distribution. For the population with $\sigma_{xx} = 4$, the modification has less effect on the variance than it does for the $\sigma_{xx} = 1$ population. Also, the $\hat{\lambda}$ adjustment to the estimator is relatively infrequent when $\sigma_{xx} = 4$.

The empirical mean and variance for the estimator with $\alpha = 2 + 2m_{XX}^{-1}\sigma_{uu}$ were 0.883 and 0.133, respectively, for the population with $\sigma_{xx} = 1$ and were 0.992 and 0.0357, respectively, for the population with $\sigma_{xx} = 4$. Therefore, the estimator with $\alpha = 2 + 2m_{XX}^{-1}\sigma_{uu}$ has a smaller mean square error than the estimator with $\alpha = 2$, a result that agrees with the conclusion in Theorem 2.5.1.

The Monte Carlo percentiles of three test statistics are compared with the percentiles of Student's t distribution and the standard normal distribution

TABLE 2.5.3. Monte Carlo percentiles for alternative test statistics for the model with $\sigma_{uu} = 1$, $\sigma_{ee} = 1$, $\beta_1 = 1$, and $n = 21$

Parameter and Variable	Percentile					
	0.01	0.05	0.10	0.90	0.95	0.99
$\sigma_{xx} = 1$						
$\hat{t}_1$ for $\hat{\beta}_1$	−2.99	−1.94	−1.43	0.72	0.83	1.06
$\tilde{t}_1$ for $\tilde{\beta}_1$	−3.20	−2.17	−1.68	0.63	0.75	0.97
$\tilde{t}_{R1}$	−2.24	−1.57	−1.27	1.28	1.59	1.97
$\sigma_{xx} = 4$						
$\hat{t}_1$ for $\hat{\beta}_1$	−2.56	−1.73	−1.29	1.18	1.44	1.80
$\tilde{t}_1$ for $\tilde{\beta}_1$	−2.71	−1.87	−1.45	1.06	1.32	1.72
$\tilde{t}_{R1}$	−2.22	−1.67	−1.30	1.34	1.69	2.33
Student's t_{19}	−2.54	−1.73	−1.33	1.33	1.73	2.54
$N(0, 1)$	−2.33	−1.64	−1.28	1.28	1.64	2.33

in Table 2.5.3. The statistics are

$$\hat{t}_1 = [\hat{V}\{\hat{\beta}_1\}]^{-1/2}(\hat{\beta}_1 - \beta_1),$$
$$\tilde{t}_1 = [\tilde{V}\{\tilde{\beta}_1\}]^{-1/2}(\tilde{\beta}_1 - \beta_1),$$
$$\tilde{t}_{R1} = [\text{sgn}(\tilde{\beta}_1 - \beta_1)]\hat{\chi},$$

where

$$\hat{V}\{\hat{\beta}_1\} = (n-1)^{-1}[\hat{\sigma}_{xx}^{-1}s_{vv} + \hat{\sigma}_{xx}^{-2}(\sigma_{uu}s_{vv} + \hat{\beta}_1^2\sigma_{uu}^2)],$$

$$\hat{\sigma}_{xx} = \begin{cases} m_{XX} - \sigma_{uu} & \text{if } \hat{\lambda} > 1 \\ m_{XX} - \hat{\lambda}\sigma_{uu} & \text{otherwise,} \end{cases}$$

$$s_{vv} = \begin{cases} (n-2)^{-1}\sum_{t=1}^{n}[Y_t - \bar{Y} - (X_t - \bar{X})\hat{\beta}_1]^2 & \text{if } \hat{\lambda} > 1 \\ \hat{\beta}_1^2\sigma_{uu} & \text{otherwise,} \end{cases}$$

$$\tilde{V}\{\tilde{\beta}_1\} = (n-1)^{-1}[\hat{H}_{xx}^{-1}\tilde{\sigma}_{vv} + \hat{H}_{xx}^{-2}(\sigma_{uu}\tilde{\sigma}_{vv} + \tilde{\beta}_1^2\sigma_{uu}^2)],$$
$$\tilde{\sigma}_{vv} = (n-2)^{-1}(n-1)(m_{YY} - 2\tilde{\beta}_1 m_{XY} + \tilde{\beta}_1^2 m_{XX}),$$

$\hat{H}_{xx}$ is defined in (2.5.3), $\text{sgn}(\tilde{\beta}_1 - \beta_1)$ is the sign of $\tilde{\beta}_1 - \beta_1$, and $\hat{\chi}$ is the square root of the likelihood ratio statistic for the hypothesis that the true value of the parameter is $\boldsymbol{\beta}_1$. See Exercise 2.37. The percentiles are given for two parametric configurations. The first is that of (2.5.7). The second parameter set is the same as the first except that $\sigma_{xx} = 4$ instead of $\sigma_{xx} = 1$. The distributions of $\hat{t}_1$ and $\tilde{t}_1$ differ considerably from that of Student's t

with 19 degrees of freedom. All tail percentiles of the statistics for the population with $\sigma_{xx} = 1$ are to the left of the tail percentiles of Student's t. At first glance, it might be surprising that the distribution of the Studentized statistic is skewed to the left when the distribution of the estimators is skewed to the right. This occurs because large estimates of β_1 are associated with small values of m_{XX}. Small values of m_{XX} produce large estimates of $V\{\hat{\beta}_1\}$ because of the small estimates for σ_{xx}. For example, the sample with $(m_{XX}, m_{XY}, m_{YY}) = (0.812, 0.058, 1.925)$ yielded the statistics

$$\hat{\beta}_1 = 33.47 \quad \text{and} \quad \hat{t}_1 = 0.005.$$

Conversely, the majority of the small estimates for β_1 are associated with large m_{XX}, and large m_{XX} produce small estimates of the variance of $\hat{\beta}_1$. For example, the sample with $(m_{XX}, m_{XY}, m_{YY}) = (3.146, 0.649, 2.122)$ yielded the statistics

$$\hat{\beta}_1 = 0.302 \quad \text{and} \quad \hat{t}_1 = -2.57.$$

Because the distribution of $\tilde{\beta}_1$ is moved left relative to that of $\hat{\beta}_1$, the distribution of $\tilde{t}_1$ is to the left of that of $\hat{t}_1$.

The percentiles of the likelihood ratio statistics are in much better agreement with percentiles suggested by large sample theory than are the percentiles of the Studentized statistics. The Monte Carlo results suggest that the normal distribution, rather than Student's t, be used to approximate the percentiles of the transformed likelihood ratio statistic. In fact, the observed percentiles for the tranformed likelihood ratio statistic for the model with $\sigma_{xx} = 1$ are closer to zero than are the percentiles of the normal distribution. We believe this is so for the same reasons that the observed variance of $\tilde{\beta}_1$ is less than that suggested by the large sample theory.

We state the following generalization of Theorem 2.5.1 without proof. The proof is of the same form as that of Theorem 2.5.1. An analogous theorem holds for $\mathbf{x}_t$ fixed.

Theorem 2.5.2. Let

$$Y_t = \beta_0 + \mathbf{x}_t\boldsymbol{\beta}_1 + e_t, \qquad \mathbf{X}_t = \mathbf{x}_t + \mathbf{u}_t,$$

where $\mathbf{x}_t$ is a k-dimensional row vector, $\mathbf{u}_t$ is a k-dimensional row vector, and $\boldsymbol{\beta}_1$ is a k-dimensional column vector. Let

$$(\mathbf{x}_t, \boldsymbol{\varepsilon}_t)' \sim \text{NI}[(\boldsymbol{\mu}_x, \mathbf{0})', \text{ block diag}(\boldsymbol{\Sigma}_{xx}, \boldsymbol{\Sigma}_{\varepsilon\varepsilon})],$$

where $\boldsymbol{\Sigma}_{xx}$ is a $k \times k$ positive definite matrix, $e_t = q_t + w_t$, and $q_j \sim \text{NI}(0, \sigma_{qq})$ independent of $(\mathbf{x}_t, w_t, \mathbf{u}_t)$ for all t and j. Let $\mathbf{S}_{aa}$ be an estimator of the covariance matrix of $\mathbf{a}_t = (w_t, \mathbf{u}_t)$ distributed as a multiple of a Wishart matrix with degrees of freedom $d_f = v^{-1}n$, where v is a fixed number. Assume $\mathbf{S}_{aa}$ is independent of $(\mathbf{x}_t, q_t, \mathbf{a}_t)$ for all t. Let

$$\tilde{\boldsymbol{\beta}}_1 = [\hat{\mathbf{H}}_{xx} + \alpha(n-1)^{-1}\mathbf{S}_{uu}]^{-1}[\hat{\mathbf{N}}_{xy} + \alpha(n-1)^{-1}\mathbf{S}_{uw}], \qquad (2.5.9)$$

where $\alpha > 0$ is a fixed real number,

$$\hat{\mathbf{H}}_{xx} = \begin{cases} \mathbf{m}_{XX} - \mathbf{S}_{uu} & \text{if } \hat{\lambda} \geq 1 + (n-1)^{-1} \\ \mathbf{m}_{XX} - [\hat{\lambda} - (n-1)^{-1}]\mathbf{S}_{uu} & \text{if } \hat{\lambda} < 1 + (n-1)^{-1}, \end{cases}$$

$$\hat{\mathbf{N}}_{xy} = \begin{cases} \mathbf{m}_{XY} - \mathbf{S}_{uw} & \text{if } \hat{\lambda} \geq 1 + (n-1)^{-1} \\ \mathbf{m}_{XY} - [\hat{\lambda} - (n-1)^{-1}]\mathbf{S}_{uw} & \text{if } \hat{\lambda} < 1 + (n-1)^{-1}, \end{cases}$$

$\hat{\lambda}$ is the smallest root of

$$|\mathbf{m}_{ZZ} - \lambda\mathbf{S}_{aa}| = 0,$$

and $\mathbf{Z}_t = (Y_t, \mathbf{X}_t)$. Then

$$E\{\tilde{\boldsymbol{\beta}}_1 - \boldsymbol{\beta}_1\} = -(n-1)^{-1}\boldsymbol{\Sigma}_{xx}^{-1}\{(k+1-\alpha) + (1+v)[\mathbf{I}\,\mathrm{tr}(\boldsymbol{\Sigma}_{uu}\boldsymbol{\Sigma}_{xx}^{-1}) + \boldsymbol{\Sigma}_{uu}\boldsymbol{\Sigma}_{xx}^{-1}]\}\boldsymbol{\Sigma}_{uv} + O(n^{-2}).$$

Furthermore, through terms of order n^{-2}, the mean square error of $\tilde{\boldsymbol{\beta}}_1$ is smaller for $\alpha = k + 4 + 2v$ than for any smaller α.

Proof. Omitted. □

The modifications of the estimators introduced in Theorems 2.5.1 and 2.5.2 guarantee that the estimator of $\boldsymbol{\Sigma}_{xx}$, denoted by $\hat{\mathbf{H}}_{xx}$, is always positive definite and that the estimator of $\boldsymbol{\beta}$ possesses finite moments. It seems clear that one should never use an α smaller than $k + 1$.

The estimators are constructed on the assumption that $\boldsymbol{\Sigma}_{xx}$ is positive definite. In practice one will wish to investigate the hypothesis that $\boldsymbol{\Sigma}_{xx}$ is positive definite. The root $\hat{\lambda}$ provides information about the validity of the model assumptions. A small root, relative to the tabular value, suggests that $\mathbf{m}_{zz}$ is singular or that the model is otherwise incorrectly specified. For the model of Theorem 2.5.2, $\mathbf{m}_{zz}$ can be singular if $\sigma_{qq} = 0$ or if $\mathbf{m}_{xx}$ is singular. Study of the sample covariance matrix of $\mathbf{X}$ helps one discriminate between the two possibilities.

When the model contains several independent variables measured with error, the smallest root, $\hat{\gamma}$, of the determinantal equation

$$|\mathbf{m}_{XX} - \gamma\mathbf{S}_{uu}| = 0 \qquad (2.5.10)$$

can be used to test the hypothesis that $|\boldsymbol{\Sigma}_{xx}| = 0$. By Theorem 2.3.2, $F = (n-k)^{-1}(n-1)\hat{\gamma}$ is approximately distributed as Snedecor's F with $n - k$ and d_f degrees of freedom when the rank of $\boldsymbol{\Sigma}_{xx}$ is $k - 1$. When F is small, the data do not support the assumption that the model is identified. One must decide if one has enough confidence in the assumption that $|\Sigma_{xx}| > 0$ to warrant constructing an estimator of $\boldsymbol{\beta}$.

Theorem 2.5.2 is for the sequence of estimators indexed by n, where n is the number of observations. The error variance, Σ_{uu}, is constant. An alternative sequence of estimators can be constructed by letting the variance of $\mathbf{u}_t$ approach zero, holding n constant. As with Theorem 2.3.2, the results of Theorem 2.5.2 hold for such sequences. See Fuller (1977).

2.5.2. No Error in the Equation

A theorem analogous to Theorem 2.5.2 can be proved for the modification of the estimator constructed for the model of Section 2.3.

Theorem 2.5.3. Assume

$$Y_t = \mathbf{x}_t\boldsymbol{\beta} + e_t, \qquad \mathbf{X}_t = \mathbf{x}_t + \mathbf{u}_t, \qquad (2.5.11)$$
$$\boldsymbol{\varepsilon}_t \sim \mathrm{NI}(\mathbf{0}, \Sigma_{\varepsilon\varepsilon}),$$

for $t = 1, 2, \ldots, n$, where $\{\mathbf{x}_t\}$ is a fixed sequence of k-dimensional row vectors and $\boldsymbol{\varepsilon}_t = (e_t, \mathbf{u}_t)$. Assume that $\{|\mathbf{x}_t|\}$ is uniformly bounded and that

$$\lim_{n\to\infty} n^{-1} \sum_{t=1}^{n} \mathbf{x}_t'\mathbf{x}_t = \bar{\mathbf{M}}_{xx}, \qquad (2.5.12)$$

where $\bar{\mathbf{M}}_{xx}$ is nonsingular. Let

$$\hat{\boldsymbol{\beta}} = [\mathbf{M}_{XX} - (\hat{\lambda} - n^{-1}\alpha)\mathbf{S}_{uu}]^{-1}[\mathbf{M}_{XY} - (\hat{\lambda} - n^{-1}\alpha)\mathbf{S}_{ue}], \qquad (2.5.13)$$

where $\hat{\lambda}$ is the smallest root of $|\mathbf{M}_{ZZ} - \lambda\mathbf{S}_{\varepsilon\varepsilon}| = 0$, $\mathbf{S}_{\varepsilon\varepsilon}$ is an unbiased estimator of $\Sigma_{\varepsilon\varepsilon}$ distributed as a multiple of a Wishart matrix with d_f degrees of freedom, d_f is proportional to n, and $\alpha > 0$ is a fixed number. Then

$$E\{\hat{\boldsymbol{\beta}} - \boldsymbol{\beta}\} = [n^{-1}(1 - \alpha)\mathbf{I} + (n^{-1} + d_f^{-1})\bar{\mathbf{M}}_{xx}^{-1}\{\Sigma_{uu} - \sigma_{vv}^{-1}\Sigma_{uv}\Sigma_{vu}\}]\bar{\mathbf{M}}_{xx}^{-1}\Sigma_{uv}$$
$$+ O(n^{-2}).$$

Furthermore, through terms of order n^{-2}, the mean square error of $\hat{\boldsymbol{\beta}}$ is, uniformly in the parameters, smaller for $\alpha = 4$ than for any smaller α.

Proof. Omitted. See Fuller (1980). □

A result analogous to Theorem 2.5.3 holds for random $\mathbf{x}_t$.

In practice one can distinguish two situations. In one the error covariance matrix is known or estimated, and in the other the covariance matrix is known only up to a multiple. If the covariance matrix is known or estimated, the estimator (2.5.13) can be used. If the covariance is given by $\mathbf{\Upsilon}_{\varepsilon\varepsilon}\sigma^2$, where $\mathbf{\Upsilon}_{\varepsilon\varepsilon}$ is known and σ^2 is unknown, the suggested estimator is

$$\hat{\boldsymbol{\beta}} = [\mathbf{M}_{XX} - \hat{\lambda}(1 - n^{-1}\alpha)\mathbf{\Upsilon}_{uu}]^{-1}[\mathbf{M}_{XY} - \hat{\lambda}(1 - n^{-1}\alpha)\mathbf{\Upsilon}_{ue}], \quad (2.5.14)$$

where $\hat{\lambda}$ is the smallest root of $|\mathbf{M}_{ZZ} - \lambda\mathbf{\Upsilon}_{\varepsilon\varepsilon}| = 0$.

Table 2.5.4 contains the sample moments and estimators for 25 samples of 21 observations generated by the model

$$Y_t = \beta_0 + \beta_1 x_t + e_t, \qquad X_t = x_t + u_t, \qquad (2.5.15)$$
$$(x_t, e_t, u_t)' \sim \mathrm{NI}(\mathbf{0}, \mathbf{I}\sigma^2),$$

TABLE 2.5.4. Maximum likelihood estimator and modified estimators for 25 samples of size 21 with $\sigma_{xx} = 1$, $\sigma_{ee} = \sigma_{uu} = 1$, and $\beta_1 = 1$

Sample Number	m_{XX}	m_{XY}	m_{YY}	$\hat{\lambda}$	$\tilde{\beta}_1$	$\tilde{\beta}$ $\alpha = 1$	$\tilde{\beta}$ $\alpha = 2 - \hat{\kappa}_{xx}$
1	1.05	0.11	1.58	1.02	4.90	1.53	0.92
2	1.24	0.53	1.20	0.69	0.96	0.91	0.87
3	1.36	1.01	2.69	0.82	1.86	1.73	1.64
4	1.45	1.02	1.81	0.59	1.19	1.15	1.13
5	1.50	0.28	1.80	1.33	1.69	1.20	1.01
6	1.53	0.82	1.75	0.81	1.14	1.08	1.04
7	1.54	0.23	2.39	1.48	3.92	1.74	1.28
8	1.62	0.52	1.75	1.16	1.13	1.00	0.94
9	1.63	0.64	1.87	1.10	1.20	1.09	1.03
10	1.67	0.56	1.40	0.97	0.79	0.74	0.71
11	1.76	1.08	2.62	1.02	1.48	1.38	1.33
12	1.78	0.81	1.51	0.82	0.85	0.81	0.80
13	1.82	1.26	2.07	0.68	1.11	1.07	1.06
14	1.83	0.24	0.94	0.88	0.26	0.25	0.24
15	1.88	1.22	1.40	0.39	0.82	0.81	0.81
16	1.90	0.98	1.79	0.87	0.94	0.90	0.89
17	2.07	1.37	1.48	0.37	0.81	0.80	0.79
18	2.12	1.11	1.74	0.80	0.84	0.82	0.80
19	2.38	1.87	3.13	0.84	1.22	1.19	1.17
20	2.40	0.93	2.14	1.33	0.87	0.82	0.80
21	2.79	1.22	2.44	1.39	0.87	0.83	0.81
22	2.88	0.43	1.09	0.99	0.23	0.22	0.22
23	3.11	1.10	1.88	0.74	0.85	0.83	0.83
24	3.30	1.42	2.13	1.18	0.67	0.65	0.65
25	3.38	1.73	2.51	1.16	0.78	0.76	0.76

where $\beta_0 = 0$, $\beta_1 = 1$, and $\sigma^2 = 1$. The set of samples is such that the mean of m_{XX} for the 25 samples is equal to 2, the population σ_{XX}. The samples are ordered on m_{XX}. Estimators of β_1 were constructed for the samples under the assumption that σ^2 is unknown. The first estimator is the maximum likelihood estimator

$$\hat{\beta}_1 = (m_{XX} - \hat{\lambda})^{-1} m_{XY}, \tag{2.5.16}$$

where $\hat{\lambda}$ is the smallest root of

$$|\mathbf{m}_{ZZ} - \lambda \mathbf{I}| = 0. \tag{2.5.17}$$

The second estimator is the modified estimator (2.5.14) with $\alpha = 1$ and the third estimator is the modified estimator (2.5.14) with

$$\alpha = 1 + m_{XX}^{-1}\sigma_{uu} = 2 - \hat{\kappa}_{xx},$$

where $\hat{\kappa}_{xx} = m_{XX}^{-1}(m_{XX} - \sigma_{uu})$. Although larger estimates tend to be associated with small values of m_{XX}, the association is not as strong as that observed for the σ_{uu} known case of Section 2.5.1. Also, the effect of the modification on the estimator is not as marked as the effect for the σ_{uu} known case. Samples 1 and 7 are examples of the kinds of samples that produce estimates for β_1 that are large in absolute value. It is for such samples that the modification produces large improvements.

The percentiles of the Monte Carlo distributions for 2000 samples of size 21 are given in Table 2.5.5 for two sets of parameters. In the first set, $\sigma_{xx} = 1$ and in the second set, $\sigma_{xx} = 4$. The remaining parameters are the same in both sets with $\sigma_{ee} = \sigma_{uu} = 1$ and $\beta_1 = 1$. The modification moves the distribution to the left with large shifts in the upper percentiles. As a result, the

TABLE 2.5.5. Monte Carlo percentiles for maximum likelihood estimator ($\hat{\beta}$) and modified estimator ($\tilde{\beta}_1$ with $\alpha = 1$), $\sigma_{ee} = \sigma_{uu} = 1$, $\beta_1 = 1$, and $n = 21$

	$\sigma_{xx} = 1$		$\sigma_{xx} = 4$	
Percentile	$\hat{\beta}_1$	$\tilde{\beta}_1$	$\hat{\beta}_1$	$\tilde{\beta}_1$
0.01	0.11	0.11	0.63	0.63
0.05	0.42	0.39	0.74	0.73
0.10	0.54	0.52	0.80	0.79
0.25	0.76	0.73	0.89	0.88
0.50	1.00	0.95	1.00	0.99
0.75	1.30	1.21	1.13	1.11
0.90	1.73	1.54	1.26	1.23
0.95	2.15	1.78	1.35	1.32
0.99	4.69	2.56	1.60	1.54

TABLE 2.5.6. Monte Carlo percentiles for alternative test statistics for the model with $\sigma_{uu} = \sigma_{ee} = 1$, $\beta_1 = 1$, and $n = 21$

Parameter and Variable	Percentile					
	0.01	0.05	0.10	0.90	0.95	0.99
$\sigma_{xx} = 1$						
$\hat{t}_1$ for $\hat{\beta}_1$	−3.75	−2.36	−1.71	0.95	1.17	1.57
$\tilde{t}_1$ for $\tilde{\beta}_1$	−3.83	−2.46	−1.82	0.85	1.08	1.50
$\tilde{t}_{R1}$	−2.45	−1.74	−1.35	1.37	1.72	2.37
$\sigma_{xx} = 4$						
$\hat{t}_1$ for $\hat{\beta}_1$	−3.06	−1.99	−1.52	1.18	1.46	1.99
$\tilde{t}_1$ for $\tilde{\beta}_1$	−3.12	−2.06	−1.58	1.11	1.41	1.93
$\tilde{t}_{R1}$	−2.44	−1.72	−1.37	1.35	1.69	2.33
Student's t_{19}	−2.54	−1.73	−1.33	1.33	1.73	2.54
$N(0, 1)$	−2.33	−1.64	−1.28	1.28	1.64	2.33

distribution of the modified estimator is more nearly symmetric about the true value. The Monte Carlo mean and variance of the modified estimator with $\alpha = 1$ are 0.998 and 0.217, respectively, for $\sigma_{xx} = 1$ and 1.004 and 0.0342, respectively, for $\sigma_{xx} = 4$. The variance of the approximating distribution is

$$V\{\hat{\beta}_1\} = (20)^{-1}[2(2) - 1] = 0.15$$

for $\sigma_{xx} = 1$ and is

$$V\{\hat{\beta}_1\} = (320)^{-1}[5(2) - 1] = 0.0281$$

for $\sigma_{xx} = 4$. The observed variance is considerably larger than that of the approximating distribution for both parameter sets, but the percentage difference for $\sigma_{xx} = 4$ is about one-half of that for $\sigma_{xx} = 1$.

The Monte Carlo mean and variance of the modified estimator with $\alpha = 1 + m_{XX}^{-1}\sigma_{uu}$ are 0.947 and 0.152 for the population with $\sigma_{xx} = 1$ and are 1.000 and 0.0329 for the population with $\sigma_{xx} = 4$. The use of an α slightly larger than one reduces the mean square error, bringing it closer to the mean square error of the approximate distribution.

Table 2.5.6 contains the percentiles of the statistics;

$$\begin{aligned} \hat{t}_1 &= [\hat{V}\{\hat{\beta}_1\}]^{-1/2}(\hat{\beta}_1 - \beta_1), \\ \tilde{t}_1 &= [\tilde{V}\{\tilde{\beta}_1\}]^{-1/2}(\tilde{\beta}_1 - \beta_1), \\ \hat{t}_{R1} &= [\text{sgn}(\tilde{\beta}_1 - \beta_1)]\hat{\chi}, \end{aligned} \tag{2.5.18}$$

where

$$\hat{V}\{\hat{\beta}_1\} = (n-1)^{-1}[\hat{\sigma}_{xx}^{-1}s_{vv} + \hat{\sigma}_{xx}^{-2}(\hat{\sigma}_{uu}s_{vv} - \hat{\beta}_1^2\hat{\sigma}_{uu}^2)], \tag{2.5.19}$$

$$s_{vv} = (n-2)^{-1}(n-1)(m_{YY} - 2\hat{\beta}_1 m_{XY} + \hat{\beta}_1^2 m_{XX}),$$

$$\tilde{V}\{\tilde{\beta}_1\} = (n-1)^{-1}[\tilde{H}_{xx}^{-1}\tilde{\sigma}_{vv} + \tilde{H}_{xx}^{-2}(\tilde{\sigma}_{vv}\hat{\sigma}_{uu} - \tilde{\beta}_1^2\hat{\sigma}_{uu}^2)], \tag{2.5.20}$$

$$\tilde{\sigma}_{vv} = (n-2)^{-1}(n-1)(m_{YY} - 2\tilde{\beta}_1 m_{XY} + \tilde{\beta}_1^2 m_{XX}),$$

$$\tilde{H}_{xx} = m_{XX} - \hat{\lambda}[1 - (n-1)^{-1}],$$

$\hat{\sigma}_{uu} = \hat{\lambda}$, $\hat{\sigma}_{xx} = m_{XX} - \hat{\lambda}$, $\text{sgn}(\tilde{\beta}_1 - \beta_1)$ is the sign of $\tilde{\beta}_1 - \beta_1$, and $\hat{\chi}$ is the square root of the likelihood ratio statistic for the hypothesis that the true value of the parameter is β_1.

The Studentized statistics are skewed to the left in much the same manner and for the same reasons that the distributions of Table 2.5.3 are skewed to the left. Generally, samples with small m_{XX} produce large estimates for β_1, but even larger estimates of the variance of the approximate distribution. The percentiles of the transformed likelihood ratio statistic are close to those of a $N(0, 1)$ random variable and display only a modest skewness to the left. This is not surprising because Student's t and the likelihood ratio statistic are closely related. See Section 1.3.4.

2.5.3. Calibration

In Section 1.6.3 we discussed the prediction of Y_t given an observation on X_t. A closely related problem is the estimation of x_t given an observation on Y_t. The estimation of x_{n+1}, given Y_{n+1} and an estimated relationship between Y_t and x_t, is the calibration problem. Consider the calibration of a new instrument using a standard instrument whose error variance is known. Let a set of objects whose true value is not known be available. We assume the true values of the objects to be fixed. For example, the objects may have been selected purposely to guarantee a wide range of true values. Our model is

$$Y_t = \beta_0 + \beta_1 x_t + e_t, \qquad X_t = x_t + u_t, \tag{2.5.21}$$

$$(e_t, u_t)' \sim \text{NI}[\mathbf{0}, \text{diag}(\sigma_{ee}, \sigma_{uu})],$$

where $\beta_1 \neq 0$, the x_t are fixed, the X_t are the readings on the standard instrument, and σ_{uu} is known. Because $\beta_1 \neq 0$, we can also write

$$x_t = \gamma_0 + \gamma_1 y_t, \tag{2.5.22}$$

where $Y_t = y_t + e_t$, $\gamma_1 = \beta_1^{-1}$, and $\gamma_0 = -\beta_1^{-1}\beta_0$. Even if the measurement error is zero, the estimation of the inverse of β_1 presents problems and the modifications of this section are appropriate. It is suggested that γ_1 be estimated by

$$\hat{\gamma}_1 = \begin{cases} \dfrac{(m_{XX} - \sigma_{uu})m_{XY} + 2(n-1)^{-1}m_{XY}\sigma_{uu}}{m_{XY}^2 + (n-1)^{-1}(m_{XX}m_{YY} - m_{XY}^2)} & \text{if } \hat{\lambda} > 1 \\ m_{YY}^{-1}m_{XY} & \text{otherwise,} \end{cases} \tag{2.5.23}$$

where $\hat{\lambda}$ is the root of

$$|\mathbf{m}_{ZZ} - \lambda \operatorname{diag}(0, \sigma_{uu})| = 0 \tag{2.5.24}$$

and $\mathbf{Z}_t = (Y_t, X_t)$. The modification of adding terms with multipliers of $(n-1)^{-1}$ in the numerator and denominator of (2.5.23) produces an estimator of γ_1 with finite mean and variance.

Given that a sample of size n has been used to calibrate the instrument whose readings are denoted by Y, the estimated equation can be used to estimate the x value for a new object. Let the x value of the new object be denoted by x_{n+1} and let the reading on the new instrument for the new object be denoted by Y_{n+1}. Only Y_{n+1} is observed. Then the estimated x value is

$$x_{n+1}^{+} = \hat{\gamma}_0 + \hat{\gamma}_1 Y_{n+1}, \tag{2.5.25}$$

where $\hat{\gamma}_0 = \bar{X} - \hat{\gamma}_1 \bar{Y}$. The error in the estimator of x_{n+1} is

$$x_{n+1}^{+} - x_{n+1} = \bar{u} - \gamma_1 \bar{e} + (\hat{\gamma}_1 - \gamma_1)(y_{n+1} - \bar{Y}) + \hat{\gamma}_1 e_{n+1}, \tag{2.5.26}$$

where we have used $x_{n+1} = \bar{x} + \gamma_1(Y_{n+1} - \bar{y}) - \gamma_1 e_{n+1}$. The estimator of γ_1 given in (2.5.23) has finite variance and, by the arguments used in Theorem 2.5.1, it can be shown that the bias is $O(n^{-2})$. It follows that

$$\begin{aligned} E\{x_{n+1}^{+} - x_{n+1}\} &= O(n^{-2}), \\ E\{(x_{n+1}^{+} - x_{n+1})^2\} &= n^{-1}(\sigma_{uu} + \gamma_1^2\sigma_{ee}) + (y_{n+1} - \bar{y})^2 V\{\hat{\gamma}_1\} \\ &\quad + [\gamma_1^2 + V\{\hat{\gamma}_1\}]\sigma_{ee} + O(n^{-2}), \end{aligned} \tag{2.5.27}$$

where we have used the fact that e_{n+1} is independent of $(\hat{\gamma}_1, \bar{e}, \bar{u})$. An estimator of the variance is

$$\hat{V}\{x_{n+1}^{+} - x_{n+1}\} = n^{-1}s_{bb} + \hat{V}\{\hat{\gamma}_1\}[(Y_{n+1} - \bar{Y})^2 - \hat{\sigma}_{ee}] + \hat{\gamma}_1^2\hat{\sigma}_{ee}, \tag{2.5.28}$$

where $\hat{\sigma}_{ee} = s_{vv} - \hat{\beta}_1^2\sigma_{uu}$, $\hat{\beta}_1 = \hat{m}_{xx}^{-1}m_{XY}$,

$$\hat{V}\{\hat{\gamma}_1\} = (n-1)^{-1}\hat{m}_{xx}^{-2}[(\hat{m}_{xx} + \sigma_{uu})s_{bb} + \sigma_{uu}^2]\hat{\gamma}_1^2, \tag{2.5.29}$$

$$\hat{m}_{xx} = \begin{cases} m_{XX} - \sigma_{uu} & \text{if } \hat{\lambda} > 1 \\ m_{XY}^{-2}m_{YY} & \text{otherwise,} \end{cases}$$

$$s_{bb} = (n-2)^{-1} \sum_{t=1}^{n} [(X_t - \bar{X}) - \hat{\gamma}_1(Y_t - \bar{Y})]^2,$$

and the difference $(Y_{n+1} - \bar{Y})^2 - \hat{\sigma}_{ee}$ is replaced with zero when the difference is negative.

We have constructed an estimator for x in the calibration problem that is nearly unbiased for the true value under the fixed model. If one is willing to treat the unobserved true value as a random drawing from the population that generated the sample, then the optimum prediction is obtained from the ordinary least squares regression of X on Y. The reader interested in alternative approaches to the calibration problem is referred to Hunter and Lamboy (1981), Brown (1982), and the discussion following those papers.

Example 2.5.1. To illustrate the calibration computations we again use the data of Example 1.2.1. We assume that we are given a yield of $Y_{14} = 114$ bushels per acre for a site and that we are asked to estimate the soil nitrogen for the site, treating the true unknown nitrogen as fixed. From Example 1.2.1 we have

$$(\bar{Y}, \bar{X}) = (97.4545, 70.6364),$$
$$(m_{YY}, m_{XY}, m_{XX}) = (87.6727, 104.8818, 304.8545).$$

By (2.5.23) our approximately unbiased estimator of γ_1 is

$$\hat{\gamma}_1 = \frac{247.8545(104.8818) + 0.2(104.8818)57}{(104.8818)^2 + 0.1[304.8545(87.6727) - (104.8818)^2]} = 2.1627$$

and $\hat{\gamma}_0 = -140.1287$. Therefore, our estimator of x_{14} is

$$x_{14}^{+} = -140.1287 + (2.1627)(114) = 106.42.$$

By (2.5.28) an estimator of the variance of x_{14}^{+} is

$$\hat{V}\{x_{14}^{+} - x_{14}\} = 416.09,$$

where $\hat{\sigma}_{ee} = 49.2343$, $s_{bb} = 290.30$, and $V\{\hat{\gamma}_1\} = 0.6983$.

The estimator of γ_1 is constructed from a relatively small sample and the error of estimation in $(\hat{\gamma}_0, \hat{\gamma}_1)$ about doubles the variance of the estimator of x_{14} relative to that one would obtain if γ_0 and γ_1 were known. If γ_0 and γ_1 were known, the estimated variance of the estimator of x_{14} is $(2.1627)^2(43.2910) = 202.48$. □ □

REFERENCES

Fuller (1977, 1980), Martinez-Garza (1970), Miller (1984).

EXERCISES

34. (Section 2.5.1) Construct estimator (2.5.3) for the data of Exercise 1.17.

35. (Section 2.5.1) Let the model of Theorem 2.5.2 hold with $k = 2$, $\mathbf{x}_t = (1, x_t)$, and σ_{uu} known. Show that the estimator of β_1 defined by (2.5.9) is the same as the estimator of (2.5.3).

36. (Section 2.5.1) Assume that two observations (Y_1, X_1) and (Y_2, X_2) are available. Assume that it is known that model (2.5.1) holds and that σ_{uu} is known.
   (a) Explain how you would estimate (β_0, β_1).
   (b) Assume that model (2.5.1) holds with $n = 2$ and that $\sigma_{uu} = 1$ is known. Assume that, unknown to the statistician, $(x_1, x_2) = (-2, 2)$ and $\beta_1 = 1$. The e_t are independently and identically distributed with

$$P\{e_t = 1\} = P\{e_t = -1\} = 0.5,$$

   u_t is independent of e_j for all t and j, and the distribution of u_t is the same as that of e_t. Give the mean square error of the estimator (2.5.9).

37. (Section 2.5.1) The likelihood ratio statistic $\tilde{l}_{R1}$ of Table 2.5.3 requires the likelihood for the model with β_1 known. Let model (2.5.1) hold with $\sigma_{xx} > 0$ and $\sigma_{ee} \geqslant 0$, let σ_{uu} be known, and let $\beta_1 = \beta_1^0$, where $\beta_1^0 \neq 0$ is known. Let $\ddot{\gamma} = (\ddot{\sigma}_{XX}, \ddot{\sigma}_{vv})'$ be a pair satisfying

$$\ddot{\sigma}_{XX}^3 - m_{XX}\ddot{\sigma}_{XX}^2 + \sigma_{Xv}^0 m_{vv}^{-1} m_{XX}(2m_{Xv} - \sigma_{Xv}^0)\ddot{\sigma}_{XX} - m_{vv}^{-1} m_{XX}^2(\sigma_{Xv}^0)^2 = 0,$$
$$\ddot{\sigma}_{vv} = m_{XX}^{-1} m_{vv}\ddot{\sigma}_{XX},$$

where $\sigma_{Xv}^0 = -\beta_1^0\sigma_{uu}$. Show that the $\ddot{\gamma}$ that maximizes the likelihood is the maximum likelihood estimator of γ, provided

$$\ddot{\sigma}_{XX} > \sigma_{uu} \quad \text{and} \quad \ddot{\sigma}_{vv} > (\beta_1^0)^2\sigma_{uu}.$$

If $\ddot{\sigma}_{XX} < \sigma_{uu}$ or $\ddot{\sigma}_{vv} < (\beta_1^0)^2\sigma_{uu}$ and if $m_{vv} < (\beta_1^0)^2 m_{XX}$, show that the supremum of the likelihood over the parameter space is given by $\tilde{\gamma} = (\tilde{\sigma}_{XX}, \tilde{\sigma}_{vv})'$, where

$$\tilde{\sigma}_{XX} = \sigma_{uu} + (\beta_1^0)^{-2} m_{YY} \quad \text{and} \quad \tilde{\sigma}_{vv} = (\beta_1^0)^2\sigma_{uu}.$$

If $\sigma_{XX} < \sigma_{uu}$ or $\ddot{\sigma}_{vv} < (\beta_1^0)^2\sigma_{uu}$ and if $m_{vv} > (\beta_1^0)^2 m_{XX}$, show that the supremum of the likelihood on the parameter space occurs for

$$\tilde{\sigma}_{XX} = \sigma_{uu} \quad \text{and} \quad \tilde{\sigma}_{vv} = (\beta_1^0)^2\sigma_{uu} + m_{YY}.$$

38. (Section 2.5.3) Let σ_{uu} be unknown and unbiasedly estimated by s_{uu}, where $d_f\sigma_{uu}^{-1}s_{uu}$ is distributed as a chi-square random variable with d_f degrees of freedom. Assume s_{uu} is independent of (Y_t, X_t), $t = 1, 2, \ldots, n$. Let the estimator $\tilde{\gamma}_1$ be the estimator (2.5.23) with σ_{uu} replaced by s_{uu}. Show that the variance of the approximate distribution of $\tilde{\gamma}_1$ is

$$(n-1)^{-1}m_{yy}^{-2}(m_{xx}\sigma_{vv} + \sigma_{uu}\sigma_{vv} + \sigma_{uv}^2) + 2d_f^{-1}m_{xy}^{-2}\sigma_{uu}^2.$$

39. (Section 2.5.3) Expand $(\hat{\gamma}_1 - \gamma_1)$, where the estimator $\hat{\gamma}_1$ is defined in (2.5.23), in a Taylor expansion through terms of $O_p(n^{-1})$. Show that the expectation of the sum of these terms is zero.

40. (Section 2.5.3) Show that if $\sigma_{uu} = 0$, the estimator (2.5.23) reduces to

$$\hat{\gamma}_1 = \hat{\beta}_{1\ell}[\hat{\beta}_{1\ell}^2 + (n-2)(n-1)^{-1}\hat{V}\{\hat{\beta}_{1\ell}\}]^{-1},$$

where $\hat{\beta}_{1\ell}$ is the ordinary least squares regression coefficient for the regression of Y_t on x_t with an intercept, and $\hat{V}\{\hat{\beta}_{1\ell}\}$ is the ordinary least squares estimator of the variance of $\hat{\beta}_{1\ell}$.

APPENDIX 2.A. LANGUAGE EVALUATION DATA

TABLE 2.A.1. A sample of 100 evaluations of essays

Observation	Developed Y	Logical X_1	Irritating X_2	Intelligent X_3	Understand W_1	Appropriate W_2	Acceptable W_3	Careful W_4
1	2	4	5	7	6	3	4	7
2	2	5	6	6	4	6	6	6
3	5	4	6	6	6	5	6	5
4	3	5	7	5	6	6	6	4
5	4	5	9	8	5	4	6	6
6	5	5	8	8	6	8	7	6
7	6	5	7	7	6	7	7	6
8	3	6	6	7	7	9	5	5
9	4	6	5	6	6	6	5	6
10	7	5	5	6	7	5	5	6
11	6	5	6	7	8	8	2	8
12	6	6	6	6	5	5	6	6
13	6	6	8	9	7	8	9	9
14	5	6	10	8	9	8	10	6
15	5	6	9	7	9	7	9	8
16	6	6	8	6	7	6	6	4
17	5	7	6	7	5	6	7	5
18	7	6	7	8	7	5	6	7
19	7	6	8	7	7	4	5	7
20	5	7	7	8	6	6	5	6
21	6	7	8	8	6	8	8	6
22	6	7	10	10	7	9	10	8
23	6	7	5	6	7	6	6	7

TABLE 2.A.1. (Continued)

Observation	Developed Y	Logical X_1	Irritating X_2	Intelligent X_3	Understand W_1	Appropriate W_2	Acceptable W_3	Careful W_4
24	6	7	8	8	7	6	6	8
25	6	7	7	8	8	6	6	8
26	7	7	7	7	7	6	7	7
27	6	7	9	8	9	7	7	8
28	7	7	7	8	8	6	7	7
29	8	7	9	10	8	9	9	8
30	6	8	7	6	6	6	6	4
31	8	7	5	8	8	3	5	5
32	6	8	7	8	8	6	7	7
33	5	8	6	7	9	8	4	6
34	7	8	8	9	8	9	8	7
35	7	8	9	8	8	9	8	8
36	6	8	6	8	10	6	8	8
37	8	8	8	5	7	8	6	2
38	8	8	8	9	8	8	8	9
39	7	8	10	9	10	9	9	8
40	7	8	10	10	10	8	8	8
41	8	8	7	8	8	5	5	7
42	8	8	10	8	10	7	9	8
43	8	9	9	9	7	8	8	8
44	7	9	8	10	9	6	6	10
45	8	9	9	10	9	7	8	9
46	9	9	8	10	9	9	9	10
47	10	9	10	10	10	10	9	8
48	9	10	10	10	10	10	10	10
49	10	10	10	10	10	10	10	10

50	10	10	10	10	10	10	10	10
51	4	5	5	8	4	4	3	6
52	3	4	6	7	3	4	4	7
53	6	8	6	7	7	4	4	7
54	5	6	6	6	6	4	5	6
55	4	5	6	7	5	5	5	5
56	6	7	8	8	8	4	4	7
57	4	4	8	8	5	6	4	5
58	5	6	8	5	4	5	5	5
59	4	5	8	7	6	5	5	6
60	6	6	7	7	5	5	6	7
61	5	5	6	8	6	6	6	7
62	6	6	6	8	6	6	6	6
63	7	9	8	8	8	5	5	6
64	8	8	7	8	8	7	4	8
65	6	4	6	8	5	6	7	5
66	6	6	8	7	6	6	6	6
67	5	6	7	7	6	7	6	6
68	7	7	5	8	6	7	7	7
69	8	8	6	9	8	6	7	10
70	6	7	8	9	8	7	6	9
71	7	7	8	8	9	8	5	8
72	7	7	7	7	7	7	7	7
73	6	7	7	7	8	7	7	5
74	8	6	8	9	6	6	8	8
75	8	8	9	9	8	6	7	5
76	5	6	9	8	9	8	6	9
77	4	7	9	8	8	6	8	6
78	6	7	6	9	6	8	8	8

TABLE 2.A.1. (Continued)

Observation	Developed Y	Logical X_1	Irritating X_2	Intelligent X_3	Understand W_1	Appropriate W_2	Acceptable W_3	Careful W_4
79	8	8	9	7	7	7	7	6
80	9	10	7	10	7	9	6	5
81	5	5	8	7	7	8	8	6
82	7	8	7	7	7	8	8	4
83	8	9	7	10	10	7	8	6
84	6	6	8	6	8	8	8	7
85	10	10	10	7	8	7	7	7
86	7	8	10	10	8	8	7	8
87	7	8	8	8	8	8	8	8
88	8	8	8	8	8	8	8	8
89	8	9	8	9	9	8	8	8
90	7	7	8	9	7	8	9	7
91	8	7	9	9	7	10	7	8
92	8	8	10	9	9	9	7	9
93	7	8	8	8	10	8	9	7
94	8	7	10	9	7	8	9	9
95	6	6	10	7	7	8	10	8
96	8	9	9	9	9	9	9	8
97	8	8	10	10	9	9	9	10
98	7	9	10	10	10	9	10	10
99	9	10	10	9	10	10	10	8
100	10	10	10	10	10	10	10	10

CHAPTER 3

Extensions of the Single Relation Model

The linear measurement error model with normally distributed errors has a considerable history. Consequently, the theory for the models of Sections 2.2–2.4 is relatively well developed. On the other hand, extensions of the model to nonnormal errors, nonlinear models, and heterogeneous error variances are currently areas of active research.

In Section 3.1 we consider the problem of heterogeneous error variances and nonnormal errors. Estimation procedures for a number of specific situations are suggested. Section 3.2 treats estimation for nonlinear models with no error in the equation. Nonlinear models with an error in the equation are considered in Section 3.3. Section 3.4 is an introduction to estimation for measurement error models containing multinomial responses.

3.1. NONNORMAL ERRORS AND UNEQUAL ERROR VARIANCES

The models of Sections 2.2 and 2.3 assumed the errors to be normally distributed and the covariance matrix of the errors to be the same for all observations. In this section the error covariance matrix is permitted to be a function of t. Estimation procedures are suggested that represent a melding of the linear errors-in-variables techniques of Sections 2.2 and 2.3 with methods for ordinary linear models containing heteroskedastic errors. Relatively distribution-free estimators of the covariance matrix of the approximate distribution of the estimators are developed. The estimators and their large sample properties are given in Section 3.1.1. Applications to several specific problems are described in the sections following.

3.1.1. Introduction and Estimators

To introduce the general problem, assume that we have conducted n experiments to measure the relationship between y_t and $\mathbf{x}_t$. Each experiment provides an estimate of $(y_t, \mathbf{x}_t)$, denoted by $(Y_t, \mathbf{X}_t)$, that contains measurement error. An estimate of the covariance matrix of the measurement error, denoted by $\hat{\boldsymbol{\Sigma}}_{aatt}$, is also obtained in each experiment. The true covariance matrix of the error, denoted by $\boldsymbol{\Sigma}_{aatt}$, may vary from experiment to experiment. In Example 3.1.7 the (Y_t, X_t) are estimated regression coefficients and $\hat{\boldsymbol{\Sigma}}_{aatt}$ is the estimated covariance matrix of the coefficients. In Example 3.1.4 the (Y_t, X_t) are the averages of two determinations and $\hat{\boldsymbol{\Sigma}}_{aatt}$ is a one degree-of-freedom estimator of $\boldsymbol{\Sigma}_{aatt}$. Let

$$y_t = \mathbf{x}_t\boldsymbol{\beta} + q_t, \qquad \mathbf{Z}_t = \mathbf{z}_t + \mathbf{a}_t, \tag{3.1.1}$$

$$(q_t, \mathbf{a}_t)' \sim \text{Ind}(\mathbf{0}, \text{block diag}\{\sigma_{qq}, \boldsymbol{\Sigma}_{aatt}\}), \tag{3.1.2}$$

where $\mathbf{Z}_t = (Y_t, \mathbf{X}_t)$, $\mathbf{a}_t = (w_t, \mathbf{u}_t)$ is the vector of measurement errors, q_t is the error in the equation, $\mathbf{x}_t$ is a k-dimensional row vector of true values, and $\text{Ind}(\mathbf{0}, \boldsymbol{\Sigma}_{\varepsilon\varepsilon tt})$ denotes a sequence of independent random variables with zero means and covariance matrices $\boldsymbol{\Sigma}_{\varepsilon\varepsilon tt}$.

We assume that estimators of $\boldsymbol{\Sigma}_{aatt}$, $t = 1, 2, \ldots, n$, are such that

$$E\{\hat{\boldsymbol{\Sigma}}_{aatt}\} = \boldsymbol{\Sigma}_{aatt} \tag{3.1.3}$$

or (and)

$$\operatorname*{plim}_{n\to\infty} n^{-1} \sum_{t=1}^{n} \hat{\boldsymbol{\Sigma}}_{aatt} = \lim_{n\to\infty} n^{-1} \sum_{t=1}^{n} \boldsymbol{\Sigma}_{aatt}. \tag{3.1.4}$$

To arrange the unique elements of $\hat{\boldsymbol{\Sigma}}_{aatt}$ and $\boldsymbol{\Sigma}_{aatt}$ in columns, we let

$$\hat{\mathbf{c}}_{at} = \text{vech } \hat{\boldsymbol{\Sigma}}_{aatt} \quad \text{and} \quad \mathbf{c}_{at} = \text{vech } \boldsymbol{\Sigma}_{aatt}, \tag{3.1.5}$$

where vech $\mathbf{A}$ is the column vector created by listing the elements of the matrix $\mathbf{A}$ on and below the diagonal in a column. See Appendix 4.A.

We have introduced several components of the estimation problem, including the specified relationship (3.1.1), the $(Y_t, \mathbf{X}_t)$ observations, and estimates of the covariance matrices of the $\mathbf{a}_t$. A final component of the problem is a set of weights, π_t, $t = 1, 2, \ldots, n$, estimated by the set of weights, $\hat{\pi}_t$, $t = 1, 2, \ldots, n$, for the observations. The weights will generally be related to the variances of the errors in the model. Often, as with ordinary generalized least squares, the estimation procedure will be an iterative one with $\pi_t = 1$ for the first iteration and an estimator of the "optimal" weight used as the $\hat{\pi}_t$ for the second and higher iterations.

To construct an estimator for the parameters of model (3.1.1) assuming that $\sigma_{qq} > 0$ is unknown and that a set of estimated weights is available, let

$$\hat{\mathbf{M}}_{z\pi z} = n^{-1} \sum_{t=1}^{n} \hat{\pi}_t(\mathbf{Z}_t'\mathbf{Z}_t - \hat{\mathbf{\Sigma}}_{aatt}). \tag{3.1.6}$$

If, for example, $\hat{\pi}_t \equiv 1$ and (3.1.3) holds, then $\hat{\mathbf{M}}_{z\pi z} \equiv \hat{\mathbf{M}}_{zz}$ is unbiased for $\mathbf{M}_{zz}$. Therefore, a natural estimator for $\boldsymbol{\beta}$ is

$$\hat{\boldsymbol{\beta}} = \hat{\mathbf{M}}_{x\pi x}^{-1}\hat{\mathbf{M}}_{x\pi y}, \tag{3.1.7}$$

where $\hat{\mathbf{M}}_{x\pi x}$ and $\hat{\mathbf{M}}_{x\pi y}$ are submatrices of $\hat{\mathbf{M}}_{z\pi z}$. In practice, one would modify estimator (3.1.7) by a procedure analogous to that of (2.5.9). We give the large sample properties of estimator (3.1.7) in Theorem 3.1.1. The specification introduced in Theorem 2.2.1, which covers both the functional and structural models, is used in the theorem. Persons interested in applications may prefer to proceed directly to Section 3.1.2.

Theorem 3.1.1. Let model (3.1.1) and (3.1.4) hold. Let $[q_t, \mathbf{a}_t, (\hat{\mathbf{c}}_{at} - \mathbf{c}_{at})', (\mathbf{x}_t - \boldsymbol{\mu}_{xt})]$, $t = 1, 2, \ldots, n$, be independent with bounded $4 + \delta$ moments ($\delta > 0$), $\sigma_{qq} > 0$, $E\{(q_t, \mathbf{a}_t, q_t\mathbf{a}_t) | \mathbf{x}_t\} = \mathbf{0}$, $E\{\mathbf{x}_t - \boldsymbol{\mu}_{xt}\} = \mathbf{0}$, and

$$\lim_{n\to\infty} n^{-1/2} \sum_{t=1}^{n} \pi_t E\{\hat{\mathbf{c}}_{at} - \mathbf{c}_{at}\} = \mathbf{0}. \tag{3.1.8}$$

Let $\{(\boldsymbol{\mu}_{xt}, \mathbf{c}_{at}', \pi_t)\}$ be a fixed sequence indexed by t, where $\{\pi_t\}$ is bounded below and above by fixed positive numbers. Let

$$\hat{\gamma}_t = [\mathbf{Z}_t, (\text{vech } \mathbf{Z}_t'\mathbf{Z}_t)', (\text{vech } \mathbf{z}_t'\mathbf{z}_t)', \hat{\mathbf{c}}_{at}']'$$

and assume that

$$\operatorname*{plim}_{n\to\infty} n^{-1} \sum_{t=1}^{n} \hat{\pi}_t^j \hat{\gamma}_t \hat{\gamma}_t' = \operatorname*{plim}_{n\to\infty} n^{-1} \sum_{t=1}^{n} \pi_t^j \hat{\gamma}_t \hat{\gamma}_t' \tag{3.1.9}$$

for $j = 1, 2$. Let

$$\operatorname*{plim}_{n\to\infty} n^{-1} \sum_{t=1}^{n} \pi_t \mathbf{x}_t'\mathbf{x}_t = \operatorname*{plim}_{n\to\infty} \mathbf{M}_{x\pi x} = \bar{\mathbf{M}}_{x\pi x}$$

$$\lim_{n\to\infty} \mathbf{G} = \bar{\mathbf{G}}$$

be positive definite, where $\mathbf{G} = n^{-1}\sum_{t=1}^{n} \pi_t^2 E\{\mathbf{d}_t'\mathbf{d}_t\}$,

$$\mathbf{d}_t' = \mathbf{X}_t' v_t - \ddot{\mathbf{\Sigma}}_{uvtt},$$

$v_t = q_t + w_t - \mathbf{u}_t\boldsymbol{\beta}$, $\ddot{\mathbf{\Sigma}}_{uvtt} = (\hat{\mathbf{\Sigma}}_{uwtt} - \hat{\mathbf{\Sigma}}_{uutt}\boldsymbol{\beta})$, and $\hat{\mathbf{\Sigma}}_{uwtt}$ and $\hat{\mathbf{\Sigma}}_{uutt}$ are submatrices of $\hat{\boldsymbol{\Sigma}}_{aatt}$. Assume

$$\operatorname*{plim}_{n\to\infty} n^{-1/2} \sum_{t=1}^{n} (\hat{\pi}_t - \pi_t)\mathbf{d}_t = \mathbf{0}, \tag{3.1.10}$$

$$\lim_{n\to\infty} n^{-1} \sum_{t=1}^{n} \pi_t \boldsymbol{\mu}_{xt}' \boldsymbol{\mu}_{xt} = \bar{\mathbf{M}}_{\mu\pi\mu}.$$

Then

$$\hat{\mathbf{V}}_{\beta\beta}^{-1/2}(\hat{\boldsymbol{\beta}} - \boldsymbol{\beta}) \xrightarrow{L} N(\mathbf{0}, \mathbf{I}), \tag{3.1.11}$$

where

$$\hat{\mathbf{V}}_{\beta\beta} = n^{-1}\hat{\mathbf{M}}_{x\pi x}^{-1}\hat{\mathbf{G}}\hat{\mathbf{M}}_{x\pi x}^{-1}, \tag{3.1.12}$$

$$\hat{\mathbf{G}} = (n-k)^{-1} \sum_{t=1}^{n} \hat{\pi}_t^2 \hat{\mathbf{d}}_t' \hat{\mathbf{d}}_t,$$

$\hat{\mathbf{d}}_t' = \mathbf{X}_t'\hat{v}_t - \hat{\boldsymbol{\Sigma}}_{uvtt}$, $\hat{\boldsymbol{\Sigma}}_{uvtt} = \hat{\boldsymbol{\Sigma}}_{aatt}(1, -\hat{\boldsymbol{\beta}}')'$, and $\hat{v}_t = Y_t - \mathbf{X}_t\hat{\boldsymbol{\beta}}$.

Proof. We have

$$\begin{aligned} n^{1/2}(\hat{\boldsymbol{\beta}} - \boldsymbol{\beta}) &= n^{1/2}\hat{\mathbf{M}}_{x\pi x}^{-1}(\hat{\mathbf{M}}_{x\pi y} - \hat{\mathbf{M}}_{x\pi x}\boldsymbol{\beta}) \\ &= n^{-1/2}\hat{\mathbf{M}}_{x\pi x}^{-1} \sum_{t=1}^{n} \hat{\pi}_t(\mathbf{X}_t'v_t - \ddot{\boldsymbol{\Sigma}}_{uvtt}) \\ &= \bar{\mathbf{M}}_{x\pi x}^{-1} n^{-1/2} \sum_{t=1}^{n} \pi_t(\mathbf{X}_t'v_t - \ddot{\boldsymbol{\Sigma}}_{uvtt}) + o_p(1), \end{aligned}$$

where we have used the distributional assumptions and assumptions (3.1.8), (3.1.9), and (3.1.10). The random variables

$$\pi_t \mathbf{d}_t' = \pi_t(\mathbf{X}_t'v_t - \ddot{\boldsymbol{\Sigma}}_{uvtt})$$

are independent random vectors and as $n \to \infty$,

$$\left[\sum_{t=1}^{n} \pi_t^2 E\{\mathbf{d}_t'\mathbf{d}_t\}\right]^{-1/2} \sum_{t=1}^{n} \pi_t \mathbf{d}_t' \xrightarrow{L} N(\mathbf{0}, \mathbf{I})$$

by Lemma 1.C.2.

To complete the proof we show that

$$\operatorname*{plim}_{n\to\infty}\left[\hat{\mathbf{G}} - n^{-1} \sum_{t=1}^{n} \pi_t^2 E\{\mathbf{d}_t'\mathbf{d}_t\}\right] = \mathbf{0}.$$

Because the random portions of the elements of $\mathbf{d}_t'\mathbf{d}_t$ have bounded $1 + \frac{1}{4}\delta$ moments and $\{\pi_t\}$ is bounded,

$$\operatorname*{plim}_{n\to\infty} n^{-1} \sum_{t=1}^{n} \pi_t^2 \mathbf{d}_t'\mathbf{d}_t = \lim_{n\to\infty} n^{-1} \sum_{t=1}^{n} \pi_t^2 E\{\mathbf{d}_t'\mathbf{d}_t\}$$

by the weak law of large numbers. Using

$$\hat{\mathbf{d}}_t' = \mathbf{X}_t'\hat{v}_t - \ddot{\Sigma}_{uvtt} - (\mathbf{X}_t'\mathbf{X}_t - \hat{\Sigma}_{uutt})(\hat{\boldsymbol{\beta}} - \boldsymbol{\beta})$$

and assumption (3.1.9), we have

$$(n - k)^{-1} \sum_{t=1}^{n} \hat{\pi}_t^2 \hat{\mathbf{d}}_t' \hat{\mathbf{d}}_t = (n - k)^{-1} \sum_{t=1}^{n} \hat{\pi}_t^2 \mathbf{d}_t' \mathbf{d}_t + O_p(n^{-1/2}).$$

Also, by assumption (3.1.9),

$$\operatorname*{plim}_{n\to\infty} n^{-1} \sum_{t=1}^{n} (\hat{\pi}_t^2 \mathbf{d}_t' \mathbf{d}_t - \pi_t^2 \mathbf{d}_t' \mathbf{d}_t) = \mathbf{0}$$

and we have the conclusion. □

Theorem 3.1.1 gives us a statistic that can be used, in large samples, to test hypotheses about $\boldsymbol{\beta}$ without completely specifying the nature of the distribution of $(\mathbf{x}_t, \mathbf{u}_t)$. If $(e_t, \mathbf{u}_t) \sim \text{NI}(\mathbf{0}, \Sigma_{\varepsilon\varepsilon})$, independent of $\mathbf{x}_t$, $\hat{\pi}_t = \pi_t = 1$, and $\hat{\Sigma}_{aatt} = \Sigma_{aa}$, the estimator of the variance of $\hat{\boldsymbol{\beta}}$ defined in (3.1.12) is estimating the same quantity as that estimated by (2.2.25) of Section 2.2. When the measurement errors are normally and identically distributed and independent of the true x values, then the estimator (2.2.25) is preferred because it has smaller variance.

The estimator (3.1.12) is usually biased downward in small samples. If the sample is not large, it may be preferable to develop a model for the error structure rather than use the distribution-free estimator of Theorem 3.1.1. As with all problems of regression type, one should plot the data to check for outliers in either $\hat{v}_t$ or $\mathbf{X}_t$. In addition to the plot of $\hat{v}_t$ against the individual X_{ti}, the plot of $\hat{v}_t$ against $[\mathbf{X}_t(n^{-1} \sum_{j=1}^{n} \mathbf{X}_j'\mathbf{X}_j)^{-1}\mathbf{X}_t']^{1/2}$ will help to identify vectors $\mathbf{X}_t$ that are separated from the main cluster of observations. Distribution-free methods of the type under study will not perform well in a situation where one or two observations are widely separated from the remaining observations. Carroll and Gallo (1982) discuss some aspects of robustness for the errors-in-variables problem.

Given the estimator of $\boldsymbol{\beta}$, one can estimate the variance of the error in the equation with

$$\hat{\sigma}_{qq} = (n - k)^{-1} \sum_{t=1}^{n} (Y_t - \mathbf{X}\hat{\boldsymbol{\beta}})^2 - n^{-1} \sum_{t=1}^{n} (1, -\hat{\boldsymbol{\beta}}')\hat{\Sigma}_{aatt}(1, -\hat{\boldsymbol{\beta}}')'. \quad (3.1.13)$$

Naturally, the estimator of σ_{qq} is taken to be the maximum of (3.1.13) and zero. If the estimator is zero, one may choose to estimate $\boldsymbol{\beta}$ by the methods of Theorem 3.1.2 below.

In some situations it is known that the q_t of the model (3.1.1) are identically zero. To construct estimators for such models, we proceed by analogy to the methods developed in Section 2.3. The suggested estimator is

$$\tilde{\beta} = \tilde{\mathbf{M}}_{x\pi x}^{-1}\tilde{\mathbf{M}}_{x\pi y}, \tag{3.1.14}$$

where

$$\tilde{\mathbf{M}}_{z\pi z} = \hat{\mathbf{M}}_{Z\pi Z} - \hat{\lambda}\hat{\boldsymbol{\Sigma}}_{a\pi a..},$$

$$(\hat{\mathbf{M}}_{Z\pi Z}, \hat{\boldsymbol{\Sigma}}_{a\pi a..}) = n^{-1} \sum_{t=1}^{n} \hat{\pi}_t(\mathbf{Z}_t'\mathbf{Z}_t, \hat{\boldsymbol{\Sigma}}_{aatt}),$$

and $\hat{\lambda}$ is the smallest root of

$$|\hat{\mathbf{M}}_{Z\pi Z} - \lambda\hat{\boldsymbol{\Sigma}}_{a\pi a..}| = 0. \tag{3.1.15}$$

The limiting process used in Theorem 3.1.2 is that introduced in Theorem 2.3.2. The error variances are assumed to be proportional to T^{-1} and the limiting behavior is obtained as $v = Tn$ becomes large. The index v becomes large as the number of $(Y_t, \mathbf{X}_t)$ observations, denoted by n, becomes large and (or) as the error variances become small.

Theorem 3.1.2. Let model (3.1.1) hold with $q_t \equiv 0$ and

$$\boldsymbol{\Sigma}_{aatt} = T^{-1}\boldsymbol{\Omega}_{aatt}$$

for $t = 1, 2, \ldots, n$. Let $[T^{1/2}\mathbf{a}_t, T(\hat{\mathbf{c}}_{at} - \mathbf{c}_{at})', (\mathbf{x}_t - \boldsymbol{\mu}_{xt})]$, $t = 1, 2, \ldots, n$, be independent with bounded $4 + \delta$ moments ($\delta > 0$), $E\{\mathbf{a}_t|\mathbf{x}_t\} = \mathbf{0}$, $E\{\mathbf{x}_t - \boldsymbol{\mu}_{xt}\} = \mathbf{0}$, and

$$\lim_{v\to\infty} Tn^{-1/2} \sum_{t=1}^{n} \pi_t E\{\hat{\mathbf{c}}_{at} - \mathbf{c}_{at}\} = \mathbf{0},$$

where $v = Tn$. Assume that

$$\operatorname*{plim}_{v\to\infty} n^{-1/2} \sum_{t=1}^{n} (\hat{\pi}_t^j - \pi_t^j)\hat{\xi}_t'(1, \hat{\xi}_t) = \mathbf{0},$$

$$n^{-1} \sum_{t=1}^{n} (\hat{\pi}_t^j - \pi_t^j)(\mathbf{z}_t'\mathbf{z}_t, T\mathbf{a}_t'\mathbf{a}_t, T\hat{\boldsymbol{\Sigma}}_{aatt}) = O_p(n^{-1/2}), \tag{3.1.16}$$

for $j = 1, 2$, where

$$\hat{\xi}_t = [T^{1/2}(\operatorname{vec} \mathbf{z}_t'\mathbf{a}_t)', T(\mathbf{c}_{at} - \operatorname{vech} \mathbf{a}_t'\mathbf{a}_t)', T(\hat{\mathbf{c}}_{at} - \mathbf{c}_{at})'].$$

Let $\{(\boldsymbol{\mu}_{xt}, T\mathbf{c}_{at}', \pi_t)\}$ be a fixed bounded sequence, where $\{\pi_t\}$ is bounded above and below by positive numbers. Let

$$\operatorname*{plim}_{v\to\infty} n^{-1} \sum_{t=1}^{n} \hat{\pi}_t\mathbf{x}_t'\mathbf{x}_t = \operatorname*{plim}_{v\to\infty} \mathbf{M}_{x\pi x} = \bar{\mathbf{M}}_{x\pi x},$$

$$\lim_{v\to\infty} \mathbf{G} = \bar{\mathbf{G}}$$

be positive definite, where $\mathbf{G} = n^{-1}\sum_{t=1}^{n} \pi_t^2 E\{\mathbf{d}_t'\mathbf{d}_t\}$,

$$\mathbf{M}_{x\pi x} = n^{-1} \sum_{t=1}^{n} \pi_t \mathbf{x}_t'\mathbf{x}_t,$$

$$\mathbf{d}_t' = \mathbf{X}_t'v_t - \ddot{\boldsymbol{\Sigma}}_{uvtt} - (v_t^2 - \ddot{\sigma}_{vvtt})\sigma_{v\pi v..}^{-1}\boldsymbol{\Sigma}_{u\pi v..},$$

$$(\sigma_{v\pi v..}, \boldsymbol{\Sigma}_{v\pi u..}, \ddot{\boldsymbol{\Sigma}}_{v\pi u..}) = n^{-1} \sum_{t=1}^{n} \pi_t(\boldsymbol{\alpha}'\boldsymbol{\Sigma}_{aatt}\boldsymbol{\alpha}, \boldsymbol{\alpha}'\boldsymbol{\Sigma}_{autt}, \boldsymbol{\alpha}'\hat{\boldsymbol{\Sigma}}_{autt}),$$

$\boldsymbol{\alpha}' = (1, -\boldsymbol{\beta}')$, $\ddot{\sigma}_{vvtt} = \boldsymbol{\alpha}'\hat{\boldsymbol{\Sigma}}_{aatt}\boldsymbol{\alpha}$, and $\ddot{\boldsymbol{\Sigma}}_{uvtt} = \hat{\boldsymbol{\Sigma}}_{uatt}\boldsymbol{\alpha}$. Assume

$$\lim_{v\to\infty} n^{-1} \sum_{t=1}^{n} \pi_t \boldsymbol{\mu}_{xt}'\boldsymbol{\mu}_{xt} = \bar{\mathbf{M}}_{\mu\pi\mu},$$

$$\operatorname*{plim}_{v\to\infty} Tn^{-1} \sum_{t=1}^{n} \hat{\pi}_t \hat{\boldsymbol{\Sigma}}_{aatt} = \operatorname*{plim}_{v\to\infty} Tn^{-1} \sum_{t=1}^{n} \pi_t \boldsymbol{\Sigma}_{aatt}.$$

Then

$$\operatorname*{plim}_{v\to\infty} \left\{\boldsymbol{\Gamma}_v^{-1/2}(\tilde{\boldsymbol{\beta}} - \boldsymbol{\beta}) - n^{-1}\boldsymbol{\Gamma}_v^{-1/2}\bar{\mathbf{M}}_{x\pi x}^{-1} \sum_{t=1}^{n} \pi_t \mathbf{d}_t'\right\} = \mathbf{0},$$

where $\boldsymbol{\Gamma}_v = n^{-1}\bar{\mathbf{M}}_{x\pi x}^{-1}\mathbf{G}\bar{\mathbf{M}}_{x\pi x}^{-1}$, $\mathbf{G} = n^{-1}\sum_{t=1}^{n} \pi_t^2 E\{\mathbf{d}_t'\mathbf{d}_t\}$, and $\tilde{\boldsymbol{\beta}}$ is defined in (3.1.14). Furthermore, if $n \to \infty$ as $v \to \infty$, then

$$\tilde{\mathbf{V}}_{\beta\beta}^{-1/2}(\tilde{\boldsymbol{\beta}} - \boldsymbol{\beta}) \xrightarrow[v\to\infty]{L} N(\mathbf{0}, \mathbf{I}),$$

where

$$\tilde{\mathbf{V}}_{\beta\beta} = n^{-1}\tilde{\mathbf{M}}_{x\pi x}^{-1}\tilde{\mathbf{G}}\tilde{\mathbf{M}}_{x\pi x}^{-1}, \tag{3.1.17}$$

$$\tilde{\mathbf{G}} = (n-k)^{-1} \sum_{t=1}^{n} \hat{\pi}_t^2 \tilde{\mathbf{d}}_t'\tilde{\mathbf{d}}_t,$$

$$\tilde{\mathbf{d}}_t' = \mathbf{X}_t'\tilde{v}_t - \tilde{\boldsymbol{\Sigma}}_{uvtt} - \tilde{\sigma}_{v\pi v..}^{-1}(\tilde{v}_t^2 - \tilde{\sigma}_{vvtt})\tilde{\boldsymbol{\Sigma}}_{u\pi v..},$$

$$(\tilde{\sigma}_{v\pi v..}, \tilde{\boldsymbol{\Sigma}}_{v\pi u..}) = n^{-1} \sum_{t=1}^{n} \hat{\pi}_t(\tilde{\sigma}_{vvtt}, \tilde{\boldsymbol{\Sigma}}_{vutt}),$$

$$(\tilde{\sigma}_{vvtt}, \tilde{\boldsymbol{\Sigma}}_{vutt}) = (\tilde{\boldsymbol{\alpha}}'\hat{\boldsymbol{\Sigma}}_{aatt}\tilde{\boldsymbol{\alpha}}, \tilde{\boldsymbol{\alpha}}'\hat{\boldsymbol{\Sigma}}_{autt}),$$

and $\tilde{\boldsymbol{\alpha}}' = (1, -\tilde{\boldsymbol{\beta}}')$.

Proof. By (3.1.16) $\hat{\mathbf{M}}_{Z\pi Z} - \mathbf{M}_{z\pi z} - \boldsymbol{\Sigma}_{a\pi a..} = O_p(n^{-1/2})$. Then, by the arguments used in the proof of Theorem 2.3.2,

$$\hat{\lambda} - 1 = O_p(n^{-1/2})$$

and

$$\begin{aligned}\tilde{\boldsymbol{\beta}} - \boldsymbol{\beta} &= \tilde{\mathbf{M}}_{x\pi x}^{-1}(\tilde{\mathbf{M}}_{x\pi y} - \tilde{\mathbf{M}}_{x\pi x}\boldsymbol{\beta}) \\ &= n^{-1}\tilde{\mathbf{M}}_{x\pi x}^{-1} \sum_{t=1}^{n} \hat{\pi}_t[(\mathbf{X}_t'v_t - \ddot{\boldsymbol{\Sigma}}_{uvtt}) - (\hat{\lambda} - 1)\boldsymbol{\Sigma}_{uvtt}] + o_p(v^{-1/2}).\end{aligned}$$

Multiplying the equation

$$\left(\sum_{t=1}^{n} \hat{\pi}_t \mathbf{Z}_t'\mathbf{Z}_t - \hat{\lambda} \sum_{t=1}^{n} \hat{\pi}_t \hat{\mathbf{\Sigma}}_{aatt}\right)\tilde{\boldsymbol{\alpha}} = \mathbf{0}$$

on the left side by $\boldsymbol{\alpha}'$ and using $\mathbf{z}_t\boldsymbol{\alpha} = 0$, we obtain

$$\sum_{t=1}^{n} \hat{\pi}_t[v_t^2 - \hat{\lambda}\boldsymbol{\alpha}'\hat{\mathbf{\Sigma}}_{aatt}\boldsymbol{\alpha} + (v_t\mathbf{Z}_t - \hat{\lambda}\boldsymbol{\alpha}'\hat{\mathbf{\Sigma}}_{aatt})(\tilde{\boldsymbol{\alpha}} - \boldsymbol{\alpha})] = 0.$$

Also,

$$n^{-1} \sum_{t=1}^{n} \pi_t v_t \mathbf{Z}_t = n^{-1} \sum_{t=1}^{n} \pi_t(\mathbf{\Sigma}_{vatt} + v_t\mathbf{z}_t) + O_p(T^{-1}n^{-1/2}),$$

$$(\ddot{\sigma}_{v\pi v..}, \ddot{\mathbf{\Sigma}}_{v\pi u..}) = (\sigma_{v\pi v..}, \mathbf{\Sigma}_{v\pi u..}) + O_p(T^{-1}n^{-1/2}).$$

It follows that

$$\hat{\lambda} - 1 = \left[n^{-1} \sum_{t=1}^{n} \hat{\pi}_t \boldsymbol{\alpha}'\hat{\mathbf{\Sigma}}_{aatt}\boldsymbol{\alpha}\right]^{-1} [M_{v\pi v} - \ddot{\sigma}_{v\pi v..} - \mathbf{M}_{v\pi x}\mathbf{M}_{x\pi x}^{-1}\mathbf{M}_{x\pi v}] + o_p(v^{-1/2}),$$

where

$$(M_{v\pi v}, \mathbf{M}_{x\pi v}) = n^{-1} \sum_{t=1}^{n} \pi_t[v_t^2, \mathbf{x}_t v_t].$$

Then we may write

$$\tilde{\boldsymbol{\beta}} - \boldsymbol{\beta} = n^{-1}\bar{\mathbf{M}}_{x\pi x}^{-1} \sum_{t=1}^{n} \pi_t \mathbf{d}_t' + o_p(v^{-1/2}).$$

The random variables $T^{1/2}\pi_t \mathbf{d}_t$ are independent and, if $n \to \infty$ as $v \to \infty$, we have

$$\mathbf{\Gamma}_v^{-1/2}(\tilde{\boldsymbol{\beta}} - \boldsymbol{\beta}) \xrightarrow[v \to \infty]{L} N(\mathbf{0}, \mathbf{I})$$

by Lemma 1.C.2 and Theorem 1.C.3. By the assumptions,

$$\operatorname*{plim}_{v\to\infty} T\tilde{\mathbf{M}}_{x\pi x}^{-1}\tilde{\mathbf{G}}\tilde{\mathbf{M}}_{x\pi x}^{-1} = \operatorname*{plim}_{v\to\infty} T\bar{\mathbf{M}}_{x\pi x}^{-1}\tilde{\mathbf{G}}\bar{\mathbf{M}}_{x\pi x}^{-1}$$

and, if $n \to \infty$ as $v \to \infty$,

$$\operatorname*{plim}_{v\to\infty} (n-k)^{-1} T \sum_{t=1}^{n} \hat{\pi}_t^2 \tilde{\mathbf{d}}_t'\tilde{\mathbf{d}}_t = \operatorname*{plim}_{v\to\infty} n^{-1} T \sum_{t=1}^{n} \pi_t^2 E\{\mathbf{d}_t'\mathbf{d}_t\}. \qquad \square$$

The normal approximation of Theorem 3.1.2 will be satisfactory when the sample size, n, is large or when the errors are normally distributed with small variances.

Using the fact that the variance of a chi-square random variable is $2b$, where b is the degrees of freedom, an approximation for the distribution of $(n-k)^{-1}n\hat{\lambda}$ is the F distribution with b and infinity degrees of freedom. The parameter b is unknown, but b can be estimated by

$$\hat{b} = 2\left[(\tilde{\alpha}'\tilde{\Sigma}_{a\pi a..}\tilde{\alpha})^{-2}(n-k)^{-2}\sum_{t=1}^{n}\hat{\pi}_t^2(\tilde{v}_t^2 - \hat{\lambda}\tilde{\alpha}'\tilde{\Sigma}_{aatt}\tilde{\alpha})^2\right]^{-1}, \qquad (3.1.18)$$

where $\tilde{\alpha}' = (1, -\tilde{\beta}')$ and $\tilde{v}_t = Y_t - \mathbf{X}_t\tilde{\beta}$.

Maximum likelihood estimation for the normal distribution model with no error in the equation is discussed in Section 3.1.6.

3.1.2. Models with an Error in the Equation

Assume that the error covariance matrices of model (3.1.1) are known. Then, a method of moments estimator for β is

$$\tilde{\beta} = \hat{\mathbf{M}}_{xx}^{-1}\hat{\mathbf{M}}_{xy}, \qquad (3.1.19)$$

where the matrices,

$$(\hat{\mathbf{M}}_{xy}, \hat{\mathbf{M}}_{xx}) = n^{-1}\sum_{t=1}^{n}[(\mathbf{X}_t'\mathbf{Y}_t - \Sigma_{uwtt}), (\mathbf{X}_t'\mathbf{X}_t - \Sigma_{uutt})],$$

are unbiased for $(\mathbf{M}_{xy}, \mathbf{M}_{xx})$. In practice one would modify estimator (3.1.19) in the manner described in Section 2.5. The modified estimator is

$$\tilde{\beta}_\alpha = \hat{\mathbf{H}}_{xx}^{-1}\hat{\mathbf{H}}_{xy}, \qquad (3.1.20)$$

where

$$\hat{\mathbf{H}}_{zz} = \begin{cases} \sum_{t=1}^{n}[\mathbf{Z}_t'\mathbf{Z}_t - (1-\alpha n^{-1})\Sigma_{aatt}] & \text{if } \hat{\lambda} > 1 + n^{-1} \\ \sum_{t=1}^{n}[\mathbf{Z}_t'\mathbf{Z}_t - (\hat{\lambda} - n^{-1} - \alpha n^{-1})\Sigma_{aatt}] & \text{if } \hat{\lambda} \leqslant 1 + n^{-1}, \end{cases}$$

$\hat{\lambda}$ is the smallest root of

$$\left|\sum_{t=1}^{n}(\mathbf{Z}_t'\mathbf{Z}_t - \lambda\Sigma_{aatt})\right| = 0, \qquad (3.1.21)$$

and $\mathbf{z}_t = (y_t, \mathbf{x}_t)$. The limiting distribution of $n^{1/2}(\tilde{\beta} - \beta)$ is the same as that of $n^{1/2}(\tilde{\beta}_\alpha - \beta)$.

Given the estimator of β, an estimator of σ_{qq} is

$$\tilde{\sigma}_{qq} = \sum_{t=1}^{n}[(n-k)^{-1}(Y_t - \mathbf{X}_t\tilde{\beta})^2 - n^{-1}(1, -\tilde{\beta}')\Sigma_{aatt}(1, -\tilde{\beta}')']. \quad (3.1.22)$$

The estimator in (3.1.22) will be positive for estimator (3.1.19) if the root $\hat{\lambda}$ defined in (3.1.21) is greater than one. If $\hat{\lambda} < 1$, the estimator of σ_{qq} is zero.

By Theorem 3.1.1, the estimator (3.1.19) is normally distributed in the limit. An estimator of the variance of the approximate distribution of $\tilde{\boldsymbol{\beta}}$ constructed under the assumption of normal errors is

$$\hat{\mathbf{V}}\{\tilde{\boldsymbol{\beta}}\} = n^{-1}\hat{\mathbf{M}}_{xx}^{-1}\hat{\mathbf{G}}\hat{\mathbf{M}}_{xx}^{-1}, \tag{3.1.23}$$

where

$$\hat{\mathbf{G}} = n^{-1} \sum_{t=1}^{n} (\mathbf{X}_t'\mathbf{X}_t\tilde{\sigma}_{vvtt} + \tilde{\mathbf{\Sigma}}_{uvtt}\tilde{\mathbf{\Sigma}}_{vutt}),$$

$$\tilde{\sigma}_{vvtt} = \tilde{\sigma}_{qq} + \sigma_{wwtt} - 2\tilde{\boldsymbol{\beta}}'\mathbf{\Sigma}_{uwtt} + \tilde{\boldsymbol{\beta}}'\mathbf{\Sigma}_{uutt}\tilde{\boldsymbol{\beta}},$$

$\tilde{\mathbf{\Sigma}}_{uvtt} = \mathbf{\Sigma}_{uwtt} - \mathbf{\Sigma}_{uutt}\tilde{\boldsymbol{\beta}}$, and $\hat{\mathbf{M}}_{xx}$ is defined in (3.1.19). The alternative estimator of the variance given in (3.1.12) is

$$\tilde{\mathbf{V}}\{\tilde{\boldsymbol{\beta}}\} = n^{-1}\hat{\mathbf{M}}_{xx}^{-1}\tilde{\mathbf{G}}\hat{\mathbf{M}}_{xx}^{-1}, \tag{3.1.24}$$

where $\tilde{\mathbf{G}} = (n-k)^{-1}\sum_{t=1}^{n}\hat{\mathbf{d}}_t'\hat{\mathbf{d}}_t$ and $\hat{\mathbf{d}}_t = \mathbf{X}_t'\hat{v}_t - \tilde{\mathbf{\Sigma}}_{uvtt}$.

The estimator $\tilde{\boldsymbol{\beta}}$ is a consistent estimator and is relatively easy to compute, but it may be possible to construct an asymptotically superior estimator using $\tilde{\boldsymbol{\beta}}$ as a preliminary estimator. Recall that we can write

$$Y_t = \mathbf{X}_t\boldsymbol{\beta} + v_t, \tag{3.1.25}$$

where $v_t = e_t - \mathbf{u}_t\boldsymbol{\beta}$ and $e_t = q_t + w_t$. The variance of v_t in (3.1.25) is analogous to the variance of the error in the equation for the tth observation of a linear model. Therefore, it is reasonable to construct an estimated generalized least squares estimator by using an estimator of σ_{vvtt} to weight the observations. Such a weighted estimator is

$$\hat{\boldsymbol{\beta}} = \left[\sum_{t=1}^{n} \tilde{\sigma}_{vvtt}^{-1}(\mathbf{X}_t'\mathbf{X}_t - \mathbf{\Sigma}_{uutt})\right]^{-1} \sum_{t=1}^{n} \tilde{\sigma}_{vvtt}^{-1}(\mathbf{X}_t'Y_t - \mathbf{\Sigma}_{uwtt}), \tag{3.1.26}$$

where $\tilde{\sigma}_{vvtt}$ is defined in (3.1.23). The estimator (3.1.26) can be modified by the method used to construct (3.1.20).

Because the error in the preliminary estimator of $\boldsymbol{\beta}$ is $O_p(n^{-1/2})$, the error in $\tilde{\sigma}_{vvtt}$ of (3.1.23) is $O_p(n^{-1/2})$. It follows that $\tilde{\sigma}_{vvtt}^{-1}$ will satisfy the conditions for $\hat{\pi}_t$ of Theorem 3.1.1. Under the assumption of normal errors, an estimator of the covariance matrix of $\hat{\boldsymbol{\beta}}$ is

$$\hat{\mathbf{V}}\{\hat{\boldsymbol{\beta}}\} = n^{-2}\hat{\mathbf{M}}_{x\pi x}^{-1}\left[\sum_{t=1}^{n} \tilde{\sigma}_{vvtt}^{-1}(\mathbf{X}_t'\mathbf{X}_t + \tilde{\sigma}_{vvtt}^{-1}\tilde{\mathbf{\Sigma}}_{uvtt}\tilde{\mathbf{\Sigma}}_{vutt})\right]\hat{\mathbf{M}}_{x\pi x}^{-1}, \tag{3.1.27}$$

where

$$\hat{\mathbf{M}}_{z\pi z} = n^{-1} \sum_{t=1}^{n} \tilde{\sigma}_{vvtt}^{-1}(\mathbf{Z}_t'\mathbf{Z}_t - \mathbf{\Sigma}_{aatt}). \tag{3.1.28}$$

If one is unwilling to assume normal errors, one can estimate the covariance matrix with expression (3.1.12).

The estimator (3.1.26) has intuitive appeal and the use of the $\tilde{\sigma}_{vvtt}^{-1}$ as weights minimizes the first part of the covariance matrix of the limiting distribution—that associated with $\mathbf{x}_t'v_t$. However, because of the contribution of the variance of $\mathbf{u}_t'v_t$ to the covariance matrix, one is not guaranteed that the large sample covariance matrix of the estimator (3.1.26) is less than that of estimator (3.1.20). See Exercise 3.1. The weights that minimize the variance of the limiting distribution depend on the unknown $\mathbf{x}_t$. Without additional assumptions, we cannot construct a best weight, because we are unable to construct a consistent estimator of each $\mathbf{x}_t$. We expect estimator (3.1.26) to be superior to estimator (3.1.20) in almost all practical situations.

One situation in which the measurement error variance will differ from observation to observation arises when the number of determinations differs from individual to individual. The different types of observations correspond to the different number of determinations. This situation occurs naturally in studies where the error variance in the explanatory variables is unknown. Then a part of the project resources must be used to estimate the error variance and this may be accomplished by making duplicate determinations on some of the study elements. Assume that it is desired to estimate the parameters of a model of the form (3.1.1) with

$$\mathbf{V}\{(q_t, \mathbf{a}_t, \mathbf{x}_t)'\} = \text{block diag}(\sigma_{qq}, \boldsymbol{\Sigma}_{aa}, \boldsymbol{\Sigma}_{xx}). \tag{3.1.29}$$

To estimate all parameters, a sample of n elements is selected from the population and duplicate determinations are made on a subset of d_f of the n elements. We assume that the two determinations are independent and identically distributed. Then we can write

$$(Y_{tj}, \mathbf{X}_{tj}) = (y_t, \mathbf{x}_t) + (w_{tj}, \mathbf{u}_{tj}), \tag{3.1.30}$$

where $(Y_{tj}, \mathbf{X}_{tj})$, $j = 1, 2$, is the jth determination on the tth element,

$$V\{(\mathbf{a}_{t1}, \mathbf{a}_{t2})'\} = \text{block diag}(\boldsymbol{\Sigma}_{aa}, \boldsymbol{\Sigma}_{aa}), \tag{3.1.31}$$

and $\mathbf{a}_{tj} = (w_{tj}, \mathbf{u}_{tj})$. An estimator of $\boldsymbol{\Sigma}_{aa}$ is

$$\mathbf{S}_{aa} = (2d_f)^{-1} \sum_{t=1}^{d_f} (\mathbf{Z}_{t1} - \mathbf{Z}_{t2})'(\mathbf{Z}_{t1} - \mathbf{Z}_{t2}), \tag{3.1.32}$$

where $\mathbf{Z}_{tj} = (Y_{tj}, \mathbf{X}_{tj})$. Given this estimator of the error covariance matrix, a consistent estimator of $\boldsymbol{\beta}$ is

$$\tilde{\boldsymbol{\beta}} = \tilde{\boldsymbol{\Sigma}}_{xx}^{-1} n^{-1} \left[\sum_{t=1}^{d_f} \bar{\mathbf{X}}_t' \bar{Y}_t + \sum_{t=d_f+1}^{n} \mathbf{X}_{t1}' Y_{t1} - (n - \tfrac{1}{2} d_f) \mathbf{S}_{uw} \right], \tag{3.1.33}$$

where

$$\tilde{\boldsymbol{\Sigma}}_{xx} = n^{-1}\left[\sum_{t=1}^{d_f} \bar{\mathbf{X}}_t'\bar{\mathbf{X}}_t + \sum_{t=d_f+1}^{n} \mathbf{X}_{t1}'\mathbf{X}_{t1} - (n - \tfrac{1}{2}d_f)\mathbf{S}_{uu}\right],$$

$$(\bar{Y}_t, \bar{\mathbf{X}}_t) = \frac{1}{2}\sum_{j=1}^{2} (Y_{tj}, \mathbf{X}_{tj}) \quad \text{for } t = 1, 2, \ldots, d_f.$$

The estimator (3.1.33) satisfies the conditions of Theorem 3.1.1 with

$$\hat{\boldsymbol{\Sigma}}_{aatt} = \begin{cases} (n - \tfrac{1}{2}d_f)(2d_f)^{-1}(\mathbf{Z}_{t1} - \mathbf{Z}_{t2})'(\mathbf{Z}_{t1} - \mathbf{Z}_{t2}), & t = 1, 2, \ldots, d_f \\ \mathbf{0}, & t = d_f + 1, \ldots, n. \end{cases}$$

Condition (3.1.4) for the estimated covariances follows from our model assumptions and the assumption that d_f is increasing. The estimator of $\boldsymbol{\beta}$ given in (3.1.33) is a simple weighted average of the estimators constructed from the first d_f observations and from the last $n - d_f$ observations. The estimator (3.1.33) is not efficient because the simple weights are not the optimum weights.

If only the first d_f observations are used to estimate $\boldsymbol{\beta}$, the variance of the approximate distribution of the estimated $\boldsymbol{\beta}$ is

$$\begin{aligned}\mathbf{V}_{\beta\beta 11} &= d_f^{-1}\boldsymbol{\Sigma}_{xx}^{-1}(\boldsymbol{\Sigma}_{xx}\sigma_{vv11} + \tfrac{1}{2}\boldsymbol{\Sigma}_{uu}\sigma_{vv11} + \tfrac{1}{4}\boldsymbol{\Sigma}_{uv}\boldsymbol{\Sigma}_{vu})\boldsymbol{\Sigma}_{xx}^{-1} \\ &\quad + (4d_f)^{-1}\boldsymbol{\Sigma}_{xx}^{-1}(\boldsymbol{\Sigma}_{uu}\sigma_{rr} + \boldsymbol{\Sigma}_{uv}\boldsymbol{\Sigma}_{vu})\boldsymbol{\Sigma}_{xx}^{-1},\end{aligned} \tag{3.1.34}$$

where $\sigma_{rr} = (1, -\boldsymbol{\beta}')\boldsymbol{\Sigma}_{aa}(1, -\boldsymbol{\beta}')'$ and $\sigma_{vv11} = \sigma_{qq} + \frac{1}{2}\sigma_{rr}$. See Theorem 2.2.1. From the expression for $\mathbf{V}_{\beta\beta 11}$, the variance of the approximate distribution of $\hat{\mathbf{M}}_{xy11}$ is

$$\mathbf{V}_{MM11} = \mathbf{V}\{\hat{\mathbf{M}}_{xy11}\} = \boldsymbol{\Sigma}_{xx}\mathbf{V}_{\beta\beta 11}\boldsymbol{\Sigma}_{xx} \tag{3.1.35}$$

because $\hat{\mathbf{M}}_{xy11} = \boldsymbol{\Sigma}_{xx}\hat{\boldsymbol{\beta}}_1 + O_p(n^{-1})$. If the last $(n - d_f)$ observations are used to estimate $\boldsymbol{\beta}$, the variance of the approximate distribution of the estimated $\boldsymbol{\beta}$ is

$$\begin{aligned}\mathbf{V}_{\beta\beta 22} &= (n - d_f)^{-1}\boldsymbol{\Sigma}_{xx}^{-1}(\boldsymbol{\Sigma}_{xx}\sigma_{vv22} + \boldsymbol{\Sigma}_{uu}\sigma_{vv22} + \boldsymbol{\Sigma}_{uv}\boldsymbol{\Sigma}_{vu})\boldsymbol{\Sigma}_{xx}^{-1} \\ &\quad + d_f^{-1}\boldsymbol{\Sigma}_{xx}^{-1}(\boldsymbol{\Sigma}_{uu}\sigma_{rr} + \boldsymbol{\Sigma}_{uv}\boldsymbol{\Sigma}_{vu})\boldsymbol{\Sigma}_{xx}^{-1},\end{aligned} \tag{3.1.36}$$

where $\sigma_{vv22} = \sigma_{qq} + \sigma_{rr}$, and the variance of the approximate distribution of $\hat{\mathbf{M}}_{xy22}$ is

$$\mathbf{V}_{MM22} = \boldsymbol{\Sigma}_{xx}\mathbf{V}_{\beta\beta 22}\boldsymbol{\Sigma}_{xx}. \tag{3.1.37}$$

Because the same estimator of the measurement error covariance matrix is used in the two estimators of $\mathbf{M}_{xy}$, the covariance of the approximate joint distribution of $\hat{\mathbf{M}}_{xy11}$ and $\hat{\mathbf{M}}_{xy22}$ is

$$\mathbf{V}_{MM12} = (2d_f)^{-1}(\boldsymbol{\Sigma}_{uu}\sigma_{rr} + \boldsymbol{\Sigma}_{uv}\boldsymbol{\Sigma}_{vu}). \tag{3.1.38}$$

Given the estimator (3.1.33), we can construct estimators of all quantities entering the covariance matrices (3.1.35), (3.1.37), and (3.1.38). Let

$$\tilde{\mathbf{V}}_{MM} = \begin{bmatrix} \tilde{\mathbf{V}}_{MM11} & \tilde{\mathbf{V}}_{MM12} \\ \tilde{\mathbf{V}}_{MM21} & \tilde{\mathbf{V}}_{MM22} \end{bmatrix}. \tag{3.1.39}$$

Then an improved estimator of $\boldsymbol{\beta}$ is

$$\begin{aligned}\hat{\boldsymbol{\beta}} &= [(\hat{\mathbf{M}}_{xx11}, \hat{\mathbf{M}}_{xx22})\tilde{\mathbf{V}}_{MM}^{-1}(\hat{\mathbf{M}}_{xx11}, \hat{\mathbf{M}}_{xx22})']^{-1} \\ &\quad \times [(\hat{\mathbf{M}}_{xx11}, \hat{\mathbf{M}}_{xx22})\tilde{\mathbf{V}}_{MM}^{-1}(\hat{\mathbf{M}}'_{xy11}, \hat{\mathbf{M}}'_{xy22})'],\end{aligned} \tag{3.1.40}$$

where

$$\hat{\mathbf{M}}_{zzjj} = \begin{cases} \mathbf{M}_{ZZ11} - \frac{1}{2}\mathbf{S}_{aa}, & j = 1 \\ \mathbf{M}_{ZZ22} - \mathbf{S}_{aa}, & j = 2. \end{cases}$$

The estimated covariance matrix of the approximate distribution of $\hat{\boldsymbol{\beta}}$ is

$$\hat{\mathbf{V}}\{\hat{\boldsymbol{\beta}}\} = [(\hat{\mathbf{M}}_{xx11}, \hat{\mathbf{M}}_{xx22})\tilde{\mathbf{V}}_{MM}^{-1}(\hat{\mathbf{M}}_{xx11}, \hat{\mathbf{M}}_{xx22})']^{-1}. \tag{3.1.41}$$

Example 3.1.1. We illustrate a situation in which duplicate observations are used to estimate the measurement error variance. Assume that the 25 observations given in Table 3.1.1 have been selected from the same population of sites as the 11 sites studied in Example 1.2.1. The data of Table 3.1.1 differ from those of Table 1.2.1 in that two determinations were made on soil nitrogen at each site for the data of Table 3.1.1. In this example we assume that the error variance is not known and must be estimated from the 25 duplicate observations. This example differs slightly from the theory presented above because it is known that $\sigma_{uw} = 0$. Therefore, no attempt is made to estimate the entire covariance matrix $\boldsymbol{\Sigma}_{aa}$.

To facilitate the computations, we let

$$\mathbf{Z}_{ij} = (Z_{ij1}, Z_{ij2}, Z_{ij3}) = (Y_{ij} - \bar{Y}_{..}, 1, N_{ij} - \bar{N}_{..}),$$

where N_{ij} is the observed nitrogen for the jth field of the ith sample ($i = 1, 2$), $\bar{N}_{..} = 68.5$ is the grand mean of observed soil nitrogen, and $\bar{Y}_{..} = 97.4444$ is the grand mean for yield. Let the data of Table 3.1.1 be sample one and let the data of Example 1.2.1 be sample two.

For the data of Table 3.1.1 the estimated error variance is

$$\hat{\sigma}_{uu} = 50^{-1} \sum_{t=1}^{25} (X_{t1} - X_{t2})^2 = 54.8,$$

and the estimator (3.1.19) is

$$\tilde{\boldsymbol{\beta}}' = (0, 0.4982),$$

TABLE 3.1.1. Additional observations on corn yield and soil nitrogen with duplicate determinations on soil nitrogen

Observation Number	Corn Yield	Soil Nitrogen		
		Determination 1	Determination 2	Average
1	106	71	70	70.5
2	119	78	66	72.0
3	87	76	77	76.5
4	100	59	58	58.5
5	105	97	87	92.0
6	98	53	69	61.0
7	98	76	63	69.5
8	97	43	45	44.0
9	99	86	81	83.5
10	88	44	58	51.0
11	105	89	71	80.0
12	91	46	66	56.0
13	90	66	53	59.5
14	94	62	54	58.0
15	95	76	69	72.5
16	83	59	57	58.0
17	94	61	76	68.5
18	101	70	69	69.5
19	78	34	47	40.5
20	115	93	87	90.0
21	80	59	62	60.5
22	93	48	40	44.0
23	91	64	48	56.0
24	111	95	103	99.0
25	118	100	97	98.5

where $\tilde{\Sigma}_{xx} = \text{diag}(1, 228.7834)$. Because σ_{eu} is known to be zero,

$$\tilde{\sigma}_{ee} = s_{vv} - (36)^{-1}(11 + 12.5)\tilde{\beta}_1^2\hat{\sigma}_{uu} = 50.4903.$$

The estimators of expressions (3.1.35), (3.1.37), and (3.1.38) for diagonal error covariance matrices are

$$\tilde{\mathbf{V}}_{MM11} = \text{diag}(2.2916, 609.4423),$$
$$\tilde{\mathbf{V}}_{MM22} = \text{diag}(5.8265, 1779.6960),$$
$$\tilde{\mathbf{V}}_{MM12} = \text{diag}(0, 29.8146).$$

where $\tilde{\sigma}_{vv11} = \tilde{\sigma}_{ee} + \frac{1}{2}\tilde{\beta}_1^2\hat{\sigma}_{uu} = 57.291$ and $\tilde{\sigma}_{vv22} = \tilde{\sigma}_{ee} + \tilde{\beta}_1^2\hat{\sigma}_{uu} = 64.0918$. The estimator of $\boldsymbol{\beta}$ defined in (3.1.40) is

$$\hat{\boldsymbol{\beta}} = (-0.0204, 0.4996)',$$

where $\hat{\mathbf{M}}_{xy11} = (-0.0044, 122.1580)'$, $\hat{\mathbf{M}}_{xy22} = (0.0101, 95.3687)'$,

$$\operatorname{vech} \hat{\mathbf{M}}_{xx11} = (1, -0.9400; 229.6100)',$$
$$\operatorname{vech} \hat{\mathbf{M}}_{xx22} = (1, 2.1364; 226.9050)',$$

and vech is the notation for the vector half of a matrix defined in Appendix 4.A. The vector half of the estimated covariance matrix of the approximate distribution of the estimator is

$$\operatorname{vech} \hat{\mathbf{V}}\{\hat{\boldsymbol{\beta}}\} = (1.63407, 0.00191; 0.00879)'.$$

In this example, the degrees of freedom for the estimated error variance and the magnitude of error variance relative to the variance of x are such that the contribution to the variance of $\hat{\boldsymbol{\beta}}$ from estimation of the error variance is small. □ □

3.1.3. Reliability Ratios Known

The model in which one has knowledge about the magnitude of the error variance relative to the variance of the observations was introduced in Section 1.1.2. We now consider the vector form of that model. Let

$$y_t = \beta_0 + \mathbf{x}_t \boldsymbol{\beta}_1 + q_t, \qquad \mathbf{Z}_t = \mathbf{z}_t + \mathbf{a}_t,$$

and let the vectors $(\mathbf{x}_t, \mathbf{a}_t, q_t)$, $t = 1, 2, \ldots$, be independently and identically distributed with mean $(\boldsymbol{\mu}_x, \mathbf{0}, 0)$, finite fourth moments, and

$$\mathbf{V}\{(\mathbf{x}_t, \mathbf{a}_t, q_t)'\} = \text{block diag}\{\boldsymbol{\Sigma}_{xx}, \boldsymbol{\Sigma}_{aa}, \sigma_{qq}\},$$

where $\boldsymbol{\Sigma}_{xx}$ is nonsingular. Let the matrix $\boldsymbol{\Lambda}_{aa}$ be known, where

$$\boldsymbol{\Lambda}_{aa} = \mathbf{D}_{ZZ}^{-1} \boldsymbol{\Sigma}_{aa} \mathbf{D}_{ZZ}^{-1},$$

and $\mathbf{D}_{ZZ}^2 = \operatorname{diag}(\sigma_{YY}, \sigma_{XX11}, \sigma_{XX22}, \ldots, \sigma_{XXkk})$. If λ_{ww} for the Y variable is not known, it is replaced with

$$\ddot{\lambda}_{ww} = \boldsymbol{\Lambda}_{wu} \boldsymbol{\Lambda}_{uu}^{\dagger} \boldsymbol{\Lambda}_{uw},$$

where $\boldsymbol{\Lambda}_{uu}^{\dagger}$ is the Moore–Penrose generalized inverse of $\boldsymbol{\Lambda}_{uu}$ and λ_{ww}, $\boldsymbol{\Lambda}_{wu}$, and $\boldsymbol{\Lambda}_{ww}$ are the submatrices of $\boldsymbol{\Lambda}_{aa}$ defined by the partition $\mathbf{a}_t = (w_t, \mathbf{u}_t)$. The diagonal elements of $\boldsymbol{\Lambda}_{aa}$, denoted in abbreviated notation by λ_{ii}, are equal to $1 - \kappa_{ii}$, where κ_{ii} is the reliability ratio for the ith variable. In many applications it is assumed that the off-diagonal elements of $\boldsymbol{\Lambda}_{aa}$ are zero.

An estimator of $\boldsymbol{\beta}_1$ can be obtained by constructing consistent estimators of $\mathbf{m}_{xx}$ and of $\mathbf{m}_{xy}$. The estimator incorporating the modifications of

Section 2.5 is

$$\tilde{\boldsymbol{\beta}}_1 = \hat{\mathbf{H}}_{xx}^{-1}\hat{\mathbf{H}}_{xy}, \tag{3.1.42}$$

where

$$\hat{\mathbf{H}}_{zz} = \begin{cases} \mathbf{m}_{ZZ} - (1 - n^{-1})\hat{\mathbf{D}}_{ZZ}\boldsymbol{\Lambda}_{aa}\hat{\mathbf{D}}_{ZZ} & \text{if } \hat{f} \geq 1 \\ \mathbf{m}_{ZZ} - (\hat{f} - n^{-1})\hat{\mathbf{D}}_{ZZ}\boldsymbol{\Lambda}_{aa}\hat{\mathbf{D}}_{ZZ} & \text{if } \hat{f} < 1, \end{cases}$$

$$\hat{\mathbf{D}}_{ZZ}^2 = \operatorname{diag}(m_{YY}, m_{XX11}, m_{XX22}, \ldots, m_{XXkk}),$$

and $\hat{f}$ is the smallest root of

$$|\mathbf{m}_{ZZ} - f\hat{\mathbf{D}}_{ZZ}\boldsymbol{\Lambda}_{aa}\hat{\mathbf{D}}_{ZZ}| = 0. \tag{3.1.43}$$

While not obvious, the estimator (3.1.42) satisfies the conditions of Theorem 3.1.1. To apply that theorem, let λ_{ij} be the ijth element of $\boldsymbol{\Lambda}_{aa}$ and express the ijth element of $\hat{\mathbf{D}}_{ZZ}\boldsymbol{\Lambda}_{aa}\hat{\mathbf{D}}_{ZZ}$ as

$$m_{ZZii}^{1/2}m_{ZZjj}^{1/2}\lambda_{ij} = n^{-1}\sum_{t=1}^{n} \tfrac{1}{2}\lambda_{ij}\sigma_{ZZii}^{1/2}\sigma_{ZZjj}^{1/2}[\sigma_{ZZii}^{-1}(Z_{ti} - \bar{Z}_i)^2 + \sigma_{ZZjj}^{-1}(Z_{tj} - \bar{Z}_j)^2] + O_p(n^{-1}). \tag{3.1.44}$$

Therefore, to the order of approximation required, Theorem 3.1.1 is applicable with the ijth element of $\hat{\boldsymbol{\Sigma}}_{aatt}$ defined by the tth element of the sum in (3.1.44). The ijth element of $\hat{\boldsymbol{\Sigma}}_{aatt}$ used to construct the vectors $\hat{\mathbf{d}}_t$ for the distribution-free variance estimator (3.1.12) is

$$\tfrac{1}{2}\lambda_{ij}m_{ZZii}^{1/2}m_{ZZjj}^{1/2}[m_{ZZii}^{-1}(Z_{ti} - \bar{Z}_i)^2 + m_{ZZjj}^{-1}(Z_{tj} - \bar{Z}_j)^2]. \tag{3.1.45}$$

If one is willing to assume that the vector $(\mathbf{x}_t, \mathbf{a}_t, q_t)$ is normally distributed and that $\boldsymbol{\Sigma}_{aa}$ is a diagonal matrix, the covariance matrix of the approximate distribution of $\hat{\boldsymbol{\beta}}_1$ can be estimated by

$$\hat{\mathbf{V}}\{\hat{\boldsymbol{\beta}}_1\} = (n - 1)^{-1}\hat{\mathbf{H}}_{xx}^{-1}\hat{\boldsymbol{\Gamma}}\hat{\mathbf{H}}_{xx}^{-1}, \tag{3.1.46}$$

where the ijth element of $\hat{\boldsymbol{\Gamma}}$ is

$$\hat{\gamma}_{ij} = m_{XXij}(s_{vv} - 2\lambda_{ii}^2\hat{\beta}_i^2 m_{XXii} - 2\lambda_{jj}^2\hat{\beta}_j^2 m_{XXjj} + 2\lambda_{ii}\lambda_{jj}\hat{\beta}_i\hat{\beta}_j m_{XXij}) + \lambda_{ii}\lambda_{jj}\hat{\beta}_i\hat{\beta}_j m_{XXii}m_{XXjj}.$$

Because of the different form of the knowledge used in computing the root $\hat{f}$, the distribution of $\hat{f}$ is different than the distribution of the root $\hat{\lambda}$ given in Theorem 2.3.2. If the entire matrix $\boldsymbol{\Lambda}_{aa}$ is known and if $q_t \equiv 0$, it can be shown that

$$[\hat{V}\{\hat{f}\}]^{-1/2}(\hat{f} - 1) \xrightarrow{L} N(0, 1),$$

where

$$\hat{V}\{\hat{f}\} = (n - k)^{-1}n^{-1}(\hat{f}\hat{\boldsymbol{\theta}}'\hat{\mathbf{D}}_{ZZ}\boldsymbol{\Lambda}_{aa}\hat{\mathbf{D}}_{ZZ}\hat{\boldsymbol{\theta}})^{-2}\sum_{t=1}^{n}(\hat{v}_t^2 - \hat{f}\hat{\boldsymbol{\theta}}'\hat{\boldsymbol{\Sigma}}_{aatt}\hat{\boldsymbol{\theta}})^2, \tag{3.1.47}$$

$\hat{v}_t = Y_t - \bar{Y} - (\mathbf{X}_t - \bar{\mathbf{X}})\tilde{\beta}_1$, $\hat{\theta}' = (1, -\tilde{\beta}_1')$, $\hat{\Sigma}_{aatt}$ is the $(k+1) \times (k+1)$ matrix with ijth element given by (3.1.45), and $\hat{\theta}$ satisfies

$$(\mathbf{m}_{ZZ} - \hat{f}\hat{\mathbf{D}}_{ZZ}\Lambda_{aa}\hat{\mathbf{D}}_{ZZ})\hat{\theta} = \mathbf{0}. \tag{3.1.48}$$

This result can be used to check the rank of Σ_{xx}. One constructs the root and vector defined by (3.1.48), replacing $\mathbf{Z}_t$ by $\mathbf{X}_t$ and Λ_{aa} by Λ_{uu}. If the rank of Σ_{xx} is $k - 1$, the normalized root will be approximately distributed as a $N(0, 1)$ random variable.

Example 3.1.2. In this example we analyze data for a sample of Iowa farm operators. The data were collected in 1977 and are a subsample of the data discussed in Abd-Ella et al. (1981). Table 3.A.1 of Appendix 3.A contains 176 observations from the sample, where the variables are the logarithm of acre size of the farm (size), the logarithm of number of years the operator has been a farm operator (experience), and the education of the operator (education). Education is the transformation of years of formal training suggested by Carter (1971). We treat the 176 observations as a simple random sample of Iowa farm operators.

To protect the confidentiality of the respondents, a random error was added to each of the variables in Table 3.A.1. Thus, the data of Table 3.A.1 contain two types of measurement error: that associated with the original responses and that added to protect confidentiality. The study of Battese, Fuller, and Hickman (1976) contains information on the reliability ratio for the original responses on farm size and that of Siegel and Hodge (1968) contains information on the reliability ratios of the original responses for education and experience. The ratio of the variance of the errors added to protect confidentiality to the sample variance is known. Combining the two sources of error, the reliability ratio for size is 0.891, the reliability ratio for experience is 0.800, and the reliability ratio for education is 0.826. The reliability ratios are treated as known in the analysis. It is assumed that the measurement errors for the three variables are uncorrelated. The model is

$$Y_t = \beta_0 + \beta_1 x_{t1} + \beta_2 x_{t2} + e_t, \qquad \mathbf{X}_t = \mathbf{x}_t + \mathbf{u}_t,$$

where Y_t is observed size, $e_t = w_t + q_t$, x_{t1} is true experience, x_{t2} is true education, X_{t1} is observed experience, and X_{t2} is observed education for the tth respondent.

The sample means are 5.5108, 2.7361, and 5.6392 for size, experience, and education, respectively. The vector half of the sample covariance matrix is

$$\text{vech } \mathbf{m}_{ZZ} = (0.91462, 0.21281, 0.07142;\ 1.00647, -0.44892;\ 1.03908)'.$$

The estimator of Σ_{xx} is the matrix of mean squares and products of $\mathbf{X}$ corrected for attenuation. Thus, vech $\hat{\boldsymbol{\Sigma}}_{xx}$ is

$$\text{vech}\{\mathbf{m}_{XX} - \hat{\mathbf{D}}_{XX}\boldsymbol{\Lambda}_{uu}\hat{\mathbf{D}}_{XX}\} = (0.80518,\ -0.44892;\ 0.85828)',$$

where $\hat{\mathbf{D}}_{XX} = \text{diag}(1.00323,\ 1.01935)$ and $\boldsymbol{\Lambda}_{uu} = \text{diag}(0.200,\ 0.174)$. The estimated equation computed from Equation (3.1.42) using the program SUPER CARP is

$$\underset{(0.79)}{\hat{Y}_t = 2.55} + \underset{(0.116)}{0.439x_{t1}} + \underset{(0.096)}{0.313x_{t2}},$$

where the numbers in parentheses are the estimated standard errors constructed with estimator (3.1.12). The vector half of the covariance matrix estimated from Equation (3.1.12) is

$$\text{vech}\ \hat{\mathbf{V}}_{\beta\beta} = (0.6237,\ -0.0765,\ -0.0713;\ 0.0134, 0.0068;\ 0.0092)'.$$

The vector half of the estimated covariance matrix (3.1.46) computed under the assumption of normality is

$$\text{vech}\ \hat{\mathbf{V}}\{\hat{\boldsymbol{\beta}}\} = (0.4768,\ -0.5025,\ -0.5931;\ 0.0087,\ 0.0047;\ 0.0082)',$$

where $s_{vv} = 0.8652$. All estimated variances are smaller under the normality assumption. It is clear that the variables cannot be exactly normally distributed because experience is reported in years and education is restricted to a few values. Also, a plot of the residuals suggests that v_t is negatively skewed. Because of these facts and because the sample is not small, the distribution-free form (3.1.12) seems the preferable variance estimation method.

The estimated squared multiple correlation between Y_t and $\mathbf{x}_t$ is

$$\hat{R}^2_{\mathbf{x}Y} = (0.91462)^{-1}[0.4386(0.21281) + 0.3126(0.07142)] = 0.126.$$

Also see Exercise 3.6. □ □

For other applications of the correction-for-attenuation model see Fuller and Hidiroglou (1978) and Hwang (1986).

3.1.4. Error Variance Functionally Related to Observations

This section is devoted to the model wherein the covariance matrices of the measurement error are known functions of observable variables or are known functions of the expectation of observable random variables. In particular, we consider estimator matrices constructed as

$$\hat{\boldsymbol{\Sigma}}_{aatt} = \sum_{i=1}^{r} \boldsymbol{\psi}'_{ti}\boldsymbol{\psi}_{ti}, \tag{3.1.49}$$

where $r \leqslant k + 1$ and ψ_{ti} are observable vectors. We treat situations in which

$$E\left\{\sum_{i=1}^{r} \psi'_{ti}\psi_{ti}\right\} = \Sigma_{aatt} \tag{3.1.50}$$

or in which

$$E\left\{\sum_{t=1}^{n}\sum_{i=1}^{r} \psi'_{ti}\psi_{ti}\right\} = \sum_{t=1}^{n} \Sigma_{aatt}. \tag{3.1.51}$$

Because any covariance matrix can be written in the form (3.1.49), the case of known error covariance matrices is automatically in the class. The model with known reliability ratios is also in the class. For example, if the Λ_{aa} of Section 3.1.3 is diagonal, then

$$\hat{\Sigma}_{aatt} = \sum_{i=1}^{k+1} \psi'_{ti}\psi_{ti},$$

where ψ_{ti} is a $(k + 1)$-dimensional row vector with $\lambda_{ii}^{1/2}(Z_{ti} - \bar{Z}_i)$ in the ith position and zeros elsewhere.

The vectors ψ_{ti} may themselves be expressed in terms of other vectors. For example, we may write

$$\psi_{ti} = \{c_{0i}\phi_{t0i}, c_{1i}\phi_{t1i}, \ldots, c_{ki}\phi_{tki}\}, \tag{3.1.52}$$

where c_{si}, $s = 0, 1, \ldots, k$, $i = 0, 1, \ldots, k$, are known constants and ϕ_{tji} are observables that may be fixed or random. Because the ϕ_{tji} are permitted to be random variables, they can be, and often are, functions of $\mathbf{Z}_t$. The class of estimator matrices (3.1.49) does not exhaust the class of possible models, but it includes many useful models and the form of the estimator (3.1.49) lends itself to computing. Also, the matrix of expression (3.1.49) is always positive semidefinite.

A situation in which the estimator (3.1.49) is applicable arises when data are available for several groups, it is desired to estimate separate slopes by group, and the explanatory variable with separate slopes is measured with error. This is one of many cases in which a problem that is straightforward in the absence of measurement error becomes relatively complex in the presence of measurement error. We consider a simple model for two groups and assume that grouping is done without error. Let f_t denote the true value of the explanatory variable of interest and let $F_t = f_t + r_t$ be the observed value, where r_t is the measurement error and $r_t \sim \mathrm{NI}(0, \sigma_{rr})$. Ordinarily there will be additional variables in the regression equation, but we restrict our attention to the vector used to estimate separate slopes for the two groups. Let

$$Y_t = \mathbf{x}_t\boldsymbol{\beta} + e_t, \qquad \mathbf{X}_t = \mathbf{x}_t + \mathbf{u}_t,$$

where $\mathbf{X}_t = (X_{t1}, X_{t2})$, $\mathbf{x}_t = (x_{t1}, x_{t2})$, $\mathbf{u}_t = (D_{t1}r_t, D_{t2}r_t)$,

$$[(X_{t1}, X_{t2}), (x_{t1}, x_{t2})] = [(D_{t1}F_t, D_{t2}F_t), (D_{t1}f_t, D_{t2}f_t)],$$

and

$$(D_{t1}, D_{t2}) = \begin{cases} (1, 0) & \text{if element } t \text{ is in group 1} \\ (0, 1) & \text{if element } t \text{ is in group 2.} \end{cases}$$

Let the variance of the measurement error in Y_t, denoted by w_t, be independent of r_t. Then the covariance matrix of $\mathbf{a}_t = (w_t, \mathbf{u}_t)$ is

$$\Sigma_{aatt} = \sum_{i=1}^{2} \psi'_{ti}\psi_{ti}, \tag{3.1.53}$$

where $\psi_{t1} = (\sigma_{ww}^{1/2}, 0, 0)$ and $\psi_{t2} = (0, \sigma_{rr}^{1/2}D_{t1}, \sigma_{rr}^{1/2}D_{t2})$. These concepts are developed further in Example 3.1.3.

Example 3.1.3. We illustrate the estimation of separate slopes by group using a constructed data set based on a survey conducted by Winakor (1975) and discussed by Hidiroglou (1974). The survey was designed to study expenditures on household textiles such as drapes, towels, and sheets. The explanatory variables for the yearly textile expenditures were the income of the household, the size of the household, and whether or not the household moved during the year. The data are given in Table 3.A.2 of Appendix 3.A, where the variables are:

- Y Log (expenditure on textiles in dollars plus 5),
- X_1 Log (income in hundreds of dollars plus 5),
- X_2 Log of number of members in the household,
- x_4 Indicator that takes the value one if the household moved and is zero otherwise

The variables in Table 3.A.2 were constructed so that the moment matrix $\mathbf{m}_{XX}$ of the explanatory variables is similar to that observed in the study. In the actual study only about one-sixth of the households moved, while in our data set about 30% are in that category. The moment matrix of the explanatory variables is similar to that observed in the study but the correlation between expenditure and the explanatory variables is much higher in our data set. It would require about 3000 of the original observations to produce standard errors for estimated coefficients of the order we obtain from our constructed data set of 100 observations.

Our model for these data is

$$Y_t = \beta_0 + \beta_1 x_{t1} + \beta_2 x_{t2} + \beta_3 x_{t1} x_{t4} + e_t,$$
$$(e_t, u_{t1}, u_{t2}) \sim \text{Ind}[\mathbf{0}, \text{diag}(\sigma_{ee}, \lambda_{11}\sigma_{XX11}, \lambda_{22}\sigma_{XX22})],$$

and $(X_{t1}, X_{t2}) = (x_{t1}, x_{t2}) + (u_{t1}, u_{t2})$. It is assumed that (e_t, u_{t1}, u_{t2}) is uncorrelated with (x_{t1}, x_{t2}) and that the vectors $(e_t, u_{t1}, u_{t2}, x_{t1}, x_{t2})$ have finite fourth moments. Note that, for the purpose of this example, it is assumed that the indicator variable for moving is measured without error. Following Hidiroglou (1974) we assume $(\lambda_{11}, \lambda_{22}) = (0.1172, 0.0485)$. The estimator of β given by expression (3.1.7) is

$$\hat{\beta} = \hat{\mathbf{M}}_{xx}^{-1}\hat{\mathbf{M}}_{xy},$$

where

$$\begin{aligned}
\mathbf{X}_t &= (1, X_{t1}, X_{t2}, X_{t1}x_{t4}),\\
\hat{\mathbf{M}}_{xy} &= \mathbf{M}_{XY},\\
\hat{\mathbf{M}}_{xx} &= n^{-1}\left(\sum_{t=1}^{n} \mathbf{X}_t'\mathbf{X}_t - \sum_{t=1}^{n}\sum_{i=1}^{3} \psi_{ti}'\psi_{ti}\right),\\
\psi_{t1} &= [0, 0, (0.1172)^{1/2}(X_{t1} - \bar{X}_1)(1 - x_{t4}), 0, 0],\\
\psi_{t2} &= [0, 0, 0, (0.0485)^{1/2}(X_{t2} - \bar{X}_2), 0],\\
\psi_{t3} &= [0, 0, (0.1172)^{1/2}(X_{t1} - \bar{X}_1)x_{t4}, 0, (0.1172)^{1/2}(X_{t1} - \bar{X}_1)x_{t4}].
\end{aligned}$$

This set of ψ-vectors differs somewhat from the set in (3.1.53). We assume we know the reliability ratios for the textile data, while the variables in (3.1.53) were constructed under the assumption that the covariance matrix of the measurement error is known.

For the textile data the sample mean vector is

$$(\bar{Y}, \bar{X}_1, \bar{X}_2, \bar{x}_4) = (3.6509, 4.6707, 0.9478, 0.28).$$

The vector half of the sample covariance matrix of $(X_{t1}, X_{t2}, X_{t1}x_{t4})$ is

$$(0.1975, 0.0584, -0.1483; 0.3255, -0.2825; 4.1985)'$$

and the corresponding portion of $n^{-1}\sum_{t=1}^{n}\sum_{i=1}^{3}\psi_{ti}'\psi_{ti}$ is

$$(0.02291, 0, 0.00749; 0.01563, 0; 0.00749)'.$$

The estimated equation is

$$\begin{aligned}
\hat{Y}_t = {} & -0.1634 + 0.6514x_{t1} + 0.3699x_{t2} + 0.3336x_{t1}x_{t4},\\
& (0.2501) \quad (0.0538) \qquad (0.0342) \qquad (0.0123)
\end{aligned}$$

where the numbers in parentheses are the standard errors obtained from the covariance matrix (3.1.12). The estimates were computed using SUPER CARP. The estimator contains a modification similar to that used in estimator (3.1.42). An estimator of σ_{ee} is

$$\hat{\sigma}_{ee} = (n - k)^{-1}\sum_{t=1}^{n}(Y_t - \mathbf{X}_t\hat{\beta})^2 - \hat{\beta}'\hat{\mathbf{\Sigma}}_{uu}\hat{\beta} = 0.02316.$$

Using the estimator of σ_{ee}, we can construct an estimator of σ_{vv} for the two types of observations. For the set of individuals that did not move ($x_{t4} = 0$), the estimated coefficient for x_{t1} is 0.6514, and for the set of individuals that moved ($x_{t4} = 1$), the estimated coefficient for x_{t1} is 0.9850. Therefore, the two estimators of σ_{vv} are

$$\hat{\sigma}_{vv(0)} = 0.03503 \quad \text{and} \quad \hat{\sigma}_{vv(1)} = 0.04753.$$

Given these two estimates of the error variances, we can construct an improved estimator of $\boldsymbol{\beta}$ by weighting the observations with the reciprocal of the appropriate $\hat{\sigma}_{vv(i)}$. To accomplish this we define the vectors

$$\hat{\sigma}_{vv(i)}^{-1/2}(Y_t, 1, X_{t1}, X_{t2}, X_{t1}x_{t4}),$$

with $i = 0, 1$ for the households that did not move and for the households that moved, respectively. The ψ vectors for the transformed problem are

$$\begin{aligned}
\psi_{t1} &= \hat{\sigma}_{vv(0)}^{-1/2}[0, 0, (0.1172)^{1/2}(X_{t1} - \bar{X})(1 - x_{t4}), 0, 0], \\
\psi_{t2} &= \hat{\sigma}_{vv(0)}^{-1/2}[0, 0, 0, (0.0485)^{1/2}(X_{t2} - \bar{X}_2)(1 - x_{t4}), 0] \\
&\quad + \hat{\sigma}_{vv(1)}^{-1/2}[0, 0, 0, (0.0485)^{1/2}(X_{t2} - \bar{X}_2)x_{t4}, 0], \\
\psi_{t3} &= \hat{\sigma}_{vv(1)}^{-1/2}[0, 0, (0.1172)^{1/2}(X_{t1} - \bar{X}_1)x_{t4}, 0, (0.1172)^{1/2}(X_{t1} - \bar{X}_1)x_{t4}],
\end{aligned}$$

and the estimated equation is

$$\begin{aligned}
\hat{Y}_t = &-0.1732 + 0.6548x_{t1} + 0.3635x_{t2} + 0.3334x_{t3}, \\
&\;(0.2362) \quad (0.0505) \qquad (0.0338) \qquad (0.0122)
\end{aligned}$$

where the numbers in parentheses are the standard errors obtained from the estimator covariance matrix (3.1.12). The estimated standard errors of the intercept and coefficient of x_{t1} are about 5% smaller than those constructed in the unweighted analysis. The estimated standard errors for x_{t2} and x_{t3} are only marginally smaller than those estimated from the unweighted analysis.

The variance of the estimator of $\boldsymbol{\beta}$ based on the data in one moving category is not strictly proportional to $\sigma_{vv(i)}$ of that category because of the term in $\Sigma_{uv}\Sigma_{vu}$ that enters the variance expression. In our example, this term is small relative to the term that is proportional to $\sigma_{vv(i)}$. Therefore, an estimated generalized least squares estimator constructed by the methods of Example 3.1.1 would be very similar to the estimator of this example that was constructed using $\hat{\sigma}_{vv(0)}$ and $\hat{\sigma}_{vv(1)}$ as weights. □□

The following example illustrates the type of economic data often collected in surveys. The variance of measurement error increases as the true value increases. The fact that duplicate observations were collected on all elements used for analysis is unusual.

Example 3.1.4. We analyze some data collected by the Statistical Laboratory of Iowa State University under contract to the Statistical Reporting Service, U.S. Department of Agriculture. The study is described in Battese, Fuller, and Hickman (1976). A sample of farmers was contacted in the first week of September 1970. The same farmers were contacted one month later and a portion of the data collected at the first interview was also obtained at the second interview. This permitted the estimation of the response variance (measurement error variance) for the repeated items. We consider two variables:

x_t Number of breeding hogs on hand September 1st

y_t Number of sows farrowing (giving birth to baby pigs) between June 1 and August 31

The data are given in Table 3.A.3 of Appendix 3.A.

Our initial model is

$$y_t = \beta_0 + \beta_1 x_t + q_t, \qquad (Y_{tj}, X_{tj}) = (y_t, x_t) + (w_{tj}, u_{tj}),$$

where Y_{tj} is the number of sows farrowing as reported in the jth interview by the tth individual, X_{tj} is the number of breeding hogs reported in the jth interview by the tth individual, and $j = 1, 2$. It is assumed that the two observations (w_{t1}, u_{t1}) and (w_{t2}, u_{t2}) are independent with zero mean vector and finite $8 + \delta$ moments. It is assumed that the q_t are independent with zero means and finite $8 + \delta$ moments, and that q_t is independent of $(w_{i1}, u_{i1}, w_{i2}, u_{i2})$ for all t and i.

The analysis of variance constructed from the two responses of 184 farmers is given in Table 3.1.2. The 184 farmers are a subset of the original data. The entries on the "Error" line are the estimates of the error variances and covariances for a single response. The covariance matrix for $(\bar{w}_t, \bar{u}_t)$ is one-half of the covariance matrix of (w_{tj}, u_{tj}) and the estimator of the covariance matrix of $\mathbf{a}_t = (\bar{w}_t, \bar{u}_t)$ has entries that are one-half of the error line of Table 3.1.2. Using the 184 means $(\bar{Y}_t, \bar{X}_t)$ and Equation (2.2.20), we obtain the following

TABLE 3.1.2. Analysis of covariance for farrowings (Y) and number of breeding hogs (X)

Source	Degrees of Freedom	Mean Squares and Products		
		YY	XY	XX
Individuals	183	304.98	444.76	1694.64
Error	184	58.23	8.72	139.09
Total	367	181.27	226.15	914.74

estimators of the parameters:

$$\tilde{\beta}_1 = (847.32 - 69.54)^{-1}(222.38 - 4.36) = 0.2803,$$
$$\tilde{\beta}_0 = 10.4321 - 0.2803(36.7745) = 0.1242,$$
$$\tilde{\sigma}_{xx} = 847.32 - 69.54 = 777.78,$$
$$\tilde{\sigma}_{qq} = 152.49 - 29.12 - (0.2803)(218.02) = 62.26,$$

where $(\bar{Y}, \bar{X}) = (10.4321, 36.7745)$. If one assumes that the response errors are normally and independently distributed and uses (2.2.25), the estimator of the covariance matrix of the approximate distribution is

$$\text{vech } \tilde{\mathbf{V}}(\tilde{\boldsymbol{\theta}}) = (1.531, -2.76, 0.099; 7.507, -0.270; 1.097)',$$

where $\tilde{\boldsymbol{\theta}}' = (\tilde{\beta}_0, 100\tilde{\beta}_1, 0.1\tilde{\sigma}_{qq})$, $s_{vv} = 94.9147$, $\tilde{\sigma}_{uv} = \hat{\sigma}_{ur} = -15.1321$, $\tilde{\sigma}_{rr} = 32.1394$, and $d_f = 184$. The estimated variance of $\tilde{\beta}_1$ is about 3% larger than it would be if the covariance matrix of the measurement error were known.

For these data we have the original duplicate observations on each individual. Therefore, we can use the duplicate observations to construct the $\boldsymbol{\psi}$ vectors of (3.1.53). Let

$$(\bar{Y}_t, \bar{\mathbf{X}}_t) = \tfrac{1}{2}[(Y_{t1} + Y_{t2}), (\mathbf{X}_{t1} + \mathbf{X}_{t2})],$$
$$\boldsymbol{\psi}_{t1} = \tfrac{1}{2}\{(Y_{t1} - Y_{t2}), (\mathbf{X}_{t1} - \mathbf{X}_{t2})\}.$$

Under the assumptions of this section, the covariance matrix of $(\bar{w}_t, \bar{\mathbf{u}}_t, q_t)$ can be a function of t. Because

$$E\{\boldsymbol{\psi}'_{t1}\boldsymbol{\psi}_{t1}\} = E\{(\bar{w}_t, \bar{\mathbf{u}}_t)'(\bar{w}_t, \bar{\mathbf{u}}_t)\},$$

assumption (3.1.3) is satisfied for $\boldsymbol{\psi}_{t1}$. The elements of

$$\text{vech}\left\{n^{-1}\sum_{t=1}^{n}\boldsymbol{\psi}'_{t1}\boldsymbol{\psi}_{t1}\right\} = (29.12, 0, 4.36; 0, 0; 69.54)'$$

are the same as the elements of the error covariance matrix $\mathbf{S}_{aa}$ constructed from the analysis of variance table. The estimated model obtained from expression (3.1.7) with the α adjustment of Section 2.5 is

$$\hat{y}_t = 0.132 + 0.2801x_t.$$

The estimated covariance matrix calculated by expression (3.1.12) is

$$\text{vech } \hat{\mathbf{V}}\{(\hat{\beta}_0, 100\hat{\beta}_1)'\} = (2.3864, -7.0196; 24.3443)'.$$

The calculations for this example were performed using the functionally related option of SUPER CARP.

The estimated covariance matrix of Theorem 3.1.1 contains much larger estimated variances than does the estimated covariance matrix constructed under the assumptions of Theorem 2.2.1. The estimate of the variance of the approximate distribution of $\hat{\beta}_0$ is 50% larger than that based on the constant-variance-normal assumptions and the estimate of the variance of the approximate distribution of $\hat{\beta}_1$ is more than three times that based on the constant-variance-normal assumptions. The farrowings data violate the assumptions of identically distributed normal errors in several respects. First, the observations are not normal because the observations are bounded below by zero and there are a number of zero observations on the farrowings variable. Second, plots of the data demonstrate that the error variances for both variables are larger for large values. In Figure 3.1.1 the mean of the two Y observations is plotted against the mean of the two X observations. It is the fact that the error variances are not constant that produces the largest increase in the estimated variance.

The variance formula for the estimator of β_1 given in Theorem 2.2.1 contains a term associated with the variance of s_{ur}, where $r_t = w_t - u_t\beta_1$. In Theorem 2.2.1 the estimated variance of s_{ur} is constructed under the assumption that the (u_t, r_t) are independent identically distributed normal vectors. The failure of these assumptions can lead to large biases in the estimated

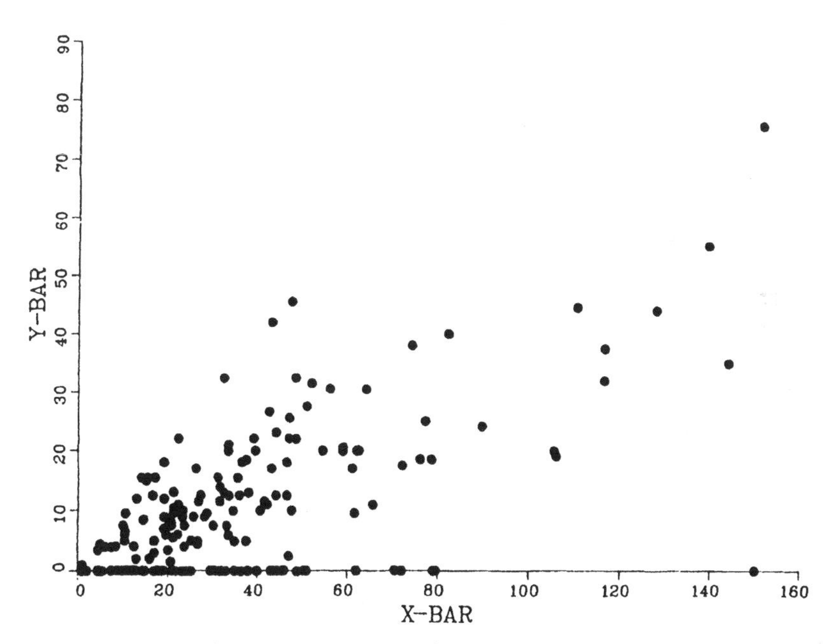

FIGURE 3.1.1. Plot of mean of two Y observations against mean of two X observations for pig farrowing data.

variance of $\tilde{\sigma}_{ur}$. For the pig data, estimators of $\sigma_{ur} = \sigma_{uv}$ and of σ_{rr} are

$$\tilde{\sigma}_{ur} = s_{uw} - s_{uu}\hat{\beta}_1 = -15.13,$$
$$\tilde{\sigma}_{rr} = s_{ww} - 2\hat{\beta}_1 s_{uw} + \hat{\beta}_1^2 s_{uu} = 32.14.$$

Therefore, under the normality assumption the estimated variance of s_{ur} is

$$\hat{V}\{s_{ur}\} = (184)^{-1}[(69.54)(32.14) + (-15.13)^2] = 13.39,$$

where $s_{uu} = 69.54$. Because the data used to estimate the error variances are available, we can construct an alternative estimator of the variance of $\tilde{\sigma}_{ur}$. The estimator $\tilde{\sigma}_{ur}$ can be written

$$\tilde{\sigma}_{ur} = (184)^{-1} \sum_{t=1}^{184} C_t = \bar{C},$$

where $C_t = (X_{t1} - X_{t2})[Y_{t1} - Y_{t2} - (X_{t1} - X_{t2})\hat{\beta}_1](0.25)$. Therefore, a direct estimator of the variance of $\tilde{\sigma}_{ur}$ is

$$\tilde{V}\{\tilde{\sigma}_{ur}\} = (184)^{-1}(183)^{-1} \sum_{t=1}^{184} (C_t - \bar{C})^2 = 36.55.$$

This consistent estimator of the variance of $\tilde{\sigma}_{ur}$ is nearly three times the estimator of variance constructed under the assumption that (w_{tj}, u_{tj}) are constant-variance-normal vectors. The contribution to the estimated variance arising from variance estimation is included in the estimation formula (3.1.12). In this example, the contribution to the total variance arising from the estimation of the error variances is modest.

Because the range in estimated error variances is so large and because the error variance seems to be related to x_t, we attempt to improve the estimator of β_1. Given our initial estimator of (β_0, β_1), we can estimate x_t with

$$\hat{x}_t = X_t - \hat{v}_t s_{vv}^{-1} \tilde{\sigma}_{vu} = X_t + 0.159\hat{v}_t,$$

where $s_{vv} = 94.9147$ and $\tilde{\sigma}_{vu} = -15.1321$. Now

$$\hat{r}_t^2 = (0.25)[Y_{t1} - \hat{\beta}_1 X_{t1} - (Y_{t2} - \hat{\beta}_1 X_{t2})]^2$$

is an estimator of $(1, -\beta_1)\Sigma_{aatt}(1, -\beta_1)'$ and

$$\hat{v}_t^2 = [\bar{Y}_t - \bar{Y} - (\bar{X}_t - \bar{X})\hat{\beta}_1]^2$$

is an estimator of $\sigma_{qq} + (1, -\beta_1)\Sigma_{aatt}(1, -\beta_1)'$. It is clear from a plot of $\hat{v}_t$ against $\hat{x}_t$ that the variance of $\hat{v}_t$ increases with $\hat{x}_t$. This plot and the plot of $\hat{r}_t$ against $\hat{x}_t$ led us to postulate

$$E\{r_t^2\} = (\alpha_0 + \alpha_1 x_t)^2,$$
$$E\{v_t^2\} = \gamma^2(\alpha_0 + \alpha_1 x_t)^2,$$

where α_0, α_1, and γ are parameters to be estimated. In this model both r_t and q_t are assumed to have variances that increase with the square of x_t. We fit the model by iterative generalized nonlinear least squares replacing (r_t^2, v_t^2, x_t) with $(\hat{r}_t^2, \hat{v}_t^2, \hat{x}_t)$. In the fitting it was assumed that the standard deviations of r_t^2 and v_t^2 were proportional to their expected values. The estimates are

$$[\hat{\alpha}_0, \hat{\alpha}_1, \hat{\gamma}^2] = \underset{(0.17,\, 0.011,\, 0.143)}{[0.68, 0.133, 1.643]},$$

where the numbers in parentheses are the estimated standard errors obtained at the last step of the generalized nonlinear least squares procedure. In this model, the variance of q_t is assumed to be a multiple of σ_{rrtt} and the multiple is estimated to be 0.643.

Let $\mathbf{Z}_t = (\bar{Y}_t, 1, \bar{X}_t)$, $\pi_t = \gamma^{-2}(\hat{\alpha}_0 + \hat{\alpha}_1 \hat{x}_t)^{-2}$, and

$$\psi_{t1} = \tfrac{1}{2}[(Y_{t1} - Y_{t2}), 0, (X_{t1} - X_{t2})].$$

Then the estimator (3.1.7) computed with $\hat{\Sigma}_{aatt}$ of (3.1.49) and the adjustment of Section 2.5 gives

$$\hat{y}_t = \underset{(0.410)}{-0.006} + \underset{(0.026)}{0.287} x_t,$$

where the numbers in parentheses are the standard errors estimated with expression (3.1.12). The use of weights produces an estimated standard error for the intercept that is less than one-third of that for the unweighted analysis. The standard error for the weighted estimator of β_1 is about one-half of that for the unweighted estimator. Figure 3.1.2 contains a plot of $\hat{\pi}_t^{1/2}\hat{v}_t$ against $\hat{x}_t$. The observations that originally fell on the X axis form the curved lower boundary of the deviations. The fact that this boundary curves away from zero and that the deviations above zero for small $\hat{x}_t$ seem somewhat smaller than the remaining positive deviations might lead one to consider alternative variance functions. However, the majority of variance inhomogeneity has been removed, and we do not pursue the issue further.

The weights are functions of the $\hat{x}_t$. Therefore, rather strong assumptions are required to apply Theorem 3.1.1. It can be demonstrated that the conditions of the theorem are satisfied if we assume the error variances decline as n increases. From the practical point of view, this means that the variance of u_t must not be too large relative to the variance of x_t. Hasabelnaby (1985) has conducted Monte Carlo studies that indicate that the approximations of Theorem 3.1.1 are satisfactory for data similar to that of this example.

□ □

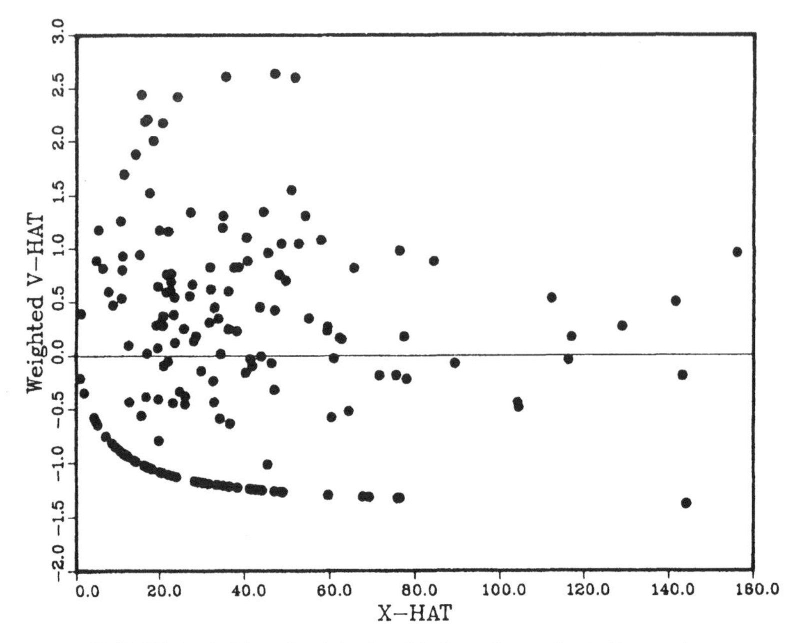

FIGURE 3.1.2. Plot of weighted residuals against estimated true values.

3.1.5. The Quadratic Model

In this section we study estimation of the parameters of the model wherein Y_t is a quadratic function of the true value of a variable observed with error. This model falls in the domain of the theory of Sections 3.1.1 and 3.1.4 because the variance of the measurement error of the square of the explanatory variable is functionally related to the true value of the explanatory variable.

Let Y_t satisfy the quadratic model,

$$Y_t = \beta_0 + \beta_1 h_t + \beta_2 h_t^2 + q_t, \qquad H_t = h_t + r_t, \tag{3.1.54}$$
$$(q_t, r_t)' \sim \text{NI}[\mathbf{0}, \text{diag}(\sigma_{qq}, \sigma_{rr})]$$

where (q_j, r_j) is independent of h_t for all t and j and (Y_t, H_t) is observed. The error in H_t^2, as an estimator of h_t^2, is $2h_t r_t + r_t^2$, where

$$E\{2h_t r_t + r_t^2\} = \sigma_{rr},$$
$$V\{2h_t r_t + r_t^2 \mid h_t\} = 4h_t^2\sigma_{rr} + 2\sigma_{rr}^2.$$

Let

$$\mathbf{X}_t = (1, H_t, H_t^2 - \sigma_{rr}) = \mathbf{x}_t + \mathbf{u}_t, \tag{3.1.55}$$

where $x_t = (1, h_t, h_t^2)$ and $\mathbf{u}_t = (0, r_t, 2h_t r_t + r_t^2 - \sigma_{rr})$. With these definitions, we can write the model (3.1.54) in the familiar form

$$Y_t = \mathbf{x}_t\boldsymbol{\beta} + q_t, \qquad \mathbf{X}_t = \mathbf{x}_t + \mathbf{u}_t,$$

where $\boldsymbol{\beta}' = (\beta_0, \beta_1, \beta_2)$. Furthermore, the conditional mean of $\mathbf{u}_t$ given h_t is the zero vector, and the conditional variance is

$$E\{\mathbf{u}_t'\mathbf{u}_t \mid h_t\} = \begin{bmatrix} 0 & 0 & 0 \\ 0 & \sigma_{rr} & 2h_t\sigma_{rr} \\ 0 & 2h_t\sigma_{rr} & 4h_t^2\sigma_{rr} + 2\sigma_{rr}^2 \end{bmatrix}. \tag{3.1.56}$$

The matrix $E\{\mathbf{u}_t'\mathbf{u}_t \mid h_t\}$ plays the role of the matrix $\boldsymbol{\Sigma}_{uutt}$ of Equation (3.1.2) of Section 3.1.1. To construct an estimator of the measurement covariance matrix, we assume σ_{rr} is known and note that the expected value of

$$\hat{\boldsymbol{\Sigma}}_{uutt} = (0, \sigma_{rr}^{1/2}, 2\sigma_{rr}^{1/2}H_t)'(0, \sigma_{rr}^{1/2}, 2\sigma_{rr}^{1/2}H_t) - \operatorname{diag}(0, 0, 2\sigma_{rr}^2) \tag{3.1.57}$$

is equal to (3.1.56).

Theorem 3.1.1 is satisfied for $\hat{\boldsymbol{\beta}}$ of (3.1.7) with $\hat{\pi}_t \equiv 1$, $\hat{\boldsymbol{\Sigma}}_{uutt}$ defined in (3.1.57), and $\mathbf{x}_t$ satisfying the conditions of the theorem. Thus, the estimator (3.1.7) can be used in large samples with large error variances.

The estimator matrix of (3.1.57) is not always positive semidefinite. It follows that there exists no set of vectors ψ_{ti} of the type described in (3.1.49) such that $\sum_{i=1}^{3} E\{\psi_{ti}'\psi_{ti} \mid h_t\}$ is exactly equal to $\boldsymbol{\Sigma}_{uutt}$. However, it is possible to construct vectors ψ_{t1} such that

$$E\left\{\sum_{t=1}^{n} \psi_{t1}'\psi_{t1}\right\} = \sum_{t=1}^{n} \boldsymbol{\Sigma}_{uutt} \tag{3.1.58}$$

and such that $E\{\psi_{t1}'\psi_{t1}\}$ is approximately equal to $\boldsymbol{\Sigma}_{uutt}$ for each t. To this end, let

$$\psi_{t1} = \{0, \sigma_{rr}^{1/2}, 2\sigma_{rr}^{1/2}[\bar{H} + \zeta^{1/2}(\bar{H}_t - \bar{H})]\}, \tag{3.1.59}$$

where

$$\zeta = \left[\sum_{t=1}^{n} (\mathrm{H}_t - H)^2\right]^{-1} \left[\sum_{t=1}^{n} (H_t - \bar{H})^2 - \tfrac{1}{2}n\sigma_{rr}\right].$$

The reader may verify that

$$n^{-1}\sum_{t=1}^{n} \psi_{t1}'\psi_{t1} = \begin{bmatrix} 0 & 0 & 0 \\ 0 & \sigma_{rr} & \bar{H}\sigma_{rr} \\ 0 & \bar{H}\sigma_{rr} & 4\sigma_{rr}M_{HH} - 2\sigma_{rr}^2 \end{bmatrix},$$

where $M_{HH} = n^{-1}\sum_{t=1}^{n} H_t^2$, and, hence, (3.1.58) is satisfied.

Example 3.1.5. A theory of the earth's structure states that the earth's surface is composed of a number of "plates" that float on the earth's mantle. At the place where two plates meet, one plate may be forced beneath the other. This movement places strain on the bedrock and produces earthquakes. The Tonga trench is a trench in the Pacific Ocean near Fiji where the Pacific plate meets the Australian plate. The data in Table 3.A.4 are the depths and locations of 43 earthquakes occurring near the Tonga trench between January 1965 and January 1966. The data are a subset of that analyzed by Sykes, Isacks, and Oliver (1969).

The variable X_1 is the perpendicular distance in hundreds of kilometers from a line that is approximately parallel to the Tonga trench. The variable X_2 is the distance in hundreds of kilometers from an arbitrary line perpendicular to the Tonga trench. The variable Y is the depth of the earthquake in hundreds of kilometers. Under the plate model, the depths of the earthquakes will increase with distance from the trench and a plot of the data shows this to be the case. The location of the earthquakes is subject to error and Sykes, Isacks, and Oliver (1969) suggest that a variance of 100 kilometers squared is a reasonable approximation for the variance in location. These same authors also explain why it is reasonable for the depth of the earthquakes to occur in a pattern that curves away from the earth's surface. Therefore, our working model is

$$y_t = \beta_0 + \beta_1 x_{t1} + \beta_2 x_{t2} + \beta_3 x_{t1}^2 + q_t,$$
$$(Y_t, X_{t1}, X_{t2}, X_{t3}) = (y_t, x_{t1}, x_{t2}, x_{t1}^2) + (w_t, r_{t1}, r_{t2}, a_{t4}),$$
$$(q_t, w_t, r_{t1}, r_{t2})' \sim \mathrm{NI}[\mathbf{0}, \mathrm{diag}(\sigma_{qq}, 0.01, 0.01, 0.01)],$$

where $X_{t3} = X_{t1}^2 - 0.01$. Because the error variance of x_{t1} is 0.01, the expected value of X_{t3} is x_{t1}^2. The sample mean of the observed vector is

$$(\bar{Y}, \bar{X}_1, \bar{X}_2, \bar{X}_3) = (1.1349, 1.5070, 1.9214, 3.5020)$$

and the matrix of mean squares and products is

$$\begin{bmatrix} 1.2129 & 1.2059 & 0.2465 & 4.3714 \\ 1.2059 & 1.2706 & 0.1633 & 4.4813 \\ 0.2465 & 0.1633 & 1.1227 & 0.6250 \\ 4.3714 & 4.4813 & 0.6250 & 16.6909 \end{bmatrix}.$$

The three vectors required for the construction of the error covariance matrix of $(Y_t, X_{t1}, X_{t2}, X_{t3})$ are

$$\psi_{t1} = \{0, 0.1, 0, 0.2[1.5070 + 0.99798(X_{t1} - 1.5070]\},$$
$$\psi_{t2} = \{0, 0, 0.1, 0\}, \quad \text{and} \quad \psi_{t3} = \{0.1, 0, 0, 0\},$$

where the ζ of Equation (3.1.59) is 0.99597. It follows that the upper left

3×3 portion of the average of the estimated error covariance matrices is $0.01\mathbf{I}$ and the last row of the matrix is

$$(0, 0.0301, 0, 0.1403).$$

The estimated equation is

$$\hat{Y}_t = \underset{(0.079)}{-0.199} + \underset{(0.113)}{0.481x_{t1}} + \underset{(0.031)}{0.077x_{t2}} + \underset{(0.033)}{0.132x_{t1}^2},$$

where the numbers in parentheses are the estimated standard errors calculated from the estimated covariance matrix of Theorem 3.1.1. The computations were performed with SUPER CARP. In this example the error variance of X_{t1} is less than 1% of the total variation. Therefore, the coefficients are very near to those obtained by ordinary least squares. In this example there are three large deviations from the fitted model, associated with observations 4, 6, and 12. Because the X_{t1} values and X_{t1}^2 values for these observations are reasonably close to their respective means, the estimated standard errors obtained from the covariance matrix (3.1.12) are smaller than those computed by the ordinary least squares formulas. □□

Example 3.1.6. In this example we illustrate the calculations for the quadratic model in which the error variances of both Y and X are known. This is an application of Theorem 3.1.2 to the quadratic model. Table 3.A.4 contains 120 observations generated to satisfy the model

$$y_t = \beta_0 + \beta_1 x_{t1} + \beta_2 x_{t1}^2, \qquad (Y_t, X_{t1}) = (y_t, x_{t1}) + (e_t, u_{t1}),$$
$$(e_t, u_{t1})' \sim \mathrm{NI}(\mathbf{0}, 0.09\mathbf{I}).$$

The sample mean vector is $\bar{\mathbf{Z}} = (\bar{Y}, \bar{X}_1, \bar{X}_2) = (0.50, 0.00, 0.50)$, where $X_{t2} = X_{t1}^2 - 0.09$. The matrix of mean squares and products corrected for the mean is

$$\mathbf{m}_{ZZ} = \begin{bmatrix} 0.41425 & 0.27227 & 0.17612 \\ 0.27227 & 0.59496 & -0.00132 \\ 0.17612 & -0.00132 & 0.36530 \end{bmatrix}.$$

The ψ vectors required for the construction of the estimated covariance matrix are

$$\psi_{t1} = (0.3, 0, 0) \quad \text{and} \quad \psi_{t2} = (0, 0.3, 0.57666X_{t1}),$$

where ζ of Equation (3.1.59) is 0.92373. The estimated mean of the error covariance matrices is

$$\hat{\boldsymbol{\Sigma}}_{aa..} = (120)^{-1} \sum_{t=1}^{120} \sum_{i=1}^{2} \psi_{ti}'\psi_{ti} = \mathrm{diag}(0.09, 0.09, 0.1962)$$

and the smallest root of

$$|\mathbf{m}_{ZZ} - \lambda\hat{\boldsymbol{\Sigma}}_{aa..}| = 0$$

is $\hat{\lambda} = 0.9771$. By the results of Section 3.1.1, the test statistic, $F = (n - k)^{-1}n\hat{\lambda} = 0.9938$, is approximately distributed as an F with b and infinity degrees of freedom, where the estimated degrees of freedom defined in (3.1.18) is $\hat{b} = 156$. One easily accepts the model because $\hat{\lambda}$ is close to one.

The modification (2.5.14) with $\alpha = 1$ of the estimator (3.1.14) is

$$\begin{aligned}\tilde{\boldsymbol{\beta}} &= [\mathbf{M}_{XX} - \hat{\lambda}(1 - n^{-1})\hat{\boldsymbol{\Sigma}}_{uu..}]^{-1}[\mathbf{M}_{XY} - \hat{\lambda}(1 - n^{-1})\hat{\boldsymbol{\Sigma}}_{ue..}] \\ &= (-0.005, 0.539, 1.009)'.\end{aligned}$$

The estimated covariance matrix of the approximate distribution of the vector of estimators is

$$\begin{bmatrix} 0.7153 & -0.0605 & -1.0350 \\ -0.0605 & 0.8920 & 0.1704 \\ -1.0350 & 0.1704 & 2.3254 \end{bmatrix} 10^{-2}.$$

This example illustrates the large effect that measurement error can have on the estimates of the parameters of the quadratic function. The ordinary least squares estimate of the quadratic equation obtained by regressing Y_{t1} on X_{t1} and X_{t1}^2 is

$$\begin{aligned} Y_t = \underset{(0.058)}{0.214} + \underset{(0.054)}{0.459}X_{t1} + \underset{(0.069)}{0.484}X_{t1}^2, \end{aligned}$$

where the numbers in parentheses are the "standard errors" computed by the ordinary least squares formulas. The ordinary least squares coefficient for the linear effect is about 85% of the consistent estimator. This agrees with the fact that the variance of the measurement error in X_{t1} is about 15% of the total variation of X_{t1}. The least squares coefficient for the quadratic effect is only about one-half of the consistent estimator. A 15% error in the original observations results in, approximately, a 50% error in the squares. See Exercise 1.7. □□

The procedures illustrated in Examples 3.1.5 and 3.1.6 are appropriate for large samples with large error variances. If the error variances are not overly large relative to the variation in x_t, it may be possible to construct more efficient estimators. This is because the variance of the deviation from the fitted function varies as the slope of the function varies. See Section 3.2.

3.1.6. Maximum Likelihood Estimation for Known Error Covariance Matrices

In this section we extend the model of Section 2.3 to permit different error covariance matrices for different observations. Let

$$y_t = \mathbf{x}_t\boldsymbol{\beta}, \qquad \mathbf{Z}_t = \mathbf{z}_t + \boldsymbol{\varepsilon}_t, \qquad (3.1.60)$$
$$\boldsymbol{\varepsilon}_t' \sim \mathrm{NI}(\mathbf{0}, \boldsymbol{\Sigma}_{\varepsilon\varepsilon tt}),$$

where $\{\mathbf{x}_t\}$ is a fixed sequence of k-dimensional row vectors and $\boldsymbol{\varepsilon}_t = (e_t, \mathbf{u}_t)$. It is assumed that the covariance matrices, $\boldsymbol{\Sigma}_{\varepsilon\varepsilon tt}$, $t = 1, 2, \ldots, n$, are known. The model can also be written

$$\mathbf{z}_t\boldsymbol{\alpha} = 0, \qquad \mathbf{Z}_t = \mathbf{z}_t + \boldsymbol{\varepsilon}_t, \qquad (3.1.61)$$

where $\boldsymbol{\alpha}' = (1, -\boldsymbol{\beta}')$.

For this model, the method of maximum likelihood produces reasonable estimators. For nonsingular $\boldsymbol{\Sigma}_{\varepsilon\varepsilon tt}$, $t = 1, 2, \ldots, n$, the logarithm of the likelihood function for a sample of n observations is

$$\log L = -\frac{1}{2}\sum_{t=1}^{n}\left[\log|(2\pi)^{k+1}\boldsymbol{\Sigma}_{\varepsilon\varepsilon tt}| + (\mathbf{Z}_t - \mathbf{z}_t)\boldsymbol{\Sigma}_{\varepsilon\varepsilon tt}^{-1}(\mathbf{Z}_t - \mathbf{z}_t)'\right]. \qquad (3.1.62)$$

We can follow the approach used in the proof of Theorem 2.3.1 to show that the maximum likelihood estimator of $\boldsymbol{\beta}$ is the $\boldsymbol{\beta}$ that minimizes

$$g_n(\boldsymbol{\beta}) = \sum_{t=1}^{n} (\boldsymbol{\alpha}'\boldsymbol{\Sigma}_{\varepsilon\varepsilon tt}\boldsymbol{\alpha})^{-1}\boldsymbol{\alpha}'\mathbf{Z}_t'\mathbf{Z}_t\boldsymbol{\alpha}. \qquad (3.1.63)$$

If we differentiate (3.1.63) with respect to $\boldsymbol{\beta}$ and set the result equal to zero, we obtain

$$\sum_{t=1}^{n} \sigma_{vvtt}^{-1}[(\mathbf{X}_t'Y_t - \mathbf{X}_t'\mathbf{X}_t\boldsymbol{\beta}) - \gamma_t(\boldsymbol{\Sigma}_{uett} - \boldsymbol{\Sigma}_{uutt}\boldsymbol{\beta})] = 0, \qquad (3.1.64)$$

where $\sigma_{vvtt} = \boldsymbol{\alpha}'\boldsymbol{\Sigma}_{\varepsilon\varepsilon tt}\boldsymbol{\alpha}$ and $\gamma_t = \sigma_{vvtt}^{-1}\boldsymbol{\alpha}'\mathbf{Z}_t'\mathbf{Z}_t\boldsymbol{\alpha}$. Because σ_{vvtt} and γ_t are functions of $\boldsymbol{\beta}$, it seems necessary to use iterative methods to obtain the $\boldsymbol{\beta}$ that minimizes (3.1.63).

One iterative procedure is suggested by (3.1.63) and (3.1.64). If σ_{vvtt} and γ_t are replaced with trial values denoted by $\sigma_{vvtt}^{(i-1)}$ and $\gamma_t^{(i-1)}$, then (3.1.64) can be solved for $\hat{\boldsymbol{\beta}}$. We suggest a slightly modified intermediate estimator of $\boldsymbol{\beta}$, defined by

$$\hat{\boldsymbol{\beta}}^{(i)} = \hat{\mathbf{H}}_{(i-1)}^{-1}\hat{\mathbf{R}}_{(i-1)}, \qquad (3.1.65)$$

where

$$\hat{\mathbf{H}}_{(i-1)} = \sum_{t=1}^{n} (\sigma_{vvtt}^{(i-1)})^{-1}(\mathbf{X}_t'\mathbf{X}_t - \hat{\lambda}\gamma_t^{(i-1)}\mathbf{\Sigma}_{uutt}),$$

$$\hat{\mathbf{R}}_{(i-1)} = \sum_{t=1}^{n} (\sigma_{vvtt}^{(i-1)})^{-1}(\mathbf{X}_t'Y_t - \hat{\lambda}\gamma_t^{(i-1)}\mathbf{\Sigma}_{uett}),$$

and $\hat{\lambda}$ is the smallest root of

$$\left|\sum_{t=1}^{n} (\sigma_{vvtt}^{(i-1)})^{-1}(\mathbf{Z}_t'\mathbf{Z}_t - \lambda\gamma_t^{(i-1)}\mathbf{\Sigma}_{\varepsilon\varepsilon tt})\right| = 0. \tag{3.1.66}$$

The estimator (3.1.65) has the advantage that the matrix to be inverted, $\hat{\mathbf{H}}_{(i-1)}^{-1}$, is positive definite (with probability one).

The iterative estimation procedure is defined by the following steps:

1. To initiate the procedure set $\sigma_{vvtt}^{(0)} = \gamma_t^{(0)} = 1$, $t = 1, 2, \ldots, n$.
2. Compute $\hat{\boldsymbol{\beta}}^{(i)}$ by (3.1.65) and go to step 3.
3. Set

$$\sigma_{vvtt}^{(i)} = (1, -\hat{\boldsymbol{\beta}}^{(i)\prime})\mathbf{\Sigma}_{\varepsilon\varepsilon tt}(1, -\hat{\boldsymbol{\beta}}^{(i)\prime})',$$
$$\gamma_t^{(i)} = (\sigma_{vvtt}^{(i)})^{-1}(Y_t - \mathbf{X}_t\hat{\boldsymbol{\beta}}^{(i)})^2.$$

Increase i by one and return to step 2.

The iterative procedure is a form of the Newton–Raphson method because $\hat{\mathbf{H}}_{(i-1)}$ is an estimator of the matrix of second partial derivatives of (3.1.63) with respect to $\boldsymbol{\beta}$ and the vector $\hat{\mathbf{R}}_{(i-1)} - \hat{\mathbf{H}}_{(i-1)}\hat{\boldsymbol{\beta}}^{(i-1)}$ is an estimator of the vector of first partial derivatives of (3.1.63) with respect to $\boldsymbol{\beta}$. It follows that the iterative procedure can be modified to guarantee convergence. It seems that the unmodified procedure described in steps 1, 2, and 3 above will almost always converge in practice.

The limiting distribution of the maximum likelihood estimator is given in Theorem 3.1.3. The theorem is proved under weaker assumptions about the distribution of $\boldsymbol{\varepsilon}_t$ than those used to derive the estimator.

Theorem 3.1.3. Let model (3.1.60) hold. Let $\{\mathbf{z}_t\}$ be a bounded sequence of fixed vectors and let the $\boldsymbol{\varepsilon}_t$ be independent $(\mathbf{0}, \mathbf{\Sigma}_{\varepsilon\varepsilon tt})$ random vectors with finite $4 + \nu$ $(\nu > 0)$ moments. Let the true parameter $\boldsymbol{\beta}$ be in the interior of the parameter space B, where B is an open bounded subset of k-dimensional Euclidean space. For all vectors $\boldsymbol{\beta}$ in B, assume

$$0 < K_L < \boldsymbol{\alpha}'\mathbf{\Sigma}_{\varepsilon\varepsilon tt}\boldsymbol{\alpha} < K_U < \infty \tag{3.1.67}$$

for all t, where K_L and K_U are fixed. Assume that the elements of the sequence $\{\boldsymbol{\Sigma}_{\varepsilon\varepsilon tt}\}$ are uniformly bounded and that the limits

$$\bar{\boldsymbol{\Sigma}}_{\varepsilon\varepsilon} = \lim_{n\to\infty} n^{-1} \sum_{t=1}^{n} \boldsymbol{\Sigma}_{\varepsilon\varepsilon tt},$$

$$\boldsymbol{\Gamma} = \lim_{n\to\infty} n^{-1}(\text{vech } \boldsymbol{\Sigma}_{\varepsilon\varepsilon tt})(\text{vech } \boldsymbol{\Sigma}_{\varepsilon\varepsilon tt})'$$

exist, where vech $\boldsymbol{\Sigma}_{\varepsilon\varepsilon tt}$ is the column vector obtained by listing the elements on and below the diagonal of $\boldsymbol{\Sigma}_{\varepsilon\varepsilon tt}$ columnwise in a single column. (See Appendix 4.A.) Assume that the limits

$$\bar{\mathbf{M}}_{xx} = \lim_{n\to\infty} n^{-1} \sum_{t=1}^{n} \mathbf{x}_t'\mathbf{x}_t,$$

$$\bar{\mathbf{M}}_{x\pi x} = \lim_{n\to\infty} n^{-1} \sum_{t=1}^{n} \sigma_{vvtt}^{-1}\mathbf{x}_t'\mathbf{x}_t$$

exist and are positive definite, where σ_{vvtt} is defined in (3.1.64). Let $\hat{\boldsymbol{\beta}}$ be the value of $\boldsymbol{\beta}$ that minimizes (3.1.63) and let $\boldsymbol{\beta}^0$ be the true value of $\boldsymbol{\beta}$. Then

$$n^{1/2}(\hat{\boldsymbol{\beta}} - \boldsymbol{\beta}^0) \xrightarrow{L} N(\mathbf{0}, \bar{\mathbf{M}}_{x\pi x}^{-1}\mathbf{G}\bar{\mathbf{M}}_{x\pi x}^{-1}),$$

where

$$\mathbf{G} = \lim_{n\to\infty} n^{-1} \sum_{t=1}^{n} E\{\mathbf{d}_t'\mathbf{d}_t\},$$

$\mathbf{d}_t' = \sigma_{vvtt}^{-1}(\mathbf{X}_t'v_t - \sigma_{vvtt}^{-1}v_t^2\boldsymbol{\Sigma}_{uvtt})$, and $\boldsymbol{\Sigma}_{uvtt} = \boldsymbol{\Sigma}_{uett} - \boldsymbol{\Sigma}_{uutt}\boldsymbol{\beta}^0$.

Proof. We first demonstrate that $\hat{\boldsymbol{\beta}}$ converges to $\boldsymbol{\beta}^0$, almost surely, as n increases. Dividing (3.1.63) by n and using $\mathbf{Z}_t = \mathbf{z}_t + \boldsymbol{\varepsilon}_t$, we obtain

$$n^{-1}g_n(\boldsymbol{\beta}) = n^{-1} \sum_{t=1}^{n} (\boldsymbol{\alpha}'\boldsymbol{\Sigma}_{\varepsilon\varepsilon tt}\boldsymbol{\alpha})^{-1}(\boldsymbol{\alpha}'\mathbf{z}_t'\mathbf{z}_t\boldsymbol{\alpha} + \boldsymbol{\alpha}'\mathbf{z}_t'\boldsymbol{\varepsilon}_t\boldsymbol{\alpha} + \boldsymbol{\alpha}'\boldsymbol{\varepsilon}_t'\mathbf{z}_t\boldsymbol{\alpha} + \boldsymbol{\alpha}'\boldsymbol{\varepsilon}_t'\boldsymbol{\varepsilon}_t\boldsymbol{\alpha}). \quad (3.1.68)$$

By assumption (3.1.67),

$$K_U^{-1}n^{-1} \sum_{t=1}^{n} \boldsymbol{\alpha}'\mathbf{z}_t'\mathbf{z}_t\boldsymbol{\alpha} \leqslant n^{-1} \sum_{t=1}^{n} (\boldsymbol{\alpha}'\boldsymbol{\Sigma}_{\varepsilon\varepsilon tt}\boldsymbol{\alpha})^{-1}\boldsymbol{\alpha}'\mathbf{z}_t'\mathbf{z}_t\boldsymbol{\alpha}$$

for all $\boldsymbol{\beta}$ in B. Because $\bar{\mathbf{M}}_{xx}$ is positive definite, given $\eta > 0$, there is an N_1 and a $\delta > 0$ such that $n > N_1$ and $|\boldsymbol{\beta} - \boldsymbol{\beta}^0|^2 > \eta$ implies $n^{-1}\sum_{t=1}^{n} \boldsymbol{\alpha}'\mathbf{z}_t\mathbf{z}_t\boldsymbol{\alpha} > K_U\delta$. Under our assumptions, for almost all sample sequences,

$$n^{-1} \sum_{t=1}^{n} (\boldsymbol{\alpha}'\boldsymbol{\Sigma}_{\varepsilon\varepsilon tt}\boldsymbol{\alpha})^{-1}(\boldsymbol{\alpha}'\boldsymbol{\varepsilon}_t\boldsymbol{\varepsilon}_t\boldsymbol{\alpha}, \boldsymbol{\alpha}'\mathbf{z}_t\boldsymbol{\varepsilon}_t\boldsymbol{\alpha}) \to (1, 0)$$

uniformly in $\boldsymbol{\beta}$ for $\boldsymbol{\beta}$ in B. Therefore, given $\delta > 0$, there is an $N_2 \geqslant N_1$ such that if $n > N_2$, the sum of the last three terms of (3.1.68) differs from one by less than $\frac{1}{2}\delta$ for all $\boldsymbol{\beta}$ in B and almost all sample sequences. The minimum value for the first term of (3.1.68) is zero because $\mathbf{z}_t\boldsymbol{\alpha}^0 = 0$. Therefore, the minimum value of (3.1.68) must be less than $1 + \frac{1}{2}\delta$ for $n > N_2$ for almost all sequences. It follows that if $n > N_2$, then the value of $\boldsymbol{\beta}$ that minimizes (3.1.68) must satisfy $|\hat{\boldsymbol{\beta}} - \boldsymbol{\beta}^0|^2 < \eta$ for almost all sample sequences.

All partial derivatives of the first two orders of $g_n(\boldsymbol{\beta})$ exist and are continuous functions of $\boldsymbol{\beta}$ on B. For $\hat{\boldsymbol{\beta}}$ in B, by Taylor's theorem,

$$\frac{\partial g_n(\hat{\boldsymbol{\beta}})}{\partial \boldsymbol{\beta}} = \frac{\partial g_n(\boldsymbol{\beta}^0)}{\partial \boldsymbol{\beta}} + \frac{\partial^2 g_n(\boldsymbol{\beta}^*)}{\partial \boldsymbol{\beta}\, \partial \boldsymbol{\beta}'}(\hat{\boldsymbol{\beta}} - \boldsymbol{\beta}^0),$$

where $\boldsymbol{\beta}^*$ is on the line segment joining $\boldsymbol{\beta}^0$ and $\hat{\boldsymbol{\beta}}$. Because $\boldsymbol{\beta}^0$ is in the interior of B, the probability that (3.1.64) is satisfied on B goes to one as n increases. Now $\hat{\boldsymbol{\beta}} \to \boldsymbol{\beta}^0$ a.s., and

$$n^{-1}\frac{\partial^2 g_n(\boldsymbol{\beta}^*)}{\partial \boldsymbol{\beta}\, \partial \boldsymbol{\beta}'} \to 2\bar{\mathbf{M}}_{x\pi x}, \text{ a.s.,}$$

where the matrix $\bar{\mathbf{M}}_{x\pi x}$ is positive definite. It follows that, given $\delta > 0$, there exists an N_δ such that

$$\hat{\boldsymbol{\beta}} - \boldsymbol{\beta} = -\left(\frac{\partial^2 g_n(\boldsymbol{\beta}^*)}{\partial \boldsymbol{\beta}\, \partial \boldsymbol{\beta}'}\right)^{-1} \frac{\partial g_n(\boldsymbol{\beta}^0)}{\partial \boldsymbol{\beta}}$$

with probability greater than $1 - \delta$ for $n > N_\delta$. From (3.1.63),

$$\frac{\partial g_n(\boldsymbol{\beta}^0)}{\partial \boldsymbol{\beta}} = 2\sum_{t=1}^{n} \mathbf{d}_t'.$$

The vectors $\mathbf{d}_t$ are independently distributed with bounded $2 + \frac{1}{2}\nu$ moments. Therefore, $n^{-1/2}\sum_{t=1}^{n} \mathbf{d}_t$ converges in distribution to a normal vector by the Liapounov central limit theorem. □

The variance of the approximate distribution of $\hat{\boldsymbol{\beta}}$ can be estimated by

$$\hat{\mathbf{V}}\{\hat{\boldsymbol{\beta}}\} = n^{-1}\hat{\mathbf{M}}_{x\pi x}^{-1}\hat{\mathbf{G}}\hat{\mathbf{M}}_{x\pi x}^{-1}, \tag{3.1.69}$$

where

$$\hat{\mathbf{M}}_{x\pi x} = n^{-1}\sum_{t=1}^{n} \hat{\sigma}_{vvtt}^{-1}(\mathbf{X}_t'\mathbf{X}_t - \hat{\lambda}\hat{\gamma}_t\boldsymbol{\Sigma}_{uutt}),$$

$$\hat{\mathbf{G}} = (n-k)^{-1}\sum_{t=1}^{n} \hat{\mathbf{d}}_t'\hat{\mathbf{d}}_t,$$

$\hat{\mathbf{d}}_t' = \hat{\sigma}_{vvtt}^{-1}\hat{x}_t\hat{v}_t$, $\hat{\sigma}_{vvtt} = (1, -\hat{\boldsymbol{\beta}}')\boldsymbol{\Sigma}_{\varepsilon\varepsilon tt}(1, -\hat{\boldsymbol{\beta}}')'$, $\hat{\boldsymbol{\Sigma}}_{vvtt} = \boldsymbol{\Sigma}_{ue} - \boldsymbol{\Sigma}_{uu}\hat{\boldsymbol{\beta}}$, $\hat{\gamma}_t = \hat{\sigma}_{vvtt}^{-1}\hat{v}_t^2$, $\hat{\mathbf{x}}_t = \mathbf{X}_t - \hat{v}_t\sigma_{vvtt}^{-1}\hat{\boldsymbol{\Sigma}}_{vutt}$, and $\hat{\lambda}$ is the smallest root of the last iterate of (3.1.66).

Theorem 3.1.3 and the estimator (3.1.69) are given for quite general error distributions. If the ε_t are normally distributed, the estimator of $\mathbf{G}$ of (3.1.69) can be replaced with

$$\hat{\mathbf{G}} = \hat{\mathbf{M}}_{x\pi x} + n^{-1} \sum_{t=1}^{n} (\hat{\sigma}_{vvtt}^{-1}\Sigma_{uutt} - \hat{\sigma}_{vvtt}^{-2}\hat{\Sigma}_{uvtt}\hat{\Sigma}_{vutt}) \tag{3.1.70}$$

where the estimators are those defined in (3.1.69).

It is also possible to construct a test of the model. Under our assumptions,

$$\hat{\chi}^2 = \sum_{t=1}^{n} \hat{\sigma}_{vvtt}^{-1}\hat{v}_t^2 \tag{3.1.71}$$

is approximately distributed as a chi-square random variable with $n - k$ degrees of freedom.

Example 3.1.7. Rust, Leventhal, and McCall (1976) analyzed the light curves of 15 Type I supernova by estimating the parameters for the relationship between apparent magnitude (a function of luminosity) and time. They were interested in studying the connection betweeen certain parameters of the relationship. To place the problem in our notation, we let (Y_i, X_i) be the estimated values of (y_i, x_i) in the equation

$$L_i(\tau) = \alpha_i \exp(-x_i^{-1}\tau) + \delta_i \exp(-y_i^{-1}\tau),$$

where $L_i(\tau)$ is the luminosity of the ith supernova at time τ and $(\alpha_i, \delta_i, y_i, x_i)$ is the vector of parameters for the ith supernova. Rust, Leventhal, and McCall constructed estimates of $(\alpha_i, \delta_i, y_i, x_i)$, $i = 1, 2, \ldots, 15$, by nonlinear least squares. The estimated values (Y_i, X_i) and the estimated covariance matrices of the estimated vector are given in Table 3.1.3. For the purposes of this example, we assume that the errors in the estimates of the parameters are normally distributed and we treat the estimates of $\Sigma_{\varepsilon\varepsilon ii}$ as if they are the true covariance matrices of the estimation (measurement) variances. The (Y_i, X_i) values are only approximately normally distributed because they are estimates constructed by nonlinear least squares.

Rust, Leventhal, and McCall (1976) explain a theory for luminosity under which

$$y_i = \beta_0 + \beta_1 x_i.$$

Minimizing (3.1.63) for the data of Table 3.1.3, we obtain the estimated relationship

$$\hat{y}_i = \underset{(16.33)}{30.04} + \underset{(3.136)}{6.116}x_i,$$

where the numbers in parentheses are the estimated standard errors. The

TABLE 3.1.3. Observations and covariance matrices for luminosity of supernova

Supernova	Y_i	X_i	σ_{eeii}	σ_{euii}	σ_{uuii}
1	67.2	6.47	46	0.60	0.024
2	102.1	9.26	8550	71.20	1.190
3	59.3	4.12	1130	28.70	1.780
4	55.6	4.14	29	0.88	0.105
5	75.9	5.70	3020	47.70	1.350
6	87.7	4.59	1470	16.20	0.483
7	47.4	4.55	71	1.99	0.188
8	112.3	10.30	1640	19.80	0.488
9	114.8	7.48	6630	80.60	1.790
10	68.1	5.21	1200	22.40	0.590
11	61.2	4.96	420	6.01	0.195
12	120.4	5.96	524	7.43	0.214
13	112.3	7.11	503	15.00	1.020
14	85.4	8.38	2000	55.30	1.850
15	47.9	8.17	922	24.40	1.000

Source: Rust, Leventhal, and McCall (1976).

estimated covariance matrix computed using Equation (3.1.70) is

$$\hat{\mathbf{V}}\{\hat{\boldsymbol{\beta}}\} = \begin{bmatrix} 266.598 & -50.126 \\ -50.126 & 9.836 \end{bmatrix}.$$

The values of $\hat{x}_i$ and $\hat{v}_i$ are given in Table 3.1.4, where

$$\hat{x}_i = X_i - \hat{v}_i \hat{\sigma}_{vvii}^{-1} \hat{\sigma}_{uvii}$$

and $\hat{\sigma}_{uvii} = \sigma_{euii} - \sigma_{uuii}\hat{\beta}_1$. An estimator of the variance of $\hat{x}_i$ is

$$\hat{V}\{\hat{x}_i - x_i\} = \sigma_{uuii} - \hat{\sigma}_{vvii}^{-1} \hat{\sigma}_{uvii}^2.$$

Because there is a considerable range in the error variance, there is a considerable range in the variance of $\hat{x}_i - x_i$.

We have

$$\hat{\chi}^2 = \sum_{i=1}^{15} \hat{\sigma}_{vvii}^{-1} \hat{v}_i^2 = 15.90.$$

Because this value is well below the 0.05 tabular value of the chi-square distribution with 13 degrees of freedom, we can accept the linear model as an adequate representation for the relationship between the true values x_i and y_i. We note that over one-third of the sum of squares is due to supernova 12. □ □

TABLE 3.1.4. Statistics for supernova luminosity

Supernova	$\hat{x}_i$	$\hat{V}\{\hat{x}_i - x_i\}$	$\hat{v}_i$	$\hat{\sigma}_{vvii}$	$\hat{\sigma}_{vvii}^{-1}\hat{v}_i^2$
1	6.50	0.019	−2.41	40	0.15
2	9.13	0.661	15.43	7724	0.03
3	4.03	1.405	4.06	846	0.02
4	4.14	0.102	0.24	22	0.00
5	5.53	0.724	11.00	2487	0.05
6	4.29	0.347	29.59	1290	0.68
7	4.71	0.175	−10.47	54	2.04
8	10.07	0.288	19.26	1416	0.26
9	7.00	0.941	39.01	5711	0.27
10	5.09	0.218	6.19	948	0.04
11	4.95	0.129	0.82	354	0.00
12	5.21	0.129	53.91	441	6.58
13	6.16	0.215	38.77	358	4.20
14	8.25	0.461	4.11	1393	0.01
15	9.06	0.494	−32.11	661	1.56

REFERENCES

Booth (1973), Chua (1983), Fuller (1980, 1984), Fuller and Hidiroglou (1978), Hasabelnaby (1985), Hidiroglou (1974), Wolter and Fuller (1982a, 1982b).

EXERCISES

1. (Section 3.1.1) Assume the model

$$Y_t = x_t\beta + e_t, \qquad X_t = x_t + u_t,$$
$$(e_t, u_t)' \sim \text{NI}[0, \text{diag}(\sigma_{ee}, \sigma_{uutt})],$$

where the σ_{uutt}, $t = 1, 2, \ldots, n$, are known and σ_{ee} is unknown. Consider the estimator

$$\ddot{\beta} = \left[\sum_{t=1}^{n} (X_t^2 - \sigma_{uutt})\gamma_t\right]^{-1} \sum_{t=1}^{n} X_t Y_t \gamma_t,$$

where the γ_t are weights.

(a) Show that the weights that minimize the large sample variance of $\ddot{\beta}$ are

$$\gamma_t = (x_t^2 + \sigma_{uutt} + \sigma_{vvtt}^{-1}\sigma_{uvtt}^2)^{-1}\sigma_{vvtt}^{-1}x_t^2.$$

(b) If we fix the error variances, it is not possible to construct consistent estimators for each optimum γ_t. Possible weights are $\gamma_{t0} = 1$, $\tilde{\gamma}_{t1} = \tilde{\sigma}_{vvtt} = \tilde{\sigma}_{ee} + \tilde{\beta}^2\sigma_{uutt}$, and

$$\tilde{\gamma}_{t2} = (\hat{M}_{xx} + \sigma_{uutt} + \tilde{\sigma}_{vvtt}^{-1}\tilde{\sigma}_{uvtt}^2)^{-1}\tilde{\sigma}_{vvtt}^{-1}\hat{M}_{xx},$$

where $\hat{M}_{xx} = n^{-1} \sum_{t=1}^{n} (X_t^2 - \sigma_{uutt})$, $\tilde{\sigma}_{uvtt} = -\tilde{\beta}\sigma_{uutt}$, and $\tilde{\beta}$ is an estimator of β satisfying $(\tilde{\beta} - \beta) = O_p(n^{-1/2})$. Ignoring the error in $\tilde{\beta}$, construct examples to show that none of the three weights dominates the other two.

(c) Consider the weights $\hat{\gamma}_t = \hat{a}_0 + \hat{a}_1 \tilde{\sigma}_{vvtt}^{-1}$, where $(\hat{a}_0, \hat{a}_1)$ are regression coefficients (restricted so that $\hat{\gamma}_t > 0$ for all t) obtained by regressing

$$(X_t^2 + \tilde{\sigma}_{uvtt}^2 \tilde{\sigma}_{vvtt}^{-1})^{-1} \tilde{\sigma}_{vvtt}^{-1} (X_t^2 - \sigma_{uutt})$$

on $\tilde{\sigma}_{vvtt}^{-1}$ with an intercept. Explain why an estimator of β constructed using $\hat{\gamma}_t$ should have smaller large sample variance than an estimator constructed with the weights of part (b).

2. (Sections 3.1.1, 3.1.4) Assume the model

$$y_t = \beta_0 + \beta_1 x_{t1} + \beta_2 x_{t2} + q_t, \qquad \mathbf{Z}_t = \mathbf{z}_t + \mathbf{a}_t,$$

where $E\{(q_t, w_t, \mathbf{u}_t)\} = \mathbf{0}$, $E\{(q_t, \mathbf{x}_t)'\mathbf{a}_t\} = \mathbf{0}$,

$$\text{vech } \mathbf{\Sigma}_{aatt} = (\gamma_{00}^2 y_t^2, 0, 0; \gamma_{11}^2 x_{t1}^2, \gamma_{12} x_{t1} x_{t2}; \gamma_{22}^2 x_{t2}^2)',$$

$\mathbf{x}_t = (x_{t1}, x_{t2})$, and $\mathbf{a}_t = (w_t, \mathbf{u}_t)$. Construct an estimator of $\mathbf{\Sigma}_{aatt}$ of the form $\sum_{i=1}^{3} \psi_{ti}' \psi_{ti}$. Are there any restrictions on γ_{12}? What is the form of the estimator if $1 + \gamma_{12} = 0$?

3. (Sections 3.1.1, 3.1.4) Compute estimates of the parameters of the textile model of Example 3.1.3 under the assumption that the covariance matrix of (u_{t1}, u_{t2}) is diag(0.0231, 0.0158). Give the estimate of the covariance matrix of the approximate distribution. Why is the estimated error covariance matrix for $(X_{t1}, X_{t2}, X_{t1}x_{t4})$ slightly different from that of Example 3.1.3?

4. (Section 3.1.6) What function of $(\boldsymbol{\beta}, \gamma_t^{(i-1)}, \sigma_{vvtt}^{(i-1)})$ does $\hat{\boldsymbol{\beta}}^{(i)}$ of (3.1.65) minimize? What is the value of $\hat{\lambda}$ if $\hat{\beta}^{(i)}$ minimizes (3.1.63)?

5. (Sections 2.2.1, 3.1.3) Assume that the data of Table 2.2.1 satisfy the model

$$y_t = \beta_0 + \sum_{i=1}^{4} x_{ti}\beta_i + q_t, \qquad \mathbf{X}_t = \mathbf{x}_t + \mathbf{u}_t,$$

where the $\mathbf{x}_t$ and $\mathbf{u}_t$ are independently distributed normal vectors with $E\{\mathbf{u}_t\} = \mathbf{0}$ and $E\{\mathbf{u}_t'\mathbf{u}_t\} = \text{diag}\{\sigma_{uu11}, \sigma_{uu22}, \sigma_{uu33}, \sigma_{uu44}\}$. Assume the ratios $\sigma_{uuii}\sigma_{XXii}^{-1}$ are known to be 0.3395, 0.3098, 0.2029, and 0.0 for $i = 1, 2, 3$, and 4, respectively. Estimate the parameters of the equation under these assumptions. Compare the estimates and estimated standard errors of the estimates with those obtained in Example 2.2.1.

6. (Sections 3.1.3, 2.2)

(a) Abd-Ella et al. (1981) assumed that the reliabilities for size, experience, and education in the original data of Example 3.1.2 (before adding error to protect confidentiality) were 0.984, 0.930, and 0.953, respectively. How much were the standard errors of the estimated coefficients of experience and size increased by the errors added to protect confidentiality?

(b) Assume the following model for the data of Example 3.1.2.

$$y_t = \beta_0 + \beta_1 x_{t1} + \beta_2 x_{t2} + q_t, \qquad (Y_t, \mathbf{X}_t) = (y_t, \mathbf{x}_t) + (w_t, \mathbf{u}_t),$$
$$(w_t, u_{t1}, u_{t2})' \sim \text{NI}(\mathbf{0}, \mathbf{\Sigma}_{aa}),$$

where $\mathbf{\Sigma}_{aa} = \text{diag}(0.0997, 0.2013, 0.1808)$ and $(\mathbf{x}_j, q_j)$ is independent of $(w_t, \mathbf{u}_t)$ for all t and j. Estimate $\boldsymbol{\beta} = (\beta_0, \beta_1, \beta_2)$ and σ_{qq}. Compare the estimate of $\boldsymbol{\beta}$ and its estimated covariance matrix with that obtained in Example 3.1.2. Test the hypothesis that $\sigma_{qq} = 0$. Estimate (y_t, x_{t1}, x_{t2}) for $t = 1, 2, \ldots, 10$, treating (x_{t1}, x_{t2}) as fixed.

7. (Sections 3.1.1, 2.4) Prove the following theorem.

Theorem. Let

$$Y_t = \beta_0 + x_t\beta_1 + e_t, \qquad X_t = x_t + u_t,$$

and let (e_t, u_t, x_t, W_t) be independently and identically distributed with finite fourth moments and

$$E\{(e_t, u_t)\} = E\{(e_t, u_t)W_t\} = (0, 0).$$

Let the instrumental variable estimator of β_1 be $\hat{\beta}_1 = m_{XW}^{-1}m_{YW}$. Then

$$\hat{V}_{\beta\beta 11}^{-1/2}(\hat{\beta}_1 - \beta_1) \xrightarrow{L} N(0, 1),$$

where

$$\hat{V}_{\beta\beta 11} = (n-1)^{-1}m_{WX}^{-1}\hat{G}m_{XW}^{-1},$$

$\hat{G} = (n-2)^{-1}\sum_{t=1}^{n}\hat{d}_t^2$, and $\hat{d}_t = (W_t - \bar{W})\hat{v}_t$.

3.2. NONLINEAR MODELS WITH NO ERROR IN THE EQUATION

3.2.1. Introduction

The regression model with fixed independent variables is called linear when the mean function is linear in the parameters. For example, the model

$$Y_t = \beta_0 + \beta_1 x_t + \beta_2 x_t^2 + e_t, \tag{3.2.1}$$

where the x_t are fixed constants known without error, is linear in the parameters $(\beta_0, \beta_1, \beta_2)$. Models such as

$$Y_t = \beta_0 + \beta_1 x_{t1} + \beta_0\beta_1 x_{t2} + e_t, \tag{3.2.2}$$

$$Y_t = \beta_0[1 - \exp(-\beta_1 x_t)] + e_t \tag{3.2.3}$$

are nonlinear in the parameters and require nonlinear methods for efficient estimation. Draper and Smith (1981) and Gallant (1975, 1986) contain discussions of nonlinear least squares procedures.

By the definition for regression models, the functional measurement error models of Sections 2.2 and 2.3 are nonlinear models. This is because the true x_t, as well as the betas, are unknown parameters. However, it is conventional to consider the measurement error model to be nonlinear only when the β parameters enter the mean function in a nonlinear manner or when the mean function is nonlinear in the explanatory variables measured with error. Let the model be

$$y_t = g(\mathbf{x}_t; \boldsymbol{\beta}), \qquad (Y_t, \mathbf{X}_t) = (y_t, \mathbf{x}_t) + (e_t, \mathbf{u}_t), \tag{3.2.4}$$

where $g(\mathbf{x}; \boldsymbol{\beta})$ is a real valued continuous function, $\{\mathbf{x}_t\}$ is a sequence of fixed p-dimensional row vectors, $\boldsymbol{\beta}$ is a k-dimensional column vector, and $\boldsymbol{\varepsilon}_t = (e_t, \mathbf{u}_t)$ is the vector of measurement errors. We assume $\boldsymbol{\varepsilon}_t$ is distributed with mean zero and covariance matrix $\boldsymbol{\Sigma}_{\varepsilon\varepsilon}$, where at least one element of $\mathbf{u}_t$ has positive variance.

Definition 3.2.1. Model (3.2.4) is nonlinear if $g(\mathbf{x}; \boldsymbol{\beta})$ is nonlinear in $\mathbf{x}$ when $\boldsymbol{\beta}$ is fixed or if $g(\mathbf{x}; \boldsymbol{\beta})$ is nonlinear in $\boldsymbol{\beta}$ when $\mathbf{x}$ is fixed.

We reverse our usual order of treatment and devote this section to the model in which $\boldsymbol{\Sigma}_{\varepsilon\varepsilon}$ is known or known up to a multiple. Section 3.2.2 contains an example of the model which is linear in the explanatory variable, while Section 3.2.3 is devoted to models nonlinear in the explanatory variables. Section 3.2.4 is devoted to modifications of the maximum likelihood estimator that improve small sample properties. These modifications are recommended for applications, but the section is easily omitted on first reading. Section 3.3 contains a treatment of the nonlinear measurement error model with an error in the equation.

3.2.2. Models Linear in x

Models that are nonlinear in $\boldsymbol{\beta}$, but linear in $\mathbf{x}$, are relatively easy to handle. Such models represent an extension of the linear errors-in-variables model to the nonlinear case that is analogous to the extension of the linear fixed-x regression model to the nonlinear fixed-x regression model. The model can be written

$$Y_t = \sum_{i=1}^{k} g_i(\boldsymbol{\beta}) x_{ti} + e_t, \qquad X_{ti} = x_{ti} + u_{ti}, \tag{3.2.5}$$

$$\boldsymbol{\varepsilon}_t' \sim \mathrm{NI}(\mathbf{0}, \boldsymbol{\Sigma}_{\varepsilon\varepsilon}),$$

where $g_i(\boldsymbol{\beta})$ are nonlinear functions of $\boldsymbol{\beta}$, $\boldsymbol{\varepsilon}_t = (e_t, \mathbf{u}_t)$, and $\boldsymbol{\Sigma}_{\varepsilon\varepsilon}$ is known. If $\sum_{t=1}^{n} \mathbf{x}_t'\mathbf{x}_t$ is nonsingular, estimators of $\{g_i(\boldsymbol{\beta}); i = 1, 2, \ldots, k\}$, ignoring the restrictions imposed by $\boldsymbol{\beta}$, can be obtained by the usual linear measurement error methods. Using these estimators of $g_i(\boldsymbol{\beta})$, improved estimators of $\boldsymbol{\beta}$ can be obtained by an adaptation of nonlinear procedures.

Example 3.2.1. The data in Table 3.2.1 were generated to satisfy the model

$$y_t = \beta_0 + \beta_1 x_{t1} + (\beta_0 + \beta_1^2) x_{t2},$$

with $(Y_t, \mathbf{X}_t) = (y_t, \mathbf{x}_t) + (e_t, \mathbf{u}_t)$ and $(e_t, \mathbf{u}_t)' \sim \mathrm{NI}(\mathbf{0}, \mathbf{I})$. In this example the x_{ti} enter the g function linearly, but there is a nonlinear restriction on the

TABLE 3.2.1. Data for Example 3.2.1

Observation	Y_t	X_{t1}	X_{t2}
1	−8.05	−1.77	−1.45
2	2.33	0.70	0.91
3	−9.35	−2.22	−3.49
4	−6.03	−1.57	−2.74
5	4.47	−1.73	2.91
6	5.58	2.14	−0.16
7	3.80	−1.60	0.92
8	−3.56	−1.11	0.41
9	−6.30	−2.74	−0.22
10	−4.11	−0.86	−1.35
11	−5.07	−0.66	−2.87
12	6.11	−0.31	1.21
13	−0.43	1.54	−0.41
14	−8.52	−1.26	−2.86
15	3.04	1.79	1.69
16	−1.67	0.87	−1.03
17	1.13	2.65	2.91
18	−1.68	−2.00	0.12
19	3.31	0.76	−1.34
20	1.32	4.47	5.99

coefficients of $(1, x_{t1}, x_{t2})$. Ignoring the fact that the coefficient of x_{t2} is a function of β_0 and β_1, we find that the coefficients of the equation

$$y_t = \beta_0 + \beta_1 x_{t1} + \beta_2 x_{t2}$$

can be estimated using the methods of Sections 2.3 and 2.5.2 under the assumption that the covariance matrix of $(e_t, \mathbf{u}_t)$ is known to be $\mathbf{I}$.

The estimated equation obtained with the program package SUPER CARP is

$$\begin{aligned}\hat{y}_t = \underset{(0.619)}{0.014} + \underset{(0.616)}{1.679}x_{t1} + \underset{(0.482)}{1.922}x_{t2},\end{aligned}$$

where the numbers in parentheses are the estimated standard errors. The vector half of the estimated covariance matrix of the estimators is

$$\text{vech}\,\hat{\mathbf{V}}_{\beta\beta} = (0.383, 0.097, 0.089; 0.379, -0.178; 0.232)'.$$

Under the original model

$$\hat{\boldsymbol{\beta}}' = (\hat{\beta}_0, \hat{\beta}_1, \hat{\beta}_2) = (\beta_0, \beta_1, \beta_0 + \beta_1^2) + (a_0, a_1, a_2), \tag{3.2.6}$$

TABLE 3.2.2. Observations for nonlinear least squares estimation

Observation	Dependent Variable $\hat{\boldsymbol{\beta}}$	Independent Variables $\boldsymbol{\Phi}_1$	$\boldsymbol{\Phi}_2$	$\boldsymbol{\Phi}_3$
1	0.014	1	0	0
2	1.679	0	1	0
3	1.922	0	0	1

where the covariance matrix of the errors of estimation (a_0, a_1, a_2) is estimated by $\hat{\mathbf{V}}_{\beta\beta}$. Using the system of equations (3.2.6), we obtain estimates of β_0 and β_1 by general nonlinear least squares. The estimates $(\tilde{\beta}_0, \tilde{\beta}_1)$ are the values of (β_0, β_1) that minimize

$$[\hat{\boldsymbol{\beta}}' - (\beta_0, \beta_1, \beta_0 + \beta_1^2)]\hat{\mathbf{V}}_{\beta\beta}^{-1}[\hat{\boldsymbol{\beta}} - (\beta_0, \beta_1, \beta_0 + \beta_1^2)'].$$

We outline one method of constructing the generalized least squares estimators. The nonlinear model may be written

$$\hat{\beta}_i = \Phi_{i1}\beta_0 + \Phi_{i2}\beta_1 + \Phi_{i3}(\beta_0 + \beta_1^2) + a_i, \qquad i = 1, 2, 3,$$

where the Φ_{ij} are displayed in Table 3.2.2. Most nonlinear regression programs are designed for observations with uncorrelated constant variance errors. To use such a program it is necessary to transform the observations. We wish to construct a transformation matrix $\mathbf{T}$ such that

$$\mathbf{T}\hat{\mathbf{V}}_{\beta\beta}\mathbf{T}' = \mathbf{I}.$$

One such transformation is that associated with Gram–Schmidt diagonalization of a covariance matrix. (See, for example, Rao (1965, p. 9).)

Table 3.2.3 contains the transformed dependent variable and the transformed matrix of independent variables. Because the original matrix of independent variables is the identity matrix, the matrix of transformed independent variables is the transformation matrix $\mathbf{T}$. The variables of Table

TABLE 3.2.3. Transformed observations for nonlinear least squares estimation

Observation	Dependent Variable $\mathbf{T}\hat{\boldsymbol{\beta}}$	Independent Variables $\mathbf{T}\boldsymbol{\Phi}_1$	$\mathbf{T}\boldsymbol{\Phi}_2$	$\mathbf{T}\boldsymbol{\Phi}_3$
1	0.023	1.615	0.000	0.000
2	2.811	−0.423	1.678	0.000
3	9.133	1.993	1.800	3.188

3.2.3 and the model (3.2.3) were entered into a standard nonlinear regression program to obtain the estimates of β_0 and β_1. These estimates are $(\tilde{\beta}_0, \tilde{\beta}_1) = (-0.080, 1.452)$ and the vector half of the estimated covariance matrix is

$$\text{vech } \hat{\mathbf{V}}\{\tilde{\boldsymbol{\beta}}\} = (0.3203, -0.0546; 0.0173)'.$$

The estimated covariance matrix of $\tilde{\boldsymbol{\beta}}$ is the usual Taylor approximation computed for nonlinear least squares. Most nonlinear regression programs compute the covariance matrix of the estimates by multiplying a residual mean square by the inverse of the matrix of sums of squares and products of the derivatives with respect to the parameters. In our problem the error mean square of the transformed variables is estimated to be one. Therefore, it is necessary to multiply the standard errors output by the program by the square root of the inverse of the residual mean square. In this example the residual mean square is 0.141 with one degree of freedom. The residual mean square is approximately distributed as an F random variable with 1 and 17 degrees of freedom when the β's satisfy the postulated restrictions. Therefore, for our example, the model is easily accepted. □ □

3.2.3. Models Nonlinear in x

In this section we derive the maximum likelihood estimator of $\boldsymbol{\beta}$ for the nonlinear model introduced in Section 3.2.1. We let

$$y_t = g(\mathbf{x}_t; \boldsymbol{\beta}), \qquad (Y_t, \mathbf{X}_t) = (y_t, \mathbf{x}_t) + (e_t, \mathbf{u}_t),$$

and assume that $\boldsymbol{\varepsilon}_t' \sim \text{NI}(\mathbf{0}, \boldsymbol{\Sigma}_{\varepsilon\varepsilon})$, where $\boldsymbol{\varepsilon}_t = (e_t, \mathbf{u}_t)$, and that $\boldsymbol{\Sigma}_{\varepsilon\varepsilon}$ is nonsingular and known (or known up to a multiple). It is assumed that $g(\mathbf{x}; \boldsymbol{\beta})$ is continuous and possesses continuous first and second derivatives with respect to both arguments for $\mathbf{x} \in A$ and $\boldsymbol{\beta} \in B$, where A and B are subsets of p-dimensional Euclidean space and k-dimensional Euclidean space, respectively. The unknown true values $(\mathbf{x}_1, \mathbf{x}_2, \ldots, \mathbf{x}_n)$ are assumed to be fixed. Under this model the density function for $(Y_t, \mathbf{X}_t)$ is proportional to

$$|\boldsymbol{\Sigma}_{\varepsilon\varepsilon}|^{-1/2} \exp\{-\tfrac{1}{2}[Y_t - g(\mathbf{x}_t; \boldsymbol{\beta}), \mathbf{X}_t - \mathbf{x}_t]\boldsymbol{\Sigma}_{\varepsilon\varepsilon}^{-1}[Y_t - g(\mathbf{x}_t; \boldsymbol{\beta}), \mathbf{X}_t - \mathbf{x}_t]'\}.$$

The maximum likelihood estimator is the $(\boldsymbol{\beta}', \mathbf{x}_1, \mathbf{x}_2, \ldots, \mathbf{x}_n)$ in $B \times A$ that minimizes the sum of squares

$$Q(\boldsymbol{\beta}', \mathbf{x}_1, \ldots, \mathbf{x}_n) = \sum_{t=1}^{n} q(\boldsymbol{\beta}', \mathbf{x}_t; \mathbf{Z}_t), \qquad (3.2.7)$$

where

$$q(\boldsymbol{\beta}', \mathbf{x}_t; \mathbf{Z}_t) = [Y_t - g(\mathbf{x}_t; \boldsymbol{\beta}), \mathbf{X}_t - \mathbf{x}_t]\boldsymbol{\Sigma}_{\varepsilon\varepsilon}^{-1}[Y_t - g(\mathbf{x}_t; \boldsymbol{\beta}), \mathbf{X}_t - \mathbf{x}_t]'.$$

The likelihood equations obtained by setting the first derivatives of (3.2.7) equal to zero are

$$\sum_{t=1}^{n} [Y_t - g(\mathbf{x}_t; \boldsymbol{\beta}), \mathbf{X}_t - \mathbf{x}_t]\boldsymbol{\Sigma}_{\varepsilon\varepsilon}^{-1}[g_{\beta(i)}(\mathbf{x}_t; \boldsymbol{\beta}), \mathbf{0}]' = 0, \tag{3.2.8}$$

$$[Y_t - g(\mathbf{x}_t; \boldsymbol{\beta}), \mathbf{X}_t - \mathbf{x}_t]\boldsymbol{\Sigma}_{\varepsilon\varepsilon}^{-1}[\mathbf{F}'_{xt}, \mathbf{I}]' = 0,$$

where the first equation holds for $i = 1, 2, \ldots, k$, the second equation holds for $t = 1, 2, \ldots, n$,

$$\mathbf{F}_{xt} = [g_{x(1)}(\mathbf{x}_t; \boldsymbol{\beta}), g_{x(2)}(\mathbf{x}_t; \boldsymbol{\beta}), \ldots, g_{x(p)}(\mathbf{x}_t; \boldsymbol{\beta})],$$

$g_{\beta(i)}(\mathbf{x}_t; \boldsymbol{\beta})$ is the partial derivative of $g(\mathbf{x}; \boldsymbol{\beta})$ with respect to β_i evaluated at $(\mathbf{x}_t; \boldsymbol{\beta})$, and $g_{x(i)}(\mathbf{x}_t; \boldsymbol{\beta})$ is the partial derivative of $g(\mathbf{x}; \boldsymbol{\beta})$ with respect to x_{ti} evaluated at $(\mathbf{x}_t; \boldsymbol{\beta})$. The maximum likelihood estimators will satisfy these equations if the solutions are in the parameter space. For the linear model, we were able to obtain explicit expressions for the estimator of the parameter vector $\boldsymbol{\beta}$ and for the estimator of $\mathbf{x}_t$. See (2.3.14) and (2.3.15). In most situations it is not possible to obtain an explicit expression for the maximum likelihood estimator of the parameter vector for the nonlinear model. In Example 3.2.2 we demonstrate how a nonlinear regression program can be used to obtain the maximum likelihood estimator. Because this procedure directly estimates the unknown $\mathbf{x}_t$, it is not computationally efficient for very large samples.

Example 3.2.2. The data in Table 3.2.4 were obtained in an experiment conducted by Frisillo and Stewart (1980b). The experiment was designed to

TABLE 3.2.4. Observations for experiment on ultrasonic absorption in Berea sandstone

Observation	Compressional Speed (Y)	Gas–Brine Saturation (X)
1	1265.0	0.0
2	1263.6	0.0
3	1258.0	5.0
4	1254.0	7.0
5	1253.0	7.5
6	1249.8	10.0
7	1237.0	16.0
8	1218.0	26.0
9	1220.6	30.0
10	1213.8	34.0
11	1215.5	34.5
12	1212.0	100.0

Source: Frisillo and Stewart (1980a).

investigate the possible use of sonic logging to detect areas with potential for natural gas production. The data are the observed compressional wave velocity (Y) of ultrasonic signals propagated through cores of Berea sandstone and the percent nitrogen gas saturation (X) in a brine solution forced into the pores of the Berea sandstone. The method used to create partial gas saturation in the brine solution could only produce saturation levels less than 35%. This explains the large gap in saturation levels between 34.5% and 100%. We assume that the data satisfy the model

$$y_t = \beta_0 + \beta_1[\exp\{\beta_2 x_t\} - 1]^2,$$

where $(Y_t, X_t) = (y_t, x_t) + (e_t, u_t)$ and $(e_t, u_t)' \sim \mathrm{NI}(\mathbf{0}, \mathbf{I}\sigma^2)$. We shall use a nonlinear least squares program to estimate the parameters of the model. Let

$$(Z_{t1}, Z_{t2}) = (Y_t, X_t) \quad \text{and} \quad (\varepsilon_{t1}, \varepsilon_{t2}) = (e_t, u_t)$$

for $t = 1, 2, \ldots, 12$. The model for nonlinear estimation can then be written

$$Z_{tj} = \beta_0 J_{tj1} + \beta_1 J_{tj1}\left[\exp\left\{\beta_2 \sum_{i=1}^{12} D_{tji} x_i\right\} - 1\right]^2 + J_{tj2} \sum_{i=1}^{12} D_{tji} x_i + \varepsilon_{tj}, \tag{3.2.9}$$

where $J_{tjl} = \delta_{jl}$, $D_{tji} = \delta_{ti}$, and δ_{ti} is Kronecker's delta. The model (3.2.9) is a nonlinear regression model with parameters $(\beta_0, \beta_1, \beta_2, x_1, x_2, \ldots, x_{12})$, explanatory variables $(D_1, D_2, \ldots, D_{12}, J_1, J_2)$, and errors ε_{tj}. Table 3.2.5 contains the variables used in the nonlinear least squares estimation of the model. The 12 pairs (Y_t, X_t) have been listed in a single column with the 12 observations on Y_t appearing first. This column of 24 observations is called the Z column. The estimates obtained by nonlinear least squares are given in Table 3.2.6. The estimated function is plotted in Figure 3.2.1.

Nonlinear regression programs generally require the user to provide start values for the parameters. In the present problem start values for $(\beta_0, \beta_1, \beta_2)$ were estimated from a plot of the data. The observed X_t values were used as start values for x_t.

The estimated standard errors in Table 3.2.6 are those output by the nonlinear regression program. The distributional properties of the estimators are investigated in Theorem 3.2.1 below.

In this experiment it is reasonable to believe that the error variances for $X = 0$ and $X = 100$ are smaller than the error variances for the mixtures. Also, one can argue that the expected value of the error given that $X = 0$ is not zero. Estimation recognizing these facts would change the estimates of $(\beta_0, \beta_1, \beta_2)$ very little because the derivative of the function is zero at $x = 0$ and the derivative is nearly zero at $x = 100$. This is reflected in the fact that

TABLE 3.2.5. Table of observations for nonlinear least squares estimation of parameters of wave velocity model

Original Observation	Indices t	Indices j	Dependent Variable Z_{tj}	D_1	D_2	D_3	D_4	D_5	D_6	D_7	D_8	D_9	D_{10}	D_{11}	D_{12}	J_1	J_2
Y_1	1	1	1265.0	1	0	0	0	0	0	0	0	0	0	0	0	1	0
Y_2	2	1	1263.6	0	1	0	0	0	0	0	0	0	0	0	0	1	0
Y_3	3	1	1258.0	0	0	1	0	0	0	0	0	0	0	0	0	1	0
Y_4	4	1	1254.0	0	0	0	1	0	0	0	0	0	0	0	0	1	0
Y_5	5	1	1253.0	0	0	0	0	1	0	0	0	0	0	0	0	1	0
Y_6	6	1	1249.8	0	0	0	0	0	1	0	0	0	0	0	0	1	0
Y_7	7	1	1237.0	0	0	0	0	0	0	1	0	0	0	0	0	1	0
Y_8	8	1	1218.0	0	0	0	0	0	0	0	1	0	0	0	0	1	0
Y_9	9	1	1220.6	0	0	0	0	0	0	0	0	1	0	0	0	1	0
Y_{10}	10	1	1213.8	0	0	0	0	0	0	0	0	0	1	0	0	1	0
Y_{11}	11	1	1215.5	0	0	0	0	0	0	0	0	0	0	1	0	1	0
Y_{12}	12	1	1212.0	0	0	0	0	0	0	0	0	0	0	0	1	1	0
X_1	1	2	0.0	1	0	0	0	0	0	0	0	0	0	0	0	0	1
X_2	2	2	0.0	0	1	0	0	0	0	0	0	0	0	0	0	0	1
X_3	3	2	5.0	0	0	1	0	0	0	0	0	0	0	0	0	0	1
X_4	4	2	7.0	0	0	0	1	0	0	0	0	0	0	0	0	0	1
X_5	5	2	7.5	0	0	0	0	1	0	0	0	0	0	0	0	0	1
X_6	6	2	10.0	0	0	0	0	0	1	0	0	0	0	0	0	0	1
X_7	7	2	16.0	0	0	0	0	0	0	1	0	0	0	0	0	0	1
X_8	8	2	26.0	0	0	0	0	0	0	0	1	0	0	0	0	0	1
X_9	9	2	30.0	0	0	0	0	0	0	0	0	1	0	0	0	0	1
X_{10}	10	2	34.0	0	0	0	0	0	0	0	0	0	1	0	0	0	1
X_{11}	11	2	34.5	0	0	0	0	0	0	0	0	0	0	1	0	0	1
X_{12}	12	2	100.0	0	0	0	0	0	0	0	0	0	0	0	1	0	1

TABLE 3.2.6. Estimated parameters and estimated standard errors obtained by nonlinear least squares

Parameter	Estimate	Estimated Standard Error
β_0	1264.65	1.03
β_1	−54.02	1.58
β_2	−0.0879	0.0063
x_1	0.00	1.54
x_2	0.00	1.54
x_3	4.93	0.74
x_4	6.73	0.71
x_5	7.17	0.71
x_6	8.69	0.74
x_7	14.67	0.93
x_8	27.46	1.30
x_9	28.81	1.33
x_{10}	34.77	1.45
x_{11}	34.58	1.44
x_{12}	100.00	1.54
σ^2	2.38	—

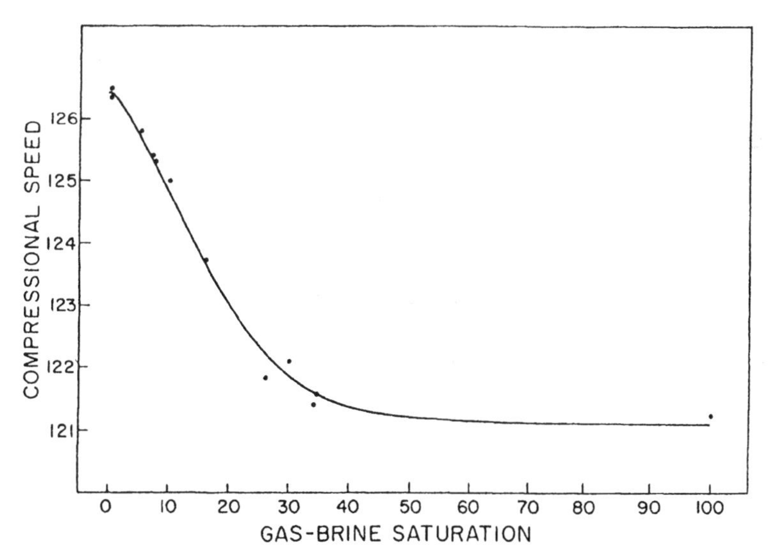

FIGURE 3.2.1. Estimated function and observed values for wave velocity.

the estimates of x_t at these points are, to two decimals, equal to X_t and the estimated standard errors for the estimated x_t values are equal to $\hat{\sigma}$. Also see Exercise 3.9 and Example 3.2.3. □ □

The nonlinear least squares calculations of Example 3.2.2 are very general and can be used in small samples for a wide range of models. To illustrate the method for models with unequal error variances and for models where the observations satisfy more than one relationship, we continue the analysis of the Frisillo and Stewart (1980) data.

Example 3.2.3. In the experiment of Frisillo and Stewart (1980b) described in Example 3.2.2, a second response variable, called quality by Frisillo and Stewart, was also determined. We let y_{t2} denote the true value of quality and y_{t1} denote the true value of compressional speed. The data are given in Table 3.2.7. For the purposes of this example, we assume the model

$$\begin{aligned}
y_{t1} &= \beta_0 + \beta_1[\exp\{\beta_2 x_t\} - 1]^2, \\
y_{t2} &= \gamma_0 + \gamma_1[\exp\{\gamma_2 x_t\} - \gamma_3]^2, \\
(Y_{t1}, Y_{t2}, X_t) &= (y_{t1}, y_{t2}, x_t) + (e_{t1}, e_{t2}, u_t), \\
(e_{t1}, e_{t2}, u_t)' &\sim \mathrm{NI}(\mathbf{0}, \mathrm{diag}\{1, 0.25, 1\}\sigma^2), \quad t = 3, 4, \ldots, 11, \\
(e_{t1}, e_{t2}, u_t)' &\sim \mathrm{NI}(\mathbf{0}, \mathrm{diag}\{1, 0.25, 0\}\sigma^2), \quad t = 1, 2, 12.
\end{aligned} \tag{3.2.10}$$

TABLE 3.2.7. Observations for experiment on ultrasonic absorption in Berea sandstone

Observation	Compressional Speed (Y_{t1})	Quality (Y_{t2})	Gas–Brine Saturation (X_t)
1	1265.0	33.0	0.0
2	1263.6	32.0	0.0
3	1258.0	21.6	5.0
4	1254.0	20.0	7.0
5	1253.0	21.0	7.5
6	1249.8	17.4	10.0
7	1237.0	14.8	16.0
8	1218.0	8.7	26.0
9	1220.6	8.7	30.0
10	1213.8	9.9	34.0
11	1215.5	10.3	34.5
12	1212.0	47.0	100.0

Source: Frisillo and Stewart (1980a).

We are assuming that the variance of e_{t2} is one-fourth that of e_{t1} for all observations. If the gas–brine saturation is greater than zero and less than 100%, we assume that the variance of e_{t1} is equal to the variance of u_t. For the zero and 100% saturation observations we assume that the gas–brine saturation is measured without error. The variables to be used in the nonlinear estimation are given in Table 3.2.8. The Z column contains the values of compressional speed, followed by the values of gas–brine saturation measured with error, followed by the values of quality multiplied by two. The gas-brine saturation for observations 1, 2, and 12 are not included in the Z column because the true values for these observations are known and need not be estimated. The values of quality are multiplied by two so that the error variance is equal to that of speed. Also, the indicator variables associated with quality (J_3 and H_3) are multiplied by two. The variable F contains the values of saturation that are measured without error for the corresponding values of speed and quality. The values of F for observations 4–11 are arbitrary because they are annihilated by the indicator variables.

The nonlinear model is written

$$Z = \beta_0 J_1 + \gamma_0 J_3 + \beta_1 H_1 \left[\exp\left\{ \beta_2 \sum_{j=3}^{11} x_j D_j \right\} - 1 \right]^2 + \sum_{j=3}^{11} x_j D_j J_2$$

$$+ \gamma_1 H_3 \left[\exp\left\{ \gamma_2 \sum_{j=3}^{11} x_j D_j \right\} - \gamma_3 \right]^2 + \beta_1 (J_1 - H_1)[\exp\{\beta_2 F\} - 1]^2$$

$$+ \gamma_1 (J_3 - H_3)[\exp\{\gamma_2 F\} - \gamma_3]^2 + \varepsilon,$$

where $(Z_{t1}, Z_{t2}, Z_{t3}) = (Y_{t1}, X_t, Y_{t2})$,

$$J_i = \begin{cases} 1 & \text{if } Z \text{ observation is for } Z_{ti} \\ 0 & \text{otherwise,} \end{cases}$$

$$D_i = D_{tji} = \begin{cases} 1 & \text{if } t = i \\ 0 & \text{otherwise,} \end{cases}$$

$$K = K_{tj} = \begin{cases} 1 & \text{if } j \neq 0, 1, 12 \\ 0 & \text{otherwise,} \end{cases}$$

$H_i = J_i K_{tj}$, $F_{tj} = x_{tj}(1 - K_{tj})$, and we have omitted the subscripts identifying the observation when no confusion will result.

The nonlinear least squares estimates of the parameters are given in Table 3.2.9. The estimates of β_0, β_1, and β_2 are very similar to those of Table 3.2.6. The addition of data on quality had little effect on the estimates of $(\beta_0, \beta_1, \beta_2)$ or on the estimated standard errors of these estimates. In fact, the estimated standard errors of $(\hat{\beta}_0, \hat{\beta}_1, \hat{\beta}_2)$ of Table 3.2.9 are slightly larger than those of Table 3.2.6. This is because $\hat{\sigma}^2 = 2.59$ of Table 3.2.9 is larger than $\hat{\sigma}^2 = 2.38$

TABLE 3.2.8. Data for nonlinear least squares estimation of multivariate wave velocity model

Original Observation	Z_{ij}	D_3	D_4	D_5	D_6	D_7	D_8	D_9	D_{10}	D_{11}	J_1	J_2	J_3	H_1	H_3	F
$Y_{1,1}$	1265.0	0	0	0	0	0	0	0	0	0	1	0	0	0	0	0
$Y_{2,1}$	1263.6	0	0	0	0	0	0	0	0	0	1	0	0	0	0	0
$Y_{3,1}$	1258.0	1	0	0	0	0	0	0	0	0	1	0	0	1	0	0
$Y_{4,1}$	1254.0	0	1	0	0	0	0	0	0	0	1	0	0	1	0	0
$Y_{5,1}$	1253.0	0	0	1	0	0	0	0	0	0	1	0	0	1	0	0
$Y_{6,1}$	1249.8	0	0	0	1	0	0	0	0	0	1	0	0	1	0	0
$Y_{7,1}$	1237.0	0	0	0	0	1	0	0	0	0	1	0	0	1	0	0
$Y_{8,1}$	1218.0	0	0	0	0	0	1	0	0	0	1	0	0	1	0	0
$Y_{9,1}$	1220.6	0	0	0	0	0	0	1	0	0	1	0	0	1	0	0
$Y_{10,1}$	1213.8	0	0	0	0	0	0	0	1	0	1	0	0	1	0	0
$Y_{11,1}$	1215.5	0	0	0	0	0	0	0	0	1	1	0	0	1	0	0
$Y_{12,1}$	1212.0	0	0	0	0	0	0	0	0	0	1	0	0	0	0	100
X_3	5.0	1	0	0	0	0	0	0	0	0	0	1	0	0	0	0
X_4	7.0	0	1	0	0	0	0	0	0	0	0	1	0	0	0	0
X_5	7.5	0	0	1	0	0	0	0	0	0	0	1	0	0	0	0

X_6	10.0	0	0	0	1	0	0	0	0	0	0	1	0	0	0	0
X_7	16.0	0	0	0	0	1	0	0	0	0	0	1	0	0	0	0
X_8	26.0	0	0	0	0	0	1	0	0	0	0	1	0	0	0	0
X_9	30.0	0	0	0	0	0	0	1	0	0	0	1	0	0	0	0
X_{10}	34.0	0	0	0	0	0	0	0	1	0	0	1	0	0	0	0
X_{11}	34.5	0	0	0	0	0	0	0	0	1	0	1	0	0	0	0
$2Y_{1,2}$	66.0	0	0	0	0	0	0	0	0	0	0	0	2	0	0	0
$2Y_{2,2}$	64.0	0	0	0	0	0	0	0	0	0	0	0	2	0	0	0
$2Y_{3,2}$	43.2	1	0	0	0	0	0	0	0	0	0	0	2	0	2	0
$2Y_{4,2}$	40.0	0	1	0	0	0	0	0	0	0	0	0	2	0	2	0
$2Y_{5,2}$	42.0	0	0	1	0	0	0	0	0	0	0	0	2	0	2	0
$2Y_{6,2}$	34.8	0	0	0	1	0	0	0	0	0	0	0	2	0	2	0
$2Y_{7,2}$	29.6	0	0	0	0	1	0	0	0	0	0	0	2	0	2	0
$2Y_{8,2}$	17.4	0	0	0	0	0	1	0	0	0	0	0	2	0	2	0
$2Y_{9,2}$	17.4	0	0	0	0	0	0	1	0	0	0	0	2	0	2	0
$2Y_{10,2}$	19.8	0	0	0	0	0	0	0	1	0	0	0	2	0	2	0
$2Y_{11,2}$	20.6	0	0	0	0	0	0	0	0	1	0	0	2	0	2	0
$2Y_{12,2}$	94.0	0	0	0	0	0	0	0	0	0	1	0	2	0	0	100

TABLE 3.2.9. Estimated parameters of multivariate wave velocity model

Parameter	Estimate	Estimated Standard Error
β_0	1265.14	1.05
β_1	-54.50	1.58
β_2	-0.0887	0.0065
γ_0	9.00	0.40
γ_1	194.45	17.71
γ_2	-0.0155	0.0018
γ_3	0.654	0.018
x_3	5.71	0.54
x_4	7.11	0.58
x_5	6.93	0.58
x_6	9.11	0.66
x_7	13.87	0.87
x_8	27.40	1.35
x_9	28.71	1.38
x_{10}	34.69	1.43
x_{11}	34.81	1.43
σ^2	2.59	

obtained in the analysis of Table 3.2.6. The conditions under which the standard errors of the estimators are appropriate are discussed in Theorem 3.2.1, following this example.

While the addition of the quality data had little effect on the estimates of $(\beta_0, \beta_1, \beta_2)$, it did have an effect on the estimates of the x values. There is a marked reduction in the estimated standard error of the estimates of x_3 through x_7. The standard errors of the estimates of x_8 through x_{11} are affected less because the slope of the quality function with respect to x is nearly zero for this range of x values. □ □

We now develop a computational algorithm for the nonlinear model that is appropriate for large samples. We consider the general model in which the true values are defined implicitly. Let the model be

$$f(\mathbf{z}_t; \boldsymbol{\beta}) = 0, \qquad \mathbf{Z}_t = \mathbf{z}_t + \boldsymbol{\varepsilon}_t,$$
$$\boldsymbol{\varepsilon}_t \sim \mathrm{NI}(\mathbf{0}, \boldsymbol{\Sigma}_{\varepsilon\varepsilon}), \tag{3.2.11}$$

where $\mathbf{z}_t$ is a p-dimensional row vector and $\boldsymbol{\beta}$ is a k-dimensional column vector. We assume $f(\mathbf{z}_t, \boldsymbol{\beta})$ is a continuous function of $\mathbf{z}_t$ and $\boldsymbol{\beta}$ with continuous first and second derivatives. We assume that the $\mathbf{z}_t$ are fixed unknown constants and that $|\boldsymbol{\Sigma}_{\varepsilon\varepsilon}| \neq 0$.

Given observations $\mathbf{Z}_t$, $t = 1, 2, \ldots, n$, the least squares (maximum likelihood) estimator is constructed by minimizing the Lagrangean

$$\sum_{t=1}^{n} (\mathbf{Z}_t - \mathbf{z}_t)\boldsymbol{\Sigma}_{\varepsilon\varepsilon}^{-1}(\mathbf{Z}_t - \mathbf{z}_t)' + \sum_{t=1}^{n} \alpha_t f(\mathbf{z}_t, \boldsymbol{\beta}) \tag{3.2.12}$$

with respect to $\boldsymbol{\beta}$ and $\mathbf{z}_t$, $t = 1, 2, \ldots, n$, where the α_t are Lagrange multipliers. Iterative methods are required to obtain the numerical solution. Britt and Luecke (1973) suggested a Gauss–Newton type of iteration that has been extended by Schnell (1983). Let $(\tilde{\boldsymbol{\beta}}, \tilde{\mathbf{z}}_1, \tilde{\mathbf{z}}_2, \ldots, \tilde{\mathbf{z}}_n)$ be initial estimates. Expand $f(\mathbf{z}_t; \boldsymbol{\beta})$ in a first-order Taylor expansion about the initial values to obtain

$$f(\mathbf{z}_t, \boldsymbol{\beta}) \doteq f(\tilde{\mathbf{z}}_t, \tilde{\boldsymbol{\beta}}) + \mathbf{f}_{\beta}(\tilde{\mathbf{z}}_t, \tilde{\boldsymbol{\beta}})(\boldsymbol{\beta} - \tilde{\boldsymbol{\beta}}) + \mathbf{f}_z(\tilde{\mathbf{z}}_t, \tilde{\boldsymbol{\beta}})(\mathbf{z}_t - \tilde{\mathbf{z}}_t)', \tag{3.2.13}$$

where $\mathbf{f}_{\beta}(\tilde{\mathbf{z}}_t, \tilde{\boldsymbol{\beta}})$ is the k-dimensional row vector containing the partial derivatives of $f(\mathbf{z}_t, \boldsymbol{\beta})$ with respect to the elements of $\boldsymbol{\beta}$ evaluated at $(\mathbf{z}_t, \boldsymbol{\beta}) = (\tilde{\mathbf{z}}_t, \tilde{\boldsymbol{\beta}})$ and $\mathbf{f}_z(\tilde{\mathbf{z}}_t, \tilde{\boldsymbol{\beta}})$ is the p-dimensional row vector of partial derivatives of $f(\mathbf{z}_t, \boldsymbol{\beta})$ with respect to the elements of $\mathbf{z}_t$ evaluated at $(\mathbf{z}_t, \boldsymbol{\beta}) = (\tilde{\mathbf{z}}_t, \tilde{\boldsymbol{\beta}})$. Replacing $f(\mathbf{z}_t, \boldsymbol{\beta})$ of (3.2.12) with (3.2.13), we have

$$\sum_{t=1}^{n} [(\mathbf{Z}_t - \tilde{\mathbf{z}}_t) - (\mathbf{z}_t - \tilde{\mathbf{z}}_t)]\boldsymbol{\Sigma}_{\varepsilon\varepsilon}^{-1}[(\mathbf{Z}_t - \tilde{\mathbf{z}}_t) - (\mathbf{z}_t - \tilde{\mathbf{z}}_t)]'$$

$$+ \sum_{t=1}^{n} \alpha_t[f(\tilde{\mathbf{z}}_t, \tilde{\boldsymbol{\beta}}) + \mathbf{f}_{\beta}(\tilde{\mathbf{z}}_t, \tilde{\boldsymbol{\beta}})(\boldsymbol{\beta} - \tilde{\boldsymbol{\beta}}) + \mathbf{f}_z(\tilde{\mathbf{z}}_t, \boldsymbol{\beta})(\mathbf{z}_t - \tilde{\mathbf{z}}_t)']. \tag{3.2.14}$$

The objective function (3.2.14) is quadratic in $(\mathbf{z}_t - \tilde{\mathbf{z}}_t)$ and $(\boldsymbol{\beta} - \tilde{\boldsymbol{\beta}})$. Differentiation produces the system of equations

$$\boldsymbol{\Sigma}_{\varepsilon\varepsilon}^{-1}[(\mathbf{Z}_t - \tilde{\mathbf{z}}_t) - (\mathbf{z}_t - \tilde{\mathbf{z}}_t)]' - \alpha_t \mathbf{f}_z'(\tilde{\mathbf{z}}_t, \tilde{\boldsymbol{\beta}}) = \mathbf{0}, \tag{3.2.15}$$

$$\sum_{t=1}^{n} \alpha_t \mathbf{f}_{\beta}'(\tilde{\mathbf{z}}_t, \tilde{\boldsymbol{\beta}}) = \mathbf{0}, \tag{3.2.16}$$

$$f(\tilde{\mathbf{z}}_t, \tilde{\boldsymbol{\beta}}) + \mathbf{f}_{\beta}(\tilde{\mathbf{z}}_t, \tilde{\boldsymbol{\beta}})(\boldsymbol{\beta} - \tilde{\boldsymbol{\beta}}) + \mathbf{f}_z(\tilde{\mathbf{z}}_t, \tilde{\boldsymbol{\beta}})(\mathbf{z}_t - \tilde{\mathbf{z}}_t)' = \mathbf{0}, \tag{3.2.17}$$

where (3.2.15) and (3.2.17) hold for $t = 1, 2, \ldots, n$. If we multiply (3.2.15) by $\mathbf{f}_z(\tilde{\mathbf{z}}_t, \tilde{\boldsymbol{\beta}})\boldsymbol{\Sigma}_{\varepsilon\varepsilon}$ and use (3.2.17), we obtain

$$\alpha_t = \tilde{\sigma}_{vvtt}^{-1}[\tilde{v}_t + f(\dot{\mathbf{z}}_t, \tilde{\boldsymbol{\beta}}) + \mathbf{f}_{\beta}(\tilde{\mathbf{z}}_t, \tilde{\boldsymbol{\beta}})(\boldsymbol{\beta} - \tilde{\boldsymbol{\beta}})], \tag{3.2.18}$$

where $\tilde{v}_t = \mathbf{f}_z(\tilde{\mathbf{z}}_t, \tilde{\boldsymbol{\beta}})(\mathbf{Z}_t - \tilde{\mathbf{z}}_t)'$ and

$$\tilde{\sigma}_{vvtt} = \mathbf{f}_z(\tilde{\mathbf{z}}_t, \tilde{\boldsymbol{\beta}})\boldsymbol{\Sigma}_{\varepsilon\varepsilon}\mathbf{f}_z'(\tilde{\mathbf{z}}_t, \tilde{\boldsymbol{\beta}}). \tag{3.2.19}$$

If we multiply (3.2.18) by $\mathbf{f}_{\beta}'(\tilde{\mathbf{z}}_t, \tilde{\boldsymbol{\beta}})$, sum, and use (3.2.16), the estimated change in $\boldsymbol{\beta}$ is given by

$$\hat{\boldsymbol{\beta}} - \tilde{\boldsymbol{\beta}} = -\left[\sum_{t=1}^{n} \mathbf{f}_{\beta}'(\tilde{\mathbf{z}}_t, \tilde{\boldsymbol{\beta}})\tilde{\sigma}_{vvtt}^{-1}\mathbf{f}_{\beta}(\tilde{\mathbf{z}}_t, \tilde{\boldsymbol{\beta}})\right]^{-1}\left\{\sum_{t=1}^{n} \mathbf{f}_{\beta}'(\tilde{\mathbf{z}}_t, \tilde{\boldsymbol{\beta}})\tilde{\sigma}_{vvtt}^{-1}[\tilde{v}_t + f(\tilde{\mathbf{z}}_t, \tilde{\boldsymbol{\beta}})]\right\}. \tag{3.2.20}$$

Therefore, the improved estimator of $\boldsymbol{\beta}$ is obtained by adding the $\hat{\boldsymbol{\beta}} - \tilde{\boldsymbol{\beta}}$ of (3.2.20) to the initial estimator $\tilde{\boldsymbol{\beta}}$. Multiplying (3.2.15) by $\boldsymbol{\Sigma}_{\varepsilon\varepsilon}$ and using (3.2.19), we have

$$\begin{aligned}\hat{\mathbf{z}}_t = \mathbf{Z}_t - \tilde{\sigma}_{vvtt}^{-1}[&\mathbf{f}_z(\tilde{\mathbf{z}}_t, \tilde{\boldsymbol{\beta}})(\mathbf{Z}_t - \tilde{\mathbf{z}}_t)' \\ &+ f(\tilde{\mathbf{z}}_t, \tilde{\boldsymbol{\beta}}) + \mathbf{f}_\beta(\tilde{\mathbf{z}}_t, \tilde{\boldsymbol{\beta}})(\hat{\boldsymbol{\beta}} - \tilde{\boldsymbol{\beta}})]\mathbf{f}_z(\tilde{\mathbf{z}}_t, \tilde{\boldsymbol{\beta}})\boldsymbol{\Sigma}_{\varepsilon\varepsilon}. \end{aligned} \tag{3.2.21}$$

The calculations are iterated using $\hat{\boldsymbol{\beta}}$ and $\hat{\mathbf{z}}_t$ from (3.2.20) and (3.2.21) as initial values $(\tilde{\boldsymbol{\beta}}, \tilde{\mathbf{z}}_t)$ for the next iteration. Modifications may be required to guarantee convergence. An alternative method of calculation is to use a nonlinear algorithm to choose $\tilde{\mathbf{z}}_t$ so that $f(\tilde{\mathbf{z}}_t, \tilde{\boldsymbol{\beta}}) = 0$ at each step.

The limiting properties of the estimator are given in Theorem 3.2.1. It is important to note that the limits in the theorem are limits as the error variances become small. Therefore, the theorem is applicable in situations where the error variance is small relative to the curvature of the function. The adequacy of this type of approximation will be explored in the examples and in Section 3.2.4. Amemiya and Fuller (1985) have given a version of Theorem 3.2.1 for sequences in which both the sample size increases and the measurement variance becomes smaller.

Theorem 3.2.1. Let

$$f(\mathbf{z}_t, \boldsymbol{\beta}) = 0, \qquad \mathbf{Z}_{rt} = \mathbf{z}_t + \boldsymbol{\varepsilon}_{rt}, \tag{3.2.22}$$
$$\boldsymbol{\varepsilon}_{rt} \sim \mathrm{NI}(\mathbf{0}, r^{-1}\boldsymbol{\Sigma}_{\varepsilon\varepsilon}),$$

where $\boldsymbol{\Sigma}_{\varepsilon\varepsilon}$ is a nonsingular fixed matrix and $\mathbf{z}_t$, $t = 1, 2, \ldots, n$, is a set of n (n fixed) p-dimensional fixed vectors. Let $f(\mathbf{z}_t, \boldsymbol{\beta})$ be a continuous function defined on $A \times B$, where A is a compact subset of p-dimensional Euclidean space and B is a compact subset of k-dimensional Euclidean space. Let $f(\mathbf{z}_t, \boldsymbol{\beta})$ have continuous first and second derivatives with respect to $(\mathbf{z}_t, \boldsymbol{\beta})$ on $A \times B$. Let A_β be the set of $\mathbf{z}_t$ in A for which $f(\mathbf{z}_t, \boldsymbol{\beta}) = 0$ and assume that for every $\xi > 0$ there exists a $\delta_\xi > 0$ such that

$$Q_n(\boldsymbol{\beta}) = n^{-1} \sum_{t=1}^{n} \inf_{\mathbf{z}_t \in A_\beta} (\mathbf{z}_t^0 - \mathbf{z}_t)\boldsymbol{\Sigma}_{\varepsilon\varepsilon}^{-1}(\mathbf{z}_t^0 - \mathbf{z}_t)' > \delta_\xi \tag{3.2.23}$$

for all $\boldsymbol{\beta}$ in B satisfying $|\boldsymbol{\beta} - \boldsymbol{\beta}^0| > \xi$. Let $\boldsymbol{\theta}' = (\boldsymbol{\beta}', \mathbf{z}_1, \mathbf{z}_2, \ldots, \mathbf{z}_n)$ and let $\boldsymbol{\theta}^0 = (\boldsymbol{\beta}^{0\prime}, \mathbf{z}_1^0, \mathbf{z}_2^0, \ldots, \mathbf{z}_n^0)'$ be the true $\boldsymbol{\theta}$, where $\boldsymbol{\beta}^0$ is an interior point of B and every $\mathbf{z}_t^0$ is in the interior of $\mathbf{A}$. Let $\hat{\boldsymbol{\theta}}$ be the value of $\boldsymbol{\theta}$ that minimizes

$$\sum_{t=1}^{n} (\mathbf{Z}_{rt} - \hat{\mathbf{z}}_t)\boldsymbol{\Sigma}_{\varepsilon\varepsilon}^{-1}(\mathbf{Z}_{rt} - \hat{\mathbf{z}}_t)', \tag{3.2.24}$$

subject to $f(\hat{\mathbf{z}}_t, \hat{\boldsymbol{\beta}}) = 0$ for $t = 1, 2, \ldots, n$. Then

$$r^{1/2}(\hat{\boldsymbol{\theta}} - \boldsymbol{\theta}^0) \xrightarrow{L} N(\mathbf{0}, \boldsymbol{\Omega}),$$

as $r \to \infty$, where

$$\boldsymbol{\Omega} = \begin{bmatrix} \boldsymbol{\Omega}_{\beta\beta} & \boldsymbol{\Omega}_{\beta z} \\ \boldsymbol{\Omega}_{z\beta} & \boldsymbol{\Omega}_{zz} \end{bmatrix}, \tag{3.2.25}$$

$$\boldsymbol{\Omega}_{\beta\beta} = \left[\sum_{t=1}^{n} \sigma_{vvtt}^{-1} \mathbf{F}'_{\beta t} \mathbf{F}_{\beta t} \right]^{-1},$$

$\boldsymbol{\Omega}_{zz}$ is an $np \times np$ matrix with $p \times p$ blocks of the form

$$\boldsymbol{\Omega}_{zztj} = \begin{cases} \sigma_{vvtt}^{-1}\sigma_{vvjj}^{-1}\mathbf{F}_{\beta t}\boldsymbol{\Omega}_{\beta\beta}\mathbf{F}'_{\beta j}\boldsymbol{\Sigma}_{\varepsilon vjj}\boldsymbol{\Sigma}_{v\varepsilon tt}, & t \neq j \\ \boldsymbol{\Sigma}_{\varepsilon\varepsilon} - \sigma_{vvtt}^{-2}(\sigma_{vvtt} - \mathbf{F}_{\beta t}\boldsymbol{\Omega}_{\beta\beta}\mathbf{F}'_{\beta t})\boldsymbol{\Sigma}_{\varepsilon vtt}\boldsymbol{\Sigma}_{v\varepsilon tt}, & t = j, \end{cases}$$

$\boldsymbol{\Omega}_{\beta z}$ is a $k \times np$ matrix composed of n blocks of the form

$$\boldsymbol{\Omega}_{\beta zt} = -\sigma_{vvtt}^{-1}\boldsymbol{\Omega}_{\beta\beta}\mathbf{F}'_{\beta t}\boldsymbol{\Sigma}_{v\varepsilon tt},$$

with $\sigma_{vvtt} = \mathbf{F}_{zt}\boldsymbol{\Sigma}_{\varepsilon\varepsilon}\mathbf{F}'_{zt}$, $\boldsymbol{\Sigma}_{\varepsilon vtt} = \boldsymbol{\Sigma}'_{v\varepsilon tt} = \boldsymbol{\Sigma}_{\varepsilon\varepsilon}\mathbf{F}'_{zt}$,

$$\mathbf{F}_{\beta t} = \left[\frac{\partial f(\mathbf{z}_t^0, \boldsymbol{\beta}^0)}{\partial \beta_1}, \frac{\partial f(\mathbf{z}_t^0, \boldsymbol{\beta}^0)}{\partial \beta_2}, \ldots, \frac{\partial f(\mathbf{z}_t^0, \boldsymbol{\beta}^0)}{\partial \beta_k} \right],$$

$$\mathbf{F}_{zt} = \left[\frac{\partial f(\mathbf{z}_t^0, \boldsymbol{\beta}^0)}{\partial z_{t1}}, \frac{\partial f(\mathbf{z}_t^0, \boldsymbol{\beta}^0)}{\partial z_{t2}}, \ldots, \frac{\partial f(\mathbf{z}_t^0, \boldsymbol{\beta}^0)}{\partial z_{tp}} \right],$$

and it is understood that all inverses are defined.

Proof. Our proofs of consistency follow those of Amemiya and Fuller (1985). Let

$$P_n(\boldsymbol{\beta}) = n^{-1} \sum_{t=1}^{n} \inf_{\mathbf{z}_t \in A_\beta} (\mathbf{Z}_{rt} - \mathbf{z}_t)\boldsymbol{\Sigma}_{\varepsilon\varepsilon}^{-1}(\mathbf{Z}_{rt} - \mathbf{z}_t)'.$$

Because $\hat{\boldsymbol{\beta}}$ minimizes $P_n(\boldsymbol{\beta})$, we have

$$P_n(\hat{\boldsymbol{\beta}}) \leqslant P_n(\boldsymbol{\beta}^0) \leqslant n^{-1} \sum_{t=1}^{n} \boldsymbol{\varepsilon}_{rt}\boldsymbol{\Sigma}_{\varepsilon\varepsilon}^{-1}\boldsymbol{\varepsilon}'_{rt} = R_n,$$

say. For any $\mathbf{z}_t$,

$$(\mathbf{z}_t^0 - \mathbf{z}_t)\boldsymbol{\Sigma}_{\varepsilon\varepsilon}^{-1}(\mathbf{z}_t^0 - \mathbf{z}_t)' \leqslant 2[(\mathbf{Z}_{rt} - \mathbf{z}_t)\boldsymbol{\Sigma}_{\varepsilon\varepsilon}^{-1}(\mathbf{Z}_{rt} - \mathbf{z}_t)' + \boldsymbol{\varepsilon}_{rt}\boldsymbol{\Sigma}_{\varepsilon\varepsilon}^{-1}\boldsymbol{\varepsilon}'_{rt}]$$

and, therefore,

$$Q_n(\hat{\boldsymbol{\beta}}) \leqslant 2[P_n(\hat{\boldsymbol{\beta}}) + R_n] \leqslant 4R_n = O_p(r^{-1}).$$

By the identifiability condition (3.2.23), $Q_n(\hat{\boldsymbol{\beta}})$ converging to zero implies that $\hat{\boldsymbol{\beta}}$ is converging to $\boldsymbol{\beta}^0$, and $\hat{\boldsymbol{\beta}}$ is a consistent estimator of $\boldsymbol{\beta}^0$.

Because

$$\sum_{t=1}^{n} (\hat{\mathbf{z}}_t - \mathbf{z}_t^0)\boldsymbol{\Sigma}_{\varepsilon\varepsilon}^{-1}(\hat{\mathbf{z}}_t - \mathbf{z}_t^0)' = nQ_n(\hat{\boldsymbol{\beta}}),$$

$(\hat{\mathbf{z}}_1, \hat{\mathbf{z}}_2, \ldots, \hat{\mathbf{z}}_n)$ is consistent for $(\mathbf{z}_1^0, \mathbf{z}_2^0, \ldots, \mathbf{z}_n^0)$.

If we expand $f(\hat{\mathbf{z}}_t, \hat{\boldsymbol{\beta}})$about $(\mathbf{z}_t^0, \boldsymbol{\beta}^0)$, the last term of the Lagrangean

$$\sum_{t=1}^{n} (\mathbf{Z}_{rt} - \hat{\mathbf{z}}_t)\boldsymbol{\Sigma}_{\varepsilon\varepsilon}^{-1}(\mathbf{Z}_{rt} - \hat{\mathbf{z}}_t)' + \sum_{t=1}^{n} \alpha_t f(\mathbf{z}_t, \boldsymbol{\beta})$$

can be written

$$\sum_{t=1}^{n} \alpha_t[\mathbf{f}_\beta(\mathbf{z}_t^*, \boldsymbol{\beta}^*)(\hat{\boldsymbol{\beta}} - \boldsymbol{\beta}^0) + \mathbf{f}_z(\mathbf{z}_t^*, \boldsymbol{\beta}^*)(\hat{\mathbf{z}}_t - \mathbf{z}_t^0)'],$$

where α_t are the Lagrange multipliers and $(\mathbf{z}_t^*, \boldsymbol{\beta}^*)$ is on the line segment joining $(\hat{\mathbf{z}}_t, \hat{\boldsymbol{\beta}})$ and $(\mathbf{z}_t^0, \boldsymbol{\beta}^0)$. Because $\hat{\boldsymbol{\theta}}$ is a consistent estimator of $\boldsymbol{\theta}^0$ and $\boldsymbol{\theta}^0$ is in the interior of the parameter space, the probability that $\hat{\boldsymbol{\theta}}$ satisfies the derivative equations associated with the Lagrangean approaches one as $r \to \infty$. Following the derivation used to obtain (3.2.20) and (3.2.21), we have

$$\hat{\boldsymbol{\beta}} - \boldsymbol{\beta}^0 = -\left[\sum_{t=1}^{n} \mathbf{f}'_\beta(\mathbf{z}_t^*, \boldsymbol{\beta}^*)(\sigma_{vvtt}^*)^{-1}\mathbf{f}_\beta(\mathbf{z}_t^*, \boldsymbol{\beta}^*)\right]^{-1}$$
$$\times \left[\sum_{t=1}^{n} \mathbf{f}'_\beta(\mathbf{z}_t^*, \boldsymbol{\beta}^*)(\sigma_{vvtt}^*)^{-1}\mathbf{f}_z(\mathbf{z}_t^*, \boldsymbol{\beta}^*)(\mathbf{Z}_{rt} - \mathbf{z}_t^0)'\right],$$

$$\hat{\mathbf{z}}_t - \mathbf{z}_t^0 = \boldsymbol{\varepsilon}_{rt} - (\sigma_{vvtt}^*)^{-1}[\boldsymbol{\varepsilon}_{rt}\mathbf{f}'_z(\mathbf{z}_t^*, \boldsymbol{\beta}^*) + \mathbf{f}_\beta(\mathbf{z}_t^*, \boldsymbol{\beta}^*)(\hat{\boldsymbol{\beta}} - \boldsymbol{\beta}^0)]\mathbf{f}_z(\mathbf{z}_t^*, \boldsymbol{\beta}^*)\boldsymbol{\Sigma}_{\varepsilon\varepsilon}$$

with probability approaching one, where

$$\sigma_{vvtt}^* = \mathbf{f}_z(\mathbf{z}_t^*, \boldsymbol{\beta}^*)\boldsymbol{\Sigma}_{\varepsilon\varepsilon}\mathbf{f}'_z(\mathbf{z}_t^*, \boldsymbol{\beta}^*).$$

Now, because $\hat{\boldsymbol{\theta}}$ is a consistent estimator,

$$[\mathbf{f}_\beta(\mathbf{z}_t^*, \boldsymbol{\beta}^*), \mathbf{f}_z(\mathbf{z}_t^*, \boldsymbol{\beta}^*), \sigma_{vvtt}^*] \xrightarrow{P} [\mathbf{F}_{\beta t}, \mathbf{F}_{zt}, \sigma_{vvtt}],$$

and it follows that

$$\hat{\boldsymbol{\beta}} - \boldsymbol{\beta}^0 = -\left[\sum_{t=1}^{n} \sigma_{vvtt}^{-1}\mathbf{F}'_{\beta t}\mathbf{F}_{\beta t}\right]^{-1} \sum_{t=1}^{n} \sigma_{vvtt}^{-1}\mathbf{F}'_{\beta t}v_{rt} + o_p(r^{-1/2}),$$

$$\hat{\mathbf{z}}_t - \mathbf{z}_t^0 = \boldsymbol{\varepsilon}_{rt} - \sigma_{vvtt}^{-1}[v_{rt} + \mathbf{F}_{\beta t}(\hat{\boldsymbol{\beta}} - \boldsymbol{\beta}^0)]\boldsymbol{\Sigma}_{v\varepsilon t} + o_p(r^{-1/2}),$$

where $v_{rt} = \mathbf{F}_{zt}\boldsymbol{\varepsilon}'_{rt}$ and $\boldsymbol{\Sigma}_{v\varepsilon t} = \mathbf{F}_{zt}\boldsymbol{\Sigma}_{\varepsilon\varepsilon}$. We obtain the distributional result because the $\boldsymbol{\varepsilon}_{rt}$ are normally distributed. □

The assumption that $\boldsymbol{\Sigma}_{\varepsilon\varepsilon}$ is nonsingular is not required in the computation of the estimates. Also, the assumption that $\boldsymbol{\Sigma}_{\varepsilon\varepsilon}$ is nonsingular can be replaced in Theorem 3.2.1 by the assumption

$$K_L < \mathbf{f}_z(\mathbf{z}_t, \boldsymbol{\beta})\boldsymbol{\Sigma}_{\varepsilon\varepsilon}\mathbf{f}'_z(\mathbf{z}_t, \boldsymbol{\beta})$$

for all $(\mathbf{z}_t, \boldsymbol{\beta})$ in $A \times B$. If $\boldsymbol{\Sigma}_{\varepsilon\varepsilon}$ is singular, one can define the quadratic form in (3.2.24) for a vector of contrasts of the $\mathbf{Z}_t$ that has a nonsingular covariance matrix. The linear functions of the $\mathbf{Z}_t$ with zero error variance are arguments of $f(\mathbf{z}_t, \boldsymbol{\beta})$ but not of the quadratic form.

The covariance matrix of the limiting distribution of $\hat{\boldsymbol{\beta}}$ is estimated by the inverse matrix of (3.2.20) computed at the last iteration. The covariance matrix for the approximate distribution of $\hat{\mathbf{z}}_t$ can be estimated by substituting estimators for the parameters in $\boldsymbol{\Omega}_{zztt}$ of (3.2.25).

In the derivation we treated $\boldsymbol{\Sigma}_{\varepsilon\varepsilon}$ as known. However, $\boldsymbol{\Sigma}_{\varepsilon\varepsilon}$ need only be known up to a multiple. If

$$\boldsymbol{\Sigma}_{\varepsilon\varepsilon} = \boldsymbol{\Upsilon}_{\varepsilon\varepsilon}\sigma^2, \tag{3.2.26}$$

where $\boldsymbol{\Upsilon}_{\varepsilon\varepsilon}$ is known and nonsingular, and the r of Theorem 3.2.1 is absorbed into σ^2, the estimator of σ^2 is

$$\begin{aligned}\hat{\sigma}^2 &= (n-k)^{-1} \sum_{t=1}^{n} (\mathbf{Z}_t - \hat{\mathbf{z}}_t)\boldsymbol{\Upsilon}_{\varepsilon\varepsilon}^{-1}(\mathbf{Z}_t - \hat{\mathbf{z}})' \\ &= (n-k)^{-1} \sum_{t=1}^{n} [\mathbf{f}_z(\hat{\mathbf{z}}_t, \hat{\boldsymbol{\beta}})\boldsymbol{\Upsilon}_{\varepsilon\varepsilon}\mathbf{f}'_z(\hat{\mathbf{z}}_t, \hat{\boldsymbol{\beta}})]^{-1}[\mathbf{f}_z(\hat{\mathbf{z}}_t, \hat{\boldsymbol{\beta}})(\mathbf{Z}_t - \hat{\mathbf{z}}_t)']^2. \end{aligned} \tag{3.2.27}$$

The second expression in (3.2.27) is also valid for singular $\boldsymbol{\Upsilon}_{\varepsilon\varepsilon}$. The approximate distribution of $\hat{\sigma}^2$ is given in Theorem 3.2.2.

Theorem 3.2.2. Let the assumptions of Theorem 3.2.1 hold with $\boldsymbol{\Sigma}_{\varepsilon\varepsilon}$ defined by (3.2.26) where $\boldsymbol{\Upsilon}_{\varepsilon\varepsilon}$ is known and nonsingular. Let the estimator of $(\boldsymbol{\beta}, \mathbf{z}_1, \mathbf{z}_2, \ldots, \mathbf{z}_n)$ be the vector that minimizes (3.2.4) with $\boldsymbol{\Upsilon}_{\varepsilon\varepsilon}$ replacing $\boldsymbol{\Sigma}_{\varepsilon\varepsilon}$. Let $\hat{\sigma}^2$ be defined by (3.2.27). Then

$$(n-k)\sigma^{-2}\hat{\sigma}^2 \xrightarrow{L} \chi^2_{n-k},$$

where χ^2_{n-k} is distributed as a chi-square random variable with $n-k$ degrees of freedom. Furthermore, the limiting distribution of $\hat{\sigma}^2$ is independent of that of $r^{1/2}(\hat{\boldsymbol{\theta}} - \boldsymbol{\theta}^0)$ defined in Theorem 3.2.1.

Proof. The estimation problem contains np unknown z_{ti} values and k unknown elements of $\boldsymbol{\beta}$. Because of the restrictions imposed by $f(\mathbf{z}_t, \boldsymbol{\beta}) = 0$, there are a total of $(p-1)n + k$ independent unknown parameters. Let γ be a $(p-1)n + k$ vector containing one set of independent parameters. Then we can write

$$\mathbf{Z}_t = \mathbf{g}_t(\gamma^0) + \boldsymbol{\varepsilon}_t, \qquad t = 1, 2, \ldots, n,$$

where the form of $\mathbf{g}_t(\gamma)$ is determined by the parameters chosen to enter γ and γ^0 is the true value of γ. Our assumptions are sufficient to guarantee the existence of functions $\mathbf{g}_t(\gamma)$ that are continuous and differentiable in the neighborhood of γ^0. By the proof of Theorem 3.2.1

$$(\mathbf{Z} - \hat{\mathbf{z}})' = \mathbf{F}_\gamma(\hat{\gamma} - \gamma^0) + \boldsymbol{\varepsilon}' + o_p(r^{-1/2}),$$
$$\hat{\gamma} - \gamma^0 = (\mathbf{F}_\gamma'\boldsymbol{\Upsilon}^{-1}\mathbf{F}_\gamma)^{-1}\mathbf{F}_\gamma'\boldsymbol{\Upsilon}^{-1}\boldsymbol{\varepsilon}' + o_p(r^{-1/2}),$$

where $\mathbf{Z} = (\mathbf{Z}_1, \ldots, \mathbf{Z}_n)$, $\hat{\mathbf{z}} = (\hat{\mathbf{z}}_1, \ldots, \hat{\mathbf{z}}_n)$, $\boldsymbol{\varepsilon} = (\boldsymbol{\varepsilon}_1, \ldots, \boldsymbol{\varepsilon}_n)$,

$$\boldsymbol{\Upsilon} = \text{block diag}(\boldsymbol{\Upsilon}_{\varepsilon\varepsilon}, \boldsymbol{\Upsilon}_{\varepsilon\varepsilon}, \ldots, \boldsymbol{\Upsilon}_{\varepsilon\varepsilon}),$$

and $\mathbf{F}_\gamma$ is the $np \times [(p-1)n + k]$ matrix of partial derivatives of $\mathbf{Z}$ with respect to γ evaluated at $\gamma = \gamma^0$. It follows that

$$\begin{aligned}\hat{\sigma}^2 &= (n-k)^{-1}r(\mathbf{Z} - \hat{\mathbf{z}})\boldsymbol{\Upsilon}^{-1}(\mathbf{Z} - \hat{\mathbf{z}})' \\ &= (n-k)^{-1}r\boldsymbol{\varepsilon}\mathbf{H}\boldsymbol{\Upsilon}^{-1}\mathbf{H}\boldsymbol{\varepsilon}' + o_p(1),\end{aligned}$$

where

$$\mathbf{H} = \mathbf{I} - \mathbf{F}_\gamma(\mathbf{F}_\gamma'\boldsymbol{\Upsilon}^{-1}\mathbf{F}_\gamma)^{-1}\mathbf{F}_\gamma'\boldsymbol{\Upsilon}^{-1}.$$

The distribution of the quadratic form in $\boldsymbol{\varepsilon}$ follows by standard regression theory. [See, e.g., Johnston (1972, Chap. 5).] By the same theory, $\hat{\sigma}^2$ and $r^{1/2}(\hat{\theta} - \theta^0)$ are independent in the limit. An alternative proof can be constructed by substituting the expressions for $\hat{\boldsymbol{\beta}} - \boldsymbol{\beta}^0$ and $\hat{\mathbf{z}}_t - \mathbf{z}_t^0$ from the proof of Theorem 3.2.1 into (3.2.27). □

Example 3.2.4. To illustrate the computations for an implicit model we use an example from Reilly and Patino-Leal (1981). The data in Table 3.2.10 are 20 observations digitized from the x-ray image of a hip prosthesis. Reilly and Patino-Leal (1981) cite Oxland, McLeod, and McNeice (1979) for the original data. The image is assumed to be that of an ellipse

$$\beta_3(y_t - \beta_1)^2 + 2\beta_4(y_t - \beta_1)(x_t - \beta_2) + \beta_5(x_t - \beta_2)^2 - 1 = 0,$$

where y_t is the true value of the vertical distance from the origin and x_t is the true horizontal distance from the origin. We assume that the observations (Y_t, X_t) of Table 3.2.7 are the sum of true values (y_t, x_t) and measurement error (e_t, u_t), where $(e_t, u_t)' \sim \text{NI}(\mathbf{0}, \mathbf{I}\sigma^2)$.

The iterative fitting method associated with (3.2.20) and (3.2.21) gives the estimated parameters:

$$\begin{aligned}&\hat{\beta}_1 = -0.9994 \quad (0.1114), && \hat{\beta}_2 = -2.9310 \quad (0.1098),\\ &\hat{\beta}_3 = 0.08757 \quad (0.00411), && \hat{\beta}_4 = 0.01623 \quad (0.00275),\\ &\hat{\beta}_5 = 0.07975 \quad (0.00350), && \hat{\sigma}^2 = 0.00588,\end{aligned}$$

TABLE 3.2.10. Observations on x-ray image of a hip prosthesis

Observation	Y_t	X_t	$\hat{y}_t$	$\hat{x}_t$	$\hat{v}_t$
1	0.50	−0.12	0.534	−0.072	−0.036
2	1.20	−0.60	1.174	−0.626	0.023
3	1.60	−1.00	1.535	−1.049	0.051
4	1.86	−1.40	1.800	−1.437	0.045
5	2.12	−2.54	2.274	−2.494	−0.099
6	2.36	−3.36	2.435	−3.354	−0.044
7	2.44	−4.00	2.427	−3.999	0.008
8	2.36	−4.75	2.268	−4.718	0.053
9	2.06	−5.25	2.030	−5.233	0.018
10	1.74	−5.64	1.738	−5.638	0.002
11	1.34	−5.97	1.358	−5.993	−0.015
12	0.90	−6.32	0.882	−6.280	−0.023
13	−0.28	−6.44	−0.278	−6.540	−0.055
14	−0.78	−6.44	−0.790	−6.508	−0.039
15	−1.36	−6.41	−1.351	−6.382	0.017
16	−1.90	−6.25	−1.868	−6.181	0.046
17	−2.50	−5.88	−2.469	−5.835	−0.034
18	−2.88	−5.50	−2.888	−5.509	−0.007
19	−3.18	−5.24	−3.174	−5.234	0.005
20	−3.44	−4.86	−3.476	−4.888	−0.029

Source: Reilly and Patino-Leal (1981).

where the numbers in parentheses are the estimated standard errors. The estimated covariance matrix of the estimators is the inverse matrix obtained at the last iteration of (3.2.20).

The estimated values of (y_t, x_t) are given in Table 3.2.10. The estimated covariance matrices of $(\hat{y}_t, \hat{x}_t)$ are given in Table 3.2.11. The covariances are estimates of the elements of the matrix Ω_{zztt} defined in (3.2.25) of Theorem 3.2.1. Note that the variance of $\hat{x}_t$ is smallest when the curve is most nearly parallel to the y axis. See Figure 3.2.2.

The deviations

$$\hat{v}_t = \mathbf{f}_z(\hat{\mathbf{z}}_t, \hat{\boldsymbol{\beta}})(\mathbf{Z}_t - \hat{\mathbf{z}}_t)' = \hat{\mathbf{F}}_{zt}(\mathbf{Z}_t - \hat{\mathbf{z}}_t)'$$

are given in Table 3.2.10. In the linear model $v_t = \boldsymbol{\varepsilon}_t(1, -\boldsymbol{\beta}')'$ is a linear function of $\boldsymbol{\varepsilon}_t$ because the vector of weights $(1, -\boldsymbol{\beta}')$ is constant over observations. In the nonlinear model the vector of weights $\mathbf{F}_{zt}$ changes from observation to observation. As a consequence, the variance of v_t will change from observation to observation and the plot of $\hat{v}_t$ against $\hat{z}_{ti}$ could be misleading for this

TABLE 3.2.11. Covariance matrices of $(\hat{y}_t, \hat{x}_t)$ and standardized deviations

Observation	$10^3\hat{V}\{\hat{y}_t\}$	$10^3\hat{C}\{\hat{y}_t, \hat{x}_t\}$	$10^3\hat{V}\{\hat{x}_t\}$	$10^2\hat{\sigma}_{vvtt}^{1/2}$	$\hat{\sigma}_{vvtt}^{-1/2}\hat{v}_t$
1	5.17	−1.00	4.48	4.77	0.76
2	3.61	−2.18	3.78	4.85	0.48
3	2.93	−2.24	4.19	4.86	1.06
4	2.54	−2.04	4.64	4.84	0.92
5	1.86	−1.21	5.52	4.70	−2.10
6	1.37	−0.34	5.86	4.52	−0.98
7	1.23	0.49	5.83	4.36	0.18
8	1.63	1.48	5.37	4.18	1.27
9	2.39	2.06	4.67	4.06	0.45
10	3.26	2.30	3.87	3.99	0.04
11	4.20	2.22	2.97	3.97	−0.38
12	5.06	1.77	2.09	4.00	0.57
13	5.88	0.08	1.29	4.24	−1.31
14	5.79	−0.63	1.40	4.37	−0.90
15	5.48	−1.30	1.66	4.51	0.38
16	5.02	−1.84	1.99	4.63	0.99
17	4.30	−2.29	2.55	4.76	0.72
18	3.82	−2.35	3.20	4.82	−0.15
19	3.66	−2.13	3.83	4.85	0.10
20	3.97	−1.51	4.69	4.86	−0.60

reason. It is suggested that $\hat{\sigma}_{vvtt}^{-1/2}\hat{v}_t$ be plotted against the elements of $\hat{\mathbf{z}}_t$. The quantities $\hat{\sigma}_{vvtt}^{-1/2}\hat{v}_t$ are given in Table 3.2.11. In this example, $\boldsymbol{\Sigma}_{\varepsilon\varepsilon} = \mathbf{I}\sigma^2$ and the quantity $\hat{\sigma}_{vvtt}^{-1/2}\hat{v}_t$ is proportional to the signed Euclidean distance from the observation, (Y_t, X_t), to the closest point, $(\hat{y}_t, \hat{x}_t)$, on the estimated function. The function and the observations are plotted in Figure 3.2.2. The runs of positive and negative values of the $\hat{v}_t$ in Figure 3.2.2 suggest that the error made in digitizing the image may be correlated, but we do not pursue that possibility further in this example.

In the linear case, every element in the vector of differences $\mathbf{Z}_t - \hat{\mathbf{z}}_t$ is a multiple of $\hat{v}_t$. Therefore, the plot of $Z_{ti} - \hat{z}_{ti}$ against $\hat{z}_{ti}$ is the same, except for a scale factor, as the plot of $\hat{v}_t$ against $\hat{z}_{ti}$. In the nonlinear case the multiple of $\hat{v}_t$ subtracted from Z_{ti} to create $\hat{z}_{ti}$ is not constant from observation to observation. However, the standardized quantities $\hat{\sigma}_{vvtt}^{-1/2}\hat{v}_t$ are equal to the differences $Z_{ti} - \hat{z}_{ti}$ divided by the estimated standard deviation of $Z_{ti} - \hat{z}_{ti}$. Therefore, a plot of $\hat{\sigma}_{vvtt}^{-1/2}\hat{v}_t$ against the ith element of $\hat{\mathbf{z}}_t$ is analogous to the plot of the standardized difference $Z_{ti} - \hat{z}_{ti}$ against $\hat{z}_{ti}$.
□□

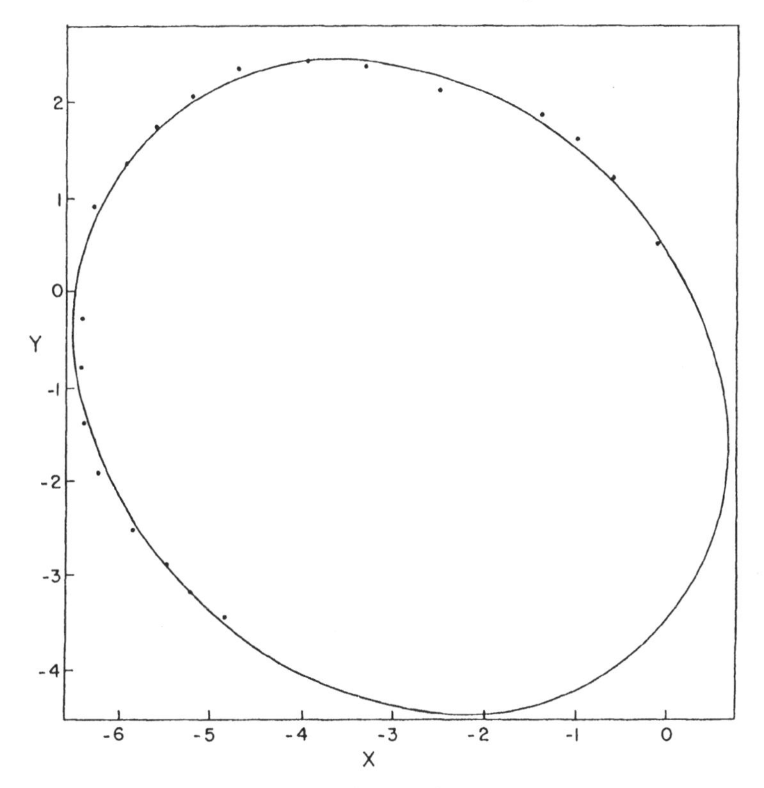

FIGURE 3.2.2. Estimated ellipse and observed data for image of hip prosthesis.

3.2.4. Modifications of the Maximum Likelihood Estimator

The conclusion of Theorem 3.2.1 requires the variance of the measurement error to be small relative to the curvature. The data analyzed in Example 3.2.2 and Example 3.2.4 are such that the approximation of Theorem 3.2.1 seems adequate. We now study maximum likelihood estimation for situations in which the variance of the measurement error is not small relative to the curvature.

Example 3.2.5. This example uses created data to illustrate some aspects of the estimation of nonlinear errors-in-variables models. Table 3.2.12 contains 120 observations generated by the model.

$$y_t = \beta_0 + \beta_1 x_t^2, \qquad (Y_t, X_t) = (y_t, x_t) + (e_t, u_t),$$
$$(e_t, u_t)' \sim \text{NI}(\mathbf{0}, 0.09\mathbf{I}),$$

TABLE 3.2.12. Data for quadratic model

Y_t	X_t	Y_t	X_t	Y_t	X_t
1.24	-1.44	0.02	-0.14	0.41	0.56
0.59	-0.90	0.04	-0.37	0.60	0.44
1.53	-0.81	0.14	-0.71	0.52	0.99
0.84	-1.14	-0.14	-0.21	-0.22	0.76
1.64	-1.36	0.29	-0.52	0.16	0.55
1.47	-0.61	0.61	-1.01	-0.10	0.53
0.84	-1.11	-0.04	-0.34	0.13	0.89
0.78	-1.46	0.58	-0.68	0.06	0.35
0.82	-0.77	-0.06	-0.41	0.46	0.33
1.40	-1.31	-0.35	-0.06	0.30	0.39
1.30	-0.98	-0.35	0.13	0.30	-0.19
1.02	-0.93	0.09	-0.18	0.09	0.55
0.65	-1.03	0.11	0.11	0.30	0.55
0.75	-1.59	0.03	-0.04	0.37	0.02
1.14	-0.45	0.26	-0.54	-0.00	0.46
1.12	-1.07	0.41	-0.10	0.98	0.84
0.64	-1.11	-0.01	0.35	1.00	1.03
0.85	-0.57	-0.45	-0.32	0.99	0.92
0.78	-0.89	0.32	-0.23	0.78	1.28
0.67	-1.25	-0.14	-0.12	0.94	1.53
0.99	-0.81	-0.06	-0.05	0.49	1.22
1.19	-1.07	0.32	-0.31	0.57	0.84
0.72	-0.51	-0.06	-0.45	0.79	1.23
1.01	-0.86	-0.22	0.65	0.89	1.36
0.45	-0.16	-0.30	0.57	1.22	0.72
0.09	-0.64	-0.26	0.17	0.62	0.71
0.42	0.21	-0.59	-0.16	1.42	1.44
0.39	-0.49	-0.31	-0.09	1.51	0.82
0.32	-0.45	0.23	0.08	1.26	0.58
-0.01	-0.71	-0.25	-0.17	1.09	0.39
-0.32	-0.69	0.59	0.09	0.38	0.63
0.65	-0.46	0.34	0.43	1.32	1.11
0.47	-0.22	-0.22	0.83	0.96	0.84
0.43	-0.98	0.24	0.70	1.50	1.07
0.72	-0.54	0.42	0.35	1.31	0.80
0.69	-0.54	-0.07	0.64	0.80	1.08
0.41	-0.81	0.81	0.52	1.06	0.85
0.06	-0.14	0.43	0.67	1.02	0.97
-0.31	-1.03	-0.10	0.50	1.06	1.00
0.05	-0.36	0.12	-0.26	1.03	1.60

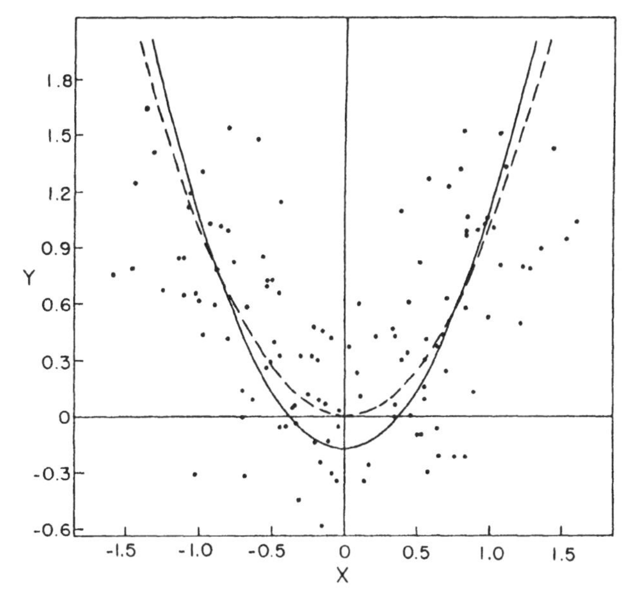

FIGURE 3.2.3. True function (dashed line) and maximum likelihood estimated function (solid line) for 120 observations generated by quadratic model.

where $(\beta_0, \beta_1) = (0, 1)$. Twenty-four observations were generated for each of the five sets of (y, x) pairs, $(1, -1)$, $(0.25, -0.50)$, $(0, 0)$, $(0.25, 0.50)$, $(1, 1)$. The errors were random normal deviates restricted so that the first and second moments for each of the five sets of 24 pairs are equal to the population moments. The data are plotted in Figure 3.2.3. Also plotted in the figure are the true function (dashed line) and the function estimated by maximum likelihood (solid line). The maximum likelihood estimate of the line is

$$\hat{y}_t = \underset{(0.059)}{-0.171} + \underset{(0.133)}{1.241}x_t^2,$$

where the numbers in parentheses are the standard errors estimated by the usual nonlinear least squares method treating σ^2 as unknown. The estimate of σ^2 is $\hat{\sigma}^2 = 0.0808$.

The estimated intercept differs by more than two estimated standard errors from the true value of zero. Also, the estimated coefficient for x_t^2 is biased upward. On the other hand, the value of the estimated function at $x = 1$ is 1.070, not greatly different from the true value of 1.000. These results illustrate the fact that choosing the function to minimize the squared distances produces a bias that is a function of the curvature. The bias moves the estimated function away from the curvature. Because the true quadratic function is concave upward at $x = 0$, the bias at that point is negative. The curvature at $x = 1$ is less than at $x = 0$ and the estimated curve is displaced less at $x = 1$. The slope of the estimated curve at $x = 1$ is biased because of the displacement at $x = 0$.

The estimation method introduced in Section 3.1.5 for the quadratic model is essentially a method of moments procedure. As such it will have a small bias for medium size samples. Using the functionally related algorithm in SUPER CARP to construct the estimator described in Example 3.1.6, we find that the estimated quadratic for the data of Table 3.2.12 is

$$\hat{y}_t = \underset{(0.074)}{0.016} + \underset{(0.134)}{1.019}x_t^2,$$

where the numbers in parentheses are the estimated standard errors computed from the estimated covariance matrix of Equation (3.1.12).

Generally, the maximum likelihood method will have a smaller variance and a larger bias than the method of moments procedure. Properties of the estimated variances based on likelihood procedures are discussed below.

□ □

Example 3.2.5 demonstrates that the maximum likelihood estimators of the parameters of nonlinear functions can be seriously biased. The maximum likelihood method chooses as the estimated surface the surface that minimizes the sum of the squared distances between the observations and the surface. When the surface is curved, the population sum of squares is minimized by a function that has been shifted away from the curvature. The bias is a function of curvature and does not decline if more observations with the same error variance and similar x values are added to the sample. Therefore, we consider modifications of the estimation procedure for samples in which the number of observations is relatively large and for models in which the measurement error in each $\mathbf{Z}$ vector is also relatively large.

Let

$$f(\mathbf{z}_t, \boldsymbol{\beta}) = 0, \qquad \mathbf{Z}_t = \mathbf{z}_t + \boldsymbol{\varepsilon}_t, \qquad \boldsymbol{\varepsilon}_t \sim \mathrm{NI}(\mathbf{0}, \boldsymbol{\Sigma}_{\varepsilon\varepsilon}), \tag{3.2.28}$$

for $t = 1, 2, \ldots, n$. The second-order Taylor expansion of $f(\mathbf{z}_t; \boldsymbol{\beta})$ about the point $(\mathbf{z}_t^0; \boldsymbol{\beta}^0)$ gives

$$\begin{aligned} f(\mathbf{z}_t, \boldsymbol{\beta}) = {} & f(\mathbf{z}_t^0, \boldsymbol{\beta}^0) + \mathbf{f}_\beta(\mathbf{z}_t^0, \boldsymbol{\beta}^0)(\boldsymbol{\beta} - \boldsymbol{\beta}^0) + \mathbf{f}_z(\mathbf{z}_t^0, \boldsymbol{\beta}^0)(\mathbf{z}_t - \mathbf{z}_t^0)' \\ & + \tfrac{1}{2}(\mathbf{z}_t - \mathbf{z}_t^0)\mathbf{F}_{zz}(\mathbf{z}_t^0, \boldsymbol{\beta}^0)(\mathbf{z}_t - \mathbf{z}_t^0)' + (\mathbf{z}_t - \mathbf{z}_t^0)\mathbf{F}_{z\beta}(\mathbf{z}_t^0, \boldsymbol{\beta}^0)(\boldsymbol{\beta} - \boldsymbol{\beta}^0) \\ & + \tfrac{1}{2}(\boldsymbol{\beta} - \boldsymbol{\beta}^0)'\mathbf{F}_{\beta\beta}(\mathbf{z}_t^0, \boldsymbol{\beta}^0)(\boldsymbol{\beta} - \boldsymbol{\beta}^0), \end{aligned} \tag{3.2.29}$$

where $\mathbf{f}_\beta(\mathbf{z}_t^0, \boldsymbol{\beta}^0)$ is the row vector of derivatives of $f(\mathbf{z}_t, \boldsymbol{\beta})$ with respect to $\boldsymbol{\beta}$, $\mathbf{f}_z(\mathbf{z}_t^0, \boldsymbol{\beta}^0)$ is the row vector of derivatives of $f(\mathbf{z}_t, \boldsymbol{\beta})$ with respect to $\mathbf{z}_t$, $\mathbf{F}_{zz}(\mathbf{z}_t^0, \boldsymbol{\beta}^0)$

is the $p \times p$ matrix of second derivatives of $f(\mathbf{z}_t, \boldsymbol{\beta})$ with respect to $\mathbf{z}_t$, $\mathbf{F}_{z\beta}(\mathbf{z}_t^0, \boldsymbol{\beta}^0)$ is the $p \times k$ matrix of second derivatives with respect to $\mathbf{z}_t$ and $\boldsymbol{\beta}$, and all derivatives are evaluated at $(\mathbf{z}_t, \boldsymbol{\beta}) = (\mathbf{z}_t^0, \boldsymbol{\beta}^0)$.

If we replace $(\mathbf{z}_t, \boldsymbol{\beta})$ by $(\hat{\mathbf{z}}_t, \hat{\boldsymbol{\beta}})$ in (3.2.29), where the $(\hat{\mathbf{z}}_t, \hat{\boldsymbol{\beta}})$ are the maximum likelihood estimators, the term in $\hat{\mathbf{z}}_t - \mathbf{z}_t^0$ is large because $\hat{\mathbf{z}}_t - \mathbf{z}_t^0$ is the same order as $\mathbf{Z}_t - \mathbf{z}_t^0$. The expected value of the term in (3.2.29) that is quadratic in $\hat{\mathbf{z}}_t - \mathbf{z}_t^0$ is

$$E\{\tfrac{1}{2}(\hat{\mathbf{z}}_t - \mathbf{z}_t^0)\mathbf{F}_{zz}(\mathbf{z}_t^0, \boldsymbol{\beta}^0)(\hat{\mathbf{z}}_t - \mathbf{z}_t^0)'\} = \tfrac{1}{2}\operatorname{tr}\{\mathbf{F}_{zz}(\mathbf{z}_t^0, \boldsymbol{\beta}^0)\boldsymbol{\Omega}_{zztt}\}, \qquad (3.2.30)$$

where $\boldsymbol{\Omega}_{zztt}$ is the covariance matrix of $\hat{\mathbf{z}}_t - \mathbf{z}_t^0$.

Let the maximum likelihood estimators of $\mathbf{z}_t$, $\boldsymbol{\beta}$, and $\boldsymbol{\Omega}_{zztt}$ be denoted by $\hat{\mathbf{z}}_t$, $\hat{\boldsymbol{\beta}}$, and $\hat{\boldsymbol{\Omega}}_{zztt}$, respectively. Then an estimator of $\boldsymbol{\beta}$ with smaller bias than the maximum likelihood estimator is obtained by minimizing

$$\sum_{t=1}^{n} (\mathbf{Z}_t - \mathbf{z}_t)\boldsymbol{\Sigma}_{\varepsilon\varepsilon}^{-1}(\mathbf{Z}_t - \mathbf{z}_t)' \qquad (3.2.31)$$

subject to the restrictions

$$f(\mathbf{z}_t, \boldsymbol{\beta}) - \tfrac{1}{2}\operatorname{tr}\{\hat{\mathbf{F}}_{zztt}\hat{\boldsymbol{\Omega}}_{zztt}\} = 0, \qquad t = 1, 2, \ldots, n, \qquad (3.2.32)$$

where $\hat{\mathbf{F}}_{zztt} = \mathbf{F}_{zz}(\hat{\mathbf{z}}_t, \hat{\boldsymbol{\beta}})$ and $\hat{\boldsymbol{\Omega}}_{zztt}$ is $\boldsymbol{\Omega}_{zztt}$ of (3.2.25) evaluated at $(\mathbf{z}_t, \boldsymbol{\beta}) = (\hat{\mathbf{z}}_t, \hat{\boldsymbol{\beta}})$.

Table 3.2.13 has been constructed to illustrate the nature of the bias in the maximum likelihood estimator of $\boldsymbol{\beta}$ and the effect of the adjustment. Sets of 200 observations of the vector (e_t, u_t) were generated, where $(e_t, u_t)' \sim NI(\mathbf{0}, \mathbf{I}\sigma^2)$. Each set of 200 observations was standardized so that the sample mean is zero and the sample covariance matrix is the identity. Then the parameters β_0 and σ^2 of the model

$$y_t = \beta_0 + cx_t^2, \qquad (Y_t, X_t) = (y_t, x_t) + (e_t, u_t)$$

were estimated for each sample. The observed (Y_t, X_t) were set equal to (e_t, u_t) so that the true value of β_0 and the true values of all x_t are zero. In the estimation, c is treated as known and the (y_t, x_t) values are treated as unknown. The estimates of β_0 are given in Table 3.2.13.

The fixed known parameter c is given in the first column of Table 3.2.13 and the maximum likelihood estimate of the intercept is given in the third column. Because the true value of β_0 is zero, the estimate of β_0 is an estimate of the bias. The fourth column contains the theoretical approximation to the bias given in (3.2.30). The empirical bias is somewhat larger than the theoretical approximation. Also, the percent difference tends to increase as the theoretical bias increases and as the curvature increases. The last column contains the modified estimator constructed to minimize (3.2.31) subject to the modified restrictions (3.2.32). The maximum likelihood estimator of σ^2 is used to construct $\hat{\boldsymbol{\Omega}}_{zztt}$. The modification removes essentially all of the bias

TABLE 3.2.13. Empirical bias of the maximum likelihood estimator of the intercept for the quadratic function with known second-degree coefficient

Quadratic Parameter c	True σ^2	M.L.E. of β_0	Approximate Bias (3.2.30)	Modified Estimator (3.2.31)
2.0000	0.25	−0.541	−0.5000	−0.428
1.0000	0.25	−0.308	−0.2500	−0.188
0.5000	0.25	−0.153	−0.1250	−0.060
0.2500	0.25	−0.068	−0.0625	−0.011
0.5000	0.5	−0.311	−0.2500	−0.157
0.2500	0.5	−0.145	−0.1250	−0.040
0.1250	0.5	−0.066	−0.0625	−0.006
0.0625	0.5	−0.032	−0.03125	−0.001
0.5000	1.0	−0.616	−0.5000	−0.376
0.2500	1.0	−0.306	−0.2500	−0.121
0.1250	1.0	−0.136	−0.1250	−0.023
0.0625	1.0	−0.064	−0.0625	−0.003
0.5000	2.0	−1.169	−1.0000	−0.823
0.2500	2.0	−0.622	−0.5000	−0.314
0.1250	2.0	−0.290	−0.2500	−0.080
0.0625	2.0	−0.132	−0.1250	−0.013
0.1250	4.0	−0.611	−0.5000	−0.242
0.0625	4.0	−0.273	−0.2500	−0.046
0.0312	4.0	−0.129	−0.1250	−0.007
0.0156	4.0	−0.063	−0.0625	−0.001

in the estimator of β_0 when the theoretical bias is less than 0.07 and the modification removes most of the bias if the theoretical bias is less than 0.15. The variance of the maximum likelihood estimator of β_0 is approximately $n^{-1}\sigma^2$. Therefore, for example, if $\sigma^2 = 0.5$ and $c = 0.25$, the squared bias of the maximum likelihood estimator of β_0 is greater than 10% of the variance whenever n is greater than 3. For $\sigma^2 = 0.5$ and $c = 0.25$, the squared bias of the modified estimator is less than 10% of the variance for n less than 59.

The modification is less effective in removing the bias when σ^2 is small and c is large. This is because the estimated variance of $\hat{\mathbf{z}}$ is more seriously biased downward in this case.

The higher-order expansions can also be used to reduce the bias in the estimator of σ^2. Expression (3.2.27) can be written

$$\hat{\sigma}^2 = (n - k)^{-1} \sum_{t=1}^{n} \hat{\Upsilon}_{vvtt}^{-1} \hat{v}_t^2, \tag{3.2.33}$$

where

$$\hat{\Upsilon}_{vvtt} = \mathbf{f}_z(\hat{\mathbf{z}}_t, \hat{\boldsymbol{\beta}})\boldsymbol{\Upsilon}_{\varepsilon\varepsilon}\mathbf{f}'_z(\hat{\mathbf{z}}_t, \hat{\boldsymbol{\beta}}), \tag{3.2.34}$$

$$\hat{v}_t = \mathbf{f}_z(\hat{\mathbf{z}}_t, \hat{\boldsymbol{\beta}})(\mathbf{Z}_t - \hat{\mathbf{z}}_t)'. \tag{3.2.35}$$

The quantity $\Upsilon_{vvtt}^{-1}v_t^2$ is the squared distance from the point $\mathbf{Z}_t$ to the point $\mathbf{z}_t$ in the metric $\boldsymbol{\Upsilon}_{\varepsilon\varepsilon}$. A source of bias in the estimator of σ^2 is the bias in $\hat{\Upsilon}_{vvtt}$ as an estimator of Υ_{vvtt}. If we replace $f_z(\hat{\mathbf{z}}_t, \hat{\boldsymbol{\beta}})$ of (3.2.34) with a Taylor series about $(\mathbf{z}_t, \boldsymbol{\beta})$ and retain only first- and second-order terms in $(\hat{\mathbf{z}}_t - \mathbf{z}_t)$, we have

$$\begin{aligned}\hat{\Upsilon}_{vvtt} \doteq \Upsilon_{vvtt} &+ 2\mathbf{f}_z(\mathbf{z}_t, \boldsymbol{\beta})\boldsymbol{\Upsilon}_{\varepsilon\varepsilon}\mathbf{F}_{zz}(\mathbf{z}_t, \boldsymbol{\beta})(\hat{\mathbf{z}}_t - \mathbf{z}_t)' \\ &+ (\hat{\mathbf{z}}_t - \mathbf{z}_t)[\mathbf{F}_{zz}(\mathbf{z}_t, \boldsymbol{\beta})\boldsymbol{\Upsilon}_{\varepsilon\varepsilon}\mathbf{F}_{zz}(\mathbf{z}_t, \boldsymbol{\beta})](\hat{\mathbf{z}}_t - \mathbf{z}_t)',\end{aligned} \tag{3.2.36}$$

where $\Upsilon_{vvtt} = \mathbf{f}_z(\mathbf{z}_t, \boldsymbol{\beta})\boldsymbol{\Upsilon}_{\varepsilon\varepsilon}\mathbf{f}'_z(\mathbf{z}_t, \boldsymbol{\beta})$. The expected value of the last term on the right side of (3.2.36) is

$$\operatorname{tr}\{\boldsymbol{\Omega}_{zztt}[\mathbf{F}_{zz}(\mathbf{z}_t, \boldsymbol{\beta})\boldsymbol{\Upsilon}_{\varepsilon\varepsilon}\mathbf{F}_{zz}(\mathbf{z}_t, \boldsymbol{\beta})]\}. \tag{3.2.37}$$

An estimator of the expression in (3.2.37) is

$$\operatorname{tr}\{\hat{\boldsymbol{\Omega}}_{zztt}\hat{\mathbf{F}}_{zztt}\boldsymbol{\Upsilon}_{\varepsilon\varepsilon}\hat{\mathbf{F}}_{zztt}\}, \tag{3.2.38}$$

where $\hat{\boldsymbol{\Omega}}_{zztt}$ and $\hat{\mathbf{F}}_{zztt}$ are the maximum likelihood estimators defined for Equation (3.2.32). It follows that an estimator of σ^2 with reduced bias is

$$\tilde{\sigma}^2 = (n-k)^{-1}\sum_{t=1}^{n}\hat{\Upsilon}_{vvtt}^{-1}[\hat{v}_t^2 + \hat{\sigma}^2\operatorname{tr}\{\hat{\boldsymbol{\Omega}}_{zztt}\hat{\mathbf{F}}_{zztt}\boldsymbol{\Upsilon}_{\varepsilon\varepsilon}\hat{\mathbf{F}}_{zztt}\}], \tag{3.2.39}$$

where $\hat{\sigma}^2$ is the estimator defined in (3.2.33).

The maximum likelihood estimates of σ^2 adjusted for degrees of freedom and the modified estimator are given in Table 3.2.14 for the data sets of Table 3.2.13. It is clear that the maximum likelihood estimator of σ^2 is biased, with the bias increasing as the curvature of the function increases and as the variance in the y direction increases. The modified estimator is less biased than the maximum likelihood estimator and displays little bias for $c\sigma^2 \leqslant 0.25$.

There is a second consideration in the use of least squares (maximum likelihood) to estimate the parameters of a nonlinear functional relationship. The estimated covariance matrix of the estimators obtained by the procedure of Example 3.2.2 is a biased estimator of the covariance matrix of the estimators. This can be seen by considering the linear model. Let

$$Y_t = \beta_0 + \beta_1 x_t + e_t = g(x_t, \boldsymbol{\beta}) + e_t, \tag{3.2.40}$$

$X_t = x_t + u_t$, and assume $(e_t, u_t)' \sim \mathrm{NI}(\mathbf{0}, \mathbf{I}\sigma^2)$. Then the estimator of β_1 given by the least squares method of Example 3.2.2 is equal to the estimator derived in Section 1.3.3. That is,

$$\hat{\beta}_1 = (m_{XX} - \hat{\lambda})^{-1}m_{XY},$$

TABLE 3.2.14. Empirical bias of the maximum likelihood estimator of the error variance for the quadratic function with known second-degree coefficient

Quadratic Parameter c	True σ^2	M.L.E. of σ^2	Modified Estimator (3.2.39)
2.0000	0.25	0.126	0.158
1.0000	0.25	0.171	0.216
0.5000	0.25	0.215	0.247
0.2500	0.25	0.240	0.253
0.5000	0.5	0.389	0.472
0.2500	0.5	0.462	0.505
0.1250	0.5	0.490	0.504
0.0625	0.5	0.499	0.503
0.5000	1.0	0.683	0.865
0.2500	1.0	0.861	0.989
0.1250	1.0	0.960	1.012
0.0625	1.0	0.991	1.006
0.5000	2.0	1.174	1.512
0.2500	2.0	1.556	1.888
0.1250	2.0	1.846	2.018
0.0625	2.0	1.961	2.018
0.1250	4.0	3.443	3.954
0.0625	4.0	3.842	4.047
0.0312	4.0	3.966	4.025
0.0156	4.0	4.003	4.018

where $\hat{\lambda}$ is the smallest root of $|\mathbf{m}_{ZZ} - \lambda \mathbf{I}| = 0$. The consistent estimator of the variance of the limiting distribution of $\hat{\beta}_1$ is given in Theorem 1.3.2 of Section 1.3.3,

$$\hat{V}\{\hat{\beta}_1\} = (n-1)^{-1}\hat{m}_{xx}^{-2}[m_{XX}s_{vv} - \hat{\sigma}_{uv}^2], \tag{3.2.41}$$

where $\hat{\sigma}_{uv} = -\hat{\beta}_1\hat{\sigma}_{uu} = -\hat{\beta}_1\hat{\lambda}$,

$$s_{vv} = (n-2)^{-1}\sum_{t=1}^{n}[Y_t - \bar{Y} - \hat{\beta}_1(X_t - \bar{X})]^2,$$

$$\hat{m}_{xx} = \tfrac{1}{2}[(m_{YY} - m_{XX})^2 + 4m_{XY}^2]^{1/2} - (m_{YY} - m_{XX}).$$

If one uses the method of Example 3.2.2 to construct an estimator of β_1 of (3.2.15), the matrix of partial derivatives at the last iteration are those displayed in Table 3.2.15. It follows that the estimated variance obtained from

TABLE 3.2.15. Partial derivatives associated with nonlinear least squares estimation of the linear errors-in-variables model

Original Observation	Index i	Dependent Variable	$\frac{\partial g(\hat{x}_i; \hat{\boldsymbol{\beta}})}{\partial \beta_0}$	$\frac{\partial g(\hat{x}_i; \hat{\boldsymbol{\beta}})}{\partial \beta_1}$	$\frac{\partial g(\hat{x}_i; \hat{\boldsymbol{\beta}})}{\partial x_1}$	$\frac{\partial g(\hat{x}_i; \hat{\boldsymbol{\beta}})}{\partial x_2}$	$\cdots$	$\frac{\partial g(\hat{x}_i; \hat{\boldsymbol{\beta}})}{\partial x_n}$
Y_1	1	Z_{11}	1	$\hat{x}_1$	$\hat{\beta}_1$	0		0
Y_2	2	Z_{21}	1	$\hat{x}_2$	0	$\hat{\beta}_1$		0
$\vdots$	$\vdots$	$\vdots$	$\vdots$	$\vdots$	$\vdots$	$\vdots$		$\vdots$
Y_n	n	Z_{n1}	1	$\hat{x}_n$	0	0		$\hat{\beta}_1$
X_1	1	Z_{12}	0	0	1	0		0
X_2	2	Z_{22}	0	0	0	1		0
$\vdots$	$\vdots$	$\vdots$	$\vdots$	$\vdots$	$\vdots$	$\vdots$		$\vdots$
X_n	n	Z_{n2}	0	0	0	0		1

the nonlinear least squares fit is

$$\hat{V}_m\{\hat{\beta}_1\} = \left[\sum_{t=1}^{n} (\hat{x}_t - \bar{X})^2\right]^{-1} s_{vv}, \tag{3.2.42}$$

where $s_{vv} = (1 + \hat{\beta}_1^2)\hat{\sigma}^2$,

$$\hat{\sigma}^2 = (2n - n - 2)^{-1} \sum_{t=1}^{n} \sum_{j=1}^{2} (Z_{tj} - \hat{z}_{tj})^2,$$

and $\hat{z}_{tj}$ is the usual least squares estimated value for z_{tj}. Now, by Theorem 2.3.2, $\hat{\sigma}^2$ is a consistent estimator of σ_{uu} and $\hat{\beta}_1$ is a consistent estimator of β_1. Also

$$E\left\{\sum_{t=1}^{n} \hat{x}_t^2\right\} = \sum_{t=1}^{n} [x_t^2 + (1 + \beta_1^2)^{-1}\sigma_{uu}] + O(1). \tag{3.2.43}$$

Therefore, if we fix $\boldsymbol{\Sigma}_{\varepsilon\varepsilon} = \mathbf{I}$, assume

$$\lim_{n\to\infty} n^{-1} \sum_{t=1}^{n} (x_t - \bar{x})^2 = \bar{m}_{xx} > 0,$$

and take the limit as the number of observations, n, becomes large, we have

$$\text{plim}[n\hat{V}_m\{\hat{\beta}_1\}] = [\bar{m}_{xx} + (1 + \beta_1^2)^{-1}]^{-1}\sigma_{vv}. \tag{3.2.44}$$

Equation (3.2.44) illustrates the fact that $\hat{V}_m\{\hat{\beta}_1\}$ is a biased estimtator of the variance of the approximate distribution of $\hat{\beta}_1$ for two reasons. First, the estimator $n^{-1} \sum_{t=1}^{n} (\hat{x}_t - \bar{X})^2$ is biased for $\bar{m}_{xx}$ and, second, no estimator of the term $\bar{m}_{xx}^{-2}(\sigma_{uu}\sigma_{vv} - \sigma_{uv}^2)$ appears in $\hat{V}_m\{\hat{\beta}_1\}$.

Example 3.2.6. We illustrate the difference between the least squares estimator of the covariance matrix and the estimator (3.2.41), using the pheasant data of Example 1.3.1. The $\hat{x}$ values constructed with the estimates of Example 1.3.1 are (9.123, 6.522, 11.724, 12.661, 11.531, 11.142, 10.493, 7.651, 9.659, 10.161, 10.116, 8.142, 11.736, 11.816, 8.222) and $s_{vv} = 0.38783$. Therefore,

$$\hat{V}_m\{\hat{\beta}_1\} = \left[\sum_{t=1}^{15} (\hat{x}_t - \bar{X})^2\right]^{-1} s_{vv} = 0.00855,$$

while the estimated variance calculated in Example 1.3.1 is

$$\hat{V}\{\hat{\beta}_1\} = 0.00925.$$

In this example the error variance is not large relative to variation in the x values. As a result, $\Sigma(\hat{x}_t - \bar{X})^2$ is only about 4% larger than $(n - 1)^{-1}\hat{\sigma}_{xx}$. Also the term $(n - 1)^{-1}\hat{\sigma}_{xx}^{-2}(\sigma_{uu}s_{vv} - \hat{\sigma}_{uv}^2)$ is small. Example 3.2.8 demonstrates that the difference between $\hat{V}_m\{\hat{\beta}_1\}$ and $\hat{V}\{\hat{\beta}_1\}$ can be large. □ □

To construct an alternative estimator of the covariance matrix for the nonlinear model, we proceed by analogy to the linear model. The linear model can be written in the form (3.2.11) with

$$f(\mathbf{z}_t; \boldsymbol{\beta}) = y_t - \mathbf{x}_t\boldsymbol{\beta}. \tag{3.2.45}$$

In the linear model with known error covariance matrix, the estimator of the covariance matrix of the limiting distribution of $\hat{\boldsymbol{\beta}}$ is given in (2.2.24) as

$$\hat{\mathbf{V}}\{\hat{\boldsymbol{\beta}}\} = n^{-1}[\hat{\mathbf{M}}_{xx}^{-1}s_{vv} + \hat{\mathbf{M}}_{xx}^{-1}(\boldsymbol{\Sigma}_{uu}s_{vv} - \hat{\boldsymbol{\Sigma}}_{uv}\hat{\boldsymbol{\Sigma}}_{vu})\hat{\mathbf{M}}_{xx}^{-1}]. \tag{3.2.46}$$

By (2.2.22), the estimated variance of $\hat{\mathbf{z}}_t$ is

$$\hat{\mathbf{V}}\{\hat{\mathbf{z}}_t\} = \boldsymbol{\Sigma}_{\varepsilon\varepsilon} - s_{vv}^{-1}\hat{\boldsymbol{\Sigma}}_{\varepsilon v}\hat{\boldsymbol{\Sigma}}_{v\varepsilon} \tag{3.2.47}$$

and we can write

$$\hat{\mathbf{V}}\{\hat{\boldsymbol{\beta}}\} = n^{-1}\hat{\mathbf{M}}_{xx}^{-1}[\hat{\mathbf{M}}_{xx} + \hat{\mathbf{F}}_{\beta z}\hat{\mathbf{V}}\{\hat{\mathbf{z}}_t\}\hat{\mathbf{F}}_{z\beta}]\hat{\mathbf{M}}_{xx}^{-1}s_{vv}, \tag{3.2.48}$$

where $\hat{\mathbf{F}}_{\beta z} = \hat{\mathbf{F}}'_{z\beta} = \mathbf{F}_{\beta z}(\hat{\mathbf{z}}_t, \hat{\boldsymbol{\beta}})$ is the $k \times (k + 1)$ matrix of second derivatives with respect to $\boldsymbol{\beta}$ and $\mathbf{z}_t$ evaluated at $(\mathbf{z}_t, \boldsymbol{\beta}) = (\hat{\mathbf{z}}_t, \hat{\boldsymbol{\beta}})$. In the linear case $\hat{\mathbf{F}}_{\beta z} = (\mathbf{0}, \mathbf{I})$. A consistent estimator of $\mathbf{M}_{xx}$ is

$$\begin{aligned}\tilde{\boldsymbol{\Sigma}}_{ff} &= n^{-1}\sum_{t=1}^{n} (\hat{\mathbf{x}}_t'\hat{\mathbf{x}}_t - \hat{\mathbf{V}}\{\hat{\mathbf{x}}_t\}) \\ &= \hat{\mathbf{M}}_{ff} - n^{-1}\sum_{t=1}^{n} \hat{\mathbf{F}}_{\beta zt}\hat{\mathbf{V}}\{\hat{\mathbf{z}}_t\}\hat{\mathbf{F}}_{z\beta t},\end{aligned} \tag{3.2.49}$$

where $\hat{\mathbf{f}}_{\beta t} = \mathbf{f}_\beta(\hat{\mathbf{z}}_t, \hat{\boldsymbol{\beta}}) = \hat{\mathbf{x}}_t$,

$$\hat{\mathbf{M}}_{ff} = n^{-1}\sum_{t=1}^{n} \hat{\mathbf{f}}_{\beta t}'\hat{\mathbf{f}}_{\beta t} = n^{-1}\sum_{t=1}^{n} \hat{\mathbf{x}}_t'\hat{\mathbf{x}}_t,$$

and $\hat{\mathbf{V}}\{\hat{\mathbf{z}}_t\}$ is defined in (3.2.47). Therefore, an estimator of the variance of the approximate distribution of $\hat{\boldsymbol{\beta}}$ for the linear model is

$$\tilde{\mathbf{V}}\{\hat{\boldsymbol{\beta}}\} = n^{-1}\tilde{\boldsymbol{\Sigma}}_{ff}^{-1}\hat{\mathbf{M}}_{ff}\tilde{\boldsymbol{\Sigma}}_{ff}^{-1}s_{vv}. \tag{3.2.50}$$

Example 3.2.7. We verify formula (3.2.50) using the pheasant data and working with the corrected sum of squares. We have

$$\hat{m}_{ff} = (n-1)^{-1}\sum_{t=1}^{n}(\hat{x}_t - \bar{X})^2 = 3.2411,$$

$$\hat{V}\{\hat{x}_t\} = (\hat{\beta}_1^2 6 + 1)^{-1}\hat{\sigma}_{uu} = 0.1122,$$

where $\hat{\beta}_1 = 0.7516$ and $\hat{\sigma}_{uu} = 0.4923$. Then

$$\tilde{\sigma}_{ff} = \hat{m}_{ff} - \hat{V}\{\hat{x}_t\} = 3.1289,$$

which agrees with the estimator of σ_{xx} given in Example 1.3.1. The estimator of the variance of the approximate distribution of $\hat{\beta}_1$ is

$$\hat{V}\{\hat{\beta}_1\} = (14)^{-1}\tilde{\sigma}_{ff}^{-1}\hat{m}_{ff}\tilde{\sigma}_{ff}^{-1}s_{vv} = 0.00925$$

which agrees with $\hat{V}\{\hat{\beta}_1\}$ of Example 1.3.1. In these calculations we omitted $\hat{\beta}_0$ from the vector of estimates because the associated x value, being identically one, is measured without error. □ □

In the nonlinear case, the matrix

$$n\hat{\mathbf{M}}_{f\pi f} = \sum_{t=1}^{n}\hat{\mathbf{f}}'_{\beta t}\hat{\sigma}_{vvtt}^{-1}\hat{\mathbf{f}}_{\beta t} \tag{3.2.51}$$

obtained at the last step of the iteration is analogous to $n\hat{\mathbf{M}}_{ff}s_{vv}^{-1}$ of the linear case and

$$n\tilde{\boldsymbol{\Sigma}}_{f\pi f} = n\hat{\mathbf{M}}_{f\pi f} - \sum_{t=1}^{n}\hat{\mathbf{F}}_{\beta zt}\hat{\mathbf{V}}\{\hat{\mathbf{z}}_t\}\hat{\mathbf{F}}_{z\beta t}\hat{\sigma}_{vvtt}^{-1} \tag{3.2.52}$$

is analogous to $n\tilde{\boldsymbol{\Sigma}}_{ff}s_{vv}^{-1}$. Therefore, a reasonable estimator of the variance of the approximate distribution of the nonlinear estimator is

$$\hat{\mathbf{V}}\{\hat{\boldsymbol{\beta}}\} = n^{-1}\tilde{\boldsymbol{\Sigma}}_{f\pi f}^{-1}\hat{\mathbf{M}}_{f\pi f}\tilde{\boldsymbol{\Sigma}}_{f\pi f}^{-1}, \tag{3.2.53}$$

where $\hat{\mathbf{M}}_{f\pi f}$ is defined in (3.2.51) and $\tilde{\boldsymbol{\Sigma}}_{f\pi f}$ is defined in (3.2.52).

Example 3.2.8. In this example we continue the analysis of the data of Example 3.2.5, incorporating the modifications to the maximum likelihood procedure developed in this section. The estimated equation obtained by maximum likelihood in Example 3.2.5 is

$$\hat{y}_t = -0.171 + 1.241x_t^2$$

and the nonlinear least squares estimated covariance matrix obtained at the last iteration is defined by

$$\text{vech}\{n^{-1}\hat{\mathbf{M}}_{f\pi f}^{-1}\} = 10^{-2}(0.3518, -0.5307, 1.7632)',$$

where $\hat{\sigma}^2 = 0.0808$ is used to construct the estimated covariance matrix and vech is defined in Appendix 4.A. To place the quadratic model in the implicit form, we write

$$f(\mathbf{z}_t, \boldsymbol{\beta}) = y_t - \beta_0 - \beta_2 x_t^2 = z_{t1} - \beta_0 - \beta_2 z_{t2}^2.$$

Then $\mathbf{F}_{zz}(\mathbf{z}_t, \boldsymbol{\beta}) = \text{diag}(0, 2\beta_2)$ and

$$\mathbf{F}_{\beta zt} = \frac{\partial^2(y_t - \beta_0 - \beta_2 x_t^2)}{\partial(\beta_0, \beta_1)' \partial(y_t, x_t)} = \text{diag}(0, -2x_t).$$

Minimizing (3.2.31) subject to the restrictions (3.3.32) gives the estimated equation

$$\hat{y}_t = -0.081 + 1.133x_t.$$

The vector half of the estimate of variance defined in (3.2.53) is

$$\text{vech } \hat{\mathbf{V}}\{\hat{\boldsymbol{\beta}}\} = 10^{-2}(0.4180, -0.8010, 2.6020)'.$$

The adjustment in the estimation procedure reduced the estimated bias in the estimated parameters by about one-half. The adjustment uses a local approximation and, as such, will not remove all the bias in situations where the variance is large relative to the curvature. The modified estimate of variance for the quadratic coefficient is nearly twice that of the least squares estimator. □ □

REFERENCES

Amemiya (1982), Amemiya and Fuller (1983), Britt and Luecke (1973), Chan (1965), Clutton-Brock (1967), Dolby and Lipton (1972), Höschel (1978a, 1978b), Reilly and Patino-Leal (1981), Schnell (1983), Stefanski and Carroll (1985b), Villegas (1969), Wolter and Fuller (1982b).

EXERCISES

8. (Sections 3.1.5, 3.2) The data in the table are 10 observations on earthquake depth (Y_t) and distance from the Tonga trench (X_t) taken from Sykes, Isacks, and Oliver (1969). Assume that these data satisfy

$$Y_t = \beta_0 + \beta_1 x_t^2 + e_t, \qquad X_t = x_t + u_t,$$

where e_t, u_t, and x_t are mutually independent.

Observation	Y_t	X_t
1	571	394
2	199	175
3	572	370
4	553	412
5	506	341
6	247	232
7	0	46
8	50	42
9	622	432
10	58	5

(a) Assume that $\sigma_{uu} = 100$ and use the method of Section 3.1.5 and Example 3.1.5 to estimate the parameters of the model. Estimate the variance of the approximate distribution of your estimators.

(b) Assume that $\sigma_{uu} = \sigma_{ee} = 100$. Construct estimator (3.1.14). What do you conclude about the adequacy of the model? Using 100 $\hat{\lambda}$, where $\hat{\lambda}$ is defined in (3.1.15), as a revised estimate of $\sigma_{uu} = \sigma_{ee}$, calculate estimates of the parameters. Calculate an estimate of the covariance matrix of the approximate distribution of your estimator setting $\sigma_{uu} = \sigma_{ee} =$ 100 $\hat{\lambda}$. Wolter and Fuller (1982a) have shown that this estimator of variance is satisfactory for the model in which the covariance matrix is known up to a multiple.

9. (Section 3.2) Use the general nonlinear procedure of Section 3.2.3 to estimate the parameters of the model of Example 3.2.1. Estimate the true values and the variances of the estimated true values. Plot $\hat{\sigma}_{vvtt}^{-1/2}\hat{v}_t$ against $\hat{x}_t$. How does $\hat{\sigma}_{vvtt}$ vary from observation to observation?

10. (Section 3.2)

(a) Use the nonlinear least squares method of Example 3.2.2 to fit the linear model

$$y_t = \beta_0 + \beta_1 x_t, \qquad (Y_t, X_t) = (y_t, x_t) + (e_t, u_t),$$
$$(e_t, u_t)' \sim \mathrm{NI}(\mathbf{0}, 0.25\mathbf{I})$$

to the data of Example 1.3.2. Compare the nonlinear least squares estimated variances for the estimated x values to the estimated variances given in Example 1.3.2.

(b) Using the data and error specification of Example 1.3.2, estimate the parameters of the model

$$y_t = \alpha_0 + \alpha_1 \exp(\alpha_2 x_t), \qquad (Y_t, X_t) = (y_t, x_t) + (e_t, u_t)$$

by nonlinear least squares. Are the data consistent with this model?

11. (Section 3.2) Estimate the parameters of the model of Example 3.2.2 assuming that there is zero measurement error in X for $X = 0$ and $X = 100$. Estimate the covariance matrix of your estimators using the methods of Example 3.2.2. Note that the degrees of freedom for the residual mean square is six and that, correspondingly, the estimated variances of $(\hat{\beta}_0, \hat{\beta}_1, \hat{\beta}_2)$ are larger than those in Example 3.2.2. The estimated variances of Example 3.2.2 are biased if the error in X at $X = 0$ and $X = 100$ is zero.

12. (Section 3.2) Using the data of Example 3.2.2 estimate the parameters of the model

$$y_t = \beta_0 + \beta_1 \exp\{\beta_2 x_t\} + \beta_3[\exp\{\beta_2 x_t\}]^2$$

under the assumption $(e_t, u_t)' \sim \mathrm{NI}[\mathbf{0}, \sigma^2 \operatorname{diag}(1.44, 1)]$. Estimate the covariance matrix of the approximate distribution of your estimators.

13. (Section 3.2) A purpose of the experiment described in Examples 3.2.2 and 3.2.3 was to develop a method of estimating the gas saturation of Berea sandstone. Assume that the experiment is conducted on a sample of unknown saturation and that $(Y_{t1}, Y_{t2}) = (1225.0, 0.12)$ is obtained. Add these observations to the data of Example 3.2.3 to obtain improved estimates of the parameters and an estimate of the gas saturation for the new sample.

14. (Section 3.2) Assume that the model

$$x_t^2 + y_t^2 = \gamma^2, \qquad (Y_t, X_t) = (y_t, x_t) + (e_t, u_t)$$

holds, where $(e_t, u_t)' \sim \mathrm{NI}(\mathbf{0}, \mathbf{I}\sigma^2)$ and γ^2 is a parameter. Let (x_t, y_t) be uniformly distributed on the circle $x^2 + y^2 = 9$. Assume that on the basis of 1000 observations generated by such a model, the parameter γ is estimated by the maximum likelihood methods of Section 3.2, wherein the x_t are treated as fixed. What would be the approximate value of the estimator of γ?

15. (Section 3.2) In Example 3.2.5 the parameters of the quadratic function were estimated by least squares. Given an observation (Y_t, X_t), show that the x value of the point on the curve that is the minimum distance from (Y_t, X_t) is the solution of a cubic equation. Give the equation.

16. (Section 3.2) An expression for the variance of the nonlinear estimator of $\boldsymbol{\beta}$ was given in Theorem 3.2.1. Show that this expression is the variance expression one obtains as the inverse of the information matrix associated with the model of Theorem 3.2.1.

17. (Section 3.2) Is the covariance matrix for $r^{1/2}(\hat{\mathbf{z}}_t - \mathbf{z}_t^0)$ of Theorem 3.2.1, denoted by $\boldsymbol{\Omega}_{zztt}$, singular? Is the $(k + p) \times (k + p)$ covariance matrix of $r^{1/2}[(\hat{\boldsymbol{\beta}} - \boldsymbol{\beta}^0)'; (\hat{\mathbf{z}}_t - \mathbf{z}_t^0)]$ singular?

18. (Section 3.2) Compare expression (2.3.32) for $\tilde{\mathbf{V}}\{\hat{\mathbf{z}}_t\}$ with expression (3.2.25) for $\boldsymbol{\Omega}_{zztt}$.

19. (Section 3.2) The data below are for 27 stag-beetles. The data have been analyzed by Turner (1978) and Griffiths and Sandland (1982). The model postulated for the true values is

$$\alpha_1 y_t + \beta_1 \log y_t + \alpha_2 x_t + \beta_2 \log x_t - 1 = 0.$$

Assume $(e_t, u_t)' \sim \mathrm{NI}(\mathbf{0}, \mathbf{I}\sigma^2)$.

Stag-beetles (*Cyclommatus tarandus*)

Observation	Mandibular Length (Y_t)	Body Length (X_t)	Observation	Mandibular Length (Y_t)	Body Length (X_t)
1	3.88	16.50	15	18.83	26.68
2	5.31	18.70	16	19.19	27.13
3	6.33	20.05	17	19.92	27.36
4	7.32	20.44	18	20.79	27.61
5	8.17	21.48	19	21.53	28.51
6	9.73	22.47	20	22.54	28.96
7	10.71	22.40	21	23.25	29.25
8	11.49	23.52	22	23.96	30.27
9	12.08	24.05	23	25.38	30.63
10	12.73	24.59	24	28.49	33.37
11	14.11	24.33	25	30.69	35.37
12	14.70	24.56	26	32.00	37.00
13	15.84	25.50	27	34.50	39.50
14	17.39	25.83			

Source. Griffiths and Sandland (1982).

(a) Estimate $(\alpha_1, \beta_1, \alpha_2, \beta_2, \sigma^2)$ under the assumption that

$$(Y_t, X_t) = (y_t, x_t) + (e_t, u_t).$$

(b) Estimate $(\alpha_1, \beta_1, \alpha_2, \beta_2, \sigma^2)$ under the assumption that

$$(Y_t^{1/2}, X_t^{1/2}) = (y_t^{1/2}, x_t^{1/2}) + (e_t, u_t).$$

(c) Estimate $(\alpha_1, \beta_1, \alpha_2, \beta_2, \sigma^2)$ under the assumption that

$$(\log Y_t, \log X_t) = (\log y_t, \log x_t) + (e_t, u_t).$$

(d) Plot the standardized residuals against $\hat{x}_t$, $\hat{x}_t^{1/2}$, and $\log \hat{x}_t$ for fits (a), (b), and (c), respectively. Choose a model for the data.

3.3. THE NONLINEAR MODEL WITH AN ERROR IN THE EQUATION

This section is devoted to the nonlinear measurement error model containing an error in the equation. The model is

$$y_t = g(\mathbf{x}_t, \boldsymbol{\beta}) + q_t, \qquad \mathbf{Z}_t = \mathbf{z}_t + \mathbf{a}_t, \tag{3.3.1}$$
$$(q_t, \mathbf{a}_t)' \sim \text{NI}[\mathbf{0}, \text{block diag}(\sigma_{qq}, \boldsymbol{\Sigma}_{aa})],$$

where $\mathbf{Z}_t = (Y_t, \mathbf{X}_t)$, $\mathbf{a}_t = (w_t, \mathbf{u}_t)$ is the vector of measurement errors, and $(q_t, \mathbf{a}_t)$ is independent of $\mathbf{x}_t$. It is assumed that $\boldsymbol{\Sigma}_{aa}$ is nonsingular and known and that the form of $g(.,.)$ is known. It is also assumed that $g(\mathbf{x}, \boldsymbol{\beta})$ is continuous and possesses continuous first and second derivatives with respect to both arguments for $\mathbf{x} \in A$ and $\boldsymbol{\beta} \in B$, where A and B are subsets of p-dimensional Euclidean space and k-dimensional Euclidean space, respectively. We shall simplify our discussion by assuming $\boldsymbol{\Sigma}_{wu} = \mathbf{0}$.

We first study the structural model where the $\mathbf{x}_t$ are a random sample from a distribution with a density and finite fourth moments. If we know the form of the $\mathbf{x}$ distribution, we might apply the method of maximum likelihood to estimate $\boldsymbol{\beta}$, σ_{qq}, and the parameters of the $\mathbf{x}$ distribution. For most distributions this is not a simple computation. Therefore, we first consider the situation in which the parameters of the $\mathbf{x}$ distribution are known.

3.3.1. The Structural Model

If we know the form and the parameters of the $\mathbf{x}$ distribution and of the $\mathbf{a}$ distribution, we can evaluate the conditional distribution of $\mathbf{x}$ given $\mathbf{X}$. In particular, when $\mathbf{x}_t$ and $\mathbf{a}_t$ are normally distributed, the conditional distribution of $\mathbf{x}_t$ given $\mathbf{X}_t$ is normal and the conditional expectation and variance of $\mathbf{x}_t$ given $\mathbf{X}_t$ are

$$E\{\mathbf{x}_t \mid \mathbf{X}_t\} = \boldsymbol{\mu}_x + (\mathbf{X}_t - \boldsymbol{\mu}_x)(\boldsymbol{\Sigma}_{xx} + \boldsymbol{\Sigma}_{uu})^{-1}\boldsymbol{\Sigma}_{xx}, \tag{3.3.2}$$
$$V\{\mathbf{x}_t \mid \mathbf{X}_t\} = \boldsymbol{\Sigma}_{xx} - \boldsymbol{\Sigma}_{xx}\boldsymbol{\Sigma}_{XX}^{-1}\boldsymbol{\Sigma}_{xx},$$

respectively. Using the conditional distribution of $\mathbf{x}_t$ given $\mathbf{X}_t$, we can evaluate the conditional moments of y_t given $\mathbf{X}_t$. We have

$$E\{y|\mathbf{X}\} = \int g(\mathbf{x}, \boldsymbol{\beta}) dF_{x|X}(\mathbf{x}|\mathbf{X}), \tag{3.3.3}$$

$$V\{y|\mathbf{X}\} = \sigma_{qq} + \int [g(\mathbf{x}, \boldsymbol{\beta}) - E\{y|\mathbf{X}\}]^2 dF_{x|X}(\mathbf{x}|\mathbf{X}),$$

where $F_{x|X}(\mathbf{x}|\mathbf{X})$ is the conditional distribution of $\mathbf{x}$ given $\mathbf{X}$. The conditional moments of Y_t given $\mathbf{X}_t$ are

$$E\{Y|\mathbf{X}\} = E\{y|\mathbf{X}\},$$
$$V\{Y|\mathbf{X}\} = V\{y|\mathbf{X}\} + \sigma_{ww}.$$

The integrals (3.3.3) will often be difficult to evaluate. If the elements of $\boldsymbol{\Sigma}_{xx}^{-1}\boldsymbol{\Sigma}_{uu}$ are not large, we may approximate the function $g(\mathbf{x}, \boldsymbol{\beta})$ with the second-order Taylor expansion,

$$g(\mathbf{x}, \boldsymbol{\beta}) \doteq g(\mathbf{W}, \boldsymbol{\beta}) + \mathbf{g}_x(\mathbf{W}, \boldsymbol{\beta})(\mathbf{x} - \mathbf{W})' + \tfrac{1}{2}(\mathbf{x} - \mathbf{W})\mathbf{g}_{xx}(\mathbf{W}, \boldsymbol{\beta})(\mathbf{x} - \mathbf{W})',$$

where $\mathbf{W}_t = \mathbf{W}(\mathbf{X}_t) = E\{\mathbf{x}_t|\mathbf{X}_t\}$, the row vector $\mathbf{g}_x(\mathbf{W}, \boldsymbol{\beta})$ is the first derivative of $g(\mathbf{x}, \boldsymbol{\beta})$ with respect to $\mathbf{x}$ evaluated at $\mathbf{x} = \mathbf{W}$, and $\mathbf{g}_{xx}(\mathbf{W}, \boldsymbol{\beta})$ is the matrix of second derivatives of $g(\mathbf{x}, \boldsymbol{\beta})$ with respect to $\mathbf{x}$ evaluated at $\mathbf{x} = \mathbf{W}$. Using this approximation for $g(\mathbf{x}, \boldsymbol{\beta})$, we can approximate the integrals of (3.3.3) by

$$E\{y|\mathbf{X}\} \doteq g(\mathbf{W}, \boldsymbol{\beta}) + (0.5)\,\mathrm{tr}[\mathbf{g}_{xx}(\mathbf{W}, \boldsymbol{\beta})\mathbf{V}\{\mathbf{x}|\mathbf{X}\}], \tag{3.3.4}$$

$$V\{y|\mathbf{X}\} \doteq [\mathbf{g}_x(\mathbf{W}, \boldsymbol{\beta})]\mathbf{V}\{\mathbf{x}|\mathbf{X}\}[\mathbf{g}_x(\mathbf{W}, \boldsymbol{\beta})]' + \sigma_{qq}. \tag{3.3.5}$$

We can write

$$Y_t = h(\mathbf{W}_t, \boldsymbol{\beta}) + r_t, \tag{3.3.6}$$

where $h(\mathbf{W}_t, \boldsymbol{\beta})$ is the conditional expected value given in (3.3.4) and the approximate variance of Y_t given $\mathbf{X}_t$ is

$$\sigma_{rrtt} = \sigma_{ee} + [\mathbf{g}_x(\mathbf{W}_t, \boldsymbol{\beta})]\mathbf{V}\{\mathbf{x}_t|\mathbf{X}_t\}[\mathbf{g}_x(\mathbf{W}_t, \boldsymbol{\beta})]',$$

with $e_t = q_t + w_t$.

One can estimate $\boldsymbol{\beta}$ by applying nonlinear least squares to expression (3.3.6). Ordinary nonlinear least squares will not be efficient for $\boldsymbol{\beta}$ because the variance of r_t is not constant. Therefore, it is necessary to iterate nonlinear estimation and variance estimation in the calculations. The first step is the application of ordinary nonlinear least squares to (3.3.6). Using the ordinary nonlinear least squares estimator of $\boldsymbol{\beta}$, denoted by $\tilde{\boldsymbol{\beta}}$, and the residual mean square

$$s_{\ell\ell} = (n - k)^{-1} \sum_{t=1}^{n} \tilde{r}_t^2, \tag{3.3.7}$$

where $\tilde{r}_t = Y_t - g(\mathbf{W}_t, \tilde{\boldsymbol{\beta}}) - (0.5)\,\mathrm{tr}[\mathbf{g}_{xx}(\mathbf{W}_t, \tilde{\boldsymbol{\beta}})\mathbf{V}\{\mathbf{x}_t|\mathbf{X}_t\}]$, one can estimate σ_{ee} with

$$\hat{\sigma}_{ee} = s_{\ell\ell} - n^{-1} \sum_{t=1}^{n} \mathbf{g}_x(\mathbf{W}_t, \tilde{\boldsymbol{\beta}})\mathbf{V}\{\mathbf{x}_t|\mathbf{X}_t\}\mathbf{g}_x'(\mathbf{W}_t, \tilde{\boldsymbol{\beta}}). \tag{3.3.8}$$

A weighted estimator of σ_{ee} is

$$\tilde{\sigma}_{ee} = \left(\sum_{t=1}^{n} \hat{\sigma}_{rrtt}^{-1}\right)^{-1} \sum_{t=1}^{n} \hat{\sigma}_{rrtt}^{-1}[\tilde{r}_t^2 - \mathbf{g}_x(\mathbf{W}_t, \tilde{\boldsymbol{\beta}})\mathbf{V}\{\mathbf{x}_t|\mathbf{X}_t\}\mathbf{g}_x'(\mathbf{W}_t, \hat{\boldsymbol{\beta}})], \tag{3.3.9}$$

where

$$\hat{\sigma}_{rrtt} = \hat{\sigma}_{ee} + \mathbf{g}_x(\mathbf{W}_t, \tilde{\boldsymbol{\beta}})\mathbf{V}\{\mathbf{x}_t|\mathbf{X}_t\}\mathbf{g}_x'(\mathbf{W}_t, \tilde{\boldsymbol{\beta}})$$

and $\tilde{\sigma}_{ee}$ and $\hat{\sigma}_{ee}$ are restricted to be nonnegative.

Given preliminary estimates of $\boldsymbol{\beta}$ and σ_{ee}, denoted by $\tilde{\boldsymbol{\beta}}$ and $\tilde{\sigma}_{ee}$, a weighted nonlinear least squares estimator of $\boldsymbol{\beta}$ is the $\boldsymbol{\beta}$ that minimizes

$$\sum_{t=1}^{n} \tilde{\sigma}_{rrtt}^{-1}\{Y_t - g(\mathbf{W}_t, \boldsymbol{\beta}) - 0.5\,\mathrm{tr}[\mathbf{g}_{xx}(\mathbf{W}_t, \boldsymbol{\beta})\mathbf{V}\{\mathbf{x}_t|\mathbf{X}_t\}]\}^2, \tag{3.3.10}$$

where

$$\tilde{\sigma}_{rrtt} = \tilde{\sigma}_{ee} + \mathbf{g}_x(\mathbf{W}_t, \tilde{\boldsymbol{\beta}})\mathbf{V}\{\mathbf{x}_t|\mathbf{X}_t\}\mathbf{g}_x'(\mathbf{W}_t, \tilde{\boldsymbol{\beta}}).$$

The procedure could be iterated with a new estimator of σ_{rrtt}, but this will generally produce little change in the estimate of $\boldsymbol{\beta}$.

If the parameters of the conditional distribution of $\mathbf{x}_t$ given $\mathbf{X}_t$ are known, the variance of the approximate distribution of the estimator of $\boldsymbol{\beta}$ that minimizes (3.3.10) is given by the usual nonlinear least squares formulas. If the parameters of the $\mathbf{x}$ distribution are not known, it is necessary to estimate them to construct the conditional expected value. Estimation of the parameters of the $\mathbf{x}$ distribution introduces additional terms into the variance of the approximate distribution of $\tilde{\boldsymbol{\beta}}$. While such terms can be estimated, they are not always simple and may depend heavily on the form of the $\mathbf{x}$ distribution.

3.3.2. General Explanatory Variables

In Section 3.3.1, $(\mathbf{x}_t, \mathbf{X}_t)$ was a random vector and approximations based on a Taylor expansion of $g(\mathbf{x}_t, \boldsymbol{\beta})$ about $g(\mathbf{W}_t, \boldsymbol{\beta})$, where $\mathbf{W}_t = E\{\mathbf{x}_t|\mathbf{X}_t\}$, were employed. In this section we relax the assumption about the distribution of $\mathbf{x}_t$ and outline an estimation procedure that relies on local quadratic approximations to the nonlinear function, where the expansions are about $\mathbf{x}_t$. As such, the procedure is appropriate for fixed or random $\mathbf{x}_t$. Because of the local nature of the approximations, the measurement error should not be overly large relative to the curvature of the function.

By expanding the function $g(\mathbf{X}_t, \boldsymbol{\beta})$ about $\mathbf{x}_t$ we can write

$$\begin{aligned} Y_t &= g(\mathbf{x}_t, \boldsymbol{\beta}) + e_t \\ &= g(\mathbf{X}_t, \boldsymbol{\beta}) + B_t + v_t, \end{aligned} \tag{3.3.11}$$

where

$$B_t = E\{-0.5(\mathbf{X}_t - \mathbf{x}_t)\mathbf{g}_{xx}(\dot{\mathbf{x}}_t, \boldsymbol{\beta})(\mathbf{X}_t - \mathbf{x}_t)'\}, \tag{3.3.12}$$

$$\begin{aligned} v_t &= Y_t - g(\mathbf{X}_t, \boldsymbol{\beta}) - B_t \\ &\doteq e_t - \mathbf{g}_x(\mathbf{x}_t, \boldsymbol{\beta})\mathbf{u}_t' - 0.5 \operatorname{tr}\{(\mathbf{u}_t'\mathbf{u}_t - \boldsymbol{\Sigma}_{uu})\mathbf{g}_{xx}(\dot{\mathbf{x}}_t, \boldsymbol{\beta})\}, \end{aligned} \tag{3.3.13}$$

$\dot{\mathbf{x}}_t$ is on the line segment joining $\mathbf{x}_t$ and $\mathbf{X}_t$, and the approximation of (3.3.13) arises from the approximation of the expectation in (3.3.12). Equation (3.3.11) is a form of the model studied in Section 2.2 with $g(\mathbf{X}_t, \boldsymbol{\beta})$ replacing $\mathbf{X}_t\boldsymbol{\beta}$ and $Y_t - B_t$ replacing Y_t.

In Exercise 2.9 of Section 2.2 it is demonstrated that the estimator of $\boldsymbol{\beta}$ for the model (3.3.1) with $g(\mathbf{x}_t, \boldsymbol{\beta}) = \mathbf{x}_t\boldsymbol{\beta}$ can be defined as the $\boldsymbol{\beta}$ that minimizes the estimator of σ_{ee}. Therefore, we choose as our initial estimator of $\boldsymbol{\beta}$, the $\boldsymbol{\beta}$ that minimizes

$$\sum_{t=1}^{n} [Y_t - g(\mathbf{X}_t, \boldsymbol{\beta}) - B_t(\mathring{\mathbf{x}}_t, \boldsymbol{\beta})]^2 - \sum_{t=1}^{n} \mathbf{g}_x(\mathring{\mathbf{x}}_t, \boldsymbol{\beta})\boldsymbol{\Sigma}_{uu}\mathbf{g}_x'(\mathring{\mathbf{x}}_t, \boldsymbol{\beta}), \tag{3.3.14}$$

where

$$B_t(\mathring{\mathbf{x}}_t, \boldsymbol{\beta}) = -0.5 \operatorname{tr}\{\boldsymbol{\Sigma}_{uu}\mathbf{g}_{xx}(\mathring{\mathbf{x}}_t, \boldsymbol{\beta})\}$$

and $\mathring{\mathbf{x}}_t$ is an estimator of $\mathbf{x}_t$. One choice for $\mathring{\mathbf{x}}_t$ is the estimated conditional mean constructed as if $\mathbf{x}_t$ were normally distributed,

$$\mathring{\mathbf{x}}_t = \bar{\mathbf{X}} + (\mathbf{X}_t - \bar{\mathbf{X}})\mathbf{m}_{XX}^{-1}(\mathbf{m}_{XX} - \boldsymbol{\Sigma}_{uu}). \tag{3.3.15}$$

Note that $g(\mathbf{X}_t, \boldsymbol{\beta})$, $B_t(\mathring{\mathbf{x}}_t, \boldsymbol{\beta})$, and $\mathbf{g}_x(\mathring{\mathbf{x}}_t, \boldsymbol{\beta})$ are all functions of $\boldsymbol{\beta}$ in (3.3.14). The quantity being minimized in (3.3.14) is an estimator of $n\sigma_{ee}$, where $\sigma_{ee} = \sigma_{qq} + \sigma_{ww}$ of model (3.3.1). Given a trial value for $\boldsymbol{\beta}$, denoted by $\tilde{\boldsymbol{\beta}}$, we can approximate expression (3.3.11) by

$$Y_t \doteq g(\mathbf{X}_t, \tilde{\boldsymbol{\beta}}) + B_t(\mathring{\mathbf{x}}_t, \tilde{\boldsymbol{\beta}}) + \mathbf{g}_\beta(\mathbf{X}_t, \tilde{\boldsymbol{\beta}})(\boldsymbol{\beta} - \tilde{\boldsymbol{\beta}}) + v_t, \tag{3.3.16}$$

where $\mathbf{g}_\beta(\mathbf{X}_t, \tilde{\boldsymbol{\beta}})$ is the row vector of derivatives of $g(\mathbf{x}, \boldsymbol{\beta})$ with respect to $\boldsymbol{\beta}$ evaluated at $(\mathbf{x}, \boldsymbol{\beta}) = (\mathbf{X}_t, \tilde{\boldsymbol{\beta}})$.

It is worthwhile to compare the approximation used in (3.3.10) with that of (3.3.16). The random model of Section 3.3.1 was used to generate the approximate expectation of Y_t given $\mathbf{X}_t$. In that case we expanded the function about $\mathbf{W}_t$, where $\mathbf{W}_t$ is the conditional expected value of $\mathbf{x}_t$ given $\mathbf{X}_t$, to obtain

$$E\{Y_t \mid \mathbf{X}_t\} \doteq g(\mathbf{W}_t, \boldsymbol{\beta}) + 0.5 \operatorname{tr}[\mathbf{g}_{xx}(\mathbf{W}_t, \boldsymbol{\beta})\mathbf{V}\{\mathbf{x}_t \mid \mathbf{X}_t\}].$$

In the resulting minimization the $\mathbf{X}_t$ and, hence, the $\mathbf{W}_t$ are treated as fixed. In the approximation (3.3.16) the expansion is about the point $\mathbf{x}_t$, which is considered fixed in the approximation. Because we can observe $\mathbf{X}_t$, an approximate expression for $g(\mathbf{x}_t, \boldsymbol{\beta})$ as a function of $\mathbf{X}_t$ is desired, but $\mathbf{X}_t$ is treated as random in the minimization. Because the expansions used in (3.3.10) and (3.3.16) are about different points, the estimated bias enters the two expressions with different signs.

Using (3.3.16) and a Gauss–Newton algorithm on (3.3.14), an improved estimator of $\boldsymbol{\beta}$, denoted by $\hat{\boldsymbol{\beta}}_{(1)}$, is defined by

$$\hat{\boldsymbol{\beta}}_{(1)} - \tilde{\boldsymbol{\beta}} = \left\{\sum_{t=1}^{n} [\mathbf{g}'_{\beta}(\mathbf{X}_t, \tilde{\boldsymbol{\beta}})\mathbf{g}_{\beta}(\mathbf{X}_t, \tilde{\boldsymbol{\beta}}) - \mathbf{g}_{\beta x}(\mathring{\mathbf{x}}_t, \tilde{\boldsymbol{\beta}})\boldsymbol{\Sigma}_{uu}\mathbf{g}_{x\beta}(\mathring{\mathbf{x}}_t, \tilde{\boldsymbol{\beta}})]\right\}^{-1}$$
$$\times \left\{\sum_{t=1}^{n} [\mathbf{g}'_{\beta}(\mathbf{X}_t, \tilde{\boldsymbol{\beta}})\tilde{v}_t + \mathbf{g}_{\beta x}(\mathring{\mathbf{x}}_t, \tilde{\boldsymbol{\beta}})\boldsymbol{\Sigma}_{uu}\mathbf{g}'_x(\mathring{\mathbf{x}}_t, \tilde{\boldsymbol{\beta}})]\right\}, \qquad (3.3.17)$$

where $\tilde{v}_t = Y_t - g(\mathbf{X}_t, \tilde{\boldsymbol{\beta}}) - B_t(\mathring{\mathbf{x}}_t, \tilde{\boldsymbol{\beta}})$ and $\mathbf{g}_{\beta x}(\mathring{\mathbf{x}}_t, \tilde{\boldsymbol{\beta}})$ is the matrix of the second derivatives with respect to $\mathbf{x}$ and $\boldsymbol{\beta}$ evaluated at $(\mathbf{x}_t, \boldsymbol{\beta}) = (\mathring{\mathbf{x}}_t, \tilde{\boldsymbol{\beta}})$. Expression (3.3.17) is similar to the expression for the estimator of $\boldsymbol{\beta}$ defined in (2.2.12). The second term within the first set of curly brackets can be viewed as an estimator of the bias in $\mathbf{g}'_{\beta}(\mathbf{X}_t, \tilde{\boldsymbol{\beta}})\mathbf{g}_{\beta}(\mathbf{X}_t, \tilde{\boldsymbol{\beta}})$ as an estimator of $\mathbf{g}'_{\beta}(\mathbf{x}_t, \tilde{\boldsymbol{\beta}})\mathbf{g}_{\beta}(\mathbf{x}_t, \tilde{\boldsymbol{\beta}})$. The second term within the second set of curly brackets on the right side of (3.3.17) can be viewed as the negative of the estimated covariance between v_t and $\mathbf{g}'_{\beta}(\mathbf{X}_t, \boldsymbol{\beta})$. The procedure associated with (3.3.14) and (3.3.17) can be iterated to obtain the $\boldsymbol{\beta}$ minimizing (3.3.14). An estimator of σ_{ee} is given by the minimum of (3.3.14) divided by $n - k$. Improved estimators of the $\mathbf{x}_t$ can be constructed as

$$\hat{\mathbf{x}}_t = \mathbf{X}_t - \tilde{v}_t\mathring{\sigma}_{vvtt}^{-1}\mathring{\boldsymbol{\Sigma}}_{vutt}, \qquad (3.3.18)$$

where

$$\mathring{\boldsymbol{\Sigma}}_{vutt} = -\mathbf{g}_x(\mathring{\mathbf{x}}_t, \tilde{\boldsymbol{\beta}})\boldsymbol{\Sigma}_{uu},$$
$$\mathring{\sigma}_{vvtt} = \tilde{\sigma}_{ee} + \mathbf{g}_x(\mathring{\mathbf{x}}_t, \tilde{\boldsymbol{\beta}})\boldsymbol{\Sigma}_{uu}\mathbf{g}'_x(\mathring{\mathbf{x}}_t, \tilde{\boldsymbol{\beta}}),$$

and $\tilde{\sigma}_{ee}$ is the estimator of σ_{ee} defined by the minimum of (3.3.14).

Having obtained estimators from the unweighted analysis, we can then construct a weighted estimator of $\boldsymbol{\beta}$ by minimizing the quantity

$$\sum_{t=1}^{n} \tilde{\sigma}_{vvtt}^{-1}\{[Y_t - g(\mathbf{X}_t, \boldsymbol{\beta}) - B_t(\hat{\mathbf{x}}_t, \boldsymbol{\beta})]^2 - \mathbf{g}_x(\hat{\mathbf{x}}_t, \boldsymbol{\beta})\boldsymbol{\Sigma}_{uu}\mathbf{g}'_x(\hat{\mathbf{x}}_t, \boldsymbol{\beta})\}, \quad (3.3.19)$$

where $\tilde{\sigma}_{vvtt}$ is the $\mathring{\sigma}_{vvtt}$ defined in (3.3.18) with $\hat{\mathbf{x}}_t$ replacing $\mathring{\mathbf{x}}_t$. The theoretical justification for the estimator obtained by minimizing (3.3.19) requires that the error variances become small in the limit because $\tilde{\sigma}_{vvtt}$ is a function of $\hat{\mathbf{x}}_t$.

Under the assumption of normal errors, a reasonable estimator of the variance of the approximate distribution of the estimator defined by (3.3.19) is

$$n^{-1}\hat{\mathbf{M}}_{g\pi g}^{-1}\hat{\mathbf{G}}\hat{\mathbf{M}}_{g\pi g}^{-1}, \tag{3.3.20}$$

where

$$\hat{\mathbf{M}}_{g\pi g} = n^{-1}\sum_{t=1}^{n}\tilde{\sigma}_{vvtt}^{-1}[\mathbf{g}'_{\beta}(\mathbf{X}_t, \hat{\boldsymbol{\beta}})\mathbf{g}_{\beta}(\mathbf{X}_t, \hat{\boldsymbol{\beta}}) - \mathbf{g}_{\beta x}(\hat{\mathbf{x}}_t, \hat{\boldsymbol{\beta}})\boldsymbol{\Sigma}_{uu}\mathbf{g}_{x\beta}(\hat{\mathbf{x}}_t, \hat{\boldsymbol{\beta}})],$$

$$\hat{\mathbf{G}} = n^{-1}\left[\sum_{t=1}^{n}\tilde{\sigma}_{vvtt}^{-1}\mathbf{g}'_{\beta}(\mathbf{X}_t, \hat{\boldsymbol{\beta}})\mathbf{g}_{\beta}(\mathbf{X}_t, \hat{\boldsymbol{\beta}}) + \sum_{t=1}^{n}\tilde{\sigma}_{vvtt}^{-2}\mathbf{g}_{\beta x}(\hat{\mathbf{x}}_t, \hat{\boldsymbol{\beta}})\boldsymbol{\Sigma}_{uu}\mathbf{g}_{x}(\hat{\mathbf{x}}_t, \hat{\boldsymbol{\beta}})\mathbf{g}'_{x}(\hat{\mathbf{x}}_t, \hat{\boldsymbol{\beta}})\boldsymbol{\Sigma}_{uu}\mathbf{g}_{x\beta}(\hat{\mathbf{x}}_t, \hat{\boldsymbol{\beta}})\right],$$

and $\hat{\boldsymbol{\beta}}$ is the final estimator of $\boldsymbol{\beta}$. For nonnormal errors, an estimator of $\mathbf{G}$ analogous to that given in (3.1.12) is

$$\hat{\mathbf{G}} = (n-k)^{-1}\sum_{t=1}^{n}\tilde{\sigma}_{vvtt}^{-2}\hat{\mathbf{d}}'_t\hat{\mathbf{d}}_t, \tag{3.3.21}$$

where

$$\hat{\mathbf{d}}'_t = \mathbf{g}'_{\beta}(\mathbf{X}_t, \hat{\boldsymbol{\beta}})\hat{v}_t + \mathbf{g}_{\beta x}(\hat{\mathbf{x}}_t, \hat{\boldsymbol{\beta}})\boldsymbol{\Sigma}_{uu}\mathbf{g}'_{x}(\hat{\mathbf{x}}_t, \hat{\boldsymbol{\beta}})$$

and $\hat{v}_t = Y_t - g(\mathbf{X}_t, \hat{\boldsymbol{\beta}}) - B_t(\hat{\mathbf{x}}_t, \hat{\boldsymbol{\beta}})$.

Example 3.3.1. We apply the procedures of this section to the data of Table 3.A.5. The data were generated using the model

$$y_t = \beta_0 + \beta_1 x_{t1} + \beta_2 x_{t1}^2, \qquad (Y_t, X_{t1}) = (y_t, x_{t1}) + (e_t, u_{t1})$$

and $(e_t, u_{t1})' \sim \mathrm{NI}[\mathbf{0}, \mathrm{diag}(\sigma_{ee}, \sigma_{uu})]$. We analyze the data under the assumption that $\sigma_{uu} = 0.09$ is known and σ_{ee} is unknown.

The estimator (3.1.6) computed with SUPER CARP is

$$[\tilde{\beta}_0, \tilde{\beta}_1, \hat{\beta}_2] = [-0.005, 0.540, 1.007],$$
$$(0.108, 0.098, 0.230)$$

where $\psi_{t1} = (0, 0.3, 0.57666X_t)$, $\mathbf{X}_t = (1, X_{t1}, X_{t1}^2 - 0.09)$, $\hat{\boldsymbol{\Sigma}}_{uutt} = \psi'_{t1}\psi_{t1}$, and $\hat{\pi}_t = 1$. The standard errors are computed from the estimated covariance matrix (3.1.12). The foundation for the ψ_{t1} vector is given in Example 3.1.5.

The computations for the quadratic model are relatively simple because the model is linear in the β parameters. Also, the second derivative, $\mathbf{g}_{xx}(\mathbf{x}_t, \boldsymbol{\beta})$, entering the bias expression, $B_t(\mathbf{x}_t, \boldsymbol{\beta})$, is a function of β_2 only. Hence,

$$g(\mathbf{X}_t, \boldsymbol{\beta}) + B_t(\dot{\mathbf{x}}_t, \boldsymbol{\beta}) = \beta_0 + \beta_1 X_{t1} + \beta_2(X_{t1}^2 - \sigma_{uu})$$

and it follows that the estimates $(\tilde{\beta}_0, \tilde{\beta}_1, \tilde{\beta}_2)$ defined by (3.1.6) minimize (3.3.14).

Estimators of the x_{t1} were constructed as

$$\hat{x}_{t1} = X_{t1} - \dot{\sigma}_{vvtt}^{-1}\dot{\sigma}_{uvtt}\tilde{v}_t,$$

where

$$(\dot{\sigma}_{vvtt}, \dot{\sigma}_{uvtt}) = [\tilde{\sigma}_{ee} + 0.09(\tilde{\beta}_1 + 2\tilde{\beta}_2\dot{x}_{t1})^2, -0.09(\tilde{\beta}_1 + 2\tilde{\beta}_2\dot{x}_{t1})],$$

$$\dot{x}_{t1} = \bar{X}_1 + 0.8489(X_{t1} - \bar{X}_1), \qquad \tilde{v}_t = Y_t - \tilde{\beta}_0 - \tilde{\beta}_1\dot{x}_{t1} - \tilde{\beta}_2(X_{t1}^2 - 0.09),$$

and $m_{XX}^{-1}(m_{XX} - 0.09) = 0.8489$. The estimator of σ_{ee} obtained by evaluating (3.3.14) at $(\tilde{\beta}_0, \tilde{\beta}_1, \tilde{\beta}_2)$ is $\tilde{\sigma}_{ee} = 0.090$. These estimates were used to construct

$$\tilde{\sigma}_{vvtt} = \tilde{\sigma}_{ee} + 0.09(\tilde{\beta}_1 + 2\tilde{\beta}_2\hat{x}_{t1})^2.$$

The values of $\hat{x}_{t1}$, $\tilde{v}_t$, and $\tilde{\sigma}_{vvtt}$ for selected observations are given in Table 3.3.1. Second round estimates of $(\beta_0, \beta_1, \beta_2)$ were obtained by minimizing

$$\sum_{t=1}^{n} \tilde{\sigma}_{vvtt}^{-1}\{[Y_t - \beta_0 - \beta_1 X_{t1} - \beta_2(X_{t1}^2 - 0.09)]^2 - [(\beta_0, \beta_1, \beta_2)\psi'_{t1}]^2\}.$$

TABLE 3.3.1. Selected observations and statistics for quadratic example

Observation	Y_t	X_{t1}	$\tilde{v}_t$	$\hat{x}_{t1}$	$\tilde{\sigma}_{vvtt}$
1	0.70	−1.44	−0.52	−1.23	0.43
2	0.05	−0.90	−0.18	−0.81	0.20
3	0.99	−0.81	0.86	−1.23	0.43
4	0.30	−1.14	−0.30	−1.00	0.29
5	1.10	−1.36	0.07	−1.39	0.55
61	−0.06	−0.05	0.06	−0.03	0.11
62	0.32	0.31	0.15	0.38	0.25
63	−0.06	−0.45	0.07	−0.47	0.10
64	−0.22	0.65	−0.90	0.25	0.19
65	−0.30	0.57	−0.84	0.18	0.16
111	0.92	0.63	0.28	0.75	0.47
112	1.86	1.11	0.11	1.15	0.83
113	1.50	0.84	0.43	1.01	0.69
114	2.04	1.07	0.40	1.21	0.89
115	1.85	0.80	0.87	1.16	0.83

This expression differs from (3.3.19) only in that a function of X_{t1} is used in the second term in place of a function of $\hat{x}_{t1}$. The second round estimates obtained by using weighted observations in SUPER CARP are

$$[\hat{\beta}_0, \hat{\beta}_1, \hat{\beta}_2] = [0.012, 0.527, 0.967],$$
$$(0.070, 0.101, 0.194)$$

where the estimated standard errors are obtained from the estimated covariance matrix (3.1.12). The estimated standard error for $\hat{\beta}_1$ is essentially the same as that for $\tilde{\beta}_1$, but the estimated standard errors for $\hat{\beta}_0$ and $\hat{\beta}_2$ are smaller than those of $\tilde{\beta}_0$ and $\tilde{\beta}_2$. The estimated variance of the approximate distribution of $\hat{\beta}_0$ is about one-half that of $\tilde{\beta}_0$.

For the approximations to be adequate, the error variances should not be too large. Limited Monte Carlo simulation indicates that the approximations hold reasonably well for data similar to that of this example.

□ □

Example 3.3.2. In this example we fit a moisture response model to some data adapted from the experiment discussed in Example 2.3.2. The yield and soil moisture data are given in the second and third columns of Table 3.3.2. We will treat the data as averages for a pair of plots. The data are not the original yields, but yields adjusted for weather conditions so that the deviations from the fitted function are smaller than actually observed.

It is assumed that yield is given by

$$Y_t = \beta_1[1 + \exp\{\beta_2 + \beta_3 x_t\}]^{-1} + e_t,$$

where $X_t = x_t + u_t$, X_t is observed soil moisture, and e_t and u_t are independent. The variance of u_t is assumed to be known and equal to 0.3313.

The first step in our estimation procedure is the fitting of the model by nonlinear least squares using a prediction of the true x_t as the explanatory variable. The estimated linear predictor of x_t given X_t is

$$\dot{x}_t = 4.612 - (6.7736)^{-1}(6.7736 - 0.3313)(X_t - 4.612),$$

where $\bar{X} = 4.612$ and $m_{XX} = 6.7736$. The nonlinear least squares estimates using $\dot{x}_t$ as the explanatory variable are given in the third column of Table 3.3.3 and the nonlinear least squares estimates constructed with X_t as the explanatory variable are given in the second column. An estimate of the bias in the function that is due to the curvature is

$$B_t(\dot{x}_t, \tilde{\beta}) = -0.5(0.3313)g_{xx}(\dot{x}_t, \tilde{\beta}),$$

where $\tilde{\beta}$ is the vector of estimates in the third column of Table 3.3.3. The fourth column of Table 3.3.3 gives the estimates obtained by nonlinear least

TABLE 3.3.2. Soil moisture and corn yield statistics

Observation	Yield	Soil Moisture	$B_t(\hat{x}_t, \hat{\beta})$	$\hat{x}_t$	$\dfrac{\partial g(\hat{x}_t, \hat{\beta})}{\partial x}$	$\tilde{\sigma}_{vvtt}^{-1/2}\hat{v}_t$
1	2.1	0.6	−0.09	0.72	1.03	0.45
2	2.0	1.3	−0.10	1.23	1.31	−0.21
3	3.1	1.8	−0.09	1.81	1.65	0.03
4	4.4	2.0	−0.08	2.24	1.88	0.63
5	6.9	2.2	−0.04	2.95	2.18	1.92
6	5.2	2.5	−0.06	2.70	2.08	0.49
7	3.6	2.9	−0.07	2.50	2.00	−0.95
8	4.8	3.0	−0.04	2.84	2.12	−0.38
9	6.2	3.4	−0.00	3.36	2.19	−0.09
10	4.7	3.6	−0.03	3.09	2.17	−1.23
11	5.3	3.7	−0.01	3.28	2.19	−1.01
12	8.4	4.0	0.06	4.17	2.04	0.41
13	11.2	4.1	0.09	4.90	1.68	1.96
14	6.2	4.1	0.03	3.68	2.17	−1.02
15	6.0	4.1	0.02	3.64	2.18	−1.14
16	8.7	4.1	0.07	4.29	1.99	0.47
17	12.4	5.8	0.08	6.15	0.97	1.22
18	9.8	5.9	0.09	5.69	1.22	−0.77
19	11.2	6.1	0.08	6.13	0.98	0.12
20	10.8	6.5	0.08	6.40	0.85	−0.46
21	10.9	7.0	0.06	6.89	0.65	−0.67
22	13.5	7.9	0.03	8.02	0.32	1.14
23	10.7	8.7	0.02	8.61	0.22	−1.36
24	13.9	9.8	0.01	9.84	0.10	1.20
25	11.6	10.2	0.01	10.18	0.08	−0.76

TABLE 3.3.3. Estimates for the moisture model

	Estimate and (Standard Error)				
Parameter	LS(X)	LS($\dot{x}$)	Bias Adjustment LS($\dot{x}$)	Bias Adjustment GLS($\dot{x}$)	Estimate (3.3.19)
β_1	12.860 (0.980)	12.859 (0.980)	12.846 (0.921)	12.741 (0.726)	12.606 (0.837)
β_2	2.138 (0.402)	2.285 (0.432)	2.426 (0.456)	2.455 (0.445)	2.363 (0.421)
β_3	−0.617 (0.140)	−0.649 (0.147)	−0.689 (0.152)	−0.708 (0.144)	−0.694 (0.152)
RMS	2.36	2.36	2.36	0.99	1.03
σ_{ee}	—	—	1.40	—	1.41

squares using $\dot{x}_t$ as the explanatory variable and $Y_t - B_t(\dot{x}_t, \tilde{\boldsymbol{\beta}})$ as the dependent variable.

The estimate of σ_{ee} calculated from the deviations associated with the fit of column four is

$$\tilde{\sigma}_{ee} = (n-3)^{-1} \sum_{t=1}^{n} \tilde{r}_t^2 - n^{-1} \sum_{t=1}^{n} 0.3313 g_x^2(\dot{x}_t, \tilde{\boldsymbol{\beta}}) = 1.40.$$

This estimate of σ_{ee} was used to construct the estimates

$$\tilde{\sigma}_{vvtt} = \tilde{\sigma}_{ee} + 0.3313 g_x^2(\dot{x}_t, \tilde{\boldsymbol{\beta}})$$

of the variances of the v_t. The weighted nonlinear least squares estimate of $\boldsymbol{\beta}$ constructed with $\tilde{\sigma}_{vvtt}$ as weight, $Y_t - B_t(\dot{x}_t, \tilde{\boldsymbol{\beta}})$ as dependent variable, and $\dot{x}_t$ as explanatory variable is given in the fifth column of Table 3.3.3.

For columns two to five of Table 3.3.3 the estimated standard errors in the parentheses are the standard errors output by a conventional nonlinear least squares program. All are biased estimates because they are constructed under the assumption that the explanatory variable is measured without error. The estimated conditional expectation $\dot{x}_t$ contains an estimation error because some parameters of the distribution are estimated, and this estimation error is not reflected in the estimated standard errors. The estimated standard errors of columns two to four are also biased because the v_t do not have constant variance. The smaller estimated standard error for the estimate of β_1 of the fifth column, relative to that of the fourth column, is due to the fact that the variance of v_t is smaller for large x_t and the large x_t are influential observations for the estimation of the limiting yield.

The last column of Table 3.3.3 contains the estimates obtained by approximately minimizing (3.3.19). Only two steps of the interative procedure were calculated starting with the estimates of column five. The estimator $\tilde{\sigma}_{vvtt}$ was used to weight the observations. Also $B_t(\hat{x}_t, \tilde{\boldsymbol{\beta}}_{(i)})$, where $\tilde{\boldsymbol{\beta}}_{(i)}$ is the estimate from the ith step, was used in the (i + 1)st step of the iteration, rather than computing the bias estimate as a function of the current estimate of $\boldsymbol{\beta}$. The estimated bias is given in Table 3.3.2. The estimates of $\boldsymbol{\beta}$ in the fifth and sixth columns of Table 3.3.3 are similar for this example. The estimated standard errors for the final column were computed with the $\hat{\mathbf{G}}$ of (3.3.21). The standardized residuals $\tilde{\sigma}_{vvtt}^{-1/2}\hat{v}_t$ are given in the last column of Table 3.3.2. No obvious anomalies are present in the plot of these residuals against $\hat{x}_t$. □□

REFERENCES

Section 3.3.1. Armstrong (1985), Stefanski (1985), Stefanski and Carroll (1985b).

EXERCISES

20. (Section 3.3.1) Let $Y_t = \mathbf{x}_t\boldsymbol{\beta} + e_t$, where e_t is independent of $(\mathbf{x}_t, \mathbf{X}_t)$ and $\mathbf{X}_t$ is observed. Let $\mathbf{W}_t = \mathbf{W}(\mathbf{X}_t) = E\{\mathbf{x}_t | \mathbf{X}_t\}$. What is the expected value of Y_t conditional on $\mathbf{X}_t$? What is the conditional variance of Y_t given $\mathbf{X}_t$, if the e_t are $\text{Ind}(0, \sigma_{ee})$ random variables?
21. (Section 3.3.1) Assume that for the model and data for Example 1.2.1 it is known that

$$E\{x | X\} = 70.64 + 0.813(X - 70.64).$$

Using this information, estimate (β_0, β_1) and estimate the covariance matrix of the estimator. If only $\sigma_{XX}^{-1}\sigma_{xx} = 0.813$ were known, how would your estimates and estimated covariance matrix change? Compare your estimated covariance matrices to that of Example 1.2.1.
22. (Section 3.3.2) Using the last two columns of Table 3.3.2 of Example 3.3.2, estimate σ_{ee}. Construct a test of the hypothesis that the variance of e_t is constant against the alternative that the variance is a linear function of x_t.

3.4. MEASUREMENT ERROR CORRELATED WITH TRUE VALUE

3.4.1. Introduction and Estimators

In certain situations the expected value of the measurement error may be a function of the true values of the variable. For example, consider the binomial random variable x that takes the values zero and one. The two possible values for the measurement error are one and zero when the true value is zero, and negative one and zero when the true value is one. Therefore, the expected value of nontrivial measurement error is a function of the true value. Typically, in situations where the expected value of the measurement error is a function of the true value, the variance of the measurement error is also related to the true value of the variable.

If the functional relationship between the mean of the error variable and the mean of the true variable is known, the model can be transformed into the model of Section 3.1.1. Assume that the vector of true values $\mathbf{z}_t$ satisfy the linear model

$$\mathbf{z}_t\boldsymbol{\alpha} = q_t, \qquad \mathbf{A}_t = \mathbf{z}_t + \boldsymbol{\xi}_t, \tag{3.4.1}$$

where $\boldsymbol{\alpha}' = (1, -\boldsymbol{\beta}')$, q_t is the error in the equation, $\mathbf{A}_t$ is the observed vector, and $\boldsymbol{\xi}_t$ is the measurement error. Let

$$E\{\boldsymbol{\xi}_t | \mathbf{z}_t\} = \mathbf{z}_t\mathbf{C}, \tag{3.4.2}$$

where $\mathbf{C}$ is known. It follows from (3.4.2) that $E\{\mathbf{A}_t | \mathbf{z}_t\} = \mathbf{z}_t(\mathbf{I} + \mathbf{C})$. If we let

$$\mathbf{Z}_t = \mathbf{A}_t(\mathbf{I} + \mathbf{C})^{-1} \quad \text{and} \quad \boldsymbol{\varepsilon}_t = \mathbf{A}_t(\mathbf{I} + \mathbf{C})^{-1} - \mathbf{z}_t \tag{3.4.3}$$

we have $\mathbf{Z}_t = \mathbf{z}_t + \boldsymbol{\varepsilon}_t$, where $E\{\boldsymbol{\varepsilon}_t | \mathbf{z}_t\} = \mathbf{0}$. Therefore, the vectors $\mathbf{Z}_t$, $\mathbf{z}_t$, and $\boldsymbol{\varepsilon}_t$ have the properties that have been associated with these symbols in earlier sections. If estimators of the error covariance matrices, $\Sigma_{\varepsilon\varepsilon tt}$, $t = 1, 2, \ldots, n$, are available, the theory of Section 3.1.1 is applicable to the transformed variables of (3.4.3). An important example of measurement error whose distribution is related to the true values is the multinomial observed subject to error.

3.4.2. Measurement Error Models for Multinomial Random Variables

We consider measurement error models for populations where the observation process consists of assigning each member of a sample of n elements to one, and only one, of r categories. We adopt the convention of writing the observation as an r vector. If the tth sample element is placed in the first category of the A classification, we write

$$\mathbf{A}_t = (1, 0, 0, \ldots, 0); \tag{3.4.4}$$

if the tth element is placed in the second category, we write

$$\mathbf{A}_t = (0, 1, 0, \ldots, 0), \quad \text{and so on.}$$

The jth element of the $\mathbf{A}_t$ vector, denoted by A_{tj}, is a binomial random variable. There are alternative ways in which the measurement error process can be formalized for such populations. The *latent structure model* assumes there exists a population of responses for each element of the population. The mean of the responses for the A classification for the tth individual is denoted by π_{At}, where

$$\pi_{At} = E\{\mathbf{A}_i \mid i = t\} \tag{3.4.5}$$

and the symbolism means that the average is for the population of possible responses for the tth individual. The jth element of the vector π_{At}, denoted by π_{Atj}, is the probability that the tth individual is placed in category j. The observation for the tth individual is

$$\mathbf{A}_t = \pi_{At} + \varepsilon_{At}, \tag{3.4.6}$$

where ε_{At} is the measurement error. By construction, the mean of the error vector for the tth individual is the zero vector. The covariance matrix of the measurement error for the tth individual is

$$E\{\varepsilon'_{At}\varepsilon_{At}\} = \Sigma_{AAtt}, \tag{3.4.7}$$

where $\Sigma_{AAtt} = \operatorname{diag} \pi_{At} - \pi'_{At}\pi_{At}$ is the covariance matrix for the multinomial distribution with probability vector π_{At} and $\operatorname{diag} \pi_{At}$ is the diagonal matrix with the elements of π_{At} on the diagonal. The latent structure model was developed by Lazarsfeld (1950, 1954) and is discussed by Andersen (1982). The latent structure model is sometimes called the average-in-repeated-trials model. When the individuals fall into a fixed number of distinct classes, with response probability $\pi_{A(j)}$ for individuals in class j, the model is called the *latent class model*. The latent class model has been used when there are responses to a number of different items. In the latent class model the number of distinct classes is often much smaller than the number of possible item response combinations.

We call the second model for classification error the *right–wrong* model. The right–wrong model assumes that every element truly belongs to one of the r categories. The response error for the population is characterized by a set of response probabilities κ_{Aij}, where κ_{Aij} is the probability that an element whose true A category is j responds as category i. Because the categories are mutually exclusive and exhaustive, the sum of the κ_{Aij} is one for each column. That is, every element in true category j is placed in one of the available categories.

There are two submodels of the right–wrong model. In the first submodel every element in true category j has the same vector of response probabilities. This submodel reduces formally to the latent class model in which there are exactly r types of individuals. Only the set of parameters used to describe the two models differ.

In the second submodel of the right–wrong model, different elements are permitted to have different response probabilities. For example, some elements might always be reported correctly under this model. The second submodel of the right–wrong model can also be parameterized as a latent structure model.

Under both submodels of the right–wrong model, the observed distribution of $\mathbf{A}_t$ obtained by making a single determination on each element of a random sample of elements is multinomial. Likewise, the distribution, over the population of elements, for the multinomial of dimension r^2 obtained by making two determinations on each element of the population is the same for the two submodels. More than two determinations per element are required to discriminate between the two submodels.

The mean vector for the observed proportions is

$$\boldsymbol{\mu}'_A = \boldsymbol{\kappa}_A \boldsymbol{\pi}'_A, \tag{3.4.8}$$

where $\boldsymbol{\pi}_A = (\pi_{A1}, \pi_{A2}, \ldots, \pi_{Ar})$ is the vector of proportions for the true classification and κ_{Aij} is the ijth element of the matrix $\boldsymbol{\kappa}_A$. The vector $\boldsymbol{\mu}_A$ is the mean over the population of elements of the observed vector $\mathbf{A}_t$. If the population mean of the observation $\mathbf{A}_t$ is equal to the vector of true proportions,

$$\boldsymbol{\mu}'_A = \boldsymbol{\pi}'_A = \boldsymbol{\kappa}_A \boldsymbol{\pi}'_A, \tag{3.4.9}$$

then the measurement error is said to be unbiased. Unbiased measurement error for the multinomial is analogous to zero mean measurement error for continuous random variables. Hence, it is an important model for applications.

For the binomial population, the matrix of response probabilities for the simple right–wrong model contains four elements. The elements of each column must sum to one and this imposes two restrictions on possible entries for the matrix. If the measurement error is unbiased, a third restriction is imposed on the elements of the matrix. It then follows that the elements of the matrix can be expressed as functions of the population proportions and one additional parameter. Under the unbiased measurement error model we can write

$$\kappa_{Aij} = \begin{cases} 1 - \alpha + \alpha\pi_{Ai}, & i = j \\ \alpha\pi_{Ai}, & j \neq i, \end{cases} \tag{3.4.10}$$

where π_{Ai} is the fraction of the population that is truly in class i and α is the parameter of the response error distribution. The two-by-two matrix of response probabilities is

$$\boldsymbol{\kappa}_A = \begin{bmatrix} 1 - \alpha + \alpha\pi_{A1} & \alpha\pi_{A1} \\ \alpha\pi_{A2} & 1 - \alpha + \alpha\pi_{A2} \end{bmatrix}. \tag{3.4.11}$$

Using (3.4.10), the expectation of the observed proportion in class one is

$$\mu_{A1} = (1 - \alpha + \alpha\pi_{A1})\pi_{A1} + \alpha\pi_{A1}\pi_{A2} = \pi_{A1},$$

which demonstrates that the response error is unbiased under the specified model. One can view the postulated structure in the following way. There is a two-level response mechanism. At the first level, the probability is $1 - \alpha$ that the correct response is obtained and the probability is α that one proceeds to the second level. At the second level the response is given with probabilities proportional to the population proportions.

These error models can be applied to the estimation of the entries in a two-way table. Assume that individuals are classified with respect to characteristic A and with respect to characteristic B, where characteristic A has r classes and characteristic B has p classes. Assume that response error in the two classifications is independent. Using the right–wrong model it is desired to estimate the true proportions in the cells of the two-way table for the AB classification. Under the model, the expected fraction in the ijth cell of the observed AB table is

$$\mu_{ABij} = \sum_{s=1}^{p} \sum_{\ell=1}^{r} \kappa_{Ai\ell}\kappa_{Bjs}\pi_{AB\ell s}, \tag{3.4.12}$$

where $\kappa_{Ai\ell}$ is the probability that an element in true category ℓ of characteristic A is observed in category i, κ_{Bjs} is the probability that an element in true category s of characteristic B is observed in category j, and $\pi_{AB\ell s}$ is the fraction of the population that is in cell ℓs of the AB table. The set of equations (3.4.12) can be written in matrix format as

$$\boldsymbol{\mu}_{AB} = \boldsymbol{\kappa}_A \boldsymbol{\pi}_{AB} \boldsymbol{\kappa}'_B, \tag{3.4.13}$$

where $\boldsymbol{\mu}_{AB}$ is the $r \times p$ matrix of expected values of the observed proportions and $\boldsymbol{\pi}_{AB}$ is the $r \times p$ matrix of population proportions for the true values. One can also write Equation (3.4.13) as

$$\text{vec}\ \boldsymbol{\mu}_{AB} = (\boldsymbol{\kappa}_B \otimes \boldsymbol{\kappa}_A)\ \text{vec}\ \boldsymbol{\pi}_{AB}, \tag{3.4.14}$$

where vec $\boldsymbol{\mu}_{AB}$ is the column vector obtained by listing the columns of $\boldsymbol{\mu}_{AB}$ one below the other, and $\boldsymbol{\kappa}_B \otimes \boldsymbol{\kappa}_A$ is the Kronecker product of $\boldsymbol{\kappa}_B$ and $\boldsymbol{\kappa}_A$. The vec notation and Kronecker products are discussed in Appendix 4.A. One can solve (3.4.14) for $\boldsymbol{\pi}_{AB}$ to obtain

$$\text{vec}\ \boldsymbol{\pi}_{AB} = (\boldsymbol{\kappa}_B^{-1} \otimes \boldsymbol{\kappa}_A^{-1})\ \text{vec}\ \boldsymbol{\mu}_{AB} \tag{3.4.15}$$

or

$$\boldsymbol{\pi}_{AB} = \boldsymbol{\kappa}_A^{-1} \boldsymbol{\mu}_{AB} \boldsymbol{\kappa}_B^{-1\prime}.$$

Let $\hat{\boldsymbol{\mu}}_{AB}$ be the observed two-way table. Then, if $\boldsymbol{\kappa}_B$ and $\boldsymbol{\kappa}_A$ are known, an estimator of $\boldsymbol{\pi}_{AB}$ is

$$\hat{\boldsymbol{\pi}}_{AB} = \boldsymbol{\kappa}_A^{-1} \hat{\boldsymbol{\mu}}_{AB} \boldsymbol{\kappa}_B^{-1\prime}. \tag{3.4.16}$$

If the original observations were obtained by multinomial sampling,

$$\hat{V}\{\text{vec}\ \hat{\boldsymbol{\mu}}_{AB}\} = n^{-1}[\text{diag}\{\text{vec}\ \hat{\boldsymbol{\mu}}_{AB}\} - (\text{vec}\ \hat{\boldsymbol{\mu}}_{AB})(\text{vec}\ \hat{\boldsymbol{\mu}}_{AB})'],$$

where n is the sample size. It follows that an estimator of the covariance matrix of vec $\hat{\boldsymbol{\pi}}_{AB}$ is

$$\hat{V}\{\text{vec}\ \hat{\boldsymbol{\pi}}_{AB}\} = (\boldsymbol{\kappa}_B^{-1} \otimes \boldsymbol{\kappa}_A^{-1})\hat{V}\{\text{vec}\ \hat{\boldsymbol{\mu}}_{AB}\}(\boldsymbol{\kappa}_B^{-1} \otimes \boldsymbol{\kappa}_A^{-1})', \tag{3.4.17}$$

where $\boldsymbol{\kappa}_A$ and $\boldsymbol{\kappa}_B$ are assumed known.

Example 3.4.1. In this example we study a model for measurement error, also called response error, in the reporting of employment status. Bershad (1967) identified the effect of response error on estimates of month-to-month change in employment status. Our model is the response error model for the change in employment status developed by Battese and Fuller (1973) and by Fuller and Chua (1983). We use constructed data that are generally consistent with the data discussed by Bershad. We assume that 40,000 individuals are interviewed in each of two months to obtain the data given in Table 3.4.1. It is assumed that the response errors in the two months are independent and unbiased. It is assumed that the α of model (3.4.10) is $\alpha = 0.166$. This value of α is consistent with the interview–reinterview studies

TABLE 3.4.1. Constructed data for two interviews on employment

First Month (A)	Second Month (B)		
	Unemployed	Employed	Total
Unemployed	0.0325	0.0275	0.0600
Employed	0.0325	0.9075	0.9400
Total	0.0650	0.9350	1.0000

of the U.S. Bureau of the Census cited by Bershad. With these assumptions and the data of Table 3.4.1, it is desired to estimate the fraction of persons that were unemployed in both month one and month two.

Under our model, the response probabilities for month A are

$$(\kappa_{A11}, \kappa_{A12}) = (1 - \alpha + \alpha\pi_{A1.}, \alpha\pi_{A1.}),$$
$$(\kappa_{A21}, \kappa_{A22}) = (\alpha\pi_{A2.}, 1 - \alpha + \alpha\pi_{A2.}),$$

where, for example, κ_{A21} is the probability that a respondent in category one is reported in category two in month A and $\pi_{A1.}$ is the fraction of the population in category one in month A. An analogous expression holds for month B.

Let $\hat{\mu}_{ABij}$ denote the observed fraction in category i for the first month (month A) and in category j for the second month (month B). Let π_{ABij} denote the corresponding population parameter. The $\hat{\mu}_{ABij}$ are given in Table 3.4.1. Under the unbiased response model with $\alpha = 0.166$, the estimated $\boldsymbol{\kappa}$ matrix for the first month is obtained from (3.4.11) by setting $\pi_{A1} = \hat{\pi}_{A1.}$ and $\pi_{A2} = \hat{\pi}_{A2.}$. Thus,

$$\hat{\boldsymbol{\kappa}}_A = \begin{bmatrix} 0.844 & 0.010 \\ 0.156 & 0.990 \end{bmatrix}$$

and the corresponding estimated $\boldsymbol{\kappa}$ matrix for the second month is

$$\hat{\boldsymbol{\kappa}}_B = \begin{bmatrix} 0.845 & 0.011 \\ 0.155 & 0.989 \end{bmatrix}.$$

Therefore, the estimated vector of π values is

$$\begin{aligned} \text{vec}\, \hat{\boldsymbol{\pi}}_{AB} &= (\hat{\boldsymbol{\kappa}}_B^{-1} \otimes \hat{\boldsymbol{\kappa}}_A^{-1})\, \text{vec}\, \hat{\mu}_{AB} \\ &= (0.0450, 0.0200, 0.0150, 0.9200)'. \end{aligned}$$

In this example measurement error is very important. The estimated fraction of individuals shifting from unemployed to employed is only slightly more than one-half of the observation subject to measurement error.

Because the sample proportions are consistent estimators of their expectations, the estimator of $\boldsymbol{\pi}_{AB}$ is consistent. The estimated covariance matrix of (3.4.17) is not applicable in this example because we have not assumed the entire matrices $\boldsymbol{\kappa}_A$ and $\boldsymbol{\kappa}_B$ to be known. We have only assumed the parameter α to be known. To construct an estimator of the covariance matrix of the limiting distribution of the estimator, we express the estimator as a function of the sample proportions. Because the sum of the probabilities is one, the two-by-two tables are determined by three parameters. If we choose $\mathbf{r}' = [\hat{\mu}_{A1.}, \hat{\mu}_{B1.}, \hat{\mu}_{AB11}]$ as our vector of observations and $\boldsymbol{\theta} = [\pi_{A1.}, \pi_{B1.}, \pi_{AB11}]$ as our vector of parameters, we can obtain explicit expressions for the π estimates in terms of the μ estimates. From (3.4.11) and (3.4.13) we have

$$\hat{\pi}_{AB11} = (1 - \alpha)^{-2}\{\hat{\mu}_{AB11} - \hat{\mu}_{A1.}\hat{\mu}_{B1.}[1 - (1 - \alpha)^2]\}, \qquad (3.4.18)$$

$\hat{\pi}_{A1.} = \hat{\mu}_{A1.}$, and $\hat{\pi}_{B1.} = \hat{\mu}_{B1.}$, where $\hat{\mu}_{A1.}$ is the observed unemployment rate in month A.

If we assume simple random sampling, the estimated covariance matrix of $\mathbf{r}$ of Table 3.4.1 is

$$\hat{\mathbf{V}}\{\mathbf{r}\} = (39{,}999)^{-1}\begin{bmatrix} 0.056400 & 0.028600 & 0.030550 \\ 0.028600 & 0.060775 & 0.030388 \\ 0.030550 & 0.030388 & 0.031444 \end{bmatrix}.$$

Let the vector of π estimates be denoted by $\hat{\boldsymbol{\theta}}$. Using Corollary 1.A.1 and expression (3.4.18), we obtain the estimated covariance matrix of $\hat{\boldsymbol{\theta}}$,

$$\hat{\mathbf{V}}\{\hat{\boldsymbol{\theta}}\} = 10^{-6}\begin{bmatrix} 1.410 & 0.715 & 1.039 \\ 0.715 & 1.519 & 1.032 \\ 1.039 & 1.032 & 1.508 \end{bmatrix}.$$

A critical assumption in the model of this example is the assumption that the probability of a correct response depends only on the current status of the individual. For example, it is assumed that the probability an employed person reports correctly is the same whether or not the person was employed last month. The effect of previous employment status could be investigated by conducting a reinterview study in which the previously reported unemployment status is used to classify respondents. □□

Example 3.4.1 illustrates the use of response probabilities to construct improved estimators of two-way tables. Dummy variables based on multinomial data are also used as explanatory variables in regression equations. Let

$$Y_t = \mathbf{x}_t\boldsymbol{\beta} + e_t, \qquad (3.4.19)$$

where $\mathbf{x}_t$ is an r-dimensional vector with a one in the jth position and zeros elsewhere when element t is in the jth category. Assume that e_t is independent of $\mathbf{x}_t$ and independent of the error made in determining $\mathbf{x}_t$. Let $\mathbf{A}_t$, as defined in (3.4.4), denote the observed classification.

Under the right–wrong model, the response error for the multinomial variable $\mathbf{A}_t$, as an estimator of $\mathbf{x}_t$, is correlated with the true value and has a variance that depends on the true value. If the matrix $\boldsymbol{\kappa}_A$ of response probabilities is known, methods introduced in Section 3.4.1 can be used to transform the observations to make the model conform to that of Section 3.1.1. Let $\boldsymbol{\kappa}_A$ denote the matrix of response probabilities and assume that a column of the matrix $\boldsymbol{\kappa}_A$ holds for every member of the population. Then, if we let

$$\mathbf{X}_t' = \boldsymbol{\kappa}_A^{-1}\mathbf{A}_t', \tag{3.4.20}$$

we can write $\mathbf{X}_t = \mathbf{x}_t + \mathbf{u}_t$, where $E\{\mathbf{u}_t\} = \mathbf{0}$ and $E\{\mathbf{X}_i \mid i = t\} = \mathbf{x}_t$ for all t. Now

$$\mathbf{V}\{\mathbf{A}_t' \mid \mathbf{x}_t = \ell_j\} = \boldsymbol{\Omega}_{jj} = \operatorname{diag}(\boldsymbol{\kappa}_{A.j}) - \boldsymbol{\kappa}_{A.j}'\boldsymbol{\kappa}_{A.j}, \tag{3.4.21}$$

where $\boldsymbol{\kappa}_{A.j}$ is the jth column of $\boldsymbol{\kappa}_A$, $\operatorname{diag}(\boldsymbol{\kappa}_{A.j})$ is the diagonal matrix with the elements of the vector $\boldsymbol{\kappa}_{A.j}$ on the diagonal, and ℓ_j is an r-dimensional vector with a one in the jth position and zeros elsewhere. It follows that

$$\boldsymbol{\Sigma}_{uutt} = \mathbf{V}\{\mathbf{u}_t' \mid \mathbf{x}_t = \ell_j\} = \boldsymbol{\kappa}_A^{-1}\boldsymbol{\Omega}_{jj}\boldsymbol{\kappa}_A^{-1\prime} \tag{3.4.22}$$

and the unconditional variance of $\mathbf{u}_t'$ is

$$\boldsymbol{\Sigma}_{uu} = \sum_{j=1}^{r} \mu_{xj}\boldsymbol{\kappa}_A^{-1}\boldsymbol{\Omega}_{jj}\boldsymbol{\kappa}_A^{-1\prime}, \tag{3.4.23}$$

where $\boldsymbol{\mu}_x = (\mu_{x1}, \mu_{x2}, \ldots, \mu_{xr}) = E\{\mathbf{x}_t\}$. We can also write

$$\boldsymbol{\Sigma}_{XX} = \boldsymbol{\Sigma}_{xx} + \boldsymbol{\Sigma}_{uu},$$

where

$$\begin{aligned}
\boldsymbol{\Sigma}_{uu} &= \boldsymbol{\kappa}_A^{-1}\boldsymbol{\Sigma}_{AA}\boldsymbol{\kappa}_A^{-1\prime} - \boldsymbol{\Sigma}_{xx}, \\
\boldsymbol{\Sigma}_{AA} &= \operatorname{diag}(\mu_{A1}, \mu_{A2}, \ldots, \mu_{Ar}) - \boldsymbol{\mu}_A'\boldsymbol{\mu}_A, \\
\boldsymbol{\Sigma}_{xx} &= \operatorname{diag}(\mu_{x1}, \mu_{x2}, \ldots, \mu_{xr}) - \boldsymbol{\mu}_x'\boldsymbol{\mu}_x.
\end{aligned} \tag{3.4.24}$$

Assume that we estimate $\boldsymbol{\Sigma}_{uutt}$ with

$$\hat{\boldsymbol{\Sigma}}_{uutt} = [P\{\mathbf{A}_t = \ell_j\}]^{-1}P\{\mathbf{x}_t = \ell_j\}\boldsymbol{\kappa}_A^{-1}\boldsymbol{\Omega}_{jj}\boldsymbol{\kappa}_A^{-1\prime} \tag{3.4.25}$$

when $\mathbf{A}_t = \ell_j$. Then an estimator of $\boldsymbol{\beta}$ is

$$\hat{\boldsymbol{\beta}} = \left[\sum_{t=1}^{n} (\mathbf{X}_t'\mathbf{X}_t - \hat{\boldsymbol{\Sigma}}_{uutt})\right]^{-1} \sum_{t=1}^{n} \mathbf{X}_t'Y_t, \tag{3.4.26}$$

where $\mathbf{X}_t$ is defined in (3.4.20) and $\hat{\boldsymbol{\Sigma}}_{uutt}$ is defined in (3.4.25).

If $\boldsymbol{\kappa}_A$ and the population probabilities are known, the estimator (3.4.26) satisfies the conditions of Theorem 3.1.1 and the variance of the approximate distribution of $\hat{\boldsymbol{\beta}}$ can be estimated using that theorem. Because the $\mathbf{A}_t$ determine r categories, it is possible to estimate the variance of $Y_t - \mathbf{X}_t\boldsymbol{\beta}$ for each category and to construct a generalized least squares estimator.

In practice we may have an external estimate of $\boldsymbol{\kappa}_A$ or of parameters such as the α of (3.4.11) but be unwillling to assume that the vector of population probabilities is known. Given $\boldsymbol{\kappa}_A$, an estimator of the vector of probabilities $\boldsymbol{\mu}_x$ is given by

$$\hat{\boldsymbol{\mu}}_x' = \boldsymbol{\kappa}_A^{-1}\bar{\mathbf{A}}',$$

where $\bar{\mathbf{A}} = n^{-1}\sum_{t=1}^{n}\mathbf{A}_t$, the jth element of $\boldsymbol{\mu}_x$ is $P\{\mathbf{x}_t = \ell_j\}$, and the jth element of $\bar{\mathbf{A}}$ is the estimator of $P\{\mathbf{A}_t = \ell_j\}$. If the $\boldsymbol{\mu}_x$ must be estimated, the $\hat{\boldsymbol{\Sigma}}_{uutt}$ do not satisfy the assumptions of Theorem 3.1.1 because the $\hat{\boldsymbol{\Sigma}}_{uutt}$ based on estimated $\boldsymbol{\mu}_x$ are not independent. In practice one may be willing to ignore the error in $\hat{\boldsymbol{\Sigma}}_{uutt}$ due to estimating $\hat{\boldsymbol{\mu}}_x$ and to use (3.1.12) as an estimator of the variance of the estimator of $\hat{\boldsymbol{\beta}}$.

The operation of transforming the observed vector to obtain errors with zero mean furnishes another verification for the method of constructing estimates of the $\boldsymbol{\pi}_{AB}$ table given by Equation (3.4.16). Assume that elements are classified according to two characteristics, A and B. Assume that there are r classes for A and p classes for B. Let the response error for the two classifications be independent. If element t belongs to category i of A and to category j of B, then

$$\mathbf{x}_t = \ell_{ri} \quad \text{and} \quad \mathbf{y}_t = \ell_{pj},$$

where ℓ_{ri} is an r-dimensional vector with a one in the ith position and zeros elsewhere and ℓ_{pj} is a p-dimensional vector with a one in the jth position and zeros elsewhere. The observed AB table can be written

$$\hat{\boldsymbol{\mu}}_{AB} = n^{-1}\sum_{t=1}^{n}\mathbf{A}_t'\mathbf{B}_t \qquad (3.4.27)$$

and the population table of true proportions can be written

$$\boldsymbol{\pi}_{AB} = E\{\mathbf{x}_t'\mathbf{y}_t\}. \qquad (3.4.28)$$

The entries in the μ_{AB} table are probabilities. If we divide the entries in the table by the marginal probabilities, we create a table of conditional probabilities. Thus,

$$P_{B|A.ij} = P\{\mathbf{y}_t = \ell_{pj} \mid \mathbf{x}_t = \ell_{ri}\} = \mu_{ABij}\mu_{Ai.}^{-1}. \qquad (3.4.29)$$

The entries in the table of conditional probabilities for B given A can be thought of as a table of population regression coefficients obtained by regressing $\mathbf{y}$ on $\mathbf{x}$. That is, the $\mathbf{P}_{B|A}$ table is

$$\mathbf{P}_{B|A} = [E\{\mathbf{x}_t'\mathbf{x}_t\}]^{-1}E\{\mathbf{x}_t'\mathbf{y}_t\}. \qquad (3.4.30)$$

Now the vector

$$(\mathbf{Y}_t, \mathbf{X}_t) = (\mathbf{B}_t\kappa_B^{-1\prime}, \mathbf{A}_t\kappa_A^{-1\prime}) \tag{3.4.31}$$

satisfies $(\mathbf{Y}_t, \mathbf{X}_t) = (\mathbf{y}_t, \mathbf{x}_t) + (\mathbf{e}_t, \mathbf{u}_t)$, where the expected value of $(\mathbf{e}_t, \mathbf{u}_t)$ is the zero vector. Under the assumption that the response errors in **A** and **B** are independent and that the matrices κ_A and κ_B are known, an estimator of the $P_{B|A}$ table can be constructed with estimator (3.4.26), where $(\mathbf{Y}_t, \mathbf{X}_t)$ is defined in (3.4.31).

REFERENCES

Andersen (1982), Battese and Fuller (1973), Bershad (1967), Chua (1983), Fuller (1984), Fuller and Chua (1984), Goodman (1974), Haberman (1977), Hansen, Hurwitz, and Bershad (1961), Lazarsfeld and Henry (1968).

EXERCISES

23. (Section 3.4) Let $\mathbf{B}_t$ be an observed r-dimensional vector with a one in the ith position and zeros elsewhere when the tth element of a population is placed in the ith category. Let κ_B denote the $r \times r$ matrix with the ijth element giving the probability that an element truly in the jth category will be placed in the ith category. Then

$$E\{\mathbf{B}_t' \mid \mathbf{b}_t'\} = \kappa_B \mathbf{b}_t',$$

where $\mathbf{b}_t$ is the r-dimensional vector with a one in the ith position and zeros elsewhere when the tth element truly belongs in the ith category.

(a) Assume $\mathbf{P} = \kappa_B \mathbf{P}$, where $\mathbf{P}' = (P_1, P_2, \ldots, P_r)$. Show that

$$E\{\mathbf{b}_t' \mid \mathbf{B}_t'\} = \mathbf{D}\kappa_B'\mathbf{D}^{-1}\mathbf{B}_t',$$

where $\mathbf{D} = \operatorname{diag}(P_1, P_2, \ldots, P_r)$ and P_i is the population proportion in category i, and it is assumed that $\mathbf{B}_t$ is the value for a randomly chosen individual. Give the general expression for the conditional probability.

(b) Assume that the elements of κ_B are given by

$$\kappa_{Bij} = \begin{cases} 1 - \alpha + \alpha P_i, & i = j \\ \alpha P_i, & j \neq i. \end{cases}$$

Show that $E\{\mathbf{b}_t' \mid \mathbf{B}_t'\} = \kappa_B \mathbf{B}_t'$.

(c) For the model of part (b) show that

$$E\{(\kappa_B^{-1}\mathbf{B}_t' - \mathbf{b}_t')(\mathbf{B}_t\kappa_B^{-1\prime} - \mathbf{b}_t) \mid \mathbf{B}_t'\} = (\kappa_B^{-1} - \kappa_B)\mathbf{B}_t'\mathbf{B}_t(\kappa_B^{-1} - \kappa_B)' + \operatorname{diag}(\kappa_B\mathbf{B}_t') - \kappa_B\mathbf{B}_t'\mathbf{B}_t\kappa_B',$$

where $\operatorname{diag}(\kappa_B\mathbf{B}_t')$ is a diagonal matrix with $\kappa_B\mathbf{B}_t'$ as the diagonal.

24. (Section 3.4) Assume that it is possible to make repeated independent determinations on the elements of a population. Assume that there are two competing submodels of the right–wrong model of Section 3.4.2. In one submodel the probability of an incorrect response is the same

for all elements of the population. In the second submodel it is assumed that there are different types of elements with different probabilities of incorrect response. Is it possible to discriminate between the two submodels with two independent determinations per element on a sample of elements? Is it possible to discriminate with three independent determinations per element?

25. (Section 3.4) Let $\mathbf{A}_t$ be a 2 vector representing the observed classification of a binomial random variable. Let $\mathbf{x}_t$ be the 2 vector representing the true classification. Let

$$E\{\mathbf{A}_t' \mid \mathbf{x}_t'\} = \boldsymbol{\kappa}_A \mathbf{x}_t',$$

where $\boldsymbol{\kappa}_A$ is the matrix of response probabilities. Assume that the response error is unbiased. Let $\mathbf{d}_t = \mathbf{A}_t - \mathbf{x}_t$. Show that

$$C\{d_{ti}, x_{ti}\} = -\tfrac{1}{2}V\{d_{ti}\}, \qquad i = 1, 2,$$
$$C\{d_{t1}, x_{t2}\} = \tfrac{1}{2}V\{d_{t1}\} = -\tfrac{1}{2}C\{d_{t1}, d_{t2}\}.$$

26. (Section 3.4) In Example 3.4.1 the model assumed only the parameter α to be known. The population proportions were estimated under the unbiased response error constraint. Assume that the matrix of response probabilities

$$\boldsymbol{\kappa} = \begin{bmatrix} 0.990 & 0.156 \\ 0.010 & 0.844 \end{bmatrix}$$

is known to hold for both months. Estimate the $\boldsymbol{\pi}_{AB}$ table and estimate the covariance matrix of your estimators.

27. (Section 3.4) Verify the expression for the variance $\hat{V}\{\hat{\theta}\}$ given in Example 3.4.1.

APPENDIX 3.A. DATA FOR EXAMPLES

TABLE 3.A.1. Data on Iowa farm size

Logarithm of Farm Size	Logarithm of Experience	Education	Logarithm of Farm Size	Logarithm of Experience	Education
5.613	2.079	5.5	5.352	1.386	6.0
3.807	3.951	3.5	5.464	3.178	5.5
6.436	2.565	5.0	5.580	3.091	6.5
6.509	3.258	4.5	5.268	2.944	5.0
4.745	2.773	6.0	5.537	1.099	7.0
5.037	0.0	6.0	5.283	3.638	6.0
5.283	2.565	6.5	5.481	3.555	5.5
5.981	0.0	7.5	5.288	1.386	6.5
5.252	2.944	6.5	5.656	3.555	5.0
6.815	3.258	6.5	5.273	3.091	4.5
5.421	3.367	6.0	6.667	2.079	8.5
5.460	1.099	5.5	5.999	3.434	5.5
4.263	3.401	4.5	6.550	3.045	4.5
6.426	2.565	7.5	5.869	2.485	4.0

TABLE 3.A.1. (Continued)

Logarithm of Farm Size	Logarithm of Experience	Education	Logarithm of Farm Size	Logarithm of Experience	Education
4.466	3.135	5.0	6.122	2.996	7.5
4.997	1.386	4.5	5.063	3.332	4.5
6.080	0.0	5.5	3.466	2.890	4.5
1.609	0.693	7.0	3.664	1.386	5.5
4.394	3.091	6.0	3.761	3.526	4.0
5.429	2.996	7.0	5.293	2.708	5.0
6.746	2.197	6.0	5.670	1.386	6.5
5.081	1.099	7.5	5.030	3.497	7.0
4.382	3.466	5.5	3.951	0.0	6.5
4.771	1.609	5.5	4.844	3.850	3.5
5.584	3.091	6.5	6.392	3.584	4.0
5.159	2.485	7.5	5.236	3.045	5.5
5.743	1.099	6.0	5.278	2.833	5.0
6.439	3.091	5.5	6.666	3.497	5.0
7.487	2.833	6.0	6.165	3.807	5.5
4.771	0.0	5.0	6.612	2.773	6.0
4.949	3.714	6.0	6.252	3.584	6.0
5.268	4.249	3.0	4.990	2.708	6.0
4.654	2.890	5.5	4.956	4.533	4.0
6.627	2.773	5.5	5.308	1.609	5.5
6.914	2.833	6.0	6.938	2.485	6.0
6.609	3.689	6.0	5.958	3.296	6.0
6.599	2.833	5.0	6.465	3.555	5.5
5.584	3.332	6.0	5.328	3.584	3.5
6.560	4.007	5.5	7.017	2.833	6.5
2.565	3.219	5.5	5.476	3.219	4.0
6.351	3.178	6.0	5.897	2.708	5.5
2.303	2.079	6.0	4.663	2.708	6.5
5.768	2.996	3.5	4.543	3.555	4.0
3.951	2.833	6.5	5.298	4.304	5.0
6.433	2.996	5.5	4.949	3.761	5.0
6.223	4.317	6.5	6.613	3.970	5.5
6.654	3.135	5.5	6.692	3.850	5.5
5.673	2.303	6.5	5.293	3.367	4.5
5.768	3.401	5.0	7.731	3.829	5.0
5.278	2.565	5.0	5.730	2.708	4.5
5.142	2.708	5.5	5.505	2.485	7.0
5.505	4.060	4.0	3.332	0.0	6.0
6.282	1.792	6.0	4.970	2.773	5.5
5.425	2.944	3.5	5.017	1.386	5.5
5.568	2.079	4.0	5.974	2.833	5.5

TABLE 3.A.1. (Continued)

Logarithm of Farm Size	Logarithm of Experience	Education	Logarithm of Farm Size	Logarithm of Experience	Education
4.898	0.693	7.0	5.313	1.099	8.0
6.852	3.638	4.5	5.464	3.258	5.5
6.784	2.485	6.5	6.468	3.219	6.0
6.829	3.829	5.0	5.900	3.367	6.5
6.201	3.045	6.0	6.535	3.045	8.0
5.595	1.609	6.5	5.011	1.946	7.0
5.971	3.091	5.0	6.346	3.367	5.5
5.505	1.609	6.5	5.561	2.890	5.5
6.006	4.007	3.5	5.118	3.738	5.0
5.537	2.773	6.0	2.079	1.792	5.0
3.555	2.996	6.5	5.707	2.773	4.5
5.187	3.638	4.5	4.913	1.386	7.0
5.407	3.045	5.5	4.956	2.197	7.5
6.054	1.099	7.0	5.835	3.178	6.5
4.820	4.220	5.0	6.532	2.079	6.0
7.299	3.045	6.0	4.654	3.466	3.5
5.293	0.693	6.0	6.031	3.434	5.0
5.136	3.219	5.5	6.344	3.555	5.0
5.501	2.565	5.5	5.905	3.258	5.5
5.900	1.792	5.5	5.635	2.398	6.5
6.236	3.258	6.0	5.537	4.094	5.5
6.380	1.792	5.5	5.900	2.944	6.0
5.724	3.401	6.0	4.844	1.099	6.0
5.598	3.258	5.0	5.182	0.0	7.5
5.481	3.784	5.5	6.512	2.996	4.5
5.165	4.043	5.5	4.094	3.434	5.0
5.011	1.792	7.5	4.511	1.609	5.5
4.787	4.159	6.0	5.429	3.258	6.5
5.308	3.178	4.5	6.178	2.996	7.0
6.764	2.398	7.0	5.811	1.609	7.0
4.836	1.792	6.0	4.663	1.609	6.0
6.686	3.135	5.0	4.875	3.912	4.0
5.338	2.639	5.0	6.201	3.091	6.0

TABLE 3.A.2. Data on textile expenditures

Observation	Log Expenditure (Y)	Log Income (X_1)	Household Size (X_2)	Moved Indicator (x_4)
1	2.8105	4.2487	0.6931	0
2	3.1605	4.7194	0.6931	0
3	3.4564	4.8075	1.0986	0
4	3.3515	4.4777	1.0986	0
5	3.1427	4.4355	1.0986	0
6	4.9828	5.5550	0.6931	1
7	3.8726	5.5558	1.7918	0
8	3.5005	4.8463	1.0986	0
9	3.2051	4.5762	1.0986	0
10	3.1133	4.2303	1.0986	0
11	5.3818	4.7218	1.6094	1
12	3.3564	5.0702	1.3863	0
13	5.1192	4.5776	1.6094	1
14	4.2950	4.0892	0.6931	1
15	2.9487	4.5041	0.0000	0
16	4.4244	4.3690	1.0986	1
17	4.5062	4.3467	1.0986	1
18	3.3970	4.8097	1.7918	0
19	3.6608	3.8930	0.0000	1
20	4.7977	4.7120	0.0000	1
21	2.9293	4.2486	1.0986	0
22	3.4415	4.6945	1.7918	0
23	3.6255	5.0868	1.3863	0
24	3.4816	5.3184	0.6931	0
25	2.9680	4.1903	1.3863	0
26	3.7426	4.7963	1.6094	0
27	3.3659	4.3260	1.3863	0
28	3.6916	4.9496	1.3863	0
29	2.6986	4.0787	0.6931	0
30	3.0485	4.7928	0.6931	0
31	3.9057	5.6423	1.0986	0
32	4.8882	5.1164	0.6931	1
33	3.1042	4.2790	0.6931	0
34	3.1020	4.3619	0.6931	0
35	2.8285	4.3596	0.6931	0
36	3.6743	5.4968	0.6931	0
37	3.2908	4.6981	0.6931	0
38	2.9464	4.5562	1.0986	0
39	3.1480	4.5094	1.3863	0
40	3.4629	4.8516	1.0986	0
41	3.3155	5.2429	1.0986	0

TABLE 3.A.2. (Continued)

Observation	Log Expenditure (Y)	Log Income (X_1)	Household Size (X_2)	Moved Indicator (x_4)
42	2.9139	4.5952	0.0000	0
43	4.5707	4.1656	1.6094	1
44	3.9754	5.1503	1.6094	0
45	3.4212	4.4454	1.6094	0
46	5.1222	4.6148	1.6094	1
47	3.7129	3.6932	0.6931	1
48	2.9538	4.6957	0.6931	0
49	3.5649	4.5040	1.0986	0
50	4.1190	5.6678	1.6094	0
51	3.6449	4.8324	1.7918	0
52	2.9474	4.0822	0.6931	0
53	3.3074	4.4091	1.0986	0
54	2.9999	4.2238	0.6931	0
55	3.0784	4.2821	1.7918	0
56	3.9149	4.3624	0.0000	1
57	4.5158	4.5258	0.0000	1
58	3.2703	5.0018	0.6931	0
59	5.7206	5.1572	1.6094	1
60	3.5917	4.7557	1.3863	0
61	3.2740	4.3027	1.7918	0
62	3.8237	4.0133	0.0000	1
63	4.9905	4.6147	1.0986	1
64	4.0648	5.8566	1.6094	0
65	3.4285	3.6468	0.0000	1
66	4.5406	4.6024	0.6931	1
67	4.0845	5.3116	1.6094	0
68	3.1397	4.5930	0.6931	0
69	2.8922	4.2982	0.6931	0
70	5.1510	4.6137	1.6094	1
71	3.9937	4.8630	1.9459	0
72	3.8553	3.8508	0.0000	1
73	2.5924	3.8723	1.3863	0
74	4.1913	4.6414	0.6931	1
75	3.3528	4.5509	1.6094	0
76	5.0442	5.1408	0.0000	1
77	3.3928	5.0257	1.0986	0
78	2.8359	4.8217	0.0000	0
79	2.6669	4.7405	0.0000	0
80	3.4381	5.3911	0.0000	0
81	4.6850	4.9156	0.6931	1
82	3.2266	4.3740	1.3863	0

TABLE 3.A.2. (Continued)

Observation	Log Expenditure (Y)	Log Income (X_1)	Household Size (X_2)	Moved Indicator (x_4)
83	3.0413	4.8804	1.0986	0
84	3.4214	4.7378	0.6931	0
85	3.7571	4.8238	1.0986	0
86	3.4723	4.8093	0.6931	0
87	4.3134	4.4191	0.0000	1
88	3.3162	4.6917	0.6931	0
89	3.2280	5.1155	1.6094	0
90	3.1080	4.8888	0.0000	0
91	3.9145	5.0968	1.9459	0
92	4.3283	4.4723	0.6931	1
93	2.5766	4.3229	0.0000	0
94	2.6888	4.2901	0.6931	0
95	3.4757	5.4455	0.0000	0
96	3.4902	5.1324	1.3863	0
97	2.9934	4.1855	0.6931	0
98	4.7598	4.9826	0.6931	1
99	4.4542	4.4554	0.6931	1
100	3.5973	4.9747	1.3863	0

TABLE 3.A.3. Data on pig farrowing

Observation	Y_{t1}	Y_{t2}	X_{t1}	X_{t2}	Observation	Y_{t1}	Y_{t2}	X_{t1}	X_{t2}
1	29	24	41	44	17	0	0	11	11
2	0	20	42	53	18	8	0	9	8
3	27	34	56	56	19	0	0	38	103
4	29	12	61	57	20	20	18	87	125
5	0	0	32	32	21	0	0	5	5
6	0	0	51	50	22	0	12	16	28
7	0	0	25	0	23	47	44	51	44
8	22	22	22	22	24	20	16	21	17
9	29	15	39	39	25	0	19	21	21
10	8	0	17	8	26	0	10	11	23
11	17	20	75	77	27	22	20	21	46
12	11	12	32	31	28	25	38	52	52
13	0	0	46	46	29	0	0	51	51
14	24	24	92	87	30	0	0	38	38
15	0	0	15	0	31	0	0	81	77
16	0	0	10	15	32	12	0	12	9

TABLE 3.A.3. (Continued)

Observation	Y_{t1}	Y_{t2}	X_{t1}	X_{t2}	Observation	Y_{t1}	Y_{t2}	X_{t1}	X_{t2}
33	30	31	64	64	76	0	0	10	10
34	15	16	18	16	77	0	0	0	4
35	0	0	12	12	78	12	13	38	29
36	0	0	18	18	79	10	10	26	20
37	17	17	43	43	80	15	0	38	28
38	20	20	36	31	81	0	0	32	29
39	0	0	46	41	82	0	0	2	57
40	5	6	21	21	83	25	0	27	27
41	0	0	5	20	84	0	0	150	150
42	0	0	18	20	85	0	0	15	0
43	5	3	6	6	86	10	0	20	30
44	11	25	36	37	87	9	10	45	78
45	32	5	82	75	88	0	4	18	14
46	0	34	63	59	89	0	0	35	35
47	0	0	6	3	90	0	15	15	45
48	20	20	83	42	91	0	0	36	51
49	0	0	84	75	92	0	12	22	17
50	0	15	21	26	93	12	12	19	19
51	0	0	13	13	94	0	10	11	10
52	0	0	9	9	95	15	15	15	15
53	13	13	23	19	96	12	12	13	13
54	24	65	113	108	97	0	25	35	53
55	16	15	31	31	98	43	3	44	44
56	13	0	21	0	99	12	8	24	45
57	8	0	23	24	100	11	11	79	52
58	0	0	56	46	101	5	10	10	10
59	10	15	36	36	102	4	0	15	11
60	29	26	51	51	103	27	24	45	49
61	16	15	17	17	104	20	20	37	42
62	15	29	50	44	105	20	20	42	67
63	9	10	24	33	106	0	0	30	40
64	0	7	20	20	107	0	0	64	60
65	0	0	0	11	108	60	28	185	71
66	5	10	15	25	109	16	18	25	27
67	11	11	22	22	110	16	9	17	16
68	20	20	61	57	111	0	0	1	1
69	0	0	25	55	112	31	13	50	47
70	0	2	0	2	113	20	3	36	47
71	4	3	5	4	114	32	32	104	129
72	18	18	51	42	115	0	0	44	54
73	60	50	152	127	116	18	13	14	14
74	0	0	19	24	117	12	9	21	21
75	0	0	21	38	118	40	35	130	103

TABLE 3.A.3. (Continued)

Observation	Y_{t1}	Y_{t2}	X_{t1}	X_{t2}	Observation	Y_{t1}	Y_{t2}	X_{t1}	X_{t2}
119	18	10	32	31	152	10	0	30	40
120	5	4	5	5	153	20	50	159	129
121	0	0	19	18	154	0	0	44	29
122	0	0	52	72	155	0	0	23	23
123	4	4	8	7	156	0	18	19	19
124	20	0	26	55	157	0	0	0	2
125	0	0	10	0	158	42	42	43	43
126	11	12	27	26	159	0	0	22	12
127	0	0	20	15	160	0	0	40	51
128	15	0	25	16	161	11	10	23	21
129	82	69	140	163	162	0	0	9	0
130	10	4	13	25	163	9	10	10	11
131	0	0	27	2	164	0	0	44	42
132	7	2	36	17	165	9	9	28	28
133	12	14	43	33	166	0	40	62	62
134	18	0	23	23	167	8	8	32	9
135	0	5	32	62	168	4	2	17	17
136	0	0	21	21	169	0	3	21	20
137	15	16	37	34	170	9	9	28	23
138	14	12	25	40	171	0	0	54	35
139	40	40	82	82	172	25	40	43	54
140	0	0	30	20	173	30	35	30	35
141	9	8	14	15	174	0	35	71	73
142	11	11	42	42	175	5	5	26	27
143	20	0	22	22	176	13	12	45	48
144	0	0	31	32	177	0	0	24	24
145	0	0	66	0	178	0	10	39	36
146	20	17	50	25	179	0	12	40	27
147	46	30	65	83	180	0	0	8	22
148	31	0	31	0	181	30	20	82	72
149	0	0	49	31	183	0	0	21	21
150	0	0	72	72	183	0	0	21	21
151	0	40	125	86	184	0	0	38	38

TABLE 3.A.4. Depths of earthquakes occurring near the Tonga trench

Earthquake	Depth Y_t	Distance Perpendicular to Trench X_{t1}	Distance Parallel to Trench X_{t2}
1	0.60	1.38	1.19
2	2.10	2.72	1.42
3	2.61	2.76	1.74
4	1.20	1.03	0.99
5	0.00	0.49	0.67
6	0.00	1.05	0.91
7	0.27	0.52	0.82
8	0.69	1.14	3.11
9	0.68	0.86	3.03
10	0.35	0.73	0.78
11	0.08	0.68	0.40
12	0.17	1.26	0.53
13	0.81	1.51	2.14
14	0.64	0.88	1.41
15	3.06	3.58	1.86
16	2.21	2.81	1.62
17	2.88	3.22	3.10
18	2.66	2.89	3.94
19	2.47	2.97	1.24
20	2.58	2.94	1.17
21	0.30	0.35	0.46
22	0.00	0.19	2.71
23	0.72	0.74	3.06
24	1.96	2.42	3.73
25	0.07	0.76	0.63
26	1.99	2.40	2.71
27	1.33	1.88	1.88
28	0.00	0.31	0.01
29	2.84	3.24	2.08
30	2.09	2.35	3.84
31	0.97	1.53	1.96
32	0.58	0.70	1.22
33	3.66	3.76	2.57
34	0.31	0.75	3.12
35	2.08	2.56	1.40
36	0.00	0.13	3.16
37	0.42	0.53	2.51
38	0.39	0.36	2.41
39	0.00	0.26	2.20
40	0.50	1.07	3.34
41	2.53	2.94	0.55
42	0.00	0.12	2.82
43	0.00	0.05	2.18

Source: Sykes, Isacks, and Oliver (1969).

TABLE 3.A.5. Observations for quadratic model

Observation	Y_t	X_{t1}	Observation	Y_t	X_{t1}
1	0.70	−1.44	44	−0.41	−0.21
2	0.05	−0.90	45	0.02	−0.52
3	0.99	−0.81	46	0.34	−1.01
4	0.30	−1.14	47	−0.31	−0.34
5	1.10	−1.36	48	0.31	−0.68
6	0.93	−0.61	49	−0.06	−0.40
7	0.30	−1.11	50	0.35	−0.06
8	0.24	−1.46	51	−0.35	0.13
9	0.28	−0.77	52	0.09	−0.18
10	0.86	−1.31	53	0.11	0.11
11	0.76	−0.98	54	0.03	−0.04
12	0.48	−0.93	55	0.26	−0.54
13	0.11	−1.02	56	0.41	−0.10
14	0.21	−1.60	57	−0.01	0.35
15	0.60	−0.45	58	−0.45	−0.32
16	0.58	−1.07	59	0.32	−0.23
17	0.10	−1.11	60	−0.14	−0.12
18	0.31	−0.57	61	−0.06	−0.05
19	0.24	−0.90	62	0.32	0.31
20	0.13	−1.25	63	−0.06	−0.45
21	0.45	−0.81	64	−0.22	0.65
22	0.66	−1.07	65	−0.30	0.57
23	0.18	−0.51	66	−0.26	0.17
24	0.47	−0.86	67	−0.59	−0.16
25	0.18	−0.16	68	−0.31	−0.02
26	−0.18	−0.64	69	0.23	0.08
27	0.15	0.21	70	−0.25	−0.17
28	0.12	−0.56	71	0.59	0.09
29	0.05	−0.45	72	0.34	0.43
30	−0.28	−0.71	73	0.05	0.83
31	−0.59	−0.70	74	0.51	0.70
32	0.38	−0.46	75	0.69	0.35
33	0.20	−0.22	76	0.20	0.64
34	0.16	−0.98	77	1.08	0.52
35	0.45	−0.54	78	0.70	0.67
36	0.42	−0.54	79	0.17	0.50
37	0.14	−0.81	80	0.39	−0.26
38	−0.21	−0.14	81	0.68	0.56
39	−0.59	−1.03	82	0.87	0.44
40	−0.22	−0.36	83	0.79	0.99
41	−0.25	−0.14	84	0.05	0.76
42	−0.23	−0.37	85	0.43	0.55
43	−0.13	−0.71	86	0.17	0.53

TABLE 3.A.5. (Continued)

Observation	Y_t	X_{t1}	Observation	Y_t	X_{t1}
87	0.40	0.89	104	1.43	1.36
88	0.33	0.35	105	1.76	0.72
89	0.73	0.33	106	1.16	0.71
90	0.57	0.39	107	1.96	1.44
91	0.57	−0.19	108	2.05	0.82
92	0.36	0.55	109	1.80	0.58
93	0.57	0.55	110	1.63	0.39
94	0.64	0.02	111	0.92	0.63
95	0.27	0.46	112	1.86	1.11
96	1.25	0.84	113	1.50	0.84
97	1.54	1.03	114	2.04	1.07
98	1.53	0.92	115	1.85	0.80
99	1.32	1.28	116	1.34	1.08
100	1.48	1.53	117	1.60	0.85
101	1.03	1.22	118	1.56	0.97
102	1.11	0.84	119	1.60	1.00
103	1.33	1.23	120	1.57	1.60

CHAPTER 4

Multivariate Models

The models studied in Section 1.2, Section 1.3, and Chapter 2 can be characterized by the fact that the true values of the observed variables satisfy a single linear equation. The models fell into two classes; those with an error in the equation and those with only measurement errors. In the models containing an error in the equation (e.g., the models of Sections 1.2 and 2.2), one variable could be identified as the "dependent" or y variable. In models with only measurement error (e.g., the models of Sections 1.3 and 2.3), the variables entered the model in a symmetric manner from a distributional point of view, but we generally chose to identify one variable as the y variable.

In this chapter we extend our treatment of measurement error models with no error in the equation to the situation in which the true variables satisfy more than one linear relation. The extension of the model with an error in the equation to the multivariate case is relatively straightforward and is not discussed.

4.1. THE CLASSICAL MULTIVARIATE MODEL

The model introduced in Section 1.3 assumes that independent information on the covariance matrix of the measurement errors is available and that the only source of error is measurement error. This section is devoted to multivariate models of that type.

4.1.1. Maximum Likelihood Estimation

We derive the maximum likelihood estimators for two models: the model with fixed $\mathbf{x}_t$ and the model with random $\mathbf{x}_t$. One representation for the fixed model is

$$\mathbf{y}_t = \mathbf{x}_t \boldsymbol{\beta}, \qquad \mathbf{Z}_t = \mathbf{z}_t + \boldsymbol{\varepsilon}_t, \qquad \boldsymbol{\varepsilon}_t \sim \mathrm{NI}(\mathbf{0}, \boldsymbol{\Sigma}_{\varepsilon\varepsilon}), \tag{4.1.1}$$

where $\{\mathbf{x}_t\}$ is a fixed sequence, $\boldsymbol{\varepsilon}_t = (\mathbf{e}_t, \mathbf{u}_t)$, $\mathbf{Z}_t = (\mathbf{Y}_t, \mathbf{X}_t)$ is observed, $\mathbf{y}_t$ is an r-dimensional row vector, $\mathbf{x}_t$ is a k-dimensional row vector, $\mathbf{z}_t = (\mathbf{y}_t, \mathbf{x}_t)$ is a p-dimensional row vector, $\boldsymbol{\beta}$ is a $k \times r$ matrix of unknown coefficients, and $p = r + k$. The r-dimensional vector $\mathbf{v}_t = \boldsymbol{\varepsilon}_t(\mathbf{I}_r, -\boldsymbol{\beta}')'$ is the vector of population deviations, where

$$\mathbf{Y}_t = \mathbf{X}_t \boldsymbol{\beta} + \mathbf{v}_t.$$

The maximum likelihood estimators for the case in which

$$\boldsymbol{\Sigma}_{\varepsilon\varepsilon} = \boldsymbol{\Upsilon}_{\varepsilon\varepsilon}\sigma^2 \tag{4.1.2}$$

and $\boldsymbol{\Upsilon}_{\varepsilon\varepsilon}$ is known are derived in Theorem 4.1.1 and are direct extensions of the estimators of Theorem 2.3.1. The estimator of $\boldsymbol{\beta}$ is expressed as a function of the characteristic vectors of $\mathbf{M}_{ZZ}$ in the metric $\boldsymbol{\Sigma}_{\varepsilon\varepsilon}$, where the vectors are defined in (4.A.21) and (4.A.22) of Appendix 4.A.

Theorem 4.1.1. Let model (4.1.1) hold. Assume that $\boldsymbol{\Sigma}_{vv}$ and $\mathbf{M}_{xx}$ are positive definite. Then the maximum likelihood estimators are

$$\hat{\boldsymbol{\beta}} = -\hat{\mathbf{B}}_{kr}\hat{\mathbf{B}}_{rr}^{-1}, \qquad \hat{\sigma}_m^2 = (p - \ell)^{-1} \sum_{i=k+1}^{p} \hat{\lambda}_i, \tag{4.1.3}$$

$$\hat{\mathbf{z}}_t = \mathbf{Z}_t(\mathbf{I} - \hat{\mathbf{B}}\hat{\mathbf{B}}'\boldsymbol{\Upsilon}_{\varepsilon\varepsilon}) = \mathbf{Z}_t - \hat{\mathbf{v}}_t \hat{\boldsymbol{\Upsilon}}_{vv}^{-1}\hat{\boldsymbol{\Upsilon}}_{v\varepsilon},$$

where $\mathbf{M}_{ZZ} = n^{-1} \sum_{t=1}^{n} \mathbf{Z}_t'\mathbf{Z}_t$, $p - \ell$ is the rank of $\boldsymbol{\Upsilon}_{\varepsilon\varepsilon}$, $\hat{\mathbf{v}}_t = \mathbf{Z}_t(\mathbf{I}_r, -\hat{\boldsymbol{\beta}}')'$,

$$(\hat{\boldsymbol{\Upsilon}}_{vv}, \hat{\boldsymbol{\Upsilon}}_{v\varepsilon}) = (\mathbf{I}_r, -\hat{\boldsymbol{\beta}}')\boldsymbol{\Upsilon}_{\varepsilon\varepsilon}[(\mathbf{I}_r, -\hat{\boldsymbol{\beta}}')', \mathbf{I}_p],$$

$\hat{\mathbf{B}}_{.i}$ are the characteristic vectors of $\mathbf{M}_{ZZ}$ in the metric $\boldsymbol{\Upsilon}_{\varepsilon\varepsilon}$, $\hat{\mathbf{B}}_{.i}$, $i = 1, 2, \ldots, r$, are the columns of $\hat{\mathbf{B}} = (\hat{\mathbf{B}}_{rr}', \hat{\mathbf{B}}_{kr}')'$, and $\hat{\lambda}_p \leqslant \hat{\lambda}_{p-1} \leqslant \cdots \leqslant \hat{\lambda}_{p-r+1}$ are the r smallest roots of

$$|\mathbf{M}_{ZZ} - \lambda\boldsymbol{\Upsilon}_{\varepsilon\varepsilon}| = 0.$$

Proof. The proof parallels the proof of Theorem 2.3.1. To simplify the proof we assume $\boldsymbol{\Upsilon}_{\varepsilon\varepsilon}$ to be nonsingular, but the conclusion is true for singular $\boldsymbol{\Upsilon}_{\varepsilon\varepsilon}$. Twice the logarithm of the likelihood for a sample of n observations is

$$2 \log L = -n \log|2\pi\boldsymbol{\Upsilon}_{\varepsilon\varepsilon}\sigma^2| - \sigma^{-2} \sum_{t=1}^{n} (\mathbf{Z}_t - \mathbf{z}_t)\boldsymbol{\Upsilon}_{\varepsilon\varepsilon}^{-1}(\mathbf{Z}_t - \mathbf{z}_t)'. \tag{4.1.4}$$

The $\mathbf{z}_t'$ that maximizes (4.1.4) for a given $\boldsymbol{\beta}$ is

$$\begin{aligned}\ddot{\mathbf{z}}_t' &= (\boldsymbol{\beta}, \mathbf{I}_k)'[(\boldsymbol{\beta}, \mathbf{I}_k)\boldsymbol{\Upsilon}_{\varepsilon\varepsilon}^{-1}(\boldsymbol{\beta}, \mathbf{I}_k)']^{-1}(\boldsymbol{\beta}, \mathbf{I}_k)\boldsymbol{\Upsilon}_{\varepsilon\varepsilon}^{-1}\mathbf{Z}_t' \\ &= \mathbf{Z}_t' - \boldsymbol{\Upsilon}_{\varepsilon\varepsilon}\mathbf{C}(\mathbf{C}'\boldsymbol{\Upsilon}_{\varepsilon\varepsilon}\mathbf{C})^{-1}\mathbf{C}'\mathbf{Z}_t', \end{aligned} \tag{4.1.5}$$

where $\mathbf{C}' = (\mathbf{I}_r, -\boldsymbol{\beta}')$. Substituting this expression for $\mathbf{z}_t'$ into (4.1.4) we have

$$2 \log L = -n \log|2\pi\boldsymbol{\Upsilon}_{\varepsilon\varepsilon}\sigma^2| - \sigma^{-2} \sum_{t=1}^{n} \mathbf{Z}_t\mathbf{C}(\mathbf{C}'\boldsymbol{\Upsilon}_{\varepsilon\varepsilon}\mathbf{C})^{-1}\mathbf{C}'\mathbf{Z}_t'. \tag{4.1.6}$$

Therefore, the $\boldsymbol{\beta}$ that maximizes the likelihood is the $\boldsymbol{\beta}$ that minimizes

$$\operatorname{tr}\{\mathbf{C}'\mathbf{M}_{ZZ}\mathbf{C}(\mathbf{C}'\boldsymbol{\Upsilon}_{\varepsilon\varepsilon}\mathbf{C})^{-1}\}. \tag{4.1.7}$$

By Corollary 4.A.10, the minimum is attained by choosing the columns of $\mathbf{C}$ to be linear combinations of the r characteristic vectors of $\mathbf{M}_{ZZ}$ in the metric $\boldsymbol{\Upsilon}_{\varepsilon\varepsilon}$ that are associated with the r smallest roots. The submatrix $\hat{\mathbf{B}}_{rr}$ is nonsingular with probability one and the estimator of $\boldsymbol{\beta}$ is given by (4.1.3).

Using the characteristic vectors in the metric $\boldsymbol{\Upsilon}_{\varepsilon\varepsilon}$, the maximum likelihood estimators of $\mathbf{z}_t$ defined by (4.1.5) are

$$\hat{\mathbf{z}}_t = \mathbf{Z}_t - \sum_{i=1}^{r} \mathbf{Z}_t\hat{\mathbf{B}}_{.i}\hat{\mathbf{B}}_{.i}'\boldsymbol{\Upsilon}_{\varepsilon\varepsilon} \tag{4.1.8}$$

$$= \mathbf{Z}_t - \hat{\mathbf{v}}_t\hat{\boldsymbol{\Upsilon}}_{vv}^{-1}\hat{\boldsymbol{\Upsilon}}_{v\varepsilon}. \tag{4.1.9}$$

Setting the derivative of (4.1.4) with respect to σ^2 equal to zero, we obtain the maximum likelihood estimator,

$$\begin{aligned}\hat{\sigma}_m^2 &= (np)^{-1} \sum_{t=1}^{n} (\mathbf{Z}_t - \hat{\mathbf{z}}_t)\boldsymbol{\Upsilon}_{\varepsilon\varepsilon}^{-1}(\mathbf{Z}_t - \hat{\mathbf{z}}_t)' \\ &= p^{-1} \operatorname{tr}\{\hat{\mathbf{B}}'\mathbf{M}_{ZZ}\hat{\mathbf{B}}\} = p^{-1} \sum_{i=k+1}^{p} \hat{\lambda}_i. \end{aligned}$$

□

Alternative derivations of the maximum likelihood estimators for fixed $\mathbf{x}_t$ are given in Healy (1975) and Gleser (1981). Also see Exercises 4.11 and 4.12.

As in the univariate model, the maximum likelihood estimator of σ^2 is not consistent for σ^2, but the estimator

$$\hat{\sigma}^2 = nr^{-1}(n-k)^{-1} \sum_{i=k+1}^{p} \hat{\lambda}_i \tag{4.1.10}$$

is consistent for σ^2. The estimator (4.1.10) can also be written

$$\hat{\sigma}^2 = (n-k)^{-1}r^{-1} \sum_{t=1}^{n} \hat{\mathbf{v}}_t\hat{\boldsymbol{\Upsilon}}_{vv}^{-1}\hat{\mathbf{v}}_t', \tag{4.1.11}$$

where $\hat{\mathbf{v}}_t = \mathbf{Z}_t(\mathbf{I}, -\hat{\boldsymbol{\beta}}')'$ and $\hat{\boldsymbol{\Upsilon}}_{vv} = (\mathbf{I}, -\hat{\boldsymbol{\beta}}')\boldsymbol{\Upsilon}_{\varepsilon\varepsilon}(\mathbf{I}, -\hat{\boldsymbol{\beta}}')'$.

We now derive the maximum likelihood estimators for the model with estimated measurement error covariance matrix. We assume that the vector of true values is randomly chosen from a normal distribution and that the true values are known to satisfy r linear equations. We let

$$\mathbf{z}_t\mathbf{C} = \boldsymbol{\beta}_0, \qquad \mathbf{Z}_t = \mathbf{z}_t + \boldsymbol{\varepsilon}_t, \tag{4.1.12}$$
$$(\boldsymbol{\varepsilon}_t, \mathbf{x}_t)' \sim \mathrm{NI}[(\mathbf{0}, \boldsymbol{\mu}_x)', \text{block diag}(\boldsymbol{\Sigma}_{\varepsilon\varepsilon}, \boldsymbol{\Sigma}_{xx})],$$

where $\mathbf{z}_t = (\mathbf{y}_t, \mathbf{x}_t)$ is a p-dimensional vector, $\mathbf{Z}_t = (\mathbf{Y}_t, \mathbf{X}_t)$, $\boldsymbol{\varepsilon}_t = (\mathbf{e}_t, \mathbf{u}_t)$, $\mathbf{C}' = (\mathbf{I}_r, -\boldsymbol{\beta}')$, $\mathbf{x}_t$ is a k-dimensional row vector of true values, $\mathbf{X}_t$ is a k-dimensional row vector of observed values, $\mathbf{Y}_t$ is an r-dimensional row vector of observations, $\boldsymbol{\beta}_0$ is an r-dimensional row vector of unknown coefficients, $\boldsymbol{\beta}$ is a $k \times r$ matrix of unknown coefficients, and $p = k + r$. The vectors $\mathbf{Z}_t = (\mathbf{Y}_t, \mathbf{X}_t)$, $t = 1, 2, \ldots, n$, are observed. It is assumed that $\boldsymbol{\Sigma}_{xx}$ is nonsingular. Under the model,

$$\mathbf{Z}_t' \sim \mathrm{NI}(\boldsymbol{\mu}_Z', \boldsymbol{\Sigma}_{ZZ}), \tag{4.1.13}$$

where

$$\boldsymbol{\Sigma}_{ZZ} = (\boldsymbol{\beta}, \mathbf{I}_k)'\boldsymbol{\Sigma}_{xx}(\boldsymbol{\beta}, \mathbf{I}_k) + \boldsymbol{\Sigma}_{\varepsilon\varepsilon} = \boldsymbol{\Sigma}_{zz} + \boldsymbol{\Sigma}_{\varepsilon\varepsilon} \tag{4.1.14}$$

and $\boldsymbol{\mu}_Z = (\boldsymbol{\beta}_0 + \boldsymbol{\mu}_x\boldsymbol{\beta}, \boldsymbol{\mu}_x)$. There are $\frac{1}{2}k(k+1)$ unknown parameters in $\boldsymbol{\Sigma}_{xx}$ and we can reparameterize the problem by defining a $k \times k$ nonsingular matrix $\mathbf{G}_2$ with $\frac{1}{2}k(k+1)$ parameters such that $\boldsymbol{\Sigma}_{xx}$ is equal to the product $\mathbf{G}_2'\mathbf{G}_2$. One possible choice for $\mathbf{G}_2$ is a lower triangular matrix, but we need not specify the nature of $\mathbf{G}_2$ at this point. If we let $\mathbf{f}_t$ denote the row vector defined by

$$\mathbf{f}_t = \mathbf{x}_t\mathbf{G}_2^{-1},$$

then it follows that $\boldsymbol{\Sigma}_{ff} = \mathbf{I}$. Model (4.1.1) specifies $\mathbf{y}_t = (z_{t1}, z_{t2}, \ldots, z_{tr})$ to be a linear function of $\mathbf{x}_t$ and, hence, of $(f_{t1}, f_{t2}, \ldots, f_{tk})$. We may write

$$\mathbf{z}_t = (\boldsymbol{\beta}_0, \mathbf{0}) + \mathbf{f}_t\mathbf{G}_2(\boldsymbol{\beta}, \mathbf{I}) = (\boldsymbol{\beta}_0, \mathbf{0}) + \mathbf{f}_t\mathbf{A}',$$

where $\mathbf{A}' = \mathbf{G}_2(\boldsymbol{\beta}, \mathbf{I})$ is a $k \times p$ matrix. It follows that

$$\boldsymbol{\Sigma}_{ZZ} = \mathbf{A}\boldsymbol{\Sigma}_{ff}\mathbf{A}' + \boldsymbol{\Sigma}_{\varepsilon\varepsilon} = \mathbf{A}\mathbf{A}' + \boldsymbol{\Sigma}_{\varepsilon\varepsilon}, \tag{4.1.15}$$

where there are $\frac{1}{2}k(k+1) + kr$ independent parameters in $\mathbf{A}$. We shall use both the parameterization associated with (4.1.15) and the parameterization (4.1.12).

Let a sample of n vectors $\mathbf{Z}_t$ be available and let

$$\mathbf{m}_{ZZ} = (n-1)^{-1} \sum_{t=1}^{n} (\mathbf{Z}_t - \bar{\mathbf{Z}})'(\mathbf{Z}_t - \bar{\mathbf{Z}}),$$

where $\bar{\mathbf{Z}} = n^{-1} \sum_{t=1}^{n} \mathbf{Z}_t$. We assume that an independent unbiased estimator of $\boldsymbol{\Sigma}_{\varepsilon\varepsilon}$, denoted by $\mathbf{S}_{\varepsilon\varepsilon}$, is available. It is assumed that $d_f\mathbf{S}_{\varepsilon\varepsilon}$ is distributed as a Wishart matrix with d_f degrees of freedom and matrix parameter $\Sigma_{\varepsilon\varepsilon}$. We first obtain the maximum likelihood estimators of the parameters of the model. Because $\Sigma_{\varepsilon\varepsilon}$ is unknown, an estimator of this covariance matrix is also constructed.

In Theorem 4.1.2 we define the estimator of Σ_{zz} in terms of $\mathbf{m}_{ZZ}$. As in earlier chapters, we call the estimator constructed with $\mathbf{m}_{ZZ}$ the maximum likelihood estimator adjusted for degrees of freedom. Matrix theory used in the development is summarized in Appendix 4.A.

Theorem 4.1.2. Let model (4.1.12) hold. Let Σ_{xx} and $\Sigma_{\varepsilon\varepsilon}$ be positive definite. Let $\hat{\lambda}_1 \geqslant \hat{\lambda}_2 \geqslant \cdots \geqslant \hat{\lambda}_p$ be the roots of

$$|\mathbf{m}_{ZZ} - \lambda \mathbf{S}_{\varepsilon\varepsilon}| = 0. \tag{4.1.16}$$

Let the latent vectors of $\mathbf{m}_{ZZ}$ in the metric $\mathbf{S}_{\varepsilon\varepsilon}$ be defined by

$$(\mathbf{m}_{ZZ} - \hat{\lambda}_i \mathbf{S}_{\varepsilon\varepsilon})\mathbf{T}_{.i} = 0, \qquad i = 1, 2, \ldots, p,$$

where $\mathbf{T}'_{.i}\mathbf{S}_{\varepsilon\varepsilon}\mathbf{T}_{.j} = \delta_{ij}$ and δ_{ij} is Kronecker's delta. Let

$$\mathbf{T} = (\mathbf{T}_{.1}, \mathbf{T}_{.2}, \ldots, \mathbf{T}_{.p}) \tag{4.1.17}$$

be the matrix whose columns are the latent vectors of $\mathbf{m}_{ZZ}$ in the metric $\mathbf{S}_{\varepsilon\varepsilon}$ arranged so that the corresponding roots are ordered from largest to smallest. Then, if $\hat{\lambda}_k > 1$, the maximum likelihood estimators adjusted for degrees of freedom are

$$\begin{aligned} (\hat{\Sigma}_{\varepsilon\varepsilon}, \hat{\Sigma}_{zz}) &= (\mathbf{S}_{\varepsilon\varepsilon}\mathbf{T}\hat{\Sigma}_{T\varepsilon\varepsilon}\mathbf{T}'\mathbf{S}_{\varepsilon\varepsilon}, \mathbf{S}_{\varepsilon\varepsilon}\mathbf{T}\hat{\Sigma}_{Tzz}\mathbf{T}'\mathbf{S}_{\varepsilon\varepsilon}), \\ \hat{\boldsymbol{\beta}} &= \hat{\Sigma}_{xx}^{-1}\hat{\Sigma}_{xy}, \\ (\hat{\beta}_0, \hat{\boldsymbol{\mu}}_x) &= (\bar{\mathbf{Y}} - \bar{\mathbf{X}}\hat{\boldsymbol{\beta}}, \bar{\mathbf{X}}), \end{aligned} \tag{4.1.18}$$

where

$$\begin{aligned} \hat{\Sigma}_{T\varepsilon\varepsilon} &= \operatorname{diag}(1, 1, \ldots, 1, \hat{\sigma}_{T\varepsilon\varepsilon,k+1,k+1}, \hat{\sigma}_{T\varepsilon\varepsilon,k+2,k+2}, \ldots, \hat{\sigma}_{T\varepsilon\varepsilon pp}), \\ \hat{\sigma}_{T\varepsilon\varepsilon ii} &= (n - 1 + d_f)^{-1}[(n-1)\hat{\lambda}_i + d_f], \qquad i = k+1, k+2, \ldots, p, \\ \hat{\Sigma}_{Tzz} &= \operatorname{diag}(\hat{\lambda}_1 - 1, \hat{\lambda}_2 - 1, \ldots, \hat{\lambda}_k - 1, 0, 0, \ldots, 0). \end{aligned}$$

If $\hat{\lambda}_k \leqslant 1$, the maximum likelihood estimator of $\hat{\Sigma}_{xx}$ is singular and $\hat{\boldsymbol{\beta}}$ is not defined. The supremum of the likelihood function is the function evaluated at $\hat{\Sigma}_{T\varepsilon\varepsilon}$ and $\hat{\Sigma}_{Tzz}$ with τ replacing k, where τ is the number of $\hat{\lambda}_i$ exceeding one.

Proof. A proof has been given by Amemiya and Fuller (1984), but our proof is closer to that given in Anderson, Anderson, and Olkin (1986). For a p-variable model in which k variables are X variables and r variables are Y variables, the logarithm of the likelihood adjusted for degrees of freedom is

$$\log L_c(\boldsymbol{\theta}) = -\tfrac{1}{2}p(n-1+d_f)\log 2\pi - \tfrac{1}{2}(n-1)\log|\boldsymbol{\Sigma}_{ZZ}|$$
$$-\tfrac{1}{2}(n-1)\operatorname{tr}\{\mathbf{m}_{ZZ}\boldsymbol{\Sigma}_{ZZ}^{-1}\} - \tfrac{1}{2}d_f\log|\boldsymbol{\Sigma}_{\varepsilon\varepsilon}| - \tfrac{1}{2}d_f\operatorname{tr}\{\mathbf{S}_{\varepsilon\varepsilon}\boldsymbol{\Sigma}_{\varepsilon\varepsilon}^{-1}\},$$

where $\boldsymbol{\theta}$ is the vector of unknown parameters. Let $\mathbf{T}$ be the matrix such that

$$\mathbf{T}'\mathbf{S}_{\varepsilon\varepsilon}\mathbf{T} = \mathbf{I}, \qquad \mathbf{T}'\mathbf{m}_{ZZ}\mathbf{T} = \hat{\boldsymbol{\Lambda}}, \tag{4.1.19}$$

where $\hat{\boldsymbol{\Lambda}} = \operatorname{diag}(\hat{\lambda}_1, \hat{\lambda}_2, \ldots, \hat{\lambda}_p)$. Define new parameters

$$\boldsymbol{\Sigma}_{T\varepsilon\varepsilon} = \mathbf{T}'\boldsymbol{\Sigma}_{\varepsilon\varepsilon}\mathbf{T}, \qquad \boldsymbol{\Sigma}_{TZZ} = \mathbf{T}'\boldsymbol{\Sigma}_{ZZ}\mathbf{T}, \tag{4.1.20}$$

where $\boldsymbol{\Sigma}_{TZZ} - \boldsymbol{\Sigma}_{T\varepsilon\varepsilon}$ is a nonnegative definite matrix of rank k. Then, employing the transformation,

$$\log L(\boldsymbol{\theta}) = -\tfrac{1}{2}(n-1+d_f)(p\log 2\pi + \log|\mathbf{S}_{\varepsilon\varepsilon}|) - \tfrac{1}{2}f(\boldsymbol{\theta}),$$

where $\boldsymbol{\theta}$ contains the elements of vech $\boldsymbol{\Sigma}_{T\varepsilon\varepsilon}$ and the free parameters in vech $\boldsymbol{\Sigma}_{TZZ}$ and

$$f(\boldsymbol{\theta}) = (n-1)[\log|\boldsymbol{\Sigma}_{TZZ}| + \operatorname{tr}\{\hat{\boldsymbol{\Lambda}}\boldsymbol{\Sigma}_{TZZ}^{-1}\}] + d_f[\log|\boldsymbol{\Sigma}_{T\varepsilon\varepsilon}| + \operatorname{tr}\{\boldsymbol{\Sigma}_{T\varepsilon\varepsilon}^{-1}\}]. \tag{4.1.21}$$

The likelihood will be maximized if $f(\boldsymbol{\theta})$ of (4.1.21) is minimized.

There exists a $\mathbf{K}$ such that

$$\boldsymbol{\Sigma}_{T\varepsilon\varepsilon} = \mathbf{K}\mathbf{G}^{-1}\mathbf{K}', \qquad \boldsymbol{\Sigma}_{TZZ} = \mathbf{K}\mathbf{K}',$$

where $\mathbf{G} = \operatorname{diag}(g_1, g_2, \ldots, g_p)$ and $g_1^{-1} \leqslant g_2^{-1} \leqslant \cdots \leqslant g_p^{-1}$ are the roots of

$$|\boldsymbol{\Sigma}_{T\varepsilon\varepsilon} - g^{-1}\boldsymbol{\Sigma}_{TZZ}| = 0.$$

The roots g_i^{-1} are the population analogues of the statistics $\hat{\lambda}_i^{-1}$ defined by (4.1.16). By the definition of $\boldsymbol{\Sigma}_{T\varepsilon\varepsilon}$ and $\boldsymbol{\Sigma}_{TZZ}$, $g_i^{-1} = 1$ for $i = k+1, k+2, \ldots, p$ and $g_i^{-1} < 1$ for $i = 1, 2, \ldots, k$. The quantity to be minimized given in expression (4.1.21) can be written

$$\begin{aligned}
f(\boldsymbol{\theta}) &= (n-1+d_f)\log|\boldsymbol{\Sigma}_{TZZ}| + d_f\log|\mathbf{G}^{-1}| + (n-1)\operatorname{tr}\{\boldsymbol{\Sigma}_{TZZ}^{-1}\hat{\boldsymbol{\Lambda}}\} \\
&\quad + d_f\operatorname{tr}\{\mathbf{K}^{-1}\mathbf{K}^{-1\prime}\mathbf{G}\} \\
&\geqslant (n-1+d_f)\sum_{i=1}^{p}\log\gamma_{Zi} + d_f\sum_{i=1}^{p}\log g_i^{-1} \\
&\quad + (n-1)\sum_{i=1}^{p}\gamma_{Zi}^{-1}\hat{\lambda}_i + d_f\sum_{i=1}^{p}\gamma_{Zi}^{-1}g_i \\
&= (n-1+d_f)\sum_{i=1}^{p}\log\gamma_{Zi} + \sum_{i=1}^{k}\gamma_{Zi}^{-1}[(n-1)\hat{\lambda}_i + d_f g_i] \\
&\quad + \sum_{i=k+1}^{p}\gamma_{Zi}^{-1}[(n-1)\hat{\lambda}_i + d_f] - d_f\sum_{i=1}^{k}\log g_i,
\end{aligned} \tag{4.1.22}$$

where $\gamma_{Z1} \geqslant \gamma_{Z2} \geqslant \cdots \geqslant \gamma_{Zp}$ are the roots of $\boldsymbol{\Sigma}_{TZZ}$, we have used the fact that the roots of $\mathbf{KK}'$ are equal to the roots of $\mathbf{K}'\mathbf{K}$, and the inequality follows by Corollary 4.A.14. The unknowns in (4.1.22) are γ_{Zi}, $i = 1, 2, \ldots, p$, and g_i, $i = 1, 2, \ldots, k$. If we first minimize (4.1.22) with respect to γ_{Zi}, we have

$$\begin{aligned} \hat{\gamma}_{Zi} &= (n - 1 + d_f)^{-1}[(n-1)\hat{\lambda}_i + d_f g_i], \qquad i = 1, 2, \ldots, k, \\ \hat{\gamma}_{Zi} &= (n - 1 + d_f)^{-1}[(n-1)\hat{\lambda}_i + d_f], \qquad i = k+1, \ldots, p, \end{aligned} \tag{4.1.23}$$

where $\hat{\gamma}_{Zi} = \hat{\sigma}_{T\varepsilon\varepsilon ii}$ for $i = k+1, \ldots, p$. Substituting the expressions of (4.1.23) for the γ_{Zi} in expression (4.1.22), we obtain

$$f(\boldsymbol{\theta}) \geqslant (n - 1 + d_f)\left[p + \sum_{i=k+1}^{p} \log \hat{\sigma}_{T\varepsilon\varepsilon ii}\right] + \sum_{i=1}^{k} h(g_i; \hat{\lambda}_i), \tag{4.1.24}$$

where

$$h(g; \lambda) = (n - 1 + d_f)\log\{(n - 1 + d_f)^{-1}[(n-1)\lambda + d_f g]\} - d_f \log g.$$

Because

$$\inf_{g > 1} h(g; \lambda) = \begin{cases} h(\lambda; \lambda) & \text{if } \lambda > 1 \\ h(1; \lambda) & \text{if } 0 < \lambda \leqslant 1, \end{cases}$$

it follows from (4.1.24) that

$$f(\boldsymbol{\theta}) \geqslant (n - 1 + d_f)\left[p + \sum_{i=q+1}^{p} \log \hat{\sigma}_{T\varepsilon\varepsilon ii}\right] + (n-1)\sum_{i=1}^{q} \log \hat{\lambda}_i, \tag{4.1.25}$$

where $q = \min\{\tau, k\}$ and the $\hat{\sigma}_{T\varepsilon\varepsilon ii}$ are defined in (4.1.23). The right side of (4.1.25) is the infimum of $f(\boldsymbol{\theta})$ for $\boldsymbol{\theta}$ in the parameter space. This infimum is attained if we evaluate (4.1.21) at

$$\begin{aligned} \boldsymbol{\Sigma}_{TZZ} &= \operatorname{diag}\{\hat{\lambda}_1, \hat{\lambda}_2, \ldots, \hat{\lambda}_q, \hat{\sigma}_{T\varepsilon\varepsilon, q+1, q+1}, \ldots, \hat{\sigma}_{T\varepsilon\varepsilon pp}\}, \\ \boldsymbol{\Sigma}_{T\varepsilon\varepsilon} &= \operatorname{diag}\{1, 1, \ldots, 1, \hat{\sigma}_{T\varepsilon\varepsilon, q+1, q+1}, \ldots, \hat{\sigma}_{T\varepsilon\varepsilon pp}\}. \end{aligned} \tag{4.1.26}$$

The estimators in (4.1.18) are obtained by applying the inverse of the transform in (4.1.20) to the estimators in (4.1.26) and then using (4.1.19). □

The estimator of $\boldsymbol{\beta}$ derived in Theorem 4.1.2 can be obtained from that of Theorem 4.1.1 by replacing $\boldsymbol{\Upsilon}_{\varepsilon\varepsilon}$ with $\mathbf{S}_{\varepsilon\varepsilon}$, provided the estimator of Theorem 4.1.2 does not fall on the boundary of the parameter space. As in the univariate case, the estimators of $\boldsymbol{\beta}$ for the fixed and random models differ only in the treatment of boundary cases.

In Theorem 4.1.2 it is assumed that $\boldsymbol{\Sigma}_{\varepsilon\varepsilon}$ is positive definite. We shall find it worthwhile to have an expression for the maximum likelihood estimator that holds for singular $\mathbf{S}_{\varepsilon\varepsilon}$. This is accomplished by using the roots and vectors of $\mathbf{S}_{\varepsilon\varepsilon}$ in the metric $\mathbf{m}_{ZZ}$. Let $\hat{\lambda}_1^{-1} \leqslant \hat{\lambda}_2^{-1} \leqslant \cdots \leqslant \hat{\lambda}_p^{-1}$ be the values of

λ^{-1} that satisfy

$$|\mathbf{S}_{\varepsilon\varepsilon} - \lambda^{-1}\mathbf{m}_{ZZ}| = 0, \tag{4.1.27}$$

where the first $\ell(\ell \geq 0)$ of the $\hat{\lambda}_i^{-1}$ are zero and $p - \ell$ is the rank of $\mathbf{S}_{\varepsilon\varepsilon}$. Let

$$\mathbf{R} = (\mathbf{R}_{.1}, \mathbf{R}_{.2}, \ldots, \mathbf{R}_{.p}) \tag{4.1.28}$$

be the matrix of characteristic vectors of $\mathbf{S}_{\varepsilon\varepsilon}$ in the metric $\mathbf{m}_{ZZ}$, where the first ℓ vectors are associated with zero roots of (4.1.27). Let $\hat{\lambda}_k^{-1} < 1$. Then the maximum likelihood estimators of $\boldsymbol{\beta}$ and $\boldsymbol{\Sigma}_{zz}$ are

$$\hat{\boldsymbol{\beta}} = (\hat{\mathbf{A}}_{rk}\hat{\mathbf{A}}_{kk}^{-1})', \qquad \hat{\boldsymbol{\Sigma}}_{zz} = \hat{\mathbf{A}}\hat{\mathbf{A}}', \tag{4.1.29}$$

where $\hat{\mathbf{A}} = [\hat{\mathbf{A}}'_{rk}, \hat{\mathbf{A}}'_{kk}]'$ is the $p \times k$ matrix

$$\hat{\mathbf{A}} = \mathbf{m}_{ZZ}[(1 - \hat{\lambda}_1^{-1})^{1/2}\mathbf{R}_{.1}, (1 - \hat{\lambda}_2^{-1})^{1/2}\mathbf{R}_{.2}, \ldots, (1 - \hat{\lambda}_k^{-1})^{1/2}\mathbf{R}_{.k}].$$

If $\mathbf{S}_{\varepsilon\varepsilon}$ is positive definite, the matrix $\hat{\mathbf{A}}$ can be expressed as

$$\hat{\mathbf{A}} = \mathbf{S}_{\varepsilon\varepsilon}[(\hat{\lambda}_1 - 1)^{1/2}\mathbf{T}_{.1}, (\hat{\lambda}_2 - 1)^{1/2}\mathbf{T}_{.2}, \ldots, (\hat{\lambda}_k - 1)^{1/2}\mathbf{T}_{.k}], \tag{4.1.30}$$

where $\mathbf{T}_{.i} = \hat{\lambda}_i^{1/2}\mathbf{R}_{.i}$ and the $\mathbf{T}_{.i}$ are defined in Theorem 4.1.2. See Result 4.A.11 of Appendix 4.A. That estimators of the form (4.1.29) maximize the likelihood for known singular $\boldsymbol{\Sigma}_{\varepsilon\varepsilon}$ is proved in Appendix 4.C.

The estimator of $\boldsymbol{\Sigma}_{\varepsilon\varepsilon}$ given in Theorem 4.1.2 can also be written

$$\hat{\boldsymbol{\Sigma}}_{\varepsilon\varepsilon} = (n - 1 + d_f)^{-1}[d_f\mathbf{S}_{\varepsilon\varepsilon} + (n - 1)(\mathbf{m}_{ZZ} - \hat{\boldsymbol{\Sigma}}_{zz})]. \tag{4.1.31}$$

In this form we see that the estimator of $\boldsymbol{\Sigma}_{\varepsilon\varepsilon}$ is obtained by "pooling" an estimator from the $\mathbf{Z}$ sample with the original estimator $\mathbf{S}_{\varepsilon\varepsilon}$. The $\mathbf{Z}$ sample can furnish information only on certain covariances, the covariances that define the covariance matrix of the v variables. In the transformed version of the problem defined in (4.1.20), the v variables correspond to the last $p - k$ variables of $\boldsymbol{\varepsilon}_t\mathbf{T}$.

In the model with a single equation, the estimated slope parameters are the elements of the vector $\hat{\boldsymbol{\alpha}}$ that minimize the ratio

$$(\boldsymbol{\alpha}'\mathbf{S}_{\varepsilon\varepsilon}\boldsymbol{\alpha})^{-1}\boldsymbol{\alpha}'\mathbf{m}_{ZZ}\boldsymbol{\alpha}, \tag{4.1.32}$$

where $\boldsymbol{\alpha}' = (1, -\boldsymbol{\beta}')$. The ratio (4.1.32), evaluated at $\hat{\boldsymbol{\alpha}}' = (1, -\hat{\boldsymbol{\beta}}')$, can be interpreted as the ratio of the mean square of the residuals $\hat{v}_t$ divided by the estimator of the variance of v_t. In the multivariate case we have r equations and there is a corresponding r-dimensional vector of deviations, $\mathbf{v}_t$. A generalization of the ratio (4.1.32) for the r-dimensional case is

$$(n - 1)^{-1} \sum_{t=1}^{n} (\mathbf{Z}_t - \bar{\mathbf{Z}})\mathbf{C}[\mathbf{C}'\mathbf{S}_{\varepsilon\varepsilon}\mathbf{C}]^{-1}\mathbf{C}'(\mathbf{Z}_t - \bar{\mathbf{Z}})', \tag{4.1.33}$$

where $\mathbf{C}' = (\mathbf{I}, -\boldsymbol{\beta}')$. It follows from the proof of Theorem 4.1.1 that the estimator obtained in Theorem 4.1.2 minimizes the ratio (4.1.33) and that the minimum value of the ratio is

$$\operatorname{tr}\{\hat{\mathbf{C}}'\mathbf{m}_{ZZ}\hat{\mathbf{C}}[\hat{\mathbf{C}}'\mathbf{S}_{\varepsilon\varepsilon}\hat{\mathbf{C}}]^{-1}\} = \sum_{i=k+1}^{p} \hat{\lambda}_i. \tag{4.1.34}$$

See Exercise 4.9.

Another generalization of (4.1.32) is the ratio of determinants

$$|\mathbf{C}'\mathbf{S}_{\varepsilon\varepsilon}\mathbf{C}|^{-1}|\mathbf{C}'\mathbf{m}_{ZZ}\mathbf{C}|. \tag{4.1.35}$$

This ratio is also minimized by the estimator obtained in Theorem 4.1.2 and the minimum value of the ratio is the product of the r smallest $\hat{\lambda}_i$.

If we know $\mathbf{v}_t$ and $\mathbf{Z}_t$, the best estimator of $\mathbf{x}_t$, treating $\mathbf{x}_t$ as fixed, is

$$\ddot{\mathbf{x}}_t = \mathbf{X}_t - \mathbf{v}_t\boldsymbol{\Sigma}_{vv}^{-1}\boldsymbol{\Sigma}_{vu}. \tag{4.1.36}$$

See, for example, Equations (1.2.34) and (1.2.20). When the parameters are estimated, the estimator of $\mathbf{x}_t$ is

$$\hat{\mathbf{x}}_t = \bar{\mathbf{X}} + (\mathbf{Z}_t - \bar{\mathbf{Z}})\hat{\mathbf{H}}_2, \tag{4.1.37}$$

where

$$\hat{\mathbf{H}}_2 = (\mathbf{0}, \mathbf{I}_k)' - (\mathbf{I}_r, -\hat{\boldsymbol{\beta}}')'\hat{\boldsymbol{\Sigma}}_{vv}^{-1}\hat{\boldsymbol{\Sigma}}_{vu}, \tag{4.1.38}$$

$\hat{\boldsymbol{\Sigma}}_{vv} = (\mathbf{I}_r, -\hat{\boldsymbol{\beta}}')\hat{\boldsymbol{\Sigma}}_{\varepsilon\varepsilon}(\mathbf{I}_r, -\hat{\boldsymbol{\beta}}')'$, and $\hat{\boldsymbol{\Sigma}}_{vu}' = \hat{\boldsymbol{\Sigma}}_{eu}' - \hat{\boldsymbol{\Sigma}}_{uu}\hat{\boldsymbol{\beta}}$. For nonsingular $\hat{\boldsymbol{\Sigma}}_{\varepsilon\varepsilon}$,

$$\hat{\mathbf{H}}_2 = \hat{\boldsymbol{\Sigma}}_{\varepsilon\varepsilon}^{-1}(\hat{\boldsymbol{\beta}}, \mathbf{I}_k)'[(\hat{\boldsymbol{\beta}}, \mathbf{I}_k)\hat{\boldsymbol{\Sigma}}_{\varepsilon\varepsilon}^{-1}(\hat{\boldsymbol{\beta}}, \mathbf{I}_k)']^{-1}. \tag{4.1.39}$$

The estimator of the first-order approximation to the variance of $\hat{\mathbf{x}}_t$ is

$$\hat{\mathbf{V}}\{\hat{\mathbf{x}}_t - \mathbf{x}_t\} = \hat{\boldsymbol{\Sigma}}_{uu} - \hat{\boldsymbol{\Sigma}}_{uv}\hat{\boldsymbol{\Sigma}}_{vv}^{-1}\hat{\boldsymbol{\Sigma}}_{vu}. \tag{4.1.40}$$

We demonstrate that the first k columns of $\hat{\mathbf{A}}$ of (4.1.29) define functions of the $\mathbf{Z}_t$ that are linear combinations of $\hat{\mathbf{x}}_t$ and the last r columns define functions of the $\mathbf{Z}_t$ that are linear combinations of $\hat{\mathbf{v}}_t$. The matrix $\mathbf{R}$ is the matrix of characteristic vectors of $\mathbf{S}_{\varepsilon\varepsilon}$ in the metric $\mathbf{m}_{ZZ}$ ordered so that the roots $\hat{\lambda}_i^{-1}$ are in increasing order and we can write

$$\mathbf{R} = (\mathbf{R}_k, \mathbf{R}_r), \tag{4.1.41}$$

where $\mathbf{R}_r = (\mathbf{R}_{rr}', \mathbf{R}_{kr}')' = (\mathbf{I}, -\hat{\boldsymbol{\beta}}')'\mathbf{R}_{rr}$. Therefore, the estimated contrasts

$$(\mathbf{Z}_t - \bar{\mathbf{Z}})\mathbf{R}_r = \hat{\mathbf{v}}_t\mathbf{R}_{rr} \tag{4.1.42}$$

are linear combinations of $\hat{\mathbf{v}}_t = \mathbf{Y}_t - \hat{\boldsymbol{\beta}}_0 - \mathbf{X}_t\hat{\boldsymbol{\beta}}$.

Direct multiplication will verify that

$$\mathbf{R}_r'\mathbf{S}_{\varepsilon\varepsilon}\hat{\mathbf{H}}_2 = \mathbf{0}. \tag{4.1.43}$$

where $\hat{\mathbf{H}}_2$ is defined in (4.1.38). Because $\mathbf{R}$ is of full rank and because

$$\mathbf{R}'\mathbf{S}_{\varepsilon\varepsilon}\mathbf{R} = \text{diag}(\hat{\lambda}_1^{-1}, \hat{\lambda}_2^{-1}, \ldots, \hat{\lambda}_p^{-1}), \tag{4.1.44}$$

it follows that the columns of $\mathbf{R}_k$ are linear combinations of the columns of $\hat{\mathbf{H}}_2$ and that the elements of $(\mathbf{Z}_t - \bar{\mathbf{Z}})\mathbf{R}_k$ are linear functions of the elements of $(\hat{\mathbf{x}}_t - \bar{\mathbf{X}})$. Therefore, the transformation $\mathbf{R}$ partitions $(\mathbf{Z}_t - \bar{\mathbf{Z}})$ into a part corresponding to $\hat{\mathbf{x}}_t$ and a part corresponding to $\hat{\mathbf{v}}_t$. Furthermore, the transformed variables are standardized so that the observed moment matrix of $(\mathbf{Z}_t - \bar{\mathbf{Z}})\mathbf{R}$ is the identity matrix.

The model (4.1.12) contains p mean parameters, $[(p-k)k + \frac{1}{2}k(k+1)]$ parameters for the covariance matrix of $\mathbf{Z}_t$, and the parameters for the covariance matrix of $\boldsymbol{\varepsilon}_t$. Because there are $\frac{1}{2}p(p+1)$ unique elements in the sample covariance matrix of $\mathbf{Z}_t$ and $\frac{1}{2}p(p+1)$ exceeds the number of parameters defining the covariance matrix of $\mathbf{Z}_t$, it is possible to construct a goodness-of-fit test of the model (4.1.12). The alternative model for $\mathbf{Z}_t$ is taken to be the normal distribution with unconstrained mean and unconstrained positive definite covariance matrix.

The likelihood ratio criterion for testing model (4.1.12) against the unconstrained alternative is given in Theorem 4.1.3. The test does not require a positive definite $\mathbf{S}_{\varepsilon\varepsilon}$.

Theorem 4.1.3. Let the null model be the model (4.1.12) and let the alternative model be the unconstrained model. Then

$$2[\log L_c(\mathbf{m}_{ZZ}, \mathbf{S}_{\varepsilon\varepsilon}) - \log L_c(\hat{\boldsymbol{\theta}})] = (n-1) \sum_{i=q+1}^{p} (\log \hat{\sigma}_{T\varepsilon\varepsilon ii} - \log \hat{\lambda}_i)$$

$$+ d_f \sum_{i=q+1}^{p} \log \hat{\sigma}_{T\varepsilon\varepsilon ii},$$

where $q = \min\{k, \tau\}$, τ is the number of $\hat{\lambda}_i$ that are greater than one, $L_c(\mathbf{m}_{ZZ}, \mathbf{S}_{\varepsilon\varepsilon})$ is the logarithm of the likelihood of the sample adjusted for degrees of freedom evaluated at $(\boldsymbol{\Sigma}_{ZZ}, \boldsymbol{\Sigma}_{\varepsilon\varepsilon}) = (\mathbf{m}_{ZZ}, \mathbf{S}_{\varepsilon\varepsilon})$, and $L_c(\hat{\boldsymbol{\theta}})$ is the logarithm of the likelihood evaluated at $(\boldsymbol{\Sigma}_{ZZ}, \boldsymbol{\Sigma}_{\varepsilon\varepsilon}) = (\hat{\boldsymbol{\Sigma}}_{ZZ}, \hat{\boldsymbol{\Sigma}}_{\varepsilon\varepsilon})$ with $\hat{\boldsymbol{\Sigma}}_{ZZ}$ and $\hat{\boldsymbol{\Sigma}}_{\varepsilon\varepsilon}$ defined in Theorem 4.1.2.

Under the null model the test statistic converges in distribution to a chi-square random variable with $\frac{1}{2}r(r+1)$ degrees of freedom as $n \to \infty$ and $d_f^{-1} = O(n^{-1})$.

Proof. Under the unconstrained model, the estimator of $\boldsymbol{\Sigma}_{TZZ}$ is $\hat{\boldsymbol{\Lambda}}$ and the estimator of $\boldsymbol{\Sigma}_{T\varepsilon\varepsilon}$ is $\mathbf{I}$. Therefore, the expression for the test statistic follows from (4.1.24). For the unconstrained parameter space, $\boldsymbol{\Sigma}_{ZZ}$ is any symmetric positive definite matrix. We transform the parameters defining $\boldsymbol{\Sigma}_{ZZ}$ on the constrained space into the set $(\boldsymbol{\beta}, \mathbf{H}_{xx}, \mathbf{H}_{yy.x})$ by the one-to-one transformation,

$$\boldsymbol{\Sigma}_{ZZ} = \mathbf{H}_{zz} + \boldsymbol{\Sigma}_{\varepsilon\varepsilon},$$
$$\mathbf{H}_{zz} = (\boldsymbol{\beta}, \mathbf{I})'\mathbf{H}_{xx}(\boldsymbol{\beta}, \mathbf{I}) + (\mathbf{I}, \mathbf{0})'\mathbf{H}_{yy.x}(\mathbf{I}, \mathbf{0}),$$

where $\boldsymbol{\beta}$ is a $k \times r$ matrix, $\mathbf{H}_{xx}$ is a $k \times k$ symmetric matrix, and $\mathbf{H}_{yy.x}$ is an $r \times r$ symmetric matrix such that $\mathbf{H}_{zz} + \boldsymbol{\Sigma}_{\varepsilon\varepsilon}$ is positive definite. Under the null model, $\mathbf{H}_{xx} = \boldsymbol{\Sigma}_{xx}$ is positive definite and $\mathbf{H}_{yy.x} = \mathbf{0}$. Since there are $2^{-1}r(r+1)$ distinct elements in $\mathbf{H}_{yy.x}$, the limiting null distribution follows from the standard likelihood theory. □

Because the test statistic depends only on $\hat{\mathbf{v}}_t$, the limiting distribution of Theorem 4.1.3 holds for any reasonable $\mathbf{x}_t$ if the $\boldsymbol{\varepsilon}_t$ are normally distributed. See Miller (1984) and Amemiya (1985b).

The degrees of freedom for the likelihood ratio statistic is the difference between the number of unique elements in the covariance matrix of $\mathbf{Z}_t$ and the number of parameters estimated. The difference is the number of unique elements in the covariance matrix of $\mathbf{v}_t$. Therefore, the likelihood ratio test is a measure of the difference between the sample covariance matrix of $\hat{\mathbf{v}}_t$ and the estimator of $\boldsymbol{\Sigma}_{vv}$ constructed as $(\mathbf{I}, -\hat{\boldsymbol{\beta}}')\mathbf{S}_{\varepsilon\varepsilon}(\mathbf{I}, -\hat{\boldsymbol{\beta}}')'$. The $\hat{v}_{ti}$ entering the likelihood ratio expression are analogous to the residuals from a regression in the sense that k coefficients are estimated in order to estimate the $\hat{v}_{ti}$ for each i. Therefore, by analogy to regression, we suggest that the test statistic be constructed as

$$\hat{F} = 2[r(r+1)]^{-1}\left[(n-k-1)\sum_{i=q+1}^{p}(\log \hat{\sigma}_{T\varepsilon ii} - \log \hat{\lambda}_i) + d_f \sum_{i=q+1}^{p} \log \hat{\sigma}_{T\varepsilon ii}\right] \tag{4.1.45}$$

and that the distribution be approximated by that of the central F distribution with $\frac{1}{2}r(r+1)$ and d_f degrees of freedom.

If $\boldsymbol{\Sigma}_{\varepsilon\varepsilon}$ is known, the likelihood ratio statistic

$$\hat{\chi}^2 = (n-k-1)\left[\sum_{j=q+1}^{p}(\hat{\lambda}_j - \log \hat{\lambda}_j) - (p-q)\right] \tag{4.1.46}$$

is approximately distributed as a chi-square random variable with $\frac{1}{2}r(r+1)$ degrees of freedom.

Another function of the roots that is also a useful diagnostic tool is the sum

$$\sum_{i=k+1}^{p} \hat{\lambda}_i = \operatorname{tr}\{\hat{\mathbf{C}}'\mathbf{m}_{ZZ}\hat{\mathbf{C}}[\hat{\mathbf{C}}'\mathbf{S}_{\varepsilon\varepsilon}\hat{\mathbf{C}}]^{-1}\}, \tag{4.1.47}$$

introduced in (4.1.34). The sum of the r smallest $\hat{\lambda}_i$ is a monotone function of individual $\hat{\lambda}_i$, while the likelihood ratio increases as an individual $\hat{\lambda}_i$ moves away from one in either direction. If the assumptions of Theorem 4.1.2 hold,

$$\sum_{i=k+1}^{p} \hat{\lambda}_i = \operatorname{tr}\{(\mathbf{m}_{vv} - \mathbf{m}_{vx}\mathbf{m}_{xx}^{-1}\mathbf{m}_{xv})\ddot{\mathbf{S}}_{vv}^{-1}\} + O_p(n^{-1}),$$

where $\ddot{\mathbf{S}}_{vv} = (1, -\boldsymbol{\beta}')\mathbf{S}_{\varepsilon\varepsilon}(1, -\boldsymbol{\beta}')'$. It follows that

$$F = (n-1)[r(n-k-1)]^{-1} \sum_{i=k+1}^{p} \hat{\lambda}_i \tag{4.1.48}$$

is approximately distributed as a central F with $r(n-k-1)$ and d_f degrees of freedom.

4.1.2. Properties of Estimators

We give the limiting distribution of the maximum likelihood estimators under weaker conditions than used in the derivation. We begin by demonstrating the strong consistency of the estimators.

Theorem 4.1.4. Let model (4.1.12) hold without the normal distribution assumption. Assume that $\boldsymbol{\varepsilon}_t$ are independently and identically distributed with mean zero and positive definite covariance matrix $\boldsymbol{\Sigma}_{\varepsilon\varepsilon}$. Let $\mathbf{S}_{\varepsilon\varepsilon}$ be an unbiased estimator of $\boldsymbol{\Sigma}_{\varepsilon\varepsilon}$ distributed as a multiple of a Wishart matrix with d_f degrees of freedom, where $d_f^{-1} = O(n^{-1})$. Assume that $\mathbf{x}_t - \boldsymbol{\mu}_{xt}$ are independently and identically distributed with mean zero and covariance matrix $\boldsymbol{\Sigma}_{xx}$, independent of $\boldsymbol{\varepsilon}_j$ for all t and j. Assume the fixed sequence $\{\boldsymbol{\mu}_{xt}\}$ satisfies

$$\lim_{n\to\infty} n^{-1} \sum_{t=1}^{n} \boldsymbol{\mu}_{xt} = \bar{\boldsymbol{\mu}}_{x.},$$

$$\lim_{n\to\infty} n^{-1} \sum_{t=1}^{n} (\boldsymbol{\mu}_{xt} - \bar{\boldsymbol{\mu}}_{x})'(\boldsymbol{\mu}_{xt} - \bar{\boldsymbol{\mu}}_{x}) = \bar{\mathbf{m}}_{\mu\mu},$$

where $\bar{\mathbf{m}}_{xx} = \boldsymbol{\Sigma}_{xx} + \bar{\mathbf{m}}_{\mu\mu}$ is positive definite. Then, as $n \to \infty$,

$$(\hat{\boldsymbol{\beta}}_0, \hat{\boldsymbol{\beta}}, \hat{\boldsymbol{\Sigma}}_{xx}) \to (\boldsymbol{\beta}_0, \boldsymbol{\beta}, \bar{\mathbf{m}}_{xx}) \quad \text{a.s.},$$

where $(\hat{\boldsymbol{\beta}}_0, \hat{\boldsymbol{\beta}}, \hat{\boldsymbol{\Sigma}}_{xx})$ is defined in Theorem 4.1.2.

Proof. Our proof follows arguments given by Gleser (1981) and by Amemiya and Fuller (1984). It can be shown that, as $n \to \infty$,

$$\mathbf{m}_{ZZ} \to \boldsymbol{\Sigma}_{ZZ}, \quad \text{a.s.},$$

where $\boldsymbol{\Sigma}_{ZZ} = \boldsymbol{\Sigma}_{zz} + \boldsymbol{\Sigma}_{\varepsilon\varepsilon}$, $\boldsymbol{\Sigma}_{zz} = (\boldsymbol{\beta}, \mathbf{I})'\bar{\mathbf{m}}_{xx}(\boldsymbol{\beta}, \mathbf{I})$, and thus,

$$\mathbf{S}_{\varepsilon\varepsilon}^{-1/2}\mathbf{m}_{ZZ}\mathbf{S}_{\varepsilon\varepsilon}^{-1/2} \to \boldsymbol{\Sigma}_{\varepsilon\varepsilon}^{-1/2}\boldsymbol{\Sigma}_{zz}\boldsymbol{\Sigma}_{\varepsilon\varepsilon}^{-1/2} + \mathbf{I}_p, \quad \text{a.s.}$$

If we let λ_i be the eigenvalues of $\boldsymbol{\Sigma}_{\varepsilon\varepsilon}^{-1/2}\boldsymbol{\Sigma}_{ZZ}\boldsymbol{\Sigma}_{\varepsilon\varepsilon}^{-1/2}$, then $\lambda_i > 1$ for $i = 1, 2, \ldots, k$ and $\lambda_i = 1$ for $i = k + 1, k + 2, \ldots, p$. A matrix of orthonormal eigenvectors corresponding to the r roots that equal one is

$$\mathbf{Q}_{(r)} = \boldsymbol{\Sigma}_{\varepsilon\varepsilon}^{1/2}(\mathbf{I}, -\boldsymbol{\beta}')'\boldsymbol{\Sigma}_{vv}^{-1/2}\mathbf{G},$$

where $\mathbf{G}$ is an $r \times r$ orthogonal matrix. Let $\hat{\lambda}_1 \geqslant \hat{\lambda}_2 \geqslant \ldots \geqslant \hat{\lambda}_p$ be the roots of (4.1.16) and let

$$\hat{\boldsymbol{\Lambda}} = \text{diag}(\hat{\lambda}_1, \hat{\lambda}_2, \ldots, \hat{\lambda}_p) = \text{block diag}(\hat{\boldsymbol{\Lambda}}_{(k)}\hat{\boldsymbol{\Lambda}}_{(r)}),$$

where $\hat{\boldsymbol{\Lambda}}_{(r)}$ contains the r smallest roots. The eigenvalues $\hat{\lambda}_i$ are locally continuous functions of the elements of $\mathbf{S}_{\varepsilon\varepsilon}^{-1/2}\mathbf{m}_{ZZ}\mathbf{S}_{\varepsilon\varepsilon}^{-1/2}$. Thus,

$$\hat{\boldsymbol{\Lambda}}_{(r)} \to \mathbf{I}_r, \quad \text{a.s.}$$

Let ω be a point in the probability space of all sequences of observations. Fix an ω such that $\mathbf{m}_{ZZ}(\omega) \to \boldsymbol{\Sigma}_{ZZ}$, $\mathbf{S}_{\varepsilon\varepsilon}(\omega) \to \boldsymbol{\Sigma}_{\varepsilon\varepsilon}$, and $\hat{\boldsymbol{\Lambda}}_{(r)}(\omega) \to \mathbf{I}_r$. The set of such ω has probability one. Let $\hat{\mathbf{Q}} = (\hat{\mathbf{Q}}_{(k)}, \hat{\mathbf{Q}}_{(r)})$ be the matrix of characteristic vectors of $\mathbf{S}_{\varepsilon\varepsilon}^{-1/2}\mathbf{m}_{ZZ}\mathbf{S}_{\varepsilon\varepsilon}^{-1/2}$ associated with the roots $\hat{\lambda}_1 \geqslant \hat{\lambda}_2 \geqslant \cdots \geqslant \hat{\lambda}_p$, where the partition corresponds to the k largest and r smallest roots. Since $\hat{\mathbf{Q}}'_{(r)}(\omega)\hat{\mathbf{Q}}_{(r)}(\omega) = \mathbf{I}$ for all n, each element of $\hat{\mathbf{Q}}_{(r)}(\omega)$ is bounded. Thus, for every subsequence of $\{\hat{\mathbf{Q}}_{(r)}(\omega)\}_{n=1}^{\infty}$, there exists a convergent subsubsequence. Taking the limit over such subsubsequences on both sides of the equation

$$[\mathbf{S}_{\varepsilon\varepsilon}(\omega)]^{-1/2}\mathbf{m}_{ZZ}(\omega)[\mathbf{S}_{\varepsilon\varepsilon}(\omega)]^{-1/2}\hat{\mathbf{Q}}_{(r)}(\omega) = \hat{\mathbf{Q}}_{(r)}(\omega)\hat{\boldsymbol{\Lambda}}_{(r)}(\omega),$$

we find that the limit of a convergent subsubsequence of $\hat{\mathbf{Q}}_{(r)}(\omega)$ is of the form $\mathbf{Q}_{(r)}$ for some orthogonal $\mathbf{G}(\omega)$, where $\mathbf{G}(\omega)$ depends on the subsequence and ω. Hence, the limit of $\mathbf{T}_{(r)}(\omega) = \mathbf{S}_{\varepsilon\varepsilon}^{-1/2}\hat{\mathbf{Q}}_{(r)}(\omega)$ for such a subsubsequence is $(\mathbf{I}, -\boldsymbol{\beta}')'\boldsymbol{\Sigma}_{vv}^{-1/2}\mathbf{G}(\omega)$ and the limit of $\hat{\boldsymbol{\beta}}(\omega) = -\mathbf{T}_{kr}(\omega)\mathbf{T}_{rr}^{-1}(\omega)$ is $\boldsymbol{\beta}$, where the partition of $\mathbf{T}$ is as defined for $\mathbf{R}$ in (4.1.41). Since the limit is the same for every subsequence, $\hat{\boldsymbol{\beta}}(\omega)$ converges to $\boldsymbol{\beta}$ and $\hat{\boldsymbol{\beta}}$ converges to $\boldsymbol{\beta}$ a.s. It follows that

$$\hat{\mathbf{H}}_2 = \mathbf{S}_{\varepsilon\varepsilon}^{-1}(\hat{\boldsymbol{\beta}}, \mathbf{I})'\hat{\boldsymbol{\Sigma}}_{\rho\rho} \to \boldsymbol{\Sigma}_{\varepsilon\varepsilon}^{-1}(\boldsymbol{\beta}, \mathbf{I})'\boldsymbol{\Sigma}_{\rho\rho}, \quad \text{a.s.,}$$

where $\hat{\boldsymbol{\Sigma}}_{\rho\rho} = [(\hat{\boldsymbol{\beta}}, \mathbf{I}_k)\mathbf{S}_{\varepsilon\varepsilon}^{-1}(\hat{\boldsymbol{\beta}}, \mathbf{I}_k)']^{-1}$ and $\boldsymbol{\Sigma}_{\rho\rho} = [(\boldsymbol{\beta}, \mathbf{I}_k)\boldsymbol{\Sigma}_{\varepsilon\varepsilon}^{-1}(\boldsymbol{\beta}, \mathbf{I}_k)']^{-1}$. Therefore,

$$\hat{\boldsymbol{\Sigma}}_{xx} = \hat{\mathbf{H}}_2'(\mathbf{m}_{ZZ} - \mathbf{S}_{\varepsilon\varepsilon})\hat{\mathbf{H}}_2 \to \boldsymbol{\Sigma}_{xx}, \quad \text{a.s.}$$

Also $\hat{\boldsymbol{\beta}}_0 \to \boldsymbol{\beta}_0$ a.s., because, $\bar{\mathbf{X}} \to \boldsymbol{\mu}_x$, a.s., and $\bar{\mathbf{Y}} \to \boldsymbol{\beta}_0 + \boldsymbol{\mu}_x\boldsymbol{\beta}$, a.s. □

Theorem 4.1.4 was stated and proved for positive definite $\Sigma_{\varepsilon\varepsilon}$, but the result also holds for singular $\Sigma_{\varepsilon\varepsilon}$.

The limiting distribution of the vector of estimators is derived in Theorem 4.1.5. The statement and proof are given in considerable detail for reference purposes. An alternative proof of asymptotic normality is given in Theorem 4.B.2 with a representation of the covariance matrix given in Corollary 4.B.2.

Theorem 4.1.5. Let

$$\mathbf{y}_t = \mathbf{x}_t \boldsymbol{\beta}, \qquad \mathbf{Z}_t = \mathbf{z}_t + \boldsymbol{\varepsilon}_t, \qquad \boldsymbol{\varepsilon}_t \sim \mathrm{NI}(\mathbf{0}, \boldsymbol{\Sigma}_{\varepsilon\varepsilon}),$$

where Σ_{vv} is positive definite. Let $d_f \mathbf{S}_{\varepsilon\varepsilon}$ be distributed as a Wishart matrix with matrix parameter $\Sigma_{\varepsilon\varepsilon}$ and d_f degrees of freedom independent of $\mathbf{Z}_t$. Let

$$c = \lim_{n \to \infty} d_f^{-1} n, \qquad 0 \leqslant c < \infty.$$

Let $\hat{\boldsymbol{\beta}}$ and $\hat{\boldsymbol{\Sigma}}_{zz}$ be defined by (4.1.29) with $\mathbf{R}$ the matrix of characteristic vectors of $\mathbf{S}_{\varepsilon\varepsilon}$ in the metric $\mathbf{M}_{ZZ}$. Let

$$\hat{\boldsymbol{\Sigma}}_{\varepsilon\varepsilon} = (n - 1 + d_f)^{-1}[d_f \mathbf{S}_{\varepsilon\varepsilon} + (n-1)(\mathbf{M}_{ZZ} - \hat{\boldsymbol{\Sigma}}_{zz})].$$

Assume that the $\mathbf{x}_t - \boldsymbol{\mu}_{xt}$ are independently and identically distributed with mean zero and covariance matrix Σ_{xx}, independent of $\boldsymbol{\varepsilon}_j$ for all t and j. Assume that the fixed sequence $\{\boldsymbol{\mu}_{xt}\}$ satisfies

$$\lim_{n \to \infty} n^{-1} \sum_{t=1}^{n} \boldsymbol{\mu}_{xt} = \bar{\boldsymbol{\mu}}_{x\cdot},$$

$$\lim_{n \to \infty} n^{-1} \sum_{t=1}^{n} \boldsymbol{\mu}'_{xt} \boldsymbol{\mu}_{xt} = \bar{\mathbf{M}}_{\mu\mu},$$

where $\bar{\mathbf{M}}_{xx} = \bar{\mathbf{M}}_{\mu\mu} + \Sigma_{xx}$ is positive definite.

Let $\hat{\boldsymbol{\theta}} = [(\mathrm{vec}\ \hat{\boldsymbol{\beta}})', (\mathrm{vech}\ \hat{\boldsymbol{\Sigma}}_{\varepsilon\varepsilon})']'$ and let $\boldsymbol{\theta}$ be the corresponding parameter vector. Then $n^{1/2}(\hat{\boldsymbol{\theta}} - \boldsymbol{\theta})$ converges in distribution to a normal random vector with zero mean and covariance matrix

$$\boldsymbol{\Gamma}_{\theta\theta} = \begin{bmatrix} \boldsymbol{\Gamma}_{\beta\beta} & \boldsymbol{\Gamma}_{\beta\varepsilon} \\ \boldsymbol{\Gamma}'_{\beta\varepsilon} & \boldsymbol{\Gamma}_{\varepsilon\varepsilon} \end{bmatrix}, \tag{4.1.49}$$

where

$$\boldsymbol{\Gamma}_{\beta\beta} = \boldsymbol{\Sigma}_{vv} \otimes [\bar{\mathbf{M}}_{xx}^{-1} + (1+c)\bar{\mathbf{M}}_{xx}^{-1}\boldsymbol{\Sigma}_{\rho\rho}\bar{\mathbf{M}}_{xx}^{-1}],$$
$$\boldsymbol{\Gamma}_{\beta\varepsilon} = -2c\{\boldsymbol{\Sigma}_{v\varepsilon} \otimes [\bar{\mathbf{M}}_{xx}^{-1}\boldsymbol{\Sigma}_{\rho\rho}(\boldsymbol{\beta}, \mathbf{I})]\}\boldsymbol{\psi}'_p,$$
$$\boldsymbol{\Gamma}_{\varepsilon\varepsilon} = 2\boldsymbol{\psi}_p[c(\boldsymbol{\Sigma}_{\varepsilon\varepsilon} \otimes \boldsymbol{\Sigma}_{\varepsilon\varepsilon}) - c^2(1+c)^{-1}(\boldsymbol{\Sigma}_{\varepsilon v}\boldsymbol{\Sigma}_{vv}^{-1}\boldsymbol{\Sigma}_{v\varepsilon}) \otimes (\boldsymbol{\Sigma}_{\varepsilon v}\boldsymbol{\Sigma}_{vv}^{-1}\boldsymbol{\Sigma}_{v\varepsilon})]\boldsymbol{\psi}'_p,$$

$\boldsymbol{\rho}_t = \mathbf{u}_t - \boldsymbol{\xi}_t$, $\boldsymbol{\xi}_t = \mathbf{v}_t \boldsymbol{\Sigma}_{vv}^{-1}\boldsymbol{\Sigma}_{vu}$, $\boldsymbol{\Sigma}_{\rho\rho} = \boldsymbol{\Sigma}_{uu} - \boldsymbol{\Sigma}_{\xi\xi}$, and $\boldsymbol{\psi}_p$ is defined in Appendix 4.A.

If, in addition, $\mathbf{x}_t = (1, \mathbf{x}_{t1})$, where $\mathbf{x}_{t1} \sim \mathrm{NI}(\boldsymbol{\mu}_x, \boldsymbol{\Sigma}_{xx})$ and $\boldsymbol{\mu}_x$ is a fixed vector, then

$$n^{1/2}(\hat{\gamma} - \gamma) \xrightarrow{L} N(\mathbf{0}, \boldsymbol{\Gamma}_{\gamma\gamma}),$$

where $\hat{\gamma} = [(\text{vec } \hat{\boldsymbol{\beta}})', (\text{vech } \hat{\boldsymbol{\Sigma}}_{\varepsilon\varepsilon})', (\text{vech } \hat{\boldsymbol{\Sigma}}_{xx})']$, γ is the corresponding parameter vector, $\hat{\boldsymbol{\Sigma}}_{xx}$ is defined in Theorem 4.1.2, the upper left portion of $\boldsymbol{\Gamma}_{\gamma\gamma}$ is $\boldsymbol{\Gamma}_{\theta\theta}$, and the remaining submatrices of $\boldsymbol{\Gamma}_{\gamma\gamma}$ are

$$\begin{aligned}
\boldsymbol{\Gamma}_{\beta x} &= 2\{\boldsymbol{\Sigma}_{vu} \otimes (\bar{\mathbf{M}}_{xx}^{-1}[\bar{\mathbf{M}}_{xx} + (1 + c)\boldsymbol{\Sigma}_{\rho\rho}])\}\boldsymbol{\psi}'_k,\\
\boldsymbol{\Gamma}_{\varepsilon x} &= 2c\boldsymbol{\psi}_p[(\boldsymbol{\Sigma}_{\varepsilon v}\boldsymbol{\Sigma}_{vv}^{-1}\boldsymbol{\Sigma}_{vu}) \otimes (\boldsymbol{\Sigma}_{\varepsilon v}\boldsymbol{\Sigma}_{vv}^{-1}\boldsymbol{\Sigma}_{vu}) - (\boldsymbol{\Sigma}_{\varepsilon u} \otimes \boldsymbol{\Sigma}_{\varepsilon u})]\boldsymbol{\psi}'_k,\\
\boldsymbol{\Gamma}_{xx} &= 2\boldsymbol{\psi}_k[(\boldsymbol{\Sigma}_{XX} \otimes \boldsymbol{\Sigma}_{XX}) + c(\boldsymbol{\Sigma}_{uu} \otimes \boldsymbol{\Sigma}_{uu}) - (1 + c)(\boldsymbol{\Sigma}_{\xi\xi} \otimes \boldsymbol{\Sigma}_{\xi\xi})]\boldsymbol{\psi}'_k.
\end{aligned}$$

Proof. Under the assumptions, the probability that $\hat{\lambda}_k^{-1}$ is less than one is coverging to one as n increases and we may restrict our attention to expression (4.1.29). By the definition of $\mathbf{R}$,

$$\mathbf{R}'_r = \mathbf{R}'_{rr}(\mathbf{I}, -\hat{\boldsymbol{\beta}}'), \tag{4.1.50}$$

$$\mathbf{R}'_{rr}\hat{\boldsymbol{\Sigma}}_{vv}\mathbf{R}_{rr} = \hat{\boldsymbol{\Lambda}}_r^{-1}, \tag{4.1.51}$$

$$\mathbf{R}'_{rr}\hat{\mathbf{M}}_{vv}\mathbf{R}_{rr} = \mathbf{I}_r, \tag{4.1.52}$$

where

$$\hat{\boldsymbol{\Sigma}}_{vv} = (\mathbf{I}, -\hat{\boldsymbol{\beta}}')\mathbf{S}_{\varepsilon\varepsilon}(\mathbf{I}, -\hat{\boldsymbol{\beta}}')', \qquad \hat{\mathbf{M}}_{vv} = (\mathbf{I}, -\hat{\boldsymbol{\beta}}')\mathbf{M}_{ZZ}(\mathbf{I}, -\hat{\boldsymbol{\beta}}')',$$
$$\hat{\boldsymbol{\Lambda}}_r^{-1} = \text{diag}(\hat{\lambda}_{k+1}^{-1}, \hat{\lambda}_{k+2}^{-1}, \ldots, \hat{\lambda}_p^{-1}).$$

By the definition of $\hat{\boldsymbol{\beta}}$ and $\hat{\boldsymbol{\Sigma}}_{zz}$,

$$\hat{\boldsymbol{\Sigma}}_{zz}\hat{\mathbf{C}} = \mathbf{0}, \tag{4.1.53}$$

where $\hat{\mathbf{C}}' = (\mathbf{I}, -\hat{\boldsymbol{\beta}}')$ and $\hat{\boldsymbol{\Sigma}}_{zz} = \hat{\mathbf{A}}\hat{\mathbf{A}}'$. It follows that

$$\hat{\boldsymbol{\beta}} - \boldsymbol{\beta} = \hat{\boldsymbol{\Sigma}}_{xx}^{-1}(\hat{\boldsymbol{\Sigma}}_{xy} - \hat{\boldsymbol{\Sigma}}_{xx}\boldsymbol{\beta}). \tag{4.1.54}$$

Now

$$\begin{aligned}
\mathbf{M}_{ZZ} - \hat{\boldsymbol{\Sigma}}_{zz} - \mathbf{S}_{\varepsilon\varepsilon} &= \mathbf{M}_{ZZ}\mathbf{R}_r(\mathbf{I} - \hat{\boldsymbol{\Lambda}}_r^{-1})\mathbf{R}'_r\mathbf{M}_{ZZ}\\
&= \hat{\boldsymbol{\Sigma}}_{\varepsilon v}\mathbf{R}_{rr}(\hat{\boldsymbol{\Lambda}}_r^2 - \hat{\boldsymbol{\Lambda}}_r)\mathbf{R}'_{rr}\hat{\boldsymbol{\Sigma}}_{v\varepsilon}\\
&= \hat{\boldsymbol{\Sigma}}_{\varepsilon v}\hat{\boldsymbol{\Sigma}}_{vv}^{-1}(\hat{\mathbf{M}}_{vv} - \hat{\boldsymbol{\Sigma}}_{vv})\hat{\boldsymbol{\Sigma}}_{vv}^{-1}\hat{\boldsymbol{\Sigma}}_{v\varepsilon},
\end{aligned} \tag{4.1.55}$$

where $\hat{\boldsymbol{\Sigma}}_{\varepsilon v} = S_{\varepsilon\varepsilon}(1, -\hat{\boldsymbol{\beta}}')'$ and we have used $\mathbf{M}_{ZZ}\mathbf{R}_r = \mathbf{S}_{\varepsilon\varepsilon}\mathbf{R}_r\hat{\Lambda}_r$, (4.1.50), (4.1.51), (4.1.52), and the spectral representation of $\mathbf{S}_{\varepsilon\varepsilon}$ in the metric $\mathbf{M}_{ZZ}$. It follows that

$$\hat{\boldsymbol{\Sigma}}_{zz} = \mathbf{M}_{ZZ} - \mathbf{S}_{\varepsilon\varepsilon} - \hat{\boldsymbol{\Sigma}}_{\varepsilon v}\hat{\boldsymbol{\Sigma}}_{vv}^{-1}(\hat{\mathbf{M}}_{vv} - \hat{\boldsymbol{\Sigma}}_{vv})\hat{\boldsymbol{\Sigma}}_{vv}^{-1}\hat{\boldsymbol{\Sigma}}_{v\varepsilon} \tag{4.1.56}$$

and, from (4.1.54),

$$\hat{\beta} - \beta = \hat{\Sigma}_{xx}^{-1}[\mathbf{M}_{Xv} - \mathbf{S}_{uv} - \hat{\Sigma}_{uv}\hat{\Sigma}_{vv}^{-1}(\hat{\mathbf{M}}_{vv} - \hat{\Sigma}_{vv})\hat{\Sigma}_{vv}^{-1}(\mathbf{I}, -\hat{\beta}')\mathbf{S}_{\varepsilon\varepsilon}(\mathbf{I}, -\beta)']. \tag{4.1.57}$$

Using equality (4.1.57) and the fact that $\hat{\beta}$ is consistent, we have

$$\begin{aligned} \hat{\beta} - \beta &= \bar{\mathbf{M}}_{xx}^{-1}[\mathbf{M}_{Xv} - \mathbf{S}_{uv} - \Sigma_{uv}\Sigma_{vv}^{-1}(\mathbf{M}_{vv} - \mathbf{S}_{vv})] + o_p(n^{-1/2}) \\ &= \bar{\mathbf{M}}_{xx}^{-1}[\mathbf{M}_{\eta v} - \mathbf{S}_{\rho v}] + o_p(n^{-1/2}), \end{aligned} \tag{4.1.58}$$

where $\mathbf{M}_{vv} = (\mathbf{I}, -\beta')\mathbf{M}_{ZZ}(\mathbf{I}, -\beta')'$, $\boldsymbol{\eta}_t = \mathbf{X}_t - \boldsymbol{\xi}_t$ and $\mathbf{S}_{vv} = (\mathbf{I}, -\beta')\mathbf{S}_{\varepsilon\varepsilon}(\mathbf{I}, -\beta')'$.

If $\mathbf{x}_t = (1, \mathbf{x}_{t1})$ and $\mathbf{x}_{t1}$ is normally distributed, then

$$\hat{\Sigma}_{xx} - \Sigma_{xx} = \mathbf{m}_{XX} - \Sigma_{XX} - (\mathbf{S}_{uu} - \Sigma_{uu}) - (\mathbf{m}_{\xi\xi} - \mathbf{S}_{\xi\xi}) + o_p(n^{-1/2}), \tag{4.1.59}$$

where $\mathbf{m}_{\xi\xi} = \Sigma_{uv}\Sigma_{vv}^{-1}\mathbf{m}_{vv}\Sigma_{vv}^{-1}\Sigma_{vu}$ and $\mathbf{S}_{\xi\xi} = \Sigma_{uv}\Sigma_{vv}^{-1}\mathbf{S}_{vv}\Sigma_{vv}^{-1}\Sigma_{vu}$. We have

$$\begin{aligned} \mathbf{T}_{rr}(\hat{\Lambda}_r - \mathbf{I}_r)\mathbf{T}_{rr}' &= \Sigma_{vv}^{-1}(\mathbf{m}_{vv} - \mathbf{S}_{vv})\Sigma_{vv}^{-1} + o_p(n^{-1/2}) \\ \mathbf{T}_{kr}\mathbf{T}_{rr}^{-1} &= -\beta + O_p(n^{-1/2}), \end{aligned}$$

where $\hat{\Lambda}_r = \operatorname{diag}(\hat{\lambda}_{k+1}, \hat{\lambda}_{k+2}, \ldots, \hat{\lambda}_p)$ and the partition of the $\mathbf{T}$ of (4.1.19) is

$$\mathbf{T} = (\mathbf{T}_k, \mathbf{T}_r), \quad \mathbf{T}_k = (\mathbf{T}_{rk}', \mathbf{T}_{kk}')', \quad \text{and} \quad \mathbf{T}_r = (\mathbf{T}_{rr}', \mathbf{T}_{kr}')'.$$

It follows that

$$\begin{aligned} \hat{\Sigma}_{\varepsilon\varepsilon} &= (n - 1 + d_f)^{-1}\mathbf{S}_{\varepsilon\varepsilon}\mathbf{T}[d_f\mathbf{I} + (n-1)\text{ block diag}(\mathbf{I}, \hat{\Lambda}_r)]\mathbf{T}'\mathbf{S}_{\varepsilon\varepsilon} \\ &= \mathbf{S}_{\varepsilon\varepsilon} + (n - 1 + d_f)^{-1}(n-1)\Sigma_{\varepsilon v}\Sigma_{vv}^{-1}(\mathbf{m}_{vv} - \mathbf{S}_{vv})\Sigma_{vv}^{-1}\Sigma_{v\varepsilon} + o_p(n^{-1/2}), \end{aligned} \tag{4.1.60}$$

which is a function only of $\mathbf{S}_{\varepsilon\varepsilon}$ and the sample covariance matrix of $\mathbf{v}_t$. For normal $\mathbf{x}_{t1}$, by Lemma 4.A.1,

$$n^{1/2}\{(\operatorname{vec}\mathbf{m}_{\eta v})', [\operatorname{vech}(\mathbf{m}_{XX} - \Sigma_{XX})]', [\operatorname{vech}(\mathbf{m}_{\xi\xi} - \Sigma_{\xi\xi})]'\}' \xrightarrow{L} N(\mathbf{0}, \boldsymbol{\Omega}),$$

where

$$\boldsymbol{\Omega} = \begin{bmatrix} \boldsymbol{\Omega}_{11} & \boldsymbol{\Omega}_{12} & \mathbf{0} \\ \boldsymbol{\Omega}_{12}' & \boldsymbol{\Omega}_{22} & \boldsymbol{\Omega}_{23} \\ \mathbf{0} & \boldsymbol{\Omega}_{23}' & \boldsymbol{\Omega}_{33} \end{bmatrix}, \tag{4.1.61}$$

$$\begin{aligned} \boldsymbol{\Omega}_{11} &= \Sigma_{vv} \otimes \Sigma_{\eta\eta}, & \boldsymbol{\Omega}_{12} &= 2(\Sigma_{vu} \otimes \Sigma_{\eta\eta})\psi_k'. \\ \boldsymbol{\Omega}_{22} &= 2\psi_k(\Sigma_{XX} \otimes \Sigma_{XX})\psi_k', & \boldsymbol{\Omega}_{23} &= \boldsymbol{\Omega}_{33} = 2\psi_k(\Sigma_{\xi\xi} \otimes \Sigma_{\xi\xi})\psi_k'. \end{aligned}$$

The limiting normality of $n^{1/2}(\hat{\gamma} - \gamma)$ follows for normal $\mathbf{x}_{t1}$ because the leading terms of the estimators are expressions in the moments of normal random vectors and $\mathbf{S}_{\varepsilon\varepsilon}$ is independent of $\mathbf{Z}_t$. For other $\mathbf{x}_t$ the limiting normality of $n^{1/2}(\hat{\boldsymbol{\beta}} - \boldsymbol{\beta})$ follows from expression (4.1.58) and the arguments of Theorem 2.1.1. The covariance matrix of the limiting distribution is obtained by evaluating the covariances of the leading terms in the expansions. □

The quantities needed to estimate the covariance matrix can be calculated from the expressions defined in Theorem 4.1.5, with

$$\tilde{\boldsymbol{\Sigma}}_{vv} = (\mathbf{I}, -\hat{\boldsymbol{\beta}}')\hat{\boldsymbol{\Sigma}}_{\varepsilon\varepsilon}(\mathbf{I}, -\hat{\boldsymbol{\beta}}')', \tag{4.1.62}$$

$$\hat{\boldsymbol{\Sigma}}_{\xi\xi} = \tilde{\boldsymbol{\Sigma}}_{uv}\tilde{\boldsymbol{\Sigma}}_{vv}^{-1}\tilde{\boldsymbol{\Sigma}}_{vu}, \tag{4.1.63}$$

$$\hat{\boldsymbol{\Sigma}}_{\rho\rho} = \hat{\boldsymbol{\Sigma}}_{uu} - \tilde{\boldsymbol{\Sigma}}_{uv}\tilde{\boldsymbol{\Sigma}}_{vv}^{-1}\tilde{\boldsymbol{\Sigma}}_{vu}, \tag{4.1.64}$$

$$\tilde{\boldsymbol{\Sigma}}_{uv} = (\mathbf{0}, \mathbf{I})\hat{\boldsymbol{\Sigma}}_{\varepsilon\varepsilon}(\mathbf{I}, -\hat{\boldsymbol{\beta}}')'. \tag{4.1.65}$$

In constructing the maximum likelihood estimator of $\boldsymbol{\Sigma}_{T\varepsilon\varepsilon}$ of Theorem 4.1.2 we used the linear combination

$$(n - 1 + d_f)^{-1}[(n - 1)\hat{\lambda}_i + d_f]$$

to estimate the diagonal elements. An alternative estimator that makes an additional degrees-of-freedom correction is

$$(n - 1 - k + d_f)^{-1}[(n - 1)\hat{\lambda}_i + d_f]. \tag{4.1.66}$$

The basis for this estimator is the expansion of (4.1.60) and our usual approximation for $\hat{\mathbf{m}}_{vv}$.

Example 4.1.1. This example is taken from Reilly and Patino-Leal (1981). The data were constructed to simulate a mixing operation. In the mixing operation the concentration of five components is determined. On the basis of material balances the true values associated with the observations of Table 4.1.1 satisfy the two equations

$$y_{t1} = \beta_{11}x_{t1}, \qquad y_{t2} = \beta_{22}x_{t2}. \tag{4.1.67}$$

The model is (4.1.67) together with

$$\mathbf{Z}_t = \mathbf{z}_t + \boldsymbol{\varepsilon}_t, \qquad \boldsymbol{\varepsilon}_t \sim \mathrm{NI}(\mathbf{0}, \boldsymbol{\Sigma}_{\varepsilon\varepsilon}),$$
$$\operatorname{vech} \boldsymbol{\Sigma}_{\varepsilon\varepsilon} = (0.01111)(2, -1, 1, -1; 2, 0, 1; 2, 0; 2)',$$

where $\mathbf{Z}_t = (Y_{t1}, Y_{t2}, X_{t1}, X_{t2})$, $\boldsymbol{\varepsilon}_t = (e_{t1}, e_{t2}, u_{t1}, u_{t2})$, $\boldsymbol{\varepsilon}_t$ is independent of $\mathbf{x}_t$, and the covariance matrix of $\boldsymbol{\varepsilon}_t$ is known. On the basis of the model specification, it seems reasonable to treat the unknown x variables as fixed. For illustrative purposes we shall estimate the covariance matrix of $\mathbf{x}_t$ as if $\mathbf{x}_t$ were a normal vector.

TABLE 4.1.1. Data for mixing example (observations are 10 times the difference in the fractions)

Observation	Y_1	Y_2	X_1	X_2
1	0.572	0.625	1.419	0.905
2	0.687	0.649	1.369	0.609
3	−0.508	−0.739	−0.863	−1.202
4	−0.067	0.448	−0.500	0.833
5	−0.088	−0.686	−0.180	−0.680
6	−0.405	0.233	−0.915	0.340
7	0.438	−0.167	0.896	−0.495
8	−0.672	−0.229	−1.546	−0.259
9	0.066	−0.065	0.045	0.017
10	−0.023	−0.069	0.275	−0.068

Source: Reilly and Patino-Leal (1981).

The sample covariance matrix for $\mathbf{Z}_t$ is

$$\mathbf{m}_{ZZ} = \begin{bmatrix} 0.2091 & 0.1325 & 0.4487 & 0.1526 \\ 0.1325 & 0.2414 & 0.2535 & 0.3237 \\ 0.4487 & 0.2535 & 1.0025 & 0.2766 \\ 0.1526 & 0.3237 & 0.2766 & 0.4693 \end{bmatrix}.$$

We first estimate the coefficients of the full model

$$y_{t1} = x_{t1}\beta_{11} + x_{t2}\beta_{21},$$
$$y_{t2} = x_{t1}\beta_{12} + x_{t2}\beta_{22}.$$

The roots of $\mathbf{m}_{ZZ}^{-1/2}\mathbf{\Sigma}_{\varepsilon\varepsilon}\mathbf{m}_{ZZ}^{-1/2}$ are $(\hat{\lambda}_1^{-1}, \hat{\lambda}_2^{-1}, \hat{\lambda}_3^{-1}, \hat{\lambda}_4^{-1}) = (0.0170, 0.0368, 1.1706, 3.2019)$. The likelihood ratio test statistic of (4.1.46) is

$$(n-k-1)\left[\sum_{i=q+1}^{p} (\hat{\lambda}_i - \log \hat{\lambda}_i) - (p-q)\right] = 3.41,$$

where $n = 10$, $q = 2$, and $p = 4$. Because the value of the test statistic is to be compared with the tabular value of the chi-square distribution with three degrees of freedom, the hypothesis that the model holds with $k = 2$ is easily accepted. The likelihood ratio test statistic for the model with $k = 1$ is 186.88 and the model with $k = 1$ is rejected.

The smallest root of

$$|\mathbf{m}_{XX} - \lambda_x \mathbf{\Sigma}_{uu}| = 0,$$

where $\boldsymbol{\Sigma}_{uu} = (90)^{-1}$ diag(2, 2), is $\hat{\lambda}_x = 15.8281$. Because $(n-1)\hat{\lambda}_x$ is approximately distributed as a chi-square random variable with $n - 2 = 8$ degrees of freedom when the rank of $\mathbf{m}_{xx}$ is one, we are comfortable with the assumption that $\mathbf{m}_{xx}$ is nonsingular and we accept the postulated model containing two x variables, where the two x variables are the last two variables of Table 4.1.1.

The matrix $\mathbf{A}'$ is estimated by

$$\hat{\mathbf{A}}' = \begin{bmatrix} 0.4111 & 0.4212 & 0.8600 & 0.5455 \\ 0.1739 & -0.2238 & 0.4908 & -0.3938 \end{bmatrix}.$$

The estimated parameters are

$$(\hat{\beta}_{11}, \hat{\beta}_{21}, \hat{\beta}_{12}, \hat{\beta}_{22}) = (0.4234, 0.0861, 0.0722, 0.6584),$$
$$\text{vech}\,\hat{\boldsymbol{\Sigma}}_{xx} = (0.9804, 0.2759, 0.4526)'.$$

The estimated covariance matrix of the estimators is given in Table 4.1.2. The estimated covariance matrix is constructed by substituting the estimators (4.1.62)–(4.1.65) into the expressions of Theorem 4.1.5 with $c = 0$.

The original model specified $\beta_{21} = \beta_{12} = 0$. An approximate test of this hypothesis can be constructed using the covariance matrix of Table 4.1.2. We have

$$\tfrac{1}{2}(0.0860, 0.0722)\hat{\mathbf{V}}^{-1}\{(\hat{\beta}_{21}, \hat{\beta}_{12})'\}(0.0860, 0.0722)' = 0.96.$$

If $\beta_{21} = \beta_{12} = 0$, the computed quantity is approximately distributed as a central F with 2 and $n - 1$ degrees of freedom. Therefore, the hypothesis is easily accepted by this test. We call the model with $\beta_{21} = \beta_{12} = 0$ the reduced (or restricted) model. One must use numerical methods to compute the maxi-

TABLE 4.1.2. Estimated covariance matrix of estimators (entries are covariances multiplied by 1000)

$\hat{\beta}_{11}$	$\hat{\beta}_{21}$	$\hat{\beta}_{12}$	$\hat{\beta}_{22}$	$\hat{\sigma}_{xx11}$	$\hat{\sigma}_{xx12}$	$\hat{\sigma}_{xx22}$
2.661	−1.656	−0.510	0.317	0.39	−1.49	0.02
−1.656	5.776	0.317	−1.101	−0.01	0.22	−2.98
−0.510	0.317	2.445	−1.522	−0.37	−0.40	0.01
0.317	−1.101	−1.522	5.308	0.00	−0.18	−0.80
0.388	−0.006	−0.366	0.005	223.41	61.45	16.90
−1.487	0.216	−0.400	−0.178	61.45	61.35	29.10
0.021	−2.977	0.006	−0.805	16.90	29.10	50.07

TABLE 4.1.3. Estimated covariance matrix for the restricted estimators (entries are covariances multiplied by 1000)

$\tilde{\beta}_{11}$	$\tilde{\beta}_{22}$	$\tilde{\sigma}_{xx11}$	$\tilde{\sigma}_{xx12}$	$\tilde{\sigma}_{xx22}$
1.8821	−0.2134	0.2151	−1.3120	−0.7222
−0.2134	3.9776	−0.0001	−0.2917	−1.4876
0.2151	−0.0001	223.3692	64.4332	18.5847
−1.3120	−0.2917	64.4332	66.4931	33.1766
−0.7222	−1.4876	18.5847	33.1766	57.4485

mum likelihood estimator of the reduced model because an explicit expression for the estimator does not exist. At the time this was written, two computer programs were available for such problems. The program LISREL VI by Jöreskog and Sörbom was available in SPSS or as an independent program. The program ISU Factor by Pantula (1983) could be used with SAS. The maximum likelihood estimates of the restricted model are

$$(\tilde{\beta}_{11}, \tilde{\beta}_{22}, \tilde{\sigma}_{xx11}, \tilde{\sigma}_{xx12}, \tilde{\sigma}_{xx22}) = (0.4574, 0.7158, 0.9804, 0.2892, 0.4924).$$

The estimated covariance matrix is given in Table 4.1.3. The output from the computer programs may differ slightly from that of Table 4.1.3 because different divisors may be used at certain points in the computation. The estimated variances for the estimated coefficients are smaller under the reduced model, but the estimated variances of the $\hat{\sigma}_{xxij}$ are changed very little. In fact, some of the estimated variances of the $\hat{\sigma}_{xxij}$ are larger for the reduced model. This is because the estimated parameters used to estimate the variances are different. If the reduced model is true, all variances for the estimators of the reduced model must be no greater than those for the full model.

Under the null hypothesis, twice the difference of the log likelihoods for the two models is approximately distributed as a chi-square random variable with two degrees of freedom. The difference is 3.33, and the hypothesis that the two coefficients are zero is easily accepted. Note that the likelihood ratio statistic gave a marginally larger value of the test statistic than the corresponding value of 2(0.96) obtained using the approximate covariance matrix of the estimators of the full model.

Table 4.1.4 contains the estimated x_{ti} values and the residuals for the restricted model, where the $\hat{\mathbf{x}}_t$ are computed by (4.1.37). The estimated covariance matrices used in the construction of $\hat{\mathbf{x}}_t$ are

$$(\hat{\boldsymbol{\Sigma}}_{vv}, \hat{\boldsymbol{\Sigma}}_{vu}) = \begin{bmatrix} 1.671 & 0.316 & 0.095 & -1.111 \\ 0.316 & 1.770 & 0.000 & -0.480 \end{bmatrix} 10^{-2}.$$

TABLE 4.1.4. Estimated true values and residuals for restricted model

Observation	$\hat{x}_{t1}$	$\hat{x}_{t2}$	$\hat{v}_{t1}$	$\hat{v}_{t2}$
1	1.424	0.839	−0.077	−0.023
2	1.363	0.740	0.061	0.213
3	−0.858	−1.237	−0.113	0.121
4	−0.508	0.893	0.162	−0.148
5	−0.178	−0.765	−0.006	−0.199
6	−0.916	0.346	0.014	−0.010
7	0.892	−0.398	0.028	0.187
8	−1.548	−0.251	0.035	−0.044
9	0.043	0.196	0.045	−0.077
10	0.284	−0.186	−0.149	−0.020

Fuller (1985) has given expansions for multivariate residuals analogous to those of Section 2.3.3. On the basis of Fuller's results and those of Miller (1986), the behavior of the $\hat{\mathbf{v}}_t$ can be approximated by the behavior of ordinary least squares residuals obtained by regressing $\mathbf{v}_t$ on $\ddot{\mathbf{x}}_t$, where

$$\ddot{\mathbf{x}}_t = \mathbf{X}_t - \mathbf{v}_t \boldsymbol{\Sigma}_{vv}^{-1} \boldsymbol{\Sigma}_{vu}.$$

Approximations for the covariance matrix of the $\hat{\mathbf{v}}_t$ are given in Fuller (1985).

When the full model estimators are used to construct $\hat{\mathbf{x}}_t$ and $\hat{\mathbf{v}}_t$ the equations

$$\sum_{t=1}^{n} \hat{v}_{ti}\hat{x}_{tj} = 0$$

hold for $i = 1, 2, \ldots, r$ and $j = 1, 2, \ldots, k$. For our restricted model the estimates satisfy

$$\sum_{t=1}^{n} (\hat{x}_{t1}, 0)\hat{\boldsymbol{\Sigma}}_{vv}^{-1}(\hat{v}_{t1}, \hat{v}_{t2})' = 0,$$

$$\sum_{t=1}^{n} (0, \hat{x}_{t2})\hat{\boldsymbol{\Sigma}}_{vv}^{-1}(\hat{v}_{t1}, \hat{v}_{t2})' = 0.$$

Because only two parameters (β_{11} and β_{22}) are estimated, only two restrictions are imposed on the residuals. It follows that a plot of $\hat{v}_{ti}$ against $\hat{x}_{tj}$ for the restricted model may exhibit some correlation. This is true in the present example, particularly for the plot of $\hat{v}_{t1}$ against $\hat{x}_{t2}$. The observed correlation is within the range of what can easily occur in random samples of size eight selected from a population with zero correlation. The deviations $\hat{v}_{ti}$ are not large when compared to their estimated standard deviations, and there are no apparent anomalies in the plots that would give one reason to reject the model. □ □

Example 4.1.2. In this example we study some data obtained in an animal breeding experiment. Animal breeders attempt to effect genetic change by selective breeding. A breeder is handicapped in this attempt because the breeder is unable to observe directly the genetic characteristics of an animal. Rather, the breeder observes measurable characteristics, called the phenotypic values, of an animal.

The phenotypic value may be divided into components attributable to the influence of genotype and to the influence of the environment. The genotype is the particular assemblage of genes possessed by an individual, and the environment is all the nongenetic circumstances that influence the phenotype value. The genotype represents the component of phenotype which can be "controlled" through selective breeding, while the environment may be regarded, from a statistical point of view, as the "error of measurement" which masks the true genotype.

The analysis of this example is based on the work of Dahm, Melton, and Fuller (1983). The basic data for this example were gathered in an experiment initiated in 1957 at the U.S.D.A. Beef Cattle Research Station in Fort Robinson, Nebraska. Detailed descriptions of this experiment can be found in Gregory et al. (1965, 1966a, 1966b). Measurements were taken on 372 steer calves, sired by 51 bulls. The following data were collected for each calf:

FC $\frac{1}{100}$ of the weight (kg) of total digestible nutrients consumed by an animal during the 252 day feeding period

ADG 10 times the average daily weight gain (kg) for the 252 day feeding period

WW $\frac{1}{30}$ of the weaning weight (kg) of an animal at a standardized 200 days of age at weaning

RP retail product = one-tenth of the weight (kg) of closely trimmed, nearly boneless meat from the right-hand side of the carcass of an animal

Let

$$\mathbf{Z}_{ts} = (Z_{ts1}, Z_{ts2}, Z_{ts3}, Z_{ts4}) = (\text{FC}_{ts}, \text{ADG}_{ts}, \text{WW}_{ts}, \text{RP}_{ts})$$

denote the phenotypic values recorded for the sth offspring of the tth sire (father). We write

$$\mathbf{Z}_{ts} = \mathbf{z}_t + \boldsymbol{\varepsilon}_{ts},$$

where $\mathbf{z}_t$ is the contribution to the genotype of the individual that is obtained from the tth sire and ε_{ts} is the sum of environmental effects and the contribution of the dam (mother) to the genotype. Throughout this example we assume all variables are coded to have zero means. Thus, no intercept terms appear in the equations. We assume $\mathbf{z}_t$ and ε_{ts} to be independent so that

$$\Sigma_{ZZ} = \Sigma_{zz} + \Sigma_{\varepsilon\varepsilon}.$$

By calculating a multivariate analysis of variance, we can obtain an unbiased estimator of the covariance matrix of ε_{ts}. The estimator of $\Sigma_{\varepsilon\varepsilon}$ is the within-sire mean square from such an analysis. The among-sire mean square in the analysis of variance is an estimator of

$$7.264\Sigma_{zz} + \Sigma_{\varepsilon\varepsilon},$$

where the constant 7.264 is determined by the number of offspring per sire. While the notation is not totally consistent with our description of the variables, we let $\mathbf{m}_{ZZ}$ denote the among-sires mean square matrix. We do this so that the among-sires matrix will play the role of $\mathbf{m}_{ZZ}$ of Theorem 4.1.2. The matrix $\mathbf{m}_{ZZ}$ does not exactly satisfy the conditions of Theorem 4.1.2 because there are different numbers of observations for the different sires. By Theorem 4.1.5, the only statistic that may be seriously affected by this fact is the estimator of the covariance matrix of the estimator of the elements of Σ_{xx}. The among-sires mean squares and products matrix for FC, ADG, WW, and RP is

$$\mathbf{m}_{ZZ} = \begin{bmatrix} 147.33 & 103.04 & 64.35 & 74.43 \\ 103.04 & 126.00 & 37.93 & 82.05 \\ 64.35 & 37.93 & 95.92 & 47.93 \\ 74.43 & 82.05 & 47.93 & 99.88 \end{bmatrix}.$$

The within-sires mean squares and products matrix is

$$\mathbf{S}_{\varepsilon\varepsilon} = \begin{bmatrix} 135.59 & 64.18 & 38.83 & 32.40 \\ 64.18 & 93.50 & 1.97 & 33.44 \\ 38.83 & 1.97 & 67.33 & 26.20 \\ 32.40 & 33.44 & 26.20 & 47.56 \end{bmatrix}.$$

The degrees of freedom are $n - 1 = 50$ and $d_f = 321$ for $\mathbf{m}_{ZZ}$ and $\mathbf{S}_{\varepsilon\varepsilon}$, respectively. In the remainder of this example we let $E\{\mathbf{m}_{ZZ}\} = \Sigma_{zz} + \Sigma_{\varepsilon\varepsilon}$. That is, we replace $7.264\Sigma_{zz}$ by Σ_{zz}. The characteristic roots of $\mathbf{S}_{\varepsilon\varepsilon}^{-1/2}\mathbf{m}_{ZZ}\mathbf{S}_{\varepsilon\varepsilon}^{-1/2}$ are

$$(\hat{\lambda}_1, \hat{\lambda}_2, \hat{\lambda}_3, \hat{\lambda}_4) = (2.1418, 1.6729, 0.7988, 0.6521),$$

and the matrix $\mathbf{T}$ containing the characteristic vectors of $\mathbf{m}_{ZZ}$ in the metric $\mathbf{S}_{\varepsilon\varepsilon}$ is

$$\mathbf{T} = \begin{bmatrix} 0.0603 & 0.3328 & -0.5432 & -1.0116 \\ 0.0777 & -0.8578 & -0.6132 & 1.0980 \\ -0.2142 & -1.5249 & 0.4641 & 0.1236 \\ 1.4455 & 1.0786 & 0.7719 & -0.2329 \end{bmatrix} (0.1).$$

While we observe four phenotypic characteristics, it does not follow that we will be able to explain the phenotypic characteristics in terms of four genotypic characteristics. Using the roots and the expression of Theorem 4.1.2, the test defined in (4.1.45) for the hypothesis that the data satisfy model (4.1.12) with k equal to one (i.e. genetic variation arises from variation in a single factor) against the unconstrained alternative is $\hat{F} = 1.82$. Under the null hypothesis, the test statistic is approximately distributed as Snedecor's F with 6 and 321 degrees of freedom. Therefore, the hypothesis of model (4.1.12) with the rank of $\boldsymbol{\Sigma}_{zz}$ equal to one is rejected at the 0.10 level of significance but would be accepted at the 0.05 level of significance.

The likelihood ratio test of the hypothesis of model (4.1.12) with k equal to two against the unrestricted alternative is $\hat{F} = 1.15$, which is approximately distributed as Snedecor's F with 3 and 321 degrees of freedom under the null. For the purposes of this example, we accept model (4.1.12) with the rank of $\boldsymbol{\Sigma}_{zz}$ equal to two. That is, we proceed with the model in which genetic variation arises from variation in two factors.

The average of the two smallest roots of $\mathbf{m}_{ZZ}$ in the metric $\mathbf{S}_{\varepsilon\varepsilon}$ is 0.7254. Under the two-factor model the statistic of (4.1.47),

$$(48)^{-1}50(2^{-1}) \sum_{i=3}^{4} \hat{\lambda}_i = 0.756,$$

is approximately distributed as an F random variable with 96 and 321 degrees of freedom. About 4% of the F distribution is to the left of the observed value. The likelihood ratio statistic for the model with $k = 2$ is larger than the statistic computed from the sum of the roots because the likelihood ratio is large if the individual roots are too small relative to one or if the roots are too large relative to one. In this example there seems little doubt that genetic variation can be explained by variation in two factors. The only cause for concern is that the variation in the error contrasts of the model, as represented by the two smallest roots, is rather small relative to the estimated variance of such contrasts as computed from the within-sire covariances. In the absence of any evidence calling for a modification of the model we proceed with the estimation.

For the two-factor model we choose weaning weight and retail product as the x variables and write

$$(z_{t1}, z_{t2}, z_{t3}, z_{t4}) = (x_{t1}, x_{t2})\begin{bmatrix} \beta_{11} & \beta_{12} & 1 & 0 \\ \beta_{21} & \beta_{22} & 0 & 1 \end{bmatrix}.$$

In the absence of prior information, the choice of x variables requires some care. For these data, the hypothesis that average daily gain, feed consumption, and retail product satisfy a model in one factor is accepted. See Exercise 4.7. Therefore, one should not use two of these three variables as the two x variables. The smallest root of the equation

$$\left|\begin{bmatrix} 95.92 & 47.93 \\ 47.93 & 99.88 \end{bmatrix} - \lambda \begin{bmatrix} 67.33 & 26.20 \\ 26.20 & 47.56 \end{bmatrix}\right| = 0$$

associated with weaning weight and retail product is $\hat{\lambda} = 1.36$ and the corresponding F value is 1.39. The value 1.39 is approximately equal to the 5% point of the F distribution with 49 and 321 degrees of freedom. The root 1.36 is the largest of the smallest roots of the determinantal equations for the six possible sets of two x variables.

Using the roots $\hat{\lambda}_i$, we have

$$\hat{\Sigma}_{Tzz} = \text{diag}(1.1418, 0.6729, 0, 0),$$
$$\hat{\Sigma}_{T\varepsilon\varepsilon} = \text{diag}(1.0000, 1.0000, 0.9782, 0.9583),$$

where we used (4.1.66). By (4.1.18) we have

$$\text{vec}\, \hat{\Sigma}_{xx} = (34.5400, 22.8352, 22.8352, 52.6057)',$$

$$\hat{\Sigma}_{\varepsilon\varepsilon} = \begin{bmatrix} 132.5120 & 64.0557 & 38.2644 & 32.4514 \\ 64.0557 & 92.1855 & 2.8257 & 33.6427 \\ 38.2644 & 2.8257 & 66.6366 & 26.0748 \\ 32.4514 & 33.6427 & 26.0748 & 47.5284 \end{bmatrix},$$

$$\text{vec}\, \hat{\boldsymbol{\beta}} = (0.4811, 0.5792, 0.3307, 0.7457).$$

The estimated covariance matrix for $\text{vec}\, \hat{\boldsymbol{\beta}}$ and $\text{vech}\, \hat{\Sigma}_{xx}$ is given in Table 4.1.5. The entries are computed from the expressions of Theorem 4.1.5 and the estimators (4.1.62)–(4.1.65). Matrices used in constructing the estimated covariance matrix include

$$\text{vech}\, \tilde{\Sigma}_{vv} = (104.002, 51.837, 86.719)',$$
$$\text{vech}\, \hat{\Sigma}_{\xi\xi} = (19.996, 4.376, 1.280)',$$
$$\text{vech}\, \hat{\Sigma}_{\eta\eta} = (81.180, 44.534, 98.855)',$$

TABLE 4.1.5. Estimated covariance matrix of estimated model parameters

	Parameter						
Parameter	β_{11}	β_{21}	β_{12}	β_{22}	σ_{xx11}	σ_{xx12}	σ_{xx22}
β_{11}	0.326	−0.139	0.163	−0.069	−0.056	−0.012	0.020
β_{21}	−0.139	0.184	−0.069	0.092	0.024	−0.006	−0.027
β_{12}	0.163	−0.069	0.273	−0.116	−0.243	0.019	0.027
β_{22}	−0.069	0.092	−0.116	0.154	0.103	−0.054	−0.037
σ_{xx11}	−0.056	0.024	−0.243	0.103	56.901	38.330	24.209
σ_{xx12}	−0.012	−0.006	0.019	−0.054	38.330	57.719	55.513
σ_{xx22}	0.020	−0.027	0.027	−0.037	24.209	55.513	124.693

Table 4.1.6 contains the estimated covariance matrix computed under the assumption that $\boldsymbol{\Sigma}_{\varepsilon\varepsilon}$ is known and equal to $\mathbf{S}_{\varepsilon\varepsilon}$. The effect of using an estimator of $\boldsymbol{\Sigma}_{\varepsilon\varepsilon}$ based on 321 degrees of freedom has a modest effect on the variances of the estimator of $\boldsymbol{\beta}$. The effect on the variances of the estimator of $\boldsymbol{\Sigma}_{xx}$ is more marked. The estimated variance of $\hat{\sigma}_{xx11}$ is nearly doubled when the estimator is constructed using the estimator $\mathbf{S}_{\varepsilon\varepsilon}$ of $\boldsymbol{\Sigma}_{\varepsilon\varepsilon}$ in place of the known $\boldsymbol{\Sigma}_{\varepsilon\varepsilon}$. As previously stated, the results for $\boldsymbol{\Sigma}_{xx}$ are only approximate because of the unequal number of observations per sire.

The portion of the estimated covariance matrix of the estimators associated with $\hat{\boldsymbol{\Sigma}}_{\varepsilon\varepsilon}$ is given in Table 4.1.7. The estimated variances of $\hat{\sigma}_{\varepsilon\varepsilon ij}$ are only slightly smaller than the estimated variances of $s_{\varepsilon\varepsilon ij}$ because $d_f = 321$ dominates $(n - 1 + d_f)$. The correlations between the estimates of the elements of $\boldsymbol{\beta}$ and the estimates of the elements of $\boldsymbol{\Sigma}_{\varepsilon\varepsilon}$ are small. The negative correlation between estimates such as $\hat{\sigma}_{\varepsilon\varepsilon 44}$ and $\hat{\sigma}_{xx22}$ reflects the fact that the sum is an estimator of σ_{XX22}. □ □

TABLE 4.1.6. Estimated covariance matrix of estimated parameters computed under assumption that $\boldsymbol{\Sigma}_{\varepsilon\varepsilon}$ is known

	Parameter						
Parameter	β_{11}	β_{21}	β_{12}	β_{22}	σ_{xx11}	σ_{xx12}	σ_{xx22}
β_{11}	0.294	−0.125	0.147	−0.062	−0.050	−0.011	0.018
β_{21}	−0.125	0.167	−0.062	0.083	0.021	−0.005	−0.024
β_{12}	0.147	−0.062	0.245	−0.104	−0.219	0.017	0.025
β_{22}	−0.062	0.083	−0.104	0.139	0.093	−0.050	−0.033
σ_{xx11}	−0.050	0.021	−0.219	0.093	31.726	28.049	20.092
σ_{xx12}	−0.011	−0.005	0.017	−0.050	28.049	45.874	47.826
σ_{xx22}	0.018	−0.024	0.025	−0.033	20.092	47.826	110.629

TABLE 4.1.7. Estimated covariances of estimates of $\sigma_{\varepsilon\varepsilon ij}$ with other estimates

	Parameter									
Parameter	$\sigma_{\varepsilon\varepsilon 11}$	$\sigma_{\varepsilon\varepsilon 21}$	$\sigma_{\varepsilon\varepsilon 31}$	$\sigma_{\varepsilon\varepsilon 41}$	$\sigma_{\varepsilon\varepsilon 22}$	$\sigma_{\varepsilon\varepsilon 32}$	$\sigma_{\varepsilon\varepsilon 42}$	$\sigma_{\varepsilon\varepsilon 33}$	$\sigma_{\varepsilon\varepsilon 43}$	$\sigma_{\varepsilon\varepsilon 44}$
$\sigma_{\varepsilon\varepsilon 11}$	102.0	50.9	31.3	27.2	25.1	15.2	13.1	9.1	7.7	6.5
$\sigma_{\varepsilon\varepsilon 21}$	50.9	48.6	9.9	20.7	35.7	11.8	16.2	0.8	4.2	6.8
$\sigma_{\varepsilon\varepsilon 31}$	31.3	9.9	31.3	14.5	1.7	13.5	5.4	15.8	9.9	5.3
$\sigma_{\varepsilon\varepsilon 41}$	27.2	20.7	14.5	22.8	13.6	9.2	12.9	6.2	8.3	9.6
$\sigma_{\varepsilon\varepsilon 22}$	25.1	35.7	1.7	13.6	50.8	2.8	19.7	−0.6	0.4	7.0
$\sigma_{\varepsilon\varepsilon 32}$	15.2	11.8	13.5	9.2	2.8	18.4	7.6	1.7	7.3	5.5
$\sigma_{\varepsilon\varepsilon 42}$	13.1	16.2	5.4	12.9	19.7	7.6	17.1	0.6	3.2	10.0
$\sigma_{\varepsilon\varepsilon 33}$	9.1	0.8	15.8	6.2	−0.6	1.7	0.6	27.3	10.8	4.2
$\sigma_{\varepsilon\varepsilon 43}$	7.7	4.2	9.9	8.3	0.4	7.3	3.2	10.8	12.0	7.7
$\sigma_{\varepsilon\varepsilon 44}$	6.5	6.8	5.3	9.6	7.0	5.5	10.0	4.2	7.7	14.1
$10\beta_{11}$	−3.7	−2.1	−3.6	0.1	−1.2	−1.6	0.1	0.7	0.3	0.0
$10\beta_{21}$	−2.8	−2.6	0.5	−2.6	−1.7	0.3	−1.1	−0.1	0.2	0.4
$10\beta_{12}$	−1.1	−1.7	−0.3	0.2	−1.8	−2.1	0.1	3.0	0.4	0.0
$10\beta_{22}$	−0.8	−1.5	0.7	−0.6	−2.7	1.0	−1.7	−0.3	1.1	0.6
σ_{xx11}	−9.1	−1.3	−15.5	−6.1	5.1	−4.8	−1.2	−25.2	−10.3	−4.1
σ_{xx21}	−7.8	−3.9	−10.1	−8.4	0.8	−8.1	−3.4	−10.3	−11.8	−7.7
σ_{xx22}	−6.4	−6.6	−5.4	−9.6	−6.7	−5.7	−10.0	−4.1	−7.7	−14.1

REFERENCES

Amemiya (1982, 1985b), Amemiya and Fuller (1984), Anderson (1951b, 1963, 1983, 1984), Bhargava (1977), Dahm, Melton, and Fuller (1983), Gleser (1981), Gleser and Watson (1973), Lawley (1940), Lawley and Maxwell (1971), Miller (1984), Theobald (1975b), Tintner (1946, 1952).

EXERCISES

1. (Section 4.1.1) Derive expression (4.1.46) under the assumption that $\boldsymbol{\Sigma}_{\varepsilon\varepsilon}$ is nonsingular. In the proof give the spectral representation of $\mathbf{m}_{ZZ}$ in the metric $\boldsymbol{\Sigma}_{\varepsilon\varepsilon}$ and show that

$$|\hat{\boldsymbol{\Sigma}}_{ZZ}| = |\boldsymbol{\Sigma}_{\varepsilon\varepsilon}| \prod_{i=1}^{k} \hat{\lambda}_i,$$

where $\hat{\lambda}_1 \geq \hat{\lambda}_2 \geq \cdots \geq \hat{\lambda}_p$ are the roots of

$$|\mathbf{m}_{ZZ} - \lambda\boldsymbol{\Sigma}_{\varepsilon\varepsilon}| = 0.$$

2. (Section 4.1.1.) In (4.1.29) the vectors $\hat{\mathbf{A}}_{.i}$ are defined so that $\hat{\mathbf{A}}'\mathbf{m}_{ZZ}^{-1}\hat{\mathbf{A}}$ is a diagonal matrix. Show that the estimator of $\boldsymbol{\beta}$ given in terms of $\hat{\mathbf{A}}$ of (4.1.29) remains unchanged if $\hat{\mathbf{A}}$ is replaced by $\hat{\mathbf{A}}\mathbf{G}$, where $\mathbf{G}$ is a $k \times k$ nonsingular matrix.

3. (Section 4.1.1) Assume that $\boldsymbol{\Sigma}_{\varepsilon\varepsilon}$ is nonsingular and let $\mathbf{T}$ be defined by (4.1.19). Show that the elements of $(\mathbf{Z}_t - \bar{\mathbf{Z}})\mathbf{T}_r$ are linear combinations of $\hat{\mathbf{v}}_t$. Show that $\mathbf{T}_r'\boldsymbol{\Sigma}_{\varepsilon\varepsilon}\hat{\mathbf{H}}_2 = \mathbf{0}$ and, hence, that the elements of $(\mathbf{Z}_t - \bar{\mathbf{Z}})\mathbf{T}_k$ are linear functions of the elements of $(\hat{\mathbf{x}}_t - \bar{\mathbf{X}})$.

4. (Section 4.1.1) Show that $\hat{\sigma}^2$ of (4.1.11) can be written

$$\hat{\sigma}^2 - \sigma^2 = r^{-1}\operatorname{tr}\{(\mathbf{M}_{vv} - \boldsymbol{\Sigma}_{vv})[(\mathbf{I}, -\boldsymbol{\beta}')\boldsymbol{\Upsilon}_{\varepsilon\varepsilon}(\mathbf{I}, -\boldsymbol{\beta}')']^{-1} + O_p(n^{-1})$$

by using

$$\begin{aligned}\hat{\sigma}^2 - \sigma^2 = {} & r^{-1}n(n-k)^{-1}\operatorname{tr}\{(\mathbf{I}, -\hat{\boldsymbol{\beta}}')\mathbf{M}_{ZZ}(\mathbf{I}, -\hat{\boldsymbol{\beta}}')'[(\mathbf{I}, -\hat{\boldsymbol{\beta}}')\boldsymbol{\Upsilon}_{\varepsilon\varepsilon}(\mathbf{I}, -\hat{\boldsymbol{\beta}}')']^{-1}\} \\ & - r^{-1}\operatorname{tr}\{(\mathbf{I}, -\boldsymbol{\beta}')(\mathbf{M}_{zz} + \boldsymbol{\Upsilon}_{\varepsilon\varepsilon}\sigma^2)(\mathbf{I}, -\boldsymbol{\beta}')'[(\mathbf{I}, -\boldsymbol{\beta}')\boldsymbol{\Upsilon}_{\varepsilon\varepsilon}(\mathbf{I}, -\boldsymbol{\beta}')']^{-1}\}\end{aligned}$$

and expanding this expression in a first-order Taylor series in $\hat{\boldsymbol{\beta}}$ about the true value of $\boldsymbol{\beta}$.

5. (Section 4.1.1) Let $\boldsymbol{\Sigma}_{\varepsilon\varepsilon}$ be a positive definite $p \times p$ matrix and let $\boldsymbol{\beta}$ be a $k \times r$ matrix. Show that

$$[(\boldsymbol{\beta}, \mathbf{I}_k)\boldsymbol{\Sigma}_{\varepsilon\varepsilon}^{-1}(\boldsymbol{\beta}, \mathbf{I}_k)']^{-1} = \boldsymbol{\Sigma}_{uu} - \boldsymbol{\Sigma}_{uv}\boldsymbol{\Sigma}_{vv}^{-1}\boldsymbol{\Sigma}_{vu},$$

where $\boldsymbol{\Sigma}_{vv} = (\mathbf{I}_r, -\boldsymbol{\beta}')\boldsymbol{\Sigma}_{\varepsilon\varepsilon}(\mathbf{I}_r, -\boldsymbol{\beta}')'$, $\boldsymbol{\Sigma}_{uv} = (\mathbf{0}, \mathbf{I}_k)\boldsymbol{\Sigma}_{\varepsilon\varepsilon}(\mathbf{I}_r, -\boldsymbol{\beta}')'$, $\boldsymbol{\Sigma}_{uu} = (\mathbf{0}, \mathbf{I}_k)\boldsymbol{\Sigma}_{\varepsilon\varepsilon}(\mathbf{0}, \mathbf{I}_k)'$, and $p = r + k$.

6. (Sections 4.1.2, 2.3) Using the formulas of Theorem 4.1.5, estimate the covariance matrix of the approximate distribution of the estimators obtained in Example 2.3.2. Estimate the covariance matrix of the approximate distribution of $(\hat{\sigma}_{pp}, \hat{\sigma}_{mm})$ of Example 2.3.2.

7. (Section 4.1.2) Using the data for [FC, ADG, RP] $= [Z_{ts1}, Z_{ts2}, Z_{ts3}]$ of Example 4.1.2, estimate the parameters of the model

$$Z_{ts1} = \beta_{11}x_{ts} + \varepsilon_{ts1}, \qquad Z_{ts2} = \beta_{12}x_{ts} + \varepsilon_{ts2},$$
$$Z_{ts3} = x_{ts} + \varepsilon_{ts3}, \qquad \boldsymbol{\varepsilon}_{ts} \sim \mathrm{NI}(\mathbf{0}, \boldsymbol{\Sigma}_{\varepsilon\varepsilon}),$$

where $\boldsymbol{\Sigma}_{\varepsilon\varepsilon}$ is estimated by $\mathbf{S}_{\varepsilon\varepsilon}$ and $x_{ts} \sim \mathrm{NI}(0, \sigma_{xx})$, independent of ε_{ij} for all ij and ts. Give the estimated covariance matrix of your estimates.

8. (Section 4.1.1) Plot the function

$$f(\lambda) = \lambda - 1 - \log \lambda$$

for $\lambda \in [0.05, 2.0]$. Compare $f(\lambda)$ with the second-order Taylor approximation, $g(\lambda) = \frac{1}{2}(\lambda - 1)^2$.

9. (Section 4.1.1 and Appendix 4.A)
   (a) Using Corollary 4.A.10, or otherwise, prove that the estimator of Theorem 4.1.2 minimizes (4.1.33) and that the minimum value of (4.1.33) is $\Sigma_{i=k+1}^{p} \hat{\lambda}_i$.
   (b) Prove that the estimator of Theorem 4.1.2 minimizes (4.1.35) and that the minimum value is $\prod_{i=k+1}^{p} \hat{\lambda}_i$.

10. (Section 4.1.1) Using the approach of Section 1.2.3 show that the two expressions for $\ddot{\mathbf{z}}_t$ given in (4.1.5) are equivalent.

11. (Section 4.1.1) Prove the following.

Theorem 4.1.6. Let

$$y_t = \beta_0 + \mathbf{x}_{t1}\boldsymbol{\beta}_1, \qquad \mathbf{Z}_t = \mathbf{z}_t + \boldsymbol{\varepsilon}_t$$
$$(\boldsymbol{\varepsilon}_t, \mathbf{x}_{t1})' \sim \mathrm{NI}[(\mathbf{0}, \boldsymbol{\mu}_x)', \text{block diag}(\boldsymbol{\Upsilon}_{\varepsilon\varepsilon}\sigma^2, \boldsymbol{\Sigma}_{xx})]$$

where $\boldsymbol{\varepsilon}_t' = (\mathbf{e}_t, \mathbf{u}_t)$, $\boldsymbol{\Sigma}_{\varepsilon\varepsilon} = \boldsymbol{\Upsilon}_{\varepsilon\varepsilon}\sigma^2$, $\boldsymbol{\Upsilon}_{\varepsilon\varepsilon}$ is known, and $\boldsymbol{\Sigma}_{xx}$ and $\boldsymbol{\Sigma}_{vv}$ are positive definite. Then the maximum likelihood estimators adjusted for degrees of freedom are

$$\hat{\boldsymbol{\beta}}_1 = (\hat{\mathbf{A}}_{rk}\hat{\mathbf{A}}_{kk}^{-1})', \qquad \hat{\beta}_0 = \bar{Y} - \bar{\mathbf{X}}\hat{\boldsymbol{\beta}}, \qquad \hat{\boldsymbol{\Sigma}}_{zz} = \hat{\mathbf{A}}\hat{\mathbf{A}}',$$

$$\tilde{\sigma}_m^2 = r^{-1} \sum_{i=k+1}^{p} \hat{v}_i,$$

where

$$\hat{\mathbf{A}} = (\hat{\mathbf{A}}_{rk}', \hat{\mathbf{A}}_{kk}')'$$
$$= \mathbf{m}_{ZZ}[(1 - \hat{v}_1^{-1}\tilde{\sigma}^2)^{1/2}\mathbf{R}_{.1}, (1 - \hat{v}_2^{-1}\tilde{\sigma}^2)^{1/2}\mathbf{R}_{.2}, \ldots, (1 - \hat{v}_k^{-1}\tilde{\sigma}^2)^{1/2}\mathbf{R}_{.k}],$$

$\hat{v}_1^{-1} \leqslant \hat{v}_2^{-1} \leqslant \cdots \leqslant \hat{v}_p^{-1}$, are the roots of

$$|\boldsymbol{\Upsilon}_{\varepsilon\varepsilon} - v_i^{-1}\mathbf{m}_{ZZ}| = 0,$$

and $\mathbf{R} = (\mathbf{R}_{.1}, \mathbf{R}_{.2}, \ldots, \mathbf{R}_{.p})$ is the matrix of the characteristic vectors of $\boldsymbol{\Upsilon}_{\varepsilon\varepsilon}$ in the metric $\mathbf{m}_{ZZ}$.

12. (Section 4.1.1) Prove the following theorem. The portion of the result associated with $\hat{\boldsymbol{\beta}}$ and $\tilde{\sigma}_m^2$ also holds for fixed $\mathbf{x}_t$.

Theorem 4.1.7. Let the assumptions of Theorem 4.1.6 of Exercise 4.11 hold. Let $\hat{\boldsymbol{\theta}}' = [\hat{\beta}_0, (\text{vec}\,\hat{\boldsymbol{\beta}}_1)', (\text{vech}\,\hat{\boldsymbol{\Sigma}}_{xx})', \tilde{\sigma}_m^2]$, where the estimators are defined in Theorem 4.1.6 of Exercise 4.11. Let $\boldsymbol{\theta}' = [(\text{vech}\,\boldsymbol{\beta})', (\text{vech}\,\boldsymbol{\Sigma}_{xx})', \sigma^2]$, where $\boldsymbol{\beta} = (\boldsymbol{\beta}_0', \boldsymbol{\beta}_1)$. Then

$$n^{1/2}(\hat{\boldsymbol{\theta}} - \boldsymbol{\theta}) \xrightarrow{L} N(\mathbf{0}, \boldsymbol{\Gamma}),$$

where

$$\boldsymbol{\Gamma} = \begin{bmatrix} \boldsymbol{\Gamma}_{\beta\beta} & \boldsymbol{\Gamma}_{\beta x} & \mathbf{0} \\ \boldsymbol{\Gamma}_{\beta x}' & \boldsymbol{\Gamma}_{xx} & \boldsymbol{\Gamma}_{x\sigma} \\ \mathbf{0} & \boldsymbol{\Gamma}_{x\sigma}' & \boldsymbol{\Gamma}_{\sigma\sigma} \end{bmatrix},$$

$\Gamma_{\beta\beta}$ and $\Gamma_{\beta x}$ are defined in Theorem 4.1.5,

$$\Gamma_{xx} = 2\psi_k[(\Sigma_{XX} \otimes \Sigma_{XX}) - (\Sigma_{\xi\xi} \otimes \Sigma_{\xi\xi})]\psi_k' + 2r^{-1}[\text{vech}(\Sigma_{uu} - \Sigma_{\xi\xi})][\text{vech}(\Sigma_{uu} - \Sigma_{\xi\xi})]',$$

$$\Gamma_{x\sigma} = -2r^{-1}\sigma^2 \text{ vech}(\Sigma_{uu} - \Sigma_{\xi\xi}), \quad \text{and} \quad \Gamma_{\sigma\sigma} = 2r^{-1}\sigma^4.$$

4.2. LEAST SQUARES ESTIMATION OF THE PARAMETERS OF A COVARIANCE MATRIX

The multivariate model of Section 4.1 and the factor model of Section 1.5 are two examples of a large class of models in which the entries in a covariance matrix are functions of a more elemental vector of parameters. One method of estimating the vector of parameters is the method of maximum likelihood. A closely related method is the method of least squares. We demonstrated in Section 1.3.3 that the maximum likelihood estimator of β could be obtained by minimizing a sum of squares function of the original observations. In this section we consider estimators obtained by applying the method of least squares to the elements of the sample covariance matrix. Least squares has been applied to covariance matrices by Anderson (1969, 1973), Jöreskog and Goldberger (1972), Browne (1974), Dahm (1979), Lee and Jennrich (1979), Lee and Bentler (1980), Dahm and Fuller (1986), and Bentler and Dijkstra (1985). Matrix results relevant to our study are given in Appendix 4.A.

4.2.1. Least Squares Estimation

Assume that the random vectors $\mathbf{Z}_t$, $t = 1, 2, \ldots$, are independently and identically distributed with finite fourth moments. Assume that the $p \times p$ covariance matrix of the $\mathbf{Z}_t$ can be expressed as a function of a q-dimensional parameter vector $\boldsymbol{\theta}$, and let

$$\Sigma_{ZZ}(\boldsymbol{\theta}) = E\{(\mathbf{Z}_t - \boldsymbol{\mu}_Z)'(\mathbf{Z}_t - \boldsymbol{\mu}_Z)\}, \tag{4.2.1}$$

where $\boldsymbol{\mu}_Z = E\{\mathbf{Z}_t\}$. In our treatment we assume $\boldsymbol{\mu}_Z$ is not a function of $\boldsymbol{\theta}$. The extension to the case where $\boldsymbol{\mu}_Z$ is a function of $\boldsymbol{\theta}$ is straightforward and is illustrated in Example 4.2.1. Let $\sigma_{ZZij}(\boldsymbol{\theta})$ denote the ijth element of $\Sigma_{ZZ}(\boldsymbol{\theta})$ and let

$$\begin{aligned} \boldsymbol{\sigma} = \boldsymbol{\sigma}(\boldsymbol{\theta}) &= \text{vech } \Sigma_{ZZ}(\boldsymbol{\theta}), \\ &= [\sigma_{ZZ11}(\boldsymbol{\theta}), \ldots, \sigma_{ZZp1}(\boldsymbol{\theta}), \sigma_{ZZ22}(\boldsymbol{\theta}), \ldots, \sigma_{ZZp2}(\boldsymbol{\theta}), \ldots, \sigma_{ZZpp}(\boldsymbol{\theta})]', \end{aligned} \tag{4.2.2}$$

where vech $\Sigma_{ZZ}(\boldsymbol{\theta})$ [also written as vech$\{\Sigma_{ZZ}(\boldsymbol{\theta})\}$] is the $\frac{1}{2}p(p + 1)$ – dimensional column vector of elements on and below the diagonal of $\Sigma_{ZZ}(\boldsymbol{\theta})$. Let the sample covariance matrix constructed from a sample of n observations be

$$\mathbf{m}_{ZZ} = (n-1)^{-1} \sum_{t=1}^{n} (\mathbf{Z}_t - \bar{\mathbf{Z}})'(\mathbf{Z}_t - \bar{\mathbf{Z}}), \tag{4.2.3}$$

where $\bar{\mathbf{Z}} = n^{-1} \sum_{t=1}^{n} \mathbf{Z}_t$. Because the elements of $\mathbf{m}_{ZZ}$ are unbiased estimators of the elements of $\boldsymbol{\Sigma}_{ZZ}(\boldsymbol{\theta})$, we can write

$$\mathbf{s} = \boldsymbol{\sigma}(\boldsymbol{\theta}) + \mathbf{a}, \tag{4.2.4}$$

where $\mathbf{s} = \text{vech } \mathbf{m}_{ZZ}$ and the vector of errors, denoted by $\mathbf{a}$, has zero expectation. If $\boldsymbol{\sigma}(\boldsymbol{\theta})$ is a linear function of $\boldsymbol{\theta}$, Equation (4.2.4) becomes

$$\mathbf{s} = \mathbf{F}\boldsymbol{\theta} + \mathbf{a}, \tag{4.2.5}$$

where the elements of $\mathbf{F}$ are known constants.

To illustrate the representation (4.2.5), assume that

$$\boldsymbol{\Sigma}_{ZZ} = \mathbf{B}\boldsymbol{\Sigma}_{xx}\mathbf{B}' + \boldsymbol{\Sigma}_{\varepsilon\varepsilon}, \tag{4.2.6}$$

where the $p \times k$ matrix $\mathbf{B}$ is known, the $p \times p$ matrix $\boldsymbol{\Sigma}_{\varepsilon\varepsilon}$ is known, and the $k \times k$ matrix $\boldsymbol{\Sigma}_{xx}$ is unknown. Then the vector of unknown parameters is $\boldsymbol{\theta} = \text{vech } \boldsymbol{\Sigma}_{xx}$ and the elements of $\boldsymbol{\Sigma}_{ZZ}$ are linear functions of the unknown parameters. That is,

$$\sigma_{ZZij} - \sigma_{\varepsilon\varepsilon ij} = \sum_{\ell=1}^{k} \sum_{t=1}^{k} b_{it}\sigma_{xxt\ell}b_{j\ell} = \mathbf{F}_{ij}\boldsymbol{\theta}, \tag{4.2.7}$$

where b_{ij} is the ijth element of $\mathbf{B}$ and

$$\mathbf{F}_{ij} = (b_{i1}b_{j1}, b_{i1}b_{j2} + b_{i2}b_{j1}, \ldots, b_{ik}b_{jk}).$$

Therefore, we can write

$$\text{vech } \mathbf{m}_{ZZ} - \text{vech } \boldsymbol{\Sigma}_{\varepsilon\varepsilon} = \mathbf{F}\boldsymbol{\theta} + \mathbf{a}, \tag{4.2.8}$$

where the row of $\mathbf{F}$ corresponding to m_{ZZij} is $\mathbf{F}_{ij}$ and $\mathbf{a}$ is the error vector.

If the expectations of the elements of $\mathbf{m}_{ZZ}$ are linear functions of the unknown parameters, if $\mathbf{F}$ has full column rank, and if

$$\mathbf{V}_{aa} = E\{\mathbf{a}'\mathbf{a}\} \tag{4.2.9}$$

is known and nonsingular, the best linear unbiased estimator for the $\boldsymbol{\theta}$ of model (4.2.5) is

$$\ddot{\boldsymbol{\theta}} = (\mathbf{F}'\mathbf{V}_{aa}^{-1}\mathbf{F})^{-1}\mathbf{F}'\mathbf{V}_{aa}^{-1}\mathbf{s}. \tag{4.2.10}$$

For models of the type (4.2.8) one replaces $\mathbf{s}$ in (4.2.10) with $\text{vech } \mathbf{m}_{ZZ} - \text{vech } \boldsymbol{\Sigma}_{\varepsilon\varepsilon}$.

In most situations the covariance matrix $\mathbf{V}_{aa}$ will not be known, but it will be possible to find a consistent estimator of $\mathbf{V}_{aa}$. If the vectors $\mathbf{Z}_t$ are normally and independently distributed, then

$$\mathbf{V}_{aa} = 2(n-1)^{-1}\psi_p(\mathbf{\Sigma}_{ZZ} \otimes \mathbf{\Sigma}_{ZZ})\psi_p'.$$

See Appendix 1.B and Lemma 4.A.1 of Appendix 4.A. Furthermore,

$$\hat{\mathbf{V}}_{aa} = 2(n-1)^{-1}\psi_p(\mathbf{m}_{ZZ} \otimes \mathbf{m}_{ZZ})\psi_p' \tag{4.2.11}$$

is an estimator of $\mathbf{V}_{aa}$ constructed with a consistent estimator of $\mathbf{\Sigma}_{ZZ}$. Because a consistent estimator of $(n-1)\mathbf{V}_{aa}$ is available, it is natural to consider the estimator

$$\hat{\theta} = (\mathbf{F}'\hat{\mathbf{V}}_{aa}^{-1}\mathbf{F})^{-1}\mathbf{F}'\hat{\mathbf{V}}_{aa}^{-1}\mathbf{s} \tag{4.2.12}$$

obtained from (4.2.10) by replacing $\mathbf{V}_{aa}$ with $\hat{\mathbf{V}}_{aa}$. A consistent distribution-free estimator of $\mathbf{V}_{aa}$ is given in Example 4.2.3.

If the elements of $\boldsymbol{\sigma}(\boldsymbol{\theta})$ are continuous nonlinear functions of $\boldsymbol{\theta}$ with continuous first derivatives, a nonlinear least squares algorithm can be used to find the $\hat{\boldsymbol{\theta}}$ that minimizes

$$Q_c(\boldsymbol{\theta}; \hat{\mathbf{V}}_{aa}) = [\mathbf{s} - \boldsymbol{\sigma}(\boldsymbol{\theta})]'\hat{\mathbf{V}}_{aa}^{-1}[\mathbf{s} - \boldsymbol{\sigma}(\boldsymbol{\theta})]. \tag{4.2.13}$$

If we let $\mathbf{F} = \mathbf{F}(\boldsymbol{\theta})$ be the $\frac{1}{2}p(p+1) \times q$ matrix of partial derivatives of $\boldsymbol{\sigma}(\boldsymbol{\theta})$ evaluated at $\boldsymbol{\theta}$ and let $\hat{\mathbf{F}} = \mathbf{F}(\hat{\boldsymbol{\theta}})$ be the $\frac{1}{2}p(p+1) \times q$ matrix of partial derivatives evaluated at $\hat{\boldsymbol{\theta}}$, the least squares estimators satisfy

$$\hat{\mathbf{F}}'\hat{\mathbf{V}}_{aa}^{-1}[\mathbf{s} - \boldsymbol{\sigma}(\hat{\boldsymbol{\theta}})] = \mathbf{0}, \tag{4.2.14}$$

provided $\hat{\boldsymbol{\theta}}$ is in the parameter space for $\boldsymbol{\theta}$. Should $\hat{\boldsymbol{\theta}}$ fall outside the parameter space (e.g., a negative estimate of variance), the function (4.2.13) can be minimized subject to the parameter constraints.

The fact that we can often express the expected values of the elements of a covariance matrix in terms of the parameters of interest, as in (4.2.8), and that we can find an estimator of $\mathbf{V}_{aa}$, as in (4.2.11), together with the availability of computer software for linear and nonlinear regression, makes the least squares approach attractive.

We first give the limiting behavior of the estimated generalized least squares estimator for the structural model.

Theorem 4.2.1. Let $\mathbf{Z}_t$ be independently and identically distributed random vectors with finite fourth moments. Let $\mathbf{m}_{ZZ}$ be the sample covariance matrix for a sample of $\mathbf{Z}_t$ vectors. Let the elements of $\boldsymbol{\sigma}(\boldsymbol{\theta})$ be continuous functions of $\boldsymbol{\theta}$ with continuous first and second derivatives at the true parameter value. Let A, the parameter space for $\boldsymbol{\theta}$, be a convex, compact subset of q-dimensional Euclidean space such that $\mathbf{\Sigma}_{ZZ}(\boldsymbol{\theta})$ is positive definite and such that if $\boldsymbol{\theta}_1$ and $\boldsymbol{\theta}_2$ are in A and $\boldsymbol{\theta}_1 \neq \boldsymbol{\theta}_2$, then $\boldsymbol{\Sigma}_{ZZ}(\boldsymbol{\theta}_1) \neq \boldsymbol{\Sigma}_{ZZ}(\boldsymbol{\theta}_2)$. Assume that the true parameter value is an interior point of A and that $\mathbf{F}$, the matrix of partial derivatives of $\boldsymbol{\sigma}(\boldsymbol{\theta})$ evaluated at the true $\boldsymbol{\theta}$, is of full column rank. Let $\hat{\mathbf{V}}_{aa}$ be a symmetric positive definite matrix such that

$$\text{plim}\{(n-1)(\hat{\mathbf{V}}_{aa} - \mathbf{V}_{aa})\} = \mathbf{0},$$

where $\mathbf{V}_{aa}$ is defined in (4.2.9). Let $\hat{\boldsymbol{\theta}}$ be the value of $\boldsymbol{\theta}$ in A that minimizes (4.2.13). Then

(i)
$$n^{1/2}(\hat{\boldsymbol{\theta}} - \boldsymbol{\theta}) \xrightarrow{L} N\{\mathbf{0}, \boldsymbol{\Gamma}_{\theta\theta}\},$$

where $\boldsymbol{\Gamma}_{\theta\theta} = \lim(n-1)(\mathbf{F}'\mathbf{V}_{aa}^{-1}\mathbf{F})^{-1}$.

(ii)
$$Q_\ell(\hat{\boldsymbol{\theta}}; \hat{\mathbf{V}}_{aa}) \xrightarrow{L} \chi^2,$$

where $Q_\ell(\hat{\boldsymbol{\theta}}; \hat{\mathbf{V}}_{aa}^{-1})$, is the minimum value of (4.2.13), and χ^2 is a chi-square random variable with $\frac{1}{2}p(p+1) - q$ degrees of freedom, provided $\frac{1}{2}p(p+1) > q$.

(iii)
$$(n-1)\hat{\mathbf{V}}\{\hat{\boldsymbol{\theta}}\} \xrightarrow{P} \boldsymbol{\Gamma}_{\theta\theta},$$

where $\hat{\mathbf{V}}\{\hat{\boldsymbol{\theta}}\} = (\hat{\mathbf{F}}'\tilde{\mathbf{V}}_{aa}^{-1}\hat{\mathbf{F}})^{-1}$, $\hat{\mathbf{F}}$ is the matrix of derivatives of $\boldsymbol{\sigma}(\boldsymbol{\theta})$ with respect to $\boldsymbol{\theta}'$ evaluated at $\hat{\boldsymbol{\theta}}$, and $(n-1)\tilde{\mathbf{V}}_{aa}$ is a consistent estimator of $(n-1)\mathbf{V}_{aa}$.

Proof. Result (i) is proved as Theorem 4.B.4 in Appendix 4.B. To prove result (ii), we note that, by assumption,

$$(n-1)^{-1}\hat{\mathbf{V}}_{aa}^{-1} = (n-1)^{-1}\mathbf{V}_{aa}^{-1} + o_p(1),$$

and from result (i)

$$n^{1/2}(\hat{\boldsymbol{\theta}} - \boldsymbol{\theta}) = n^{1/2}(\mathbf{F}'\mathbf{V}_{aa}^{-1}\mathbf{F})^{-1}\mathbf{F}'\mathbf{V}_{aa}^{-1}\mathbf{a} + o_p(1).$$

Therefore,

$$\boldsymbol{\sigma}(\hat{\boldsymbol{\theta}}) = \boldsymbol{\sigma}(\boldsymbol{\theta}) + \mathbf{F}(\hat{\boldsymbol{\theta}} - \boldsymbol{\theta}) + o_p(n^{-1/2})$$

and

$$Q_\ell(\hat{\boldsymbol{\theta}}; \hat{\mathbf{V}}_{aa}) = \mathbf{a}'\mathbf{V}_{aa}^{-1}\mathbf{a} - \mathbf{a}'\mathbf{V}_{aa}^{-1}\mathbf{F}(\mathbf{F}'\mathbf{V}_{aa}^{-1}\mathbf{F})^{-1}\mathbf{F}'\mathbf{V}_{aa}^{-1}\mathbf{a} + o_p(1).$$

Now $(n-1)^{1/2}\,\mathbf{a}$ is converging in distribution to a normal random vector with zero mean and covariance matrix $(n-1)\mathbf{V}_{aa}$. The quadratic form in $\mathbf{a}$ that approximates $Q_\ell(\hat{\boldsymbol{\theta}}; \hat{\mathbf{V}}_{aa})$ is of rank $\frac{1}{2}p(p+1) - q$. The result follows from standard linear least squares theory.

Result (iii) follows from the consistency of $(n-1)\tilde{\mathbf{V}}_{aa}$, the fact that the error in $\hat{\boldsymbol{\theta}}$ is $O_p(n^{-1/2})$, and the assumption that $\mathbf{F}$ is a continuous function of $\boldsymbol{\theta}$. □

Example 4.2.1. Least squares provides a procedure for estimating the parameters of a model using several sets of data. In Example 1.4.1 we introduced a data set containing three measures of the magnitudes of Alaskan earthquakes. In addition to the sample of earthquakes for which three measures are available, there are also 261 earthquakes for which two measures, amplitude of surface waves and logarithm of seismogram amplitude, are available. In this analysis we use 61 of the 62 earthquakes of Example 1.4.1, omitting observation number 54, and the second sample of 261 earthquakes. The sample mean vector is

$$(\bar{Y}_1, \bar{X}_1, \bar{W}_1) = (5.0836, 5.2328, 5.2436)$$

and the covariance matrix is

$$\mathbf{m}_{11} = \begin{bmatrix} m_{YY11} & m_{YX11} & m_{YW11} \\ m_{XY11} & m_{XX11} & m_{XW11} \\ m_{WY11} & m_{WX11} & m_{WW11} \end{bmatrix} = \begin{bmatrix} 0.6301 & 0.2702 & 0.4119 \\ 0.2702 & 0.1946 & 0.2305 \\ 0.4119 & 0.2305 & 0.4105 \end{bmatrix}$$

for the sample of 61 earthquakes. The mean vector is $(\bar{Y}_2, \bar{X}_2) = (5.0318, 5.2452)$ and the covariance matrix is

$$\mathbf{m}_{22} = \begin{bmatrix} m_{YY22} & m_{YX22} \\ m_{XY22} & m_{XX22} \end{bmatrix} = \begin{bmatrix} 0.5304 & 0.2280 \\ 0.2280 & 0.1595 \end{bmatrix}$$

for the sample of 261 earthquakes with observations only on (Y, X). We have used the subscript 1 to identify the first sample and the subscript 2 to identify the second sample. We postulate a model in which the structural lines are the same for the two sets of data. The model permits the variance of the earthquake magnitudes to be different in the two populations from which the two samples are selected. We assume the measurement errors to be independent and to have the same variances in the two populations. Thus, our model is

$$Y_{ti} = \beta_0 + \beta_1 x_{ti} + e_{ti}, \qquad X_{ti} = x_{ti} + u_{ti}, \qquad i = 1, 2,$$

$$W_{t1} = \gamma_0 + \gamma_1 x_{t1} + a_{t1},$$

$$(x_{t1}, e_{t1}, u_{t1}, a_{t1})' \sim \mathrm{NI}[(\mu_{x1}, \mathbf{0})', \mathrm{diag}(\sigma_{xx11}, \sigma_{ee}, \sigma_{uu}, \sigma_{aa})],$$

$$(x_{t2}, e_{t2}, u_{t2})' \sim \mathrm{NI}[(\mu_{x2}, \mathbf{0})', \mathrm{diag}(\sigma_{xx22}, \sigma_{ee}, \sigma_{uu})].$$

The observation vector in our analysis is

$$\mathbf{s} = [(\mathrm{vech}\ \mathbf{m}_{11})', (\mathrm{vech}\ \mathbf{m}_{22})', (\bar{Y}_1, \bar{X}_1, \bar{W}_1), (\bar{Y}_2, \bar{X}_2)]'.$$

The covariance matrix of **s** is block diagonal with four blocks corresponding to the four subvectors of **s**. By Lemma 4.A.1 the covariance matrix of the vector of covariances for the first population is estimated by

$$\hat{\mathbf{\Omega}}_{11} = \hat{\mathrm{V}}\{\mathrm{vech}\ \mathbf{m}_{11}\} = 2(60)^{-1}\psi_3(\mathbf{m}_{11} \otimes \mathbf{m}_{11})\psi_3'$$

and the covariance matrix of the vector of covariances for the second population is estimated by

$$\hat{\boldsymbol{\Omega}}_{22} = \hat{\mathbf{V}}\{\text{vech } \mathbf{m}_{22}\} = 2(260)^{-1}\boldsymbol{\psi}_2(\mathbf{m}_{22} \otimes \mathbf{m}_{22})\boldsymbol{\psi}_2'.$$

To compute the estimates we use the estimated covariance matrix of $\mathbf{s}$ to construct a transformation such that the estimated covariance matrix of the transformed variables is the identity. Then we apply ordinary nonlinear least squares to the transformed problem. Given the matrices $\hat{\boldsymbol{\Omega}}_{11}$, $\hat{\boldsymbol{\Omega}}_{22}$, $\mathbf{m}_{11}$, and $\mathbf{m}_{22}$ we can find matrices $\mathbf{P}_1$, $\mathbf{P}_2$, $\mathbf{P}_3$, and $\mathbf{P}_4$ such that

$$\mathbf{P}_1'\mathbf{P}_1 = \hat{\boldsymbol{\Omega}}_{11}^{-1}, \qquad \mathbf{P}_2'\mathbf{P}_2 = \hat{\boldsymbol{\Omega}}_{22}^{-1},$$
$$\mathbf{P}_3'\mathbf{P}_3 = 61\mathbf{m}_{11}^{-1}, \qquad \mathbf{P}_4'\mathbf{P}_4 = 261\mathbf{m}_{22}^{-1}.$$

One method of computing the $\mathbf{P}$ matrices is to find the triangular matrix defined by the Cholesky decomposition. If we let $\mathbf{P} = \text{block diag}(\mathbf{P}_1, \mathbf{P}_2, \mathbf{P}_3, \mathbf{P}_4)$, then

$$\mathbf{P}'\mathbf{P} = \text{block diag}(\hat{\boldsymbol{\Omega}}_{11}^{-1}, \hat{\boldsymbol{\Omega}}_{22}^{-1}, \mathbf{m}_{11}^{-1}, \mathbf{m}_{22}^{-1})$$

is an estimator of the inverse of the covariance matrix of $\mathbf{s}$. It follows that the estimated covariance matrix of $\mathbf{Ps}$ is the identity matrix. Therefore, we can use any of the standard nonlinear regression programs to estimate the parameters of the model by the method of nonlinear least squares applied to the transformed vector of observations $\mathbf{Ps}$. For our example the nonlinear model can be written

$$\begin{aligned}\mathbf{Ps} = {} & (\beta_1^2\sigma_{xx11} + \sigma_{ee})\mathbf{P}_{.1} + \beta_1\sigma_{xx11}\mathbf{P}_{.2} + \beta_1\gamma_1\sigma_{xx11}\mathbf{P}_{.3} \\ & + (\sigma_{xx11} + \sigma_{uu})\mathbf{P}_{.4} + \gamma_1\sigma_{xx11}\mathbf{P}_{.5} + (\gamma_1^2\sigma_{xx11} + \sigma_{uu})\mathbf{P}_{.6} \\ & + (\beta_1^2\sigma_{xx22} + \sigma_{ee})\mathbf{P}_{.7} + \beta_1\sigma_{xx22}\mathbf{P}_{.8} + (\sigma_{xx22} + \sigma_{uu})\mathbf{P}_{.9} \\ & + (\beta_0 + \beta_1\mu_{x1})\mathbf{P}_{.10} + \mu_{x1}\mathbf{P}_{.11} + (\gamma_0 + \gamma_1\mu_{x1})\mathbf{P}_{.12} \\ & + (\beta_0 + \beta_1\mu_{x2})\mathbf{P}_{.13} + \mu_{x2}\mathbf{P}_{.14} + \mathbf{P}(\mathbf{s} - \boldsymbol{\sigma}),\end{aligned} \tag{4.2.15}$$

where $\mathbf{P}_{.i}$ is the ith column of $\mathbf{P}$, $\boldsymbol{\sigma} = E\{\mathbf{s}\}$, and $\mathbf{P}(\mathbf{s} - \boldsymbol{\sigma})$ is the error in the equation.

The nonlinear least squares estimates of the model are given in Table 4.2.1. The residual sum of squares for the regression is 2.20 with $14 - 11 = 3$ degrees of freedom. Because 2.20 is less than the expected value of the chi-square distribution with three degrees of freedom, the residual sum of squares gives no reason to reject the model.

The sample moments and the model estimates of the moments are given in Table 4.2.2. The variances of the differences between the observed and estimated moments were estimated by

$$\hat{\mathbf{V}}\{\mathbf{s} - \hat{\boldsymbol{\sigma}}\} = \mathbf{P}^{-1}[\mathbf{I} - \hat{\mathbf{K}}(\hat{\mathbf{K}}'\hat{\mathbf{K}})^{-1}\hat{\mathbf{K}}']\mathbf{P}^{-1\prime},$$

TABLE 4.2.1. Least squares estimates for the earthquake model

Parameter	Least Squares Estimate	Least Squares Standard Error
β_0	−4.456	0.701
β_1	1.811	0.134
γ_0	−2.595	0.681
γ_1	1.493	0.130
μ_{x1}	5.251	0.053
μ_{x2}	5.242	0.025
σ_{xx11}	1.533	0.333
σ_{xx22}	1.253	0.156
σ_{ee}	1.205	0.283
$10\sigma_{uu}$	0.350	0.086
$10\sigma_{aa}$	0.686	0.201

where $\hat{\mathbf{K}}$ is the 14×11 matrix of derivatives associated with the nonlinear function (4.2.15) evaluated at the least squares estimates. The matrix $(\hat{\mathbf{K}}'\hat{\mathbf{K}})^{-1}$ is equivalent to the matrix $(\hat{\mathbf{F}}'\hat{\mathbf{V}}_{aa}^{-1}\hat{\mathbf{F}})^{-1}$ of Theorem 4.2.1. The entries in the column titled "Standard Error of Difference" are the square roots of the diagonal elements of $\hat{\mathbf{V}}\{\mathbf{s} - \hat{\boldsymbol{\sigma}}\}$. The model seems to fit the data quite well.

TABLE 4.2.2. Sample moments and least squares estimates

Sample Moment	Value of Sample Moment	Model Estimate	Standard Error of Difference	Standardized Difference
m_{YY11}	0.6301	0.6233	0.0471	0.15
m_{XY11}	0.2702	0.2776	0.0069	−1.08
m_{WY11}	0.4119	0.4145	0.0148	−0.17
m_{XX11}	0.1946	0.1883	0.0153	0.41
m_{WX11}	0.2305	0.2289	0.0097	0.17
m_{WW11}	0.4105	0.4103	0.0012	0.15
m_{YY22}	0.5304	0.5315	0.0090	−0.12
m_{XY22}	0.2280	0.2269	0.0010	1.10
m_{XX22}	0.1595	0.1603	0.0026	−0.32
$\bar{Y}_1$	5.0836	5.0531	0.0308	0.99
$\bar{X}_1$	5.2328	5.2506	0.0180	−0.99
$\bar{W}_1$	5.2436	5.2448	0.0012	−0.99
$\bar{Y}_2$	5.0318	5.0377	0.0060	−0.99
$\bar{X}_2$	5.2452	5.2421	0.0031	0.99

The standard errors of Tables 4.2.1 and 4.2.2 are constructed under the assumption that the original observations are normally distributed. Deviations from normality have modest effects on the distribution of $\hat{\beta}_1$ and $\hat{\gamma}_1$ but have a larger effect on the distribution of estimators of variance such as $\hat{\sigma}_{xxii}$ and $\hat{\sigma}_{ee}$. Therefore, any analysis of a sample moment matrix should include normality checks on the original data. Ganse, Amemiya, and Fuller (1983) report that the hypothesis of normality is rejected for these data, but they retained the normal assumption in their computations on the basis that the deviation from normality was not large in absolute terms. □ □

The least squares estimator has an explicit form when the covariance matrix is a linear function of the unknown parameters. In addition, the Kronecker product form of the estimated covariance matrix for normal covariances leads to a compact expression for the estimator in certain important normal distribution cases. Let a $p \times p$ covariance matrix $\boldsymbol{\Sigma}_{ZZ}$, the expectation of an observed covariance matrix $\mathbf{m}_{ZZ}$, be defined in terms of the $q \times q$ covariance matrix $\boldsymbol{\Sigma}_{ff}$ by

$$\boldsymbol{\Sigma}_{ZZ} = \mathbf{C}'\boldsymbol{\Sigma}_{ff}\mathbf{C}, \tag{4.2.16}$$

where the $q \times p$ matrix $\mathbf{C}$ is known. Assume that the elements of $\boldsymbol{\Sigma}_{ff}$ are defined in terms of a vector $\boldsymbol{\theta}$ by

$$\operatorname{vec} \boldsymbol{\Sigma}_{ff} = \mathbf{R}\boldsymbol{\theta},$$

where $\mathbf{R}$ is a fixed matrix. Then, by Result 4.A.4, the system of equations

$$\operatorname{vech} \mathbf{m}_{ZZ} = \boldsymbol{\psi}_p(\mathbf{C}' \otimes \mathbf{C}')\mathbf{R}\boldsymbol{\theta} + \mathbf{a} \tag{4.2.17}$$

defines the vector half of the observed covariance matrix in terms of a linear function of $\boldsymbol{\theta}$ and an error vector $\mathbf{a}$. Assume that $\boldsymbol{\psi}_p(\mathbf{C}' \otimes \mathbf{C}')\mathbf{R}$ is of full column rank and let $\mathbf{W}$ be an estimator of $\boldsymbol{\Sigma}_{ZZ}$. Then the generalized least squares estimator of $\boldsymbol{\theta}$ is

$$\begin{aligned}\hat{\boldsymbol{\theta}} = {} & [\mathbf{R}'\boldsymbol{\kappa}_q\{(\mathbf{C}\mathbf{W}^{-1}\mathbf{C}') \otimes (\mathbf{C}\mathbf{W}^{-1}\mathbf{C}')\}\boldsymbol{\kappa}_q\mathbf{R}]^{-1} \\ & \times [\mathbf{R}'\boldsymbol{\kappa}_q\{(\mathbf{C}\mathbf{W}^{-1}) \otimes (\mathbf{C}\mathbf{W}^{-1})\} \operatorname{vec} \mathbf{m}_{ZZ}],\end{aligned} \tag{4.2.18}$$

where $\boldsymbol{\kappa}_q = \boldsymbol{\Phi}_q\boldsymbol{\psi}_q$, $2(n-1)^{-1}\, \boldsymbol{\psi}_p(\mathbf{W} \otimes \mathbf{W})\boldsymbol{\psi}_p'$ is an estimator of $\mathbf{V}_{aa}$, and we have used Result 4.A.1(i) and Result 4.A.2. If $\boldsymbol{\theta}$ is a vector composed of some elements of $\boldsymbol{\Sigma}_{ff}$, then the columns of $\mathbf{R}$ are in the column space of $\boldsymbol{\Phi}_q$ and

$$\boldsymbol{\kappa}_q\mathbf{R} = \boldsymbol{\Phi}_q(\boldsymbol{\Phi}_q'\boldsymbol{\Phi}_q)^{-1}\boldsymbol{\Phi}_q'\mathbf{R} = \mathbf{R}.$$

For such cases

$$\hat{\boldsymbol{\theta}} = [\mathbf{R}'\{(\mathbf{C}\mathbf{W}^{-1}\mathbf{C}') \otimes (\mathbf{C}\mathbf{W}^{-1}\mathbf{C}')\}\mathbf{R}]^{-1}\mathbf{R}' \operatorname{vec}(\mathbf{C}\mathbf{W}^{-1}\mathbf{m}_{ZZ}\mathbf{W}^{-1}\mathbf{C}'). \tag{4.2.19}$$

If $\boldsymbol{\Sigma}_{ff}$ is a $q \times q$ diagonal matrix and the elements of $\boldsymbol{\theta}$ are those diagonal elements, then the matrix $\mathbf{R}$ is equal to $\mathbf{L}$, where $\mathbf{L}$ is a $q^2 \times q$ matrix with ones in the $[1, 1]$, $[(q + 2), 2]$, $[2q + 3, 3], \ldots, [q^2, q]$ positions and zeros elsewhere. The least squares results for diagonal $\boldsymbol{\Sigma}_{ff}$ are developed in Theorem 4.2.2.

Theorem 4.2.2. Let $\mathbf{m}_{ZZ}$ be an unbiased estimator of the nonsingular covariance matrix $\boldsymbol{\Sigma}_{ZZ}$, where $\mathbf{m}_{ZZ}$ is distributed as a multiple of a Wishart matrix with $n - 1$ degrees of freedom. Let $\boldsymbol{\Sigma}_{ZZ}$ satisfy (4.2.16), where $\boldsymbol{\Sigma}_{ff} = \text{diag}(\sigma_{ff11}, \sigma_{ff22}, \ldots, \sigma_{ffqq})$ and $\mathbf{C}$ is a known matrix. Let $\psi_p(\mathbf{C}' \otimes \mathbf{C}')\mathbf{L}$ be of full column rank. Let $\psi_p(\mathbf{W} \otimes \mathbf{W})\psi_p'$ be used as an estimator of a multiple of $\mathbf{V}_{aa}$ in constructing the least squares estimator. Then the least squares estimator of the vector of diagonal elements of $\boldsymbol{\Sigma}_{ff}$ is

$$\hat{\boldsymbol{\theta}} = \mathbf{B}^{-1}\mathbf{d}, \tag{4.2.20}$$

where $\boldsymbol{\theta} = (\sigma_{ff11}, \sigma_{ff22}, \ldots, \sigma_{ffqq})'$, $\mathbf{B}$ is a matrix whose ijth element is the square of the ijth element of $\mathbf{CW}^{-1}\mathbf{C}'$, and $\mathbf{d}$ is a column vector whose ith element is the ith diagonal element of

$$\mathbf{CW}^{-1}\mathbf{m}_{ZZ}\mathbf{W}^{-1}\mathbf{C}'. \tag{4.2.21}$$

If $\mathbf{W}$ is a consistent estimator of $\boldsymbol{\Sigma}_{ZZ}$, a consistent estimator of the covariance matrix of the approximate distribution of $\hat{\boldsymbol{\theta}}$ is $2(n - 1)^{-1}\mathbf{B}^{-1}$. Also, if $\mathbf{W}$ is a consistent estimator of $\boldsymbol{\Sigma}_{ZZ}$, the residual sum of squares

$$\hat{\chi}^2 = \tfrac{1}{2}(n - 1)\,\text{tr}\{[(\mathbf{m}_{ZZ} - \hat{\boldsymbol{\Sigma}}_{ZZ})\mathbf{W}^{-1}]^2\} = \tfrac{1}{2}(n - 1)[\text{tr}\{(\mathbf{m}_{ZZ}\mathbf{W}^{-1})^2\} - \mathbf{d}'\hat{\boldsymbol{\theta}}],$$

where $\hat{\boldsymbol{\Sigma}}_{ZZ} = \mathbf{C}'\boldsymbol{\Sigma}_{ff}(\hat{\boldsymbol{\theta}})\mathbf{C}$, converges in distribution to a chi-square random variable with $\frac{1}{2}p(p + 1) - q$ degrees of freedom.

Proof. If $\boldsymbol{\Sigma}_{ff}$ is diagonal, the matrix $\mathbf{L}$ has ones in the rows corresponding to the diagonal elements. These are the$[(i - 1)q + i, i]$ positions or, in terms of the double subscript notation, the $[ii, ii]$ positions. Therefore, the ijth element of

$$\mathbf{L}'[(\mathbf{CW}^{-1}\mathbf{C}') \otimes (\mathbf{CW}^{-1}\mathbf{C}')]\mathbf{L}$$

is the square of the ijth element of $\mathbf{CW}^{-1}\mathbf{C}'$. Because $\mathbf{L}'$ selects the diagonal elements from the vector representation of the matrix, the ith element of

$$\mathbf{d} = \mathbf{L}'\,\text{vec}(\mathbf{CW}^{-1}\mathbf{m}_{ZZ}\mathbf{W}^{-1}\mathbf{C}')$$

is the ith diagonal element of $\mathbf{CW}^{-1}\mathbf{m}_{ZZ}\mathbf{W}^{-1}\mathbf{C}'$.

If $\mathbf{W}$ is a consistent estimator of $\boldsymbol{\Sigma}_{ZZ}$, then $\mathbf{CW}^{-1}\mathbf{C}'$ converges to $\mathbf{C}\boldsymbol{\Sigma}_{ZZ}^{-1}\mathbf{C}'$ in probability and the residual sum of squares converges in distribution to a chi-square random variable by arguments of the type used in the proof of result (ii) of Theorem 4.2.1. □

Example 4.2.2. To illustrate the least squares computations for a model of the form given in Theorem 4.2.2 we use the 37 observations on hectares of corn introduced in Example 1.5.1. We begin the analysis by assuming that the data satisfy the model

$$Y_{ti} = \beta_{0i} + x_t + e_{ti}, \qquad X_t = x_t + u_t, \tag{4.2.22}$$

for $i = 1, 2$, where Y_{t1} is the photographic determination, Y_{t2} is the satellite determination, and X_t is the interview determination. We provided evidence in Example 1.5.1 that the assumption of constant variance normal errors is not appropriate for this data set. However, to illustrate the computations, we carry out the analysis of this example under the assumption

$$(x_t, e_{t1}, e_{t2}, u_t)' \sim \mathrm{NI}[(\mu_x, \mathbf{0}), \mathrm{diag}(\sigma_{xx}, \sigma_{ee11}, \sigma_{ee22}, \sigma_{uu})].$$

Under model (4.2.22), the covariance matrix of $\mathbf{Z}_t = (Y_{t1}, Y_{t2}, X_t)$ is of the form

$$\mathbf{\Sigma}_{ZZ} = \mathbf{C}'[\mathrm{diag}(\sigma_{xx}, \sigma_{ee11}, \sigma_{ee22}, \sigma_{uu})]\mathbf{C},$$

where $\mathrm{vec}\,\mathbf{C} = (1, 1, 1; 1, 0, 0; 0, 1, 0; 0, 0, 1)'$.

The sample covariance matrix for $\mathbf{Z}_t$ is given in Example 1.5.1 and

$$\mathbf{m}_{ZZ}^{-1} = \begin{bmatrix} 2.9131 & -0.1669 & -2.9171 \\ -0.1669 & 0.3223 & -0.0839 \\ -2.9171 & -0.0839 & 3.2170 \end{bmatrix} 10^{-2}.$$

If we use $\mathbf{m}_{ZZ}$ as the $\mathbf{W}$ in constructing the estimator (4.2.20), we have

$$\mathbf{B} = \begin{bmatrix} 1.36 & 2.92 & 0.51 & 4.67 \\ 2.92 & 848.64 & 2.79 & 850.92 \\ 0.51 & 2.79 & 10.39 & 0.70 \\ 4.67 & 850.92 & 0.76 & 1034.90 \end{bmatrix} 10^{-6},$$

$$\mathbf{d}' = (0.1166, 2.9131, 0.3223, 3.2170)10^{-2}$$

and the estimated parameters are

$$\underset{(206.1)}{\hat{\sigma}_{xx} = 672.1}, \qquad \underset{(19.4)}{\hat{\sigma}_{ee11} = 18.1}, \qquad \underset{(74.0)}{\hat{\sigma}_{ee22} = 271.5}, \qquad \underset{(17.6)}{\hat{\sigma}_{uu} = 13.0},$$

where the numbers in parentheses are the estimated standard errors calculated as the square roots of the diagonal elements of $2(n-1)^{-1}\mathbf{B}^{-1}$. The $\hat{\chi}^2$ value defined in Theorem 4.2.2 is 7.13. The value 7.13 falls between the tabular 0.05 and 0.025 points of the chi-square distribution with two degrees of freedom. Therefore, unless there was strong external evidence for the model, we would reject the fitted model for these data.

Although we would reject the model, we continue the analysis in order to illustrate some other computational aspects of the problem. The vector half of the estimated covariance matrix of the observations based on the least squares estimates is

$$\text{vech } \hat{\mathbf{\Sigma}}_{ZZ} = (690.2, 672.1, 672.1; 943.6, 672.1; 685.1)'.$$

If we use this matrix as the $\mathbf{W}$ matrix in a second iteration of least squares, the second round least squares estimates are

$$\hat{\sigma}_{xx} = 1089.2, \qquad \hat{\sigma}_{ee11} = 45.3, \qquad \hat{\sigma}_{ee22} = 349.9, \qquad \hat{\sigma}_{uu} = -6.9.$$

Variances must be nonnegative and the estimator of σ_{uu} is outside the parameter space. Because the system of equations used to obtain the estimates is linear, it is easy to impose the nonnegativity condition on the estimates. We do this by removing the row and column of the $\mathbf{B}$ matrix that correspond to the negative estimate. In our case the last row and last column are removed. The corresponding element of the $\mathbf{d}$ vector is removed and the reduced system of equations is solved for the remaining parameters. The reduced system of three equations gives

$$\hat{\sigma}_{xx} = 1088.0, \qquad \hat{\sigma}_{ee11} = 39.16, \qquad \hat{\sigma}_{ee22} = 348.62.$$

These estimates, together with $\hat{\sigma}_{uu} = 0$, can be used to construct another round of least squares estimates, and so on. After four rounds of calculation the estimates are

$$\underset{(250.0)}{\hat{\sigma}_{xx} = 1058.6}, \qquad \underset{(18.8)}{\hat{\sigma}_{ee11} = 37.8}, \qquad \underset{(86.6)}{\hat{\sigma}_{ee22} = 361.0}, \qquad \underset{(16.6)}{\hat{\sigma}_{uu} = 0.0}.$$

The standard errors were computed from the inverse of the 4×4 matrix $\mathbf{B}$, but it was necessary to use the 3×3 submatrix of $\mathbf{B}$ to obtain nonnegative estimates. The associated estimate of the covariance matrix of $\mathbf{Z}_t$ is

$$\text{vech } \hat{\mathbf{\Sigma}}_{ZZ} = (1096.4, 1058.6, 1058.6; 1419.6, 1058.6; 1058.6)'.$$

The least squares estimates after four iterations are equal to the maximum likelihood estimates to the accuracy reported here. See Section 4.2.2. The roots of the matrix $\mathbf{m}_{ZZ}^{-1/2}\hat{\mathbf{\Sigma}}_{ZZ}\mathbf{m}_{ZZ}^{-1/2}$ are

$$(\hat{\tau}_1, \hat{\tau}_2, \hat{\tau}_3) = (1.842, 0.935, 0.720).$$

It follows that the likelihood ratio test of the model against the unconstrained normal model is

$$2[\log L_c(\mathbf{m}_{ZZ}) - \log L_c(\hat{\mathbf{\Sigma}}_{ZZ})] = (n-1) \sum_{i=1}^{3} \{\log \hat{\tau}_i + \hat{\tau}_i^{-1} - 1\} = 7.79.$$

Because the value is beyond the 0.05 tabular value of the chi-square distribution with two degrees of freedom, the model is rejected at the 0.05 level by the likelihood ratio test. □ □

Example 4.2.3. In this example we present alternative least squares estimates of the parameters of model (4.2.22) for the corn hectares data. We replace the assumption of homoskedastic normal errors with the weaker distributional assumption that the vectors $(x_t, e_{t1}, e_{t2}, u_t)$ are independently and identically distributed with finite fourth moments and

$$E\{(e_{t1}, e_{t2}, u_t)\} = E\{x_t(e_{t1}, e_{t2}, u_t)\} = (0, 0, 0).$$

Let $\mathbf{b}_t = \text{vech}\{(\mathbf{Z}_t - \bar{\mathbf{Z}})'(\mathbf{Z}_t - \bar{\mathbf{Z}})\}$. Then

$$\hat{\mathbf{V}}_{bb} = (n-2)^{-1}(n-1)^{-1} \sum_{t=1}^{n} (\mathbf{b}_t - \bar{\mathbf{b}})(\mathbf{b}_t - \bar{\mathbf{b}})'$$

is a consistent estimator of the covariance matrix of vech $\mathbf{m}_{ZZ}$. The matrix $\hat{\mathbf{V}}_{bb}$ is given in Table 4.2.3. The elements of the matrix of Table 4.2.3 are similar to the elements of the matrix estimated under the normal assumption. For example, the diagonal elements of $2(n-1)^{-1}\psi_3(\mathbf{m}_{ZZ} \otimes \mathbf{m}_{ZZ})\psi_3'$ are

$$(79{,}549,\ 56{,}231,\ 69{,}335,\ 55{,}786,\ 49{,}530,\ 62{,}260).$$

Under our assumptions, model (4.2.22) specifies

$$E\{\text{vech } \mathbf{m}_{ZZ}\} = \mathbf{F}\boldsymbol{\theta},$$

where $\boldsymbol{\theta}' = (\sigma_{xx}, \sigma_{ee11}, \sigma_{ee22}, \sigma_{uu})$ and

$$\mathbf{F}' = \begin{bmatrix} 1 & 1 & 1 & 1 & 1 & 1 \\ 1 & 0 & 0 & 0 & 0 & 0 \\ 0 & 0 & 0 & 1 & 0 & 0 \\ 0 & 0 & 0 & 0 & 0 & 1 \end{bmatrix}.$$

TABLE 4.2.3. Distribution-free estimate of the covariance matrix of the sample moments

Moment					
m_{YY11}	m_{YY12}	m_{YX11}	m_{YY22}	m_{YX21}	m_{XX}
80097	63963	71119	48472	57672	64300
63963	58856	57943	47655	53820	53380
71119	57943	65340	44404	53919	61058
48472	47655	44404	43963	43895	41188
57672	53820	53919	43895	50592	51176
64300	53380	61058	41188	51176	58826

The estimated generalized least squares estimator of $\boldsymbol{\theta}$ based on $\hat{\mathbf{V}}_{bb}$ is

$$\hat{\boldsymbol{\theta}} = (\mathbf{F}'\hat{\mathbf{V}}_{bb}^{-1}\mathbf{F})^{-1}\mathbf{F}'\hat{\mathbf{V}}_{bb}^{-1}(\text{vech } \mathbf{m}_{ZZ}) = \underset{(210.6, \quad 14.7, \quad 73.5, \ 12.2)}{[687.9, \ -16.5, \ 226.8, \ 31.1]'},$$

where the standard errors are the square roots of the diagonal elements of $(\mathbf{F}'\hat{\mathbf{V}}_{bb}^{-1}\mathbf{F})^{-1}$. By Theorem 4.2.1, $n^{1/2}(\hat{\boldsymbol{\theta}} - \boldsymbol{\theta})$ converges in distribution to a normal random vector. Therefore, in large samples, the standard errors can be used for approximate tests and confidence intervals. The negative estimate for σ_{ee11} is unsatisfactory and, in practice, one would modify the estimator. A method of estimation in which $\hat{\sigma}_{ee11}$ is set equal to zero is illustrated in Example 4.2.2. We do not modify the estimator because our objective is to compare the estimates and estimated standard errors with those obtained using the estimated covariance matrix of vech $\mathbf{m}_{ZZ}$ based on the normality assumption. The vector of estimates obtained in Example 4.2.2. is

$$\hat{\boldsymbol{\theta}}_N = (\mathbf{F}'\hat{\mathbf{V}}_{aa}^{-1}\mathbf{F})^{-1}\mathbf{F}'\hat{\mathbf{V}}_{aa}^{-1}(\text{vech } \mathbf{m}_{ZZ}) = \underset{(206.1, \ 19.4, \quad 74.0, \ 17.6)}{[672.1, \ 18.1, \ 271.5, \ 13.0]'},$$

where $\hat{\mathbf{V}}_{aa}^{-1} = 2(n-1)^{-1}\boldsymbol{\psi}_3(\mathbf{m}_{ZZ} \otimes \mathbf{m}_{ZZ})\boldsymbol{\psi}_3'$ and the standard errors given in parentheses are the square roots of the diagonal elements of $(\mathbf{F}'\hat{\mathbf{V}}_{aa}^{-1}\mathbf{F})^{-1}$. The two estimates of σ_{ee11} differ by about two standard errors and the two estimates of σ_{uu} differ by about one standard error. In this example the estimated standard errors for $\hat{\sigma}_{ee11}$ and $\hat{\sigma}_{uu}$ are smaller when constructed with the distribution-free estimator of the covariance matrix than when constructed with the normal distribution estimator.

The residual sum of squares is

$$(\text{vech } \mathbf{m}_{ZZ})'\hat{\mathbf{V}}_{bb}^{-1}(\text{vech } \mathbf{m}_{ZZ}) - \hat{\boldsymbol{\theta}}'\mathbf{F}'\hat{\mathbf{V}}_{bb}^{-1}(\text{vech } \mathbf{m}_{ZZ}) = 8.91.$$

When the model is true the residual sum of squares is approximately distributed as a chi-square random variable with two degrees of freedom. Therefore, the distribution-free procedures of this example, as well as the normal distribution procedures of Example 4.2.2, lead to the rejection of model (4.2.22) at the 5% level. □ □

4.2.2. Relationships Between Least Squares and Maximum Likelihood

The generalized least squares estimator and the maximum likelihood estimator are closely related for the normal distribution model. In this section we demonstrate that the two estimators have the same limiting distribution under mild conditions and that, for most samples, the generalized least squares estimator computed using the maximum likelihood estimator of the covariance matrix of the sample moments is the maximum likelihood estimator.

If $\mathbf{Z}_t \sim \mathrm{NI}(\boldsymbol{\mu}_Z, \boldsymbol{\Sigma}_{ZZ})$ and if $\boldsymbol{\Sigma}_{ZZ} = \boldsymbol{\Sigma}_{ZZ}(\boldsymbol{\theta})$ is a function of the parameter vector $\boldsymbol{\theta}$, then the likelihood adjusted for degrees of freedom for a sample of n observations is

$$\begin{aligned}\log L_c(\boldsymbol{\mu}_Z, \boldsymbol{\theta}) = & -\tfrac{1}{2}(n-1)[p \log 2\pi + \log|\boldsymbol{\Sigma}_{ZZ}| + \operatorname{tr}\{\mathbf{m}_{ZZ}\boldsymbol{\Sigma}_{ZZ}^{-1}\}] \\ & -\tfrac{1}{2}n(\bar{\mathbf{Z}} - \boldsymbol{\mu}_Z)\boldsymbol{\Sigma}_{ZZ}^{-1}(\bar{\mathbf{Z}} - \boldsymbol{\mu}_Z)'. \qquad (4.2.23)\end{aligned}$$

If $\boldsymbol{\mu}_Z$ is not functionally related to $\boldsymbol{\theta}$, the maximum likelihood estimator of $\boldsymbol{\mu}_Z$ is $\bar{\mathbf{Z}}$. Because $|\mathbf{m}_{ZZ}|$ is not a function of $\boldsymbol{\theta}$, the maximum likelihood estimator also minimizes

$$Q_m^*(\boldsymbol{\theta}) = -\log|\mathbf{m}_{ZZ}\boldsymbol{\Sigma}_{ZZ}^{-1}| + \operatorname{tr}\{\mathbf{m}_{ZZ}\boldsymbol{\Sigma}_{ZZ}^{-1}\}. \qquad (4.2.24)$$

The least squares estimator is the value of $\boldsymbol{\theta}$ that minimizes

$$Q_\ell(\boldsymbol{\theta}; \hat{\mathbf{V}}_{aa}) = [\mathbf{s} - \boldsymbol{\sigma}(\boldsymbol{\theta})]'\hat{\mathbf{V}}_{aa}^{-1}[\mathbf{s} - \boldsymbol{\sigma}(\boldsymbol{\theta})], \qquad (4.2.25)$$

where $\mathbf{s} = \operatorname{vech} \mathbf{m}_{ZZ}$ and $\boldsymbol{\sigma}(\boldsymbol{\theta}) = \operatorname{vech} \boldsymbol{\Sigma}_{ZZ}(\boldsymbol{\theta})$. It is demonstrated in Lemma 4.A.1 that the covariance matrix of vech $\mathbf{m}_{ZZ}$ is

$$\mathbf{V}_{aa} = (n-1)^{-1}\boldsymbol{\Omega} = 2(n-1)^{-1}\boldsymbol{\psi}_p(\boldsymbol{\Sigma}_{ZZ} \otimes \boldsymbol{\Sigma}_{ZZ})\boldsymbol{\psi}_p' \qquad (4.2.26)$$

and that

$$\mathbf{V}_{aa}^{-1} = \tfrac{1}{2}(n-1)\boldsymbol{\Phi}_p'(\boldsymbol{\Sigma}_{ZZ}^{-1} \otimes \boldsymbol{\Sigma}_{ZZ}^{-1})\boldsymbol{\Phi}_p.$$

Let $\hat{\mathbf{V}}_{aa}$ be the estimator of $\mathbf{V}_{aa}$ obtained from (4.2.26) by replacing $\boldsymbol{\Sigma}_{ZZ}$ with $\mathbf{m}_{ZZ}$. Then (4.2.25) becomes

$$\begin{aligned}Q_\ell(\boldsymbol{\theta}; \hat{\mathbf{V}}_{aa}) &= \tfrac{1}{2}(n-1)[\operatorname{vec}(\mathbf{m}_{ZZ} - \boldsymbol{\Sigma}_{ZZ})]'(\mathbf{m}_{ZZ}^{-1} \otimes \mathbf{m}_{ZZ}^{-1})[\operatorname{vec}(\mathbf{m}_{ZZ} - \boldsymbol{\Sigma}_{ZZ})] \\ &= \tfrac{1}{2}(n-1) \operatorname{tr}\{[(\mathbf{m}_{ZZ} - \boldsymbol{\Sigma}_{ZZ})\mathbf{m}_{ZZ}^{-1}]^2\}, \qquad (4.2.27)\end{aligned}$$

where we have used Result 4.A.6, assumed $\mathbf{m}_{ZZ}$ to be nonsingular, and, for notational convenience, suppressed the dependency of $\boldsymbol{\Sigma}_{ZZ}$ on $\boldsymbol{\theta}$.

Both $Q_\ell(\boldsymbol{\theta}; \hat{\mathbf{V}}_{aa})$ and $Q_m^*(\boldsymbol{\theta})$ can be written in terms of the roots of $\boldsymbol{\Sigma}_{ZZ}$ in the metric $\mathbf{m}_{ZZ}$. Let $\hat{\tau}_i$ be the roots of

$$|\mathbf{m}_{ZZ}^{-1/2}\boldsymbol{\Sigma}_{ZZ}\mathbf{m}_{ZZ}^{-1/2} - \tau\mathbf{I}| = 0. \qquad (4.2.28)$$

Then

$$\begin{aligned}Q_\ell(\boldsymbol{\theta}; \hat{\mathbf{V}}_{aa}) &= \tfrac{1}{2}(n-1) \operatorname{tr}\{[\mathbf{I} - \mathbf{m}_{ZZ}^{-1/2}\boldsymbol{\Sigma}_{ZZ}\mathbf{m}_{ZZ}^{-1/2}]^2\} \\ &= \tfrac{1}{2}(n-1) \sum_{i=1}^{p} (\hat{\tau}_i - 1)^2\end{aligned}$$

and

$$Q_m^*(\boldsymbol{\theta}) = \sum_{i=1}^{p} \{\log \hat{\tau}_i + \hat{\tau}_i^{-1}\}.$$

If we treat $Q_m^*(\boldsymbol{\theta})$ as a function of $(\hat{\tau}_1, \hat{\tau}_2, \ldots, \hat{\tau}_p)$ and expand the function in a Taylor series about the point $(1, 1, \ldots, 1)$, we have

$$Q_m^*(\boldsymbol{\theta}) = p + \frac{1}{2} \sum_{i=1}^{p} (\hat{\tau}_i - 1)^2 + \text{remainder}.$$

The remainder is of order $(\hat{\tau}_i - 1)^3$ and will be small in large samples because the $\hat{\tau}_i$ are continuous functions of the elements of $\mathbf{m}_{ZZ}$ and $\mathbf{m}_{ZZ}$ is converging to $\boldsymbol{\Sigma}_{ZZ}$ in probability. Therefore, the two estimation methods are minimizing similar quantities and the estimators will be "close" in that sense.

To investigate the relationship between the two estimators further, we consider minimization of the quantity

$$\begin{aligned} Q_\ell(\boldsymbol{\theta}; \mathbf{W}) &= [\mathbf{s} - \boldsymbol{\sigma}(\boldsymbol{\theta})]'[\boldsymbol{\psi}_p(\mathbf{W} \otimes \mathbf{W})\boldsymbol{\psi}_p']^{-1}[\mathbf{s} - \boldsymbol{\sigma}(\boldsymbol{\theta})] \\ &= \operatorname{tr}\{[(\mathbf{m}_{ZZ} - \boldsymbol{\Sigma}_{ZZ})\mathbf{W}^{-1}]^2\}, \end{aligned} \tag{4.2.29}$$

where $\mathbf{W}$ is a symmetric positive definite matrix that is treated as fixed in the minimization. As usual, we assume the elements of $\boldsymbol{\Sigma}_{ZZ}(\boldsymbol{\theta})$ to be continuous functions of $\boldsymbol{\theta}$ with continuous first derivatives. Let $\hat{\boldsymbol{\theta}}_m$ be the maximum likelihood estimator of $\boldsymbol{\theta}$ obtained by minimizing (4.2.24).

The derivative of $Q_\ell(\boldsymbol{\theta}; \mathbf{W})$ with respect to the ith element of $\boldsymbol{\theta}$ is

$$\frac{\partial Q_\ell(\boldsymbol{\theta}; \mathbf{W})}{\partial \theta_i} = -2 \operatorname{tr}\left\{\mathbf{W}^{-1}(\mathbf{m}_{ZZ} - \boldsymbol{\Sigma}_{ZZ})\mathbf{W}^{-1} \frac{\partial \boldsymbol{\Sigma}_{ZZ}(\boldsymbol{\theta})}{\partial \theta_i}\right\}. \tag{4.2.30}$$

The derivative of (4.2.24) with respect to θ_i is

$$\begin{aligned} \frac{\partial Q_m^*(\boldsymbol{\theta})}{\partial \theta_i} &= \operatorname{tr}\left\{\boldsymbol{\Sigma}_{ZZ}^{-1} \frac{\partial \boldsymbol{\Sigma}_{ZZ}(\boldsymbol{\theta})}{\partial \theta_i}\right\} - \operatorname{tr}\left\{\mathbf{m}_{ZZ}\boldsymbol{\Sigma}_{ZZ}^{-1} \frac{\partial \boldsymbol{\Sigma}_{ZZ}(\boldsymbol{\theta})}{\partial \theta_i} \boldsymbol{\Sigma}_{ZZ}^{-1}\right\} \\ &= \operatorname{tr}\left\{\boldsymbol{\Sigma}_{ZZ}^{-1}(\boldsymbol{\Sigma}_{ZZ} - \mathbf{m}_{ZZ})\boldsymbol{\Sigma}_{ZZ}^{-1} \frac{\partial \boldsymbol{\Sigma}_{ZZ}(\boldsymbol{\theta})}{\partial \theta_i}\right\}, \end{aligned} \tag{4.2.31}$$

where we have used Result 4.A.8 and sometimes suppressed the dependency of $\boldsymbol{\Sigma}_{ZZ}$ on $\boldsymbol{\theta}$. Therefore, if $\hat{\boldsymbol{\theta}}_m$ satisfies the likelihood equations, then $\hat{\boldsymbol{\theta}}_m$ is also a solution of the first derivative equations associated with $Q_\ell[\boldsymbol{\theta}; \boldsymbol{\Sigma}_{ZZ}(\hat{\boldsymbol{\theta}}_m)]$ of (4.2.29), where $\boldsymbol{\Sigma}_{ZZ}(\hat{\boldsymbol{\theta}}_m)$ is the covariance matrix of $\mathbf{Z}_t$ evaluated at $\boldsymbol{\theta} = \hat{\boldsymbol{\theta}}_m$.

The fact that the maximum likelihood estimator satisfies the least squares equations does not guarantee that the maximum likelihood estimator is equal to the least squares estimator constructed with the estimated covariance matrix $\boldsymbol{\Sigma}(\hat{\boldsymbol{\theta}}_m)$, because of the second-order conditions. Browne (1974) has demonstrated that the probability that the least squares estimator constructed with $\mathbf{W} = \boldsymbol{\Sigma}_{ZZ}(\hat{\boldsymbol{\theta}}_m)$ equals the maximum likelihood estimator approaches one as the sample size increases. Also see Exercise 4.21.

Equations (4.2.30) and (4.2.31) suggest an iterative scheme for constructing maximum likelihood estimators. The iteration can be described as follows:

1. Initiate the iteration by setting
$$\mathbf{W}^{(0)} = \mathbf{m}_{ZZ}. \tag{4.2.32}$$
Set $i = 0$ and go to step 2.
2. Using $\mathbf{W}^{(i)}$, minimize $Q_\ell(\boldsymbol{\theta}; \mathbf{W}^{(i)})$ defined in (4.2.29). Let the $\boldsymbol{\theta}$-value giving the minimum value be denoted by $\boldsymbol{\theta}^{(i+1)}$. Go to step 3.
3. Using $\boldsymbol{\theta}^{(i+1)}$ from step 2, construct
$$\mathbf{W}^{(i+1)} = \boldsymbol{\Sigma}_{ZZ}(\boldsymbol{\theta}^{(i+1)}). \tag{4.2.33}$$
Set i equal to $i + 1$ and go to step 2.

This iterative procedure, as it stands, is not guaranteed to converge to the maximum likelihood estimator. However, it is possible to modify the procedure so that convergence is guaranteed. Critical to the modification is the fact that, if $\boldsymbol{\theta}^{(i)}$ is not the maximum likelihood estimator, there is a direction for which the directional derivative of the likelihood function at $\boldsymbol{\theta}^{(i)}$ is positive.

Lemma 4.2.1. Let the iterative scheme be as described in steps 1, 2, and 3. Let $\boldsymbol{\Sigma}_{ZZ}(\boldsymbol{\theta})$ be a continuous function of $\boldsymbol{\theta}$ with continuous first derivatives on a set B. Let $\boldsymbol{\theta}^{(i)}$ be in the interior of B and assume $\boldsymbol{\theta}^{(i)}$ is not a stationary point of $Q_\ell(\boldsymbol{\theta}; \mathbf{W}^{(i)})$. Let the directional derivative of $Q_\ell(\boldsymbol{\theta}; \mathbf{W}^{(i)})$ evaluated at $\boldsymbol{\theta}^{(i)}$ be a minimum in the direction ξ. Then the directional derivative of the likelihood function in the direction ξ evaluated at $\boldsymbol{\theta}^{(i)}$ is positive.

Proof. The partial derivative of $Q_\ell(\boldsymbol{\theta}; \mathbf{W})$ with respect to θ_i defined in (4.2.30) evaluated at $(\boldsymbol{\theta}; \mathbf{W}) = (\boldsymbol{\theta}^{(i)}; \mathbf{W}^{(i)})$ is equal to twice the partial derivative of $Q_m^*(\boldsymbol{\theta})$ with respect to θ_i defined in (4.2.31) evaluated at $\boldsymbol{\theta} = \boldsymbol{\theta}^{(i)}$. Because $\boldsymbol{\theta}^{(i)}$ is not a stationary point of $Q_\ell(\boldsymbol{\theta}; \mathbf{W}^{(i)})$, the directional derivative of $Q_\ell(\boldsymbol{\theta}; \mathbf{W}^{(i)})$ in the direction ξ evaluated at $\boldsymbol{\theta}^{(i)}$ is negative. Therefore, the directional derivative of the likelihood function evaluated at $\boldsymbol{\theta}^{(i)}$ is positive in the direction ξ. □

By Lemma 4.2.1 there is some point in the direction ξ from $\boldsymbol{\theta}^{(i)}$ such that the likelihood is larger than at $\boldsymbol{\theta}^{(i)}$. Therefore, by choosing such a point at each step, one can guarantee that the likelihood increases at each step initiated at a point that is not the maximum point. The resulting sequence of likelihoods is monotone increasing and bounded. Hence, the modified iterative scheme converges. The modification is seldom needed to achieve convergence in practice.

Example 4.2.4. To illustrate the iteration of the least squares estimation procedure, we continue with the earthquake data of Example 4.2.1. The estimates of the covariance matrices for the two populations obtained in Example 4.2.1 are

$$\operatorname{vech} \hat{\boldsymbol{\Sigma}}_{11} = (0.6233, 0.2776, 0.4145; 0.1883, 0.2289; 0.4103)',$$
$$\operatorname{vech} \hat{\boldsymbol{\Sigma}}_{22} = (0.5315, 0.2269; 0.1603)'.$$

These two matrices can be used to create the estimated covariance matrices of the sample moments for a second iteration of the least squares procedure. The estimates obtained for the nonlinear model using the $\mathbf{P}$ matrix constructed from the estimated covariance matrices $\hat{\boldsymbol{\Sigma}}_{11}$ and $\hat{\boldsymbol{\Sigma}}_{22}$ are given in the "Iteration 2" column of Table 4.2.4, where the $\mathbf{P}$ matrix is defined in Example 4.2.1. The estimates obtained in Example 4.2.1 are given as "Iteration 1" of the table and two additional iterations are given in the table. In this example the model fits well and iteration produces only modest changes in the estimates.

The bottom line of Table 4.2.4 contains the likelihood ratio statistic computed with the unconstrained two-population normal model as the alternative. The statistic is

$$\begin{aligned}
&60[-\log|\hat{\boldsymbol{\Sigma}}_{11}^{-1}\mathbf{m}_{11}| + \operatorname{tr}\{\mathbf{m}_{11}\hat{\boldsymbol{\Sigma}}_{11}^{-1}\}] + 61(\bar{\mathbf{Z}}_1 - \hat{\boldsymbol{\mu}}_1)\hat{\boldsymbol{\Sigma}}_{11}^{-1}(\bar{\mathbf{Z}}_1 - \hat{\boldsymbol{\mu}}_1)' \\
&\quad + 260[-\log|\hat{\boldsymbol{\Sigma}}_{22}^{-1}\mathbf{m}_{22}| + \operatorname{tr}\{\mathbf{m}_{22}\hat{\boldsymbol{\Sigma}}_{22}^{-1}\}] \\
&\quad + 261(\bar{\mathbf{Z}}_2 - \hat{\boldsymbol{\mu}}_2)\hat{\boldsymbol{\Sigma}}_{22}^{-1}(\bar{\mathbf{Z}}_2 - \hat{\boldsymbol{\mu}}_2)' - [60(3) + 260(2)],
\end{aligned}$$

TABLE 4.2.4. Iterated least squares estimates for parameters of earthquake model

Parameter	Iteration 1	2	3	4	Standard Error
β_0	−4.4564	−4.4411	−4.4420	−4.4418	0.652
β_1	1.8111	1.8087	1.8089	1.8088	0.124
γ_0	−2.5952	−2.6035	−2.6038	−2.6038	0.678
γ_1	1.4932	1.4950	1.4950	1.4950	0.129
μ_{x1}	5.2506	5.2489	5.2490	5.2490	0.054
μ_{x2}	5.2421	5.2414	5.2414	5.2414	0.025
$10\sigma_{xx11}$	1.5327	1.5329	1.5333	1.5333	0.330
$10\sigma_{xx22}$	1.2529	1.2516	1.2515	1.2515	0.153
$10\sigma_{ee}$	1.2051	1.2255	1.2252	1.2252	0.261
$10\sigma_{uu}$	0.3503	0.3560	0.3561	0.3561	0.078
$10\sigma_{aa}$	0.6859	0.6793	0.6793	0.6793	0.201
Residual sum of squares	2.1986	2.8745	2.7989	2.7992	
Likelihood ratio	2.7217	2.6435	2.6429	2.6429	

where $\bar{\mathbf{Z}}_1 = (\bar{Y}_1, \bar{X}, \bar{W}_1)$, $\bar{\mathbf{Z}}_2 = (\bar{Y}_2, \bar{X}_2)$, and, for example,

$$\hat{\boldsymbol{\mu}}_1 = (\hat{\beta}_0 + \hat{\beta}_1\hat{\mu}_{x1}, \hat{\gamma}_0 + \hat{\gamma}_1\hat{\mu}_{x1}).$$

In this example the likelihood ratio decreases (the likelihood increases) with each iteration. The residual sum of squares, given in the next-to-last line of Table 4.2.4, does not necessarily decline with iteration because a different quadratic form is being minimized at each step.

Both the residual sum of squares and the likelihood ratio statistic of the last iteration are approximately distributed as chi-square random variables with three degrees of freedom under the null model. Both statistics indicate good agreement with the model.

The last column of Table 4.2.4 contains the standard errors of the estimates computed at the fourth iteration. The standard errors for the fourth iteration are very similar to those obtained at the first iteration. The first iteration standard errors are given in Table 4.2.1. □ □

4.2.3. Least Squares Estimation for the Multivariate Functional Model

In this section we apply the least squares method to the matrix of sample mean squares and products generated by the functional model. Let

$$\mathbf{z}_t = \boldsymbol{\beta}_0 + \mathbf{x}_t(\boldsymbol{\beta}, \mathbf{I}), \qquad \mathbf{Z}_t = \mathbf{z}_t + \boldsymbol{\varepsilon}_t, \tag{4.2.34}$$

for $t = 1, 2, \ldots, n$, where $\mathbf{Z}_t = (\mathbf{Y}_t, \mathbf{X}_t)$ is a p-dimensional vector, $\boldsymbol{\varepsilon}_t = (\mathbf{e}_t, \mathbf{u}_t)$ are $\text{NI}(\mathbf{0}, \boldsymbol{\Sigma}_{\varepsilon\varepsilon})$ random vectors, and $\{\mathbf{x}_t\}$ is a sequence of fixed k-dimensional vectors. Because the $\mathbf{x}_t$ are fixed,

$$E\{\mathbf{m}_{ZZ}\} = (\boldsymbol{\beta}, \mathbf{I}_k)'\mathbf{m}_{xx}(\boldsymbol{\beta}, \mathbf{I}_k) + \boldsymbol{\Sigma}_{\varepsilon\varepsilon}. \tag{4.2.35}$$

We employ generalized least squares to estimate the unknown parameters, using the unique elements of $\mathbf{m}_{ZZ}$ as the vector of observations. It is assumed that the model is identified and that the elements of $\boldsymbol{\beta}$, $\boldsymbol{\Sigma}_{\varepsilon\varepsilon}$, and $\mathbf{m}_{xx}$ can be expressed as continuous differentiable functions of a parameter vector. Because $\mathbf{m}_{xx}$ is a function of n, those elements of the parameter vector defining $\mathbf{m}_{xx}$ are functions of n. In many situations the elements of $\boldsymbol{\beta}$, $\boldsymbol{\Sigma}_{\varepsilon\varepsilon}$, and $\mathbf{m}_{xx}$ are defined by separate sets of parameters and the parameter vector can be partitioned as

$$\boldsymbol{\theta}_n = [\boldsymbol{\theta}'_\beta, \boldsymbol{\theta}'_\varepsilon, (\text{vech } \mathbf{m}_{xx})']',$$

where $\boldsymbol{\beta}$ is a function of $\boldsymbol{\theta}_\beta$, $\boldsymbol{\Sigma}_{\varepsilon\varepsilon}$ is a function of $\boldsymbol{\theta}_\varepsilon$, and $(\boldsymbol{\theta}'_\beta, \boldsymbol{\theta}_\varepsilon)$ is free of n. As in Section 4.2.1, we write

$$\mathbf{s} = \boldsymbol{\sigma}(\boldsymbol{\theta}_n) + \mathbf{a}_n, \tag{4.2.36}$$

where $\mathbf{s} = \text{vech } \mathbf{m}_{ZZ}$ and $E\{\mathbf{a}_n\} = \mathbf{0}$. To simplify the notation we will often omit the subscript n from $\boldsymbol{\theta}_n$ and $\mathbf{a}_n$. Under the conditions of Theorem 1.C.2 or Theorem 1.C.3,

$$(n-1)^{1/2}\mathbf{a}_n \xrightarrow{L} N(\mathbf{0}, \bar{\boldsymbol{\Gamma}}_{aa}), \tag{4.2.37}$$

where

$$\bar{\boldsymbol{\Gamma}}_{aa} = \lim_{n\to\infty} (n-1)E\{[\text{vech}(\mathbf{m}_{ZZ} - \mathbf{m}_{zz} - \boldsymbol{\Sigma}_{\varepsilon\varepsilon})][\text{vech}(\mathbf{m}_{ZZ} - \mathbf{m}_{zz} - \boldsymbol{\Sigma}_{\varepsilon\varepsilon})]'\}.$$

If the ε_t are normally distributed,

$$\bar{\boldsymbol{\Gamma}}_{aa} = 2\psi_p\{2[(\boldsymbol{\beta}, \mathbf{I})'\bar{\mathbf{m}}_{xx}(\boldsymbol{\beta}, \mathbf{I})] \otimes \boldsymbol{\Sigma}_{\varepsilon\varepsilon} + (\boldsymbol{\Sigma}_{\varepsilon\varepsilon} \otimes \boldsymbol{\Sigma}_{\varepsilon\varepsilon})\}\psi_p', \tag{4.2.38}$$

where $\lim \mathbf{m}_{xx} = \bar{\mathbf{m}}_{xx}$ and we have used

$$\mathbf{V}\{\text{vec } \mathbf{m}_{z\varepsilon}\} = (n-1)^{-1}(\boldsymbol{\Sigma}_{\varepsilon\varepsilon} \otimes \mathbf{m}_{zz}), \tag{4.2.39}$$

$$\mathbf{V}\{\text{vech}(\mathbf{m}_{z\varepsilon} + \mathbf{m}_{\varepsilon z})\} = 4(n-1)^{-1}\psi_p(\mathbf{m}_{zz} \otimes \boldsymbol{\Sigma}_{\varepsilon\varepsilon})\psi_p', \tag{4.2.40}$$

$$\mathbf{V}\{\psi_p \text{ vec } \mathbf{m}_{\varepsilon\varepsilon}\} = 2(n-1)^{-1}\psi_p(\boldsymbol{\Sigma}_{\varepsilon\varepsilon} \otimes \boldsymbol{\Sigma}_{\varepsilon\varepsilon})\psi_p'. \tag{4.2.41}$$

Note that the variance of $\mathbf{a}_n$ under the functional model differs from the variance of $\mathbf{a}$ under the random model of Section 4.2.1.

Let $\mathbf{G}_{aan}$ be a symmetric positive definite matrix of the same dimension as the covariance matrix of $\mathbf{a}_n$. Often $\mathbf{G}_{aan}$ is an estimator of the inverse of the covariance matrix of $\mathbf{a}_n$. The generalized least squares estimator of $\boldsymbol{\theta}$ computed in the metric $\mathbf{G}_{aan}$ is the $\boldsymbol{\theta}$ that minimizes the quadratic form

$$Q_c(\boldsymbol{\theta}, \mathbf{s}; \mathbf{G}_{aan}) = [\mathbf{s} - \boldsymbol{\sigma}(\boldsymbol{\theta})]'\mathbf{G}_{aan}[\mathbf{s} - \boldsymbol{\sigma}(\boldsymbol{\theta})]. \tag{4.2.42}$$

The generalized least squares estimator of $\boldsymbol{\theta}$ is denoted by $\tilde{\boldsymbol{\theta}}$.

We have defined a parameter vector to be identified if knowledge of the distribution function of the observable random variables uniquely determines the value of the parameter vector. In this section we are only considering models in which the covariance matrix is a function of $\boldsymbol{\theta}$. Thus, for the structural model, $\boldsymbol{\theta}$ is said to be identified if $\boldsymbol{\Sigma}_{ZZ}(\boldsymbol{\theta}_1) = \boldsymbol{\Sigma}_{ZZ}(\boldsymbol{\theta}_2)$ implies $\boldsymbol{\theta}_1 = \boldsymbol{\theta}_2$. By analogy to the structural case, we say that $\boldsymbol{\theta}_n$ is identified for the functional model if $\boldsymbol{\sigma}(\boldsymbol{\theta}_{1n}) = \boldsymbol{\sigma}(\boldsymbol{\theta}_{2n})$ implies $\boldsymbol{\theta}_{1n} = \boldsymbol{\theta}_{2n}$ in Equation (4.2.36). For identified models we can obtain the limiting distribution of the least squares estimators.

Theorem 4.2.3. Assume model (4.2.34) with normal ε_t and suppose that $\mathbf{m}_{xx}$ converges to a positive definite matrix $\bar{\mathbf{m}}_{xx}$. Let

$$\lim_{n\to\infty} \boldsymbol{\theta}_n = \boldsymbol{\theta}_\infty = [\boldsymbol{\theta}_\beta', \boldsymbol{\theta}_\varepsilon', (\text{vech } \bar{\mathbf{m}}_{xx})']' \quad \text{and} \quad \operatorname*{plim}_{n\to\infty} \mathbf{G}_{aan} = \bar{\mathbf{G}}_{aa},$$

where $\bar{\mathbf{G}}_{aa}$ is positive definite. Let A, the parameter space for $\boldsymbol{\theta}_n$, be a convex, compact subset of q-dimensional Euclidean space such that $\boldsymbol{\sigma}(\boldsymbol{\theta}) \neq \boldsymbol{\sigma}(\boldsymbol{\theta}_\infty)$ for any $\boldsymbol{\theta}$ in A with $\boldsymbol{\theta} \neq \boldsymbol{\theta}_\infty$. Assume that $\boldsymbol{\theta}_\infty$ is an interior point of A. Assume that $\boldsymbol{\sigma}(\boldsymbol{\theta})$ is a continuous function of $\boldsymbol{\theta}$ with continuous first and second derivatives in a neighborhood of $\boldsymbol{\theta}_\infty$. Let $\mathbf{F}$ be of full column rank, where

$$\mathbf{F} = \frac{\partial \boldsymbol{\sigma}(\boldsymbol{\theta}_\infty)}{\partial \boldsymbol{\theta}'} \tag{4.2.43}$$

is the matrix of partial derivatives evaluated at $\boldsymbol{\theta}_\infty$. Let $\tilde{\boldsymbol{\theta}}$ be the estimator of $\boldsymbol{\theta}_n$ that minimizes (4.2.42) over A. Then

$$n^{1/2}(\tilde{\boldsymbol{\theta}} - \boldsymbol{\theta}_n) \xrightarrow{L} N(\mathbf{0}, \boldsymbol{\Gamma}_{\theta\theta}),$$

where

$$\boldsymbol{\Gamma}_{\theta\theta} = (\mathbf{F}'\bar{\mathbf{G}}_{aa}\mathbf{F})^{-1}\mathbf{F}'\bar{\mathbf{G}}_{aa}\bar{\boldsymbol{\Gamma}}_{aa}\bar{\mathbf{G}}_{aa}\mathbf{F}(\mathbf{F}'\bar{\mathbf{G}}_{aa}\mathbf{F})^{-1} \tag{4.2.44}$$

and $\bar{\Gamma}_{aa}$ is defined in (4.2.38). If $\bar{\mathbf{G}}_{aa} = \bar{\boldsymbol{\Gamma}}_{aa}^{-1}$ and $\frac{1}{2}p(p+1) - q > 0$, then

$$\hat{\chi}^2 = [\mathbf{s} - \boldsymbol{\sigma}(\tilde{\boldsymbol{\theta}})]'\mathbf{G}_{aan}[\mathbf{s} - \boldsymbol{\sigma}(\tilde{\boldsymbol{\theta}})]$$

converges in distribution to a chi-square random variable with $\frac{1}{2}p(p+1) - q$ degrees of freedom.

Proof. The function $Q_c(\boldsymbol{\theta}, \mathbf{s}; \mathbf{G}_{aan})$ in (4.2.42) is continuous in $\mathbf{s}$, $\mathbf{G}_{aan}$, and $\boldsymbol{\theta}$. By the identifiability assumptions, $Q_c[\boldsymbol{\theta}, \boldsymbol{\sigma}(\boldsymbol{\theta}_\infty); \bar{\mathbf{G}}_{aa}]$ has a unique minimum of zero at $\boldsymbol{\theta} = \boldsymbol{\theta}_\infty$. Because $\mathbf{s}$ and $\mathbf{G}_{aan}$ converge to $\boldsymbol{\sigma}(\boldsymbol{\theta}_\infty)$ and $\bar{\mathbf{G}}_{aa}$ in probability, the arguments of the proof of Theorem 4.B.1 can be used to show that the least squares estimator $\tilde{\boldsymbol{\theta}}$ converges to $\boldsymbol{\theta}_\infty$ in probability. Let

$$\mathbf{h}(\boldsymbol{\theta}) = 2\frac{\partial \boldsymbol{\sigma}'(\boldsymbol{\theta})}{\partial \boldsymbol{\theta}}\mathbf{G}_{aan}[\mathbf{s} - \boldsymbol{\sigma}(\boldsymbol{\theta})].$$

By Taylor's theorem,

$$\mathbf{h}(\tilde{\boldsymbol{\theta}}) = \mathbf{h}(\boldsymbol{\theta}_n) + \mathbf{A}_n(\tilde{\boldsymbol{\theta}} - \boldsymbol{\theta}_n),$$

where the ijth element of $\mathbf{A}_n$ is

$$a_{ij} = \frac{\partial h_i(\boldsymbol{\theta}_{*i})}{\partial \theta_j}, \tag{4.2.45}$$

$h_i(\boldsymbol{\theta})$ is the ith element of $\mathbf{h}(\boldsymbol{\theta})$, θ_j is the jth element of $\boldsymbol{\theta}$, and the point $\boldsymbol{\theta}_{*i}$ is on the line segment joining $\boldsymbol{\theta}_n$ and $\tilde{\boldsymbol{\theta}}$. Now

$$\frac{\partial h_i(\boldsymbol{\theta}_{*i})}{\partial \theta_j} = 2\frac{\partial^2 \boldsymbol{\sigma}'(\boldsymbol{\theta}_{*i})}{\partial \theta_i \partial \theta_j}\mathbf{G}_{aan}[\mathbf{s} - \boldsymbol{\sigma}(\boldsymbol{\theta}_{*i})] - 2\frac{\partial \boldsymbol{\sigma}'(\boldsymbol{\theta}_{*i})}{\partial \theta_i}\mathbf{G}_{aan}\frac{\partial \boldsymbol{\sigma}(\boldsymbol{\theta}_{*i})}{\partial \theta_j}.$$

The elements of $\mathbf{s} - \boldsymbol{\sigma}(\boldsymbol{\theta}_{*i})$ and $\boldsymbol{\theta}_{*i} - \boldsymbol{\theta}_n$ converge to zero in probability, $\mathbf{G}_{aan}$ converges to $\bar{\mathbf{G}}_{aa}$ in probability, and the partial derivatives are bounded in a neighborhood of $\boldsymbol{\theta}_\infty$. It follows that

$$\operatorname*{plim}_{n\to\infty} \mathbf{A}_n = -2\mathbf{F}'\bar{\mathbf{G}}_{aa}\mathbf{F}.$$

By assumption, $\mathbf{F}$ is of full column rank, implying that $2\mathbf{F}'\bar{\mathbf{G}}_{aa}\mathbf{F}$ is nonsingular.

Because (4.2.42) is continuously differentiable in a neighborhood of $\boldsymbol{\theta}_\infty$ and because $\tilde{\boldsymbol{\theta}}$ converges to $\boldsymbol{\theta}_\infty$ in probability, it follows that the probability that $\mathbf{h}(\tilde{\boldsymbol{\theta}}) = \mathbf{0}$ converges to one. Therefore,

$$\begin{aligned}\tilde{\boldsymbol{\theta}} - \boldsymbol{\theta}_n &= -\mathbf{A}_n^{-1}\mathbf{h}(\boldsymbol{\theta}_n) + O_p(n^{-1}).\\ &= (\mathbf{F}'\bar{\mathbf{G}}_{aa}\mathbf{F})^{-1}\mathbf{F}'\bar{\mathbf{G}}_{aa}[\mathbf{s} - \boldsymbol{\sigma}(\boldsymbol{\theta}_n)] + o_p(n^{-1/2}).\end{aligned}$$

By Theorem 1.C.2, $n^{1/2}[\mathbf{s} - \boldsymbol{\sigma}(\boldsymbol{\theta}_n)]$ converges in distribution to a normal vector with covariance matrix $\bar{\boldsymbol{\Gamma}}_{aa}$ of (4.2.38). Therefore, the limiting distribution of $n^{1/2}(\tilde{\boldsymbol{\theta}} - \boldsymbol{\theta}_n)$ is multivariate normal with zero mean vector and covariance matrix given in (4.2.44).

If $\bar{\mathbf{G}}_{aa} = \bar{\boldsymbol{\Gamma}}_{aa}^{-1}$, then

$$[\mathbf{s} - \boldsymbol{\sigma}(\tilde{\boldsymbol{\theta}})]'\mathbf{G}_{aan}[\mathbf{s} - \boldsymbol{\sigma}(\tilde{\boldsymbol{\theta}})] = \mathbf{a}'\mathbf{K}'\bar{\boldsymbol{\Gamma}}_{aa}^{-1}\mathbf{K}\mathbf{a} + o_p(1),$$

where $\mathbf{K} = \mathbf{I} - \mathbf{F}(\mathbf{F}'\bar{\boldsymbol{\Gamma}}_{aa}^{-1}\mathbf{F})^{-1}\mathbf{F}'\bar{\boldsymbol{\Gamma}}_{aa}^{-1}$. The limiting distribution follows from the fact that $n^{1/2}\mathbf{a}_n$ is converging in distribution to a normal random vector with covariance matrix $\bar{\boldsymbol{\Gamma}}_{aa}$. □

Once one has obtained estimators of $\boldsymbol{\beta}$ and $\boldsymbol{\Sigma}_{\varepsilon\varepsilon}$, estimators of the individual $\mathbf{x}_t$ are

$$\hat{\mathbf{x}}_t = \mathbf{X}_t - \tilde{\mathbf{v}}_t\tilde{\boldsymbol{\Sigma}}_{vv}^{-1}\tilde{\boldsymbol{\Sigma}}_{vu}, \tag{4.2.46}$$

where this estimator is that of (4.1.37) with the generalized least squares estimators of $\boldsymbol{\beta}$ and $\boldsymbol{\Sigma}_{\varepsilon\varepsilon}$ replacing the maximum likelihood estimators.

To obtain the most efficient estimator of $\boldsymbol{\theta}$, one should use a consistent estimator of $\bar{\boldsymbol{\Gamma}}_{aa}^{-1}$ as the matrix $\mathbf{G}_{aan}$. Generally, this matrix cannot be constructed until one has an estimator of $\mathbf{m}_{xx}$. We shall demonstrate that an estimator of $(\boldsymbol{\theta}_\beta, \boldsymbol{\theta}_\varepsilon)$ and an estimator of the covariance matrix of the approximate distribution of $[\boldsymbol{\theta}'_\beta, \boldsymbol{\theta}'_\varepsilon]'$ can be obtained by using the matrix that is appropriate for normally distributed $\mathbf{x}_t$ as the matrix $\mathbf{G}_{aan}$. We first present some matrix and derivative results that will be used in the proof of the theorem.

Result 4.2.1. Let model (4.2.34) hold and let $\boldsymbol{\sigma}(\boldsymbol{\theta})$ be defined by (4.2.36). Then the partial derivative of the vector $\boldsymbol{\sigma}(\boldsymbol{\theta})$ with respect to vech $\mathbf{m}_{xx} = \boldsymbol{\theta}_x$ is

$$\frac{\partial\boldsymbol{\sigma}(\boldsymbol{\theta})}{\partial\boldsymbol{\theta}'_x} = \mathbf{F}_x = \boldsymbol{\psi}_p[(\boldsymbol{\beta}, \mathbf{I})' \otimes (\boldsymbol{\beta}, \mathbf{I})']\boldsymbol{\Phi}_k, \tag{4.2.47}$$

where $\boldsymbol{\psi}_p$ and $\boldsymbol{\Phi}_k$ are defined in Appendix 4.A.

Proof. From (4.2.35), we have

$$\text{vech}[E\{\mathbf{m}_{ZZ}\}] = \boldsymbol{\psi}_p[(\boldsymbol{\beta}, \mathbf{I})' \otimes (\boldsymbol{\beta}, \mathbf{I})']\boldsymbol{\Phi}_k \text{ vech } \mathbf{m}_{xx} + \text{vech } \boldsymbol{\Sigma}_{\varepsilon\varepsilon},$$

where we have used Result 4.A.4. Because the expression for $\text{vech}[E\{\mathbf{m}_{ZZ}\}]$ is linear in vech $\mathbf{m}_{xx}$, the derivative result follows. □

Result 4.2.2. Let model (4.2.34) hold. Then

$$\boldsymbol{\psi}_p(\mathbf{m}_{zz} \otimes \mathbf{m}_{zz})\boldsymbol{\psi}'_p = \mathbf{F}_x\boldsymbol{\psi}_k(\mathbf{m}_{xx} \otimes \mathbf{m}_{xx})\boldsymbol{\psi}'_k\mathbf{F}'_x, \tag{4.2.48}$$

where $\mathbf{F}_x$ is defined in (4.2.47).

Proof. The matrix $\boldsymbol{\psi}_p(\mathbf{m}_{zz} \otimes \mathbf{m}_{zz})\boldsymbol{\psi}'_p$ is equal to

$$\begin{aligned}
&\boldsymbol{\psi}_p[(\boldsymbol{\beta}, \mathbf{I})' \otimes (\boldsymbol{\beta}, \mathbf{I})'](\mathbf{m}_{xx} \otimes \mathbf{m}_{xx})[(\boldsymbol{\beta}, \mathbf{I}) \otimes (\boldsymbol{\beta}, \mathbf{I})]\boldsymbol{\psi}'_p \\
&\quad = \boldsymbol{\psi}_p[(\boldsymbol{\beta}, \mathbf{I})' \otimes (\boldsymbol{\beta}, \mathbf{I})']\boldsymbol{\Phi}_k\boldsymbol{\psi}_k(\mathbf{m}_{xx} \otimes \mathbf{m}_{xx})\boldsymbol{\psi}'_k\boldsymbol{\Phi}'_k[(\boldsymbol{\beta}, \mathbf{I}) \otimes (\boldsymbol{\beta}, \mathbf{I})]\boldsymbol{\psi}'_p
\end{aligned}$$

and the conclusion follows by application of Result 4.A.2. □

Result 4.2.3. Let

$$\begin{aligned}
\boldsymbol{\Omega} &= 2\boldsymbol{\psi}_p[\mathbf{m}_{zz} \otimes \mathbf{m}_{zz} + 2\mathbf{m}_{zz} \otimes \boldsymbol{\Sigma}_{\varepsilon\varepsilon} + \boldsymbol{\Sigma}_{\varepsilon\varepsilon} \otimes \boldsymbol{\Sigma}_{\varepsilon\varepsilon}]\boldsymbol{\psi}'_p, \\
\boldsymbol{\Gamma}_{aa} &= 2\boldsymbol{\psi}_p[2\mathbf{m}_{zz} \otimes \boldsymbol{\Sigma}_{\varepsilon\varepsilon} + \boldsymbol{\Sigma}_{\varepsilon\varepsilon} \otimes \boldsymbol{\Sigma}_{\varepsilon\varepsilon}]'\boldsymbol{\psi}'_p,
\end{aligned}$$

where it is understood that $\boldsymbol{\Omega}$ and $\boldsymbol{\Gamma}_{aa}$ are evaluated at the same value of $\boldsymbol{\theta}$. Then

$$\boldsymbol{\Omega}^{-1} = \boldsymbol{\Gamma}_{aa}^{-1} - 2\boldsymbol{\Gamma}_{aa}^{-1}\mathbf{F}_x\boldsymbol{\psi}_k(\mathbf{m}_{xx} \otimes \mathbf{m}_{xx})\boldsymbol{\psi}'_k\mathbf{F}'_x\boldsymbol{\Omega}^{-1}. \tag{4.2.49}$$

Proof. From Result 4.2.2 we have

$$\boldsymbol{\Omega} = \boldsymbol{\Gamma}_{aa} + 2\mathbf{F}_x\boldsymbol{\psi}_k(\mathbf{m}_{xx} \otimes \mathbf{m}_{xx})\boldsymbol{\psi}'_k\mathbf{F}'_x$$

and the expression for $\boldsymbol{\Omega}^{-1}$ follows. □

We are now in a position to demonstrate that the random model covariance matrix for $\mathbf{s}$ can be used to construct the least squares estimator of the parameter vector for the fixed model.

Theorem 4.2.4. Let the model assumptions of Theorem 4.2.3 hold. Let

$$\ddot{\mathbf{F}} = (\ddot{\mathbf{F}}_\beta, \ddot{\mathbf{F}}_\varepsilon, \ddot{\mathbf{F}}_x)$$

denote the $\frac{1}{2}p(p+1) \times q$ matrix of first derivatives of $\boldsymbol{\sigma}(\boldsymbol{\theta})$ evaluated at $\boldsymbol{\theta} = \ddot{\boldsymbol{\theta}}$, where the partition corresponds to the partition $[\boldsymbol{\theta}'_\beta, \boldsymbol{\theta}'_\varepsilon, (\text{vech } \mathbf{m}_{xx})']$. Let

$$\ddot{\mathbf{a}} = \text{vech}\{\mathbf{m}_{ZZ}\} - \boldsymbol{\sigma}(\ddot{\boldsymbol{\theta}}),$$

where $\sigma(\ddot{\theta}) = \text{vech}\{(\ddot{\beta}, \mathbf{I})'\ddot{\mathbf{m}}_{xx}(\ddot{\beta}, \mathbf{I}) + \ddot{\Sigma}_{\varepsilon\varepsilon}\}$, and let $\ddot{\Omega}$ and $\ddot{\Gamma}_{aa}$ be the Ω and Γ_{aa} of Result 4.2.3 evaluated at $\theta = \ddot{\theta}$. Then

(i) $\delta_{(f)} = \delta_{(r)}$, where

$$\delta_{(f)} = (\ddot{\mathbf{F}}'\ddot{\Gamma}_{aa}^{-1}\ddot{\mathbf{F}})^{-1}\ddot{\mathbf{F}}'\ddot{\Gamma}_{aa}^{-1}\ddot{\mathbf{a}},$$
$$\delta_{(r)} = (\ddot{\mathbf{F}}'\ddot{\Omega}^{-1}\ddot{\mathbf{F}})^{-1}\ddot{\mathbf{F}}'\ddot{\Omega}^{-1}\ddot{\mathbf{a}};$$

(ii) the portion of $(\ddot{\mathbf{F}}'\ddot{\Gamma}_{aa}^{-1}\ddot{\mathbf{F}})^{-1}$ associated with $(\theta_\beta', \theta_\varepsilon')$ is equal to the corresponding portion of $(\ddot{\mathbf{F}}'\ddot{\Omega}^{-1}\ddot{\mathbf{F}})^{-1}$;

(iii) $(\ddot{\mathbf{F}}'\ddot{\Omega}^{-1}\ddot{\mathbf{F}})^{-1}\ddot{\mathbf{F}}'\ddot{\Omega}^{-1}\ddot{\Gamma}_{aa}\ddot{\Omega}^{-1}\ddot{\mathbf{F}}(\ddot{\mathbf{F}}'\ddot{\Omega}^{-1}\ddot{\mathbf{F}})^{-1} = (\ddot{\mathbf{F}}'\ddot{\Gamma}_{aa}^{-1}\ddot{\mathbf{F}})^{-1}$.

Proof. Replacing $\ddot{\Omega}^{-1}$ in the definition of $\delta_{(r)}$ by the expression in Result 4.2.3, we have

$$\{\ddot{\mathbf{F}}'\ddot{\Gamma}_{aa}^{-1}\ddot{\mathbf{F}} - 2\ddot{\mathbf{F}}'\ddot{\Gamma}_{aa}^{-1}\ddot{\mathbf{F}}_x[\psi_k(\ddot{\mathbf{m}}_{xx} \otimes \ddot{\mathbf{m}}_{xx})\psi_k']\ddot{\mathbf{F}}_x'\ddot{\Omega}^{-1}\ddot{\mathbf{F}}\}\delta_{(r)}$$
$$= \ddot{\mathbf{F}}'\ddot{\Gamma}_{aa}^{-1}\ddot{\mathbf{a}} - 2\ddot{\mathbf{F}}'\ddot{\Gamma}_{aa}^{-1}\ddot{\mathbf{F}}_x[\psi_k(\ddot{\mathbf{m}}_{xx} \otimes \ddot{\mathbf{m}}_{xx})\psi_k']\ddot{\mathbf{F}}_x'\ddot{\Omega}^{-1}\ddot{\mathbf{a}}.$$

Using the definition of $\delta_{(r)}$, we have

$$\ddot{\mathbf{F}}_x'\ddot{\Omega}^{-1}\ddot{\mathbf{F}}\delta_{(r)} = \ddot{\mathbf{F}}_x'\ddot{\Omega}^{-1}\ddot{\mathbf{F}}(\ddot{\mathbf{F}}'\ddot{\Omega}^{-1}\ddot{\mathbf{F}})^{-1}\ddot{\mathbf{F}}'\ddot{\Omega}^{-1}\ddot{\mathbf{a}} = \ddot{\mathbf{F}}_x'\ddot{\Omega}^{-1}\ddot{\mathbf{a}}.$$

Therefore, $(\ddot{\mathbf{F}}'\ddot{\Gamma}_{aa}^{-1}\ddot{\mathbf{F}})\delta_{(r)} = \ddot{\mathbf{F}}'\ddot{\Gamma}_{aa}^{-1}\ddot{\mathbf{a}}$ and $\delta_{(r)} = \delta_{(f)}$. Now

$$(\ddot{\mathbf{F}}'\ddot{\Omega}^{-1}\ddot{\mathbf{F}})^{-1} = (\ddot{\mathbf{F}}'\ddot{\Gamma}_{aa}^{-1}\ddot{\mathbf{F}})^{-1} + (\ddot{\mathbf{F}}'\ddot{\Gamma}_{aa}^{-1}\ddot{\mathbf{F}})^{-1}\ddot{\mathbf{K}}(\ddot{\mathbf{F}}'\ddot{\Omega}^{-1}\ddot{\mathbf{F}})^{-1} \quad (4.2.50)$$
$$= (\ddot{\mathbf{F}}'\ddot{\Gamma}_{aa}^{-1}\ddot{\mathbf{F}})^{-1} + 2(\mathbf{0}, \mathbf{I}_x)'[(\psi_k(\ddot{\mathbf{m}}_{xx} \otimes \ddot{\mathbf{m}}_{xx})\psi_k'](\mathbf{0}, \mathbf{I}_x),$$

where $\ddot{\mathbf{K}} = 2\ddot{\mathbf{F}}'\ddot{\Gamma}_{aa}^{-1}\ddot{\mathbf{F}}_x[\psi_k(\ddot{\mathbf{m}}_{xx} \otimes \ddot{\mathbf{m}}_{xx})\psi_k']\ddot{\mathbf{F}}_k'\ddot{\Omega}^{-1}\ddot{\mathbf{F}}'$ and $\mathbf{I}_x$ is an identity matrix with dimension equal to $\frac{1}{2}k(k+1)$, the number of columns in $\mathbf{F}_x$. Because the upper left portion of the second term of the last expression of (4.2.50) is zero, conclusion (ii) is proved.

Using the expression for $\ddot{\Omega}^{-1}$ given in Result 4.2.3, we have

$$\ddot{\mathbf{F}}'\ddot{\Omega}^{-1}\ddot{\Gamma}_{aa}\ddot{\Omega}^{-1}\ddot{\mathbf{F}} = \ddot{\mathbf{F}}'\{\ddot{\Omega}^{-1} - 2\ddot{\Omega}^{-1}\ddot{\mathbf{F}}_x[\psi_k(\ddot{\mathbf{m}}_{xx} \otimes \ddot{\mathbf{m}}_{xx})\psi_k']\ddot{\mathbf{F}}_x'\ddot{\Omega}^{-1}\}\ddot{\mathbf{F}}.$$

Post- and premultiplying by $(\ddot{\mathbf{F}}'\ddot{\Omega}^{-1}\ddot{\mathbf{F}})^{-1}$ makes the right side equal to

$$(\ddot{\mathbf{F}}'\ddot{\Omega}^{-1}\ddot{\mathbf{F}})^{-1} - 2(\mathbf{0}, \mathbf{I}_x)'[\psi_k(\ddot{\mathbf{m}}_{xx} \otimes \ddot{\mathbf{m}}_{xx})\psi_k'](\mathbf{0}, \mathbf{I}_x)$$

and conclusion (iii) follows from expression (4.2.50). □

By part (i) of Theorem 4.2.4, the least squares estimator computed using Ω is equal to that computed using Γ_{aa} at each step of an iterative Gauss–Newton procedure that updates the covariance matrix at each step.

It follows from Theorem 4.2.4 that the limiting distribution of $\tilde{\theta}$, obtained by using $\psi_p(\mathbf{m}_{ZZ} \otimes \mathbf{m}_{ZZ})\psi_p'$ for the estimated covariance matrix of $\mathbf{s}$, is the same as the limiting distribution obtained using Γ_{aa}. That is, the estimators constructed using the consistent estimator of the random covariance matrix are, asymptotically, the best generalized least squares estimators for the fixed model. Also, by Theorem 4.2.4 the estimated covariance matrix of the estimators computed using $\psi_p(\mathbf{m}_{ZZ} \otimes \mathbf{m}_{ZZ})\psi'_p$ is appropriate for the estimators of $\boldsymbol{\beta}$ and $\boldsymbol{\Sigma}_{\varepsilon\varepsilon}$ for the fixed model. The theorem does require that no functional relation exist between the elements of $\mathbf{m}_{xx}$ and the elements of $\boldsymbol{\beta}$ and $\boldsymbol{\Sigma}_{\varepsilon\varepsilon}$. If one is interested in tests or confidence statements about $\mathbf{m}_{xx}$, one must compute the estimated covariance matrix $(\tilde{\mathbf{F}}'\tilde{\boldsymbol{\Gamma}}_{aa}^{-1}\tilde{\mathbf{F}})^{-1}$ using, for example,

$$\tilde{\boldsymbol{\Gamma}}_{aa} = 2\psi_p\{2[(\tilde{\boldsymbol{\beta}}, \mathbf{I})'\tilde{\mathbf{m}}_{xx}(\tilde{\boldsymbol{\beta}}, \mathbf{I})] \otimes \tilde{\boldsymbol{\Sigma}}_{\varepsilon\varepsilon} + (\tilde{\boldsymbol{\Sigma}}_{\varepsilon\varepsilon} \otimes \tilde{\boldsymbol{\Sigma}}_{\varepsilon\varepsilon})\}\psi'_p, \tag{4.2.51}$$

where $\tilde{\boldsymbol{\beta}}$, $\tilde{\boldsymbol{\Sigma}}_{\varepsilon\varepsilon}$, and $\tilde{\mathbf{m}}_{xx}$ are the least squares estimators.

Example 4.2.5. We use the data on corn hectares of Example 1.5.1 to illustrate some of the results of this section. For the purposes of this example we postulate the reduced model,

$$Y_{t1} = \beta_{11}x_t + e_{t1}, \qquad Y_{t2} = \beta_{02} + \beta_{12}x_t + e_{t2},$$
$$(e_{t1}, e_{t2}, u_t)' \sim \mathrm{NI}[\mathbf{0}, \mathrm{diag}(\sigma_{uu}, \sigma_{ee22}, \sigma_{uu})],$$

and $X_t = x_t + u_t$, for $t = 1, 2, \ldots, 37$. Under this model the personal interview (Y_{t1}) and aerial photography (X_t) procedures have equal error variance and the systematic parts of these two procedures differ only by a multiple. The systematic part of the satellite determination (Y_{t2}) is given by an affine transformation of the systematic part of the interview determination.

We will compute estimates under both the random and fixed models for x_t. The sample means are

$$(\bar{Y}_1, \bar{Y}_2, \bar{X}) = (123.28, 133.83, 120.32)$$

and the vector half of the sample covariance matrix is

$$\mathrm{vech}\, \mathbf{m}_{ZZ} = (1196.61, 908.43, 1108.74; 1002.07, 849.87; 1058.62)'.$$

To initiate our estimation procedure we first construct simple consistent estimates of the parameters. Under the model, $\bar{X}^{-1}\bar{Y}_1 = 1.025$ is a consistent estimate of β_{11} and

$$\ddot{\sigma}_{xx} = \bar{Y}_1^{-1}\bar{X}m_{YX11} = (1.025)^{-1}(1108.74) = 1081.70$$

is a consistent estimate of σ_{xx} for the random model or for $\mathbf{m}_{xx}$ of the fixed model. Also,

$$\ddot{\beta}_{12} = 2^{-1}(1081.70)^{-1}[(1.025)^{-1}908.43 + 849.87] = 0.8025,$$
$$\ddot{\sigma}_{uu} = 2^{-1}[1196.61 - (1.025)^2(1081.70) + 1058.62 - 1081.70] = 18.53,$$
$$\ddot{\sigma}_{ee22} = 1002.07 - (0.8025)^2 1081.10 = 305.45$$

are consistent estimates of β_{12}, σ_{uu}, and σ_{ee22}, respectively. If we construct an estimate of $\mathbf{\Sigma}_{ZZ}$ for random $\mathbf{Z}_t$, using these preliminary estimates, we obtain

$$\text{vech } \ddot{\mathbf{\Sigma}}_{ZZ} = (1154.99, 889.77, 1108.74; 1002.07, 868.06; 1100.23)',$$
$$\text{vech } \ddot{\mathbf{\Sigma}}_{zz} = (1136.46, 889.77, 1108.74; 696.62, 868.06; 1081.70)'.$$

Based on these preliminary estimates, the estimated covariance matrix of

$$[(\text{vech } \mathbf{m}_{ZZ})', (\bar{Y}, \bar{X}, \bar{W})]'$$

constructed under the random normal model is

$$\ddot{\mathbf{\Omega}} = \text{block diag}\{(18)^{-1}\psi_3(\ddot{\mathbf{\Sigma}}_{ZZ} \otimes \ddot{\mathbf{\Sigma}}_{ZZ})\psi_3', (37)^{-1}\ddot{\mathbf{\Sigma}}_{ZZ}\}.$$

Under the fixed-x model with normal error, the estimated covariance matrix is

$$\ddot{\mathbf{\Gamma}}_{aa} = \text{block diag}\{(18)^{-1}\psi_3[2(\ddot{\mathbf{\Sigma}}_{\varepsilon\varepsilon} \otimes \ddot{\mathbf{\Sigma}}_{zz}) + (\ddot{\mathbf{\Sigma}}_{\varepsilon\varepsilon} \otimes \ddot{\mathbf{\Sigma}}_{\varepsilon\varepsilon})]\psi_3', (37)^{-1}\ddot{\mathbf{\Sigma}}_{\varepsilon\varepsilon}\},$$

where $\ddot{\mathbf{\Sigma}}_{\varepsilon\varepsilon} = \text{diag}\{18.53, 305.45, 18.53\}$.

If we compute one step of the Gauss–Newton iteration starting with our initial consistent estimators, we obtain the estimates of Table 4.2.5. The one-step estimates are the same for the two covariance matrices because $\delta_{(f)} = \delta_{(r)}$ by Theorem 4.2.4. Also, as stated in Theorem 4.2.4, the estimated standard errors are the same for the estimates of $\boldsymbol{\beta}$ and $\mathbf{\Sigma}_{\varepsilon\varepsilon}$ for the two models. It is only the standard errors for the estimates associated with x that differ. Under the random model the variance of the distribution of x is being estimated and the estimated variance of $\hat{\sigma}_{xx}$ reflects the fact that the values of x change from sample to sample. Under the fixed model the set of x values is the same for the collection of samples for which the variance is being estimated. For example, the variance of the estimator of $\bar{x}$ is close to $n^{-1}\hat{\sigma}_{uu}$ for the fixed model, and the variance of the estimator of μ_x is close to $n^{-1}\hat{\sigma}_{XX}$ for the random model.

TABLE 4.2.5. One-step estimates for random and fixed normal models

Parameter			Standard Error	
Random x	Fixed x	Estimate	Random x	Fixed x
$10\beta_{11}$	$10\beta_{11}$	10.272	0.082	0.082
β_{02}	β_{02}	37.332	11.174	11.174
β_{12}	β_{12}	0.803	0.090	0.090
σ_{uu}	σ_{uu}	17.510	4.368	4.368
$10^{-1}\sigma_{ee22}$	$10^{-1}\sigma_{ee22}$	30.536	7.338	7.338
$10^{-2}\sigma_{xx}$	$10^{-2}m_{xx}$	10.803	2.567	0.342
$10^{-1}\mu_x$	$10^{-1}\bar{x}$	12.016	0.545	0.070

TABLE 4.2.6. Final estimates for random and fixed normal models

Parameter			Standard Error	
Random x	Fixed x	Estimate	Random x	Fixed x
$10\beta_{11}$	$10\beta_{11}$	10.272	0.079	0.079
β_{02}	β_{02}	37.338	11.169	11.169
β_{12}	β_{12}	0.803	0.090	0.090
σ_{uu}	σ_{uu}	17.422	4.106	4.106
$10^{-1}\sigma_{ee22}$	$10^{-1}\sigma_{ee22}$	30.538	7.328	7.328
$10^{-2}\sigma_{xx}$	$10^{-2}m_{xx}$	10.804	2.568	0.331
$10^{-1}\mu_x$	$10^{-1}\bar{x}$	12.016	0.545	0.067

Estimates for the model obtained after iteration are given in Table 4.2.6. Because the initial estimates were quite good, the final estimates differ little from the one-step estimates. The estimated standard errors changed more on iteration than did the estimates. The residual sum of squares from the weighted regression is 1.58 and is the same for the fixed and random regression models. Under the null model the residual sum of squares is distributed as a chi-square random variable with two degrees of freedom. The model of this example is easily accepted for these data, under the assumption of normal homoskedastic errors. □ □

We have seldom discussed the efficiency of our estimators, except to note that maximum likelihood estimators of the normal structural model are efficient by the properties of maximum likelihood estimators. The least squares results provide us some additional information on efficiency. We have the following results:

(i) The vector of sample covariances is converging in distribution to a normal vector under very weak conditions.

(ii) The limiting distribution of the least squares estimator constructed with a consistent estimator for the covariance matrix of vech $\mathbf{m}_{ZZ}$ is the same as the limiting distribution of the least squares estimator constructed with the true covariance matrix of vech $\mathbf{m}_{ZZ}$.

(iii) The distribution-free estimator of the variance of vech $\mathbf{m}_{ZZ}$ converges to the covariance matrix of the sample covariances under the fourth moment assumption.

Because least squares is asymptotically efficient for normal random vectors, it follows from (i), (ii), and (iii) that the least squares estimator is asymptotically efficient in the class of estimators based only on $\mathbf{m}_{ZZ}$.

In least squares estimation we are able to construct a consistent estimator of the covariance matrix of the distribution of the sample covariances without knowing the distribution of the $\mathbf{x}_t$. An extension of this result has been used by Bickel and Ritov (1985) to construct efficient estimators in a more general setting. For other results on efficiency in measurement error models, see Nussbaum (1977, 1984), Gleser (1983), and Stefanski and Carroll (1985a).

Because the normal distribution maximum likelihood procedures and least squares procedures retain desirable asymptotic properties for a wide range of distributions, because the covariance matrix of the approximate distribution of the estimators is easily computed for nonnormal distributions, because the estimators are relatively easy to compute, because the estimators have been subjected to considerable Monte Carlo study, and because many distributions encountered in practice are approximately normal, normal distribution maximum likelihood and least squares are the estimators most heavily used in practice.

REFERENCES

Anderson (1969, 1973), Bentler (1983), Browne (1974, 1984), Dahm (1979), Dahm and Fuller (1986), Fuller and Pantula (1982), Ganse, Amemiya, and Fuller (1983), Jöreskog and Goldberger (1972), Lee and Bentler (1980), Shapiro (1983).

EXERCISES

13. (Sections 4.2.1, 4.2.2) Let the sample covariance matrix for 11 observations be

$$\text{vech } \mathbf{m}_{ZZ} = (70.8182, 13.8545, 48.5636)'.$$

Assume that the data satisfy the model

$$Y_{ti} = \beta_0 + \beta_1 x_t + e_t, \qquad X_t = x_t + u_t,$$
$$(x_t, e_t, u_t) \sim \text{NI}[\mathbf{0}, \text{diag}(\sigma_{xx}, \sigma_{ee}, 57)].$$

Let it be hypothesized that $\beta_1 = 0.5$. This implies that $\sigma_{Xv} = \sigma_{uv} = -28.5$.

(i) Construct the least squares estimator of $(\sigma_{XX}, \sigma_{vv})$ using each of the two matrices
  (a) $\text{vech } \ddot{\Sigma} = (m_{vv}, m_{Xv}, m_{XX})' = (69.1046, -10.4270, 48.5636)'$
  (b) $\text{vech } \tilde{\Sigma} = (m_{vv}, \sigma_{Xv}, m_{XX})' = (69.1046, -28.500, 48.5636)'$
  in the estimated covariance matrix, $\hat{\mathbf{V}}_{aa} = 2(n-1)\psi_p(\ddot{\Sigma} \otimes \ddot{\Sigma})\psi_p'$.

(ii) Complete one additional iteration using the first step least squares estimates obtained with the matrix of (b) to construct $\hat{\mathbf{V}}_{aa}$.

(iii) Compare the least squares chi-square test of the model computed under the procedures of part (ii) with the likelihood ratio statistic computed by the method of Exercise 2.37.

14. (Sections 4.2.1, 4.2.2) Use the least squares method of (4.2.20) to estimate the parameters of the model

$$Z_{ti} = \beta_{0i} + z_t + \varepsilon_{ti}, \qquad i = 1, 2, 3,$$
$$(z_t, \varepsilon_{t1}, \varepsilon_{t2}, \varepsilon_{t3})' \sim \text{NI}(\mathbf{0}, \text{diag}\{\sigma_{zz}, \sigma_{\varepsilon\varepsilon 11}, \sigma_{\varepsilon\varepsilon 22}, \sigma_{\varepsilon\varepsilon 33}\}),$$

for the Grubbs data of Exercise 1.41, where Z_{ti}, i = 1, 2, 3, are the fuse burning times reported by the three observers.

(a) Use $\mathbf{m}_{ZZ}$ for $\mathbf{W}$. Is the model accepted by the data?

(b) Repeat the least squares procedure replacing $\mathbf{W}$ by the estimator of $\hat{\mathbf{\Sigma}}_{ZZ}$ obtained in part (a).

(c) Estimate the parameters of the reduced model in which $\sigma_{\varepsilon\varepsilon 11} = \sigma_{\varepsilon\varepsilon 22} = \sigma_{\varepsilon\varepsilon 33}$ using $\mathbf{m}_{ZZ}$ for $\mathbf{W}$. Compare the model with equal error variances to the model of part (a).

15. (Section 4.2.2) Assume that $E\{\mathbf{m}_{ZZ}\} = \alpha\mathbf{\Sigma}$, where $\mathbf{\Sigma}$ is a known $p \times p$ nonsingular matrix and α is a parameter to be estimated. Show that if the least squares method is used with $\mathbf{W}$ of (4.2.29) equal to $\mathbf{m}_{ZZ}$, the least squares estimator is

$$\hat{\alpha} = [\text{tr}\{(\mathbf{\Sigma}\mathbf{m}_{ZZ}^{-1})^2\}]^{-1} \text{tr}\{\mathbf{\Sigma}\mathbf{m}_{ZZ}^{-1}\}.$$

Show that if the least squares method is used with $\mathbf{W} = \mathbf{\Sigma}$, the least squares estimator is

$$\hat{\alpha} = p^{-1} \text{tr}\{\mathbf{m}_{ZZ}\mathbf{\Sigma}^{-1}\}.$$

Show that the estimator constructed with $\mathbf{W} = \mathbf{\Sigma}$ is the maximum likelihood estimator under the assumption of normality.

16. (Sections 4.2.1, 4.2.2) The following data are for three measurements on 10 units as reported by Hahn and Nelson (1970). The first measurement is made by the standard device and the last two measurements are replicates of measurements made by a new device. For the purpose of this example we assume

$$Z_{t1} = x_t + \varepsilon_{t1}, \qquad Z_{ti} = \gamma_0 + \gamma_1 x_t + \varepsilon_{ti}, \quad \text{for } i = 2, 3$$

and

$$(x_t, \varepsilon_{t1}, \varepsilon_{t2}, \varepsilon_{t3}) \sim \text{NI}[(\mu_x, \mathbf{0}), \text{diag}(\sigma_{xx}, \sigma_{\varepsilon\varepsilon 11}, \sigma_{\varepsilon\varepsilon 22}, \sigma_{\varepsilon\varepsilon 22})].$$

		Device 2	
Unit	Device 1	Determination 1	Determination 2
1	71	77	80
2	108	105	96
3	72	71	74
4	140	152	146
5	61	88	83
6	97	117	120
7	90	93	103
8	127	130	119
9	101	112	108
10	114	105	115

Source: Hahn and Nelson (1970).

(a) Estimate the parameters of the model by least squares applied to the covariance matrix with

$$\hat{\mathbf{V}}_{aa} = 2(n-1)^{-1}\psi(\mathbf{m}_{ZZ} \otimes \mathbf{m}_{ZZ})\psi'.$$

(b) Let $(Y_{t1}, Y_{t2}, Y_{t3}) = [Z_{t1}, \frac{1}{2}(Z_{t2} + Z_{t3}), \frac{1}{2}(Z_{t2} - Z_{t3})]$. Show that under the model of this exercise

$$C\{Y_{t1}, Y_{t3}\} = C\{Y_{t2}, Y_{t3}\} = 0.$$

Also show that the covariances between (m_{YY13}, m_{YY23}) and all other sample covariances of $\mathbf{Y}_t$ are zero. Therefore, m_{YY13} and m_{YY23} need not be included in the least squares fit. Use these facts to derive the least squares estimators (equivalent to the maximum likelihood estimators) of the parameters of the model. Compute the estimated covariance matrix of your estimates and compute the lack of fit statistic. Note that the model is a reparameterization of the model of Section 2.2.

(c) Compute the covariance matrix for $(\hat{\gamma}_0, \hat{\gamma}_1, \hat{m}_{xx}, \hat{\sigma}_{\varepsilon\varepsilon 11}, \hat{\sigma}_{\varepsilon\varepsilon 22})$ under the assumption that the x values are fixed. Use (4.2.51) and the estimates obtained in part (b).

17. (Section 4.2.2) The estimates of μ_{x1} and μ_{x2} of Table 4.2.4 are nearly equal and the estimates of σ_{xx11} and σ_{xx22} are very similar. Using the estimates of Table 4.2.4 and $\sigma_{xx11} = \sigma_{xx22} = 1.31$ to construct an estimate of the covariance matrix of $\mathbf{s}$, compute least squares estimates of the model with $\mu_{x1} = \mu_{x2} = \mu_x$ and $\sigma_{xx11} = \sigma_{xx22} = \sigma_{xx}$. Is the restricted model acceptable?

18. (Sections 4.2.3, 1.4) Assume that, for a sample of 25 observations on $\mathbf{Z}_t = (Y_{t1}, Y_{t2}, X_t)$,

$$\text{vech } \mathbf{m}_{ZZ} = (3.45, 2.49, 1.80; 6.42, 3.90; 4.18)'$$

and the vector of sample means is $(\bar{Y}_1, \bar{Y}_2, \bar{X}) = (1.51, 3.37, 2.63)$. Assume that the data satisfy the model

$$Y_{ti} = \beta_{0i} + \beta_{1i}x_t + e_{ti}, \quad \text{for } i = 1, 2,$$
$$X_t = x_t + u_t,$$

and

$$(x_t, e_{t1}, e_{t2}, u_t)' \sim \text{NI}[(\mu_x, \mathbf{0})', \text{block diag}(\sigma_{xx}, \boldsymbol{\Sigma}_{ee}, \sigma_{uu})].$$

(a) Assume that $\sigma_{uu} = 1$. Estimate the remaining parameters of the model and estimate the covariance matrix of your estimates.

(b) Assume that $\sigma_{uu} = 1$ and that $\sigma_{ee12} = 0$. Apply least squares to the covariance matrix to estimate the remaining parameters of the model. Use $2(n-1)^{-1}\psi_3(\mathbf{m}_{ZZ} \otimes \mathbf{m}_{ZZ})\psi_3'$ as the estimated covariance matrix for vech $\mathbf{m}_{ZZ}$.

(c) Complete a second iteration of the least squares procedure of part (b) using

$$2(n-1)^{-1}\psi_3(\hat{\boldsymbol{\Sigma}}_{ZZ} \otimes \hat{\boldsymbol{\Sigma}}_{ZZ})\psi_3'$$

as the estimated covariance matrix for vech $\mathbf{m}_{ZZ}$, where $\hat{\boldsymbol{\Sigma}}_{ZZ}$ is the estimated covariance matrix constructed from the estimates of part (b).

(d) Beginning with the estimates of part (b), complete a second iteration of the least squares procedure using the fixed-x covariance matrix (4.2.51) for vech $\mathbf{m}_{ZZ}$.

(e) Compare the estimated variance for your estimator of β_{11} of part (c) with the estimated variance of the single equation estimator (1.2.3) of Section 1.2 that uses only the knowledge of σ_{uu} and with the variance of the instrumental variable estimator (1.4.12) of Section 1.4 that uses only the knowledge that $\sigma_{ee12} = 0$.

19. (Sections 4.2.1, 4.2.2) Use the method of least squares to combine the two sets of corn yield–soil nitrogen data analyzed in Example 3.1.1. Assume the model of Example 3.1.1. Include

the sample means in your vector of observations. For the first iteration use the sample covariances and Lemma 4.A.1 to estimate the covariance matrices of the sample covariances. Then compute an additional iteration using estimated covariance matrices for the sample moments constructed from your first round parameter estimates. Does the model seem appropriate for these data?

20. (Sections 4.2.3, 1.5)
    (a) Use the nonparametric estimate of the covariance matrix of the sample moments of the corn hectares data constructed in Example 4.2.3 to estimate the covariance matrix of the estimated parameters of the factor model constructed in Example 1.5.1. Because the factor model of Example 1.5.1 is just identified, the least squares estimators are equal to the estimators presented in Example 1.5.1.
    (b) Compute an estimator of the covariance matrix of $(\bar{Y}_1, \bar{Y}_2, \bar{X}, m_{YY11}, m_{YY12}, m_{YX11}, m_{YY22}, m_{YX21}, m_{XX})$ of the form of $\hat{\mathbf{V}}_{bb}$ of Example 4.2.3 for the corn hectares data. Use this covariance matrix to estimate the parameters of the model of Example 4.2.5 by nonlinear generalized least squares. Construct a test for the model.

21. (Section 4.2.2) Prove the following.

Theorem. Let the assumptions of Theorem 4.2.1 hold. Let $\hat{\boldsymbol{\theta}}_m$ be the value of $\boldsymbol{\theta}$ that minimizes

$$Q_m(\boldsymbol{\theta}) = \log|\boldsymbol{\Sigma}_{ZZ}(\boldsymbol{\theta})| + \operatorname{tr}\{\mathbf{m}_{ZZ}\boldsymbol{\Sigma}_{ZZ}^{-1}(\boldsymbol{\theta})\}.$$

and let $\hat{\boldsymbol{\theta}}_\ell$ be the value of $\boldsymbol{\theta}$ that minimizes

$$Q_\ell(\boldsymbol{\theta}; \hat{\mathbf{V}}_{aa}) = [\mathbf{s} - \boldsymbol{\sigma}(\boldsymbol{\theta})]'\hat{\mathbf{V}}_{aa}^{-1}[\mathbf{s} - \boldsymbol{\sigma}(\boldsymbol{\theta})],$$

where $\mathbf{s} = \operatorname{vech} \mathbf{m}_{ZZ}$, $\boldsymbol{\sigma}(\boldsymbol{\theta}) = \operatorname{vech} \boldsymbol{\Sigma}_{ZZ}(\boldsymbol{\theta})$, and $\hat{\mathbf{V}}_{aa}$ is given in (4.2.11). Then

$$n^{1/2}(\hat{\boldsymbol{\theta}}_\ell - \hat{\boldsymbol{\theta}}_m) \xrightarrow{L} \mathbf{0}.$$

4.3. FACTOR ANALYSIS

Factor analysis has a rich history in psychology beginning with the work of Charles Spearman (1904). The statistical development of the subject began with the work in least squares of Adcock (1878) and Pearson (1901). Harman (1976) describes the development of factor analysis and cites a number of applications in the physical and social sciences.

The application of the method of maximum likelihood to factor analysis is primarily due to Lawley (1940, 1941, 1943) and is well described in Lawley and Maxwell (1971). Anderson and Rubin (1955) contains a detailed investigation of the factor model. The applications of the method often involve large quantities of data and there has been a continuing interest in efficient computational methods.

4.3.1. Introduction and Model

The factor model was introduced in Section 1.5. In the model of (1.5.1) the elements of an observation vector are specified to be the sum of a linear function of a single unobserved factor x_t and an error vector. Specification of the nature of the covariance matrix of the error vector and the independence of x_t and the error vector permitted the estimation of the parameters of the model. In this chapter we extend the model to higher dimensions. In the simple model of Section 1.5, the number of parameters to be estimated was equal to the number of sample moments. As a result, we were able to obtain explicit expressions for the estimators. This is generally not possible for models of higher dimension.

Let a p-dimensional vector of observations on the tth individual be denoted by

$$\mathbf{Z}_t = (Z_{t1}, Z_{t2}, \ldots, Z_{tp}). \tag{4.3.1}$$

It is assumed that the vector can be expressed as the sum of a linear function of a k-dimensional factor vector $\mathbf{f}_t$ and a p-dimensional error vector $\boldsymbol{\varepsilon}_t$. It is often assumed that the unobserved factor vector is distributed as a normal random vector with mean zero and positive definite covariance matrix. Under the normal distribution assumption,

$$\mathbf{Z}_t = \boldsymbol{\mu}_Z + \mathbf{f}_t\mathbf{A}' + \boldsymbol{\varepsilon}_t, \tag{4.3.2}$$

where

$$(\mathbf{f}_t, \boldsymbol{\varepsilon}_t)' \sim \mathrm{NI}[\mathbf{0}, \text{block diag}(\boldsymbol{\Sigma}_{ff}, \boldsymbol{\Sigma}_{\varepsilon\varepsilon})],$$

$\mathbf{A}'$ is a $k \times p$ matrix, and $\boldsymbol{\mu}_Z$ is the mean of the $\mathbf{Z}$ vector. The matrix $\mathbf{A}'$ is sometimes called the matrix of *factor loadings*. We also assume that $\boldsymbol{\Sigma}_{\varepsilon\varepsilon}$ is a diagonal matrix. Because $\mathbf{f}_t$ and $\boldsymbol{\varepsilon}_t$ are normal random vectors, the vector $\mathbf{Z}_t$ is normal with mean $\boldsymbol{\mu}_Z$ and covariance matrix

$$\mathrm{V}\{\mathbf{Z}_t'\} = \mathbf{A}\boldsymbol{\Sigma}_{ff}\mathbf{A}' + \boldsymbol{\Sigma}_{\varepsilon\varepsilon}. \tag{4.3.3}$$

In the terminology common in factor analysis that was introduced in Section 1.5, ε_{ti} is called the unique factor for the variable Z_{ti} and the elements of $\mathbf{f}_t$ are called the common factors. The fraction of the variance of Z_{ti} that is due to the variation in the common factors is called the communality. The communality for the ith observed variable is

$$\mathbf{A}_{i.}\boldsymbol{\Sigma}_{ff}\mathbf{A}_{i.}'(\sigma_{ZZii})^{-1},$$

where $\mathbf{A}_{i.}$ is the ith row of $\mathbf{A}$ and σ_{ZZii} is the variance of Z_{ti}. The variance $\sigma_{\varepsilon\varepsilon ii}$ is sometimes called the uniqueness of the variable Z_{ti}. Uniqueness is also often expressed as a fraction of total variance. The model (4.3.2) contains p means and

$$pk + \tfrac{1}{2}k(k+1) + p \tag{4.3.4}$$

other parameters, where pk is the number of elements in $\mathbf{A}$, $\frac{1}{2}k(k+1)$ is the number of unique elements in $\boldsymbol{\Sigma}_{ff}$, and p is the number of diagonal elements in $\boldsymbol{\Sigma}_{\varepsilon\varepsilon}$.

We can transform both $\mathbf{f}_t$ and $\mathbf{A}$ without altering the content of the model. That is, if we let $\mathbf{G}$ be an arbitrary nonsingular $k \times k$ matrix, then

$$\begin{aligned}\mathbf{Z}_t &= \boldsymbol{\mu}_Z + \mathbf{f}_t\mathbf{G}\mathbf{G}^{-1}\mathbf{A}' + \boldsymbol{\varepsilon}_t \\ &= \boldsymbol{\mu}_Z + \mathbf{g}_t\mathbf{B}' + \boldsymbol{\varepsilon}_t, \end{aligned} \tag{4.3.5}$$

where $\mathbf{g}_t = \mathbf{f}_t\mathbf{G}$ and $\mathbf{B}' = \mathbf{G}^{-1}\mathbf{A}'$. Because of (4.3.5), the model is often parameterized by assuming $\boldsymbol{\Sigma}_{ff} = \mathbf{I}$.

While there are $pk + \frac{1}{2}k(k + 1) + p$ parameters entering the covariance matrix of model (4.3.2), the existence of the arbitrary transformation matrix $\mathbf{G}$ means that there are only

$$pk + \tfrac{1}{2}k(k + 1) + p - k^2 = k(p - k) + \tfrac{1}{2}k(k + 1) + p \tag{4.3.6}$$

independent parameters associated with the model for $\boldsymbol{\Sigma}_{ZZ}$. The sample covariance matrix $\mathbf{m}_{ZZ}$ contains $\frac{1}{2}p(p + 1)$ unique elements and a necessary condition for the model to be identified is that the number of unique elements equal or exceed the number of parameters. By (4.3.6) the condition is

$$(p - k)^2 \geqslant p + k. \tag{4.3.7}$$

If the inequality (4.3.7) is not satisfied, it is possible to identify the model by imposing additional restrictions on $\mathbf{A}$ or (and) $\boldsymbol{\Sigma}_{ff}$ or (and) $\boldsymbol{\Sigma}_{\varepsilon\varepsilon}$.

We can also consider the model expressed in a form consistent with the errors-in-variables development of earlier chapters. To this end the vector $\mathbf{Z}_t$ is divided into two parts,

$$\mathbf{Z}_t = (\mathbf{Y}_t, \mathbf{X}_t), \tag{4.3.8}$$

where $\mathbf{X}_t$ is a k-dimensional vector, $\mathbf{Y}_t$ is an r-dimensional vector, and $r = p - k$. We assume

$$\mathbf{Y}_t = \boldsymbol{\beta}_0 + \mathbf{x}_t\boldsymbol{\beta} + \mathbf{e}_t, \qquad \mathbf{X}_t = \mathbf{x}_t + \mathbf{u}_t, \tag{4.3.9}$$

or, equivalently,

$$\mathbf{Z}_t = (\boldsymbol{\beta}_0, \mathbf{0}) + \mathbf{x}_t(\boldsymbol{\beta}, \mathbf{I}) + \boldsymbol{\varepsilon}_t,$$

where $\boldsymbol{\varepsilon}_t = (\mathbf{e}_t, \mathbf{u}_t)$. The assumptions that $\boldsymbol{\Sigma}_{\varepsilon\varepsilon}$ is diagonal, that $\boldsymbol{\Sigma}_{xx}$ is positive definite, and that $\mathbf{x}_t$ is independent of $\boldsymbol{\varepsilon}_t$ are retained. The model contains p unknown means and

$$(p - k)k + p + \tfrac{1}{2}k(k + 1) \tag{4.3.10}$$

unknown parameters in the covariance matrix, where $(p - k) \times k$ is the dimension of $\boldsymbol{\beta}$, p is the number of diagonal elements in $\boldsymbol{\Sigma}_{\varepsilon\varepsilon}$, and $\frac{1}{2}k(k + 1)$ is the number of unique elements in $\boldsymbol{\Sigma}_{xx}$. Note that the number in (4.3.10) is the number of unique parameters obtained in (4.3.6) for the alternative parameterization.

The specification that Σ_{xx} is positive definite together with (4.3.7) are necessary conditions for identification of $\boldsymbol{\beta}$ and $\Sigma_{\varepsilon\varepsilon}$, but they are not sufficient. To understand this, recall the one-factor model in three variables. Identification of all parameters of that model required that the coefficients of the factor (β_{11} and β_{21} of (1.5.1)) be nonzero. In the present model, if a column of $\boldsymbol{\beta}$ is composed of all zeros, it may not be possible to estimate all parameters of the model. In practice, care should also be taken in the specification of the $\mathbf{X}_t$ vector, to guarantee that Σ_{xx} is nonsingular.

Before proceeding to other methods of estimation, we observe that the columns of $\boldsymbol{\beta}$ can often be estimated by the method of instrumental variables. Without loss of generality, consider the first equation of (4.3.9)

$$Y_{t1} = \beta_{01} + \mathbf{x}_t \boldsymbol{\beta}_{.1} + e_{t1}, \tag{4.3.11}$$

where $\boldsymbol{\beta}_{.1}$ is the first column of $\boldsymbol{\beta}$. The model specifies $\mathbf{x}_t$ to be independent of $\boldsymbol{\varepsilon}_t$ and $\Sigma_{\varepsilon\varepsilon}$ to be diagonal. This means that $Y_{t2}, Y_{t3}, \ldots, Y_{t,p-k}$ are independent of e_{t1}. Therefore, if

$$p \geqslant 2k + 1, \tag{4.3.12}$$

it may be possible to use the method of instrumental variables to estimate $\boldsymbol{\beta}_{.1}$. The condition (4.3.12) is stronger than condition (4.3.7). However, for many problems the instrumental variable estimators, which are easy to compute by two-stage least squares or limited information maximum likelihood, can furnish preliminary information about the parameters of the model. Section 2.4 contains a description of the instrumental variable estimators and Example 2.4.1 is an application of the method to the factor model.

4.3.2. Maximum Likelihood Estimation

We now obtain the maximum likelihood estimators of the parameters for the normal model. Our development follows closely that of Lawley and Maxwell (1971). In Section 4.1 we obtained an explicit expression for the estimator of $\boldsymbol{\beta}$ under the assumption that an independent estimator of $\Sigma_{\varepsilon\varepsilon}$ is available. In the factor model, $\Sigma_{\varepsilon\varepsilon}$ is a matrix containing unknown parameters to be estimated from the $\mathbf{Z}_t$ data. This complicates the estimation procedure and no explicit expression for the maximum likelihood estimator has been obtained.

We develop some properties of the likelihood function and outline methods for numerical maximization of the likelihood function in this section. Material following Result 4.3.1 can be easily omitted by persons not interested in computational procedures.

Two parameterizations of the factor model have been developed. The first is that given in (4.3.2), which we write as

$$\mathbf{Z}_t = \boldsymbol{\mu}_Z + \mathbf{f}_t\mathbf{A}' + \boldsymbol{\varepsilon}_t, \tag{4.3.13}$$

where the matrix of factors satisfies

$$E\{\mathbf{f}_t\} = 0, \qquad E\{\mathbf{f}_t'\mathbf{f}_t\} = \mathbf{I}$$

and there are $\frac{1}{2}k(k+1) + k(p-k)$ independent parameters in $\mathbf{A}$. With this form of the model, the covariance matrix of $\mathbf{Z}_t$ can be written

$$\boldsymbol{\Sigma}_{ZZ} = \mathbf{AA}' + \boldsymbol{\Sigma}_{\varepsilon\varepsilon}. \tag{4.3.14}$$

The second form of the model is that given in (4.3.9). In the parameterization of (4.3.9)

$$\boldsymbol{\Sigma}_{ZZ} = (\boldsymbol{\beta}, \mathbf{I})'\boldsymbol{\Sigma}_{xx}(\boldsymbol{\beta}, \mathbf{I}) + \boldsymbol{\Sigma}_{\varepsilon\varepsilon}.$$

Under the normal model, the density of $\mathbf{Z}_t$ is

$$(2\pi)^{-p/2}|\boldsymbol{\Sigma}_{ZZ}|^{-1/2} \exp\{-\tfrac{1}{2}(\mathbf{Z}_t - \boldsymbol{\mu}_Z)\boldsymbol{\Sigma}_{ZZ}^{-1}(\mathbf{Z}_t - \boldsymbol{\mu}_Z)'\}, \tag{4.3.15}$$

where we assume that the elements of $\boldsymbol{\Sigma}_{\varepsilon\varepsilon}$, $\boldsymbol{\Sigma}_{xx}$, and $\boldsymbol{\beta}$ are such that $|\boldsymbol{\Sigma}_{ZZ}| > 0$. The logarithm of the likelihood for a sample of n observations is

$$\begin{aligned}\log L = &-\tfrac{1}{2}np \log 2\pi - \tfrac{1}{2}n \log|\boldsymbol{\Sigma}_{ZZ}| - \tfrac{1}{2}(n-1)\operatorname{tr}\{\mathbf{m}_{ZZ}\boldsymbol{\Sigma}_{ZZ}^{-1}\} \\ &-\tfrac{1}{2}n(\bar{\mathbf{Z}} - \boldsymbol{\mu}_Z)\boldsymbol{\Sigma}_{ZZ}^{-1}(\bar{\mathbf{Z}} - \boldsymbol{\mu}_Z)'.\end{aligned} \tag{4.3.16}$$

It is clear that the maximum likelihood estimator of $\boldsymbol{\mu}_Z$ is $\bar{\mathbf{Z}}$,

$$\hat{\boldsymbol{\mu}}_Z = \bar{\mathbf{Z}} = (\hat{\beta}_0 + \hat{\boldsymbol{\mu}}_x\hat{\boldsymbol{\beta}}, \hat{\boldsymbol{\mu}}_x), \tag{4.3.17}$$

because replacing $\boldsymbol{\mu}_Z$ by $\hat{\boldsymbol{\mu}}_Z$ reduces the last term on the right side of (4.3.16) to zero.

As we did in Section 4.1, we define the logarithm of the reduced likelihood function adjusted for degrees of freedom by

$$\log L_c = -\tfrac{1}{2}(n-1)(p \log 2\pi + \log|\boldsymbol{\Sigma}_{ZZ}| + \operatorname{tr}\{\mathbf{m}_{ZZ}\boldsymbol{\Sigma}_{ZZ}^{-1}\}). \tag{4.3.18}$$

The maximum likelihood estimators of the parameters in $\boldsymbol{\Sigma}_{ZZ}$ adjusted for degrees of freedom are described in Result 4.3.1.

Result 4.3.1. Let

$$\mathbf{Y}_t = \beta_0 + \mathbf{x}_t\boldsymbol{\beta} + \mathbf{e}_t, \qquad \mathbf{X}_t = \mathbf{x}_t + \mathbf{u}_t,$$
$$(\boldsymbol{\varepsilon}_t, \mathbf{x}_t)' \sim \mathrm{NI}[(\mathbf{0}, \boldsymbol{\mu}_x)', \text{block diag}(\boldsymbol{\Sigma}_{\varepsilon\varepsilon}, \boldsymbol{\Sigma}_{xx})],$$

where $\boldsymbol{\Sigma}_{\varepsilon\varepsilon}$ is a diagonal matrix with positive elements, $\boldsymbol{\Sigma}_{ZZ}$ and $\boldsymbol{\Sigma}_{xx}$ are positive definite, and $(p-k)^2 \geqslant p + k$. Then the maximum likelihood estimators adjusted for degrees of freedom satisfy the equations

$$\hat{\beta} = \hat{\Sigma}_{xx}^{-1}\hat{\Sigma}_{xy}, \tag{4.3.19}$$

$$\hat{\Sigma}_{ZZ}^{-1}(\mathbf{m}_{ZZ} - \hat{\Sigma}_{ZZ})\hat{\Sigma}_{ZZ}^{-1}\hat{\mathbf{A}} = \mathbf{0}, \tag{4.3.20}$$

$$\operatorname{diag}(\mathbf{m}_{ZZ} - \hat{\Sigma}_{ZZ}) = \mathbf{0}, \tag{4.3.21}$$

where $\hat{\Sigma}_{ZZ} = \hat{\Sigma}_{zz} + \hat{\Sigma}_{\varepsilon\varepsilon}$ and $\hat{\Sigma}_{zz} = \hat{\mathbf{A}}\hat{\mathbf{A}}'$, provided the maximum of the likelihood occurs in the interior of parameter space.

Proof. The derivatives of the likelihood function with respect to the elements of $\mathbf{A}$ are evaluated in Theorem 4.C.1 of Appendix 4.C and Equation (4.3.20) is the same as Equation (4.C.6). The derivative of (4.3.18) with respect to the ith diagonal element of $\Sigma_{\varepsilon\varepsilon}$ is

$$\frac{\partial \log L_c}{\partial \sigma_{\varepsilon\varepsilon ii}} = \tfrac{1}{2}(n-1)\operatorname{tr}\left\{\Sigma_{ZZ}^{-1}(\mathbf{m}_{ZZ} - \Sigma_{ZZ})\Sigma_{ZZ}^{-1}\frac{\partial \Sigma_{ZZ}}{\partial \sigma_{\varepsilon\varepsilon ii}}\right\}.$$

Because the partial derivative of Σ_{ZZ} with respect to $\sigma_{\varepsilon\varepsilon ii}$ is a matrix with a one in the ii position and zeros elsewhere, we have the p equations,

$$\operatorname{diag}\{\Sigma_{ZZ}^{-1}(\mathbf{m}_{ZZ} - \Sigma_{ZZ})\Sigma_{ZZ}^{-1}\} = \mathbf{0}, \tag{4.3.22}$$

where diag $\mathbf{m}$ is the diagonal matrix composed of the diagonal elements of $\mathbf{m}$. If $\Sigma_{\varepsilon\varepsilon}$ is positive definite,

$$\Sigma_{ZZ}^{-1} = (\mathbf{A}\mathbf{A}' + \Sigma_{\varepsilon\varepsilon})^{-1} = \Sigma_{\varepsilon\varepsilon}^{-1} - \Sigma_{\varepsilon\varepsilon}^{-1}\mathbf{A}(\mathbf{I} + \mathbf{A}'\Sigma_{\varepsilon\varepsilon}^{-1}\mathbf{A})^{-1}\mathbf{A}'\Sigma_{\varepsilon\varepsilon}^{-1}. \tag{4.3.23}$$

Multiplying Equation (4.3.23) on the right side by $\mathbf{A}$, we find that Equation (4.3.20), as an expression in Σ_{ZZ} and $\mathbf{A}$, can also be written

$$(\mathbf{m}_{ZZ} - \Sigma_{ZZ})\Sigma_{\varepsilon\varepsilon}^{-1}\mathbf{A}(\mathbf{I} + \mathbf{A}'\Sigma_{\varepsilon\varepsilon}^{-1}\mathbf{A})^{-1} = \mathbf{0}. \tag{4.3.24}$$

Using (4.3.23) and (4.3.24), we can write (4.3.22) as

$$\operatorname{diag}\{\Sigma_{\varepsilon\varepsilon}^{-1}(\mathbf{m}_{ZZ} - \Sigma_{ZZ})\Sigma_{\varepsilon\varepsilon}^{-1}\} = \mathbf{0} \tag{4.3.25}$$

and, because $\Sigma_{\varepsilon\varepsilon}$ is diagonal, we obtain (4.3.21). □

We now give some results that suggest an iterative method of computing the estimators.

Result 4.3.2. Let $\hat{\Sigma}_{\varepsilon\varepsilon}$ be the maximum likelihood estimator of $\Sigma_{\varepsilon\varepsilon}$ adjusted for degrees of freedom. Assume that $\hat{\Sigma}_{\varepsilon\varepsilon}$ is in the interior of parameter space. Then the maximum likelihood estimator of $\boldsymbol{\beta}$ adjusted for degrees of freedom is given by

$$\hat{\boldsymbol{\beta}}' = \hat{\mathbf{A}}_{rk}\hat{\mathbf{A}}_{kk}^{-1}, \tag{4.3.26}$$

where $\hat{\mathbf{A}} = (\hat{\mathbf{A}}'_{rk}, \hat{\mathbf{A}}'_{kk})'$,

$$\hat{\mathbf{A}} = \mathbf{m}_{ZZ}[(1 - \hat{\lambda}_1^{-1})^{1/2}\hat{\mathbf{R}}_{.1}, \ldots, (1 - \hat{\lambda}_k^{-1})^{1/2}\hat{\mathbf{R}}_{.k}], \tag{4.3.27}$$

$\hat{\mathbf{R}}_{.j}$ are the characteristic vectors of $\hat{\mathbf{\Sigma}}_{\varepsilon\varepsilon}$ in the metric $\mathbf{m}_{ZZ}$, and $\hat{\lambda}_1^{-1} \leqslant \hat{\lambda}_2^{-1} \leqslant \cdots \leqslant \hat{\lambda}_p^{-1}$ are the p roots of $\hat{\mathbf{\Sigma}}_{\varepsilon\varepsilon}$ in the metric $\mathbf{m}_{ZZ}$. The maximum likelihood estimator of the covariance matrix of $\mathbf{z}_t$ adjusted for degrees of freedom is

$$\hat{\mathbf{\Sigma}}_{zz} = \hat{\mathbf{A}}\hat{\mathbf{A}}'. \tag{4.3.28}$$

Proof. It can be shown that if $\hat{\mathbf{\Sigma}}_{\varepsilon\varepsilon}$ satisfies the equations in Result 4.3.1, then $\hat{\lambda}_k^{-1} < 1$. Thus, the result follows from Theorem 4.C.1. The maximum likelihood estimator of $\mathbf{A}$ adjusted for degrees of freedom was obtained in that theorem for known $\mathbf{\Sigma}_{\varepsilon\varepsilon}$. If $\mathbf{A}$ maximizes the adjusted likelihood conditional on $\hat{\mathbf{\Sigma}}_{\varepsilon\varepsilon}$ and if $\hat{\mathbf{\Sigma}}_{\varepsilon\varepsilon}$ is associated with the global maximum, then $(\hat{\mathbf{A}}, \hat{\mathbf{\Sigma}}_{\varepsilon\varepsilon})$ give the global maximum of the adjusted likelihood. □

Result 4.3.3. Let the model of Result 4.3.1 hold. Let the vector of parameters be

$$\gamma = (\gamma'_\beta, \gamma'_\varepsilon, \gamma'_x)',$$

where $\gamma_\beta = \operatorname{vec} \boldsymbol{\beta}$, $\gamma'_\varepsilon = (\sigma_{\varepsilon\varepsilon 11}, \sigma_{\varepsilon\varepsilon 22}, \ldots, \sigma_{\varepsilon\varepsilon pp})$, and $\gamma_x = \operatorname{vech} \mathbf{\Sigma}_{xx}$. Consider an iterative Gauss–Newton nonlinear least squares procedure in which a step of the procedure is the application of generalized least squares to the approximating linear system

$$\tilde{\mathbf{g}} = \tilde{\mathbf{F}}(\gamma - \tilde{\gamma}) + \mathbf{a}, \tag{4.3.29}$$

where $\tilde{\gamma}$ is the value of γ obtained at the preceding step, $\tilde{\mathbf{g}} = \mathbf{g}(\tilde{\gamma}) = \operatorname{vech}[\mathbf{m}_{ZZ} - \mathbf{\Sigma}_{ZZ}(\tilde{\gamma})]$, $\tilde{\mathbf{F}} = \mathbf{F}(\tilde{\gamma})$ is the matrix of partial derivatives of $\operatorname{vech} \mathbf{\Sigma}_{ZZ} = \operatorname{vech} \mathbf{\Sigma}_{ZZ}(\gamma)$ with respect to γ' evaluated at $\gamma = \tilde{\gamma}$, and $E\{\mathbf{aa}'\}$ is approximated with

$$(n-1)^{-1}\tilde{\mathbf{\Omega}} = 2(n-1)^{-1}\psi_p[\mathbf{\Sigma}_{ZZ}(\tilde{\gamma}) \otimes \mathbf{\Sigma}_{ZZ}(\tilde{\gamma})]\psi'_p.$$

Let the estimate at the next step be $\hat{\gamma}$, where

$$\hat{\gamma} - \tilde{\gamma} = (\tilde{\mathbf{F}}'\tilde{\mathbf{\Omega}}^{-1}\tilde{\mathbf{F}})^{-1}\tilde{\mathbf{F}}'\tilde{\mathbf{\Omega}}^{-1}\tilde{\mathbf{g}}. \tag{4.3.30}$$

Then the estimate of γ_ε defined in (4.3.30) is

$$\hat{\gamma}_\varepsilon = \tilde{\mathbf{B}}^{-1}\tilde{\mathbf{d}}, \tag{4.3.31}$$

where $\tilde{\mathbf{B}}$ is the matrix whose ijth element is the square of the ijth element of $\tilde{\mathbf{C}}\tilde{\mathbf{\Sigma}}_{vv}^{-1}\tilde{\mathbf{C}}'$, $\tilde{\mathbf{d}}$ is a column vector whose ith element is the ith diagonal element of

$$\tilde{\mathbf{C}}\tilde{\mathbf{\Sigma}}_{vv}^{-1}\tilde{\mathbf{m}}_{vv}\tilde{\mathbf{\Sigma}}_{vv}^{-1}\tilde{\mathbf{C}}',$$

with $\tilde{\boldsymbol{\Sigma}}_{vv} = \tilde{\mathbf{C}}'\tilde{\boldsymbol{\Sigma}}_{\varepsilon\varepsilon}\tilde{\mathbf{C}}$, $\tilde{\mathbf{C}}' = (\mathbf{I}, -\tilde{\boldsymbol{\beta}}')$, $\tilde{\mathbf{m}}_{vv} = \tilde{\mathbf{C}}'\mathbf{m}_{ZZ}\tilde{\mathbf{C}}$, and

$$\tilde{\boldsymbol{\Sigma}}_{\varepsilon\varepsilon} = \operatorname{diag}(\tilde{\sigma}_{\varepsilon\varepsilon 11}, \tilde{\sigma}_{\varepsilon\varepsilon 22}, \ldots, \tilde{\sigma}_{\varepsilon\varepsilon pp}).$$

Proof. Let $\mathbf{F} = \mathbf{F}(\gamma) = (\mathbf{F}_\beta, \mathbf{F}_\varepsilon, \mathbf{F}_x)$, where

$$\mathbf{F}_\beta = 2\psi_p[(\mathbf{I}_r, \mathbf{0}_{rk})' \otimes (\boldsymbol{\beta}, \mathbf{I}_k)'\boldsymbol{\Sigma}_{xx}],$$
$$\mathbf{F}_x = \psi_p[(\boldsymbol{\beta}, \mathbf{I}_k)' \otimes (\boldsymbol{\beta}, \mathbf{I}_k)']\boldsymbol{\Phi}_k,$$

$\mathbf{F}_\varepsilon = \psi_p\mathbf{L}$, and $\mathbf{L}$ is the $p^2 \times p$ matrix defined in Result 4.A.3.2. Note that the partition of $\mathbf{F}$ conforms to the partition of γ. To obtain an explicit expression for the estimate of γ_ε at the next step in the least squares procedure, we shall apply a linear transformation to the system (4.3.30). The objectives of the transformation are to isolate the coefficients of γ_β and γ_x and to reduce the covariance matrix of the error vector to a block diagonal. Let

$$(\mathbf{v}_t, \boldsymbol{\eta}_t) = \mathbf{Z}_t(\mathbf{C}, \mathbf{H}_2) = \mathbf{Z}_t\mathbf{H},$$

where $\mathbf{H} = (\mathbf{C}, \mathbf{H}_2)$ and $\mathbf{H}_2 = (\mathbf{0}_{kr}, \mathbf{I}_k)' - \mathbf{C}\boldsymbol{\Sigma}_{vv}^{-1}\boldsymbol{\Sigma}_{vu}$. Then,

$$V\{(\mathbf{v}_t, \boldsymbol{\eta}_t)'\} = \mathbf{H}'\boldsymbol{\Sigma}_{ZZ}\mathbf{H} = \text{block diag}\{\boldsymbol{\Sigma}_{vv}, \boldsymbol{\Sigma}_{\eta\eta}\}.$$

If the $\mathbf{m}_{ZZ}$ of (4.3.29) is replaced by the moment matrix of $\mathbf{Z}_t\mathbf{H}$, the transformed error vector is

$$\mathbf{Q}'\mathbf{a} = \operatorname{vech}\{\mathbf{H}'\mathbf{m}_{ZZ}\mathbf{H}\} - \operatorname{vech}\{\mathbf{H}'\boldsymbol{\Sigma}_{ZZ}\mathbf{H}\},$$

where $\mathbf{Q}' = \psi_p(\mathbf{H}' \otimes \mathbf{H}')\boldsymbol{\Phi}_p$ and the variance of $(n-1)^{1/2}\mathbf{Q}'\mathbf{a}$ is

$$2\psi_p[\text{block diag}\{\boldsymbol{\Sigma}_{vv}, \boldsymbol{\Sigma}_{\eta\eta}\}) \otimes (\text{block diag}\{\boldsymbol{\Sigma}_{vv}, \boldsymbol{\Sigma}_{\eta\eta}\})]\psi_p'.$$

If we rearrange the elements of $\mathbf{Q}'\mathbf{a}$ and write

$$\begin{aligned}\mathbf{a}_* &= [(\operatorname{vech}\mathbf{m}_{vv})', (\operatorname{vec}\mathbf{m}_{\eta v})', (\operatorname{vech}\mathbf{m}_{\eta\eta})']' \\ &\quad -[(\operatorname{vech}\boldsymbol{\Sigma}_{vv})', (\operatorname{vec}\mathbf{0}_{kr})', (\operatorname{vech}\boldsymbol{\Sigma}_{\eta\eta})']',\end{aligned} \tag{4.3.32}$$

then the variance of $(n-1)^{1/2}\mathbf{a}_*$ is

$$\text{block diag}\{\mathbf{V}_1, \mathbf{V}_2, \mathbf{V}_3\} = \boldsymbol{\Omega}_*, \tag{4.3.33}$$

where

$$\mathbf{V}_1 = 2\psi_r(\boldsymbol{\Sigma}_{vv} \otimes \boldsymbol{\Sigma}_{vv})\psi_r', \qquad \mathbf{V}_2 = \boldsymbol{\Sigma}_{vv} \otimes \boldsymbol{\Sigma}_{\eta\eta}, \qquad \mathbf{V}_3 = 2\psi_k(\boldsymbol{\Sigma}_{\eta\eta} \otimes \boldsymbol{\Sigma}_{\eta\eta})\psi_k'.$$

The transformed derivatives are

$$\mathbf{Q}'\mathbf{F}_\beta = 2\psi_p[(\mathbf{I}_r, -\boldsymbol{\Sigma}_{vv}^{-1}\boldsymbol{\Sigma}_{vu})' \otimes (\mathbf{0}_{kr}, \boldsymbol{\Sigma}_{xx})'],$$
$$\mathbf{Q}'\mathbf{F}_\varepsilon = \psi_p(\mathbf{H}' \otimes \mathbf{H}')\mathbf{L},$$
$$\mathbf{Q}'\mathbf{F}_x = \psi_p[(\mathbf{0}_{kr}, \mathbf{I}_k)' \otimes (\mathbf{0}_{kr}, \mathbf{I}_k)'].$$

Arranging the transformed observations based on the estimated transformation in the same order as $\mathbf{a}_*$, we have the transformed version of (4.3.29):

$$\tilde{\mathbf{g}}_* = \tilde{\mathbf{F}}_*(\gamma - \tilde{\gamma}) + \mathbf{a}_*, \tag{4.3.34}$$

where $\tilde{\mathbf{g}}_* = (\tilde{\mathbf{g}}'_{1*}, \tilde{\mathbf{g}}'_{2*}, \tilde{\mathbf{g}}'_{3*})'$ and

$$\tilde{\mathbf{F}}_* = \begin{bmatrix} \mathbf{0} & \psi_r(\tilde{\mathbf{C}}' \otimes \tilde{\mathbf{C}}')\mathbf{L} & \mathbf{0} \\ \mathbf{I}_r \otimes \tilde{\boldsymbol{\Sigma}}_{xx} & (\tilde{\mathbf{C}}' \otimes \tilde{\mathbf{H}}'_2)\mathbf{L} & \mathbf{0} \\ -2\psi_k(\tilde{\boldsymbol{\Sigma}}_{uv}\tilde{\boldsymbol{\Sigma}}_{vv}^{-1} \otimes \tilde{\boldsymbol{\Sigma}}_{xx}) & \psi_k(\tilde{\mathbf{H}}'_2 \otimes \tilde{\mathbf{H}}'_2)\mathbf{L} & \mathbf{I} \end{bmatrix}.$$

Because $\tilde{\boldsymbol{\Omega}}_*$ is block diagonal, because $(\gamma'_\beta, \gamma'_x)$ does not appear in the equations for $\tilde{\mathbf{g}}'_{1*}$, and because the dimension of $(\gamma'_\beta, \gamma'_x)$ is the same as that of $(\tilde{\mathbf{g}}'_{2*}, \tilde{\mathbf{g}}'_{3*})$, it follows that the generalized least squares estimator of $(\gamma - \tilde{\gamma})$ obtained at the next step is

$$\begin{bmatrix} \hat{\gamma}_\beta - \tilde{\gamma}_\beta \\ \hat{\gamma}_x - \tilde{\gamma}_x \end{bmatrix} = \begin{bmatrix} \mathbf{I}_r \otimes \tilde{\boldsymbol{\Sigma}}_{xx}^{-1} & \mathbf{0} \\ 2\psi_k(\tilde{\boldsymbol{\Sigma}}_{uv}\tilde{\boldsymbol{\Sigma}}_{vv}^{-1} \otimes \mathbf{I}_k) & \mathbf{I} \end{bmatrix} \begin{bmatrix} \tilde{\mathbf{g}}_{2*} - (\tilde{\mathbf{C}}' \otimes \tilde{\mathbf{H}}'_2)\mathbf{L}(\hat{\gamma}_\varepsilon - \tilde{\gamma}_\varepsilon) \\ \tilde{\mathbf{g}}_{3*} - \psi_k(\tilde{\mathbf{H}}'_2 \otimes \tilde{\mathbf{H}}'_2)\mathbf{L}(\hat{\gamma}_\varepsilon - \tilde{\gamma}_\varepsilon) \end{bmatrix}, \tag{4.3.35}$$

$$\hat{\gamma}_\varepsilon - \tilde{\gamma}_\varepsilon = [\mathbf{L}'(\tilde{\mathbf{C}} \otimes \tilde{\mathbf{C}})\psi'_r\tilde{\mathbf{V}}_1^{-1}\psi_r(\tilde{\mathbf{C}}' \otimes \tilde{\mathbf{C}}')\mathbf{L}]^{-1}\mathbf{L}'(\tilde{\mathbf{C}} \otimes \tilde{\mathbf{C}})\psi'_r\tilde{\mathbf{V}}_1^{-1}\tilde{\mathbf{g}}_{1*}.$$

Using $\tilde{\mathbf{g}}_{1*} = \operatorname{vech}\{\tilde{\mathbf{m}}_{vv} - \tilde{\boldsymbol{\Sigma}}_{vv}\}$, $\mathbf{L}\tilde{\gamma}_\varepsilon = \operatorname{vec} \tilde{\boldsymbol{\Sigma}}_{\varepsilon\varepsilon}$, the expression for $\tilde{\mathbf{V}}_1$ given in (4.3.33), and Result 4.A.2, we have

$$\hat{\gamma}_\varepsilon = [\mathbf{L}'\{(\tilde{\mathbf{C}}\tilde{\boldsymbol{\Sigma}}_{vv}^{-1}\tilde{\mathbf{C}}') \otimes (\tilde{\mathbf{C}}\tilde{\boldsymbol{\Sigma}}_{vv}^{-1}\tilde{\mathbf{C}}')\}\mathbf{L}]^{-1}\mathbf{L}' \operatorname{vec}(\tilde{\mathbf{C}}\tilde{\boldsymbol{\Sigma}}_{vv}\tilde{\mathbf{m}}_{vv}\tilde{\boldsymbol{\Sigma}}_{vv}^{-1}\tilde{\mathbf{C}}'). \tag{4.3.36}$$

Expression (4.3.31) follows from (4.3.36) and Theorem 4.2.2. □

The importance of Result 4.3.3 is that, at every step in an iterative procedure using the current estimate of the covariance matrix, the estimated generalized least squares estimator of γ_ε can be computed from Equation (4.3.36), ignoring the remaining portion of the large system. The estimator of γ_ε is given in (4.3.31).

Results 4.3.2 and 4.3.3 suggest an iterative method for obtaining the maximum likelihood estimators. Given an initial estimator of $\boldsymbol{\Sigma}_{\varepsilon\varepsilon}$, one uses Result 4.3.2 to obtain an estimator of $\mathbf{A}$ (or $\boldsymbol{\beta}$). Given the estimator of $\mathbf{A}$, an estimator of $\boldsymbol{\Sigma}_{\varepsilon\varepsilon}$ is computed by the method of Result 4.3.3.

One way to initiate the procedure is to use the residual mean square of the regression of each Z_i on all other Z_i's as the estimator of $\sigma_{\varepsilon\varepsilon ii}$. This procedure of constructing start values requires n to be greater than p and the estimator of $\sigma_{\varepsilon\varepsilon ii}$ will be too large on the average. An efficient method of computing this start vector is to use the reciprocals of the diagonal elements of $\mathbf{m}_{ZZ}^{-1}$.

Because of the form of $\hat{\gamma}_\varepsilon$ in Result 4.3.3, it is not necessary to compute $\hat{\mathbf{A}}$ of Result 4.3.2 at each step in the application of the suggested iterative method. Let $\tilde{\boldsymbol{\Sigma}}_{\varepsilon\varepsilon}$ be the current trial value for $\boldsymbol{\Sigma}_{\varepsilon\varepsilon}$. Let $\tilde{\lambda}_1^{-1} \leqslant \tilde{\lambda}_2^{-1} \leqslant \cdots \leqslant \tilde{\lambda}_p^{-1}$ be the p roots of

$$|\tilde{\boldsymbol{\Sigma}}_{\varepsilon\varepsilon} - \lambda^{-1}\mathbf{m}_{ZZ}| = 0, \tag{4.3.37}$$

and let $\tilde{\mathbf{R}} = (\tilde{\mathbf{R}}_{(k)}, \tilde{\mathbf{R}}_{(r)})$ be the matrix of characteristic vectors of $\tilde{\boldsymbol{\Sigma}}_{\varepsilon\varepsilon}$ in the metric $\mathbf{m}_{ZZ}$. Then by (4.1.50), (4.1.51), and (4.1.52),

$$\tilde{\mathbf{C}}\tilde{\boldsymbol{\Sigma}}_{vv}^{-1}\tilde{\mathbf{C}}' = \tilde{\mathbf{R}}_{(r)}\tilde{\boldsymbol{\Lambda}}_{(r)}\tilde{\mathbf{R}}_{(r)}', \tag{4.3.38}$$

$$\tilde{\mathbf{C}}\tilde{\boldsymbol{\Sigma}}_{vv}^{-1}\tilde{\mathbf{m}}_{vv}\tilde{\boldsymbol{\Sigma}}_{vv}^{-1}\tilde{\mathbf{C}}' = \tilde{\mathbf{R}}_{(r)}\tilde{\boldsymbol{\Lambda}}_{(r)}^2\tilde{\mathbf{R}}_{(r)}', \tag{4.3.39}$$

where $\tilde{\boldsymbol{\Lambda}}_{(r)} = \text{diag}\{\tilde{\lambda}_{k+1}, \tilde{\lambda}_{k+2}, \ldots, \tilde{\lambda}_p\}$. Thus, the next-step estimate of γ_ε defined in (4.3.31) requires only $\tilde{\boldsymbol{\Lambda}}_{(r)}$ and $\tilde{\mathbf{R}}_{(r)}$.

If $\tilde{\boldsymbol{\Sigma}}_{\varepsilon\varepsilon}$ is the maximum likelihood estimator, then

$$\text{diag}\{\tilde{\mathbf{R}}_{(r)}(\tilde{\boldsymbol{\Lambda}}_{(r)}^2 - \tilde{\boldsymbol{\Lambda}}_{(r)})\tilde{\mathbf{R}}_{(r)}'\} = \text{diag}\{\tilde{\boldsymbol{\Sigma}}_{\varepsilon\varepsilon}^{-1}(\mathbf{m}_{ZZ} - \tilde{\boldsymbol{\Sigma}}_{ZZ})\tilde{\boldsymbol{\Sigma}}_{\varepsilon\varepsilon}^{-1}\} = \mathbf{0}$$

and

$$\hat{\gamma}_\varepsilon - \tilde{\gamma}_\varepsilon = \tilde{\mathbf{B}}^{-1}\mathbf{L}' \text{vec}[\tilde{\mathbf{R}}_{(r)}(\tilde{\boldsymbol{\Lambda}}_{(r)}^2 - \tilde{\boldsymbol{\Lambda}}_{(r)})\tilde{\mathbf{R}}_{(r)}'] = \mathbf{0}.$$

Hence, the maximum likelihood estimator is a stationary point for the method.

Convergence of the iterative procedure is not guaranteed. However, the generalized sum of squares associated with (4.3.34) decreases in the direction from $\tilde{\gamma}$ to $\hat{\gamma}$ when $\tilde{\gamma}$ is not a stationary point. Therefore, by Lemma 4.2.1, it is possible to find a point between $\hat{\gamma}$ and $\tilde{\gamma}$ such that the likelihood is increased and the iterative procedure can be modified so that it will converge to a local maximum of the likelihood function.

Fuller and Pantula (1982) suggested a further modification of the method. They approximated the expressions (4.3.38) and (4.3.39) by $\tilde{\mathbf{R}}_{(r)}\tilde{\mathbf{R}}_{(r)}'$ and $\tilde{\mathbf{R}}_{(r)}\tilde{\boldsymbol{\Lambda}}_{(r)}\tilde{\mathbf{R}}_{(r)}'$, respectively. Their modification is based on the fact that if $\tilde{\boldsymbol{\Sigma}}_{\varepsilon\varepsilon}$ is the true $\boldsymbol{\Sigma}_{\varepsilon\varepsilon}$, then $\tilde{\boldsymbol{\Lambda}}_{(r)}$ estimates $\mathbf{I}_r$.

The construction of the least squares estimator of $\boldsymbol{\Sigma}_{\varepsilon\varepsilon}$ is relatively simple if some of the estimates of $\sigma_{\varepsilon\varepsilon ii}$ are zero. Should $\mathbf{B}^{-1}\mathbf{d}$ produce a negative estimate for $\sigma_{\varepsilon\varepsilon ii}$, the estimate for that element is set equal to zero. Then the associated row of the system of equations

$$\mathbf{B}\hat{\gamma}_\varepsilon = \mathbf{d}$$

and the corresponding column of $\mathbf{B}$ are removed. The reduced system of equations is solved for the remaining elements of γ_ε.

From the derivation of the estimators it is clear that the maximum likelihood estimator of $\boldsymbol{\Sigma}_{\varepsilon\varepsilon}$ is a function of

$$\hat{\mathbf{m}}_{vv} = (n-1)^{-1} \sum_{t=1}^{n} \hat{\mathbf{v}}_t'\hat{\mathbf{v}}_t,$$

where $\hat{\mathbf{v}}_t = (\mathbf{Z}_t - \bar{\mathbf{Z}})(\mathbf{I}, -\hat{\boldsymbol{\beta}}')'$. On the basis of expansions such as that of (4.1.70) it seems that

$$(n - k - 1)^{-1}(n - 1)\hat{\boldsymbol{\Sigma}}_{\varepsilon\varepsilon} \tag{4.3.40}$$

will be a less biased estimator of $\boldsymbol{\Sigma}_{\varepsilon\varepsilon}$ than the maximum likelihood estimator. If the estimator (4.3.40) is used, it is suggested that the covariance matrix of the estimators be estimated by

$$(n - k - 1)^{-2}(n - 1)^2\hat{\boldsymbol{\Gamma}}_{\varepsilon\varepsilon},$$

where $\boldsymbol{\Gamma}_{\varepsilon\varepsilon}$ is defined in Theorem 4.3.1 of the next section.

The computational algorithm we have outlined is described by Fuller and Pantula (1982). It is closely related to an algorithm given by Jöreskog (1977). Jöreskog and Lawley (1968) (see also Jöreskog (1967)) and Jennrich and Robinson (1969) have given computational algorithms for the factor model.

4.3.3. Limiting Distribution of Factor Estimators

In this section we derive an expression for the covariance matrix of the limiting distribution of the maximum likelihood estimators of the parameters of the factor model.

Theorem 4.3.1. Let

$$\mathbf{Z}_t = (\mathbf{Y}_t, \mathbf{X}_t) = (\boldsymbol{\beta}_0, \mathbf{0}) + \mathbf{x}_t(\boldsymbol{\beta}, \mathbf{I}) + \boldsymbol{\varepsilon}_t,$$
$$(\boldsymbol{\varepsilon}_t, \mathbf{x}_t)' \sim \mathrm{NI}[(\mathbf{0}, \boldsymbol{\mu}_x)', \text{block diag}(\boldsymbol{\Sigma}_{\varepsilon\varepsilon}, \boldsymbol{\Sigma}_{xx})],$$

where $\boldsymbol{\Sigma}_{\varepsilon\varepsilon}$ is a positive definite diagonal matrix, $\boldsymbol{\Sigma}_{xx}$ is positive definite, and the model is identified. Let

$$\hat{\gamma}' = (\hat{\gamma}'_\beta, \hat{\gamma}'_\varepsilon, \hat{\gamma}'_x) = [(\text{vec}\ \hat{\boldsymbol{\beta}})', (\hat{\sigma}_{\varepsilon\varepsilon 11}, \ldots, \hat{\sigma}_{\varepsilon\varepsilon pp}), (\text{vech}\ \hat{\boldsymbol{\Sigma}}_{xx})']$$

be the maximum likelihood estimator adjusted for degrees of freedom and let γ' denote the corresponding vector of parameters. Let $\hat{\boldsymbol{\beta}}_0 = \bar{\mathbf{Y}} - \bar{\mathbf{X}}\hat{\boldsymbol{\beta}}$. Assume that A, the parameter space for γ, is a convex, compact set. Assume that the true parameter value of γ is an interior point of A, and that $\boldsymbol{\Sigma}_{ZZ}$ evaluated at any value of γ in A other than the true value differs from $\boldsymbol{\Sigma}_{ZZ}$ evaluated at the true value. Then

$$n^{1/2}[(\hat{\boldsymbol{\beta}}_0, \hat{\gamma}')' - (\boldsymbol{\beta}_0, \gamma')'] \xrightarrow{L} N(\mathbf{0}, \boldsymbol{\Gamma}),$$

where

$$\boldsymbol{\Gamma} = \begin{bmatrix} \boldsymbol{\Gamma}_{00} & \boldsymbol{\Gamma}_{0\beta} & \boldsymbol{\Gamma}_{0\varepsilon} & \boldsymbol{\Gamma}_{0x} \\ \boldsymbol{\Gamma}_{\beta 0} & \boldsymbol{\Gamma}_{\beta\beta} & \boldsymbol{\Gamma}_{\beta\varepsilon} & \boldsymbol{\Gamma}_{\beta x} \\ \boldsymbol{\Gamma}_{\varepsilon 0} & \boldsymbol{\Gamma}_{\varepsilon\beta} & \boldsymbol{\Gamma}_{\varepsilon\varepsilon} & \boldsymbol{\Gamma}_{\varepsilon x} \\ \boldsymbol{\Gamma}_{x0} & \boldsymbol{\Gamma}_{x\beta} & \boldsymbol{\Gamma}_{x\varepsilon} & \boldsymbol{\Gamma}_{xx} \end{bmatrix},$$

$$\Gamma_{0*} = (\Sigma_{vv}, \mathbf{0}) - (\mathbf{I}_r \otimes \boldsymbol{\mu}_x)[-\Gamma_{\beta\beta}(\mathbf{I}_r \otimes \boldsymbol{\mu}'_x), \Gamma_{\beta\beta}, \Gamma_{\beta\varepsilon}, \Gamma_{\beta x}],$$
$$\Gamma_{\beta\beta} = \Sigma_{vv} \otimes (\Sigma_{xx}^{-1}\Sigma_{\eta\eta}\Sigma_{xx}^{-1}) + \mathbf{P}_1\Gamma_{\varepsilon\varepsilon}\mathbf{P}'_1,$$
$$\Gamma_{\beta x} = 2(\Sigma_{vu} \otimes \Sigma_{xx}^{-1}\Sigma_{\eta\eta})\psi'_k + \mathbf{P}_1\Gamma_{\varepsilon\varepsilon}\mathbf{P}'_2,$$
$$\Gamma_{xx} = 2\psi_k(\Sigma_{XX} \otimes \Sigma_{XX})\psi'_k - 2\psi_k(\Sigma_{\xi\xi} \otimes \Sigma_{\xi\xi})\psi'_k + \mathbf{P}_2\Gamma_{\varepsilon\varepsilon}\mathbf{P}'_2,$$
$$\Gamma_{\varepsilon\varepsilon} = 2[\mathbf{L}'\{(\mathbf{C}\Sigma_{vv}^{-1}\mathbf{C}') \otimes (\mathbf{C}\Sigma_{vv}^{-1}\mathbf{C}')\}\mathbf{L}]^{-1},$$
$$\Gamma_{\beta\varepsilon} = -\mathbf{P}_1\Gamma_{\varepsilon\varepsilon}, \qquad \Gamma_{x\varepsilon} = -\mathbf{P}_2\Gamma_{\varepsilon\varepsilon},$$
$$\mathbf{P}_1 = \{\mathbf{C}' \otimes (\Sigma_{xx}^{-1}\mathbf{H}'_2)\}\mathbf{L}, \qquad \mathbf{P}_2 = \psi_k\{(2\Sigma_{uv}\Sigma_{vv}^{-1}\mathbf{C}' + \mathbf{H}'_2) \otimes \mathbf{H}'_2\}\mathbf{L},$$
$$\mathbf{H}_2 = (\mathbf{0}_{kr}, \mathbf{I}_k)' - \mathbf{C}\Sigma_{vv}^{-1}\Sigma_{vu}, \qquad \Sigma_{\eta\eta} = \Sigma_{XX} - \Sigma_{uv}\Sigma_{vv}^{-1}\Sigma_{vu},$$

$\Sigma_{\xi\xi} = \Sigma_{uv}\Sigma_{vv}^{-1}\Sigma_{vu}$, $\Gamma_{0*} = (\Gamma_{00}, \Gamma_{0\beta}, \Gamma_{0\varepsilon}, \Gamma_{0x})$, $\mathbf{I}_k$ is the k-dimensional identity matrix, $\mathbf{C}' = (\mathbf{I}_r, \boldsymbol{\beta}')$, $\mathbf{0}_{kr}$ is a $k \times r$ matrix of zeros, and $\mathbf{L}$ is the $p^2 \times p$ matrix defined in Result 4.A.3.2 of Appendix 4.A.

Proof. By Corollary 4.B.2,

$$n^{1/2}(\hat{\gamma} - \gamma) \xrightarrow{L} N[\mathbf{0}, (\mathbf{F}'\boldsymbol{\Omega}^{-1}\mathbf{F})^{-1}],$$

where $\boldsymbol{\Omega} = 2\psi_p(\Sigma_{ZZ} \otimes \Sigma_{ZZ})\psi'_p$ is the covariance matrix of $(n-1)^{1/2}$ vech $\mathbf{m}_{ZZ}$ and $\mathbf{F}$ is the matrix of partial derivatives

$$\mathbf{F} = \frac{\partial \text{ vech } \Sigma_{ZZ}(\gamma)}{\partial \gamma'}.$$

Therefore, it suffices to obtain expressions for the elements of $(\mathbf{F}'\boldsymbol{\Omega}^{-1}\mathbf{F})^{-1}$. The matrix $(\mathbf{F}'\boldsymbol{\Omega}^{-1}\mathbf{F})^{-1}$ can be considered to be the covariance matrix of the generalized least squares estimator $\hat{\gamma}$ of γ computed from the system

$$\mathbf{g} = \mathbf{F}\gamma + \mathbf{a}, \tag{4.3.41}$$

where $\mathbf{g}$ is a $2^{-1}p(p+1)$-dimensional vector of observations,

$$\mathbf{a} = \text{vech}[(\mathbf{Z}_t - \boldsymbol{\mu}_Z)(\mathbf{Z}_t - \boldsymbol{\mu}_Z)' - \Sigma_{ZZ}],$$

and, by the properties of the normal distribution, $E\{\mathbf{aa}'\} = \boldsymbol{\Omega}$. In the proof of Result 4.3.3 we gave expressions for the matrix $\mathbf{F}$ and demonstrated how the transformation

$$\mathbf{Q} = \psi_p(\mathbf{H} \otimes \mathbf{H})\boldsymbol{\Phi}_p$$

transformed the system (4.3.41) into a system

$$\mathbf{g}_* = \mathbf{F}_*\gamma + \mathbf{a}_* \tag{4.3.42}$$

with a block diagonal error covariance matrix. Therefore, we evaluate the elements of

$$(\mathbf{F}'_*\boldsymbol{\Omega}_*^{-1}\mathbf{F}_*)^{-1},$$

where $\mathbf{F}_*$ is defined following (4.3.34) and $\boldsymbol{\Omega}_*$ is defined in (4.3.33). Using (4.3.35) we have

$$\begin{aligned}\hat{\gamma}_\beta - \gamma_\beta &= (\mathbf{I}_r \otimes \boldsymbol{\Sigma}_{xx}^{-1}) \operatorname{vech} \mathbf{m}_{\eta v} - \mathbf{P}_1(\hat{\gamma}_\varepsilon - \gamma_\varepsilon) + o_p(n^{-1/2}),\\ \hat{\gamma}_x - \gamma_x &= 2\psi_k(\boldsymbol{\Sigma}_{uv}\boldsymbol{\Sigma}_{vv}^{-1} \otimes \mathbf{I}_k) \operatorname{vech} \mathbf{m}_{\eta v} + \operatorname{vech}(\mathbf{m}_{\eta\eta} - \boldsymbol{\Sigma}_{\eta\eta})\\ &\quad - \mathbf{P}_2(\hat{\gamma}_\varepsilon - \gamma_\varepsilon) + o_p(n^{-1/2}),\end{aligned}$$

where

$$\begin{aligned}\{2\psi_k(\boldsymbol{\Sigma}_{uv}\boldsymbol{\Sigma}_{vv}^{-1} \otimes \mathbf{I}_k)(\mathbf{C}' \otimes \mathbf{H}_2') + \psi_k(\mathbf{H}_2' \otimes \mathbf{H}_2')\}\mathbf{L}&\\ = \psi_k[(2\boldsymbol{\Sigma}_{uv}\boldsymbol{\Sigma}_{vv}^{-1}\mathbf{C}' + \mathbf{H}_2') \otimes \mathbf{H}_2']\mathbf{L} &= \mathbf{P}_2.\end{aligned}$$

Hence, using (4.3.33), (4.3.42), and the fact that vech $\mathbf{m}_{vv}$ is uncorrelated with $(\mathbf{g}_{2*}, \mathbf{g}_{3*})$,

$$\begin{aligned}\boldsymbol{\Gamma}_{\beta\beta} &= (\mathbf{I}_r \otimes \boldsymbol{\Sigma}_{xx}^{-1})\mathbf{V}_2(\mathbf{I}_r \otimes \boldsymbol{\Sigma}_{xx}^{-1}) + \mathbf{P}_1\boldsymbol{\Gamma}_{\varepsilon\varepsilon}\mathbf{P}_1',\\ \boldsymbol{\Gamma}_{\beta x} &= 2(\mathbf{I}_r \otimes \boldsymbol{\Sigma}_{xx}^{-1})\mathbf{V}_2(\boldsymbol{\Sigma}_{vv}^{-1}\boldsymbol{\Sigma}_{vu} \otimes \mathbf{I}_k)\psi_k' + \mathbf{P}_1\boldsymbol{\Gamma}_{\varepsilon\varepsilon}\mathbf{P}_2',\\ \boldsymbol{\Gamma}_{xx} &= 4\psi_k(\boldsymbol{\Sigma}_{uv}\boldsymbol{\Sigma}_{vv}^{-1} \otimes \mathbf{I}_k)\mathbf{V}_2(\boldsymbol{\Sigma}_{vv}^{-1}\boldsymbol{\Sigma}_{vu} \otimes \mathbf{I}_k)\psi_k' + \mathbf{V}_3 + \mathbf{P}_2\boldsymbol{\Gamma}_{\varepsilon\varepsilon}\mathbf{P}_2'\\ &= 4\psi_k(\boldsymbol{\Sigma}_{\xi\xi} \otimes \boldsymbol{\Sigma}_{\eta\eta})\psi_k' + 2\psi_k(\boldsymbol{\Sigma}_{\eta\eta} \otimes \boldsymbol{\Sigma}_{\eta\eta})\psi_k' + \mathbf{P}_2\boldsymbol{\Gamma}_{\varepsilon\varepsilon}\mathbf{P}_2',\\ \boldsymbol{\Gamma}_{\varepsilon\varepsilon} &= 2[\mathbf{L}'(\mathbf{C} \otimes \mathbf{C})\psi_r'\boldsymbol{\Phi}_r'(\boldsymbol{\Sigma}_{vv}^{-1} \otimes \boldsymbol{\Sigma}_{vv}^{-1})\boldsymbol{\Phi}_r\psi_r(\mathbf{C}' \otimes \mathbf{C}')\mathbf{L}]^{-1}\\ &= 2[\mathbf{L}'\{(\mathbf{C}\boldsymbol{\Sigma}_{vv}^{-1}\mathbf{C}') \otimes (\mathbf{C}\boldsymbol{\Sigma}_{vv}^{-1}\mathbf{C}')\}\mathbf{L}]^{-1},\end{aligned}$$

where $\mathbf{V}_2$ and $\mathbf{V}_3$ are defined in (4.3.33) and we have used

$$\boldsymbol{\Phi}_r\psi_r(\mathbf{C}' \otimes \mathbf{C}')\mathbf{L} = (\mathbf{C}' \otimes \mathbf{C}')\mathbf{L}.$$

Algebraic substitution gives the expressions of the theorem statement. The covariance matrices for $\hat{\boldsymbol{\beta}}_0$ follow from the definition of $\hat{\boldsymbol{\beta}}_0$. □

The expressions of Theorem 4.3.1 are rather complicated, but all expressions are straightforward matrix functions of the parameters. Because the parameters can be estimated consistently, estimators of the covariances can be constructed by substituting the corresponding estimators for the unknown parameters.

The variance expressions for $\hat{\boldsymbol{\beta}}$ and $\hat{\boldsymbol{\Sigma}}_{xx}$ of Theorem 4.3.1 warrant comparison with the variance expressions of estimators of the same parameters given in Theorem 4.1.1. The expressions of Theorem 4.3.1 each contain an additional term that is introduced because the sample is used to estimate $\boldsymbol{\Sigma}_{\varepsilon\varepsilon}$. In both cases it is possible to express this term as a function of the covariance matrix of the estimators of $\boldsymbol{\Sigma}_{\varepsilon\varepsilon}$.

The expressions for the covariance matrix of $\hat{\gamma}_\beta$ and $\hat{\gamma}_\varepsilon$ are appropriate for a wide range of specifications on the $\mathbf{x}_t$. In particular, the expressions remain valid for fixed $\mathbf{x}_t$ and for nonnormal, but random $\mathbf{x}_t$. The normality of the ε_t is required for all variance expressions, but modest deviations from normality should have modest effects on the covariance matrix of $\hat{\gamma}_\beta$. Deviations from normality will have larger effects on the covariance matrices of the estimators of Σ_{xx} and $\Sigma_{\varepsilon\varepsilon}$. It is clear that the estimators will converge in distribution to normal random variables for any distribution for which $n^{1/2}\,\text{vech}(\mathbf{m}_{ZZ} - \Sigma_{ZZ})$ converges to a normal vector.

The likelihood ratio test of the adequacy of the factor model relative to the unconstrained model has the same form as that obtained in Theorem 4.1.3 of Section 4.1. If we let the null model be the identified normal factor model (4.3.9) and let the alternative model be the unconstrained normal distribution model, the negative of twice the logarithm of the likelihood ratio computed from the likelihood adjusted for degrees of freedom is

$$\begin{aligned} 2[\log L_c(\mathbf{m}_{ZZ}) - \log L_c(\hat{\Sigma}_{ZZ})] &= (n-1)\sum_{j=k+1}^{p} (\hat{\lambda}_j - \log \hat{\lambda}_j - 1) \\ &= (n-1)\sum_{j=k+1}^{p} \log \hat{\lambda}_j^{-1}, \end{aligned} \tag{4.3.43}$$

where $\hat{\lambda}_1^{-1} \leqslant \hat{\lambda}_2^{-1} \leqslant \cdots \leqslant \hat{\lambda}_p^{-1}$ are the roots of

$$|\hat{\Sigma}_{\varepsilon\varepsilon} - \lambda^{-1}\mathbf{m}_{ZZ}| = 0.$$

The limiting distribution of the statistic is that of a chi-square random variable with $\frac{1}{2}r(r+1) - p$ degrees of freedom. See Exercises 4.21 and 4.22.

Bartlett (1950) suggested that the distribution of the statistic

$$\hat{\chi}^2 = [n - k - 1 - 6^{-1}(2r+5)] \sum_{j=k+1}^{p} \log \hat{\lambda}_j^{-1} \tag{4.3.44}$$

is better approximated by that of a chi-square random variable than is the distribution of the unmodified likelihood ratio statistic. The test statistic is generally computed with the multiplier of (4.3.44).

We have given the covariance matrices of the estimators for the parameterization in which the factors are a subset of the z variables. In some situations one may be interested in redefining the factors in terms of a linear function of the vector of true values $\mathbf{z}_t$. We consider the class of linear transformations

$$\mathbf{f}_t = \mathbf{z}_t \mathbf{K}, \tag{4.3.45}$$

where $\mathbf{K}$ is a fixed $p \times k$ matrix of rank k. Using the original x parameterization, we can write

$$\begin{aligned} \mathbf{f}_t &= [(\boldsymbol{\beta}_0, \mathbf{0}) + \mathbf{x}_t(\boldsymbol{\beta}, \mathbf{I}_k)]\mathbf{K} \\ &= (\boldsymbol{\beta}_0, \mathbf{0})\mathbf{K} + \mathbf{x}_t \mathbf{G}, \end{aligned}$$

where $\mathbf{G} = (\boldsymbol{\beta}, \mathbf{I}_k)\mathbf{K}$. The transformed system becomes

$$\begin{aligned} \mathbf{z}_t &= \boldsymbol{\mu}_z + (\mathbf{f}_t - \boldsymbol{\mu}_f)\mathbf{A}' \\ &= \mathbf{A}_0' + \mathbf{f}_t\mathbf{A}', \end{aligned} \tag{4.3.46}$$

where $\mathbf{A}' = \mathbf{G}^{-1}(\boldsymbol{\beta}, \mathbf{I}_k)$ and $\mathbf{A}_0' = \boldsymbol{\mu}_z[\mathbf{I}_p - \mathbf{K}\mathbf{A}']$. The estimators are

$$\begin{aligned} \hat{\mathbf{A}}' &= [(\hat{\boldsymbol{\beta}}, \mathbf{I}_k)\mathbf{K}]^{-1}(\hat{\boldsymbol{\beta}}, \mathbf{I}_k), \\ \hat{\mathbf{A}}_0' &= \bar{\mathbf{Z}}[\mathbf{I}_p - \mathbf{K}\hat{\mathbf{A}}']. \end{aligned} \tag{4.3.47}$$

It follows that, to the first order of approximation,

$$\begin{aligned} \operatorname{vec}(\hat{\mathbf{A}}' - \mathbf{A}') &\doteq [(\mathbf{I}_p - \mathbf{A}\mathbf{K}') \otimes \mathbf{G}^{-1}] \operatorname{vec}(\hat{\boldsymbol{\beta}} - \boldsymbol{\beta}, \mathbf{0}_{kk}), \\ \hat{\mathbf{A}}_0' - \mathbf{A}_0' &\doteq (\bar{\mathbf{Z}} - \boldsymbol{\mu}_z)(\mathbf{I}_p - \mathbf{K}\mathbf{A}') - \boldsymbol{\mu}_z\mathbf{K}(\hat{\mathbf{A}}' - \mathbf{A}'). \end{aligned} \tag{4.3.48}$$

Because the estimated covariance matrix for the approximate distribution of $(\bar{\mathbf{Z}}, [\operatorname{vec}(\hat{\boldsymbol{\beta}} - \boldsymbol{\beta})]')$ is available, one can compute the estimated covariance matrix for the approximate distribution of $[\hat{\mathbf{A}}_0', (\operatorname{vec} \hat{\mathbf{A}}')']$ using expressions (4.3.48).

The estimator of the vector of true values, treating the true values as fixed, is given by expression (4.1.9), which can be written

$$\hat{\mathbf{z}}_t = \mathbf{Z}_t - \hat{\mathbf{v}}_t\hat{\boldsymbol{\Sigma}}_{vv}^{-1}\hat{\boldsymbol{\Sigma}}_{v\varepsilon}, \tag{4.3.49}$$

where $\hat{\boldsymbol{\Sigma}}_{vv} = \hat{\mathbf{C}}'\hat{\boldsymbol{\Sigma}}_{\varepsilon\varepsilon}\hat{\mathbf{C}}$ and $\hat{\boldsymbol{\Sigma}}_{v\varepsilon} = \hat{\mathbf{C}}'\hat{\boldsymbol{\Sigma}}_{\varepsilon\varepsilon}$. As in previous sections, the estimator of the first-order approximation to the variance is

$$\hat{V}\{\hat{\mathbf{z}}_t - \mathbf{z}_t\} = \hat{\boldsymbol{\Sigma}}_{\varepsilon\varepsilon} - \hat{\boldsymbol{\Sigma}}_{\varepsilon v}\hat{\boldsymbol{\Sigma}}_{vv}^{-1}\hat{\boldsymbol{\Sigma}}_{v\varepsilon}. \tag{4.3.50}$$

The predictor of the vector of true values constructed under the normal random model is

$$\tilde{\mathbf{z}}_t = \bar{\mathbf{Z}} + (\hat{\mathbf{x}} - \bar{\mathbf{x}})\hat{\boldsymbol{\Sigma}}_{\eta\eta}^{-1}\hat{\boldsymbol{\Sigma}}_{xz}, \tag{4.3.51}$$

where $\hat{\boldsymbol{\Sigma}}_{\eta\eta} = \hat{\boldsymbol{\Sigma}}_{xx} + \hat{\boldsymbol{\Sigma}}_{uu} - \hat{\boldsymbol{\Sigma}}_{uv}\hat{\boldsymbol{\Sigma}}_{vv}^{-1}\hat{\boldsymbol{\Sigma}}_{vu}$, $\hat{\boldsymbol{\Sigma}}_{xz} = (\hat{\boldsymbol{\Sigma}}_{xy}, \hat{\boldsymbol{\Sigma}}_{xx})$, and $\hat{\boldsymbol{\Sigma}}_{zz}$ is defined in Result 4.3.1. The estimator of the first-order approximation to the variance of the error in the random predictor is

$$\tilde{V}\{\tilde{\mathbf{z}}_t - \mathbf{z}_t\} = \hat{\boldsymbol{\Sigma}}_{zz} - \hat{\boldsymbol{\Sigma}}_{zx}\hat{\boldsymbol{\Sigma}}_{\eta\eta}^{-1}\hat{\boldsymbol{\Sigma}}_{xz}. \tag{4.3.52}$$

In factor analysis, the estimated true values of the factors of model (4.3.2) are usually called *factor scores*. The estimated values of the factors computed under the fixed factor assumption are often called Bartlett's factor scores and the predicted values computed under the random factor assumption are often called Thomsons' factor scores.

Example 4.3.1. In this example we analyze the Bekk smoothness tests for 14 paper materials conducted by six laboratories. The data for six labora-

TABLE 4.3.1. Transformed Bekk smoothness for 14 materials as determined by six laboratories

	Laboratory					
Material	1	2	3	4	5	6
1	8.04	6.03	6.05	6.15	6.46	5.99
2	8.29	7.69	6.75	7.20	7.47	6.76
3	10.84	10.69	9.65	9.83	11.06	9.97
4	11.59	11.27	9.95	10.66	11.26	10.68
5	11.60	11.04	10.18	11.54	11.66	10.51
6	12.69	12.08	11.17	11.87	12.32	11.68
7	16.23	16.09	14.22	15.62	16.34	15.51
8	16.60	16.20	14.70	15.74	16.77	15.71
9	19.38	19.60	17.62	19.13	19.64	18.97
10	20.43	20.02	17.39	19.99	20.61	19.63
11	21.88	22.23	19.18	21.79	22.37	21.20
12	21.57	23.16	19.83	22.07	23.04	21.79
13	22.17	23.16	20.08	22.21	23.30	22.34
14	22.82	23.01	20.10	22.61	23.38	22.70
Mean	16.01	15.88	14.06	15.46	16.12	15.25

Source: Mandel and Lashof (1959).

tories are a subset of data for 14 laboratories analyzed by Mandel and Lashof (1959). The smoothness data are given in Table 4.3.1 with the materials arranged so that the magnitudes tend to increase as one reads down the table. We assume that the data satisfy the model

$$Z_{tj} = \beta_{0j} + \beta_{1j}x_t + \varepsilon_{tj}, \qquad t = 1, 2, \ldots, 14, \qquad j = 1, 2, \ldots, 6,$$
$$\boldsymbol{\varepsilon}_t' = (\varepsilon_{t1}, \varepsilon_{t2}, \ldots, \varepsilon_{t6})' \sim \mathrm{NI}(\mathbf{0}, \boldsymbol{\Sigma}_{\varepsilon\varepsilon}), \tag{4.3.53}$$

where $\boldsymbol{\Sigma}_{\varepsilon\varepsilon} = \mathrm{diag}(\sigma_{\varepsilon\varepsilon 11}, \sigma_{\varepsilon\varepsilon 22}, \ldots, \sigma_{\varepsilon\varepsilon 66})$ and $(\beta_{06}, \beta_{16}) = (0, 1)$. Given that the test materials were specifically chosen for the laboratory experiment, the x_t values are considered fixed.

The sample moment matrix $\mathbf{m}_{ZZ}$ is given in Table 4.3.2. At the time this was written, most major statistical packages provided algorithms for computing factor estimates, but there were few programs available for the computation of the estimated covariance matrix of the approximate distribution of the estimates. Two programs that provided estimated covariance matrices of the estimates were LISREL VI by Jöreskog and Sörbom and ISU Factor by Pantula. Our estimates are computed using Pantula's program. The

TABLE 4.3.2. Sample moment matrix for Bekk smoothness of 14 materials tested at six laboratories

Laboratory					
1	2	3	4	5	6
29.295	32.702	27.340	31.867	32.667	32.117
32.702	36.803	30.695	35.714	36.671	36.002
27.340	30.695	25.675	29.832	30.625	30.081
31.867	35.714	29.832	34.803	35.652	35.006
32.667	36.671	30.625	35.652	36.609	35.929
32.117	36.002	30.081	35.006	35.929	35.312

estimated equations are

$$\hat{z}_{t1} = \underset{(0.278)}{2.144} + \underset{(0.017)}{0.909}x_t, \qquad \hat{z}_{t2} = \underset{(0.236)}{0.321} + \underset{(0.015)}{1.020}x_t,$$

$$\hat{z}_{t3} = \underset{(0.184)}{1.068} + \underset{(0.011)}{0.852}x_t, \qquad \hat{z}_{t4} = \underset{(0.238)}{0.333} + \underset{(0.014)}{0.992}x_t,$$

$$\hat{z}_{t5} = \underset{(0.169)}{0.597} + \underset{(0.010)}{1.018}x_t,$$

where the numbers in parentheses are the estimated standard errors. The estimate of $\Sigma_{\varepsilon\varepsilon}$ is

$$\hat{\Sigma}_{\varepsilon\varepsilon} = \text{diag}\,\{0.107, 0.068, 0.039, 0.071, 0.024, 0.021\},$$
$$(0.046, 0.031, 0.018, 0.032, 0.014, 0.013)$$

where the standard errors of the estimates are given below the estimates. In computing the estimated variances, estimators are substituted for parameters in the expressions of Theorem 4.3.1. The divisor $n - k - 1$ is used in place of $n - 1$ when computing the estimated covariance matrices. The standard errors of the estimated error variances are large relative to the estimates because there are only 14 observations. Recall that the standard error of a sample variance based on 12 degrees of freedom is about 0.41 of the true variance, a ratio not greatly different from the ratios for the elements of $\hat{\Sigma}_{\varepsilon\varepsilon}$.

Because the number of observations is quite small, we expect the estimator (4.3.40), $(12)^{-1}(13)\hat{\Sigma}_{\varepsilon\varepsilon}$, to be a less biased estimator of $\Sigma_{\varepsilon\varepsilon}$ than $\hat{\Sigma}_{\varepsilon\varepsilon}$. Thus, our preferred estimator of $\Sigma_{\varepsilon\varepsilon}$ for this problem is

$$\tilde{\Sigma}_{\varepsilon\varepsilon} = \text{diag}\{0.116, 0.073, 0.042, 0.076, 0.026, 0.023\}.$$
$$(0.050, 0.034, 0.020, 0.035, 0.015, 0.014)$$

TABLE 4.3.3. Deviations between observed and estimated values

	Deviation						
Material	$Z_1 - \hat{z}_1$	$Z_2 - \hat{z}_2$	$Z_3 - \hat{z}_3$	$Z_4 - \hat{z}_4$	$Z_5 - \hat{z}_5$	$Z_6 - \hat{z}_6$	$\hat{z}_6$
1	0.55	−0.29	−0.03	−0.01	−0.12	0.12	5.87
2	−0.05	0.42	−0.12	0.11	−0.06	−0.05	6.81
3	−0.43	0.13	0.03	−0.46	0.25	−0.06	10.03
4	−0.15	0.19	−0.11	−0.14	−0.08	0.13	10.55
5	−0.28	−0.21	−0.01	0.58	0.16	−0.20	10.71
6	−0.03	−0.10	0.19	0.00	−0.12	0.05	11.63
7	0.02	−0.01	−0.03	−0.06	−0.01	0.04	15.47
8	0.11	−0.22	0.19	−0.24	0.11	−0.07	15.78
9	0.01	−0.04	0.41	0.01	−0.24	0.03	18.94
10	0.47	−0.28	−0.37	0.23	0.07	0.04	19.59
11	0.31	0.12	−0.09	−0.27	0.03	−0.16	21.36
12	−0.53	0.45	0.06	−0.03	0.10	−0.16	21.94
13	−0.24	0.10	0.02	−0.23	0.01	0.06	22.28
14	0.23	−0.25	−0.13	−0.03	−0.11	0.21	22.49

The largest root of the determinantal equation associated with (4.3.43) is $\hat{\lambda}_1 = 5149$ and the five smallest $\hat{\lambda}_i$ are

$$(1.9552, 1.3069, 0.8251, 0.6561, 0.2567).$$

The test statistic given in (4.3.44) is $\hat{\chi}^2 = 9.84$. Under the model the distribution of the test statistic is approximately that of a chi-square random variable with nine degrees of freedom. Hence, the one-factor model is easily accepted.

The deviations $Z_{ti} - \hat{z}_{ti}$, for $i = 2, \ldots, 6$, and $\hat{x}_t = \hat{z}_{t6}$ are given in Table 4.3.3. Because there is only one factor, the $\hat{z}_{ti}$, for the first five variables, are linear functions of $\hat{z}_{t6}$. By the nature of the estimators

$$\sum_{t=1}^{n} \hat{v}_{t\ell} \hat{x}_t = 0$$

for $\ell = 1, 2, \ldots, 5$. Because $Z_{ti} - \hat{z}_{ti}$ is a linear function of the $\hat{v}_{t\ell}$ and because $\hat{z}_{ti}$ is a linear function of $\hat{x}_t$,

$$\sum_{t=1}^{n} (Z_{ti} - \hat{z}_{ti})\hat{z}_{tj} = 0$$

for all i and j. Therefore, the plot of $Z_{ti} - \hat{z}_{ti}$ against $\hat{z}_{ti}$ is one form of a residual plot for the factor model. In Figure 4.3.1 $Z_{t1} - \hat{z}_{t1}$ is plotted against

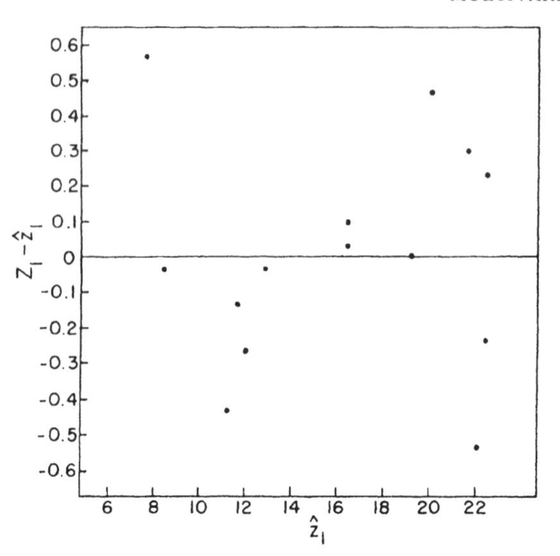

FIGURE 4.3.1. Deviation $Z_{t1} - \hat{z}_{t1}$ plotted against $\hat{z}_{t1}$.

$\hat{z}_{t1}$. There are only 14 observations in this example but plots such as Figure 4.3.1 display no anomalies.

In this example the selection of Z_6 as the x variable was an arbitrary choice. In such situations one may wish to treat the x variables in a symmetric manner and express each of the six variables in terms of the average of the six true values. To this end we define

$$f_t = 6^{-1} \sum_{i=1}^{6} z_{ti}.$$

It follows that

$$z_{ti} = \alpha_{0i} + \alpha_{1i} f_t,$$

where $\hat{\alpha}_{1i} = 6(\Sigma_{j=1}^{6} \hat{\beta}_{1j})^{-1} \hat{\beta}_{1i}$, $\hat{\alpha}_{0i} = \bar{Z}_i - \hat{\alpha}_{1i} \bar{Z}$, and $\bar{Z}$ is the average of the $\bar{Z}_i$. In terms of this factor definition, the estimated equations are

$$\hat{z}_{t1} = \underset{(0.235)}{1.442} + \underset{(0.0144)}{0.9421} f_t, \qquad \hat{z}_{t2} = \underset{(0.191)}{-0.464} + \underset{(0.0117)}{1.0568} f_t,$$

$$\hat{z}_{t3} = \underset{(0.153)}{0.411} + \underset{(0.0094)}{0.8829} f_t, \qquad \hat{z}_{t4} = \underset{(0.195)}{-0.431} + \underset{(0.0119)}{1.0276} f_t,$$

$$\hat{z}_{t5} = \underset{(0.132)}{-0.187} + \underset{(0.0081)}{1.0547} f_t, \qquad \hat{z}_{t6} = \underset{(0.127)}{-0.771} + \underset{(0.0078)}{1.0359} f_t.$$

The standard errors were computed using the covariance matrix for $(\hat{\boldsymbol{\beta}}_0, \hat{\boldsymbol{\beta}}_1)$ and the Taylor expansion of the estimators given in (4.3.48). The entire covariance matrix is singular because the sum of the $\hat{\alpha}_{0i}$ is zero and the average of the $\hat{\alpha}_{1i}$ is one. □ □

Example 4.3.2. To illustrate the computations associated with maximum likelihood factor analysis with more than one factor, we use a part of the language data of Example 2.4.1. Assume that observations on the first six items of that example satisfy the model in two factors. We rearrange the data letting

$$\mathbf{Z}_t = (Y_{t1}, Y_{t2}, Y_{t3}, Y_{t4}, X_{t1}, X_{t2}),$$

where Y_{t1} = developed, Y_{t2} = understand, Y_{t3} = appropriate, Y_{t4} = acceptable, X_{t1} = logical, and X_{t2} = irritating. The mean and covariance matrix for 100 observations are given in Table 2.4.1. The model may be written in the identified form

$$\mathbf{Z}_t = (\boldsymbol{\beta}_0, \mathbf{0}) + \mathbf{x}_t(\boldsymbol{\beta}, \mathbf{I}) + \boldsymbol{\varepsilon}_t,$$

or as

$$\mathbf{Z}_t = \boldsymbol{\mu}_z + \mathbf{f}_t\mathbf{A}' + \boldsymbol{\varepsilon}_t,$$

where $\mathbf{Z}_t = (\mathbf{Y}_t, \mathbf{X}_t)$, $\boldsymbol{\varepsilon}_t = (\mathbf{e}_t, \mathbf{u}_t)$, $E\{\mathbf{f}_t\} = \mathbf{0}$, $\boldsymbol{\Sigma}_{ff} = \mathbf{I}$, and $\mathbf{A}'\boldsymbol{\Sigma}_{\varepsilon\varepsilon}^{-1}\mathbf{A}$ is a diagonal matrix. Because the observations on the Z_{ti} are limited to the nine integers from 2 to 10, it is clear that the data are not normally distributed. Nonetheless, we first construct all statistics as if we were dealing with normal data. The maximum likelihood estimates of the equations obtained by the iterative computational procedure are

$$\begin{aligned}
\hat{Y}_{t1} &= \underset{(0.745)}{-1.139} + \underset{(0.171)}{0.779}x_{t1} + \underset{(0.178)}{0.275}x_{t2}, \\
\hat{Y}_{t2} &= \underset{(0.814)}{-0.166} + \underset{(0.141)}{0.537}x_{t1} + \underset{(0.165)}{0.490}x_{t2}, \\
\hat{Y}_{t3} &= \underset{(1.087)}{-2.189} - \underset{(0.140)}{0.085}x_{t1} + \underset{(0.198)}{1.102}x_{t2}, \\
\hat{Y}_{t4} &= \underset{(1.262)}{-3.122} - \underset{(0.163)}{0.080}x_{t1} + \underset{(0.240)}{1.367}x_{t2},
\end{aligned}$$

where the numbers in parentheses are estimated standard errors. The estimated error covariance matrix is

$$\hat{\boldsymbol{\Sigma}}_{\varepsilon\varepsilon} = \text{diag}\{0.963, 1.034, 1.022, 0.864, 0.073, 1.893\}$$
$$(0.203, 0.170, 0.203, 0.246, 0.250, 0.174)$$

and the vector half of the estimated covariance matrix of $\mathbf{x}$ is

$$\operatorname{vech} \hat{\boldsymbol{\Sigma}}_{xx} = \underset{(0.427 \quad 0.266 \quad 0.327)}{[2.336, 1.151; 1.412]'}.$$

If the method of normal maximum likelihood is used to construct the estimator of the matrix of factor loadings $\mathbf{A}$, subject to the restrictions that $\hat{\mathbf{A}}'\hat{\boldsymbol{\Sigma}}_{\varepsilon\varepsilon}^{-1}\hat{\mathbf{A}}$ is a diagonal matrix and that $\hat{\boldsymbol{\Sigma}}_{ff} = \mathbf{I}$, we obtain

$$\hat{\mathbf{A}}' = \begin{bmatrix} 1.417 & 1.233 & 1.068 & 1.045 & 1.519 & 0.853 \\ 0.092 & 0.311 & 0.897 & 1.144 & -0.174 & 0.828 \end{bmatrix}.$$

The likelihood ratio test of the hypothesis that the observed covariance matrix is a sample matrix generated by a two-factor model against the general covariance matrix alternative is

$$[n - k - 1 - 6^{-1}(2r + 5)]\left\{\sum_{i=k+1}^{p} (\hat{\lambda}_i - \log \hat{\lambda}_i) - (p - k)\right\} = 1.40,$$

where $\hat{\lambda}_1 \geqslant \hat{\lambda}_2 \geqslant \cdots \geqslant \hat{\lambda}_p$ are the roots of

$$|\mathbf{m}_{ZZ} - \lambda\hat{\boldsymbol{\Sigma}}_{\varepsilon\varepsilon}| = 0$$

and the test statistic is that given in (4.3.44). The six roots are (39.381, 4.588, 1.112, 1.032, 0.980, 0.877). There are 21 unique elements in $\mathbf{m}_{ZZ}$ and $8 + 6 + 3 = 17$ parameters are estimated. Therefore, if the model is correct, the likelihood ratio statistic is approximately distributed as a chi-square random variable with four degrees of freedom. For our sample, the hypothesis of the two-factor model is easily accepted.

The estimated covariance matrix of the vector $\mathbf{v}$ is

$$\begin{aligned} \tilde{\boldsymbol{\Sigma}}_{vv} &= (\mathbf{I}, -\hat{\boldsymbol{\beta}}')\hat{\boldsymbol{\Sigma}}_{\varepsilon\varepsilon}(\mathbf{I}, -\hat{\boldsymbol{\beta}}')' \\ &= \begin{bmatrix} 1.075 & 0.151 & 0.275 & 0.331 \\ 0.151 & 1.269 & 0.485 & 0.594 \\ 0.275 & 0.485 & 2.107 & 1.344 \\ 0.331 & 0.594 & 1.344 & 2.533 \end{bmatrix}. \end{aligned}$$

The estimated covariance matrix of the elements in $\hat{\boldsymbol{\Sigma}}_{\varepsilon\varepsilon}$ is obtained as output from the last step of the iterative procedure used to obtain the maximum likelihood estimator. The communalities and their standard errors are given in Table 4.3.4.

The estimated coefficients suggest that X_{t1}, Y_{t3}, and Y_{t4}—logical, appropriate, and acceptable—are three measures of the same factor. To investigate this conjecture further, we estimate the model with β_{13} and β_{14} set equal to

TABLE 4.3.4 The estimated communalities and their estimated standard errors

	Variables					
	Z_1	Z_2	Z_3	Z_4	Z_5	Z_6
Communalities	0.676	0.610	0.656	0.735	0.970	0.613
Standard errors	0.073	0.069	0.074	0.080	0.104	0.080

zero. The estimated equations for the reduced model are

$$\begin{aligned}
\hat{Y}_{t1} &= \underset{(0.746)}{-1.132} + \underset{(0.177)}{0.771}x_{t1} + \underset{(0.184)}{0.280}x_{t2}, \\
\hat{Y}_{t2} &= \underset{(0.817)}{-0.191} + \underset{(0.140)}{0.527}x_{t1} + \underset{(0.164)}{0.501}x_{t2}, \\
\hat{Y}_{t3} &= \underset{(1.130)}{-2.270} \qquad\qquad\quad + \underset{(0.144)}{1.189}x_{t2}, \\
\hat{Y}_{t4} &= \underset{(1.188)}{-2.969} \qquad\qquad\quad + \underset{(0.152)}{1.274}x_{t2},
\end{aligned}$$

the estimated error covariance matrix is

$$\hat{\Sigma}_{\varepsilon\varepsilon} = \operatorname{diag}\{0.970, 1.031, 0.966, 0.969, 0.060, 0.891\},$$
$$(0.208, 0.170, 0.200, 0.214, 0.266, 0.167)$$

and the estimated covariance matrix of the factors is

$$\operatorname{vech} \hat{\Sigma}_{xx} = [\underset{(0.436)}{2.349}, \ \underset{(0.251)}{1.155}; \ \underset{(0.324)}{1.415}]'.$$

The likelihood ratio test that the observed covariance matrix is a sample matrix generated by the restricted two-factor model against the general covariance matrix alternative is $\hat{\chi}_6^2 = 2.76$. Under the null, the limiting distribution of the test statistic is that of a chi-square random variable with six degrees of freedom. Also the difference between the chi-square tests for the full and reduced models,

$$\hat{\chi}_6^2 - \hat{\chi}_4^2 = 2.76 - 1.40 = 1.36,$$

is approximately distributed as a chi-square random variable with two degrees of freedom when the restrictions are true. Therefore, the reduced model is easily accepted. The further restricted model in which β_{21}, β_{13}, and β_{14} are all set equal to zero would also be accepted.

To investigate the effect of nonnormality on the estimated standard errors, we compute the covariance matrix of the unrestricted vector of estimated parameters using a distribution-free estimator of the covariance matrix of vech $\mathbf{m}_{ZZ}$. By (4.3.34), the maximum likelihood estimator can be considered to be a generalized least squares estimator using vech $\mathbf{m}_{ZZ}$ as the vector of observations and $2(n-1)^{-1}\psi_4(\hat{\Sigma}_{ZZ} \otimes \hat{\Sigma}_{ZZ})\psi_4'$ as the estimator of the covariance matrix of vech $\mathbf{m}_{ZZ}$. Applying (4.B.9), an estimator of the covariance matrix of the maximum likelihood estimator for nonnormal distributions is

$$(\hat{\mathbf{F}}'\hat{\boldsymbol{\Omega}}_{ZZ}^{-1}\hat{\mathbf{F}})^{-1}\hat{\mathbf{F}}'\hat{\boldsymbol{\Omega}}_{ZZ}^{-1}\hat{\mathbf{V}}_{bb}\hat{\boldsymbol{\Omega}}_{ZZ}^{-1}\hat{\mathbf{F}}(\hat{\mathbf{F}}'\boldsymbol{\Omega}_{ZZ}\hat{\mathbf{F}})^{-1}, \tag{4.3.54}$$

where $\hat{\mathbf{F}}$ is the matrix $\mathbf{F}$ of (4.B.9) evaluated at $\gamma = \hat{\gamma}$,

$$\begin{aligned}
\gamma' &= [\boldsymbol{\beta}_0, \boldsymbol{\mu}_x, (\text{vec } \boldsymbol{\beta})', \gamma_\varepsilon', (\text{vech } \boldsymbol{\Sigma}_{xx})'], \\
\gamma_\varepsilon' &= (\sigma_{\varepsilon\varepsilon 11}, \sigma_{\varepsilon\varepsilon 22}, \ldots, \sigma_{\varepsilon\varepsilon 66}), \\
\hat{\boldsymbol{\Omega}}_{ZZ} &= (100)^{-1} \text{ block diag}\{\boldsymbol{\Sigma}_{ZZ}, 2\psi_4(\hat{\boldsymbol{\Sigma}}_{ZZ} \otimes \hat{\boldsymbol{\Sigma}}_{ZZ})\psi_4'\}, \\
\hat{\mathbf{V}}_{bb} &= (99)^{-1}(100)^{-1} \sum_{t=1}^{100} (\mathbf{b}_t - \bar{\mathbf{b}})'(\mathbf{b}_t - \bar{\mathbf{b}}), \\
\mathbf{b}_t &= \{\mathbf{Z}_t, [\text{vech}(\mathbf{Z}_t - \bar{\mathbf{Z}})'(\mathbf{Z}_t - \bar{\mathbf{Z}})]'\}.
\end{aligned}$$

Table 4.3.5 contains the estimated variances computed by the two methods.

For a normal random variable Z_t the variance of the ratio

$$\tfrac{1}{2}\left[(n-1)^{-1} \sum_{t=1}^{n} (Z_t - \bar{Z})^2\right]^{-2} (n-1)^{-1} \sum_{t=1}^{n} [(Z_t - \bar{Z})^2 - m_{ZZ}]^2$$

is about $6(n-1)^{-1}$ and the variance of the square root of the ratio is about $1.5(n-1)^{-1}$. See Kendall and Stuart (Vol. 1, 1977, p. 258). Using this property as a crude guide for estimators that are functions of the second moments, about 95% of the ratios of estimated standard errors computed by the two methods should fall in the interval of (0.75, 1.25). Of the 17 ratios in the table that are functions of second moments only, five fall outside this interval. The average of the ratios for the slope parameters and the $\sigma_{\varepsilon\varepsilon}$'s is about 1.27. This suggests that the standard errors of these estimators based on normality are biased downward. □ □

For models discussed in Chapter 1, Chapter 2, and Section 4.1 we were able to provide tests for model identification. The construction of such tests is more difficult for the factor model.

Some information about identification is provided by the estimated covariance matrix of the estimates. If the model is not identified, the covariance matrix of the limiting distribution is singular and, in finite samples, the estimated variance of estimators of certain functions of parameters will often be very large.

The following result provides an alternative method of investigating the identification status of the model. See Anderson and Rubin (1956).

TABLE 4.3.5. Estimated standard errors computed by alternative methods

Parameter	Standard Errors Multiplied by 10 Normal	Moment	Ratio Moment/Normal
β_{01}	5.44	5.93	1.09
β_{02}	6.51	6.76	1.04
β_{03}	1.16	1.28	1.10
β_{04}	1.56	1.30	0.83
μ_{x1}	0.24	0.24	1.00
μ_{x2}	0.23	0.23	1.00
β_{11}	0.29	0.43	1.51
β_{21}	0.31	0.48	1.56
β_{12}	0.20	0.24	1.24
β_{22}	0.27	0.32	1.20
β_{13}	0.19	0.23	1.20
β_{23}	0.38	0.48	1.24
β_{14}	0.26	0.30	1.16
β_{24}	0.56	0.59	1.05
$\sigma_{\varepsilon\varepsilon 11}$	0.40	0.47	1.15
$\sigma_{\varepsilon\varepsilon 22}$	0.28	0.20	0.69
$\sigma_{\varepsilon\varepsilon 33}$	0.41	0.60	1.47
$\sigma_{\varepsilon\varepsilon 44}$	0.59	1.06	1.79
$\sigma_{\varepsilon\varepsilon 55}$	0.61	0.89	1.45
$\sigma_{\varepsilon\varepsilon 66}$	0.30	0.32	1.07
σ_{xx11}	1.78	1.65	0.93
σ_{xx12}	0.69	0.58	0.84
σ_{xx22}	1.05	0.79	0.76

Theorem 4.3.2. Let the factor model (4.3.2) hold. Suppose that there exists a set of factors $\mathbf{f}_t = (f_{t1}, f_{t2}, \ldots, f_{tk})$ satisfying $\mathbf{\Sigma}_{ff} = \mathbf{I}$, such that

$$\mathbf{z}_t - \boldsymbol{\mu}_z = \mathbf{f}_t\mathbf{A}',$$

where at least one column of $\mathbf{A}$ contains no more than two nonzero elements. Then the model is not identified.

Proof. By assumption, there exists an $\mathbf{A}$ and a diagonal $\mathbf{\Sigma}_{\varepsilon\varepsilon}$ with diagonal elements $\{\sigma_{\varepsilon\varepsilon 11}, \sigma_{\varepsilon\varepsilon 22}, \ldots, \sigma_{\varepsilon\varepsilon pp}\}$ such that

$$\mathbf{\Sigma}_{ZZ} = \mathbf{AA}' + \mathbf{\Sigma}_{\varepsilon\varepsilon} \tag{4.3.55}$$

and such that a column of $\mathbf{A}$ contains no more than two nonzero elements. With no loss of generality we let the first column satisfy the restriction and assume that $A_{i1} = 0$, $i = 3, 4, \ldots, p$. Among the equations in (4.3.55), the parameters A_{11}, A_{21}, $\sigma_{\varepsilon\varepsilon 11}$, and $\sigma_{\varepsilon\varepsilon 22}$ appear only in the three equations

$$\begin{aligned} A_{11}^2 + \sigma_{\varepsilon\varepsilon 11} &= \sigma_{ZZ11} - \sum_{\ell=2}^{k} A_{1\ell}^2, \\ A_{11}A_{21} &= \sigma_{ZZ12} - \sum_{\ell=2}^{k} A_{1\ell}A_{2\ell}, \\ A_{21}^2 + \sigma_{\varepsilon\varepsilon 22} &= \sigma_{ZZ22} - \sum_{\ell=2}^{k} A_{2\ell}^2. \end{aligned} \tag{4.3.56}$$

Because the four parameters A_{11}, A_{21}, $\sigma_{\varepsilon\varepsilon 11}$, and $\sigma_{\varepsilon\varepsilon 22}$ appear only in three equations, there exist A_{11}^*, A_{21}^*, $\sigma_{\varepsilon\varepsilon 11}^* > 0$, and $\sigma_{\varepsilon\varepsilon 22}^* > 0$ with $\sigma_{\varepsilon\varepsilon 11}^* \neq \sigma_{\varepsilon\varepsilon 11}$ that also satisfy the equations in (4.3.56). Let $\mathbf{A}^*$ and $\mathbf{\Sigma}_{\varepsilon\varepsilon}^*$ denote the $\mathbf{A}$ and $\mathbf{\Sigma}_{\varepsilon\varepsilon}$ with A_{11}^*, A_{21}^*, $\sigma_{\varepsilon\varepsilon 11}^*$, and $\sigma_{\varepsilon\varepsilon 22}^*$ replacing A_{11}, A_{21}, $\sigma_{\varepsilon\varepsilon 11}$, and $\sigma_{\varepsilon\varepsilon 22}$. Then

$$\mathbf{AA}' + \mathbf{\Sigma}_{\varepsilon\varepsilon} = \mathbf{A}^*\mathbf{A}^{*\prime} + \mathbf{\Sigma}_{\varepsilon\varepsilon}^*,$$

but there is no orthogonal matrix $\mathbf{G}$ such that

$$(\mathbf{A}^*, \mathbf{\Sigma}_{\varepsilon\varepsilon}^*) = (\mathbf{AG}, \mathbf{\Sigma}_{\varepsilon\varepsilon}).$$

Hence, the model is not identified. □

One sees from the proof of Theorem 4.3.2 that it is possible to take either $\sigma_{\varepsilon\varepsilon 11}$ or $\sigma_{\varepsilon\varepsilon 22}$ to be zero and to obtain unique solutions for the remaining parameters when only Z_1 and Z_2 are correlated with f_1. Thus, if the estimation procedure produces an estimate of zero for an element of $\sigma_{\varepsilon\varepsilon ii}$ or an estimate of $\sigma_{\varepsilon\varepsilon ii}$ with a large variance relative to other estimates of the error variances, the possibility that the model is not identified should be considered.

One can estimate the model with the offending $\sigma_{\varepsilon\varepsilon ii}$ set equal to zero and with the true value of the corresponding Z variable designated to be one of the x variables. If only one of the β coefficients associated with the designated x variable is significantly different from zero, the hypothesis that the model is not identified is not rejected.

The factor model in k orthogonal factors with only the first variable loading on one of the factors is observationally indistinguishable from the $(k-1)$-factor model. This is because only the sum $A_{1k}^2 + \sigma_{\varepsilon\varepsilon 11}$ of model (4.3.2) is identified for such a configuration.

Example 4.3.3. To illustrate some of the identification concepts we analyze all eight of the language items introduced in Example 2.4.1. If we fit a two-factor model to the data for the eight items, the test statistic of (4.3.44) is $\hat{\chi}_{13}^2 = 31.66$ with 13 degrees of freedom. Therefore, the two-factor model is rejected. The test statistic for the three-factor model is $\hat{\chi}_7^2 = 2.33$ with seven degrees of freedom. It seems that three factors provide an adequate description of the systematic variation in the eight variables. In this example we identify the variables by the first three letters of the description. Thus, DEV is the variable developed, UND is the variable understand, and so on. With this coding, the estimated equations are

$$
\begin{aligned}
\text{DEV} &= \underset{(0.73)}{-1.23} + \underset{(0.178)}{0.755}\ \text{log} + \underset{(0.181)}{0.245}\ \text{irr} + \underset{(0.099)}{0.068}\ \text{car},\\
\text{UND} &= \underset{(0.76)}{-0.32} + \underset{(0.144)}{0.499}\ \text{log} + \underset{(0.176)}{0.367}\ \text{irr} + \underset{(0.165)}{0.195}\ \text{car},\\
\text{APP} &= \underset{(1.10)}{-2.16} + \underset{(0.141)}{0.092}\ \text{log} + \underset{(0.235)}{1.149}\ \text{irr} - \underset{(0.131)}{0.062}\ \text{car},\\
\text{ACC} &= \underset{(1.27)}{3.10} - \underset{(0.161)}{0.070}\ \text{log} + \underset{(0.283)}{1.414}\ \text{irr} - \underset{(0.147)}{0.065}\ \text{car},\\
\text{INT} &= \underset{(0.65)}{1.83} + \underset{(0.103)}{0.190}\ \text{log} + \underset{(0.193)}{0.303}\ \text{irr} + \underset{(0.252)}{0.347}\ \text{car},
\end{aligned}
$$

where the numbers in parentheses below the estimates are the estimated standard errors. The estimate of $\Sigma_{\varepsilon\varepsilon}$ is

$$
\hat{\Sigma}_{\varepsilon\varepsilon} = \text{diag}\{\underset{(0.20)}{0.96},\ \underset{(0.16)}{0.97},\ \underset{(0.21)}{1.01},\ \underset{(0.25)}{0.86},\ \underset{(0.17)}{0.67},\ \underset{(0.27)}{0.06},\ \underset{(0.17)}{0.90},\ \underset{(1.19)}{0.00}\}
$$

and the variables are in the order DEV, UND, APP, ACC, INT, LOG, IRR, CAR. The estimator of vech Σ_{xx} is

$$
\text{vech}\ \hat{\Sigma}_{xx} = [\underset{(0.44)}{2.35},\ \underset{(0.27)}{1.15},\ \underset{(0.28)}{1.13};\ \underset{(0.33)}{1.41},\ \underset{(0.28)}{1.15};\ \underset{(1.25)}{2.64}]'.
$$

The distinctive aspect of the set of estimates is the estimate of zero for the error variance for the variable careful. Not only is the estimate zero, but the estimated standard error is about five times the standard errors of the other estimated error variances. Also, none of the estimated β coefficients for careful exceeds twice the standard error in absolute value.

An option available in ISU Factor was used as an aid in selecting the variables designated as the x variables in the fitting. This option selects a set of variables to approximately maximize the determinant of the standardized $\hat{\Sigma}_{xx}$. With the selected set of x variables, and the fact that the two-factor model was rejected, we believe that $|\Sigma_{xx}| \neq 0$. Therefore, it is reasonable to conclude that large standard errors of estimators are due to causes other than a singular Σ_{xx}.

If we reestimate the model, fixing the error variance of careful at zero, the standard errors of the estimated coefficients for careful, in the order of the original five equations, are

$$(0.080, 0.093, 0.066, 0.080, 0.103).$$

The large change in the magnitudes of the standard errors suggests that the zero error variance maximized the likelihood because the model is not identified, rather than because the error variance is small for an identified model. The coefficient of intelligent on careful is about five times the standard error, where the standard error is constructed on the basis of zero error variance. This evidence suggests that the data configuration can be explained by a model in three factors in which only intelligent and careful are correlated with one of the factors. The fact that the coefficient of understand on careful is more than twice its standard error, in the model with the careful error variance fixed equal to zero, clouds interpretation of the statistics. We are guided in our model building by the grouping of the variables given in Example 2.4.1. In that grouping, the study directors placed intelligent and careful in a group of two items pertaining to the writer.

To investigate the conjecture of a three-factor nonidentified model further, we fit the three-factor model subject to the restrictions

$$\sigma_{\varepsilon\varepsilon 88} = \beta_{31} = \beta_{32} = \beta_{33} = \beta_{34} = 0, \tag{4.3.57}$$

where $\sigma_{\varepsilon\varepsilon 88}$ is the error variance for careful and all coefficients for careful except that for intelligent are set equal to zero. Either the program ISU Factor or LISREL VI can be used to estimate the reduced model.

Under the restrictions (4.3.57) the restricted model is identified when the two-factor model in the first six variables is identified. If one attempts to fit the model with zero restrictions for $r - 1$ of the elements of a row of $\boldsymbol{\beta}$, but without the zero restriction for $\sigma_{\varepsilon\varepsilon 88}$, one will obtain a singular matrix for

$$\mathbf{L}'\{(\tilde{\mathbf{C}}\tilde{\boldsymbol{\Sigma}}_{vv}^{-1}\tilde{\mathbf{C}}') \otimes (\tilde{\mathbf{C}}\tilde{\boldsymbol{\Sigma}}_{vv}^{-1}\tilde{\mathbf{C}}')\}\mathbf{L}$$

of (4.3.36). Therefore, when investigating the hypothesis that only one Y variable has a nonzero coefficient for a particular x variable, one should always set the error variance of that x variable equal to zero, or otherwise impose restrictions to identify the model for estimation purposes. The estimated equations of the restricted model are

$$\begin{aligned}
\text{DEV} &= \underset{(0.74)}{-1.22} + \underset{(0.166)}{0.837}\,\text{log} + \underset{(0.176)}{0.233}\,\text{irr},\\
\text{UND} &= \underset{(0.80)}{-0.29} + \underset{(0.138)}{0.564}\,\text{log} + \underset{(0.162)}{0.480}\,\text{irr},\\
\text{APP} &= \underset{(1.07)}{-2.10} + \underset{(0.150)}{0.096}\,\text{log} + \underset{(0.200)}{1.080}\,\text{irr},
\end{aligned}$$

$$\begin{aligned} \text{ACC} &= -2.98 - 0.075 \log + 1.343 \text{ irr}, \\ &\quad\ (1.22)\quad (0.173) \qquad (0.239) \\ \text{INT} &= \ \ 1.84 + 0.211 \log + 0.284 \text{ irr} + 0.344 \text{ car}. \\ &\quad\ (0.62)\quad (0.092) \qquad (0.127) \qquad (0.067) \end{aligned}$$

The chi-square value for the test of the restricted factor model against the model with unrestricted covariance matrix ($\hat{\boldsymbol{\Sigma}}_{ZZ} = \mathbf{m}_{ZZ}$) is 9.94. If the restricted model is correct, the approximate distribution of the statistic is that of a chi-square random variable with 12 degrees of freedom. Therefore, the restricted three-factor model is easily accepted by the data. The difference between the chi-square values for the full three-factor model and the reduced three-factor model is 7.61, which is well below the 0.10 point of the chi-square distribution with five degrees of freedom. If the original model is not identified, the limiting distribution of the difference of the two test statistics is not that of a chi-square random variable. This is because the information matrix for the full model is singular. Nonetheless, it seems reasonable to compute the difference and to question the validity of the restrictions if the difference is large relative to the tabulated chi-square distribution for the difference in the degrees of freedom.

The description of the data by a nonidentified three-factor model is also supported by the estimation of the two-factor model for two data sets each containing seven variables. One data set is obtained by deleting intelligent and one set is obtained by deleting careful from the full set of eight variables. When intelligent is omitted from the data set, the likelihood ratio statistic of (4.3.44) for the two-factor model is $\hat{\chi}^2 = 8.54$ with eight degrees of freedom. When careful is omitted from the data set, the likelihood ratio statistic of (4.3.44) is $\hat{\chi}^2 = 2.96$ with eight degrees of freedom.

Thus, it is reasonable to conclude that the original full three-factor model in eight variables is not identified because only intelligent and careful load on one factor. This conclusion is consistent with the description of the original variables because the variables intelligent and careful are the only two variables that are directly concerned with the characteristics of the writer.

Our method of setting the error variance for careful, $\sigma_{\varepsilon\varepsilon 88}$, equal to zero to obtain an identified model was an arbitrary one. Exactly the same likelihood would be obtained for the reduced model if one set the error variance of intelligent, $\sigma_{\varepsilon\varepsilon 55}$, equal to zero and estimated $\sigma_{\varepsilon\varepsilon 88}$, or if one set $\sigma_{\varepsilon\varepsilon 55} = \sigma_{\varepsilon\varepsilon 88}$. The two-factor model fails because there is a correlation between intelligent and careful that is not explained by the two factors, logical and irritating. Another model for such behavior is the model in two factors that permits a nonzero value for $\sigma_{\varepsilon\varepsilon 58}$, the covariance between the errors for intelligent and careful. The two-factor model with $\sigma_{\varepsilon\varepsilon 58} \neq 0$ will give the same likelihood as the three-factor model with diagonal $\boldsymbol{\Sigma}_{\varepsilon\varepsilon}$ and $\sigma_{\varepsilon\varepsilon 88} = 0$.

We complete our study of the language data by fitting the model subject to the restrictions imposed in Example 4.3.2. The estimates of the reduced equations are

$$\begin{aligned}
\text{DEV} &= \underset{(0.74)}{-1.22} + \underset{(0.171)}{0.835}\ \text{log} + \underset{(0.181)}{0.234}\ \text{irr}, \\
\text{UND} &= \underset{(0.81)}{-0.31} + \underset{(0.135)}{0.556}\ \text{log} + \underset{(0.161)}{0.491}\ \text{irr}, \\
\text{APP} &= \underset{(1.11)}{-2.16} \qquad\qquad\quad + \underset{(0.142)}{1.175}\ \text{irr}, \\
\text{ACC} &= \underset{(1.17)}{-2.88} \qquad\qquad\quad + \underset{(0.149)}{1.263}\ \text{irr}, \\
\text{INT} &= \underset{(0.62)}{1.84} + \underset{(0.090)}{0.208}\ \text{log} + \underset{(0.126)}{0.288}\ \text{irr} + \underset{(0.067)}{0.343}\ \text{car}.
\end{aligned}$$

The estimated error covariance matrix is

$$\hat{\boldsymbol{\Sigma}}_{\varepsilon\varepsilon} = \text{diag}\{\underset{(0.19)}{0.89},\ \underset{(0.16)}{0.98},\ \underset{(0.20)}{1.00},\ \underset{(0.21)}{0.99},\ \underset{(0.10)}{0.67},\ \underset{(0.21)}{0.18},\ \underset{(0.16)}{0.88},\ \underset{(-)}{0}\},$$

where the error variance of careful is fixed at zero. The estimated covariance matrix of the x variables is

$$\text{vech}\ \hat{\boldsymbol{\Sigma}}_{xx} = [\underset{(0.40)}{2.23},\ \underset{(0.25)}{1.16},\ \underset{(0.28)}{1.17};\ \underset{(0.32)}{1.43},\ \underset{(0.26)}{1.14};\ \underset{(0.38)}{2.64}]'.$$

The estimates for the portion of the model that was estimated in Example 4.3.2 are very similar to the estimates obtained in that example. On average, the standard errors for corresponding estimates are slightly smaller for the estimates obtained in this example because the additional two variables provide some information for the original model. □ □

REFERENCES

Amemiya, Fuller, and Pantula (1987), Anderson (1984), Anderson and Rubin (1956), Fuller and Pantula (1982), Harman (1976), Jennrich and Robinson (1969), Jöreskog (1967, 1970b, 1973), Jöreskog and Lawley (1968), Lawley (1940, 1941, 1943), Lawley and Maxwell (1971).

EXERCISES

22. (Section 4.3.2)
   (a) Show that

$$\log|\mathbf{m}_{ZZ}\hat{\boldsymbol{\Sigma}}_{ZZ}^{-1}| - \text{tr}\{\mathbf{m}_{ZZ}\hat{\boldsymbol{\Sigma}}_{ZZ}^{-1}\} = \log|\hat{\mathbf{m}}_{vv}\hat{\boldsymbol{\Sigma}}_{vv}^{-1}| - \text{tr}\{\hat{\mathbf{m}}_{vv}\hat{\boldsymbol{\Sigma}}_{vv}^{-1}\} - k,$$

where $\hat{\mathbf{m}}_{vv} = (\mathbf{I}, -\hat{\boldsymbol{\beta}}')\mathbf{m}_{ZZ}(\mathbf{I}, -\hat{\boldsymbol{\beta}}')'$, $\hat{\boldsymbol{\Sigma}}_{vv} = (\mathbf{I}, -\hat{\boldsymbol{\beta}}')\hat{\boldsymbol{\Sigma}}_{\varepsilon\varepsilon}(\mathbf{I}, -\hat{\boldsymbol{\beta}}')'$, and $\hat{\boldsymbol{\Sigma}}_{ZZ}$, $\hat{\boldsymbol{\beta}}$, and $\hat{\boldsymbol{\Sigma}}_{\varepsilon\varepsilon}$ are the maximum likelihood estimators defined in Result 4.3.1.

(b) Use the equations of Result 4.3.1 to show that the diagonal of $\hat{\boldsymbol{\Sigma}}_{ZZ}^{-1}\mathbf{m}_{ZZ}$ is $\mathbf{I}_p$ when the estimates of the factor model are in the interior of the parameter space.

23. (Sections 4.3.3, 4.1.1) For the factor model show that if the estimates are in the interior of the parameter space, the likelihood ratio criterion of (4.3.43) is

$$-2[\log L_c(\hat{\boldsymbol{\Sigma}}_{ZZ}) - \log L_c(\mathbf{m}_{ZZ})] = (n-1) \sum_{j=k+1}^{p} (\hat{\lambda}_j - \log \hat{\lambda}_j - 1),$$

where $\hat{\lambda}_j^{-1}$ are the roots of

$$|\mathbf{m}_{ZZ}^{-1/2}\hat{\boldsymbol{\Sigma}}_{\varepsilon\varepsilon}\mathbf{m}_{ZZ}^{-1/2} - \lambda^{-1}\mathbf{I}| = 0.$$

Also see Equation (4.1.46). Then, using Exercise 4.22, show that the criterion can also be written

$$(n-1) \sum_{j=k+1}^{p} \log \hat{\lambda}_j^{-1}.$$

24. (Sections 4.3.1, 4.2.2) Let (Z_{t1}, Z_{t2}) be distributed as a bivariate normal vector with mean (μ_1, μ_2) and covariance matrix defined by

$$\operatorname{vech} \boldsymbol{\Sigma}_{ZZ} = (\sigma_{11}, \rho\sigma_{11}, \sigma_{22.1} + \rho^2\sigma_{11})'.$$

Let ρ be known.

(a) Let the covariance matrix of vech $\mathbf{m}_{ZZ}$ be estimated by $2(n-1)^{-1}\psi_2(\ddot{\boldsymbol{\Sigma}} \otimes \ddot{\boldsymbol{\Sigma}})\psi_2'$, where $\ddot{\boldsymbol{\Sigma}}$ is any matrix of the form $\boldsymbol{\Sigma}_{ZZ}$ with $\sigma_{11} > 0$ and $\sigma_{22.1} > 0$. Prove that the least squares estimator of $(\sigma_{11}, \sigma_{22.1})$ based on the sample covariances is

$$(\hat{\sigma}_{11}, \hat{\sigma}_{22.1}) = (m_{11}, m_{22} - 2\rho m_{12} + \rho^2 m_{11}),$$

where $m_{ii} = m_{ZZii} = (n-1)^{-1} \sum_{t=1}^{n} (Z_{ti} - \bar{Z}_i)^2$.

(b) Show that the estimator of part (a) is the maximum likelihood estimator adjusted for degrees of freedom.

25. (Section 4.3)

(a) Assume that

$$\mathbf{Z}_t = (Z_{t1}, Z_{t2}, \ldots, Z_{t6}) = (Y_{t1}, Y_{t2}, Y_{t3}, Y_{t4}, X_{t1}, X_{t2})$$

satisfies the factor model

$$\mathbf{Y}_t = \boldsymbol{\beta}_0 + \mathbf{x}_t \begin{bmatrix} \beta_{11} & 0 & 0 & 0 \\ 0 & \beta_{22} & \beta_{23} & \beta_{24} \end{bmatrix} + \boldsymbol{\varepsilon}_t,$$

where $\mathbf{x}_t \sim NI(\mathbf{0}, \boldsymbol{\Sigma}_{xx})$. Is β_{11} identified for all positive definite $\boldsymbol{\Sigma}_{xx}$? Explain.

(b) Assume $\mathbf{Z}_t \sim NI(\boldsymbol{\mu}_Z, \boldsymbol{\Sigma}_{ZZ})$, with

$$\boldsymbol{\Sigma}_{ZZ} = \begin{bmatrix} 1.00 & 0.90 & 0.90 & 0.10 & 0.20 \\ 0.90 & 1.00 & 0.90 & 0.10 & 0.20 \\ 0.90 & 0.90 & 1.00 & 0.10 & 0.20 \\ 0.10 & 0.10 & 0.10 & 1.00 & 0.70 \\ 0.20 & 0.20 & 0.20 & 0.70 & 1.00 \end{bmatrix}.$$

Does $\mathbf{Z}_t$ satisfy the factor model in two factors? Are the parameters of the model identified?

26. (Section 4.3.3) The data in the table are observations on the vital capacity of the human lung as measured by two instruments, each operated by a skilled and an unskilled operator. The observation identified as X is the standard instrument operated by a skilled operator, Y_1

is the standard instrument operated by an unskilled operator, Y_2 is the experimental instrument operated by a skilled operator and Y_3 is the experimental instrument operated by an unskilled operator. The data are described in Barnett (1969). Assume that the data satisfy the factor model

$$Y_{tj} = \beta_{0j} + x_t\beta_{1j} + e_{tj}, \qquad j = 1, 2, 3,$$
$$X_t = x_t + u_t, \qquad \varepsilon_t' \sim \mathrm{NI}(\mathbf{0}, \boldsymbol{\Sigma}_{\varepsilon\varepsilon}),$$
$$\boldsymbol{\Sigma}_{\varepsilon\varepsilon} = \mathrm{diag}(\sigma_{ee11}, \sigma_{ee22}, \sigma_{ee33}, \sigma_{uu}),$$

Readings of lung vital capacity for 72 patients on four instrument–operator combinations

X	Y_1	Y_2	Y_3	X	Y_1	Y_2	Y_3
345	353	403	372	106	100	85	60
131	132	161	160	200	180	127	170
382	372	415	370	228	228	238	235
211	288	274	252	194	180	167	158
186	142	154	169	258	270	285	211
194	178	202	180	140	144	168	148
236	226	243	235	126	110	100	103
288	292	265	286	232	242	236	236
198	172	180	166	200	194	198	198
312	318	325	304	240	190	147	174
176	163	139	120	288	298	324	314
148	176	170	164	342	315	320	320
184	166	140	165	100	113	65	84
358	348	368	396	140	140	135	138
188	200	209	207	188	171	160	135
240	232	255	248	128	126	116	133
222	212	229	227	312	300	311	325
254	250	262	196	377	334	390	370
92	120	64	103	342	322	312	329
224	216	230	230	274	288	285	288
224	213	203	214	284	292	271	275
226	251	240	245	380	374	344	340
386	418	398	368	210	168	165	193
278	210	189	200	182	140	106	105
222	140	184	136	140	132	135	110
188	182	190	184	220	168	164	111
94	96	106	100	194	190	182	127
248	222	215	215	326	320	325	327
166	178	176	180	196	194	189	192
404	418	400	377	132	126	114	100
254	256	208	225	284	306	365	351
178	170	139	120	206	184	172	178
128	130	80	113	220	197	190	227
194	206	203	188	126	115	86	115
176	200	186	186	304	284	285	267
204	166	147	116	214	218	256	272

Source: Barnett (1969).

where $x_t \sim \mathrm{NI}(0, \sigma_{xx})$ and x_t is independent of ε_i for all i and t.

(a) Estimate the parameters of the model and estimate the covariance matrix of your estimates. Plot $\hat{v}_{tj}$ against $\hat{x}_t$ for $j = 1, 2, 3$ and plot $X_t - \hat{x}_t = \hat{v}_{t4}$ against $\hat{x}_t$.

(b) Estimate the reduced model in which $\beta_{11} = 1$ and $\beta_{12} = \beta_{13}$. Are the data compatible with this model?

(c) Estimate the reduced model with $\beta_{01} = 0$, $\beta_{11} = 1$, $\beta_{02} = \beta_{03}$, and $\beta_{12} = \beta_{13}$. Are the data compatible with this model?

27. (Sections 4.1, 4.2, 4.3) The data in the table are observations made at three stages in a production process.

(a) Fit the model of Example 4.2.3 to the data. That is, assume that the observations, denoted by Y_{tj}, satisfy

$$Y_{tj} = \beta_{0j} + x_t + e_{tj},$$

where $e_{tj} \sim \mathrm{NI}(0, \sigma_{jj})$, e_{tj} is independent of e_{ij} for $t \neq i$, and e_{tj} is independent of x_i for all t and i. Do you feel the unit coefficient model is appropriate for these data?

(b) Fit the unrestricted factor model to the data. Fit the factor model with $\sigma_{\varepsilon\varepsilon 11} = \sigma_{\varepsilon\varepsilon 22}$.

(c) Fit the model of Section 4.1,

$$Z_{ti} = \beta_{0i} + \beta_{1i} z_{t3} + \varepsilon_{ti}, \quad \text{for } i = 1, 2,$$

$$Z_{t3} = z_{t3} + \varepsilon_{t3},$$

under the assumption $\boldsymbol{\varepsilon}_t \sim \mathrm{NI}(\mathbf{0}, \mathbf{I}\sigma^2)$.

(d) What model would you choose for these data?

Observation	Stage 1	Stage 2	Stage 3	Observation	Stage 1	Stage 2	Stage 3
1	34	48	63	21	42	54	69
2	34	58	63	22	41	57	68
3	26	44	53	23	35	52	68
4	29	56	62	24	44	54	69
5	24	53	62	25	40	49	65
6	29	46	58	26	40	50	67
7	34	45	59	27	41	54	69
8	27	41	53	28	37	55	70
9	29	45	55	29	32	46	66
10	40	40	59	30	42	47	63
11	30	45	62	31	27	47	53
12	29	52	60	32	36	52	63
13	38	53	63	33	37	48	69
14	34	50	60	34	37	51	62
15	32	49	58	35	32	40	60
16	38	53	63	36	34	45	63
17	30	49	56	37	24	47	58
18	41	54	69	38	34	50	59
19	40	52	63	39	40	53	68
20	39	48	63	40	36	49	67

28. (Section 4.3.3) Estimate the factor model in two factors for the seven language items of Table 2.4.1 obtained by deleting careful. Use logical and irritating as the variables defining the factors. Compare the estimates to those of Example 4.3.3.

APPENDIX 4.A. MATRIX–VECTOR OPERATIONS

In this appendix we present notation for performing various operations on the elements of matrices. When interested in functions of elements of a matrix, it is often more convenient to arrange the elements in a vector.

Definition 4.A.1. Let $\mathbf{A} = (a_{ij})$ be a $p \times q$ matrix and let $\mathbf{A}_{.j}$ denote the jth column of $\mathbf{A}$. Then

$$\begin{aligned}\text{vec}\,\mathbf{A} &= (a_{11}, a_{21}, \ldots, a_{p1}, a_{12}, a_{22}, \ldots, a_{p2}, \ldots, a_{1q}, a_{2q}, \ldots, a_{pq})' \\ &= (\mathbf{A}'_{.1}, \mathbf{A}'_{.2}, \ldots, \mathbf{A}'_{.q})'.\end{aligned}$$

Sometimes, to identify the elements of the original matrix more clearly, we will separate the elements of different columns with a semicolon and write

$$\text{vec}\,\mathbf{A} = (a_{11}, a_{21}, \ldots, a_{p1}; a_{12}, \ldots, a_{p2}; \ldots; a_{1q}, \ldots, a_{pq})'.$$

Note that vec $\mathbf{A}$, also written vec$\{\mathbf{A}\}$, is the vector obtained by listing the columns of $\mathbf{A}$ one beneath the other beginning with the leftmost column. It follows that

$$\begin{aligned}\text{vec}\{\mathbf{A}'\} &= (a_{11}, a_{12}, \ldots, a_{1q}, a_{21}, a_{22}, \ldots, a_{2q}, \ldots, a_{p1}, a_{p2}, \ldots, a_{pq})' \\ &= (\mathbf{A}_{1.}, \mathbf{A}_{2.}, \ldots, \mathbf{A}_{p.})',\end{aligned}$$

where $\mathbf{A}_{i.}$ in the ith row of $\mathbf{A}$, is a listing of the rows of $\mathbf{A}$ as column vectors one below the other commencing with the top row.

If the matrix $\mathbf{A}$ is a symmetric $p \times p$ matrix, vec $\mathbf{A}$ will contain $\frac{1}{2}p(p-1)$ pairs of identical elements. In some situations one will wish to retain only one element of each pair. This can be accomplished by listing the elements in each column that are on and below the diagonal.

Definition 4.A.2. Let $\mathbf{A} = (a_{ij})$ be a $p \times p$ matrix. Then

$$\text{vech}\,\mathbf{A} = (a_{11}, a_{21}, \ldots, a_{p1}, a_{22}, a_{32}, \ldots, a_{p2}, a_{33}, a_{43}, \ldots, a_{p3}, \ldots, a_{pp})'.$$

The vector vech $\mathbf{A}$, also written vech$\{\mathbf{A}\}$, is sometimes called the vector half of $\mathbf{A}$. As with vec $\mathbf{A}$, we will sometimes separate the elements of different columns with a semicolon and write

$$\text{vech}\,\mathbf{A} = (a_{11}, a_{21}, \ldots, a_{p1}; a_{22}, a_{32}, \ldots, a_{p2}; \ldots; a_{pp})'.$$

For symmetric $\mathbf{A}$, vech $\mathbf{A}$ contains the unique elements of $\mathbf{A}$. Therefore, it is possible to recreate vec $\mathbf{A}$ from vech $\mathbf{A}$.

Definition 4.A.3. Let $\mathbf{A} = (a_{ij})$ be a $p \times p$ symmetric matrix. Let $\mathbf{\Phi}_p$ be the $p^2 \times \frac{1}{2}p(p+1)$ matrix such that

$$\text{vec } \mathbf{A} = \mathbf{\Phi}_p \text{ vech } \mathbf{A}, \tag{4.A.1}$$

and define ψ_p by

$$\psi_p = (\mathbf{\Phi}_p'\mathbf{\Phi}_p)^{-1}\mathbf{\Phi}_p'. \tag{4.A.2}$$

Note that $\mathbf{\Phi}_p$ is unique and of full column rank and that

$$\text{vech } \mathbf{A} = \psi_p \text{ vec } \mathbf{A} = (\mathbf{\Phi}_p'\mathbf{\Phi}_p)^{-1}\mathbf{\Phi}_p'\mathbf{\Phi}_p \text{ vech } \mathbf{A}.$$

If no confusion will result, we may omit the dimension subscript p from $\mathbf{\Phi}_p$. There are many linear transformations of vec $\mathbf{A}$ into vech $\mathbf{A}$, but the transformation ψ, which is the Moore–Penrose generalized inverse of $\mathbf{\Phi}$, is particularly useful.

Single subscripts could be used to denote the elements of vec $\mathbf{A}$ and vech $\mathbf{A}$, but it is more convenient to retain the original double subscript notation associated with the matrix $\mathbf{A}$. In the double subscript notation for the elements of a matrix, the first subscript of vec $\mathbf{A}$ is nested within the second because the elements of vec $\mathbf{A}$ are listed by column. If the elements of $\mathbf{\Phi}$ are defined with double subscripts to match those of vec $\mathbf{A}$ and vech $\mathbf{A}$, we have, for $i = 1, 2, \ldots, p, j = 1, 2, \ldots, p, k \geqslant \ell = 1, 2, \ldots, p$,

$$[\mathbf{\Phi}]_{ij,k\ell} = \begin{cases} 1 & \text{if } (i, j) = (k, \ell) \\ 1 & \text{if } (i, j) = (\ell, k) \\ 0 & \text{otherwise.} \end{cases}$$

If follows that the elements of $\mathbf{\Phi}'\mathbf{\Phi}$ are given by

$$[\mathbf{\Phi}'\mathbf{\Phi}]_{ij,k\ell} = \begin{cases} 1 & \text{if } (i, j) = (k, \ell), \quad i = j \\ 2 & \text{if } (i, j) = (k, \ell), \quad i \neq j \\ 0 & \text{otherwise.} \end{cases}$$

Therefore, if the elements of ψ are defined with double subscripts to match those of vec $\mathbf{A}$ and vech $\mathbf{A}$, we have

$$[\psi]_{ij,ks} = \tfrac{1}{2}(\delta_{kj}\delta_{si} + \delta_{ki}\delta_{sj}), \qquad j \leqslant i, \tag{4.A.3}$$

where δ_{ij} is Kronecker's delta,

$$\delta_{ij} = \begin{cases} 1 & \text{if } j = i \\ 0 & \text{otherwise.} \end{cases}$$

Also,

$$[\mathbf{\Phi}]_{ij,ks} = (2 - \delta_{ij})[\psi]_{ks,ij}, \qquad s \leqslant k.$$

An element of the product of two matrices identified in the double subscript notation can be written as a double sum. Let

$$\mathbf{A} = (a_{ij,ks})$$

be an $mn \times pq$ matrix and let

$$\mathbf{B} = (b_{rt,uv})$$

be a $pq \times dr$ matrix. Then

$$[\mathbf{AB}]_{ij,uv} = \sum_{k=1}^{p} \sum_{s=1}^{q} a_{ij,ks} b_{ks,uv}. \tag{4.A.4}$$

Definition 4.A.4. The Kronecker product of a $p \times q$ matrix $\mathbf{A} = (a_{ij})$ and an $m \times n$ matrix $\mathbf{B} = (b_{ij})$, denoted by $\mathbf{A} \otimes \mathbf{B}$, is the $pm \times qn$ matrix

$$\mathbf{A} \otimes \mathbf{B} = \begin{bmatrix} a_{11}\mathbf{B} & a_{12}\mathbf{B} & \dots & a_{1q}\mathbf{B} \\ a_{21}\mathbf{B} & a_{22}\mathbf{B} & \dots & a_{2q}\mathbf{B} \\ \vdots & \vdots & & \vdots \\ a_{p1}\mathbf{B} & a_{p2}\mathbf{B} & \dots & a_{pq}\mathbf{B} \end{bmatrix}.$$

In the double subscript notation the (ij, ks)th element of $\mathbf{A} \otimes \mathbf{B}$ is

$$[\mathbf{A} \otimes \mathbf{B}]_{ij,ks} = a_{js} b_{ik}. \tag{4.A.5}$$

Some elementary properties of the Kronecker product are given in Result 4.A.1.

Result 4.A.1. Let the matrices **A**, **B**, **C**, and **D** be suitably conformable. Then

(i) $$(\mathbf{A} \otimes \mathbf{B})(\mathbf{C} \otimes \mathbf{D}) = (\mathbf{AC}) \otimes (\mathbf{BD}), \tag{4.A.6}$$

(ii) $$(\mathbf{A} \otimes \mathbf{B})^{-1} = \mathbf{A}^{-1} \otimes \mathbf{B}^{-1}, \tag{4.A.7}$$

(iii) $$(\mathbf{A} \otimes \mathbf{B})' = \mathbf{A}' \otimes \mathbf{B}', \tag{4.A.8}$$

(iv) $$(\mathbf{A} + \mathbf{B}) \otimes (\mathbf{C} + \mathbf{D}) = (\mathbf{A} \otimes \mathbf{C}) + (\mathbf{A} \otimes \mathbf{D}) + (\mathbf{B} \otimes \mathbf{C}) + (\mathbf{B} \otimes \mathbf{D}). \tag{4.A.9}$$

Proof. Reserved for the reader. □

The next two results establish relationships involving Kronecker products and the matrices $\mathbf{\Phi}$ and $\boldsymbol{\psi}$.

Result 4.A.2 Let the $p^2 \times p^2$ symmetric idempotent matrix $\boldsymbol{\kappa}_p$ be defined by

$$\boldsymbol{\kappa}_p = \mathbf{\Phi}_p \boldsymbol{\psi}_p = \mathbf{\Phi}_p(\mathbf{\Phi}_p'\mathbf{\Phi}_p)^{-1}\mathbf{\Phi}_p', \tag{4.A.10}$$

where $\boldsymbol{\Phi}_p$ is defined by (4.A.1) and ψ_p is defined in (4.A.2), and let $\mathbf{A}$ be a $p \times q$ matrix. Then

$$\boldsymbol{\kappa}_p(\mathbf{A} \otimes \mathbf{A}) = (\mathbf{A} \otimes \mathbf{A})\boldsymbol{\kappa}_q. \qquad (4.A.11)$$

Proof. Note that a typical element of $\boldsymbol{\kappa}_p$ is given by

$$[\boldsymbol{\kappa}_p]_{ij,st} = \tfrac{1}{2}(\delta_{is}\delta_{jt} + \delta_{it}\delta_{js}).$$

Using this definition we have

$$\begin{aligned}
[\boldsymbol{\kappa}_p(\mathbf{A} \otimes \mathbf{A})]_{ij,km} &= \sum_{s=1}^{p} \sum_{t=1}^{p} \tfrac{1}{2}(\delta_{is}\delta_{jt} + \delta_{it}\delta_{js})(\mathbf{A} \otimes \mathbf{A})_{st,km} \\
&= \tfrac{1}{2}[(\mathbf{A} \otimes \mathbf{A})_{ij,km} + (\mathbf{A} \otimes \mathbf{A})_{ji,km}] \\
&= \tfrac{1}{2}[(\mathbf{A} \otimes \mathbf{A})_{ij,km} + a_{im}a_{jk}] \\
&= \tfrac{1}{2}[(\mathbf{A} \otimes \mathbf{A})_{ij,km} + (\mathbf{A} \otimes \mathbf{A})_{ij,mk}] \\
&= \sum_{u=1}^{q} \sum_{v=1}^{q} \tfrac{1}{2}(\mathbf{A} \otimes \mathbf{A})_{ij,uv}(\delta_{uk}\delta_{vm} + \delta_{um}\delta_{vk}) \\
&= \sum_{u=1}^{q} \sum_{v=1}^{q} (\mathbf{A} \otimes \mathbf{A})_{ij,uv}[\boldsymbol{\kappa}_q]_{uv,km} \\
&= [(\mathbf{A} \otimes \mathbf{A})\boldsymbol{\kappa}_q]_{ij,km}. \qquad \square
\end{aligned}$$

The matrix $\boldsymbol{\kappa}_p$ is a projection matrix that creates a $p^2 \times p^2$ column whose (ij, km) element is the average of the (ij, km) and (ji, km) elements of the original $p^2 \times p^2$ vector. Result 4.A.2 follows from the fact that the average of $a_{ij}a_{km}$ and $a_{ji}a_{km}$ can be obtained as the average of the ij and ji elements of column km of $\mathbf{A} \otimes \mathbf{A}$ or as the average of the km and mk elements of row ij of $\mathbf{A} \otimes \mathbf{A}$.

Result 4.A.3.1. Let $\mathbf{A}$ be a $p \times p$ nonsingular matrix. Then

$$[\psi_p(\mathbf{A} \otimes \mathbf{A})\psi'_p]^{-1} = \boldsymbol{\Phi}'_p(\mathbf{A}^{-1} \otimes \mathbf{A}^{-1})\boldsymbol{\Phi}_p. \qquad (4.A.12)$$

Proof. We verify the result by multiplication,

$$\begin{aligned}
\psi_p(\mathbf{A} \otimes \mathbf{A})\psi'_p\boldsymbol{\Phi}'_p(\mathbf{A}^{-1} \otimes \mathbf{A}^{-1})\boldsymbol{\Phi}_p &= \psi_p(\mathbf{A} \otimes \mathbf{A})\boldsymbol{\kappa}_p(\mathbf{A}^{-1} \otimes \mathbf{A}^{-1})\boldsymbol{\Phi}_p \\
&= \psi_p(\mathbf{A} \otimes \mathbf{A})(\mathbf{A} \otimes \mathbf{A})^{-1}\boldsymbol{\kappa}_p\boldsymbol{\Phi}_p \\
&= \mathbf{I}. \qquad \square
\end{aligned}$$

Result 4.A.3.2. Let $\mathbf{L}$ (sometimes denoted by $\mathbf{L}_p$) be the $p^2 \times p$ matrix defined by

$$\mathbf{L} = (\mathbf{I}_{.1} \otimes \mathbf{I}_{.1}, \mathbf{I}_{.2} \otimes \mathbf{I}_{.2}, \ldots, \mathbf{I}_{.p} \otimes \mathbf{I}_{.p}), \qquad (4.A.13)$$

where $\mathbf{I}_{.i}$ is the ith column of the $p \times p$ identity matrix. Let $\mathbf{A}$ and $\mathbf{B}$ be $p \times p$ matrices. Then

(i) $(\mathbf{A} \otimes \mathbf{B})\mathbf{L} = (\mathbf{A}_{.1} \otimes \mathbf{B}_{.1}, \mathbf{A}_{.2} \otimes \mathbf{B}_{.2}, \ldots, \mathbf{A}_{.p} \otimes \mathbf{B}_{.p})$,
(ii) the jith element of $\mathbf{L}'(\mathbf{A} \otimes \mathbf{B})\mathbf{L}$ is $a_{ji}b_{ji}$,
(iii) $\text{vec}\{\text{diag } \mathbf{d}\} = \mathbf{Ld}$,

where $\mathbf{A}_{.i}$ is the ith column of $\mathbf{A}$, $\mathbf{B}_{.i}$ is the ith column of $\mathbf{B}$, $\mathbf{a}_{ji}$ is the jith element of $\mathbf{A}$, b_{ji} is the jith element of $\mathbf{B}$, and diag $\mathbf{d}$ is a $p \times p$ diagonal matrix whose diagonal is composed of the p elements of the vector $\mathbf{d}$.

Proof. By (4.A.6),

$$\begin{aligned}(\mathbf{A} \otimes \mathbf{B})(\mathbf{I}_{.i} \otimes \mathbf{I}_{.i}) &= (\mathbf{AI}_{.i}) \otimes (\mathbf{BI}_{.i}) \\ &= \mathbf{A}_{.i} \otimes \mathbf{B}_{.i}\end{aligned}$$

and the result (i) follows. Again applying (4.A.6) we have

$$\begin{aligned}(\mathbf{I}'_{.j} \otimes \mathbf{I}'_{.j})(\mathbf{A}_{.i} \otimes \mathbf{B}_{.i}) &= (\mathbf{I}'_{.j}\mathbf{A}_{.i}) \otimes (\mathbf{I}'_{.j}\mathbf{B}_{.i}) \\ &= a_{ji}b_{ji}\end{aligned}$$

and result (ii) is established. Result (iii) follows from the fact that the column vector $\mathbf{I}_{.i} \otimes \mathbf{I}_{.i}$ has a one in the $[(i-1)p + i]$ position and zeros elsewhere. □

The ψ matrix can be used to obtain a compact expression for the covariance matrix of the sample covariances of a normal distribution.

Lemma 4.A.1. Let $\mathbf{Z}_t$ be normally distributed with mean μ_Z and covariance matrix Σ_{ZZ}. Let

$$\mathbf{m}_{ZZ} = (n-1)^{-1} \sum_{t=1}^{n} (\mathbf{Z}_t - \bar{\mathbf{Z}})'(\mathbf{Z}_t - \bar{\mathbf{Z}}).$$

Then the covariance matrix of vech $\mathbf{m}_{ZZ}$ is

$$\begin{aligned}\mathbf{V}\{\text{vech } \mathbf{m}_{ZZ}\} &= \mathbf{V}\{\psi_p \text{ vec } \mathbf{m}_{ZZ}\} \\ &= 2(n-1)^{-1}\psi_p(\Sigma_{ZZ} \otimes \Sigma_{ZZ})\psi'_p.\end{aligned} \tag{4.A.14}$$

Furthermore,

$$\mathbf{V}^{-1}\{\text{vech } \mathbf{m}_{ZZ}\} = \tfrac{1}{2}(n-1)\mathbf{\Phi}'_p(\Sigma_{ZZ}^{-1} \otimes \Sigma_{ZZ}^{-1})\mathbf{\Phi}_p. \tag{4.A.15}$$

Proof. By the moment results of Appendix 1.B

$$C\{m_{ZZij}, m_{ZZk\ell}\} = (n-1)^{-1}(\sigma_{ZZik}\sigma_{ZZj\ell} + \sigma_{ZZi\ell}\sigma_{ZZjk}).$$

Now the ij, $k\ell$ element of $\psi(\Sigma_{ZZ} \otimes \Sigma_{ZZ})\psi'$ is, by (4.A.4) and (4.A.5),

$$\sum_{u=1}^{p} \sum_{v=1}^{p} \sum_{t=1}^{p} \sum_{s=1}^{p} \psi_{ij,ts}[\Sigma_{ZZ} \otimes \Sigma_{ZZ}]_{ts,uv}\psi_{k\ell,uv}$$

$$= \sum_{u=1}^{p} \sum_{v=1}^{p} \sum_{t=1}^{p} \sum_{s=1}^{p} \tfrac{1}{4}(\delta_{tj}\delta_{si} + \delta_{ti}\delta_{sj})\sigma_{ZZsv}\sigma_{ZZtu}(\delta_{u\ell}\delta_{vk} + \delta_{uk}\delta_{v\ell})$$

$$= \sum_{u=1}^{p} \sum_{v=1}^{p} \sum_{t=1}^{p} \sum_{s=1}^{p} \tfrac{1}{4}(\delta_{tj}\delta_{si}\delta_{u\ell}\delta_{vk} + \delta_{tj}\delta_{si}\delta_{uk}\delta_{v\ell}$$

$$+ \delta_{ti}\delta_{sj}\delta_{u\ell}\delta_{vk} + \delta_{ti}\delta_{sj}\delta_{uk}\delta_{v\ell})\sigma_{ZZsv}\sigma_{ZZtu}$$

$$= \tfrac{1}{2}(\sigma_{ZZik}\sigma_{ZZj\ell} + \sigma_{ZZi\ell}\sigma_{ZZjk})$$

and the first result is established. The inverse result follows by an application of Result 4.A.3. □

The next three results involve Kronecker products and the trace and vec operators.

Result 4.A.4. Let $\mathbf{A}$, $\mathbf{B}$, and $\mathbf{C}$ be $p \times q$, $q \times m$, and $m \times n$ matrices, respectively. Then

$$\text{vec}(\mathbf{ABC}) = (\mathbf{C}' \otimes \mathbf{A}) \text{ vec } \mathbf{B}. \tag{4.A.16}$$

Proof. Let $\mathbf{C}_{.j}$ be the jth column of $\mathbf{C}$. Then the jth p-dimensional subvector of vec($\mathbf{ABC}$) is

$$(\mathbf{ABC})_{.j} = \sum_{i=1}^{m} c_{ij}\mathbf{AB}_{.i}$$

$$= (\mathbf{C}'_{.j} \otimes \mathbf{A}) \text{ vec } \mathbf{B}$$

and the conclusion follows. □

Result 4.A.5. Let $\mathbf{A}$ and $\mathbf{B}$ be $p \times q$ and $q \times p$ matrices, respectively, and let the trace of $\mathbf{C}$, denoted by tr $\mathbf{C}$ or by tr$\{\mathbf{C}\}$, be the sum of the diagonal elements of a square matrix $\mathbf{C}$. Then

$$\text{tr}\{\mathbf{AB}\} = (\text{vec } \mathbf{A})' \text{ vec } \mathbf{B}' = (\text{vec } \mathbf{A}')' \text{ vec } \mathbf{B}. \tag{4.A.17}$$

Proof. The trace is

$$\text{tr}\{\mathbf{AB}\} = \sum_{i=1}^{p} \sum_{j=1}^{q} a_{ij}b_{ji}$$

$$= \sum_{i=1}^{p} \mathbf{A}_{i.}\mathbf{B}_{.i}$$

$$= (\text{vec } \mathbf{A}')' \text{ vec } \mathbf{B}.$$

The other result follows because tr$\{\mathbf{AB}\}$ = tr$\{\mathbf{BA}\}$. □

Result 4.A.6. Let **A**, **B**, **C**, and **D** be $p \times q$, $q \times m$, $p \times n$, and $n \times m$ matrices, respectively. Then

$$\operatorname{tr}\{\mathbf{ABD}'\mathbf{C}'\} = (\operatorname{vec} \mathbf{A})'(\mathbf{B} \otimes \mathbf{C})(\operatorname{vec} \mathbf{D}). \tag{4.A.18}$$

Proof. Now

$$\begin{aligned}\operatorname{tr}\{\mathbf{ABD}'\mathbf{C}'\} &= (\operatorname{vec} \mathbf{A})'\operatorname{vec}(\mathbf{CDB}') \\ &= (\operatorname{vec} \mathbf{A})'(\mathbf{B} \otimes \mathbf{C}) \operatorname{vec} \mathbf{D},\end{aligned}$$

where we have used Result 4.A.5 and Result 4.A.4. □

We next give some results on derivatives of matrix and vector functions.

Definition 4.A.5. Let $\mathbf{A} = \mathbf{A}(\boldsymbol{\theta})$ be a $p \times q$ matrix whose typical element $a_{ij} = a_{ij}(\boldsymbol{\theta})$ is a function of the r-dimensional column vector $\boldsymbol{\theta} = (\theta_1, \theta_2, \ldots, \theta_r)'$. Then

$$\frac{\partial \mathbf{A}}{\partial \theta_i} = \begin{bmatrix} \dfrac{\partial a_{11}}{\partial \theta_i} & \dfrac{\partial a_{12}}{\partial \theta_i} & \cdots & \dfrac{\partial a_{1q}}{\partial \theta_i} \\ \dfrac{\partial a_{21}}{\partial \theta_i} & \dfrac{\partial a_{22}}{\partial \theta_i} & \cdots & \dfrac{\partial a_{2q}}{\partial \theta_i} \\ \vdots & \vdots & & \vdots \\ \dfrac{\partial a_{p1}}{\partial \theta_i} & \dfrac{\partial a_{p2}}{\partial \theta_i} & \cdots & \dfrac{\partial a_{pq}}{\partial \theta_i} \end{bmatrix}.$$

The derivatives of products of matrices follow directly from Definition 4.A.5. We have, for example,

$$\frac{\partial \mathbf{AB}}{\partial \theta_i} = \frac{\partial \mathbf{A}}{\partial \theta_i} \mathbf{B} + \mathbf{A} \frac{\partial \mathbf{B}}{\partial \theta_i}. \tag{4.A.19}$$

Definition 4.A.6. Let $\mathbf{a} = \mathbf{a}(\boldsymbol{\theta})$ be a p-dimensional column vector whose typical element $a_j = a_j(\boldsymbol{\theta})$ is a function of the r-dimensional column vector $\boldsymbol{\theta}$. Then

$$\frac{\partial \mathbf{a}}{\partial \boldsymbol{\theta}'} = \begin{bmatrix} \dfrac{\partial a_1}{\partial \theta_1} & \dfrac{\partial a_1}{\partial \theta_2} & \cdots & \dfrac{\partial a_1}{\partial \theta_r} \\ \dfrac{\partial a_2}{\partial \theta_1} & \dfrac{\partial a_2}{\partial \theta_2} & \cdots & \dfrac{\partial a_2}{\partial \theta_r} \\ \vdots & \vdots & & \vdots \\ \dfrac{\partial a_p}{\partial \theta_1} & \dfrac{\partial a_p}{\partial \theta_2} & \cdots & \dfrac{\partial a_p}{\partial \theta_r} \end{bmatrix}.$$

Definition 4.A.7. The matrix

$$\frac{\partial \mathbf{a}'}{\partial \boldsymbol{\theta}} = \left[\frac{\partial \mathbf{a}}{\partial \boldsymbol{\theta}'}\right]'$$

is the transpose of the matrix of Definition 4.A.6.

Definition 4.A.8. Let $g(\mathbf{A})$ be a scalar function of the $p \times q$ matrix $\mathbf{A}$. Then

$$\frac{\partial g(\mathbf{A})}{\partial \mathbf{A}} = \begin{bmatrix} \frac{\partial g(\mathbf{A})}{\partial a_{11}} & \frac{\partial g(\mathbf{A})}{\partial a_{12}} & \cdots & \frac{\partial g(\mathbf{A})}{\partial a_{1q}} \\ \frac{\partial g(\mathbf{A})}{\partial a_{21}} & \frac{\partial g(\mathbf{A})}{\partial a_{22}} & \cdots & \frac{\partial g(\mathbf{A})}{\partial a_{2q}} \\ \vdots & \vdots & & \vdots \\ \frac{\partial g(\mathbf{A})}{\partial a_{p1}} & \frac{\partial g(\mathbf{A})}{\partial a_{p2}} & \cdots & \frac{\partial g(\mathbf{A})}{\partial a_{pq}} \end{bmatrix}.$$

The derivative of the determinant of a $p \times p$ nonsingular matrix has a simple form.

Result 4.A.7. Let $\mathbf{A}(\boldsymbol{\theta})$ be a nonsingular $p \times p$ matrix. Then

$$\frac{\partial |\mathbf{A}|}{\partial \mathbf{A}} = |\mathbf{A}|\mathbf{A}^{-1} \quad \text{and} \quad \frac{\partial \log|\mathbf{A}|}{\partial \mathbf{A}} = \mathbf{A}^{-1}.$$

Proof. The expansion of the determinant using cofactors is

$$|\mathbf{A}| = \sum_{j=1}^{p} a_{ij} \operatorname{Cof}(a_{ij}).$$

Because $\operatorname{Cof}(a_{ij})$ does not depend on a_{ij},

$$\frac{\partial |\mathbf{A}|}{\partial a_{ij}} = \operatorname{Cof}(a_{ij}) = a^{ij}|\mathbf{A}|,$$

where a^{ij} is the ijth element of $\mathbf{A}^{-1}$. It follows by the chain rule that

$$\frac{\partial \log|\mathbf{A}|}{\partial \mathbf{A}} = \mathbf{A}^{-1}. \qquad \square$$

Result 4.A.8. Let $\mathbf{A}(\boldsymbol{\theta})$ be a $p \times p$ matrix with positive determinant. Then

$$\frac{\partial \log|\mathbf{A}|}{\partial \theta_i} = \operatorname{tr}\left\{\mathbf{A}^{-1}\frac{\partial \mathbf{A}}{\partial \theta_i}\right\}.$$

Proof. Using Result 4.A.7 and the chain rule, we obtain

$$\frac{\partial \log|\mathbf{A}|}{\partial \theta_i} = \sum_{j=1}^{p} \sum_{k=1}^{p} \frac{\partial \log|\mathbf{A}|}{\partial a_{jk}} \frac{\partial a_{jk}}{\partial \theta_i}$$

$$= \operatorname{tr}\left\{\mathbf{A}^{-1} \frac{\partial \mathbf{A}}{\partial \theta_i}\right\}. \qquad \square$$

Note that if $\mathbf{A}$ is a symmetric matrix and $a_{tt} = \theta_i$, then

$$\frac{\partial \log|\mathbf{A}|}{\partial a_{tt}} = a^{tt},$$

and if $\mathbf{A}$ is symmetric and $A_{tj} = \theta_i$ for $t \neq j$, then

$$\frac{\partial \log|\mathbf{A}|}{\partial \theta_i} = 2a^{tj},$$

where a_{tj} is the tjth element of $\mathbf{A}$ and a^{tj} is the tjth element of $\mathbf{A}^{-1}$.

Result 4.A.9. Let $\mathbf{A} = \mathbf{A}(\boldsymbol{\theta})$ be a $p \times p$ nonsingular matrix. Then

$$\frac{\partial \mathbf{A}^{-1}}{\partial \theta_i} = -\mathbf{A}^{-1} \frac{\partial \mathbf{A}}{\partial \theta_i} \mathbf{A}^{-1}.$$

Proof. If $\mathbf{B}$ is a $p \times p$ matrix, we have

$$\frac{\partial \mathbf{AB}}{\partial \theta_i} = \mathbf{A} \frac{\partial \mathbf{B}}{\partial \theta_i} + \frac{\partial \mathbf{A}}{\partial \theta_i} \mathbf{B}.$$

Letting $\mathbf{B} = \mathbf{A}^{-1}$,

$$\mathbf{0} = \mathbf{A} \frac{\partial \mathbf{A}^{-1}}{\partial \theta_i} + \frac{\partial \mathbf{A}}{\partial \theta_i} \mathbf{A}^{-1}$$

and the conclusion follows. $\square$

If $\mathbf{A}$ is symmetric, note that

$$\frac{\partial a^{rs}}{\partial a_{ij}} = \begin{cases} -a^{rj}a^{is} - a^{ri}a^{js} & \text{for } i \neq j \\ -a^{ri}a^{is} & \text{for } i = j. \end{cases}$$

We next give some theorems about the roots of symmetric matrices. These lead to alternative representations for positive semidefinite symmetric matrices.

Result 4.A.10. (Courant–Fischer min–max theorem) Let $\mathbf{A}$ be a $p \times p$ symmetric matrix. Let $\lambda_1 \geqslant \lambda_2 \geqslant \cdots \geqslant \lambda_p$ be the characteristic roots of $\mathbf{A}$. Then

$$\begin{aligned}
\lambda_1 &= \max_{\mathbf{x}} (\mathbf{x}'\mathbf{x})^{-1}\mathbf{x}'\mathbf{A}\mathbf{x}, \\
\lambda_2 &= \min_{\mathbf{y} \neq \mathbf{0}} \max_{\mathbf{x}'\mathbf{y}=0} (\mathbf{x}'\mathbf{x})^{-1}\mathbf{x}'\mathbf{A}\mathbf{x}, \\
&\vdots \\
\lambda_k &= \min_{\substack{\mathbf{y}_i \neq 0 \\ i=1,\ldots,(k-1)}} \max_{\mathbf{x}'\mathbf{y}_i=0} (\mathbf{x}'\mathbf{x})^{-1}\mathbf{x}'\mathbf{A}\mathbf{x}, \\
&\vdots \\
\lambda_p &= \min_{\substack{\mathbf{y}_i \neq 0 \\ i=1,\ldots,(p-1)}} \max_{\mathbf{x}'\mathbf{y}_i=0} (\mathbf{x}'\mathbf{x})^{-1}\mathbf{x}'\mathbf{A}\mathbf{x} \\
&= \min_{\mathbf{x} \neq 0} (\mathbf{x}'\mathbf{x})^{-1}\mathbf{x}'\mathbf{A}\mathbf{x}.
\end{aligned}$$

Proof. See Bellman (1960, p. 115). □

The Courant–Fischer min–max theorem is typically stated with the condition $\mathbf{y}_i'\mathbf{y}_i = 1$, on the vectors over which the minimum is evaluated. Such a condition does not change the minimum value of the ratio.

Let $\mathbf{m}$ be a symmetric $p \times p$ matrix and let $\boldsymbol{\Sigma}$ be a symmetric positive definite $p \times p$ matrix. Then the roots $\lambda_1 \geqslant \lambda_2 \geqslant \cdots \geqslant \lambda_p$ of

$$|\mathbf{m} - \lambda\boldsymbol{\Sigma}| = 0 \tag{4.A.20}$$

are called the roots of $\mathbf{m}$ in the metric $\boldsymbol{\Sigma}$. There exists a matrix $\mathbf{T}$ such that

$$\mathbf{T}'\boldsymbol{\Sigma}\mathbf{T} = \mathbf{I}, \tag{4.A.21}$$

$$\mathbf{T}'\mathbf{m}\mathbf{T} = \text{diag}(\lambda_1, \lambda_2, \ldots, \lambda_p). \tag{4.A.22}$$

We call the columns of $\mathbf{T}$ the characteristic vectors of $\mathbf{m}$ in the metric $\boldsymbol{\Sigma}$. Some authors call any vector $\mathbf{T}_{.i}$ satisfying

$$(\mathbf{m} - \lambda_i\boldsymbol{\Sigma})\mathbf{T}_{.i} = \mathbf{0}$$

a characteristic vector of $\mathbf{m}$ in the metric $\boldsymbol{\Sigma}$, but we reserve the term for vectors that also satisfy (4.A.21).

The matrix $\mathbf{T}$ can be constructed as

$$\mathbf{T} = \mathbf{Q}\boldsymbol{\Gamma}^{-1/2}\mathbf{P},$$

where the columns of $\mathbf{Q}$ are the characteristic vectors of $\boldsymbol{\Sigma}$,

$$\mathbf{Q}'\boldsymbol{\Sigma}\mathbf{Q} = \boldsymbol{\Gamma} = \text{diag}\{\gamma_1, \gamma_2, \ldots, \gamma_p\},$$

$\mathbf{QQ}' = \mathbf{I}$, γ_i are the characteristic roots of $\boldsymbol{\Sigma}$, the columns of $\mathbf{P}$ are the characteristic vectors of $\boldsymbol{\Gamma}^{-1/2}\mathbf{Q}'\mathbf{mQ}\boldsymbol{\Gamma}^{-1/2}$, and $\boldsymbol{\Gamma}^{-1/2} = \text{diag}\{\gamma_1^{-1/2}, \gamma_2^{-1/2}, \ldots, \gamma_p^{-1/2}\}$.

The corollary of Result 4.A.10 for the roots of $\mathbf{m}$ in the metric $\boldsymbol{\Sigma}$ is used repeatedly in the text.

Corollary 4.A.10. Let $\mathbf{m}$ and $\boldsymbol{\Sigma}$ be $p \times p$ symmetric positive definite matrices. Let $\lambda_1 \geqslant \lambda_2 \geqslant \cdots \geqslant \lambda_p$ be the roots of the determinantal equation

$$|\mathbf{m} - \lambda\boldsymbol{\Sigma}| = 0.$$

Let $\mathbf{P}$ be a $p \times r$ matrix of rank r, where $r + k = p$. Then

$$\max_{\mathbf{P}} |\mathbf{P}'\boldsymbol{\Sigma}\mathbf{P}|^{-1}|\mathbf{P}'\mathbf{mP}| = \prod_{i=1}^{r} \lambda_i,$$

$$\max_{\mathbf{P}} \text{tr}\{(\mathbf{P}'\boldsymbol{\Sigma}\mathbf{P})^{-1}(\mathbf{P}'\mathbf{mP})\} = \sum_{i=1}^{r} \lambda_i,$$

$$\min_{\mathbf{P}} |\mathbf{P}'\boldsymbol{\Sigma}\mathbf{P}|^{-1}|\mathbf{P}'\mathbf{mP}| = \prod_{i=k+1}^{p} \lambda_i,$$

$$\min_{\mathbf{P}} \text{tr}\{(\mathbf{P}'\boldsymbol{\Sigma}\mathbf{P})^{-1}(\mathbf{P}'\mathbf{mP})\} = \sum_{i=k+1}^{p} \lambda_i.$$

The maxima are attained when the ith column of $\mathbf{P}$ is proportional to the ith characteristic vector of $\mathbf{m}$ in the metric $\boldsymbol{\Sigma}$, $i = 1, 2, \ldots, r$, and the minima are attained when the ith column of $\mathbf{P}$ is proportional to the $(k + i)$th characteristic vector of $\mathbf{m}$ in the metric $\boldsymbol{\Sigma}$.

Proof. Reserved for the reader. □

Result 4.A.11. Let $\mathbf{m}$ and $\boldsymbol{\Sigma}$ be $p \times p$ positive definite symmetric matrices. Let $\hat{\lambda}_1 > \hat{\lambda}_2 > \cdots > \hat{\lambda}_p$ be the roots of $\mathbf{m}$ in the metric $\boldsymbol{\Sigma}$ and let $\hat{\gamma}_1 < \hat{\gamma}_2 < \cdots < \hat{\gamma}_p$ be the roots of $\boldsymbol{\Sigma}$ in the metric $\mathbf{m}$. Let $\mathbf{T}_{.i}$ be the vectors of $\mathbf{m}$ in the metric $\boldsymbol{\Sigma}$ and let $\mathbf{R}_{.i}$ be the vectors of $\boldsymbol{\Sigma}$ in the metric $\mathbf{m}$. Then

(a) $\hat{\lambda}_i = \hat{\gamma}_i^{-1}$,

(b) $\mathbf{T}_{.i} = \hat{\lambda}_i^{1/2}\mathbf{R}_{.i}$ (or $\mathbf{T}_{.i} = -\hat{\lambda}_i^{1/2}\mathbf{R}_{.i}$),

(c) $\text{tr}\{\mathbf{m}\boldsymbol{\Sigma}^{-1}\} = \sum_{i=1}^{p} \hat{\lambda}_i$,

(d) $|\mathbf{m}| = |\boldsymbol{\Sigma}| \prod_{i=1}^{p} \hat{\lambda}_i$.

Proof. Reserved for the reader. □

Result 4.A.12. (Spectral decomposition) Let $\mathbf{m}$ be a real symmetric $p \times p$ matrix. Let $\lambda_1, \lambda_2, \ldots, \lambda_p$ be the, not necessarily distinct, roots of $\mathbf{m}$. Let $\mathbf{Q}$ be a matrix such that

$$\mathbf{Q}'\mathbf{m}\mathbf{Q} = \mathbf{\Lambda} = \text{diag}(\lambda_1, \lambda_2, \ldots, \lambda_p)$$

and $\mathbf{Q}'\mathbf{Q} = \mathbf{Q}\mathbf{Q}' = \mathbf{I}$. Then

$$\mathbf{m} = \sum_{i=1}^{p} \lambda_i \mathbf{Q}_{.i}\mathbf{Q}'_{.i},$$

where $\mathbf{Q}_{.i}$ is the ith column of $\mathbf{Q}$.

Proof. We have $\mathbf{m} = \mathbf{Q}\mathbf{\Lambda}\mathbf{Q}'$ and

$$\mathbf{Q}\mathbf{\Lambda}\mathbf{Q}' = (\mathbf{Q}_{.1}, \mathbf{Q}_{.2}, \ldots, \mathbf{Q}_{.p})(\lambda_1\mathbf{Q}_{.1}, \lambda_2\mathbf{Q}_{.2}, \ldots, \lambda_p\mathbf{Q}_{.p})'$$

$$= \sum_{i=1}^{p} \lambda_i \mathbf{Q}_{.i}\mathbf{Q}'_{.i}. \qquad \square$$

If there are repeated roots, it is possible to group the products associated with the repeated roots to obtain

$$\mathbf{m} = \sum_{i=1}^{d} \lambda_i \mathbf{A}_i,$$

where $\mathbf{d}$ is the number of distinct roots and

$$\mathbf{A}_i = \sum_{j=1}^{r_i} \mathbf{Q}_{j.}\mathbf{Q}'_{j.}$$

is the sum of the matrices associated with the ith distinct root. Note that $\mathbf{A}_i = \mathbf{A}'_i$ and $\mathbf{A}_i^2 = \mathbf{A}_i$.

Corollary 4.A.12. Let $\mathbf{m}$ be a real symmetric $p \times p$ matrix and let $\mathbf{\Sigma}$ be a real positive definite symmetric matrix. Let $\lambda_1 \geqslant \lambda_2 \geqslant \cdots \geqslant \lambda_p$ be the roots of $\mathbf{m}$ in the metric $\mathbf{\Sigma}$. Let $\mathbf{T}$ be a matrix such that

$$\mathbf{T}'\mathbf{m}\mathbf{T} = \text{diag}(\lambda_1, \lambda_2, \ldots, \lambda_p)$$

and $\mathbf{T}'\mathbf{\Sigma}\mathbf{T} = \mathbf{I}$. Then

$$\mathbf{m} = \sum_{i=1}^{p} \lambda_i \mathbf{\Sigma}\mathbf{T}_{.i}\mathbf{T}'_{.i}\mathbf{\Sigma},$$

where $\mathbf{T}_{.i}$ is the ith column of $\mathbf{T}$.

Proof. Reserved for the reader. $\square$

Result 4.A.13. (Positive square root) Let $\mathbf{m}$ be a real symmetric positive semidefinite $p \times p$ matrix. Let $\mathbf{Q}$ be the matrix such that

$$\mathbf{Q}'\mathbf{m}\mathbf{Q} = \text{diag}(\lambda_1, \lambda_2, \ldots, \lambda_p) = \mathbf{\Lambda}$$

and $\mathbf{Q}'\mathbf{Q} = \mathbf{Q}\mathbf{Q}' = \mathbf{I}$. Define the positive square root of $\mathbf{m}$ by

$$\mathbf{m}^{1/2} = \sum_{i=1}^{p} \lambda_i^{1/2}\mathbf{Q}_{.i}\mathbf{Q}'_{.i} = \mathbf{Q}\mathbf{\Lambda}^{1/2}\mathbf{Q}',$$

where

$$\mathbf{\Lambda}^{1/2} = \text{diag}(\lambda_1^{1/2}, \lambda_2^{1/2}, \ldots, \lambda_p^{1/2}),$$

the positive square roots of λ_i are chosen, and $\mathbf{Q}_{.i}$ is the ith column of $\mathbf{Q}$. Then

$$\mathbf{m} = [\mathbf{m}^{1/2}]^2 = \mathbf{m}^{1/2}\mathbf{m}^{1/2}.$$

If $\mathbf{m}$ is positive definite, $\mathbf{I} = \mathbf{m}^{-1/2}\mathbf{m}\mathbf{m}^{-1/2}$ and $\mathbf{m}^{-1} = [\mathbf{m}^{-1/2}]^2$, where

$$\mathbf{m}^{-1/2} = [\mathbf{m}^{1/2}]^{-1} = \mathbf{Q}\mathbf{\Lambda}^{-1/2}\mathbf{Q}'$$

and $\mathbf{\Lambda}^{-1/2} = \text{diag}(\lambda_1^{-1/2}, \lambda_2^{-1/2}, \ldots, \lambda_p^{-1/2})$

Proof. Reserved for the reader. □

The following results on the traces of products of matrices appear several places in the literature. Our proofs are based on those of Theobald (1975a, 1975b).

Result 4.A.14. Let $\mathbf{M}$ and $\mathbf{S}$ be real symmetric $p \times p$ matrices. Let the characteristic roots of $\mathbf{M}$ be $\lambda_{M1} \geqslant \lambda_{M2} \geqslant \cdots \geqslant \lambda_{Mp}$ and let the characteristic roots of $\mathbf{S}$ be $\lambda_{S1} \geqslant \lambda_{S2} \geqslant \cdots \geqslant \lambda_{Sp}$. Then

$$\text{tr}\{\mathbf{MS}\} \leqslant \text{tr}\{\mathbf{\Lambda}_M\mathbf{\Lambda}_S\}, \tag{4.A.23}$$

where

$$\mathbf{\Lambda}_M = \text{diag}(\lambda_{M1}, \lambda_{M2}, \ldots, \lambda_{Mp}),$$
$$\mathbf{\Lambda}_S = \text{diag}(\lambda_{S1}, \lambda_{S2}, \ldots, \lambda_{Sp}).$$

Furthermore, $\text{tr}\{\mathbf{MS}\} = \text{tr}\{\mathbf{\Lambda}_M\mathbf{\Lambda}_S\}$ if and only if there exists an orthogonal matrix $\mathbf{Q}$ such that

$$\mathbf{Q}'\mathbf{MQ} = \mathbf{\Lambda}_M \quad \text{and} \quad \mathbf{Q}'\mathbf{SQ} = \mathbf{\Lambda}_S. \tag{4.A.24}$$

Proof. Let $\mathbf{Q}_M$ be an orthogonal matrix composed of characteristic vectors of $\mathbf{M}$ and let $\mathbf{Q}_s$ be an orthogonal matrix composed of characteristic vectors of S. We have

$$\mathrm{tr}\{\mathbf{MS}\} = \mathrm{tr}\{\mathbf{Q}_M\mathbf{\Lambda}_M\mathbf{Q}'_M\mathbf{Q}_S\mathbf{\Lambda}_S\mathbf{Q}'_S\} = \mathrm{tr}\{\mathbf{Q}'_S\mathbf{Q}_M\mathbf{\Lambda}_M\mathbf{Q}'_M\mathbf{Q}_S\mathbf{\Lambda}_S\}.$$

Let $\mathbf{B} = \mathbf{Q}'_M\mathbf{Q}_S$ and note that $\mathbf{B}$ is an orthogonal matrix. Then

$$\mathrm{tr}\{\mathbf{B}'\mathbf{\Lambda}_M\mathbf{B}\mathbf{\Lambda}_S\} = \sum_{i=1}^{p}\sum_{j=1}^{p} b_{ji}\lambda_{Mj}b_{ji}\lambda_{Si} = \sum_{i=1}^{p}\sum_{j=1}^{p} b_{ji}^2\lambda_{Mj}\lambda_{Si},$$

where $\sum_{i=1}^{p} b_{ji}^2 = \sum_{j=1}^{p} b_{ji}^2 = 1$. Let $\mathbf{C}$ be the matrix with elements $c_{ij} = b_{ij}^2$. A matrix $\mathbf{C}$ with nonnegative elements that sum to one by both rows and columns is called a doubly stochastic matrix. Let c_{ij}, c_{jk} be two elements of $\mathbf{C}$ with $i > j$ and $k > j$. Let $a_{ik(j)} = \min(c_{ij}, c_{jk})$. If we replace c_{ij}, c_{jk}, c_{jj}, and c_{ik} of $\mathbf{C}$ by

$$c_{ij} - a_{ik(j)}, \quad c_{jk} - a_{ik(j)}, \quad c_{jj} + a_{ik(j)}, \quad \text{and} \quad c_{ik} + a_{ik(j)},$$

respectively, the new matrix, denoted by $\mathbf{C}_1$, has at least one zero element, is doubly stochastic, and

$$\boldsymbol{\lambda}'_M\mathbf{C}_1\boldsymbol{\lambda}_S = \boldsymbol{\lambda}'_M\mathbf{C}\boldsymbol{\lambda}_S + a_{ik(j)}(\lambda_{Mi} - \lambda_{Mj})(\lambda_{Sk} - \lambda_{Sj}),$$

where $\boldsymbol{\lambda}'_M = (\lambda_{M1}, \lambda_{M2}, \ldots, \lambda_{Mp})$ and $\boldsymbol{\lambda}'_S = (\lambda_{S1}, \lambda_{S2}, \ldots, \lambda_{Sp})$. The operation can be repeated until the matrix $\mathbf{C}$ is reduced to the $p \times p$ identity and the inequality result follows. If $\mathbf{Q}_M = \mathbf{Q}_S$, then $\mathbf{C} = \mathbf{I}$ and the equality result is established. □

Corollary 4.A.14. Let $\mathbf{M}$ and $\mathbf{S}$ be real symmetric matrices with $\mathbf{S}$ nonsingular. Then

$$\mathrm{tr}\{\mathbf{MS}^{-1}\} \geqslant \mathrm{tr}\{\mathbf{\Lambda}_M\mathbf{\Lambda}_S^{-1}\}, \tag{4.A.25}$$

where

$$\mathbf{\Lambda}_M = \mathrm{diag}(\lambda_{M1}, \lambda_{M2}, \ldots, \lambda_{Mp}),$$
$$\mathbf{\Lambda}_S = \mathrm{diag}(\lambda_{S1}, \lambda_{S2}, \ldots, \lambda_{Sp}),$$

$\lambda_{M1} \geqslant \lambda_{M2} \geqslant \cdots \geqslant \lambda_{Mp}$ are the roots of $\mathbf{M}$, and $\lambda_{S1} \geqslant \lambda_{S2} \geqslant \cdots \geqslant \lambda_{Sp}$ are the roots of $\mathbf{S}$. Expression (4.A.25) is an equality if and only if there exists a $\mathbf{Q}$ satisfying Equations (4.A.24).

Proof. Replace $\mathbf{S}$ by $-\mathbf{S}^{-1}$ in the proof of Result 4.A.14. □

Result 4.A.15. Let $\mathbf{M}$, $\mathbf{R}$, and $\mathbf{S}$ be real symmetric $p \times p$ matrices and let $\mathbf{S}$ be positive definite. Let $\gamma_{M1} \geqslant \gamma_{M2} \geqslant \cdots \geqslant \gamma_{Mp}$ be the roots of

$$|\mathbf{M} - \gamma\mathbf{S}| = 0,$$

and let $\gamma_{R1} \geqslant \gamma_{R2} \geqslant \cdots \geqslant \gamma_{Rp}$ be the roots of

$$|\mathbf{R} - \gamma\mathbf{S}| = 0.$$

Then

$$\operatorname{tr}\{\mathbf{M}(\mathbf{R} + \mathbf{S})^{-1}\} \geqslant \operatorname{tr}\{\mathbf{\Gamma}_M(\mathbf{I} + \mathbf{\Gamma}_R)^{-1}\}, \tag{4.A.26}$$

where $\mathbf{\Gamma}_M = \operatorname{diag}(\gamma_{M1}, \gamma_{M2}, \ldots, \gamma_{Mp})$ and $\mathbf{\Gamma}_R = \operatorname{diag}(\gamma_{R1}, \gamma_{R2}, \ldots, \gamma_{Rp})$. Furthermore, expression (4.A.26) is an equality if and only if there exists a $\mathbf{T}$ such that

$$\mathbf{T}'\mathbf{M}\mathbf{T} = \mathbf{\Gamma}_M, \quad \mathbf{T}'\mathbf{R}\mathbf{T} = \mathbf{\Gamma}_R, \quad \text{and} \quad \mathbf{T}'\mathbf{S}\mathbf{T} = \mathbf{I}. \tag{4.A.27}$$

Proof. Let $\mathbf{P}_M$ be an orthogonal matrix of characteristic vectors of the matrix $\mathbf{S}^{-1/2}\mathbf{M}\mathbf{S}^{-1/2}$, and let $\mathbf{P}_R$ be an orthogonal matrix of characteristic vectors of $\mathbf{S}^{-1/2}\mathbf{R}\mathbf{S}^{-1/2}$. Then $\mathbf{P}_R$ is a matrix such that

$$\mathbf{P}_R'(\mathbf{S}^{-1/2}\mathbf{R}\mathbf{S}^{-1/2} + \mathbf{I})\mathbf{P}_R = (\mathbf{\Gamma}_R + \mathbf{I}).$$

It follows from Corollary 4.A.14 that

$$\begin{aligned}\operatorname{tr}\{\mathbf{M}(\mathbf{R} + \mathbf{S})^{-1}\} &= \operatorname{tr}\{\mathbf{S}^{-1/2}\mathbf{M}\mathbf{S}^{-1/2}(\mathbf{S}^{-1/2}\mathbf{R}\mathbf{S}^{-1/2} + \mathbf{I})^{-1}\} \\ &\geqslant \operatorname{tr}\{\mathbf{\Gamma}_M(\mathbf{\Gamma}_R + \mathbf{I})^{-1}\}\end{aligned}$$

with equality holding if and only if there exists a $\mathbf{T}$ satisfying Equations (4.A.27). □

APPENDIX 4.B. PROPERTIES OF LEAST SQUARES AND MAXIMUM LIKELIHOOD ESTIMATORS

In this appendix we present theorems giving conditions under which the least squares and maximum likelihood estimators of parameters of multivariate measurement error models are consistent and asymptotically normally distributed. A critical part of the proof is the fact that the estimator defined implicitly as the value giving the maximum of a continuous function is itself a continuous function of the remaining arguments of the original function. We begin with two lemmas that establish this result. Lemma 4.B.1 is an adaptation of Lemma 1 of Jennrich (1969).

Lemma 4.B.1. Let $g(\mathbf{x}, \mathbf{y})$ be a continuous real valued function defined on the Cartesian product $A \times B$, where A is a subset of p-dimensional Euclidean space and B is a compact subset of q-dimensional Euclidean space.

Then the function

$$h(\mathbf{x}) = \max_{\mathbf{y} \in B} g(\mathbf{x}, \mathbf{y}) \tag{4.B.1}$$

is a continuous function of $\mathbf{x}$ on the interior of A.

Proof. Let $\mathbf{x}_0$ be any interior point of A. Then there exists a $\delta_0 > 0$ such that $\mathbf{x}$ is in A if $|\mathbf{x} - \mathbf{x}_0| \leqslant \delta_0$, where $|\mathbf{x} - \mathbf{x}_0|$ is the Euclidean norm of $\mathbf{x} - \mathbf{x}_0$. Because $S = \{(\mathbf{x}, \mathbf{y}): |\mathbf{x} - \mathbf{x}_0| \leqslant \delta_0, \mathbf{y} \in B\}$ is compact, $g(\mathbf{x}, \mathbf{y})$ is uniformly continuous on S. Therefore, for any $\varepsilon > 0$, there is a $\delta_1 > 0$ such that

$$|g(\mathbf{x}_1, \mathbf{y}_1) - g(\mathbf{x}_2, \mathbf{y}_2)| < \varepsilon$$

for all $(\mathbf{x}_1, \mathbf{y}_1)$ and $(\mathbf{x}_2, \mathbf{y}_2)$ in S satisfying $|\mathbf{x}_1 - \mathbf{x}_2| < \delta_1$ and $|\mathbf{y}_1 - \mathbf{y}_2| < \delta_1$. Hence, if

$$|\mathbf{x} - \mathbf{x}_0| < \delta,$$

where $\delta = \min(\delta_0, \delta_1)$, then

$$g(\mathbf{x}_0, \mathbf{y}) - \varepsilon < g(\mathbf{x}, \mathbf{y}) < g(\mathbf{x}_0, \mathbf{y}) + \varepsilon \tag{4.B.2}$$

for all $\mathbf{y} \in B$. Therefore, for $|\mathbf{x} - \mathbf{x}_0| < \delta$

$$\max_{\mathbf{y} \in B} g(\mathbf{x}_0, \mathbf{y}) - \varepsilon \leqslant \max_{\mathbf{y} \in B} g(\mathbf{x}, \mathbf{y}) \leqslant \max_{\mathbf{y} \in B} g(\mathbf{x}_0, \mathbf{y}) + \varepsilon.$$

Because ε is arbitrary, $h(\mathbf{x})$ is continuous at $\mathbf{x} = \mathbf{x}_0$, where $h(\mathbf{x})$ is defined in (4.B.1). Because $\mathbf{x}_0$ is an arbitrary interior point of A, $h(\mathbf{x})$ is continuous on the interior of A. □

Lemma 4.B.2. Let the assumptions of Lemma 4.B.1 hold. Let $\mathbf{x}_0$ be an interior point of A. Assume that the point $\mathbf{y}_0$ is the unique point for which $\max_{\mathbf{y} \in B} g(\mathbf{x}_0, \mathbf{y})$ is attained. Let $\mathbf{y}_M(\mathbf{x})$ be a point in B such that

$$g(\mathbf{x}, \mathbf{y}_M(\mathbf{x})) = \max_{\mathbf{y} \in B} g(\mathbf{x}, \mathbf{y}).$$

Then $\mathbf{y}_M(\mathbf{x})$ is a continuous function of $\mathbf{x}$ at $\mathbf{x} = \mathbf{x}_0$.

Proof. By assumption, $\mathbf{y}_0$ is unique and $g(\mathbf{x}_0, \mathbf{y})$ is continuous on the compact set B. Hence, for any $\varepsilon > 0$ there is an $\eta(\varepsilon) > 0$ such that

$$g(\mathbf{x}_0, \mathbf{y}) < g(\mathbf{x}_0, \mathbf{y}_0) - \eta(\varepsilon) \tag{4.B.3}$$

for all $\mathbf{y}$ in B satisfying $|\mathbf{y} - \mathbf{y}_0| \geqslant \varepsilon$. By Lemma 4.B.1, $h(\mathbf{x})$ of (4.B.1) is a continuous function of $\mathbf{x}$. Therefore, given $\eta(\varepsilon) > 0$, there exists a $\delta_2 > 0$ such that for all $|\mathbf{x} - \mathbf{x}_0| < \delta_2$

$$h(\mathbf{x}) > g(\mathbf{x}_0, \mathbf{y}_0) - \tfrac{1}{4}\eta(\varepsilon). \tag{4.B.4}$$

By Equation (4.B.2) there exists a $\delta_3 > 0$, such that for all $|\mathbf{x} - \mathbf{x}_0| < \delta_3$ and all $\mathbf{y} \in B$

$$g(\mathbf{x}, \mathbf{y}) < g(\mathbf{x}_0, \mathbf{y}) + \tfrac{1}{4}\eta(\varepsilon). \tag{4.B.5}$$

Thus, for all $\mathbf{y}$ in B satisfying $|\mathbf{y} - \mathbf{y}_0| \geqslant \varepsilon$ and $\mathbf{x}$ such that $|\mathbf{x} - \mathbf{x}_0| < \delta$, where $\delta = \min(\delta_2, \delta_3)$,

$$g(\mathbf{x}, \mathbf{y}) < g(\mathbf{x}_0, \mathbf{y}) + \tfrac{1}{4}\eta(\varepsilon) < g(\mathbf{x}_0, \mathbf{y}_0) - \tfrac{3}{4}\eta(\varepsilon),$$

where we have used (4.B.3). Thus, by (4.B.4), for $|\mathbf{x} - \mathbf{x}_0| < \delta$ and all $\mathbf{y}$ in B satisfying $|\mathbf{y} - \mathbf{y}_0| \geqslant \varepsilon$,

$$g(\mathbf{x}, \mathbf{y}) < g(\mathbf{x}, \mathbf{y}_M(\mathbf{x})) - \tfrac{1}{2}\eta(\varepsilon). \tag{4.B.6}$$

Therefore, $|\mathbf{x} - \mathbf{x}_0| < \delta$ implies $|\mathbf{y}_M(\mathbf{x}) - \mathbf{y}_0| < \varepsilon$ and $\mathbf{y}_M(\mathbf{x})$ is continuous at $\mathbf{x} = \mathbf{x}_0$. □

Theorem 4.B.1 Let $\{\mathbf{s}_n\}$ be a sequence of p-dimensional statistics. Assume $\mathbf{s}_n$ converges to the p-dimensional vector $\boldsymbol{\sigma}_0$ almost surely as $n \to \infty$. Let $g(\mathbf{s}, \gamma)$ be a continuous function defined on $A \times B$, where B is a compact set and $\boldsymbol{\sigma}_0$ is in the interior of A. Assume $\max_{\gamma \in B} g(\boldsymbol{\sigma}_0, \gamma)$ is uniquely attained at $\gamma = \gamma_0$. Let

$$\hat{\gamma} = \hat{\gamma}(\mathbf{s}_n)$$

be the value of γ such that

$$g(\mathbf{s}_n, \hat{\gamma}) = \max_{\gamma \in B} g(\mathbf{s}_n, \gamma).$$

Then $\hat{\gamma} \to \gamma_0$ almost surely as $n \to \infty$.

Proof. By Lemma 4.B.2, $\hat{\gamma}(\mathbf{s}_n)$ is a continuous function of $\mathbf{s}_n$ and the result follows. □

We now derive the limiting distribution of the estimator constructed by maximizing the normal likelihood for a covariance matrix.

Theorem 4.B.2. Let $\{\mathbf{Z}_t\}$ be a sequence of independent identically distributed random vectors with mean $\boldsymbol{\mu}_Z$, covariance matrix $\boldsymbol{\Sigma}_{ZZ}(\gamma)$, and finite fourth moments. Let $\boldsymbol{\Sigma}_{ZZ}(\gamma)$ be a continuous function of γ with continuous first and second derivatives. Let A, the parameter space for γ, be a convex compact subset of q-dimensional Euclidean space such that $\boldsymbol{\Sigma}_{ZZ}(\gamma)$ is positive definite for all γ in A. Assume that the true parameter value γ_0 is in the interior of A and that $\boldsymbol{\Sigma}_{ZZ}(\gamma) \neq \boldsymbol{\Sigma}_{ZZ}(\gamma_0)$ for any γ in A with $\gamma \neq \gamma_0$. Let $\hat{\gamma}$ be the value of γ that maximizes

$$f(\gamma, \mathbf{s}) = -\log|\boldsymbol{\Sigma}_{ZZ}(\gamma)| - \operatorname{tr}\{\mathbf{m}_{ZZ}\boldsymbol{\Sigma}_{ZZ}^{-1}(\gamma)\}, \tag{4.B.7}$$

where $\mathbf{s} = \operatorname{vech} \mathbf{m}_{ZZ}$. Let $\mathbf{F}$ be of full column rank, where

$$\mathbf{F} = \frac{\partial \operatorname{vech} \boldsymbol{\Sigma}_{ZZ}(\gamma_0)}{\partial \gamma'}$$

is the matrix of partial derivatives of $\boldsymbol{\Sigma}_{ZZ}(\gamma)$ with respect to γ evaluated at $\gamma = \gamma_0$. Then

$$\mathrm{n}^{1/2}(\hat{\gamma} - \gamma_0) \xrightarrow{L} N(\mathbf{0}, \mathbf{V}_{\gamma\gamma}), \tag{4.B.8}$$

where

$$\mathbf{V}_{\gamma\gamma} = [\mathbf{F}'\boldsymbol{\Omega}^{-1}\mathbf{F}]^{-1}\mathbf{F}'\boldsymbol{\Omega}^{-1}\boldsymbol{\Sigma}_{aa}\boldsymbol{\Omega}^{-1}\mathbf{F}[\mathbf{F}'\boldsymbol{\Omega}^{-1}\mathbf{F}]^{-1}, \tag{4.B.9}$$

$$\boldsymbol{\Sigma}_{aa} = E\{\mathbf{a}\mathbf{a}'\},$$

$$\mathbf{a} = \operatorname{vech}[(\mathbf{Z}_t - \boldsymbol{\mu}_Z)'(\mathbf{Z}_t - \boldsymbol{\mu}_Z) - \boldsymbol{\Sigma}_{ZZ}(\gamma_0)], \tag{4.B.10}$$

$$\boldsymbol{\Omega} = 2\psi_p[\boldsymbol{\Sigma}_{ZZ}(\gamma_0) \otimes \boldsymbol{\Sigma}_{ZZ}(\gamma_0)]\psi_p'. \tag{4.B.11}$$

Proof. By assumption, $\boldsymbol{\Sigma}_{ZZ}(\gamma)$ is nonsingular for all γ in A. Also, all partial derivatives of the first two orders of $\boldsymbol{\Sigma}_{ZZ}(\gamma)$ exist and are continuous functions of γ on A. Hence, all partial derivatives of the first two orders of the function (4.B.7) with respect to γ exist and are continuous functions of γ and of $\operatorname{vech} \mathbf{m}_{ZZ} = \mathbf{s}$ on the Cartesian product $A \times B$, where B is a subset of $2^{-1}p \times (p + 1)$-dimensional Euclidean space such that $\mathbf{s}$ in B implies that $\mathbf{m}_{ZZ}$ is positive definite. Then, by Taylor's theorem, for γ in A,

$$\frac{\partial f(\gamma, \mathbf{s})}{\partial \gamma} = \frac{\partial f(\gamma_0, \mathbf{s})}{\partial \gamma} + \frac{\partial^2 f(\gamma_{**}, \mathbf{s})}{\partial \gamma\, \partial \gamma'}(\gamma - \gamma_0), \tag{4.B.12}$$

where γ_{**} is used to denote the fact that the elements of the matrix are evaluated at points on the line segment joining γ_0 and γ. By Theorem 4.B.1, $\hat{\gamma}$ is consistent for γ_0. Therefore, with probability approaching one as n increases, $f(\gamma, \mathbf{s})$ attains its maximum at an interior point $\hat{\gamma}$ in A and

$$\frac{\partial f(\hat{\gamma}, \mathbf{s})}{\partial \gamma} = \mathbf{0} \tag{4.B.13}$$

with probability approaching one. Using (4.B.13), we evaluate (4.B.12) at $\gamma = \hat{\gamma}$ to obtain

$$\frac{\partial^2 f(\gamma_*, \mathbf{s})}{\partial \gamma\, \partial \gamma'}(\hat{\gamma} - \gamma_0) = -\frac{\partial f(\gamma_0, \mathbf{s})}{\partial \gamma}, \tag{4.B.14}$$

with probability approaching one, where γ_* is used to denote the fact that the elements of the matrix are evaluated at points on the line segment joining

γ_0 and $\hat{\gamma}$. By Theorem 4.B.1

$$\hat{\gamma} \to \gamma_0 \quad \text{a.s. as } n \to \infty. \tag{4.B.15}$$

Because $\mathbf{m}_{ZZ}$ is, essentially, composed of the sample means of independent identically distributed random variables with finite population means,

$$\mathbf{m}_{ZZ} \to \mathbf{\Sigma}_{ZZ}(\gamma_0), \quad \text{a.s. as } n \to \infty. \tag{4.B.16}$$

Thus,

$$\frac{\partial^2 f(\gamma_*, \mathbf{s})}{\partial \gamma \, \partial \gamma'} \to \mathbf{H}_0, \quad \text{a.s.,} \tag{4.B.17}$$

where

$$\mathbf{H}_0 = \frac{\partial^2 f(\gamma_0, \boldsymbol{\sigma}_0)}{\partial \gamma \, \partial \gamma'} \tag{4.B.18}$$

and $\boldsymbol{\sigma}_0 = \text{vech } \mathbf{\Sigma}_{ZZ}(\gamma_0)$, because the second derivative is a continuous function of γ and $\mathbf{s}$. The matrix $\mathbf{H}_0$ is negative definite because the function $f(\gamma, \boldsymbol{\sigma}_0)$ has continuous first and second derivatives and attains its maximum uniquely at an interior point $\gamma = \gamma_0$. Therefore, the probability that the matrix of partial derivatives on the left side of (4.B.18) is nonsingular approaches one as $n \to \infty$. Hence, by (4.B.14)

$$\hat{\gamma} - \gamma_0 = -\left(\frac{\partial^2 f(\gamma_*, \mathbf{s})}{\partial \gamma \partial \gamma'}\right)^{-1} \frac{\partial f(\gamma_0, \mathbf{s})}{\partial \gamma} \tag{4.B.19}$$

with probability approaching one as $n \to \infty$. For an element γ_i of γ,

$$\begin{aligned} \frac{\partial f(\gamma, \text{vech } \mathbf{m}_{ZZ})}{\partial \gamma_i} &= \text{tr}\left\{\frac{\partial \mathbf{\Sigma}_{ZZ}(\gamma)}{\partial \gamma_i} \mathbf{\Sigma}_{ZZ}^{-1}(\gamma)[\mathbf{m}_{ZZ} - \mathbf{\Sigma}_{ZZ}(\gamma)]\mathbf{\Sigma}_{ZZ}^{-1}(\gamma)\right\} \\ &= \left[\frac{\partial \text{ vec } \mathbf{\Sigma}_{ZZ}(\gamma)}{\partial \gamma_i}\right]' [\mathbf{\Sigma}_{ZZ}^{-1}(\gamma) \otimes \mathbf{\Sigma}_{ZZ}^{-1}(\gamma)] \text{ vec}[\mathbf{m}_{ZZ} - \mathbf{\Sigma}_{ZZ}(\gamma)], \end{aligned}$$

where we have used Results 4.A.8, 4.A.9, and 4.A.6. It follows that

$$\begin{aligned} \frac{\partial f(\gamma_0, \text{vech } \mathbf{m}_{ZZ})}{\partial \gamma} &= \mathbf{F}'\mathbf{\Phi}_p'\{\mathbf{\Sigma}_{ZZ}^{-1}(\gamma_0) \otimes \mathbf{\Sigma}_{ZZ}^{-1}(\gamma_0)\}\mathbf{\Phi}_p \text{ vech}[\mathbf{m}_{ZZ} - \mathbf{\Sigma}_{ZZ}(\gamma_0)] \\ &= 2\mathbf{F}'\mathbf{\Omega}^{-1} \text{ vech}[\mathbf{m}_{ZZ} - \mathbf{\Sigma}_{ZZ}(\gamma_0)], \end{aligned} \tag{4.B.20}$$

where $\mathbf{\Omega}^{-1}$ is defined in terms of $\mathbf{\Sigma}_{ZZ}^{-1}$ in Lemma 4.A.1. For elements γ_i and γ_j of γ,

$$\begin{aligned} \frac{\partial^2 f(\gamma, \text{vech } \mathbf{m}_{ZZ})}{\partial \gamma_i \, \partial \gamma_j} = -\text{tr}\bigg\{&\frac{\partial \mathbf{\Sigma}_{ZZ}(\gamma)}{\partial \gamma_i} \mathbf{\Sigma}_{ZZ}^{-1}(\gamma) \frac{\partial \mathbf{\Sigma}_{ZZ}(\gamma)}{\partial \gamma_j} \mathbf{\Sigma}_{ZZ}^{-1}(\gamma)[2\mathbf{m}_{ZZ} - \mathbf{\Sigma}_{ZZ}(\gamma)]\mathbf{\Sigma}_{ZZ}^{-1}(\gamma) \\ &- \frac{\partial^2 \mathbf{\Sigma}_{ZZ}(\gamma)}{\partial \gamma_i \, \partial \gamma_j} \mathbf{\Sigma}_{ZZ}^{-1}(\gamma)[\mathbf{m}_{ZZ} - \mathbf{\Sigma}_{ZZ}(\gamma)]\mathbf{\Sigma}_{ZZ}^{-1}(\gamma)\bigg\}. \end{aligned} \tag{4.B.21}$$

Because the partial derivatives are continuous, we have, by (4.B.15), (4.B.17), and (4.B.21),

$$\frac{\partial^2 f(\gamma_*, \operatorname{vech} \mathbf{m}_{ZZ})}{\partial\gamma\,\partial\gamma'} = -2\mathbf{F}'\boldsymbol{\Omega}^{-1}\mathbf{F} + o_p(1). \tag{4.B.22}$$

It follows from (4.B.19), (4.B.20), and (4.B.22) that

$$\hat{\gamma} - \gamma_0 = [\mathbf{F}'\boldsymbol{\Omega}^{-1}\mathbf{F}]^{-1}\mathbf{F}'\boldsymbol{\Omega}^{-1} \operatorname{vech}[\mathbf{m}_{ZZ} - \boldsymbol{\Sigma}_{ZZ}(\gamma_0)] + o_p(n^{-1/2}). \tag{4.B.23}$$

The conclusion follows from (4.B.23) and the fact that $n^{1/2} \operatorname{vech}[\mathbf{m}_{ZZ} - \boldsymbol{\Sigma}_{ZZ}(\gamma_0)]$ converges in distribution to a normal vector. See Theorem 1.C.1 □

Corollary 4.B.2. Let the assumptions of Theorem 4.B.2 hold. Assume that the $\mathbf{Z}_t$ are normally distributed. Then

$$n^{1/2}(\hat{\gamma} - \gamma_0) \xrightarrow{L} N(\mathbf{0}, [\mathbf{F}'\boldsymbol{\Omega}^{-1}\mathbf{F}]^{-1}), \tag{4.B.24}$$

where $\boldsymbol{\Omega}$ and $\mathbf{F}$ are defined in Theorem 4.B.2.

Proof. The conclusion is immediate because the $\boldsymbol{\Sigma}_{aa}$ of (4.B.9) is equal to $\boldsymbol{\Omega}$ for normally distributed $\mathbf{Z}_t$. See Lemma 4.A.1. □

Theorem 4.B.3. Let

$$\mathbf{Z}_t = (\boldsymbol{\beta}_0, \mathbf{0}) + \mathbf{x}_t(\boldsymbol{\beta}, \mathbf{I}) + \boldsymbol{\varepsilon}_t, \tag{4.B.25}$$

where the $\boldsymbol{\varepsilon}_t$ are independent identically distributed random variables with zero mean vector, covariance matrix $\boldsymbol{\Sigma}_{\varepsilon\varepsilon}$, and finite fourth moments. Let $\{\mathbf{x}_t\}$ be a sequence of k-dimensional fixed vectors. Let γ_β be the vector containing the unknown portion of $\boldsymbol{\beta}$ and let γ_ε be the vector containing the unknown portion of $\boldsymbol{\Sigma}_{\varepsilon\varepsilon}$. Assume that

$$\gamma' = [\gamma'_\beta, \gamma'_\varepsilon, (\operatorname{vech} \mathbf{m}_{xx})'] = [\gamma'_\beta, \gamma'_\varepsilon, \gamma'_x]$$

is an element of a convex, compact parameter space A. Assume that for any γ in A, $\mathbf{m}_{xx}$ and

$$\boldsymbol{\Sigma}_{ZZ}(\gamma) = (\boldsymbol{\beta}, \mathbf{I})'\mathbf{m}_{xx}(\boldsymbol{\beta}, \mathbf{I}) + \boldsymbol{\Sigma}_{\varepsilon\varepsilon}$$

are positive definite. For $n > k$, let

$$\gamma'_n = [\gamma'_{\beta 0}, \gamma'_{\varepsilon 0}, (\operatorname{vech} \mathbf{m}_{xxn})'] = [\gamma'_{\beta 0}, \gamma'_{\varepsilon 0}, \gamma'_{xn}]$$

be the true value of γ in A and let

$$\lim_{n\to\infty} \mathbf{m}_{xxn} = \bar{\mathbf{m}}_{xx},$$

where $\mathbf{m}_{xxn}$ and $\bar{\mathbf{m}}_{xx}$ are positive definite. Assume that

$$\gamma'_0 = [\gamma'_{\beta 0}, \gamma'_{\varepsilon 0}, (\operatorname{vech} \bar{\mathbf{m}}_{xx})'] = [\gamma'_{\beta 0}, \gamma'_{\varepsilon 0}, \gamma'_{x0}]$$

is an interior point of A. Assume that for any γ in A, with $\gamma \neq \gamma_0$, $\Sigma_{ZZ}(\gamma) \neq \Sigma_{ZZ}(\gamma_0)$. Let

$$\Gamma_{aan} = \mathbf{V}\{(n-1)^{1/2} \operatorname{vech}(\mathbf{m}_{ZZ} - \mathbf{m}_{zzn} - \Sigma_{\varepsilon\varepsilon})\},$$

$$\bar{\Gamma}_{aa} = \lim_{n\to\infty} \Gamma_{aan}, \tag{4.B.26}$$

$$\mathbf{F} = \frac{\partial \operatorname{vech} \Sigma_{ZZ}(\gamma_0)}{\partial \gamma'}, \tag{4.B.27}$$

where $\mathbf{F}$ is the matrix of partial derivatives of vech $\Sigma_{ZZ}(\gamma)$ with respect to γ' evaluated at $\gamma = \gamma_0$ and $\mathbf{F}$ is full column rank. Then

$$\hat{\gamma} \to \gamma_0, \quad \text{a.s.},$$
$$n^{1/2}(\hat{\gamma} - \gamma_n) \xrightarrow{L} N(\mathbf{0}, \Gamma_{\gamma\gamma}),$$

where $\hat{\gamma}$ is the value of γ that maximizes (4.B.7),

$$\Gamma_{\gamma\gamma} = (\mathbf{F}'\boldsymbol{\Omega}^{-1}\mathbf{F})^{-1}\mathbf{F}'\boldsymbol{\Omega}^{-1}\bar{\Gamma}_{aa}\boldsymbol{\Omega}^{-1}\mathbf{F}(\mathbf{F}'\boldsymbol{\Omega}^{-1}\mathbf{F})^{-1}, \tag{4.B.28}$$
$$\boldsymbol{\Omega} = 2\psi_p[\Sigma_{ZZ}(\gamma_0) \otimes \Sigma_{ZZ}(\gamma_0)]\psi_p'.$$

Proof. Because $f(\gamma, \mathbf{s})$ of (4.B.7) evaluated at $\mathbf{s} = \operatorname{vech} \Sigma_{ZZ}(\gamma_0)$ has a unique maximum on A at $\gamma = \gamma_0$ and because $\mathbf{m}_{ZZ} \to \Sigma_{ZZ}(\gamma_0)$, a.s., the consistency result follows by Theorem 4.B.1. The arguments used in the proof of Theorem 4.B.2 can be used to show that

$$\hat{\gamma} - \gamma_n = (\mathbf{F}'\boldsymbol{\Omega}^{-1}\mathbf{F})^{-1}\mathbf{F}'\boldsymbol{\Omega}^{-1} \operatorname{vech}[\mathbf{m}_{ZZ} - \Sigma_{ZZ}(\gamma_n)] + o_p(n^{-1/2}). \tag{4.B.29}$$

The result then follows by Theorem 1.C.2 of Appendix 1.C. □

Corollary 4.B.3. Let the assumptions of Theorem 4.B.3 hold and, in addition, assume the ε_t to be normally distributed. Then

$$n^{1/2}[(\hat{\gamma}_\beta', \hat{\gamma}_\varepsilon')' - (\gamma_{\beta 0}', \gamma_{\varepsilon 0}')'] \xrightarrow{L} N(\mathbf{0}, \mathbf{G}_{11}), \tag{4.B.30}$$

where $\mathbf{G}_{11}$ is the upper left block of the matrix $(\mathbf{F}'\boldsymbol{\Omega}^{-1}\mathbf{F})^{-1}$ and $\mathbf{F}$ and $\boldsymbol{\Omega}$ are defined in Theorem 4.B.3.

Proof. This result follows from Theorem 4.2.4 of Section 4.2.3. □

We now give the limiting distribution of the estimator $\hat{\gamma}$, where $\hat{\gamma}$ is obtained by applying generalized least squares to the elements of the sample covariance matrix.

Theorem 4.B.4. Let $\{\mathbf{Z}_t\}$ be a sequence of independent identically distributed random vectors with mean $\boldsymbol{\mu}_Z$, covariance matrix $\Sigma_{ZZ}(\gamma)$, and finite fourth moments. Let $\mathbf{\Sigma}_{ZZ}(\gamma)$ be a continuous function of γ with continuous first and second derivatives. Let A, the parameter space for γ, be a convex, compact subset of q-dimensional Euclidean space such that $\mathbf{\Sigma}_{ZZ}(\gamma)$ is positive definite. Let γ_0, the true parameter value, be an interior point of A. Assume that for any γ in A, with $\gamma \neq \gamma_0$, $\mathbf{\Sigma}_{zz}(\gamma) \neq \mathbf{\Sigma}_{zz}(\gamma_0)$. Let $\tilde{\gamma}$ be the value of γ that minimizes

$$f_\ell(\gamma, \mathbf{s}; \mathbf{G}_{aan}) = [\mathbf{s} - \boldsymbol{\sigma}(\gamma)]'\mathbf{G}_{aan}[\mathbf{s} - \boldsymbol{\sigma}(\gamma)], \tag{4.B.31}$$

where $\mathbf{s} = \text{vech } \mathbf{m}_{ZZ}$, $\boldsymbol{\sigma}(\gamma) = \text{vech } \mathbf{\Sigma}_{ZZ}(\gamma)$, and $\mathbf{G}_{aan}$ is a positive definite random matrix that converges in probability to a positive definite matrix $\bar{\mathbf{G}}_{aa}$ as n approaches infinity. Let $\mathbf{F}$ be of full column rank, where $\mathbf{F}$ is the matrix of partial derivatives of $\boldsymbol{\sigma}(\gamma)$ with respect to γ' evaluated at $\gamma = \gamma_0$. Then

$$n^{1/2}(\tilde{\gamma} - \gamma_0) \xrightarrow{L} N(\mathbf{0}, \mathbf{V}_{\gamma\gamma}),$$

where

$$\mathbf{V}_{\gamma\gamma} = (\mathbf{F}'\bar{\mathbf{G}}_{aa}\mathbf{F})^{-1}\mathbf{F}'\bar{\mathbf{G}}_{aa}\mathbf{\Sigma}_{aa}\bar{\mathbf{G}}_{aa}\mathbf{F}(\mathbf{F}'\bar{\mathbf{G}}_{aa}\mathbf{F})^{-1}, \tag{4.B.32}$$

$\mathbf{\Sigma}_{aa} = E\{\mathbf{aa}'\}$, and $\mathbf{a} = \text{vech}[(\mathbf{Z}_t - \boldsymbol{\mu}_Z)'(\mathbf{Z}_t - \boldsymbol{\mu}_Z) - \mathbf{\Sigma}_{ZZ}(\gamma_0)]$.

Proof. Using the arguments associated with (4.B.12) and (4.B.13), we can write, with probability approaching one as $n \to \infty$,

$$\frac{\partial^2 f_\ell(\gamma_*, \mathbf{s}; \mathbf{G}_{aan})}{\partial\gamma\, \partial\gamma'}(\tilde{\gamma} - \gamma_0) = -\frac{\partial f_\ell(\gamma_0, \mathbf{s}; \mathbf{G}_{aan})}{\partial\gamma}, \tag{4.B.33}$$

where γ_* is on the line segment joining γ_0 and $\tilde{\gamma}$. Because $(\tilde{\gamma}, \mathbf{s}; \mathbf{G}_{aan})$ is converging in probability to $(\gamma_0, \boldsymbol{\sigma}(\gamma_0); \bar{\mathbf{G}}_{aa})$, because the second derivatives are continuous, and because $f_\ell(\gamma, \boldsymbol{\sigma}(\gamma_0); \bar{\mathbf{G}}_{aa})$ has a unique minimum at $\gamma = \gamma_0$,

$$\tilde{\gamma} - \gamma_0 = -\left(\frac{\partial^2 f_\ell(\gamma_*, \mathbf{s}; \mathbf{G}_{aan})}{\partial\gamma\, \partial\gamma'}\right)^{-1} \frac{\partial f_\ell(\gamma_0, s; \mathbf{G}_{aan})}{\partial\gamma}$$

with probability approaching one as $n \to \infty$. Also,

$$\frac{\partial^2 f_\ell(\gamma_*, \mathbf{s}; \mathbf{G}_{aan})}{\partial\gamma\, \partial\gamma'} \xrightarrow{P} 2\mathbf{F}'\bar{\mathbf{G}}_{aa}\mathbf{F}, \tag{4.B.34}$$

and

$$\frac{\partial f_\ell(\gamma_0, \mathbf{s}; \mathbf{G}_{aan})}{\partial\gamma} = -2\mathbf{F}'\bar{\mathbf{G}}_{aa}[\mathbf{s} - \boldsymbol{\sigma}(\gamma_0)] + o_p(n^{-1/2}). \tag{4.B.35}$$

The result follows because, by Theorem 1.C.1, $[\mathbf{s} - \boldsymbol{\sigma}(\gamma_0)]$ is converging in distribution to a multivariate normal random variable with mean zero and covariance matrix $\mathbf{\Sigma}_{aa}$. □

Corollary 4.B.4. Let the assumptions of Theorem 4.B.4 hold. In addition, assume that the $\mathbf{Z}_t$ are normally distributed and that

$$\mathbf{G}_{aan} = \tfrac{1}{2}\mathbf{\Phi}_p'(\mathbf{m}_{ZZ}^{-1} \otimes \mathbf{m}_{ZZ}^{-1})\mathbf{\Phi}_p.$$

Then

$$n^{1/2}(\tilde{\gamma} - \gamma_0) \xrightarrow{L} N(\mathbf{0}, [\mathbf{F}'\mathbf{\Omega}^{-1}\mathbf{F}]^{-1}),$$

where $\mathbf{\Omega}$ is defined in Theorem 4.B.2.

Proof. Under the assumptions, $\mathbf{\Sigma}_{aa} = \mathbf{\Omega}$ and

$$2\psi_p(\mathbf{m}_{ZZ} \otimes \mathbf{m}_{ZZ})\psi_p' \xrightarrow{P} \mathbf{\Omega}.$$

The result then follows from (4.B.32). □

APPENDIX 4.C. MAXIMUM LIKELIHOOD ESTIMATION FOR SINGULAR MEASUREMENT COVARIANCE

In this appendix we derive the maximum likelihood estimator for the multivariate model, permitting the known error covariance matrix to be singular. The formulation provides a unified treatment for models that contain both explanatory variables measured without error and explanatory variables measured with error.

Theorem 4.C.1. Let the normal model (4.1.12) hold with $\mathbf{\Sigma}_{ZZ}$ and $\mathbf{\Sigma}_{xx}$ nonsingular. Let $\hat{\lambda}_1^{-1} \leqslant \hat{\lambda}_2^{-1} \leqslant \ldots \leqslant \hat{\lambda}_p^{-1}$ be the values of λ_i^{-1} that satisfy

$$|\mathbf{\Sigma}_{\varepsilon\varepsilon} - \lambda^{-1}\mathbf{m}_{ZZ}| = 0, \tag{4.C.1}$$

where the first ℓ of the $\hat{\lambda}^{-1}$ are zero and $p - \ell$ is the rank of $\mathbf{\Sigma}_{\varepsilon\varepsilon}$. Let

$$\mathbf{R} = (\mathbf{R}_{.1}, \mathbf{R}_{.2}, \ldots, \mathbf{R}_{.p}) \tag{4.C.2}$$

be the matrix of characteristic vectors of $\mathbf{\Sigma}_{\varepsilon\varepsilon}$ in the metric $\mathbf{m}_{ZZ}$ where the first $\ell(\ell \geqslant 0)$ vectors are associated with zero roots of (4.C.1) and let $\hat{\lambda}_k^{-1} < 1$. Then the maximum likelihood estimator of $\boldsymbol{\beta}$ is

$$\hat{\boldsymbol{\beta}} = (\hat{\mathbf{A}}_{rk}\hat{\mathbf{A}}_{kk}^{-1})', \tag{4.C.3}$$

where $\hat{\mathbf{A}} = [\hat{\mathbf{A}}_{rk}', \hat{\mathbf{A}}_{kk}']'$ and

$$\hat{\mathbf{A}} = \mathbf{m}_{ZZ}[(1 - \hat{\lambda}_1^{-1})^{1/2}\mathbf{R}_{.1}, (1 - \hat{\lambda}_2^{-1})^{1/2}\mathbf{R}_{.2}, \ldots, (1 - \hat{\lambda}_k^{-1})^{1/2}\mathbf{R}_{.k}].$$

If τ, the number of $\hat{\lambda}_i^{-1}$ less than one, is less than k, the supremum of the likelihood is attained with $\hat{\mathbf{\Sigma}}_{zz} = \hat{\mathbf{A}}_\tau\hat{\mathbf{A}}_\tau'$, where

$$\hat{\mathbf{A}}_\tau = \mathbf{m}_{ZZ}[(1 - \hat{\lambda}_1^{-1})^{1/2}\mathbf{R}_{.1}, (1 - \hat{\lambda}_2^{-1})^{1/2}\mathbf{R}_{.2}, \ldots, (1 - \hat{\lambda}_\tau^{-1})^{1/2}\mathbf{R}_{.\tau}].$$

Proof. Let $\log L_c$ be the logarithm of the likelihood. Then, by Results 4.A.7, 4.A.8, and 4.A.9, we have

$$\frac{\partial \log L_c}{\partial A_{ij}} = \tfrac{1}{2}(n-1)\,\mathrm{tr}\left\{\boldsymbol{\Sigma}_{ZZ}^{-1}(\mathbf{m}_{ZZ} - \boldsymbol{\Sigma}_{ZZ})\boldsymbol{\Sigma}_{ZZ}^{-1}\frac{\partial \boldsymbol{\Sigma}_{ZZ}}{\partial A_{ij}}\right\}, \tag{4.C.4}$$

where

$$\frac{\partial \boldsymbol{\Sigma}_{ZZ}}{\partial A_{ij}} = \mathbf{A}\frac{\partial \mathbf{A}'}{\partial A_{ij}} + \frac{\partial \mathbf{A}}{\partial A_{ij}}\mathbf{A}'. \tag{4.C.5}$$

If the derivatives of (4.C.4) are to be zero, the elements of $\mathbf{A}$ must satisfy the equations

$$\boldsymbol{\Sigma}_{ZZ}^{-1}(\mathbf{m}_{ZZ} - \boldsymbol{\Sigma}_{ZZ})\boldsymbol{\Sigma}_{ZZ}^{-1}\mathbf{A} = \mathbf{0}. \tag{4.C.6}$$

Because $\boldsymbol{\Sigma}_{ZZ}$ is nonsingular, the sample moment matrix will be nonsingular with probability one for n greater than p. Therefore, premultiplying (4.C.6) by $\boldsymbol{\Sigma}_{ZZ}\mathbf{m}_{ZZ}^{-1}\boldsymbol{\Sigma}_{ZZ}$, we have

$$\mathbf{A} = (\mathbf{A}\mathbf{A}' + \boldsymbol{\Sigma}_{\varepsilon\varepsilon})\mathbf{m}_{ZZ}^{-1}\mathbf{A} \tag{4.C.7}$$

or

$$\mathbf{A}(\mathbf{I} - \mathbf{A}'\mathbf{m}_{ZZ}^{-1}\mathbf{A}) = \boldsymbol{\Sigma}_{\varepsilon\varepsilon}\mathbf{m}_{ZZ}^{-1}\mathbf{A}. \tag{4.C.8}$$

Recall that there are $k \times p$ elements in $\mathbf{A}$, but only $rk + \frac{1}{2}k(k+1)$ of these elements are free parameters. Therefore, we may impose $\frac{1}{2}k(k-1)$ restrictions on the elements of $\mathbf{A}$. One method of imposing these restrictions is to require that

$$\mathbf{A}'\mathbf{m}_{ZZ}^{-1}\mathbf{A} = \mathbf{D}, \tag{4.C.9}$$

where $\mathbf{D}$ is diagonal. Applying these restrictions to (4.C.8) we obtain

$$\boldsymbol{\Sigma}_{\varepsilon\varepsilon}\mathbf{m}_{ZZ}^{-1}\mathbf{A} = \mathbf{A}(\mathbf{I} - \mathbf{D}). \tag{4.C.10}$$

It follows that the columns of $\mathbf{A}$, denoted by $\mathbf{A}_{.i}$, must satisfy

$$[\boldsymbol{\Sigma}_{\varepsilon\varepsilon} - (1 - d_{ii})\mathbf{m}_{ZZ}]\mathbf{m}_{ZZ}^{-1}\mathbf{A}_{.i} = \mathbf{0}, \qquad i = 1, 2, \ldots, k, \tag{4.C.11}$$

where d_{ii} is the ith diagonal element of $\mathbf{D}$.

Expression (4.C.11) will be satisfied if $(1 - d_{ii}) = \hat{\lambda}_i^{-1}$ and $\mathbf{m}_{ZZ}^{-1}\mathbf{A}_{.i}$ is proportional to a characteristic vector of $\boldsymbol{\Sigma}_{\varepsilon\varepsilon}$ in the metric $\mathbf{m}_{ZZ}$. By (4.C.8) and (4.C.9) the constant of proportionality must be $(1 - \hat{\lambda}_i^{-1})^{1/2}$, where $\hat{\lambda}_i^{-1}$ is the corresponding root of $\boldsymbol{\Sigma}_{\varepsilon\varepsilon}$ in the metric $\mathbf{m}_{ZZ}$. Therefore, $\hat{\mathbf{A}}$ is of the form given in (4.C.3), but it remains to determine which $\hat{\lambda}_i^{-1}$ are to be used to construct $\hat{\mathbf{A}}$. Let c denote the set of k subscripts of the $\hat{\lambda}_i^{-1}$ used to construct $\hat{\mathbf{A}}$ and let $\bar{c}$ denote the set of subscripts not in c. Because $\hat{\mathbf{A}}$ must have rank k, c must consist of k distinct subscripts for which $\hat{\lambda}_i^{-1} < 1$. If

$\tau < k$, the likelihood has no stationary point in the open parameter space and the maximum likelihood estimator does not exist. Now assume $\tau \geqslant k$ so that k subscripts are included in c. Then

$$\hat{\mathbf{\Sigma}}_{ZZ} = \hat{\mathbf{A}}\hat{\mathbf{A}}' + \mathbf{\Sigma}_{\varepsilon\varepsilon} \tag{4.C.12}$$

$$= \sum_{i \in c} (1 - \hat{\lambda}_i^{-1})\mathbf{m}_{ZZ}\mathbf{R}_{.i}\mathbf{R}'_{.i}\mathbf{m}_{ZZ} + \mathbf{\Sigma}_{\varepsilon\varepsilon} \tag{4.C.13}$$

$$= \sum_{i \in c} \mathbf{m}_{ZZ}\mathbf{R}_{.i}\mathbf{R}'_{.i}\mathbf{m}_{ZZ} + \sum_{i \in \bar{c}} \hat{\lambda}_i^{-1}\mathbf{m}_{ZZ}\mathbf{R}_{.i}\mathbf{R}'_{.i}\mathbf{m}_{ZZ}, \tag{4.C.14}$$

where

$$\mathbf{\Sigma}_{\varepsilon\varepsilon} = \sum_{i=1}^{p} \hat{\lambda}_i^{-1}\mathbf{m}_{ZZ}\mathbf{R}_{.i}\mathbf{R}'_{.i}\mathbf{m}_{ZZ} \tag{4.C.15}$$

is the spectral representation of $\mathbf{\Sigma}_{\varepsilon\varepsilon}$ in the metric $\mathbf{m}_{ZZ}$. For $\hat{\mathbf{\Sigma}}_{ZZ}$ to be positive definite the subscripts for the zero values of $\hat{\lambda}_i^{-1}$ must be in c. It follows that the inverse of $\hat{\lambda}_i^{-1}$ is well defined for $i \in \bar{c}$. By Result 4.A.11,

$$|\hat{\mathbf{\Sigma}}_{ZZ}| = |\mathbf{m}_{ZZ}| \prod_{i \in \bar{c}} \hat{\lambda}_i^{-1} \tag{4.C.16}$$

$$\operatorname{tr}\{\mathbf{m}_{ZZ}\hat{\mathbf{\Sigma}}_{ZZ}^{-1}\} = k + \sum_{i \in \bar{c}} \hat{\lambda}_i, \tag{4.C.17}$$

and the logarithm of the likelihood evaluated at $\hat{\mathbf{\Sigma}}_{ZZ}$ is

$$\log L_c(\hat{\mathbf{\Sigma}}_{ZZ}) = -\tfrac{1}{2}(n-1)\Big[p \log 2\pi + \log|\mathbf{m}_{ZZ}| + k + \sum_{i \in \bar{c}} (-\log \hat{\lambda}_i + \hat{\lambda}_i)\Big]. \tag{4.C.18}$$

For the maximum likelihood estimators to be defined by (4.C.3), we required $\hat{\lambda}_k^{-1} < 1$. The function

$$-\log \lambda + \lambda \tag{4.C.19}$$

is increasing in λ for $\lambda > 1$ and it follows that the maximum of the likelihood over the stationary points is obtained by choosing the columns of $\hat{\mathbf{A}}$ to be the columns associated with the k smallest $\hat{\lambda}_i^{-1}$. This, together with the fact that the likelihood decreases without bound as an individual element of $\mathbf{A}$ is increased in absolute value, demonstrates that the use of the k smallest $\hat{\lambda}_i^{-1}$ to construct $\hat{\mathbf{A}}$ maximizes the likelihood when $\hat{\lambda}_k^{-1} < 1$.

The boundary of the parameter space associated with an $\mathbf{A}$ of less than full column rank is the union of the sets Ω_j, $j = \ell, \ell + 1, \ldots, k - 1$, where Ω_j is the parameter space for a model with an $\mathbf{A}$ of rank j. On each Ω_j the model (4.1.12) can be reparameterized to a model with a j-dimensional $\mathbf{x}_t$.

By the above arguments, for $j \leqslant \tau$, the maximum of the likelihood on Ω_j is

$$-\tfrac{1}{2}(n-1)\Big[p \log 2\pi + \log|\mathbf{m}_{ZZ}| + j + \sum_{i=j+1}^{p} (-\log \hat{\lambda}_i + \hat{\lambda}_i)\Big]. \quad (4.C.20)$$

Because expression (4.C.19) is increasing in λ, expression (4.C.20) is maximized over $j \leqslant \tau$ by choosing $j = \tau$. It follows that the supremum of the original likelihood for $\tau < k$ is given by the value of likelihood for $\hat{\boldsymbol{\Sigma}}_{zz} = \hat{\mathbf{A}}_\tau \hat{\mathbf{A}}'_\tau$.

□

REFERENCES

Appendix 4.A. Bellman (1960), Browne (1974), Dahm (1979), Henderson and Searle (1979), McCulloch (1982), Nel (1980), Neudecker (1969), Roth (1934), Searle and Quaas (1978), Theil (1971), Theobald (1975a, 1975b), Tracy and Dwyer (1969).

Appendix 4.B. Amemiya (1982), Amemiya, Fuller, and Pantula (1987), Anderson and Rubin (1956), Dahm and Fuller (1986), Jennrich (1969).

EXERCISES

29. (Appendix 4.A) Let $\mathbf{m}$ be a real symmetric $p \times p$ matrix and let $\mathbf{Q}$ be the matrix such that

$$\mathbf{Q}'\mathbf{Q} = \mathbf{Q}\mathbf{Q}' = \mathbf{I} \quad \text{and} \quad \mathbf{Q}'\mathbf{m}\mathbf{Q} = \text{diag}(\lambda_1, \lambda_2, \ldots, \lambda_p).$$

Show that the regression coefficient obtained by regressing the jth column of $\mathbf{m}$ on the ith column of $\mathbf{Q}$ is $\lambda_i q_{ji}$, where q_{ji} is the jith element of $\mathbf{Q}$. Hence, show that the jth column of $\mathbf{m}$ is given by

$$\mathbf{m}_{.j} = \sum_{i=1}^{p} \mathbf{Q}_{.i} \lambda_i q_{ji}$$

and that $\mathbf{m} = \Sigma_{i=1}^{p} \lambda_i \mathbf{Q}_{.i} \mathbf{Q}'_{.i}$.

30. (Appendix 4.A) Give the 9×6 matrix $\mathbf{\Phi}$ and the 6×9 matrix ψ_3 for a 3×3 matrix. Let $\boldsymbol{\Sigma} = \{\sigma_{ij}\}$ be a 3×3 covariance matrix. Compute $2\psi_3(\boldsymbol{\Sigma} \otimes \boldsymbol{\Sigma})\psi'_3$. The matrices $\mathbf{\Phi}_3$ and ψ_3 are those of Definition 4.A.3.

31. (Appendix 4.A) Let $\mathbf{A}$ and $\mathbf{B}$ be symmetric $k \times k$ matrices. Show that

$$\psi_k(\mathbf{A} \otimes \mathbf{B})\psi'_k = \psi_k(\mathbf{B} \otimes \mathbf{A})\psi'_k,$$

where ψ_k is that of Definition 4.A.3. Hence, show that $\psi_k(\mathbf{A} \otimes \mathbf{B})\psi'_k$ is symmetric.

32. (Appendix 4.A) Prove Result 4.A.1.

33. (Appendix 4.A) Let $\boldsymbol{\Sigma}$ be a $p \times p$ matrix satisfying

$$\boldsymbol{\Sigma} = \mathbf{A}\boldsymbol{\Sigma}\mathbf{B}' + \mathbf{C},$$

where $\mathbf{A}$, $\mathbf{B}$, and $\mathbf{C}$ are given $p \times p$ matrices. If $[\mathbf{I} - (\mathbf{B} \otimes \mathbf{A})]$ is nonsingular, show that

$$\text{vec}\, \boldsymbol{\Sigma} = [\mathbf{I} - (\mathbf{B} \otimes \mathbf{A})]^{-1}\, \text{vec}\, \mathbf{C}.$$

34. (Appendix 4.A) Prove Result 4.A.11.
35. (Appendix 4.A) Prove the following.

Lemma. Let **A**, **B**, and **S** be $p \times p$ symmetric positive semidefinite matrices. Let **S** be positive definite. Then

$$\operatorname{tr}\{\mathbf{AB}\} \leqslant \sum_{i=T}^{p} \theta_{Ai}\lambda_{Bi},$$

where $\theta_{A1} \geqslant \theta_{A2} \geqslant \cdots \geqslant \theta_{Ap}$ are the roots of

$$|\mathbf{A} - \theta\mathbf{S}| = 0$$

and $\lambda_{B1} \geqslant \lambda_{B2} \geqslant \cdots \geqslant \lambda_{Bp}$ are the roots of

$$|\mathbf{B} - \lambda\mathbf{S}^{-1}| = 0.$$

36. (Appendix 4.A, Section 4.2) Assume that we construct an estimator of the variance of m_{ZZ} in two different ways: as

$$\hat{V}\{m_{ZZ}\} = (n-2)^{-1}(n-1)^{-1} \sum_{t=1}^{n} [(Z_t - \bar{Z})^2 - m_{ZZ}]^2$$

and as

$$\tilde{V}\{m_{ZZ}\} = 2(n-1)^{-1}m_{ZZ}^2.$$

What is the variance of the approximate distribution of $[\tilde{V}\{m_{ZZ}\}]^{-1/2}[\hat{V}\{m_{ZZ}\}]^{1/2}$ when the Z_t are NI$(0, \sigma_{ZZ})$ random variables? Recall that the $2r$th central moment of a normal distribution with standard deviation σ is $(2^r r!)^{-1}(2r)!\sigma^{2r}$.

Bibliography

Abd-Ella, M. M., Hoiberg, E. O., and Warren, R. D. (1981), Adoption behavior in family farm systems: An Iowa study. *Rural Sociology* **46**, 42–61.

Acton, F. S. (1966), *Analysis of Straight-Line Data*. Dover, New York.

Adcock, R. J. (1877), Note on the method of least squares. *Analyst* **4**, 183–184.

Adcock, R. J. (1878), A problem in least squares. *Analyst* **5**, 53–54.

Aigner, D. J. (1973), Regression with a binary variable subject to errors of observation. *J. Econometrics* **1**, 49–60.

Aigner, D. J. and Goldberger, A. S., Eds. (1977), *Latent Variables in Socio-Economic Models*. North-Holland, Amsterdam.

Aigner, D. J., Hsiao, C., Kapteyn, A., and Wansbeek, T. (1984), Latent variable models in econometrics. In *Handbook of Econometrics*, Vol. 2, Z. Griliches and M. D. Intriligator (Eds.). North-Holland, Amsterdam.

Allen, R. G. D. (1939), The assumptions of linear regression. *Economica* **6**, 191–204.

Amemiya, Y. (1980), Likelihood techniques for the errors in variables model. Creative component for the M.S. degree. Iowa State University, Ames, Iowa.

Amemiya, Y. (1982), Estimators for the errors-in-variables model. Unpublished Ph.D. thesis. Iowa State University, Ames, Iowa.

Amemiya, Y. (1985a), Multivariate functional and structural relationship with general error covariance structure. Unpublished manuscript. Iowa State University, Ames, Iowa.

Amemiya, Y. (1985b), On the goodness-of-fit tests for linear statistical relationships. Technical Report No. 10. Econometric Workshop, Stanford University.

Amemiya, Y. (1985c), Instrumental variable estimator for the nonlinear errors-in-variables model. *J. Econometrics* **28**, 273–289.

Amemiya, Y. and Fuller, W. A. (1983), Estimation for the nonlinear functional relationship. *Proc. Business Economic Statist. Sect. Am. Statist. Assoc.*, 47–56.

Amemiya, Y. and Fuller, W. A. (1984), Estimation for the multivariate errors-in-variables model with estimated error covariance matrix. *Ann. Statist.* **12**, 497–509.

Amemiya, Y. and Fuller, W. A. (1985), Estimation for the nonlinear functional relationship. Unpublished manuscript. Iowa State University, Ames, Iowa.

Amemiya, Y., Fuller, W. A., and Pantula, S. G. (1987), The covariance matrix of estimators for the factor model. *J. Multivariate Anal.* **22**, 51–64.

Andersen, E. B. (1982), Latent structure analysis: A survey. *Scand. J. Statist.* **9**, 1–12.

Anderson, B. M., Anderson, T. W., and Olkin, I. (1986), Maximum likelihood estimators and likelihood ratio criteria in multivariate components of variance. *Ann. Statist.* **14**, 405–417.

Anderson, D. A. (1981), The circular structural model. *J. R. Statist. Soc. Ser. B* **43**, 131–141.

Anderson, P. O. (1975), Large sample and jackknife procedures for small sample orthogonal least squares inference. *Commun. Statist.* **4**, 193–202.

Anderson, R. D., Quaas, R. L., and Searle, S. R. (1977), Fourth moments in the general linear model; and the variance of translation invariant quadratic forms. Paper No. BU-630-M in the Biometric Unit, Cornell University, Ithaca, New York.

Anderson, T. W. (1948), The asymptotic distribution of the roots of certain determinantal equations. *J. R. Statist. Soc. Ser. B* **10**, 132–139.

Anderson, T. W. (1951a), The asymptotic distribution of certain characteristic roots and vectors. *Proceedings of the Second Berkeley Symposium.* University of California Press, Berkeley.

Anderson, T. W. (1951b), Estimating linear restrictions on regression coefficients for multivariate normal distributions. *Ann. Math. Statist.* **22**, 327–351.

Anderson, T. W. (1958), *Introduction to Multivariate Statistical Analysis.* Wiley, New York.

Anderson, T. W. (1963), Asymptotic theory for principal component analysis. *Ann. Math. Statist.* **34**, 122–148.

Anderson, T. W. (1966), Estimation of covariance matrices which are linear combinations or whose inverses are linear combinations of given matrices. Technical Report No. 3, Contract AF 41(609)-2653, Teachers College, Columbia University, New York.

Anderson, T. W. (1969), Statistical inference for covariance matrices with linear structure. In *Multivariate Analysis II*, P. R. Krishnaiah (Ed.). Academic Press, New York.

Anderson, T. W. (1973), Asymptotically efficient estimation of covariance matrices with linear structure. *Ann. Statist.* **1**, 135–141.

Anderson, T. W. (1974), An asymptotic expansion of the distribution of the limited information maximum likelihood estimate of a coefficient in a simultaneous equation system. *J. Am. Statist. Assoc.* **69**, 565–573.

Anderson, T. W. (1976), Estimation of linear functional relationships: Approximate distributions and connections with simultaneous equations in econometrics. *J. R. Statist. Soc. Ser. B* **38**, 1–20.

Anderson, T. W. (1977), Asymptotic expansions of the distributions of estimates in simultaneous equations for alternative parameter sequences. *Econometrica* **45**, 509–518.

Anderson, T. W. (1983), Maximum likelihood for linear structural relationships. Paper presented at the Workshop on Functional and Structural Relationships and Factor Analysis at the University of Dundee, Dundee, Scotland, September 7, 1983.

Anderson, T. W. (1984), Estimating linear statistical relationships. *Ann. Statist.* **12**, 1–45.

Anderson, T. W. and Rubin, H. (1949), Estimation of the parameters of a single equation in a complete system of stochastic equations. *Ann. Math. Statist.* **20**, 46–63.

Anderson, T. W. and Rubin, H. (1950), The asymptotic properties of estimates of the parameters of a single equation in a complete system of stochastic equations. *Ann. Math. Statist.* **21**, 570–582.

Anderson, T. W. and Rubin, H. (1956), Statistical inference in factor analysis. *Proceedings of the Third Berkeley Symposium.* Vol. V. University of California Press, Berkeley.

Anderson, T. W. and Sawa, T. (1977), Tables of the distribution of the maximum likelihood estimate of the slope coefficient and approximations. Technical Report No. 234, Economic Series, IMSSS, Stanford University.

Anderson, T. W. and Sawa, T. (1982), Exact and approximate distributions of the maximum likelihood estimator of a slope coefficient. *J. R. Statist. Soc. Ser. B* **44**, 52–62.

Armstrong, B. (1985), Measurement error in the generalized linear model. *Commun. Statist. Part B* **14**, 529–544.

Armstrong, J. S. (1967), Derivation of theory by means of factor analysis or Tom Swift and his electric factor analysis machine. *Am. Statist.* **21**, 17–21.

Barnett, V. D. (1967), A note on linear structural relationships when both residual variances are known. *Biometrika* **54**, 670–672.

Barnett, V. D. (1969), Simultaneous pairwise linear structural relationships. *Biometrics* **25**, 129–142.

Barnett, V. D. (1970), Fitting straight lines—the linear functional relationship with replicated observations. *Appl. Statist.* **19**, 135–144.

Barr, A. J., Goodnight, J. H., Sali, J. P., Blair, W. H., and Chilko, D. M. (1979), *SAS User's Guide.* SAS Institute, Raleigh, North Carolina.

Bartholomew, D. J. (1980), Factor analysis for categorical data. *J. R. Statist. Soc. Ser. B* **42**, 293–312.

Bartlett, M. S. (1948), A note on the statistical estimation of supply and demand relations from time series. *Econometrica* **16**, 323–329.

Bartlett, M. S. (1949), Fitting a straight line when both variables are subject to error. *Biometrics* **5**, 207–212.

Bartlett, M. S. (1950), Contribution to discussion of Kendall and Smith's paper on factor analysis. *J. R. Statist. Soc. Ser. B* **12**, 86.

Bartlett, M. S. (1951), The effect of standardization on an approximation in factor analysis. *Biometrika* **38**, 337–344.

Bartlett, M. S. (1954), A note on the multiplying factor for various approximations. *J. R. Statist. Soc. Ser. B* **16**, 296–298.

Basmann, R. L. (1960), On finite sample distributions of generalized classical linear identifiability test statistics. *J. Am. Statist. Assoc.* **55**, 650–659.

Basmann, R. L. (1961), A note on the exact finite sample frequency functions of generalized classical linear estimators in two leading overidentified cases. *J. Am. Statist. Assoc.* **56**, 619–636.

Battese, G. E. (1973), Parametric models for response errors in survey sampling. Unpublished Ph.D. thesis. Iowa State University, Ames, Iowa.

Battese, G. E. and Fuller, W. A. (1973), An unbiased response model for analysis of categorical data. *Proc. Soc. Statist. Sect. Am. Statist. Assoc.* 202–207.

Battese, G. E., Fuller, W. A., and Hickman, R. D. (1976), Estimation of response variances from interview–reinterview surveys. *J. Indian Soc. Agric. Statist.* **28**, 1–14.

Beaton, A. E., Rubin, D. E., and Barone, J. L. (1976), The acceptability of regression solutions: Another look at computational accuracy. *J. Am. Statist. Assoc.* **71**, 158–168.

Bekker, P. A. (1986), Comment on identification in the linear errors in variables model. *Econometrica* **54**, 215–217.

Bekker, P. A., Wansbeek, T. J., and Kapteyn, A. (1985), Errors in variables in econometrics: New developments and recurrent themes. *Statist. Neerlandica* **39**, 129–141.

Bellman, R. (1960), *Introduction to Matrix Analysis.* McGraw-Hill, New York.

Bentler, P. M. (1980), Multivariate analysis with latent variables: Casual modeling. *Ann. Rev. Psychol.* **31**, 419–456.

Bentler, P. M. (1983), Some contributions to efficient statistics in structural models: Specification and estimation of moment structures. *Psychometrika* **48**, 493–517.

Bentler, P. M. (1985), Theory and implementation of EQS, a structural equations program. BMDP Statistical Software, Los Angeles.

Bentler, P. M. and Dijkstra, T. (1985), Efficient estimation via linearization in structural models. In *Multivariate Analysis VI*, P. R. Krishnaiah (Ed.). North-Holland, New York.

Berkson, J. (1950), Are there two regressions? *J. Am. Statist. Assoc.* **45**, 164–180.

Bershad, M. A. (1967), Gross changes in the presence of response errors. Unpublished memorandum, U.S. Bureau of the Census, Washington, D.C.

Bhargava, A. K. (1977), Maximum likelihood estimation in a multivariate "errors-in-variables" regression model with unknown error covariance matrix. *Commun. Statist. Part A* **6**, 587–601.

Bhargava, A. K. (1979), Estimation of a linear transformation and an associated distributional problem. *J. Statist. Planning Inference* **3**, 19–26.

Bickel, P. J. and Ritov, Y. (1985), Efficient estimation in the errors in variables model. Unpublished manuscript. University of California, Berkeley, California.

Birch, M. W. (1964), A note on the maximum likelihood estimation of a linear structural relationship. *J. Am. Statist. Assoc.* **59**, 1175–1178.

Bock, R. D. and Bargmann, R. E. (1966), Analysis of covariance structures. *Psychometrika* **31**, 507–534.

Bock, R. D. and Peterson, A. C. (1975), A multivariate correction for attenuation. *Biometrika* **62**, 673–678.

Bohrnstedt, G. W. and Carter, T. M. (1971), Robustness in regression analysis. In *Sociological Methodology*, H. L. Costner (Ed.). Jossey-Bass, San Francisco.

Booth, G. D. (1973), The errors-in-variables model when the covariance matrix is not constant. Unpublished Ph.D. dissertation. Iowa State University, Ames, Iowa.

Box, G. E. P. (1949), A general distribution theory for a class of likelihood criteria. *Biometrika* **36**, 317–346.

Box, G. E. P. (1961), The effects of errors in factor levels and experimental design. *Bull. Int. Statist. Inst.* **38**, 339–355.

Brailovsky, V. (1981), On multiple regression for the case with error in both dependent and independent variables. *Ann. N. Y. Acad. Sci.* **373**, Fourth International Conference on Collective Phenomena, 113–135.

Britt, H. I. and Luecke, R. H. (1973), The estimation of parameters in nonlinear implicit models. *Technometrics* **15**, 233–247.

Brown, P. J. (1982), Multivariate calibration. *J. R. Statist. Soc. Ser. B* **44**, 287–321.

Brown, R. L. (1957), Bivariate structural relation. *Biometrika* **44**, 84–96.

Brown, R. L. and Fereday, F. (1958), Multivariate linear structural relations. *Biometrika* **45**, 136–153.

Browne, M. W. (1974), Generalized least squares estimators in the analysis of covariance structures. *S. Afr. Statist. J.* **8**, 1–24.

Browne, M. W. (1984), Asymptotically distribution-free methods for the analysis of covariance structures. *Br. J. Math. Statist. Psychol.* **37**, 62–83.

Burt, C. (1909), Experimental tests of general intelligence. *Br. J. Psychol.* **3**, 94–177.

Carlson, F. D., Sobel, E., and Watson, G. S. (1966), Linear relationship between variables affected by errors. *Biometrics* **22**, 252–267.

Carroll, R. J. and Gallo, P. (1982), Some aspects of robustness in functional errors-in-variables regression models. *Commun. Statist. Part A*, **11**, 2573–2585.

Carroll, R. J. and Gallo, P. (1984), Comparisons between maximum likelihood and method of moments in a linear errors-in-variables regression model. In *Design of Experiments: Ranking and Selection*, T. J. Santner and A. C. Tamhane (Eds.). Marcel Dekker, New York.

Carroll, R. J., Gallo, P., and Gleser, L. J. (1985), Comparison of least squares and errors-in-variables regression, with special reference to randomized analysis of covariance. *J. Am. Statist. Assoc.* **80**, 929–932.

Carroll, R. J., Spiegelman, C. H., Lan, K. K., Bailey, K. T., and Abbot, R. D. (1982), On errors-in-variables for binary regression models. *Biometrika* **71**, 19–25.

Carter, L. F. (1971), Inadvertent sociological theory. *Soc. Forces* **30**, 12–25.

Carter, R. L. (1976), Instrumental variable estimation of the simple errors-in-variables model. Ph.D. thesis. Iowa State University, Ames, Iowa.

Carter, R. L. (1981), Restricted maximum likelihood estimation of bias and reliability in the comparison of several measuring methods. *Biometrics* **37**, 733–742.

Carter, R. L. and Fuller, W. A. (1980), Instrumental variable estimation of the simple errors in variables model. *J. Am. Statist. Assoc.* **75**, 687–692.

Cattell, R. B. and Dickman, K. (1962), A dynamic model for physical influences demonstrating the necessity of oblique simple structure, *Psychol. Bull.* **59**, 389–400.

Chai, J. J. (1971), Correlated measurement errors and least squares estimators of the regression coefficient. *J. Am. Statist. Assoc.* **66**, 478–483.

Chamberlain, G. and Griliches, Z. (1975), Unobservables with a variance-components structure: Ability, schooling, and the economic success of brothers. *Int. Econ. Rev.* **16**, 422–449.

Chan, L. K. and Mak, T. K. (1979), Maximum likelihood estimation of a linear structural relationship with replication. *J. R. Statist. Soc. Ser. B* **41**, 263–268.

Chan, N. N. (1965), On circular functional relationships. *J. R. Statist. Soc. Ser. B* **27**, 45–56.

Chan, N. N. (1980), Estimating linear functional relationships. In *Recent Developments in Statistical Inference and Data Analysis*, K. Matusita (Ed.). North-Holland, Amsterdam.

Chan, N. N. (1982), Linear structural relationships with unknown error variances. *Biometrika* **69**, 277–279.

Chan, N. N. and Mak, T. K. (1983), Estimation of multivariate linear functional relationships. *Biometrika* **70**, 263–267.

Chan, N. N. and Mak, T. K. (1984), Heteroscedastic errors in a linear functional relationship. *Biometrika* **71**, 212–215.

Chan, N. N. and Mak, T. K. (1985), Estimation in multivariate errors-in-variables models. *Linear Algebra Appl.* **70**, 197–207.

Chua, T. C. (1983), Response errors in repeated surveys with duplicated observations. Unpublished Ph.D. thesis. Iowa States University, Ames, Iowa.

Chua, T. C. and Fuller, W. A. (1987), A model for multinomial response error applied to labor flows. *J. Am. Statist. Assoc.* **82**, 46–51.

Clutton-Brock, M. (1967), Likelihood distributions for estimating functions when both variables are subject to error. *Technometrics* **9**, 261–269.

Cochran, W. G. (1968), Errors of measurement in statistics. *Technometrics* **10**, 637–666.

Cochran, W. G. (1970), Some effects of errors of measurement on multiple correlation. *J. Am. Statist. Assoc.* **65**, 22–34.

Cochran, W. G. (1972), Some effects of errors of measurement on linear regression. *Proceedings of the Sixth Berkeley Symposium on Mathematical Statistics and Probability*, Vol. I, pp. 527–540. University of California Press, Berkeley.

Cochran, W. G. (1977), *Sampling Techniques*. Wiley, New York.

Cohen, J. E. and D'Eustachio, P. (1978), An affine linear model for the relation between two sets of frequency counts. Response to a query. *Biometrics* **34**, 514–516.

Cohen, J. E., D'Eustachio, P., and Edelman, G. M. (1977), The specific antigen-binding cell populations of individual fetal mouse spleens: Repertoire composition, size and genetic control. *J. Exp. Med.* **146**. 394–411.

Cook, W. R. (1931), On curve fitting by means of least squares. *Philos. Mag. Ser.* 7 **12**, 231–237.

Copas, J. B. (1972), The likelihood surface in the linear functional relationship problem. *J. R. Statist. Soc. Ser. B* **34**, 190–202.

Cox, N. R. (1976), The linear structural relation for several groups of data. *Biometrika* **63**, 231–237.

Cramér, H. (1946), *Mathematical Methods of Statistics*, Chap. 27. Princeton University Press, Princeton.

Creasy, M. A. (1956), Confidence limits for the gradient in the linear functional relationship. *J. R. Statist. Soc. Ser. B* **18**, 65–69.

Dahm, P. F. (1979), Estimation of the parameters of the multivariate linear errors in variables model. Unpublished Ph.D. dissertation. Iowa State University, Ames, Iowa.

Dahm, P. F. and Fuller, W. A. (1986), Generalized least squares estimation of the functional multivariate linear errors in variables model. *J. Multivariate Anal.* **19**, 132–141.

Dahm, P. F., Heady, E. O., and Sonka, S. T. (1976), Estimation and application of a production function in decision rules for swine producers. *Can. J. Agric. Econ.* **24**, 1–16.

Dahm, P. F., Melton, B. E., and Fuller, W. A. (1983), Generalized least squares estimation of a genotypic covariance matrix. *Biometrics* **39**, 587–597.

Dalenius, T. (1977a), Bibliography on nonsampling errors in surveys—I. *Int. Statist. Rev.* **45**, 71–89.

Dalenius, T. (1977b), Bibliography on nonsampling errors in surveys—II. *Int. Statist. Rev.* **45**, 181–197.

Dalenius, T. (1977c), Bibliography on nonsampling errors in surveys—III. *Int. Statist. Rev.* **45**, 303–317.

Davis, F. B. (1944), Fundamental factors of comprehension in reading. *Psychometrika* **9**, 185–197.

Deemer, W. L. and Olkin, I. (1951), Jacobians of matrix transformations useful in multivariate analysis. *Biometrika* **38**, 345–367.

DeGracie, J. S. (1968), Analysis of covariance when the concomitant variable is measured with error. Unpublished Ph.D. thesis. Iowa State University, Ames, Iowa.

DeGracie, J. S. and Fuller, W. A. (1972), Estimation of the slope and analysis of covariance when the concomitant variable is measured with error. *J. Am. Statist. Assoc.* **67**, 930–937.

Deming, W. E. (1931), The application of least squares, *Philos. Mag. Ser. 7* **11**, 146–158.

Deming, W. E. (1943), *Statistical Adjustment of Data.* Wiley, New York.

Denton, F. T. and Kuiper, J. (1965), The effects of measurement errors on parameter estimates and forecasts: A case study based on the Canadian preliminary national accounts. *Rev. Econ. Statist.* **47**, 198–206.

Dolby, G. R. (1972), Generalized least squares and maximum likelihood estimation of nonlinear functional relationships. *J. R. Statist. Soc. Ser. B* **34**, 393–400.

Dolby, G. R. (1976), The ultra-structural relation: A synthesis of the functional and structural relations. *Biometrika* **63**, 39–50.

Dolby, G. R. (1976), A note on the linear structural relation when both residual variances are known. *J. Am. Statist. Assoc.* **71**, 351–352.

Dolby, G. R. and Freeman, T. G. (1975), Functional relationships having many independent variables and errors with multivariate normal distribution. *J. Multivariate Anal.* **5**, 466–479.

Dolby, G. R. and Lipton, S. (1972), Maximum likelihood estimation of the general nonlinear functional relationship with replicated observations and correlated errors. *Biometrika* **59**, 121–129.

Dorff, M. and Gurland, J. (1961a), Estimation of the parameters of a linear functional relation *J. R. Statist. Soc. Ser. B* **23**, 160–170.

Dorff, M. and Gurland, J. (1961b), Small sample behavior of slope estimators in a linear functional relation. *Biometrics* **17**, 283–298.

Draper, N. R. and Beggs, W. J. (1971), Errors in the factor levels and experimental design. *Ann Math. Statist.* **41**, 46–58.

Draper, N. R. and Smith, H. (1981), *Applied Regression Analysis*, 2nd ed. Wiley, New York.

Driel, O. P. van (1978), On various causes of improper solutions in maximum likelihood factor analysis. *Psychometrika* **43**, 225–243.

Duncan, O. D. (1975), *Introduction to Structural Equation Models*. Academic Press, New York.

Durbin, J. (1954), Errors-in-variables. *Int. Statist. Rev.* **22**, 23–32.

Dwyer, P. S. (1967), Some applications of matrix derivatives in multivariate analysis. *J. Am. Statist. Assoc.* **62**, 607–625.

Efron, B. (1982), *The Jackknife, the Bootstrap and Other Resampling Plans*. SIAM, Philadelphia.

Farris, A. L., Klonglan, E. D., and Nomsen, R. C. (1977), *The Ring-Necked Pheasant in Iowa*. Iowa Conservation Commission, Des Moines, Iowa.

Featherman, D. L. (1971), A research note: A social structural model for the socioeconomic career. *Am. J. Soc.* **77**, 293–304.

Fedorov, V. V. (1974), Regression problems with controllable variables subject to error. *Biometrika* **61**, 49–56.

Feldstein, M. (1974), Errors in variables: A consistent estimator with smaller MSE in finite samples. *J. Am. Statist. Assoc.* **69**, 990–996.

Fieller, E. C. (1954), Some problems in interval estimation. *J. R. Statist. Soc. Ser. B* **16**, 175–185.

Fisher, R. A. (1938), The statistical utilization of multiple measurements. *Ann. Eugenics* **8**, 376–386.

Fisher, F. M. (1966), *The Identification Problem in Econometrics*. McGraw-Hill, New York.

Fletcher, R. and Powell, M. J. D. (1963), A rapidly convergent descent method for minimization. *Comput. J.* **6**, 163–168.

Florens, J. P., Mouchart, M., and Richard, F. (1974), Bayesian inference in error-in-variables models. *J. Multivariate Anal.* **4**, 419–452.

Franklin, J. N. (1968), *Matrix Theory.* Prentice-Hall, Englewood Cliffs, New Jersey.

Frisillo, A. L. and Stewart, T. J. (1980a), Effect of partial gas/brine saturation on ultrasonic absorption in sandstone. Amoco Production Company Research Center, Tulsa, Oklahoma.

Frisillo, A. L. and Stewart, T. J. (1980b), Effect of partial gas/brine saturation on ultrasonic absorption in sandstone. *J. Geophys. Res.* **85**, 5209–5211.

Fuller, W. A. (1975), Regression analysis for sample survey. *Sankhyā C* **37**, 117–132.

Fuller, W. A. (1976), *Introduction to Statistical Time Series.* Wiley, New York.

Fuller, W. A. (1977), Some properties of a modification of the limited information estimator. *Econometrica* **45**, 939–953.

Fuller, W. A. (1978), An affine linear model for the relation between two sets of frequency counts: Response to query. *Biometrics* **34**, 517–521.

Fuller, W. A. (1980), Properties of some estimators for the errors-in-variables model. *Ann. Statist.* **8**, 407–422.

Fuller, W. A. (1984), Measurement error models with heterogeneous error variances. In *Topics in Applied Statistics*, Y. P. Chaubey and T. D. Dwivedi (Eds.). Concordia University, Montreal.

Fuller, W. A. (1985), Properties of residuals from errors-in-variables analyses. (Abstract) *Inst. Math. Statist. Bull.* **14**, 203–204.

Fuller, W. A. (1986), Estimators of the factor model for survey data. In *Proceedings of the Symposia in Statistics and Festschrift in Honour of V. M. Joshi*, I. B. Mac Neill and G. J. Umphrey (Eds.). Reidel, Boston.

Fuller, W. A. and Chua, T. C. (1983), A model for multinomial response error. *Proceedings of the 44th Session of the International Statistical Institute, Contributed Papers*, Vol. 1, pp. 406–409.

Fuller, W. A. and Chua, T. C. (1984), Gross change estimation in the presence of response error. In *Proceedings of the Conference on Gross Flows in Labor Force Statistics.* Bureau of the Census and Bureau of Labor Statistics, Washington, D.C.

Fuller, W. A. and Harter, R. M. (1987), The multivariate components of variance model for small area estimation. In *Small Area Statistics: An International Symposium*, R. Platek, J. N. K. Rao, C. E. Särndal, and M. B. Singh (Eds.). Wiley, New York.

Fuller, W. A. and Hidiroglou, M. A. (1978), Regression estimation after correcting for attenuation. *J. Am. Statist. Assoc.* **73**, 99–104.

Fuller, W. A. and Pantula, S. G. (1982), A computational algorithm for the factor model. Iowa State University, Ames, Iowa.

Gallant, A. R. (1975), Nonlinear regression. *Am. Statist.* **29**, 73–81.

Gallant, A. R. (1986), *Nonlinear Statistical Models.* Wiley, New York.

Gallo, P. P. (1982), Consistency of regression estimates when some variables are subject to error. *Commun. Statist. Part A* **11**, 973–983.

Ganse, R. A., Amemiya, Y., and Fuller, W. A. (1983), Prediction when both variables are subject to error, with application to earthquake magnitude. *J. Am. Statist. Assoc.* **78**, 761–765.

Garber, S. and Klepper, S. (1980), Extending the classical normal errors-in-variables model. *Econometrics* **48**, 1541–1546.

Geary, R. C. (1942), Inherent relations between random variables. *Proc. R. Irish Acad. Sect. A* **47**, 63–76.

Geary, R. C. (1943), Relations between statistics: The general and the sampling problem when the samples are large. *Proc. R. Irish Acad Sect. A* **49**, 177–196.

Geary, R. C. (1948), Studies in relations between economic time series. *J. R. Statist. Soc. Ser. B* **10**, 140–158.

Geary, R. C. (1949), Determinations of linear relations between systematic parts of variables with errors of observation the variances of which are unknown. *Econometrica* **17**, 30–58.

Geraci, V. J. (1976), Identification of simultaneous equation models with measurement error. *J. Econometrics* **4**, 263–282.

Girshick, M. A. (1939), On the sampling theory of roots of determinantal equations. *Ann. Math. Statist.* **10**, 203–224.

Gleser, L. J. (1981), Estimation in a multivariate "errors-in-variables" regression model: Large sample results. *Ann. Statist.* **9**, 24–44.

Gleser, L. J. (1982), Confidence regions for the slope in a linear errors-in-variables regression model. Technical Report 82–23. Department of Statistics, Purdue University, Lafayette, Indiana.

Gleser, L. J. (1983), Functional, structural and ultrastructural errors-in-variables models. *Proc. Business Economic Statist. Sect. Am. Statist. Assoc.*, 57–66.

Gleser, L. J. (1985), A note on G. R. Dolby's ultrastructural model. *Biometrika* **72**, 117–124.

Gleser, L. J. and Hwang, J. T. (1985), The nonexistence of $100(1 - \alpha)\%$ confidence sets of finite expected diameter in errors-in-variables and related models. Technical Report 85–15. Department of Statistics, Purdue University, Lafayette, Indiana.

Gleser, L. J. and Watson, G. S. (1973), Estimation of a linear transformation. *Biometrika* **60**, 525–534.

Goldberger, A. S. (1972), Maximum likelihood estimation of regressions containing unobservable independent variables. *Int. Econ. Rev.* **13**, 1–15.

Goldberger, A. S. (1972), Structural equation models in the social sciences. *Econometrica* **40**, 979–1002.

Goldberger, A. S. and Duncan, O. D. (Eds.) (1973), *Structural Equation Models in the Social Sciences*. Seminar Press, New York.

Goodman, L. A. (1974), Exploratory latent structure analysis using both identifiable and unidentifiable models. *Biometrika* **61**, 215–231.

Gregory, K. E., Swiger, L. A., Koch, R. M., Sumption L. J., Rosden, W. W., and Ingalls, J. E. (1965), Heterosis in pre-weaning traits of beef cattle. *J. Animal Sci.* **24**, 21–28.

Gregory, K. E., Swiger, L. A., Sumption, L. J., Koch, R. M., Ingalls, J. E., Rowden, W. W., and Rothlisberger, J. A. (1966a), Heterosis effects on growth rate and feed efficiency of beef steers. *J. Animal Sci.* **25**, 299–310.

Gregory, K. E., Swiger, L. A., Sumption L. J., Koch, R. M., Ingalls, J. E., Rowden, W. W., and Rothlisberger, J. A. (1966b), Heterosis effects on carcass of beef steers. *J. Animal Sci.* **25**, 311–322.

Griffiths, D. A. and Sandland, R. L. (1982), Allometry and multivariate growth revisited. *Growth* **46**, 1–11.

Griliches, Z. (1974), Errors in variables and other unobservables. *Econometrica* **42**, 971–998.

Griliches, Z. and Ringstad, V. (1970), Errors in the variables bias in nonlinear contexts. *Econometrica* **38**, 368–370.

Grubbs, F. E. (1948), On testing precision of measuring instruments and product variability. *J. Am. Statist. Assoc.* **43**, 243–264.

Grubbs, F. E. (1973), Errors of measurement, precision, accuracy and the statistical comparison of measuring instruments. *Technometrics* **15**, 53–66.

Haberman, S. J. (1977), Product models for frequency tables involving indirect observation. *Ann. Statist.* **5**, 1124–1147.

Hahn, G. J. and Nelson, W. (1970), A problem in the statistical comparison of measuring devices. *Technometrics* **12**, 95–102.

Haitovsky, Y. (1972), On errors of measurement in regression analysis in economics. *Rev. Int. Statist. Inst.* **30**, 23–35.

Halperin, M. (1961), Fitting of straight lines and prediction when both variables are subject to error. *J. Am. Statist. Assoc.* **56**, 657–659.

Halperin, M. (1964), Interval estimation in linear regression when both variables are subject to error. *J. Am. Statist. Assoc.* **59**, 1112–1120.

Halperin, M. and Gurian, J. (1971), A note on estimation in straight line regression when both variables are subject to error. *J. Am. Statist. Assoc.* **66**, 587–589.

Hanamura, R. C. (1975), Estimating variances in simultaneous measurement procedures. *Am. Statist.* **29**, 108–109.

Hansen, M. H., Hurwitz, W. N., and Bershad, M. A. (1961), Measurement errors in censuses and surveys. *Bull. Int. Statist. Inst.* **38**, 359–374.

Harman, H. H. (1976), *Modern Factor Analysis*, 3rd ed. University of Chicago Press, Chicago.

Hasabelnaby, N. A. (1985), Functionally related analysis of an error-in-variables model. Creative component for the M.S. degree. Iowa State University, Ames, Iowa.

Heady, E. O., Sonka, S. T., and Dahm, P. F. (1976), Estimation and application of gain isoquants in decision rules for swine producers. *J. Agric Econ.* **27**, 235–242.

Healy, J. D. (1975), Estimation and tests for unknown restrictions in multivariate linear models. Department of Statistics, Purdue University, Mimeo Series 471.

Healy, J. D. (1980), Maximum likelihood estimation of a multivariate linear functional relationship. *J. Multivariate Anal.* **10**, 243–251.

Henderson, H. V. and Searle, S. R. (1979), Vec and Vech operators for matrices, with some uses in Jacobian and multivariate statistics. *Can. J. Statist.* **7**, 65–81.

Henrici, P. (1964), *Elements of Numerical Analysis.* Wiley, New York.

Hey, E. N. and Hey, M. H. (1960), The statistical estimation of a rectangular hyperbola. *Biometrics* **16**, 606–617.

Hidiroglou, M. A. (1974), Estimation of regression parameters for finite populations. Unpublished Ph.D. thesis. Iowa State University, Ames, Iowa.

Hidiroglou, M. A., Fuller, W. A., and Hickman, R. D. (1980), *SUPER CARP.* Department of Statistics, Iowa State University, Ames, Iowa.

Hinich, M. J. (1983), Estimating the gain of a linear filter from noisy data. In *Handbook of Statistics*, Vol. 3, D. R. Brillinger and P. R. Krishnaiah (Eds.). North-Holland, Amsterdam.

Hodges, S. D. and Moore, P. G. (1972), Data uncertainties and least squares regression. *J. Appl. Statist.* **21**, 185–195.

Höschel, H. P. (1978a), Least squares and maximum likelihood estimation of functional relations. In *Transactions of the Eighth Prague Conference on Information Theory, Statistical Decision Functions, Random Processes.* Academia, Prague.

Höschel, H. P. (1978b), Generalized least squares estimators of linear functional relations with known error covariance. *Math. Operationsforsch. Statist. Ser. Statist.* **9**, 9–26.

Hotelling, H. (1933), Analysis of a complex of statistical variables into principal components. *J. Educ. Psychol.* **24**, 417–441, 498–520.

Hotelling, H. (1957), The relation of the newer multivariate statistical methods to factor analysis. *Br. J. Statist. Psychol.* **10**, 69–79.

Hsu, P. L. (1941a), On the problem of rank and the limiting distribution of Fisher's test function. *Ann. Eugenics* **11**, 39–41.

Hsu, P. L. (1941b), Canonical reduction of the general regression problem. *Ann. Eugenics* **11**, 42–46.

Hunter, J. S. (1980), The national system of scientific measurement. *Science* **210**, 869–874.

Hunter, W. G. and Lamboy, W. F. (1981), A Bayesian analysis of the linear calibration problem. *Technometrics* **23**, 323–328.

Hwang, J. T. (1986), Multiplicative errors-in-variables models with applications to the recent data released by U.S. Department of Energy. *J. Am. Statist. Assoc.* **81**, 680–688.

Isogawa, Y. (1985), Estimating a multivariate linear structural relationship with replication. *J. R. Statist. Soc. Ser. B* **47**, 211–215.

Jaech, J. L. (1971), Further tests of significance for Grubbs' estimators. *Biometrics* **27**, 1097–1101.

Jaech, J. L. (1976), Large sample tests for Grubbs' estimators of instrument precision with more than two instruments. *Technometrics* **18**, 127–132.

James, A. T. (1954), Normal multivariate analysis and the orthogonal group. *Ann. Math. Statist.* **25**, 40–75.

Jennrich, R. I. (1969), Asymptotic properties of nonlinear least squares estimators. *Ann. Math. Statist.* **40**, 633–643.

Jennrich, R. I. (1973), Standard errors for obliquely rotated factor loadings. *Psychometrika* **38**, 593–604.

Jennrich, R. I. (1974), Simplified formulae for standard errors in maximum-likelihood factor analysis. *Br. J. Math. Statist. Psychol.* **27**, 122–131.

Jennrich, R. I. and Robinson, S. M. (1969), A Newton–Raphson algorithm for maximum likelihood factor analysis. *Psychometrika* **34**, 111–123.

Jennrich, R. I. and Thayer, D. T. (1973), A note on Lawley's formulas for standard errors in maximum likelihood factor analysis. *Psychometrika* **38**, 571–580.

Johnston, J. (1972), *Econometric Methods*, 2nd ed. McGraw-Hill, New York.

Jones, T. A. (1979), Fitting straight lines when both variables are subject to error I. Maximum likelihood and least squares estimation. *J. Int. Assoc. Math. Geol.* **11**, 1–25.

Jöreskog, K. G. (1966), Testing a simple structure hypothesis in factor analysis. *Psychometrika* **31**, 165–178.

Jöreskog, K. G. (1967), Some contributions to maximum likelihood factor analysis. *Psychometrika* **32**, 443–482.

Jöreskog, K. G. (1969), A general approach to confirmatory maximum likelihood factor analysis. *Psychometrika* **34**, 183–202.

Jöreskog, K. G. (1970a), Estimation and testing of simple models. *Br. J. Math. Statist. Psychol.* **23**, 121–145.

Jöreskog, K. G. (1970b), A general method for analysis of covariance structures. *Biometrika* **57**, 239–251.

Jöreskog, K. G. (1971), Statistical analysis of sets of congeneric tests. *Psychometrika* **36**, 109–133.

Jöreskog, K. G. (1973), A general method for estimating a linear structural equation system. In *Structural Equation Models in the Social Sciences*, A. S. Goldberger and O. D. Duncan (Eds.). Seminar Press, New York.

Jöreskog, K. G. (1977), Factor analysis by least squares and maximum likelihood methods. In *Statistical Methods for Digital Computers*, Vol. 13, K. Enslein, R. Ralston, and S. W. Wilf (Eds.). Wiley, New York.

Jöreskog, K. G. (1978), Structural analysis of covariance and correlation matrices. *Psychometrika* **43**, 443–477.

Jöreskog, K. G. (1981), Analysis of covariance structures. *Scand. J. Statist.* **8**, 65–92.

Jöreskog, K. G. and Goldberger, A. S. (1972), Factor analysis by generalized least squares. *Psychometrika* **37**, 243–260.

Jöreskog, K. G. and Lawley, D. N. (1968), New methods in maximum likelihood factor analysis. *Br. J. Math. Statist. Psychol.* **21**, 85–96.

Jöreskog, K. G. and Sörbom, D. (1981), LISREL V: Analysis of linear structural relationships by maximum likelihood and least squares methods. University of Uppsala, Uppsala, Sweden.

Kadane, J. B. (1970), Testing overidentifying restrictions when the disturbances are small. *J. Am. Statist. Assoc.* **65**, 182–185.

Kadane, J. B. (1971), Comparison of k-class estimators when the disturbances are small. *Econometrica* **39**, 723–737.

Kalbfleisch, J. D. and Sprott, D. A. (1970), Application of likelihood methods to models involving large numbers of parameters. *J. R. Statist. Soc. Ser. B* **32**, 175–208.

Kang, Y. J. (1985), Estimation for the no-intercept errors-in-variables model. Creative component for the M.S. degree. Iowa State University, Ames, Iowa.

Karni, E. and Weissman, I. (1974), A consistent estimator of the slope in a regression model with errors in the variables. *J. Am. Statist. Assoc.* **69**, 211–213, corrections 840.

Kelley, J. (1973), Causal chain models for the socioeconomic career. *Am. Soc. Rev.* **38**, 481–493.

Kelly, G. (1984), The influence function in the errors in variables problem. *Ann. Statist.* **12**, 87–100.

Kendall, M. G. (1951), Regression, structure, and functional relationship, I. *Biometrika* **38**, 11–25.

Kendall, M. G. (1952), Regression, structure, and functional relationship, II. *Biometrika* **39**, 96–108.

Kendall, M. G. and Stuart, A. (1977), *The Advanced Theory of Statistics*, Vol. 1, 4th ed. Hafner, New York.

Kendall, M. G. and Stuart, A. (1979), *The Advanced Theory of Statistics*, Vol. 2, 4th ed. Hafner, New York.

Kerrich, J. E. (1966), Fitting the line $Y = \alpha X$ when errors of observation are present in both variables. *Am. Statist.* **20**, 24.

Ketellapper, R. H. (1982), Two-stage least squares estimation in the simultaneous equation model with errors in the variables. *Rev. Econ. Statist.* **64**, 696–701.

Ketellapper, R. H. and Ronner, A. E. (1984), Are robust estimation methods useful in the structural errors-in-variables model? *Metrika* **31**, 33–41.

Kiefer, J. and Wolfowitz, J. (1956), Consistency of the maximum likelihood estimator in the presence of infinitely many incidental parameters. *Ann. Math. Statist.* **27**, 887–906.

Klepper, S. and Leamer, E. E. (1984), Consistent sets of estimates for regression with errors in all variables. *Econometrica* **55**, 163–184.

Konijn, H. S. (1981), Maximum likelihood estimator and confidence intervals for a simple errors in variables model. *Commun. Statist. Part A* **10**, 983–996.

Koopmans, T. C. (1937), *Linear Regression Analysis of Economic Time Series*. DeErven F. Bohn, Haarlem, The Netherlands.

Koopmans, T. C. and Hood, W. C. (1953), The estimation of simultaneous linear economic relationships. In *Studies in Econometric Method*, W. C. Hood and T. C.

Koopmans (Eds.). Yale University Press, New Haven.

Koopmans, T. C. and Reiersol, O. (1950), The identification of structural characteristics. *Ann. Math. Statist.* **21**, 165–181.

Korn, E. I. (1982), The asymptotic efficiency of tests using misclassified data in contingency tables. *Biometrics* **38**, 445–450.

Kummell, C. H. (1879), Reduction of observed equations which contain more than one observed quantity. *Analyst* **6**, 97–105.

Lakshminarayanan, M. Y. and Gunst, R. F. (1984), Estimation of parameters in linear structural relationships: Sensitivity to the choice of the ratio of error variances. *Biometrika* **71**, 569–573.

Lawley, D. N. (1940), The estimation of factor loadings by the method of maximum likelihood. *Proc. R. Soc. Edinburgh A* **60**, 64–82.

Lawley, D. N. (1941), Further investigations in factor estimation. *Proc. R. Soc. Edinburgh A* **61**, 176–185.

Lawley, D. N. (1943), The application of the maximum likelihood method to factor analysis. *Br. J. Psychol.* **33**, 172–175.

Lawley, D. N. (1953), A modified method of estimation in factor analysis and some large sample results. *Uppsala Symposium on Psychological Factor Analysis, Nordisk Psykologi's Monograph, No.* 3. Almqvist and Wiksell, Stockholm.

Lawley, D. N. (1956), Tests of significance for the latent roots of covariance and correlation matrices. *Biometrika* **43**, 128–136.

Lawley, D. N. (1967), Some new results in maximum likelihood factor analysis. *Proc. R. Soc. Edinburgh A* **67**, 256–264.

Lawley, D. N. (1976), The inversion of an augmented information matrix occurring in factor analysis. *Proc. R. Soc. Edinburgh A* **75**, 171–178.

Lawley, D. N. and Maxwell, A. E. (1971), *Factor Analysis as a Statistical Method*, 2nd ed. American Elsevier, New York.

Lawley, D. N. and Maxwell, A. E. (1973), Regression and factor analysis. *Biometrika* **60**, 331–338.

Lazarsfeld, P. F. (1950), The logical and mathematical foundation of latent structure analysis. In *Measurement and Prediction*, S. A. Stouffer et al. (Eds.), Princeton University Press, Princeton, NJ.

Lazarsfeld, P. F. (1954), A conceptual introduction to latent structure analysis. In *Mathematical Thinking in the Social Sciences*, P. F. Lazarsfeld (Ed.), The Free Press, Glencoe, IL.

Lazarsfeld, P. F. and Henry, N. W. (1968), *Latent Structure Analysis*. Houghton Mifflin, Boston.

Leamer, E. E. (1978), Least-squares versus instrumental variables estimation in a simple errors in variables model. *Econometrica* **46**, 961–968.

Lee, S. Y. (1985), On testing functional constraints in structural equation models. *Biometrika* **72**, 125–132.

Lee, S. Y. and Bentler, P. M. (1980), Some asymptotic properties of constrained generalized least squares estimation in covariance structure models. *S. Afr. Statist. J.* **14**, 121–136.

Lee, S. Y. and Jennrich, R. I. (1979), A study of algorithms for covariance structure analysis with specific comparisons using factor analysis. *Psychometrika* **44**, 99–113.

Levi, M. D. (1973), Errors in the variables bias in the presence of correctly measured variables. *Econometrica* **41**, 985–986.

Lindley, D. V. (1947), Regression lines and the linear functional relationship. *J. R. Statist. Soc. Suppl.* **9**, 218–244.

Lindley, D. V. (1953), Estimation of a functional relationship. *Biometrika* **40**, 47–49.

Lindley, D. V. and El Sayyad, G. M. (1968), The Bayesian estimation of a linear functional relationship. *J. R. Statist. Soc. Ser. B* **30**, 190–202.

Linssen, H. N. (1980), Functional relationships and minimum sum estimation. Doctoral dissertation. Technische Hogeschool, Eindhoven, The Netherlands.

Linssen, H. N. and Banens, P. J. A. (1983), Estimation of the radius of a circle when the coordinates of a number of points on its circumference are observed: An example of bootstrapping. *Statist. Prob. Lett.* **1**, 307–311.

Longley, J. W. (1967), An appraisal of least squares programs for the electronic computer from the point of view of the user. *J. Am. Statist. Assoc.* **62**, 819–841.

Lord, F. M. (1960), Large-sample covariance analysis when the control variable is fallible. *J. Am. Statist. Assoc.* **55**, 307–321.

Madansky, A. (1959), The fitting of straight lines when both variables are subject to error. *J. Am. Statist. Assoc.* **54**, 173–205.

Madansky, A. (1976), *Foundations of Econometrics*. North-Holland, Amsterdam.

Magnus, J. R. (1984), On differentiating eigenvalues and eigenvectors. London School of Economics, London.

Mak, T. K. (1981), Large sample results in the estimation of a linear transformation. *Biometrika* **68**, 323–325.

Mak, T. K. (1983), On Sprent's generalized least-squares estimator. *J. R. Statist. Soc. Ser. B.* **45**, 380–383.

Malinvaud, E. (1970), *Statistical Methods of Econometrics*. North-Holland, Amsterdam.

Maloney, C. J. and Rastogi, S. C. (1970), Significance test for Grubbs' estimators. *Biometrics* **26**, 671–676.

Mandel, J. (1959), The measuring process. *Technometrics* **1**, 251–267.

Mandel, J. (1976), Models, transformations of scale and weighting. *J. Quality Technol.* **8**, 86–97.

Mandel, J. (1982), The linear functional relationships with correlated errors in the variables; a heuristic approach. Unpublished manuscript. National Bureau of Standards, Washington, D.C.

Mandel, J. and Lashof, T. W. (1959), The interlaboratory evaluation of testing methods. *ASTM Bull. No. 239*, 53–61.

Mann, H. B. and Wald, A. (1943), On stochastic limit and order relationships. *Ann. Math. Statist.* **14**, 217–226.

Mariano, R. S. and Sawa, T. (1972), The exact finite-sample distribution of the limited-information maximum likelihood estimator in the case of two included endogenous variables. *J. Am. Statist. Assoc.* **67**, 159–163.

Martinez-Garza, A. (1970), Estimators for the errors in variables model. Unpublished Ph.D. dissertation. Iowa State University, Ames, Iowa.

McCulloch, C. E. (1982), Symmetric matrix derivatives with applications. *J. Am. Statist. Assoc.* **77**, 679–682.

McDonald, R. P. (1978), A simple comprehensive model for the analysis of covariance structures. *Br. J. Math. Statist. Psychol.* **31**, 59–72.

McDonald, R. P. (1980), A simple comprehensive model for the analysis of covariance structures: Some remarks on applications. *Br. J. Math. Statist. Psychol.* **33**, 161–183.

McRae, E. C. (1974), Matrix derivatives with an application to an adaptive linear decision problem. *Ann. Statist.* **1**, 763–765.

Meyers, H. and von Hake, C. A. (1976), *Earthquake Data File Summary.* National Geophysical and Solar–Terrestrial Data Center, U.S. Department of Commerce, Boulder, Colorado.

Miller, M. H. and Modigliani, F. (1966), Some estimates of the cost of capital to the electric utility industry, 1954–57. *Am. Econ. Rev.* **56**, 333–391.

Miller, R. C., Aurand, L. W., and Flack, W. R. (1950), Amino acids in high and low protein corn. *Science* **112**, 57–58.

Miller, S. M. (1984), Tests for the slope in the univariate linear errors-in-variables model. Creative component for the M.S. degree. Iowa State University, Ames, Iowa.

Miller, S. M. (1986), The limiting behavior of residuals from measurement error regressions. Unpublished Ph.D. dissertation. Iowa State University, Ames, Iowa.

Moberg, L. and Sundberg, R. (1978), Maximum likelihood estimator of a linear functional relationship when one of the departure variances is known. *Scand. J. Statist.* **5**, 61–64.

Moran, P. A. P. (1961), Path coefficients reconsidered. *Aust. J. Statist.* **3**, 87–93.

Moran, P. A. P. (1971), Estimating structural and functional relationships. *J. Multivariate Anal.* **1**, 232–255.

Morgan, W. A. (1939), A test for the significance of the difference between the two variances in a sample from a normal bivariate population. *Biometrika* **31**, 13–19.

Morgenstern, O. (1963), *On the accuracy of economic observations.* Princeton University Press, Princeton, New Jersey.

Morimune, K. and Kunitomo, N. (1980), Improving the maximum likelihood estimate in linear functional relationships for alternative parameter sequences. *J. Am. Statist. Assoc.* **75**, 230–237.

Morton, R. (1981a), Efficiency of estimating equations and the use of pivots. *Biometrika* **68**, 227–233.

Morton, R. (1981b), Estimating equations for an ultrastructural relationship. *Biometrika* **68**, 735–737.

Mowers, R. P. (1981), Effects of rotations and nitrogen fertilization on corn yields at the Northwest Iowa (Galva–Primghar) Research Center. Unpublished Ph.D. dissertation. Iowa State University, Ames, Iowa.

Mowers, R. P., Fuller, W. A., and Shrader, W. D. (1981), Effect of soil moisture on corn yields on Moody soils. Iowa Agriculture and Home Economics Experiment Station Research Bulletin 593, Ames, Iowa.

Mulaik, S. A. (1972), *The Foundations of Factor Analysis*. McGraw-Hill, New York.

Nagar, A. L. (1959), The bias and moment matrix of the general k-class estimators of the parameters in simultaneous equations. *Econometrica* **27**, 575–595.

Nair, K. R. and Banerjee, K. S. (1942–1943), A note on fitting straight lines if both variables are subject to error. *Sankhyā* **6**, 331.

Nair, K. R. and Shrivastava, M. P. (1942–1943), On a simple method of curve fitting. *Sankhyā* **6**, 121–132.

National Opinion Research Center (1947), Jobs and occupations: A population evaluation. *Opinion News* **9**, 3–13.

Nel, D. G. (1980), On matrix differentiation in statistics. *S. Afr. Statist. J.* **14**, 137–193.

Neudecker, H. (1969), Some theorems on matrix differentiation with special reference to Kronecker matrix products. *J. Am. Statist. Assoc.* **64**, 953–963.

Neyman, J. (1951), Existence of consistent estimates of the directional parameter in a linear structural relation between two variables. *Ann. Math. Statist.* **22**, 497–512.

Neyman, J. and Scott, E. L. (1948), Consistent estimates based on partially consistent observations. *Econometrica* **16**, 1–32.

Neyman, J. and Scott, E. L. (1951), On certain methods of estimating the linear structural relation. *Ann. Math. Statist.* **22**, 352–361.

Nussbaum, M. (1976), Maximum likelihood and least squares estimation of linear functional relationships. *Math. Operationsforsch. Statist. Ser. Statist.* **7**, 23–49.

Nussbaum, M. (1977), Asymptotic optimality of estimators of a linear functional relation if the ratio of the error variances is known. *Math. Operationsforsch. Statist. Ser. Statist.* **8**, 173–198.

Nussbaum, M. (1979), Asymptotic efficiency of estimators of a multivariate linear functional relation. *Math. Operationsforsch. Statist. Ser. Statist.* **10**, 505–527.

Nussbaum, M. (1984), An asymptotic minimax risk bound for estimation of a linear functional relationship. *J. Multivariate Anal.* **14**, 300–314.

Okamoto, M. (1973), Distinctness of the eigenvalues of a quadratic form in a multivariate sample. *Ann. Statist.* **1**, 763–765.

Okamoto, M. (1983), Asymptotic theory of Brown–Fereday's method in a linear structural relationship. *J. Japan Statist. Soc.* **13**. 53–56.

Okamoto, M. and Isogawa, Y. (1981), Asymptotic confidence regions for a linear structural relationship. *J. Japan Statist. Soc.* **11**, 119–126.

Okamoto, M. and Isogawa, Y. (1983), Maximum likelihood method in the Brown–Fereday model of multivariate linear structural relationship. *Math. Japanica* **28**, 173–180.

Okamoto, M. and Masamori, I. (1983), A new algorithm for the least-squares solution in factor analysis. *Psychometrika* **48**, 597–605.

O'Neill, M., Sinclair, L. G., and Smith, F. J. (1969), Polynomial curve fitting when abscissas and ordinates are both subject to error. *Comput. J.* **12**, 52–56.

Oxland, P. R., McLeod, D. S., and McNeice, G. M. (1979), An investigation of a radiographic technique for evaluating prosthetic hip performance. Technical Report, University of Waterloo, Department of Systems Design.

Pakes, A. (1982), On the asymptotic bias of the Wald-type estimators of a straight line when both variables are subject to error. *Int. Econ. Rev.* **23**, 491–497.

Pal, M. (1980), Consistent moment estimators of regression coefficients in the presence of errors in variables. *J.* Econometrics **14**, 349–364.

Pal, M. (1981), Estimation in Errors-in-Variables Models. Ph.D. thesis. Indian Statistical Institute, Calcutta.

Pantula, S. G. (1983), ISU FACTOR. Department of Statistics, North Carolina State University, Raleigh, North Caroline.

Pantula, S. G. and Fuller, W. A. (1986), Computational algorithms for the factor model. *Commun. Statist. Part B* **15**, 227–259.

Patefield, W. M. (1976), On the validity of approximate distributions arising in fitting a linear functional relationship. *J. Statist. Comput. Simul.* **5**, 43–60.

Patefield, W. M. (1977), On the information matrix in the linear functional relationship problem. *Appl. Statist.* **26**, 69–70.

Patefield, W. M. (1978), The unreplicated ultrastructural relation: Large sample properties. *Biometrika* **65**, 535–540.

Patefield, W. M. (1981), Multivariate linear relationships: Maximum likelihood estimation and regression bounds. *J. R. Statist. Soc. Ser. B* **43**, 342–352.

Patefield, W. M. (1985), Information from the maximized likelihood function. *Biometrika,* **72**, 664–668.

Pearson, K. (1901), On lines and planes of closest fit to systems of points in space. *Philos. Mag.* **2**, 559–572.

Pearson, K. (1902), On the mathematical theory of errors of judgment, with special reference to the personal equations. *Philos. Trans. R. Soc. London* **A198**, 235–299.

Pierce, D. A. (1981), Sources of error in economic time series. *J.* Econometrics **17**, 305–322.

Pitman, E. J. G. (1939), A note on normal correlation. *Biometrika* **31**, 9–12.

Prentice, R. L. (1982), Covariate measurement errors and parameter estimation in a failure time regression model. *Biometrika* **69**, 331–342.

Rao, C. R. (1955), Estimation and test of significance in factor analysis. *Psychometrika* **20**, 93–111.

Rao, C. R. (1965), *Linear Statistical Inference and Its Applications.* Wiley, New York.

Reilly, P. M. and Patino-Leal, H. (1981), A Bayesian study of the error-in-variables model. *Technometrics* **23**, 221–231.

Reilman, M. A., Gunst, R. F., and Lakshminarayanan, M. Y. (1985), Structural model estimation with correlated measurement errors. *Biometrika* **72**, 669–672.

Reiersol, O. (1950), Identifiability of a linear relation between variables which are subject to error. *Econometrica* **18**, 375–389.

Richardson, D. H. (1968), The exact distribution of a structural coefficient estimator. *J. Am. Statist. Assoc.* **63**, 1214–1226.

Riggs, D. S., Guarnieri, J. A., and Addelman, S. (1978), Fitting straight lines when both variables are subject to error. *Life Sci.* **22**, 1305–1360.

Robertson. C. A. (1974), Large sample theory for the linear structural relation. *Biometrika* **61**, 353–359.

Robinson, P. M. (1977), The estimation of a multivariate linear relation. *J. Multivariate Anal.* **7**, 409–423.

Ronner, A. E. (1986), Moment estimators in a structural regression model with outliers in the explanatory variable; theorems and proofs. *Metrika*, to be published.

Roos, C. F. (1937), A general invariant criterion of fit for lines and planes where all variates are subject to error. *Metron* **13**, 3–20.

Roth, W. E. (1934), On direct product matrices. *Bull. Am. Math. Soc.* **40**, 461–468.

Russel, T. S. and Bradley, R. A. (1958), One-way variances in a two-way classification. *Biometrika* **45**, 111–129.

Rust, B. W., Leventhal, M., and McCall, S. L. (1976), Evidence for a radioactive decay hypothesis for supernova luminosity. *Nature* (*London*) **262**, 118–120.

Sampson, A. R. (1974), A tale of two regressions. *J. Am. Statist. Assoc.* **69**, 682–689.

Sargan, J. D. (1958), The estimation of economic relationships using instrumental variables. *Econometrica* **26**, 393–415.

Sargan, J. D. and Mikhail, W. M. (1971), A general approximation to the distribution of instrumental variable estimates. *Econometrica* **39**, 131–169.

SAS Institute Inc. (1985), SAS® *User's Guide: Statistics, Version 5 Edition*. SAS Institute Inc., Cary, North Carolina.

Sawa, T. (1969), The exact sampling distribution of ordinary least squares and two-stage least squares estimators. *J. Am. Statist. Assoc.* **64**, 923–937.

Sawa, T. (1973), Almost unbiased estimator in simultaneous equations system. *Int. Econ. Rev.* **14**, 97–106.

Schafer, D. W. (1986), Combining information on measurement error in the errors-in-variables model. *J. Am. Statist. Assoc.* **81**, 181–185.

Scheffé, H. (1958), Fitting straight lines when one variable is controlled. *J. Am. Statist. Assoc.* **53**, 106–117.

Schneeweiss, H. (1976), Consistent estimation of a regression with errors in the variables. *Metrika* **23**, 101–115.

Schneeweiss, H. (1980), An efficient linear combination of estimators in a regression with errors in the variables. In *Mathematische Systeme in der Ökonomie*, M. J. Beckmann, W. Eichhorn, and W. Krelle (Eds.). Athenäum, Königstein.

Schneeweiss, H. (1982), Note on Creasy's confidence limits for the gradient in the linear functional relationship. *J. Multivariate Anal.* **12**, 155–158.

Schneeweiss, H. (1985), Estimating linear relations with errors in the variables; the merging of two approaches. In *Contributions to Econometrics and Statistics Today*, H. Schneeweiss and H. Strecker (Eds.). Springer-Verlag, Berlin.

Schneeweiss, H. and Mittag, H. J. (1986), *Lineare Modelle mit fehlerbehafteten Daten*. Physica-Verlag, Heidelberg.

Schnell, D. (1983), Maximum likelihood estimation of parameters in the implicit nonlinear errors-in-variables model. Creative component for the M.S. degree. Iowa State University, Ames, Iowa.

Schnell, D. (1987), Estimation of parameters in the nonlinear functional errors-in-variables model. Unpublished Ph.D. thesis, Iowa State University, Ames, Iowa.

Scott, E. L. (1950), Note on consistent estimates of the linear structural relation between two variables. *Ann. Math. Statist.* **21**, 284–288.

Seares, F. H. (1944), Regression lines and the functional relation. *Astrophys. J.* **100**, 255–263.

Searle, S. R. and Quaas, R. L. (1978), A notebook on variance components: A detailed description of recent methods of estimating variance components, with applications in animal breeding. Biometrics Unit Mimeo Series Paper No. BU-640-M. Cornell University, Ithaca, New York.

Selén, J. (1986), Adjusting for errors in classification and measurement in the analysis of partly and purely categorical data. *J. Am. Statist. Assoc.* **81**, 75–81.

Shapiro, A. (1983), Asymptotic distribution theory in the analysis of covariance structures. *S. Afr. Statist. J.* **17**, 33–81.

Shapiro, A. (1985), Asymptotic distribution of test statistics in the analysis of moment structures under inequality constraints. *Biometrika* **72**, 133–144.

Shaw, R. H., Nielsen, D. R., and Runkles, J. R. (1959), Evaluation of some soil moisture characteristics of Iowa soils. Iowa Agriculture and Home Economics Experiment Station Research Bulletin 465, Ames, Iowa.

Shukla, G. K. (1973), Some exact tests of hypothesis about Grubbs' estimators. *Biometrics* **29**, 373–378.

Siegel, P. M. and Hodge, R. W. (1968), A causal approach to the study of measurement error. In *Methodology in Social Research*, H. M. Blalock and A. B. Blalock (Eds.). McGraw-Hill, New York.

Smith, K. (1918), On the standard deviations of adjusted and interpolated values of an observed polynomial function and its constants and the guidance they give towards a proper choice of the distribution of observations. *Biometrika* **12**, 1–85.

Solari, M. E. (1969), The "maximum likelihood solution" to the problem of estimating a linear functional relationship. *J. R. Statist. Soc. Ser. B* **31**, 372–375.

Sonka, S. T., Heady, E. O., and Dahm, P. F. (1976), Estimation of gain isoquants and a decision model application for swine production. *Am. J. Agric. Econ.* **58**, 466–474.

Spearman, C. (1904), General intelligence, objectively determined and measured. *Am. J. Psychol.* **15**, 201–293.

Spiegelman, C. (1979), On estimating the slope of a straight line when both variables are subject to error. *Ann. Statist.* **7**, 201–206.

Spiegelman, C. (1982), A note on the behavior of least squares regression estimates when both variables are subject to error. *J. Res. Nat. Bur. Stand.* **87**, 67–70.

Sprent, P. (1966), A generalized least-squares approach to linear functional relationships. *J. R. Statist. Soc. Ser. B* **28**, 278–297.

Sprent, P. (1968), Linear relationships in growth and size studies. *Biometrics* **24**, 639–656.

Sprent, P. (1969), *Models in Regression and Related Topics*. Methuen, London.

Sprent, P. (1970), The saddle point of a likelihood surface for a linear functional relationship. *J. R. Statist. Soc. Ser. B* **32**, 432–434.

Sprent, P. (1976), Modified likelihood estimation of a linear relationship. In *Studies in Probability and Statistics*, E. J. Williams (Ed.). North-Holland, Amsterdam.

Stefanski, L. A. (1985), The effects of measurement error on parameter estimation. *Biometrika* **74**, 583–592.

Stefanski, L. A. (1985), Unbiased estimation of a nonlinear function of a normal mean with application to measurement-error models. Unpublished manuscript. Cornell University, Ithaca, New York.

Stefanski, L. A. and Carroll, R. J. (1985a), Conditional scores and optimal scores for generalized linear measurement-errors models. Unpublished manuscript. Cornell University, Ithaca, New York.

Stefanski, L. A. and Carroll, R. J. (1985b), Covariate measurement error in logistic regression. *Ann. Statist.* **12**, 1335–1351.

Stroud, T. W. F. (1972), Comparing conditional means and variances in a regression model with measurement errors of known variances. *J. Am. Statist. Assoc.* **67**, 407–412, correction (1973) **68**, 251.

Sykes, L. R., Isacks, B. L., and Oliver, J. (1969), Spatial distribution of deep and shallow earthquakes of small magnitudes in the Fiji–Tonga Region. *Bull. Seismol. Soc. Am.* **59**, 1093–1113.

Takemura, A., Momma, M., and Takeuchi, K. (1985), Prediction and outlier detection in errors-in-variables model. Unpublished manuscript. University of Tokyo.

Tenenbein, A. (1970), A double sampling scheme for estimating from binomial data with misclassifications. *J. Am. Statist. Assoc.* **65**, 1350–1361.

Theil, H. (1958), *Economic Forecasts and Policy*. North-Holland, Amsterdam.

Theil, H. (1971), *Principles of Econometrics*. Wiley, New York.

Theobald, C. M. (1975a), An inequality for the trace of the product of two symmetric matrices. *Math. Proc. Cambridge Philos. Soc.* **77**, 265–267.

Theobald, C. M. (1975b), An inequality with application to multivariate analysis. *Biometrika* **62**, 461–466.

Theobald, C. M. and Mallinson, J. R. (1978), Comparative calibration, linear structural relationships and congeneric measurements. *Biometrics* **34**, 39–45.

Thompson, W. A., Jr. (1963), Precision of simultaneous measurement procedures. *J. Am. Statist. Assoc.* **58**, 474–479.

Thomson, G. H. (1951), *The Factorial Analysis of Human Ability*. London University Press, London.

Thurstone, L. L. (1974), *Multiple Factor Analysis.* University of Chicago Press, Chicago.

Tintner, G. (1945), A note on rank, multicollinearity, and multiple regression. *Ann. Math. Statist.* **16**, 304–308.

Tintner, G. (1946), Multiple regression for systems of equations. *Econometrica* **14**, 5–36.

Tintner, G. (1952), *Econometrics.* Wiley, New York.

Tracy, D. S. and Dwyer, P. S. (1969), Multivariate maximum and minimum with matrix derivatives. *J. Am. Statist. Assoc.* **64**, 1576–1594.

Tukey, J. W. (1951), Components in regression. *Biometrics* **5**, 33–70.

Turner, M. E. (1978), Allometry and multivariate growth. *Growth* **42**, 434–450.

U.S. Department of Commerce (1975), 1970 Census of Population and Housing. Accuracy of Data for Selected Population Characteristics as measured by the 1970 CPS-Census Match. PHE(E)-11 U.S. Government Printing Office, Washington, D.C.

Vann, R. and Lorenz, F. (1984), Faculty response to writing of nonnative speakers of English. Unpublished manuscript. Department of English, Iowa State University, Ames, Iowa.

Villegas, C. (1961), Maximum likelihood estimation of a linear functional relationship. *Ann. Math. Statist.* **32**, 1048–1062.

Villegas, C. (1964), Confidence region for a linear relation. *Ann. Math. Statist.* **35**, 780–787.

Villegas, C. (1966), On the asymptotic efficiency of least squares estimators. *Ann. Math. Statist.* **37**, 1676–1683.

Villegas, C. (1969), On the least squares estimation of non-linear relations. *Ann. Math. Statist.* **40**, 462–466.

Villegas, C. (1972), Bayesian inference in linear relations. *Ann. Math. Statist.* **43**, 1767–1791.

Villegas, C. (1982), Maximum likelihood and least squares estimation in linear and affine functional models. *Ann. Statist.* **10**, 256–265.

Voss, R. E. (1969), Response by corn to NPK fertilization on Marshall and Monona soils as influenced by management and meteorological factors. Unpublished Ph.D. thesis. Iowa State University, Ames, Iowa.

Wald, A. (1940), Fitting of straight lines if both variables are subject to error. *Ann. Math. Statist.* **11**, 284–300.

Wald, A. (1943), Tests of statistical hypotheses concerning several parameters when the number of observations is large. *Trans. Am. Math. Soc.* **54**, 426–482.

Walker, H. M. and Lev, J. (1953), *Statistical Inference.* Holt, Rinehart and Winston, New York.

Ware, J. H. (1972), Fitting straight lines when both variables are subject to error and the ranks of the means are known. *J. Am. Statist. Assoc.* **67**, 891–897.

Warren, R. D., White, J. K., and Fuller, W. A. (1974), An errors-in-variables analysis of managerial role performance. *J. Am. Statist. Assoc.* **69**, 886–893.

Wax, Y. (1976), The adjusted covariance regression estimate. Unpublished Ph.D. thesis. Yale University, New Haven, Connecticut.

Werts, C. E., Rock, D. A., Linn, R. L., and Jöreskog, K. C. (1976), Testing the equality of partial correlations. *Am. Statist.* **30**, 101–102.

Whittle, P. (1952), On principal components and least square methods of factor analysis. *Skand. Aktuarietidskr.* **35**, 223–239.

Whitwell, J. C. (1951), Estimating precision of textile instruments. *Biometrics*, **7**, 101–112.

Wiley, D. E. (1973), The identification problem for structural equations models with unmeasured variables. In *Structural Equation Models in the Social Sciences*, A. S. Goldberger and O. D. Duncan (Eds.). Seminar Press, New York.

Wilks, S. S. (1963), *Mathematical Statistics*. Wiley, New York.

Willassen, Y. (1977), On identifiability of stochastic difference equations with errors-in-variables in relation to identifiability of the classical errors-in-variables (EIV) models. *Scand. J. Statist.* **4**, 119–124.

Williams, E. J. (1955), Significance tests for discriminant functions and linear functional relationships. *Biometrika* **42**, 360–381.

Williams, E. J. (1969), Regression methods in calibration problems. *Bull. Int. Statist. Inst.* **43**, 17–27.

Williams, E. J. (1973), Tests of correlation in multivariate analysis. *Bull. Int. Statist. Inst. Proc. 39th Session*. Book 4, 218–234.

Williams, J. S. (1978), A definition for the common-factor analysis model and the elimination of problems of factor score undeterminacy. *Psychometrika* **43**, 293–306.

Williams, J. S. (1979), A synthetic basis for a comprehensive factor-analysis theory. *Biometrics* **35**, 719–733.

Winakor, G. (1975), Household textiles consumption by farm and city families: assortment owned, annual expenditures, and sources. *Home Economics Research J.* **4**, 2–26.

Wolfowitz, J. (1957), The minimum distance method. *Ann. Math. Statist.* **28**, 75–88.

Wolter, K. M. (1974), Estimators for a nonlinear functional relationship. Unpublished Ph.D. dissertation. Iowa State University, Ames, Iowa.

Wolter, K. M. and Fuller, W. A. (1975), Estimating a nonlinear errors-in-variables model with singular error covariance matrix. *Proc. Business Econ. Statist. Sect. Am. Statist. Assoc.*, 624–629.

Wolter, K. M. and Fuller, W. A. (1982a), Estimation of the quadratic errors-in-variables model. *Biometrika* **69**, 175–182.

Wolter, K. M. and Fuller, W. A. (1982b), Estimation of nonlinear errors-in-variables models. *Ann. Statist.* **10**, 539–548.

Wu, D. (1973), Alternative tests of independence between stochastic regressors and disturbances. *Econometrica* **41**, 733–750.

Zellner, A. (1970), Estimation of regression relationships containing unobservable variables. *Int. Econ. Rev.* **11**, 441–454.

Author Index

Subject Index

WILEY SERIES IN PROBABILITY AND STATISTICS
ESTABLISHED BY WALTER A. SHEWHART AND SAMUEL S. WILKS

The ***Wiley Series in Probability and Statistics*** is well established and authoritative. It covers many topics of current research interest in both pure and applied statistics and probability theory. Written by leading statisticians and institutions, the titles span both state-of-the-art developments in the field and classical methods.

Reflecting the wide range of current research in statistics, the series encompasses applied, methodological and theoretical statistics, ranging from applications and new techniques made possible by advances in computerized practice to rigorous treatment of theoretical approaches.

This series provides essential and invaluable reading for all statisticians, whether in academia, industry, government, or research.

† ABRAHAM and LEDOLTER · Statistical Methods for Forecasting
AGRESTI · Analysis of Ordinal Categorical Data
AGRESTI · An Introduction to Categorical Data Analysis
AGRESTI · Categorical Data Analysis, *Second Edition*
ALTMAN, GILL, and McDONALD · Numerical Issues in Statistical Computing for the Social Scientist
AMARATUNGA and CABRERA · Exploration and Analysis of DNA Microarray and Protein Array Data
ANDĚL · Mathematics of Chance
ANDERSON · An Introduction to Multivariate Statistical Analysis, *Third Edition*
* ANDERSON · The Statistical Analysis of Time Series
ANDERSON, AUQUIER, HAUCK, OAKES, VANDAELE, and WEISBERG · Statistical Methods for Comparative Studies
ANDERSON and LOYNES · The Teaching of Practical Statistics
ARMITAGE and DAVID (editors) · Advances in Biometry
ARNOLD, BALAKRISHNAN, and NAGARAJA · Records
* ARTHANARI and DODGE · Mathematical Programming in Statistics
* BAILEY · The Elements of Stochastic Processes with Applications to the Natural Sciences
BALAKRISHNAN and KOUTRAS · Runs and Scans with Applications
BALAKRISHNAN and NG · Precedence-Type Tests and Applications
BARNETT · Comparative Statistical Inference, *Third Edition*
BARNETT · Environmental Statistics
BARNETT and LEWIS · Outliers in Statistical Data, *Third Edition*
BARTOSZYNSKI and NIEWIADOMSKA-BUGAJ · Probability and Statistical Inference
BASILEVSKY · Statistical Factor Analysis and Related Methods: Theory and Applications
BASU and RIGDON · Statistical Methods for the Reliability of Repairable Systems
BATES and WATTS · Nonlinear Regression Analysis and Its Applications

*Now available in a lower priced paperback edition in the Wiley Classics Library.
†Now available in a lower priced paperback edition in the Wiley–Interscience Paperback Series.

BECHHOFER, SANTNER, and GOLDSMAN · Design and Analysis of Experiments for Statistical Selection, Screening, and Multiple Comparisons
BELSLEY · Conditioning Diagnostics: Collinearity and Weak Data in Regression
† BELSLEY, KUH, and WELSCH · Regression Diagnostics: Identifying Influential Data and Sources of Collinearity
BENDAT and PIERSOL · Random Data: Analysis and Measurement Procedures, *Third Edition*
BERRY, CHALONER, and GEWEKE · Bayesian Analysis in Statistics and Econometrics: Essays in Honor of Arnold Zellner
BERNARDO and SMITH · Bayesian Theory
BHAT and MILLER · Elements of Applied Stochastic Processes, *Third Edition*
BHATTACHARYA and WAYMIRE · Stochastic Processes with Applications
BILLINGSLEY · Convergence of Probability Measures, *Second Edition*
BILLINGSLEY · Probability and Measure, *Third Edition*
BIRKES and DODGE · Alternative Methods of Regression
BLISCHKE AND MURTHY (editors) · Case Studies in Reliability and Maintenance
BLISCHKE AND MURTHY · Reliability: Modeling, Prediction, and Optimization
BLOOMFIELD · Fourier Analysis of Time Series: An Introduction, *Second Edition*
BOLLEN · Structural Equations with Latent Variables
BOLLEN and CURRAN · Latent Curve Models: A Structural Equation Perspective
BOROVKOV · Ergodicity and Stability of Stochastic Processes
BOULEAU · Numerical Methods for Stochastic Processes
BOX · Bayesian Inference in Statistical Analysis
BOX · R. A. Fisher, the Life of a Scientist
BOX and DRAPER · Empirical Model-Building and Response Surfaces
* BOX and DRAPER · Evolutionary Operation: A Statistical Method for Process Improvement
BOX and FRIENDS · Improving Almost Anything, *Revised Edition*
BOX, HUNTER, and HUNTER · Statistics for Experimenters: Design, Innovation, and Discovery, *Second Editon*
BOX and LUCEÑO · Statistical Control by Monitoring and Feedback Adjustment
BRANDIMARTE · Numerical Methods in Finance: A MATLAB-Based Introduction
BROWN and HOLLANDER · Statistics: A Biomedical Introduction
BRUNNER, DOMHOF, and LANGER · Nonparametric Analysis of Longitudinal Data in Factorial Experiments
BUCKLEW · Large Deviation Techniques in Decision, Simulation, and Estimation
CAIROLI and DALANG · Sequential Stochastic Optimization
CASTILLO, HADI, BALAKRISHNAN, and SARABIA · Extreme Value and Related Models with Applications in Engineering and Science
CHAN · Time Series: Applications to Finance
CHARALAMBIDES · Combinatorial Methods in Discrete Distributions
CHATTERJEE and HADI · Regression Analysis by Example, *Fourth Edition*
CHATTERJEE and HADI · Sensitivity Analysis in Linear Regression
CHERNICK · Bootstrap Methods: A Practitioner's Guide
CHERNICK and FRIIS · Introductory Biostatistics for the Health Sciences
CHILÈS and DELFINER · Geostatistics: Modeling Spatial Uncertainty
CHOW and LIU · Design and Analysis of Clinical Trials: Concepts and Methodologies, *Second Edition*
CLARKE and DISNEY · Probability and Random Processes: A First Course with Applications, *Second Edition*
* COCHRAN and COX · Experimental Designs, *Second Edition*
CONGDON · Applied Bayesian Modelling
CONGDON · Bayesian Models for Categorical Data
CONGDON · Bayesian Statistical Modelling

*Now available in a lower priced paperback edition in the Wiley Classics Library.
†Now available in a lower priced paperback edition in the Wiley–Interscience Paperback Series.

CONOVER · Practical Nonparametric Statistics, *Third Edition*
COOK · Regression Graphics
COOK and WEISBERG · Applied Regression Including Computing and Graphics
COOK and WEISBERG · An Introduction to Regression Graphics
CORNELL · Experiments with Mixtures, Designs, Models, and the Analysis of Mixture Data, *Third Edition*
COVER and THOMAS · Elements of Information Theory
COX · A Handbook of Introductory Statistical Methods
* COX · Planning of Experiments
CRESSIE · Statistics for Spatial Data, *Revised Edition*
CSÖRGŐ and HORVÁTH · Limit Theorems in Change Point Analysis
DANIEL · Applications of Statistics to Industrial Experimentation
DANIEL · Biostatistics: A Foundation for Analysis in the Health Sciences, *Eighth Edition*
* DANIEL · Fitting Equations to Data: Computer Analysis of Multifactor Data, *Second Edition*
DASU and JOHNSON · Exploratory Data Mining and Data Cleaning
DAVID and NAGARAJA · Order Statistics, *Third Edition*
* DEGROOT, FIENBERG, and KADANE · Statistics and the Law
DEL CASTILLO · Statistical Process Adjustment for Quality Control
DEMARIS · Regression with Social Data: Modeling Continuous and Limited Response Variables
DEMIDENKO · Mixed Models: Theory and Applications
DENISON, HOLMES, MALLICK and SMITH · Bayesian Methods for Nonlinear Classification and Regression
DETTE and STUDDEN · The Theory of Canonical Moments with Applications in Statistics, Probability, and Analysis
DEY and MUKERJEE · Fractional Factorial Plans
DILLON and GOLDSTEIN · Multivariate Analysis: Methods and Applications
DODGE · Alternative Methods of Regression
* DODGE and ROMIG · Sampling Inspection Tables, *Second Edition*
* DOOB · Stochastic Processes
DOWDY, WEARDEN, and CHILKO · Statistics for Research, *Third Edition*
DRAPER and SMITH · Applied Regression Analysis, *Third Edition*
DRYDEN and MARDIA · Statistical Shape Analysis
DUDEWICZ and MISHRA · Modern Mathematical Statistics
DUNN and CLARK · Basic Statistics: A Primer for the Biomedical Sciences, *Third Edition*
DUPUIS and ELLIS · A Weak Convergence Approach to the Theory of Large Deviations
EDLER and KITSOS · Recent Advances in Quantitative Methods in Cancer and Human Health Risk Assessment
* ELANDT-JOHNSON and JOHNSON · Survival Models and Data Analysis
ENDERS · Applied Econometric Time Series
† ETHIER and KURTZ · Markov Processes: Characterization and Convergence
EVANS, HASTINGS, and PEACOCK · Statistical Distributions, *Third Edition*
FELLER · An Introduction to Probability Theory and Its Applications, Volume I, *Third Edition*, Revised; Volume II, *Second Edition*
FISHER and VAN BELLE · Biostatistics: A Methodology for the Health Sciences
FITZMAURICE, LAIRD, and WARE · Applied Longitudinal Analysis
* FLEISS · The Design and Analysis of Clinical Experiments
FLEISS · Statistical Methods for Rates and Proportions, *Third Edition*
† FLEMING and HARRINGTON · Counting Processes and Survival Analysis
FULLER · Introduction to Statistical Time Series, *Second Edition*
† FULLER · Measurement Error Models

*Now available in a lower priced paperback edition in the Wiley Classics Library.
†Now available in a lower priced paperback edition in the Wiley–Interscience Paperback Series.

GALLANT · Nonlinear Statistical Models
GEISSER · Modes of Parametric Statistical Inference
GELMAN and MENG · Applied Bayesian Modeling and Causal Inference from Incomplete-Data Perspectives
GEWEKE · Contemporary Bayesian Econometrics and Statistics
GHOSH, MUKHOPADHYAY, and SEN · Sequential Estimation
GIESBRECHT and GUMPERTZ · Planning, Construction, and Statistical Analysis of Comparative Experiments
GIFI · Nonlinear Multivariate Analysis
GIVENS and HOETING · Computational Statistics
GLASSERMAN and YAO · Monotone Structure in Discrete-Event Systems
GNANADESIKAN · Methods for Statistical Data Analysis of Multivariate Observations, *Second Edition*
GOLDSTEIN and LEWIS · Assessment: Problems, Development, and Statistical Issues
GREENWOOD and NIKULIN · A Guide to Chi-Squared Testing
GROSS and HARRIS · Fundamentals of Queueing Theory, *Third Edition*
* HAHN and SHAPIRO · Statistical Models in Engineering
HAHN and MEEKER · Statistical Intervals: A Guide for Practitioners
HALD · A History of Probability and Statistics and their Applications Before 1750
HALD · A History of Mathematical Statistics from 1750 to 1930
† HAMPEL · Robust Statistics: The Approach Based on Influence Functions
HANNAN and DEISTLER · The Statistical Theory of Linear Systems
HEIBERGER · Computation for the Analysis of Designed Experiments
HEDAYAT and SINHA · Design and Inference in Finite Population Sampling
HEDEKER and GIBBONS · Longitudinal Data Analysis
HELLER · MACSYMA for Statisticians
HINKELMANN and KEMPTHORNE · Design and Analysis of Experiments, Volume 1: Introduction to Experimental Design
HINKELMANN and KEMPTHORNE · Design and Analysis of Experiments, Volume 2: Advanced Experimental Design
HOAGLIN, MOSTELLER, and TUKEY · Exploratory Approach to Analysis of Variance
* HOAGLIN, MOSTELLER, and TUKEY · Exploring Data Tables, Trends and Shapes
* HOAGLIN, MOSTELLER, and TUKEY · Understanding Robust and Exploratory Data Analysis
HOCHBERG and TAMHANE · Multiple Comparison Procedures
HOCKING · Methods and Applications of Linear Models: Regression and the Analysis of Variance, *Second Edition*
HOEL · Introduction to Mathematical Statistics, *Fifth Edition*
HOGG and KLUGMAN · Loss Distributions
HOLLANDER and WOLFE · Nonparametric Statistical Methods, *Second Edition*
HOSMER and LEMESHOW · Applied Logistic Regression, *Second Edition*
HOSMER and LEMESHOW · Applied Survival Analysis: Regression Modeling of Time to Event Data
† HUBER · Robust Statistics
HUBERTY · Applied Discriminant Analysis
HUBERTY and OLEJNIK · Applied MANOVA and Discriminant Analysis, *Second Edition*
HUNT and KENNEDY · Financial Derivatives in Theory and Practice, *Revised Edition*
HUSKOVA, BERAN, and DUPAC · Collected Works of Jaroslav Hajek—with Commentary
HUZURBAZAR · Flowgraph Models for Multistate Time-to-Event Data
IMAN and CONOVER · A Modern Approach to Statistics

*Now available in a lower priced paperback edition in the Wiley Classics Library.
†Now available in a lower priced paperback edition in the Wiley–Interscience Paperback Series.

† JACKSON · A User's Guide to Principle Components
JOHN · Statistical Methods in Engineering and Quality Assurance
JOHNSON · Multivariate Statistical Simulation
JOHNSON and BALAKRISHNAN · Advances in the Theory and Practice of Statistics: A Volume in Honor of Samuel Kotz
JOHNSON and BHATTACHARYYA · Statistics: Principles and Methods, *Fifth Edition*
JOHNSON and KOTZ · Distributions in Statistics
JOHNSON and KOTZ (editors) · Leading Personalities in Statistical Sciences: From the Seventeenth Century to the Present
JOHNSON, KOTZ, and BALAKRISHNAN · Continuous Univariate Distributions, Volume 1, *Second Edition*
JOHNSON, KOTZ, and BALAKRISHNAN · Continuous Univariate Distributions, Volume 2, *Second Edition*
JOHNSON, KOTZ, and BALAKRISHNAN · Discrete Multivariate Distributions
JOHNSON, KEMP, and KOTZ · Univariate Discrete Distributions, *Third Edition*
JUDGE, GRIFFITHS, HILL, LÜTKEPOHL, and LEE · The Theory and Practice of Econometrics, *Second Edition*
JUREČKOVÁ and SEN · Robust Statistical Procedures: Aymptotics and Interrelations
JUREK and MASON · Operator-Limit Distributions in Probability Theory
KADANE · Bayesian Methods and Ethics in a Clinical Trial Design
KADANE AND SCHUM · A Probabilistic Analysis of the Sacco and Vanzetti Evidence
KALBFLEISCH and PRENTICE · The Statistical Analysis of Failure Time Data, *Second Edition*
KARIYA and KURATA · Generalized Least Squares
KASS and VOS · Geometrical Foundations of Asymptotic Inference
† KAUFMAN and ROUSSEEUW · Finding Groups in Data: An Introduction to Cluster Analysis
KEDEM and FOKIANOS · Regression Models for Time Series Analysis
KENDALL, BARDEN, CARNE, and LE · Shape and Shape Theory
KHURI · Advanced Calculus with Applications in Statistics, *Second Edition*
KHURI, MATHEW, and SINHA · Statistical Tests for Mixed Linear Models
KLEIBER and KOTZ · Statistical Size Distributions in Economics and Actuarial Sciences
KLUGMAN, PANJER, and WILLMOT · Loss Models: From Data to Decisions, *Second Edition*
KLUGMAN, PANJER, and WILLMOT · Solutions Manual to Accompany Loss Models: From Data to Decisions, *Second Edition*
KOTZ, BALAKRISHNAN, and JOHNSON · Continuous Multivariate Distributions, Volume 1, *Second Edition*
KOVALENKO, KUZNETZOV, and PEGG · Mathematical Theory of Reliability of Time-Dependent Systems with Practical Applications
LACHIN · Biostatistical Methods: The Assessment of Relative Risks
LAD · Operational Subjective Statistical Methods: A Mathematical, Philosophical, and Historical Introduction
LAMPERTI · Probability: A Survey of the Mathematical Theory, *Second Edition*
LANGE, RYAN, BILLARD, BRILLINGER, CONQUEST, and GREENHOUSE · Case Studies in Biometry
LARSON · Introduction to Probability Theory and Statistical Inference, *Third Edition*
LAWLESS · Statistical Models and Methods for Lifetime Data, *Second Edition*
LAWSON · Statistical Methods in Spatial Epidemiology
LE · Applied Categorical Data Analysis
LE · Applied Survival Analysis
LEE and WANG · Statistical Methods for Survival Data Analysis, *Third Edition*
LEPAGE and BILLARD · Exploring the Limits of Bootstrap

*Now available in a lower priced paperback edition in the Wiley Classics Library.
†Now available in a lower priced paperback edition in the Wiley–Interscience Paperback Series.

LEYLAND and GOLDSTEIN (editors) · Multilevel Modelling of Health Statistics
LIAO · Statistical Group Comparison
LINDVALL · Lectures on the Coupling Method
LIN · Introductory Stochastic Analysis for Finance and Insurance
LINHART and ZUCCHINI · Model Selection
LITTLE and RUBIN · Statistical Analysis with Missing Data, *Second Edition*
LLOYD · The Statistical Analysis of Categorical Data
LOWEN and TEICH · Fractal-Based Point Processes
MAGNUS and NEUDECKER · Matrix Differential Calculus with Applications in Statistics and Econometrics, *Revised Edition*
MALLER and ZHOU · Survival Analysis with Long Term Survivors
MALLOWS · Design, Data, and Analysis by Some Friends of Cuthbert Daniel
MANN, SCHAFER, and SINGPURWALLA · Methods for Statistical Analysis of Reliability and Life Data
MANTON, WOODBURY, and TOLLEY · Statistical Applications Using Fuzzy Sets
MARCHETTE · Random Graphs for Statistical Pattern Recognition
MARDIA and JUPP · Directional Statistics
MASON, GUNST, and HESS · Statistical Design and Analysis of Experiments with Applications to Engineering and Science, *Second Edition*
McCULLOCH and SEARLE · Generalized, Linear, and Mixed Models
McFADDEN · Management of Data in Clinical Trials
* McLACHLAN · Discriminant Analysis and Statistical Pattern Recognition
McLACHLAN, DO, and AMBROISE · Analyzing Microarray Gene Expression Data
McLACHLAN and KRISHNAN · The EM Algorithm and Extensions
McLACHLAN and PEEL · Finite Mixture Models
McNEIL · Epidemiological Research Methods
MEEKER and ESCOBAR · Statistical Methods for Reliability Data
MEERSCHAERT and SCHEFFLER · Limit Distributions for Sums of Independent Random Vectors: Heavy Tails in Theory and Practice
MICKEY, DUNN, and CLARK · Applied Statistics: Analysis of Variance and Regression, *Third Edition*
* MILLER · Survival Analysis, *Second Edition*
MONTGOMERY, PECK, and VINING · Introduction to Linear Regression Analysis, *Fourth Edition*
MORGENTHALER and TUKEY · Configural Polysampling: A Route to Practical Robustness
MUIRHEAD · Aspects of Multivariate Statistical Theory
MULLER and STOYAN · Comparison Methods for Stochastic Models and Risks
MURRAY · X-STAT 2.0 Statistical Experimentation, Design Data Analysis, and Nonlinear Optimization
MURTHY, XIE, and JIANG · Weibull Models
MYERS and MONTGOMERY · Response Surface Methodology: Process and Product Optimization Using Designed Experiments, *Second Edition*
MYERS, MONTGOMERY, and VINING · Generalized Linear Models. With Applications in Engineering and the Sciences
† NELSON · Accelerated Testing, Statistical Models, Test Plans, and Data Analyses
† NELSON · Applied Life Data Analysis
NEWMAN · Biostatistical Methods in Epidemiology
OCHI · Applied Probability and Stochastic Processes in Engineering and Physical Sciences
OKABE, BOOTS, SUGIHARA, and CHIU · Spatial Tesselations: Concepts and Applications of Voronoi Diagrams, *Second Edition*
OLIVER and SMITH · Influence Diagrams, Belief Nets and Decision Analysis

*Now available in a lower priced paperback edition in the Wiley Classics Library.
†Now available in a lower priced paperback edition in the Wiley–Interscience Paperback Series.

PALTA · Quantitative Methods in Population Health: Extensions of Ordinary Regressions
PANJER · Operational Risk: Modeling and Analysis
PANKRATZ · Forecasting with Dynamic Regression Models
PANKRATZ · Forecasting with Univariate Box-Jenkins Models: Concepts and Cases
* PARZEN · Modern Probability Theory and Its Applications
PEÑA, TIAO, and TSAY · A Course in Time Series Analysis
PIANTADOSI · Clinical Trials: A Methodologic Perspective
PORT · Theoretical Probability for Appiications
POURAHMADI · Foundations of Time Series Analysis and Prediction Theory
PRESS · Bayesian Statistics: Principles, Models, and Applications
PRESS · Subjective and Objective Bayesian Statistics, *Second Edition*
PRESS and TANUR · The Subjectivity of Scientists and the Bayesian Approach
PUKELSHEIM · Optimal Experimental Design
PURI, VILAPLANA, and WERTZ · New Perspectives in Theoretical and Applied Statistics
† PUTERMAN · Markov Decision Processes: Discrete Stochastic Dynamic Programming
QIU · Image Processing and Jump Regression Analysis
* RAO · Linear Statistical Inference and Its Applications, *Second Edition*
RAUSAND and HØYLAND · System Reliability Theory: Models, Statistical Methods, and Applications, *Second Edition*
RENCHER · Linear Models in Statistics
RENCHER · Methods of Multivariate Analysis, *Second Edition*
RENCHER · Multivariate Statistical Inference with Applications
* RIPLEY · Spatial Statistics
* RIPLEY · Stochastic Simulation
ROBINSON · Practical Strategies for Experimenting
ROHATGI and SALEH · An Introduction to Probability and Statistics, *Second Edition*
ROLSKI, SCHMIDLI, SCHMIDT, and TEUGELS · Stochastic Processes for Insurance and Finance
ROSENBERGER and LACHIN · Randomization in Clinical Trials: Theory and Practice
ROSS · Introduction to Probability and Statistics for Engineers and Scientists
ROSSI, ALLENBY, and McCULLOCH · Bayesian Statistics and Marketing
† ROUSSEEUW and LEROY · Robust Regression and Outlier Detection
* RUBIN · Multiple Imputation for Nonresponse in Surveys
RUBINSTEIN · Simulation and the Monte Carlo Method
RUBINSTEIN and MELAMED · Modern Simulation and Modeling
RYAN · Modern Regression Methods
RYAN · Statistical Methods for Quality Improvement, *Second Edition*
SALEH · Theory of Preliminary Test and Stein-Type Estimation with Applications
* SCHEFFE · The Analysis of Variance
SCHIMEK · Smoothing and Regression: Approaches, Computation, and Application
SCHOTT · Matrix Analysis for Statistics, *Second Edition*
SCHOUTENS · Levy Processes in Finance: Pricing Financial Derivatives
SCHUSS · Theory and Applications of Stochastic Differential Equations
SCOTT · Multivariate Density Estimation: Theory, Practice, and Visualization
† SEARLE · Linear Models for Unbalanced Data
† SEARLE · Matrix Algebra Useful for Statistics
† SEARLE, CASELLA, and McCULLOCH · Variance Components
SEARLE and WILLETT · Matrix Algebra for Applied Economics
SEBER and LEE · Linear Regression Analysis, *Second Edition*
† SEBER · Multivariate Observations
† SEBER and WILD · Nonlinear Regression
SENNOTT · Stochastic Dynamic Programming and the Control of Queueing Systems
* SERFLING · Approximation Theorems of Mathematical Statistics

*Now available in a lower priced paperback edition in the Wiley Classics Library.
†Now available in a lower priced paperback edition in the Wiley–Interscience Paperback Series.

SHAFER and VOVK · Probability and Finance: It's Only a Game!
SILVAPULLE and SEN · Constrained Statistical Inference: Inequality, Order, and Shape Restrictions
SMALL and McLEISH · Hilbert Space Methods in Probability and Statistical Inference
SRIVASTAVA · Methods of Multivariate Statistics
STAPLETON · Linear Statistical Models
STAUDTE and SHEATHER · Robust Estimation and Testing
STOYAN, KENDALL, and MECKE · Stochastic Geometry and Its Applications, *Second Edition*
STOYAN and STOYAN · Fractals, Random Shapes and Point Fields: Methods of Geometrical Statistics
STYAN · The Collected Papers of T. W. Anderson: 1943–1985
SUTTON, ABRAMS, JONES, SHELDON, and SONG · Methods for Meta-Analysis in Medical Research
TAKEZAWA · Introduction to Nonparametric Regression
TANAKA · Time Series Analysis: Nonstationary and Noninvertible Distribution Theory
THOMPSON · Empirical Model Building
THOMPSON · Sampling, *Second Edition*
THOMPSON · Simulation: A Modeler's Approach
THOMPSON and SEBER · Adaptive Sampling
THOMPSON, WILLIAMS, and FINDLAY · Models for Investors in Real World Markets
TIAO, BISGAARD, HILL, PEÑA, and STIGLER (editors) · Box on Quality and Discovery: with Design, Control, and Robustness
TIERNEY · LISP-STAT: An Object-Oriented Environment for Statistical Computing and Dynamic Graphics
TSAY · Analysis of Financial Time Series, *Second Edition*
UPTON and FINGLETON · Spatial Data Analysis by Example, Volume II: Categorical and Directional Data
VAN BELLE · Statistical Rules of Thumb
VAN BELLE, FISHER, HEAGERTY, and LUMLEY · Biostatistics: A Methodology for the Health Sciences, *Second Edition*
VESTRUP · The Theory of Measures and Integration
VIDAKOVIC · Statistical Modeling by Wavelets
VINOD and REAGLE · Preparing for the Worst: Incorporating Downside Risk in Stock Market Investments
WALLER and GOTWAY · Applied Spatial Statistics for Public Health Data
WEERAHANDI · Generalized Inference in Repeated Measures: Exact Methods in MANOVA and Mixed Models
WEISBERG · Applied Linear Regression, *Third Edition*
WELSH · Aspects of Statistical Inference
WESTFALL and YOUNG · Resampling-Based Multiple Testing: Examples and Methods for p-Value Adjustment
WHITTAKER · Graphical Models in Applied Multivariate Statistics
WINKER · Optimization Heuristics in Economics: Applications of Threshold Accepting
WONNACOTT and WONNACOTT · Econometrics, *Second Edition*
WOODING · Planning Pharmaceutical Clinical Trials: Basic Statistical Principles
WOODWORTH · Biostatistics: A Bayesian Introduction
WOOLSON and CLARKE · Statistical Methods for the Analysis of Biomedical Data, *Second Edition*
WU and HAMADA · Experiments: Planning, Analysis, and Parameter Design Optimization
WU and ZHANG · Nonparametric Regression Methods for Longitudinal Data Analysis
YANG · The Construction Theory of Denumerable Markov Processes

*Now available in a lower priced paperback edition in the Wiley Classics Library.
†Now available in a lower priced paperback edition in the Wiley–Interscience Paperback Series.

YOUNG, VALERO-MORA, and FRIENDLY · Visual Statistics: Seeing Data with Dynamic Interactive Graphics
ZELTERMAN · Discrete Distributions—Applications in the Health Sciences
* ZELLNER · An Introduction to Bayesian Inference in Econometrics
ZHOU, OBUCHOWSKI, and McCLISH · Statistical Methods in Diagnostic Medicine